AERODYNAMICS, AERONAUTICS, AND FLIGHT MECHANICS

SECOND EDITION

AERODYNAMICS, AERONAUTICS, AND FLIGHT MECHANICS

Barnes W. McCormick, Ph.D.
The Pennsylvania State University
Department of Aerospace Engineering

WILEY-
INDIA

AERODYNAMICS, AERONAUTICS, AND FLIGHT MECHANICS
Second Edition

Authorized reprint by Wiley India (P.) Ltd., 4435/7, Ansari Road, Daryaganj, New Delhi 110 002.

Reprint: 2010

Printed at : S.P. Printers, Delhi,

ISBN : 978-81-265-2378-8

Dedication

This book is dedicated to my Grandchildren
Joan
Gerry
Rebecca
Tommy
Emily

PREFACE

This second edition of *Aerodynamics, Aeronautics, and Flight Mechanics* marks fifty years from the time I entered a U. S. Navy V-12 program to study Aeronautical Engineering. A lot of water has gone over the dam in those fifty years, causing a flood of profound changes in the profession. As an undergraduate, I remember bending a horizontal tail in a structures lab because someone in the group (maybe me) had slipped a decimal point on a slide rule. Who could have imagined in 1944 the invention of the "simple" handheld scientific calculator or the PC's, which do all the marvelous things that today's student takes for granted. I can only hope that the next fifty years are as exciting, interesting, and satisfying for today's student of Aerospace Engineering as the last fifty have been for me—and I hope that this text plays at least a small part in the student's career.

If you are interested in the analysis, design, and operation of aircraft, then you should find this second edition valuable. Like the first edition, this revision will probably be used as much as a reference, as it will as a text—at least this is the feedback I have received from those in the industry.

The major changes in this edition include the addition of a chapter on helicopters and V/STOL aircraft; a revision of the material on static and open-loop, dynamic stability and control; and the addition of an introductory chapter on automatic stability. The first edition contained material on the use of the analog computer; this has been removed and, instead, the use of the digital computer has been emphasized. Computer exercises are included in each chapter. Some material has been added on the subject of stealth, and throughout the text, example data has been revised and new data has been added. Also, the material on delta wings has been expanded to include nonlinear effects caused by vortex bursting.

There is too much material in this book for one course. It can be the basis for two or three courses. An introductory course in low-speed aeronautics can be formed around the material in Chapters 1, 2, 3, 4, 6, and 7, drawing selectively on Chapters 6 and 7. A second course in aeronautics can follow introducing compressible aerodynamics from Chapter 5, and the other end of the spectrum, vertical flight from Chapter 8. Finally, the material in Chapters 9, 10, and 11 are sufficient for a good introductory course in aircraft stability and control.

Again, my thanks go to the many persons in the industry and to the reviewers of the draft for their contributions and constructive criticisms. At this point, I could almost reiterate the preface to be found in the first edition. Indeed, to more fully appreciate the content of this edition, you may wish to read the preface for the first one.

Barnes W. McCormick

University Park, PA
April 5, 1994

CONTENTS

1 / INTRODUCTION

Aeronautics is defined as "the science that treats of the operation of aircraft; also, the art or science of operating aircraft." Basically, with aeronautics, one is concerned with predicting and controlling the forces and moments of an aircraft that is traveling through the atmosphere.

A BRIEF HISTORY.

Thursday, December 17, 1903

When we got up a wind of between 20 and 25 miles was blowing from the north. We got the machine out early and put out the signal for the men at the station. Before we were quite ready, John T. Daniels, W. S. Dough, A. D. Etheridge, W. C. Brinkly of Manteo, and Johnny Moore of Nags Head arrived. After running the engine and propellers a few minutes to get them in working order, I got on the machine at 10:35 for the first trial. The wind, according to our anemometers at this time, was blowing a little over 20 miles (corrected) 27 miles according to the government anemometer at Kitty Hawk. On slipping the rope the machine started off increasing in speed to probably 7 or 8 miles. The machine lifted from the truck just as it was entering the fourth rail. Mr. Daniels took a picture just as it left the tracks. I found the control of the front rudder quite difficult on account of its being balanced too near the center and thus had a tendency to turn itself when started so that the rudder was turned too far on one side and then too far on the other. As a result the machine would rise suddenly to about 10 ft. and then as suddenly, on turning the rudder, dart for the ground. A sudden dart when out about 100 feet from the end of the tracks ended the flight. Time about 12 seconds (not known exactly as watch was not promptly stopped). The level for throwing off the engine was broken, and the skid under the rudder cracked. After repairs, at 20 min. after 11 o'clock Will made the second trial.

The above, taken from Orville Wright's diary, as reported in Reference 1.1, describes mankind's first sustained, controlled, powered flight in a heavier-than-air machine. The photograph, mentioned by Orville Wright, is shown here as Figure 1.1. Three more flights were made that morning. The last one, by Wilbur Wright, began just at 12 o'clock and covered 260 m in 59 s. Shortly after this flight, a strong gust of wind struck the airplane, turning it over and over. Although the machine was severely damaged and never flew again, the Wright Brothers achieved their goal, begun approximately 4 yr earlier.

Their success was no stroke of luck. The Wright Brothers were painstaking in their research and confident of their own results. They built their own wind tunnel and tested, in a methodical manner, hundreds of different airfoil and wing platform shapes. They were anything but a "couple of bicycle mechanics." Their letters to Octave Chanute, a respected civil engineer and aviation enthusiast of the day, reveal the Wright Brothers to have been learned men well versed in basic concepts such as work, energy, statics, and dynamics. A three-view drawing of their first airplane is presented in Figure 1.2.

On September 18, 1901, Wilbur Wright was invited to deliver a lecture before the Western Society of Engineers at a meeting in Chicago, Illinois. Among the conclusions reached by him in that paper were the following:

1. *That the ratio of drift to lift in well-shaped surfaces is less at angles of incidence of five degrees to 12 degrees than at an angle of three degrees. ("Drift" is what we now call "drag.")*

Figure 1.1 The first flight, December 17, 1903. (Courtesy of the National Air and Space Museum, Smithsonian Institution)

2. *That in arched surfaces the center of pressure at 90 degrees is near the center of the surface, but moves slowly forward as the angle becomes less, till a critical angle varying with the shape and depth of the curve is reached, after which it moves rapidly toward the rear till the angle of no lift is found."*
3. *That a pair of superposed, or tandem surfaces, has less lift in proportion to drift than either surface separately, even after making allowance for weight and head resistance of the connections.*

These statements and other remarks (see Ref. 1.1) show that the Wright Brothers had a good understanding of wing and airfoil behavior well beyond that of other experimenters of the time.

Following their first successful flights at Kitty Hawk, North Carolina in 1903, the Wright Brothers returned to their home in Dayton, Ohio. Two years later they were making flights there, almost routinely, in excess of 30 km and 30 min while others were still trying to get off the ground.

Most of the success of the Wright Brothers must be attributed to their own research, which utilized their wind tunnel and numerous experiments with controlled kites and gliders. However, their work was built, to some degree, on the gliding experiments of Otto Lilienthal and Octave Chanute. Beginning in 1891, Lilienthal, working near Berlin, Germany, made approximately 2000 gliding flights over a 5-yr period. Based on measurements obtained from these experiments, he published tables of lift and drag measurements on which the Wright Brothers based their early designs. Unfortunately, Lilienthal had no means of providing direct aerodynamic control to his gliders and relied instead on kinesthetic control, whereby he shifted his weight fore and aft and side to side. On August 9, 1896, as the result of a gust, Otto Lilienthal lost control from an altitude of approximately 15 m. He died the next day. During 1896 and 1897, Octave Chanute, inspired by Lilienthal's work, designed and built several gliders that were flown by others near Miller, Indiana. Chanute recognized Lilienthal's control problems and was attempt-

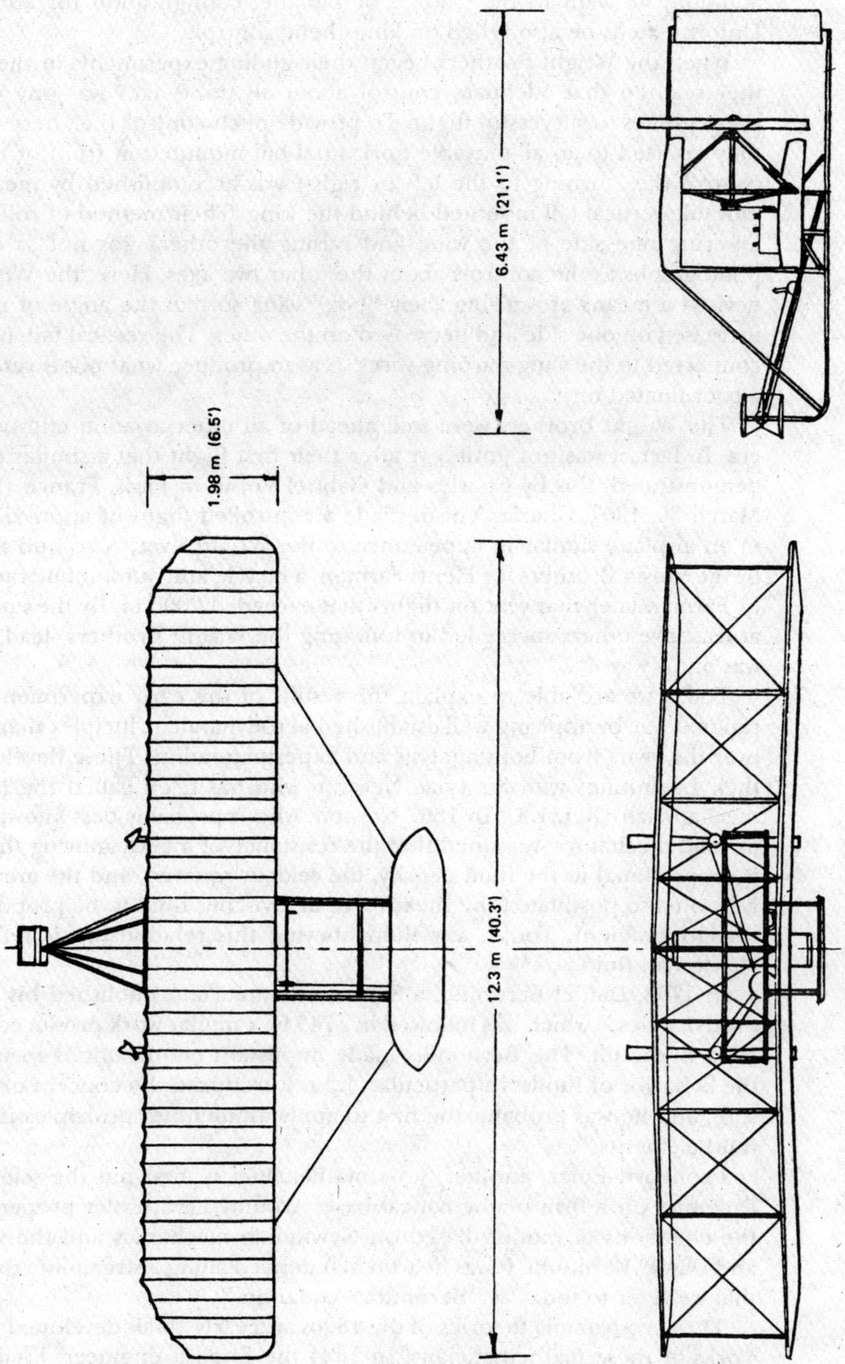

Figure 1.2 Three views of the Wright Brothers Flyer.

ing to achieve an "automatic" stability in his designs. Chanute's principal contribution was the addition of both vertical and horizontal stabilizing tail surfaces. In addition, he went to the "box," or biplane, configuration for added strength. Unfortunately, he also relied on kinesthetic control.

When the Wright Brothers began their gliding experiments in the fall of 1900, they realized that adequate control about all three axes was one of the major prerequisites to successful flight. To provide pitch control (i.e., nose up or down), they resorted to an all-movable horizontal tail mounted in front of the wing. Yaw control (i.e., turning to the left or right) was accomplished by means of an all-movable vertical tail mounted behind the wing. Their method of roll control (i.e., lowering one side of the wing and raising the other) was not as obvious from photographs as the controls about the other two axes. Here, the Wright Brothers devised a means of warping their "box" wing so that the angle of incidence was increased on one side and decreased on the other. The vertical tail, or rudder, was connected to the wing-warping wires so as to produce what pilots refer to today as a coordinated turn.

The Wright Brothers were well ahead of all other aviation enthusiasts of their era. In fact, it was not until 3 yr after their first flight that a similar capability was demonstrated, this by Charles and Gabriel Voisin in Paris, France (Ref. 1.2). On March 30, 1907, Charles Voisin made a controlled flight of approximately 100 m in an airplane similar in appearance to the Wright flyer. A second machine built by the Voisin Brothers for Henri Farman, a bicycle and automobile racer, was flown by Farman later that year on flights that exceeded 2000 m. By the end of that year at least five others succeeded in following the Wright Brothers' lead, and aviation was on its way.

Today we are able to explain the results of the early experimenters in a very rational way by applying well-established aerodynamic principles that have evolved over the years from both analysis and experimentation. These developments have their beginnings with Sir Isaac Newton, who has been called the first real fluid mechanician (Ref. 1.3). In 1687 Newton, who is probably best known for his work in solid mechanics, reasoned that the resistance of a body moving through a fluid is proportional to the fluid density, the velocity squared, and the area of the body. Newton also postulated the shear force in a viscous fluid to be proportional to the velocity gradient. Today, any fluid obeying this relationship is referred to as a Newtonian fluid.

In 1738, Daniel Bernoulli, a Swiss mathematician, published his treatise, "Hydrodynamics," which was followed in 1743 by a similar work produced by his father, John Bernoulli. The Bernoullis made important contributions to understanding the behavior of fluids. In particular, John introduced the concept of internal pressure, and he was probably the first to apply momentum principles to infinitesimal fluid elements.

Leonhard Euler, another Swiss mathematician, first put the science of hydrodynamics on a firm mathematical base. Around 1755, Euler properly formulated the equations of motion based on Newtonian mechanics and the works of John and Daniel Bernoulli. It was he who first derived along a streamline the relationship that we refer to today as "Bernoulli's equation."

The aerodynamic theories of the 1800s and early 1900s developed from the early works of these mathematicians. In 1894 the English engineer, Frederick William Lanchester, developed a theory to predict the aerodynamic behavior of wings. Unfortunately, this work was not made generally known until 1907 in a book published by Lanchester. By then the Wright Brothers had been flying for 3 yr.

Much of the knowledge that they had laboriously deduced from experiment could have been reasoned from Lanchester's theory. In 1894, Lanchester completed an analysis of airplane stability that could also have been of value to the Wrights. Again, this work was not published until 1908.

Lanchester's wing theory was somewhat intuitive in its development. In 1918, Ludwig Prandtl, a German professor of mechanics, presented a mathematical formulation of three-dimensional wing theory; today both men are credited with this accomplishment. Prandtl also made another important contribution to the science with his formalized boundary layer concept.

Around 1917 Nikolai Ergorovich Joukowski (the spelling has been anglicized), a Russian professor of rational mechanics and aerodynamics in Moscow, published a series of lectures on hydrodynamics in which the behavior of a family of airfoils was investigated analytically.

The work of these early hydro- and aerodynamicists contributed little, if any, to the progress and ultimate success of those struggling to fly. However, it was the analytical base laid by Euler and those who followed him on which the rapid progress in aviation was built.

After 1908, the list of aviators, engineers, and scientists contributing to the development of aviation grew rapidly. Quantum improvements were accomplished with the use of flaps, retractable gear, the cantilevered wing, all-metal construction, and the turbojet engine.

In 1903, the Wright Brothers Flyer had a wing span of 12.3 m (40.3 ft) and was able to carry approximately 150 lb. In 1988, the Antonov An-218, developed in the USSR, made its first flight. This airplane, the world's largest, is powered by six turbofan engines rated at 229.5 kN (51,590 lb) each. With a wing span of 88.4 m (275.6 ft), the An-218 is capable of carrying a payload of 2450 kN (551,150 lb). The span of the Antonov is almost three times as long as the distance covered by the Wright Brothers on their first flight. This tremendous growth in size and payload capacity is shown in Figures 1.3 and 1.4. The data points are taken from References 1.5 and 1.6. The one isolated point well above the others in Figure 1.3 is the famous "Spruce Goose," an all-wood flying boat built by Howard Hughes, a Hollywood

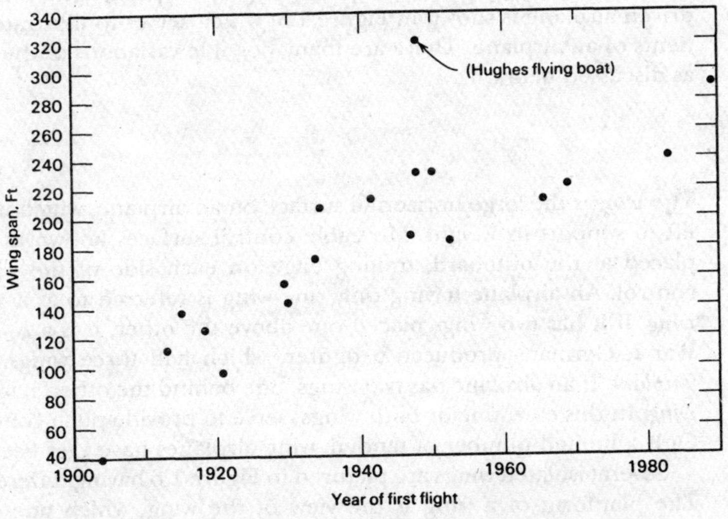

Figure 1.3 Increase in wing span since the Wright Brothers flight in 1903.

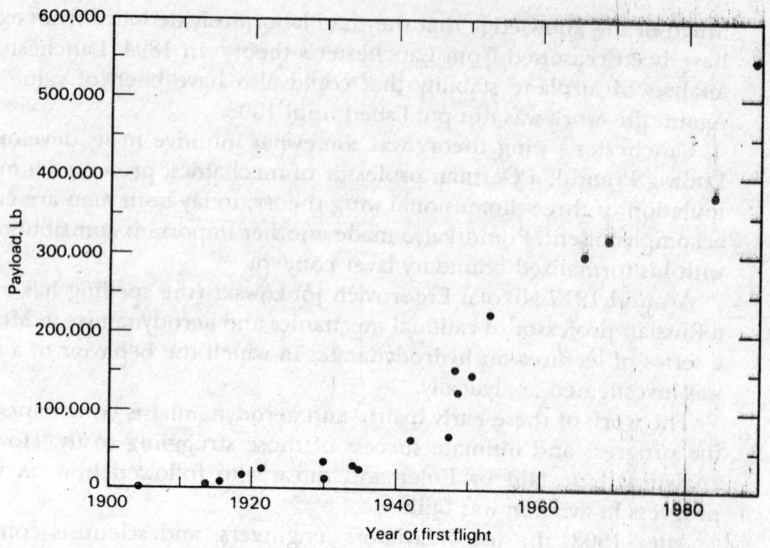

Figure 1.4 Increase in payload capability since the Wright Brothers flight in 1903.

mogul and entrepreneur who founded the Hughes Aircraft Co. This unique airplane, at one time on display in Long Beach, California, was far ahead of its time in size and payload, but its development was terminated by the ending of World War II. Its one and only flight consisted of lifting a few feet off above the water with Howard Hughes at the controls.

AIRPLANE GEOMETRY

Most readers of this book will have some knowledge of airplanes. However, to assist those who do not, this section will describe the basic components of an airplane and introduce, in an elementary way, some of the technical aspects of flight.

Airplanes come in many shapes and sizes. A typical, light, twin-engine, propeller-driven airplane is shown in Figure 1.5, which serves to illustrate the major components of an airplane. There are many possible variations of the major components as discussed below.

WING

The *wing* is the large horizontal surface on an airplane, which provides most of the lift to support its weight. Moveable control surfaces known as *ailerons* are usually placed at the outboard, trailing edge on each side of the wing to provide roll control. An airplane having only one wing is referred to as a *monoplane,* or *single-wing.* If it has two wings placed one above the other, it is a *biplane.* During World War I, Germany produced a fighter, which had three wings, called the Fokker *Triplane.* If an airplane has two wings, one behind the other, it is known as a *tandem-wing.* In this case, one or both wings, serve to provide pitch control in lieu of a tail. Only a limited number of tandem-wing airplanes have ever been produced.

Several isolated wings are pictured in Figure 1.6 having different *planform* shapes. The planform of a wing is the view of the wing, which one sees when looking

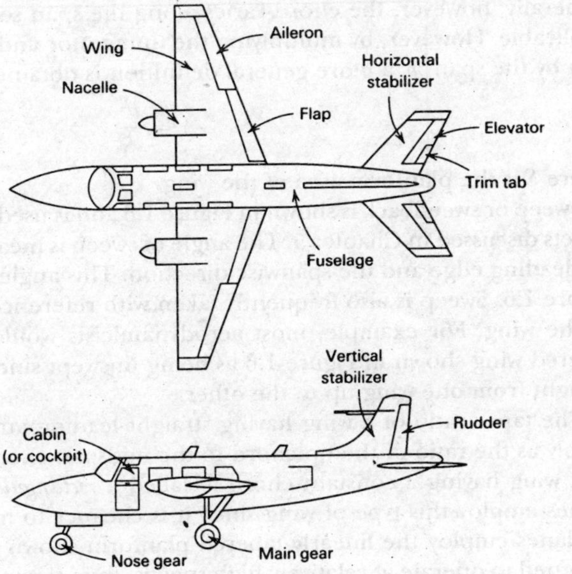

Figure 1.5 A light twin-engine, propeller-driven, monoplane with tricycle gear and conventional empennage.

directly up or down on the wing. The span is the distance from one wing tip to the other as shown in the planform view. The chord is defined as the distance from the leading edge to the trailing edge in the x direction and generally varies along the span.

There are several parameters that characterize wing geometry. The first of these is the aspect ratio, A, which is a measure of the span length relative to the chord. If the chord, c, is constant along the span, then the aspect ratio is given simply by

$$A = \frac{b}{c} \tag{1.1}$$

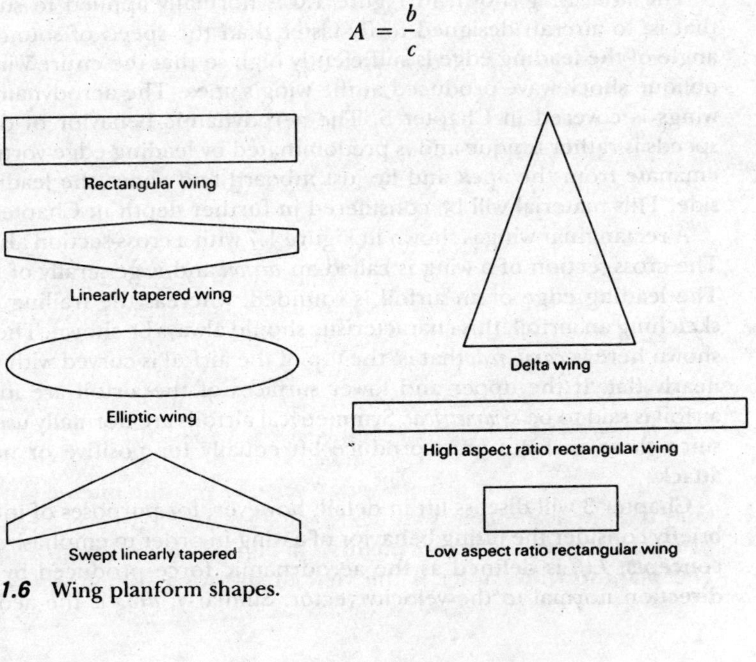

Figure 1.6 Wing planform shapes.

Generally, however, the chord varies along the span so that Equation (1.1) is not applicable. However, by multiplying the numerator and denominator of this equation by the span, b, a more general definition is obtained.

$$A = \frac{b^2}{S} \tag{1.2}$$

where S is the planform area of the wing.

Sweep or sweepback is shown in Figure 1.6 and is used to alleviate compressibility effects discussed in Chapter 5. The angle of sweep is measured as the angle between the leading edge and the spanwise direction. This angle, denoted as Λ, is shown in Figure 1.6. Sweep is also frequently taken with reference to the quarter-chord line of the wing. For example, most aerodynamicists would characterize the linearly tapered wing shown in Figure 1.6 as being unswept since the quarter-chord line is straight from one wing tip to the other.

The taper ratio of a wing having straight leading and trailing edges is defined simply as the ratio of the tip chord to the midspan chord.

A wing having a constant chord is called a *rectangular* wing. Many low-cost airplanes employ this type of wing since it is cheaper to manufacture. Most modern airplanes employ the linearly tapered planform shown in Figure 1.6. If wings are designed to operate at relatively high speeds, then they are swept as well as tapered.

From an aerodynamic efficiency standpoint, the elliptic planform shown in Figure 1.6 is nearly optimum as will be shown in Chapter 3. Recent NASA studies (Ref. 1.7) have shown that even the elliptic planform may not be quite the optimum measured on the basis of lift-to-drag ratio because of higher-order effects not considered in the classical analysis. Probably the most famous airplane to employ an elliptic planform was the Supermarine Spitfire flown by the Royal Air Force in Britain during World War II. It would probably have done as well with a linearly tapered wing since, as will be shown later, a taper ratio (tip chord to midchord) of approximately 0.3 results in a predicted efficiency close to that of an elliptic planform.

The *delta wing* shown in Figure 1.6 is normally applied to supersonic aircraft; that is, to aircraft designed to fly faster than the speed of sound. The sweepback angle of the leading edge is sufficiently high so that the entire wing lies behind the oblique shock wave produced at the wing's apex. The aerodynamics of supersonic wings is covered in Chapter 5. The aerodynamic behavior of delta wings at low speeds is rather unique and is predominated by leading edge vortices (LEV), which emanate from the apex and lie just inboard and above the leading edge on each side. This material will be considered in further depth in Chapter 3.

A rectangular wing is shown in Figure 1.7 with a cross-section also being depicted. The cross-section of a wing is called an *airfoil* and is generally of the shape shown. The leading edge of an airfoil is rounded, whereas the trailing edge is sharp. In sketching an airfoil, this characteristic should always be shown. The particular airfoil shown here is *cambered;* that is, the top of the airfoil is curved with the bottom being nearly flat. If the upper and lower surfaces of the airfoil are identical, then the airfoil is said to be *symmetrical.* Symmetrical airfoils are normally used on tail surfaces since these surfaces must produce lift equally for positive or negative angles of attack.

Chapter 3 will discuss lift in detail; however, for purposes of introduction, let us briefly consider the lifting behavior of a wing in order to emphasize some important concepts. *Lift* is defined as the aerodynamic force produced by a surface in the direction normal to the velocity vector. Similarly, *drag* is the aedrodynamic force

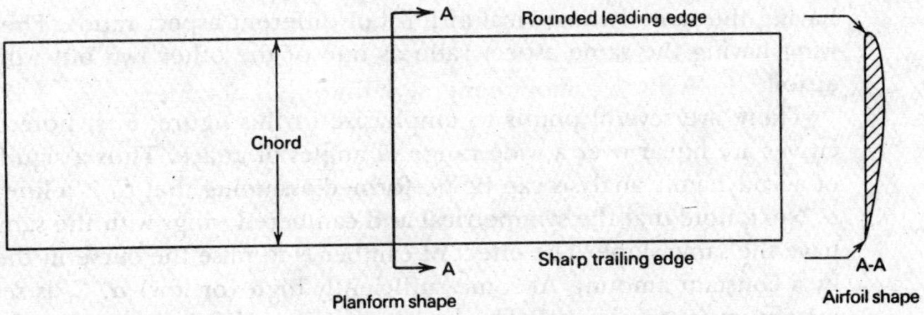

Figure 1.7 Section of a rectangular wing showing the airfoil shape.

parallel to the velocity vector. Chapter 2 will show that the lift can be expressed in terms of a dimensionless lift coefficient, C_L, which is a measure of the lift but independent (within limits to be discussed) of the size of the wing, the airspeed, or air density. At this point, it will suffice to assume that C_L is a function only of the shape of the wing and its angle of attack. Knowing these factors, one can then calculate the lift of a wing of any size operating at any altitude and speed from the definition of C_L.

$$C_L = \frac{L}{\frac{1}{2}\rho V^2 S} \tag{1.3}$$

Figure 1.8 depicts a typical variation of C_L with angle of attack for three wings having rectangular planforms. This figure shows the effect of aspect ratio and airfoil shape on the lift. The inset in this figure illustrates a cambered airfoil. The dotted line is the mean camber line midway between the upper and lower surfaces of the airfoil. The line connecting the leading and trailing edges of the mean camber line is the chord and, in this figure, the angle of attack is measured relative to the chord. There are three curves shown on the figure. Two of the curves are for wings

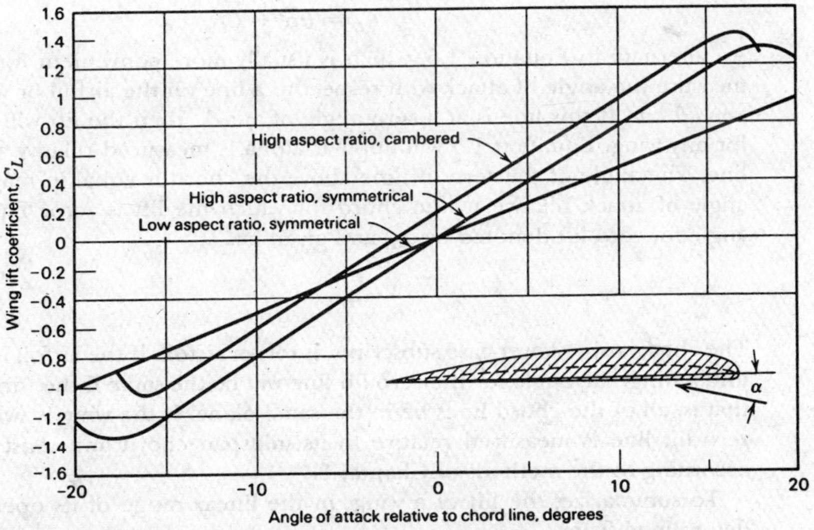

Figure 1.8 Effect of aspect ratio and camber on the lift behavior of wings.

having the same symmetrical airfoil but different aspect ratios. The third is for a wing having the same aspect ratio as one of the other two but with a cambered airfoil.

There are several points to emphasize in this figure. First note that all of the curves are linear over a wide range of angles of attack. Thus a significant amount of aerodynamic analysis can be performed assuming that C_L is a linear function of α. Next, note that the symmetrical and cambered wings with the same aspect ratio have the same slope. The effect of camber is to raise the curve in the linear range by a constant amount. At some sufficiently high (or low) α, C_L is seen to reach a maximum (or minimum). This maximum value of C_L is denoted as $C_{L_{max}}$. Generally, the effect of camber is to increase $C_{L_{max}}$ positively but to decrease its magnitude for negative angles of attack.

Increasing the aspect ratio is seen to increase the lift curve slope. As the aspect ratio approaches infinity, the theoretical value of the slope approaches 2π/rad. This is called the 2-D, or airfoil or section (referring to a section of the wing), lift curve slope. This follows since, for an infinite aspect ratio, the flow in any plane normal to the span will be the same. Dimensionless coefficients for airfoils are denoted by lower-case letters. Since the planform area for the 2-D airfoil will be given by the product of the chord, c, and a unit span, the section lift coefficient will be

$$C_l = \frac{L}{\frac{1}{2}\rho V^2 c} \tag{1.4}$$

ρ is the mass density of the air and can be found in Appendix B. The airspeed, wing area, lift, and ρ are in consistent units.

If a denotes the slope of the lift curve for a symmetrical wing, since C_L is zero at an α of zero, the wing lift coefficient can be obtained from

$$C_L = a\alpha \tag{1.5}$$

However, for a cambered wing, the above equation will not hold. Instead, one must add C_{L_0}, the wing lift coefficient at a zero angle of attack.

$$C_L = a\alpha + C_{L_0} \tag{1.6}$$

An alternate to Equation 1.6, which is usually more convenient for analysis, is to measure the angle of attack with respect to a line on the airfoil or wing called the *zero lift line.* If this line is at a zero angle of attack, then the lift will be zero. Thus for any wing, Equation 1.5 will apply if alpha is measured relative to the zero lift line. The angle of the zero lift line above the chord is equal in magnitude to the angle of attack relative to the chord for which the lift is zero. This angle is the angle for zero lift denoted by α_{0l} and given by

$$\alpha_{0l} = -\frac{C_{l_0}}{a} \tag{1.7}$$

The above, using lower-case subscripts, is for an airfoil. If the airfoil is incorporated into a wing, the angle for the zero lift line will be the same if the wing if untwisted, that is, all of the chord lines lie in the same plane. If the wing is twisted, then the zero lift line is measured relative to its midspan chord and must be calculated according to the methods of Chapter 3.

To summarize, the lift of a wing, in the linear range of its operation, can be determined from

$$C_L = a\alpha + C_{L0} \quad (\alpha \text{ relative to chord line}) \tag{1.8}$$

$$C_L = a(\alpha - \alpha_{0l}) \quad (\alpha \text{ relative to chord line}) \tag{1.9}$$

$$C_L = a\alpha \quad (\alpha \text{ relative to zero lift line}) \tag{1.10}$$

EMPENNAGE

The tail assembly of an airplane consisting of the horizontal tail and the vertical tail is known as the *empennage.* The purpose of the empennage is to provide for stability and control of the airplane in both pitch and yaw. If the horizontal tail is ahead of the wing, it is known as a *canard.* From Figure 1.2, The Wright Brothers Flyer is seen to have the canard configuration. From shortly after the Wright Brothers' success until the 1970s, most airplane designs have favored the conventional tail. However, the canard configuration is being adopted more and more, partly in response to its chief proponent, Elburt (Burt) Rutan, designer of the Voyager, the only airplane to fly around the earth nonstop without refueling. Canards are also finding application with high-performance airplanes in combination with a conventional horizontal tail. For these applications, the canards provide added maneuverability, spin resistance, and may interact favorably with the main wing to increase maximum lift. The first all-composite airplane to be certified by the FAA is a canard configuration. This airplane, the Beechcraft Starship, shown in Figure 1.9, is capable of cruising at 7620 m (25,000 ft) at a speed of 335 kts. Powered by two Pratt & Whitney Canada PT6A-67A turboprop engines, each rated at 895 KW (1200 shp), the Starship has a range of 2544 km (1394 n. mi.) with a payload of 9408 N (2115 lb).

A perturbation on the conventional empennage is the so-called vee-tail. Here, the vertical tail is removed and each side of the horizontal tail inclined upward approximately 45° to form a vee. The resulting vee-shaped surface produces aerodynamic forces having both vertical and horizontal components which can provide the same degree of stability and control as the conventional empennage. Only a few airplanes have been built incorporating the vee-tail, the best known of which is the vee-tail Beechcraft Bonanza. First produced in 1959, many vee-tail Bonanzas are still flyng although in later models the vee-tail was replaced in favor of a conventional empennage.

Figure 1.9 The Beech Starship. (Courtesy of the Beech Aircraft Corporation)

As shown in Figure 1.5, the horizontal tail is composed of a forward, fixed part known as the *horizontal stabilizer* (or tailplane) and a moveable part, the *elevator*. The entire horizontal tail contributes to the stability of the airplane but only the moveable elevator provides control to *pitch* the airplane about an axis parallel to the wing; that is, to move the nose up or down. Some airplanes employ an all-moveable, or *flying*, tail where the entire tail rotates to provide pitch control. Another configuration is the *stabilator*, which looks like a conventional horizontal tail but differs considerably in its operation. In this case the "horizontal stabilizer" rotates in response to pilot control input and the "elevator" is mechanically linked to move also. Thus, for example, as the forward part of the stabilator rotates nose-up, the aft part of the stabilator rotates tail-down at a greater angle so as to produce, in effect, a cambered airfoil. The resulting increase in the tail lift is greater than that which would be obtained from a flying tail for the same rotation. The conventional and stabilator configurations will be considered in more detail later in Chapter 9.

The vertical tail is composed of a forward, fixed part known as the *vertical stabilizer* (or *fin*) and a moveable part, the *rudder*. The action of the vertical tail is similar to that of the horizontal tail. The total tail, both the fin and the rudder, contributes to the airplane's stability but only the rudder provides *yaw* control about a vertical axis through the airplane's center-of-gravity (CG). For example, if the trailing edge of the rudder is deflected to the left, this will produce a lift on the vertical tail to the right, which, in turn, will produce a moment about the vertical axis tending to turn (yaw) the airplane to the left.

Figure 1.5 shows another small, moveable surface on the elevator designated as a *trim tab*. Trim tabs can be placed on any control surface; that is, ailerons, elevator, or rudder. The trim tab can be adjusted to maintain a fixed deflection angle between the tab and the control surface to which it is attached, which does not change as the control surface is moved. By changing the angle of the trim tab, the pilot produces primarily a moment about the hinge line of the control surface, which relieves the amount of force required of the pilot to deflect the control surface. For example, at a given operating condition, the pilot can adjust the elevator trim tab so that the airplane can fly straight and level without any force being applied to the wheel. This relieves the fatigue on a long trip, which the pilot would suffer if he or she found it necessary to apply a force continuously to the wheel.

LANDING GEAR

The landing gear pictured in Figure 1.5 is a *tri-cycle* gear having a single nose-gear forward and two main gear aft. This configuration is now used more universally than the *conventional* gear where the main gear is forward with a small tail wheel at the rear. The conventional gear configuration, being a relic of the past is affectionately dubbed a "tail dragger" by pilots. This configuration is unstable, and, while rolling along the runway with this type of gear, the pilot must exercise prudent brake and rudder steering to prevent excessive yaw. Otherwise an uncontrollable motion will occur, known as a *ground loop*, where the front and rear of the airplane swap positions. If the landing gear does not collapse, the least favorable circumstance will be embarrassment to the pilot as he or she rolls down the runway backward. Your author speaks from personal experience in a Cessna L-19, which was borrowed from the Army for a research project. Conversely, the tri-cycle gear is stable and will tend to roll straight down the runway without yawing.

Instrument Panel

Finally, before leaving the description of airplane components, it should be noted that the panel in front of a pilot is not a "dashboard"; it is an *instrument panel.* Instruments typical of a private aircraft to be found on an instrument panel include:

Altimeter An altimeter measures the pressure altitude. Set to the proper barometric pressure, it measures approximately the altitude of the airplane above sea level.

Airspeed Indicator As the name implies, an airspeed indicator is the counterpart of a speedometer in an automobile. Its operation is covered in detail in Chapter 2.

Tachometer A tachometer measures the propeller RPM. For turbine-powered aircraft, meters are also found, which measure engine RPM. Also included on most tachometers is a digital counter, which records engine time based on an average RPM. This is referred to as "tach time."

Chronometer A chronometer is a clock that is used, not simply for telling the time, but for navigational and instrument-flying purposes as well.

Artificial Horizon An artificial horizon contains gyroscopes, which measure roll and pitch (see the next section for definition of roll and pitch). This artificial horizon, driven by the engine suction pump, provides the pilot with an earth-fixed reference system for instrument flight.

Turn-Bank Indicator A turn-bank indicator has a needle driven by an electrically powered gyroscope, which senses the yawing; that is, a turn to the left or right, of the airplane. It also contains a small ball, which rolls in a sealed U-shaped glass tube. When the ball is centered while making a turn, the turn is *coordinated,* meaning that the resultant force on the airplane (and pilot) is parallel to the plane of symmetry of the airplane. An uncoordinated turn results in the uncomfortable feeling that the occupants are sliding to one side or the other of their seats. As will be seen later in Chapter 10, the rolling and yawing motions of an airplane are coupled so that the wings will remain level if the airplane is not turning. Thus, the Turn-Bank Indicator provides a backup to the Artificial Horizon for instrument flight and is therefore powered from a separate source.

DME DME is an acronym for "distance measuring equipment." This is an electronic instrument, which receives a signal from a ground-based, omnidirectional, VHF radio beacon and converts it to a distance of the airplane from the station. It will also differentiate the signal to obtain the groundspeed and then provide the pilot with the time to reach the station.

ADF ADF is an acronym for Automatic Direction Finder, which is sometimes referred to as a radio compass. When turned to a low-frequency radio station, the needle of the ADF will point to the station.

Magnetic Compass A magnetic compass, which is installed in all airplanes, is used to determine the heading of the airplane relative to magnetic north. However, it is difficult to maneuver an airplane by reference to the magnetic compass because the movement of the compass will lag appreciably the motion of the airplane. Therefore, the main usefulness of the magnetic compass is to set the gryo compass, described next, while flying straight and level.

DG DG is an acronym for "directional gyroscope" and is frequently referred to as a "gyro compass." This instrument consists of a single gyroscope, which drives a compass needle, which responds immediately to the turning of the airplane. Actually, the needle remains fixed relative to the earth, and it is the turning of the airplane that results in a change in the indicated heading. Over time, the directional gyroscope will precess so that it is necessary to reset it to agree with the magnetic compass.

NavCom NavCom is an acronym for "navigation and communication" and is a VHF voice radio combined with an instrument, a *VOR* (very high frequency, omnidirectional, radio), which receives signals from a VOR ground beacon. The pilot rotates a compass face until an indicator needle is centered and reads the direction to or from the station off the face. A window on the instrument will read either "to" or "from." An *ILS* (instrument landing system) receiver is frequently incorporated into the navcom package. It consists of a localizer needle, which moves left or right in the manner of a VOR needle and a glide slope needle, which moves up or down. The pilot sets the heading of the localizer course on the compass face of the VOR. Then he or she proceeds to a low-frequency, vertical radio beacon, called the *outer marker* (OM), which is typically 5 miles away from the runway threshold. As the airplane passes over the OM at an altitude prescribed on an approach plate, both the localizer and glide slope needles will be centered with visual and aural signal indicating the location. The pilot then flies down the path defined by the needles keeping them centered. If either needle drifts off center, the pilot simply changes the direction of the airplane toward either needle.

It should be noted that, at the time of this writing, the FAA is beginning a transition to the global positioning system (GPS). This system uses signals from satellites, which can accurately define a position on the earth within a few meters. It is assumed that in the near future, the many airways presently defined by the location of VOR beacons will be converted to GPS. Amazingly, a modification of this system (DGPS) using a differentiation technique reduces the position error to centimeters. In this case, one wonders if the system designers are talking about the front or rear of the airplane. The system is so accurate that it promises instrument landings under zero visibility at any airport providing the location of the runway threshold, its heading, and altitude are known.

Transponder A transponder is a radar transmitter, which responds to an interrogation from a ground-based radar. Like the radar detector found in some automobiles, the transponder detects the signal from the ground-based radar and transmits back a response. Unlike the automobile radar detector, the transponder is required by law for most operations within the air traffic system and is not a device intended to circumvent the law. It is an active system, as opposed to a passive radar, which provides a strong signal on the ground radar screen. By pushing an "ident" button at the air traffic controller's request, the pilot will cause the transponder to transmit a unique code that provides positive identification to the controller.

In addition to these instruments, a bank of gauges is on the instrument panel, which provides information on fuel quantity, fuel pressure, oil pressure, oil temperature, manifold pressure, and battery charging similar to automobile gauges. In military and commercial airplanes, one will find many more specialized instruments designed to facilitate navigation, improve instrument landings, monitor systems, and detect other aircraft.

AIRPLANE PERFORMANCE

Our state of knowledge is now such that one can predict with some certainty the performance of an airplane before it is ever flown. Where analytical or numerical techniques are insufficient, sophisticated experimental facilities are utilized to investigate areas such as high-lift devices or aerothermodynamics. At this time, as an introduction, we will consider some first principles regarding the prediction of airplane performance. To begin, refer to Figure 1.10, which shows a right-handed coordinate system defining positive directions for the forces and moments acting on an airplane as well as its linear and angular velocity vectors at the CG. The coordinate system, velocity components, forces, and moments are easily memorized because of the cyclic relationships associated with the right-handed system. If you point your right thumb along the *x*-axis, your fingers will curl in a direction so as to rotate the *y*-axis into the *z*-axis. Similarly, with the thumb along the *y*-axis, the *z*-axis rotates into the *x*-axis, and with the thumb along the *z*-axis, the *x*-axis rotates into the *y*-axis. A similar scheme, using your thumb, will indicate the positive direction of rotation about the three axes for angular displacements, angular velocities, and moments. To summarize:

Axes:	$x,y,z,$
Resultant aerodynamic forces along axes:	X,Y,Z
Linear velocities along axes:	u,v,w
Angular velocities about axes:	P,Q,R
Aerodynamic moments about axes:	L,M,N
Angular displacements about axes:	ϕ,θ,ψ

One has to be careful in interpreting these, or any other symbols, since there is some duplication of definitions like, for example, *L* for lift and *L* for rolling moment. Usually, the definition is clear from the way in which the symbol is being used. The angular rates about the *x*-, *y*-, and *z*-axes are called *rolling, pitching,* and *yawing,* respectively, with similar labels for the moments about these axes. Motion in the airplane's plane of symmetry is called *longitudinal* motion. This encompasses linear motion along the *x*- and *z*-axes and rotation about the *y*-axis. Motion of the

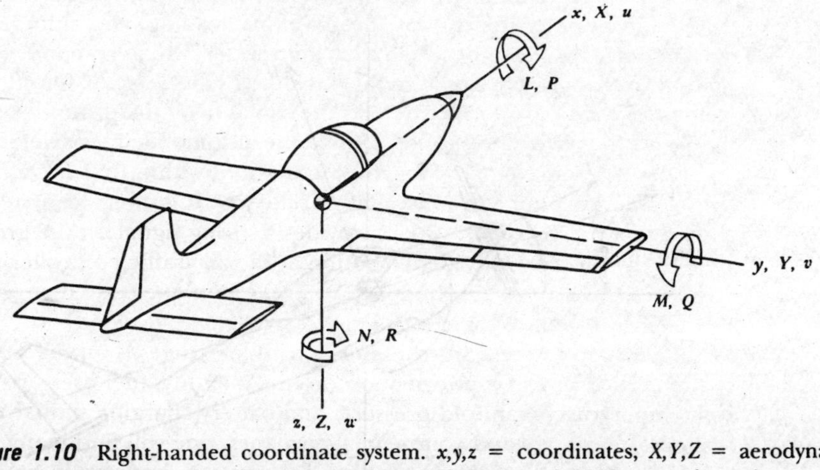

Figure 1.10 Right-handed coordinate system. x,y,z = coordinates; X,Y,Z = aerodynamic forces along axes; U,V,W = velocity components along axes; L,M,N = aerodynamic moments about axes; P,Q,R = angular velocities about axes.

plane of symmetry is called *lateral-directional* motion and encompasses linear motion along the y-axis and rotation about the x- and z-axes.

Now consider a view of the airplane in the x-z plane as shown in Figure 1.11. The airplane is in a *steady* climb, meaning that the airplane is not accelerating. The vector sum of all of the aerodynamic and gravitational forces and moments on the airplane are equal to zero; that is, the airplane is in equilibrium. The aerodynamic and gravitational forces and moments are shown in this figure at the CG. By definition, the weight acts at the CG but the aerodynamic forces generally do not. Of course, in trim, since the weight must act through the CG, the resultant aerodynamic force vectors must also act through the CG.

In Figure 1.11, V represents the velocity of the airplane's center of gravity. This vector is shown inclined upward from the horizontal through the angle of climb, θ_c. The angle between the horizontal and the thrust line is denoted as θ. If this line is taken as the reference line for the airplane, then the airplane is said to be pitched at the angle θ. The angle of attack, measured relative to the thrust line, is the angle between the velocity vector and the reference line given by

$$\alpha = \theta - \theta_c \tag{1.11}$$

The thrust, T, is the propelling force that balances mainly the aerodynamic drag on the airplane. T can be produced by a propeller, a turbojet, or a rocket engine. The total lift on the airplane is the sum of the lifts on the various components such as the wing, the tail, fuselage, nacelles, and propellers. In level flight, the lift is mainly the vertical force upward on the wing. However, most airplanes have a nose-down aerodynamic moment about the CG, which must be balanced by a download on the horizontal tail. The reason for this will become clear later in Chapter 9. It follows therefore that the lift on the wing for trimmed flight must be in excess of the airplane's weight in order to compensate for the tail download. In landing, when flaps are lowered, the nose-down pitching moment can increase appreciably making it mandatory that the tail download be considered when calculating the landing speed.

Similar to the lift, the drag, D, is defined as the component of all aerodynamic forces generated by the airplane in the direction opposite to the velocity vector, V.

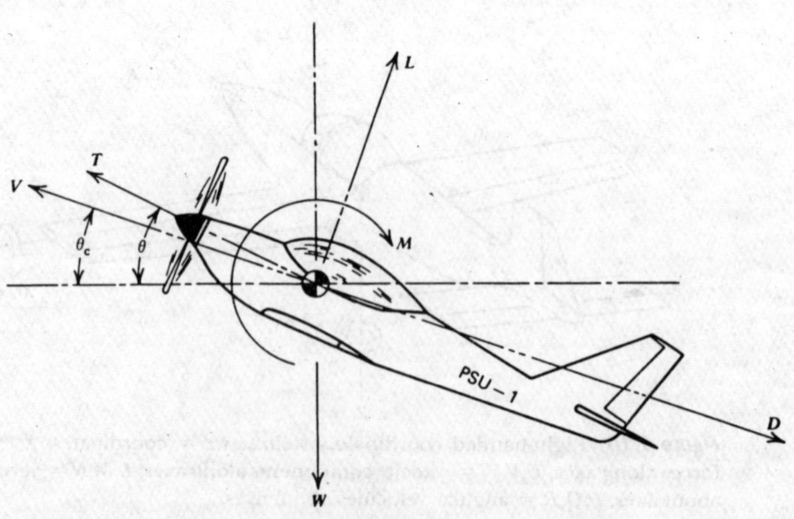

Figure 1.11 Forces and moments on an airplane in a steady climb.

This force is composed of two principal parts; the parasite drag and the induced drag. The induced drag is generated as a result of producing lift; the parasite drag is the drag of the fuselage, landing gear, struts, and other surfaces exposed to the air. There is a fine point concerning the drag of the wing to be mentioned here that will be elaborated on later. Part of the wing drag contributes to the parasite drag and is sometimes referred to as profile drag. The profile drag is closely equal to the drag of the wing at zero lift; however, it does increase with increasing lift. This increase is therefore usually included as part of the induced drag. In a strict sense this is incorrect, as will become clearer later on.

W is the gross weight of the airplane and, by definition, acts at the center of gravity of the airplane and is directed vertically downward. It is composed of the empty weight of the airplane and its useful load. This latter weight includes the payload (passengers and cargo) and the fuel weight.

The pitching moment, M, is defined as positive in the nose-up direction (clockwise in Figure 1.11) and results from the distribution of aerodynamic forces on the wing, tail, fuselage, engine nacelles, and other surfaces exposed to the flow. Obviously, if the airplane is in trim, the sum of these moments about the center of gravity must be zero.

We know today that the aerodynamic forces on an airplane are the same whether we move the airplane through still air or fix the airplane and move the air past it. In other words, it is the relative motion between the air and airplane and not the absolute motion of either that determines the aerodynamic forces. This statement was not always so obvious. When he learned of the Wright Brothers' wind tunnel tests, Octave Chanute wrote to them on October 12, 1901 (Ref. 1.1) and referred to "natural wind." Chanute conjectured in his letter:

> *It seems to me that there may be a difference in the result whether the air is impinged upon by a moving body or whether the wind impinges upon the same body at rest. In the latter case each molecule, being driven from behind, tends to transfer more of its energy to the body than in the former case when the body meets each molecule successively before it has time to react on its neighbors.*

Fortunately, Wilbur and Orville Wright chose to believe their own wind tunnel results.

Returning to Figure 1.11, we may equate the vector sum of all forces to zero, since the airplane is in equilibrium. Hence, in the direction of flight,

$$T\cos(\theta - \theta_c) - D - W\sin\theta_c = 0 \tag{1.12}$$

Normal to this direction,

$$W\cos\theta_c - L - T\sin(\theta - \theta_c) = 0 \tag{1.13}$$

These equations can be solved for the angle of climb to give

$$\theta_c = \tan^{-1}\frac{T\cos(\theta - \theta_c) - D}{L + T\sin(\theta - \theta_c)} \tag{1.14}$$

In this form, θ_c appears on both sides of the equation. However, let us assume *a priori* that θ_c and $(\theta - \theta_c)$ are small angles. Also, except for very high performance and V/STOL (vertical or short takeoff and landing) airplanes, the thrust for most airplanes is only a fraction of the weight. Thus, Equation 1.13 becomes

$$\theta_c \simeq \frac{T - D}{W} \tag{1.15}$$

For airplanes propelled by turbojets or rockets, Equation 1.14 is in the form that one would normally use for calculating the angle of climb. However, in the case of airplanes with shaft engines, this equation is modified so that we can deal with power instead of thrust.

First, consider a thrusting propeller that moves a distance S in time t at a constant velocity, V. The work that the propeller performs during this time is, obviously,

$$\text{work} = TS$$

Power is the rate at which work is performed; hence,

$$\text{power} = T\frac{S}{t}$$

But S/t is equal to the velocity at which the airplane is traveling. Hence, the power available from the propeller, P_A, is given by

$$P_A = TV \tag{1.16}$$

If P_E is the power delivered by the engine to the propeller shaft, P_A and P_E are related by

$$P_A = \eta P_E \tag{1.17}$$

where η is the propeller efficiency. As will be seen later in Chapter 6, propellers are efficient devices for converting from engine power (BHP in English units) to thrust power (THP). In cruise, η can be from 85 to 90% for a well-designed propeller. Similar to the derivation of Equation 1.14, we can obtain the power required to overcome the drag of the airplane as

$$P_R = DV \tag{1.18}$$

Thus, returning to Equation 1.13, by multiplying through by WV, we get

$$W(V\theta_c) = \eta P_E - P_R \tag{1.19}$$

The quantity $V\theta_c$ is the rate of climb, R/C. The difference between the power that is required and that available, $P_A - P_R$, is referred to as the *excess power*. Thus, Equation 1.17 shows that the power expended in raising the weight of the airplane at the rate of climb is equal to the excess power. In operating an airplane, this means the following. A pilot is flying at a given speed with the engine throttle only partially open. If the pilot advances the throttle, he or she can choose to accelerate or climb. If the pilot pulls back on the control column as the throttle is advanced so as to keep the airspeed constant, and hence the required power constant, the airplane will climb according to Equation 1.8. If, on the other hand, the pilot maintains a constant altitude, the airplane will accelerate because of the increased thrust, attaining a higher speed where the new required power equals the increased power available from the propeller and engine. When a wide open throttle (WOT) condition is reached, the maximum power available is equal to the power required. This is the condition for maximum airspeed, "straight and level."

In addition to performance, the area of *flying qualities* is very important to the acceptance of an airplane by the customer. Flying qualities refers primarily to stability and control including *maneuverability* and *agility*. Maneuverability relates to a steady operating state of an airplane such as a turn or a pull-up. For example, an airplane is more maneuverable the shorter its turning radius or the more quickly it can complete a loop. Agility is a modern term, which relates to how quickly an airplane can transition from one steady maneuver to another. Agility is obviously important for a military airplane, which might engage in air-to-air combat.

Chapters 9, 10, and 11 will treat the matter of stability and control in some depth; for introductory purposes, let us consider only briefly the *longitudinal static stability* of an airplane. The aerodynamic pitching moment, M, about the CG of an airplane at a fixed speed and altitude depends only on the angle of attack, α. This moment, shown in Figure 1.11, will vary qualitatively with α as shown in Figure

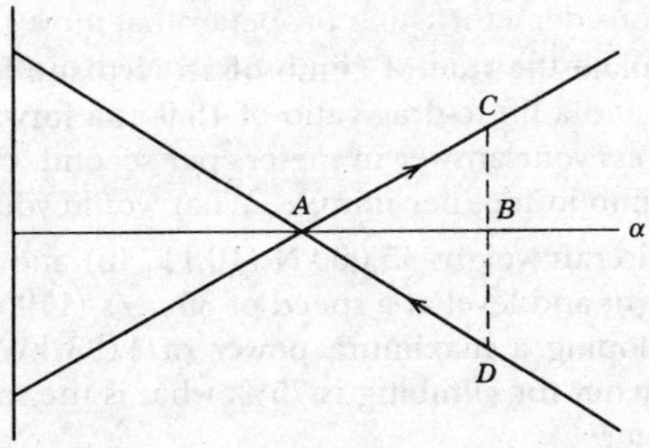

Figure 1.12 Variation of pitching moment with angle of attack for stable and unstable airplanes.

1.12. Here, we have chosen to measure α relative to the zero lift line of the airplane. Remember, this is a line such that the lift on the total airplane is zero if this line is at a zero angle of attack. Referring to this figure, consider the point A, which represents the point at which the airplane is trimmed. Since the lift must equal the weight of the airplane in trimmed flight, obviously the angle of attack must be positive at this point. Now suppose a disturbance occurs, such as an atmospheric gust, to cause the angle of attack to increase suddenly to point B. With the concomitant increase in lift, the airplane will begin to depart from its flight path. If the pitching moment, M, increases with α so that the moment is now positive as shown by point C, this positive moment will tend to increase α even more, causing a further departure from the flight path. Obviously, this is an unstable situation. Conversely, if the increase in α results in a negative M to point D, this nose-down pitching moment will tend to decrease α returning the airplane to its original trimmed state. Thus, the requirement for longitudinal static stability is that the rate of change of pitching moment with angle of attack be negative.

$$\frac{dM}{d\alpha} < 0 \tag{1.20}$$

This is about as far as we can go without considering in detail the generation of aerodynamic forces and moments on an airplane and its components. The preceding discussion has shown the importance of being able to predict these quantities from both performance and flying qualities viewpoints. The following chapters will present detailed analytical, numerical, and experimental material sufficient to determine the performance and stability and control characteristics of an airplane. The material will use both the SI and English system of units. Students should become familiar with both since the metric system is used almost exclusively outside of the United States, whereas the English system is still used extensively by engineers within this country. "Becoming familiar" means more than simply knowing the conversion factors from one system of units to the other. One should develop a feeling for orders of magnitude in both systems. For example, what is a high or low pressure in pascals or in pounds per square foot?

As you study the following material, keep in mind that it took the early aviation pioneers a lifetime to accumulate only a fraction of the knowledge that is yours to gain with a few months of study. Also, to the student using this text, remember—*You have not really learned the material until you can duplicate the derivations in the book and apply them to solving a practical problem.*

PROBLEMS

1.1 Calculate the rate of climb of an airplane having a thrust-to-weight ratio of 0.25 and a lift-to-drag ratio of 15.0 at a forward velocity of 70 m/s (230 fps). Express your answer in meters per second. Current practice is to express rate of climb in feet per minute. What would your answer be in these units?

1.2 An aircraft weighs 45,000 N (10,117 lb) and requires 597 kW (800 thp) to fly straight and level at a speed of 80 m/s (179 mph). If the engine is capable of developing a maximum power of 1193 kW (1600 bhp) and the propeller efficiency for climbing is 75%, what is the rate of climb for maximum power in fpm?

1.3 An airplane has a mass moment of inertia about its y-axis through the CG of 1300 slug-ft^2. The lift of the horizontal tail acts at a distance of 4.5 m behind the airplane's CG. The tail has a planform area of 3 sq. m., and the slope of its lift coefficient curve, $dC_L/d\alpha$, is equal to 0.08 per degree. The airplane is trimmed at an altitude of 6000 ft and a true airspeed of 110 kts when the pilot suddenly pulls back on the wheel causing the tail to nose down 5°. What will be the instantaneous pitching acceleration in rad/s/s?

1.4 An airplane has a lift-to-drag ratio of 15. It is at an altitude of 1500 (4921 ft) when the engine fails. An airport is 16 km (9.94 mi) ahead. Will the pilot be able to glide far enough to reach it?

1.5 The lift curve slope for a rectangular wing is given by

$$\frac{dC_L}{d\alpha} = 0.1 \frac{A}{A + 2}$$

in C_L per degree, and the lift acts at a quarter of the chord back from the leading edge. Assume, for the Wright Brothers Flyer shown in Figure 1.2, that there is no interaction of the two wings with each other or with the canard tail and use the figure to estimate the wing and canard geometry.

(a) Calculate how far back the CG could have been before the airplane became statically unstable. Do you think the airplane was statically stable?

(b) Orville Wright weighed approximately 145 lb and 5 lb of gasoline was used for his short flight. The empty weight of the airplane was 450 lb. His last flight covered 260 m in 59 s. What would the lift coefficient have been for this flight for one of the wings, assuming the lift to be distributed equally between the wings?

1.6 A rectangular wing weighs 5.0 lb. It has a span of 12 in. and a chord of 2.5 in. The wing is attached to one end of a slender rod 10 ft long, which is free to pivot at the other end as shown in the figure. The apparatus is placed in a wind tunnel at the University of Denver (the "mile-high" city). The C_L curve for the wing corresponds to the curve labeled "HIGH ASPECT RATIO-SYMMETRICAL" in Figure 1.8. What must the airspeed be in the test section for the wing to operate at its maximum lift coefficient?

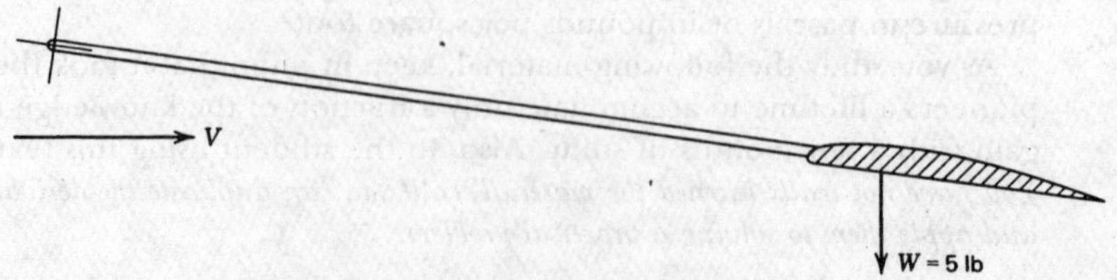

1.7 A drag coefficient, C_D, can be defined similar to the lift coefficient with the drag force replacing the lift and the planform area replaced by a reference area. A flat plate normal to the flow has a C_D of approximately 1.0 based on the projected area of the plate. In a hurricane-force wind, derive the relationship between weight, size, and wind speed at standard sea level (SSL) conditions, which could lead to the overturning of a mobile home.

1.8 A wing with a symmetrical airfoil develops a lift of 10,000 N at an altitude of 10 km at an angle of attack of 10° and a certain speed. How much lift will a geometrically similar wing having half the area develop at an angle of attack of 5° at standard sea level conditions at twice the speed?

1.9 A wing has a leading edge sweep of 35°. The midspan chord equals 10 ft and the taper ratio is 0.5. If the aspect ratio equals 8.0, what is the value of the wing span?

1.10 An airplane weighing 10,000 lb is climbing at a rate of 1500 fpm at a true airspeed of 200 kts at an altitude of 10,000 ft. How much power is being expended in order to climb?

REFERENCES

1.1 McFarland, Marvin W., editor, *The Papers of Wilbur and Orville Wright, Including the Chaunute-Wright Letters,* McGraw-Hill, New York, 1953.

1.2 Harris, Sherwood, *The First to Fly, Aviation's Pioneer Days,* Simon and Schuster, New York, 1970.

1.3 Robertson, James M., *Hydrodynamics in Theory and Application,* Prentice-Hall, Englewood Cliffs, NJ, 1965.

1.4 Mechtly, E. A., *The International System of Units, Physical Constants and Conversion Factors,* NASA SP-7012, U.S. Government Printing Office, Washington, D.C., 1969.

1.5 Cleveland, F. A., "Size Effects in Conventional Aircraft Design," *AIAA J. of Aircraft, 7(6),* November–December 1970 (33d Wright Brothers Lecture).

1.6 Anonymous, *Jane's All the World's Aircraft,* Pilot Press Ltd., published annually.

1.7 van Dam, C. P., Vijgen, P. M. H. W., Holmes, B. J., "Aerodynamic characteristics of Crescent and Elliptic Wings at High Angles of Attack," *AIAA J. of Aircraft, 28(4),* April 1991.

2 FLUID MECHANICS

This chapter will stress the principles in fluid mechanics that are especially important to the study of aerodynamics. For the reader whose preparation does not include fluid mechanics, the material in this chapter should be sufficient to understand the developments in succeeding chapters. For a more complete treatment, see any of the many available texts on fluid mechanics (e.g., Refs. 2.1 and 2.2).

Unlike solid mechanics, one normally deals with a continuous medium in the study of fluid mechanics. An airplane in flight does not experience a sudden change in the properties of the air surrounding it. The stream of water from a firehose exerts a steady force on the side of a burning building, unlike the impulse on a swinging bat as it connects with the discrete mass of the baseball.

In solid mechanics, one is concerned with the behavior of a given, finite system of solid masses under the influence of force and moment vectors acting on the system. In fluid mechanics, one generally deals not with a finite system, but with the flow of a continuous fluid mass under the influence of distributed pressures and shear stresses.

The term *fluid* should not be confused with the term *liquid*, since the former includes not only the latter, but gases as well. Generally, a fluid is defined as any substance that will readily deform under the influence of shearing forces. Thus, a fluid is the antonym of a solid. Since both liquids and gases satisfy this definition, they are both known as fluids. A liquid is distinguished from a gas by the fact that the former is nearly incompressible. Unlike a gas, the volume of a given mass of liquid remains nearly constant, independent of the pressure imposed on the mass.

FLUID STATICS AND THE ATMOSPHERE

Before treating the more difficult case of a fluid in motion, let us consider a fluid at rest in static equilibrium. The mass per unit volume of a fluid is defined as the mass density, usually denoted by ρ. The mass density is a constant for liquids, but it is a function of temperature, T, and pressure, p, for gases. Indeed, for a gas, p, ρ, and T are related by the equation of state

$$p = \rho RT \tag{2.1}$$

R is referred to as the universal gas constant. Its value can be found in Appendix A1 for both the English and SI systems. T, in Equation 2.1, is the absolute temperature measured in degrees Kelvin in the SI system and degrees Rankine in the English system.

A vertical cylinder of fluid is shown in Figure 2.1 having a unit cross-sectional area and a differential height, dh. The weight of the fluid contained within the cylinder is given by $\rho g dh$ and is shown in the figure as a gravitational force acting downward. A normal pressure, p, is shown acting upward on the bottom of the cylinder and a pressure $p + (dp/dh)\,dh$ acting downward on the top where dp/dh is the gradient of the static pressure with increasing height. Since the liquid element is at rest, all of the forces acting on it must be in equilibrium. Thus, summing forces in the vertical direction results in an expression for the pressure gradient.

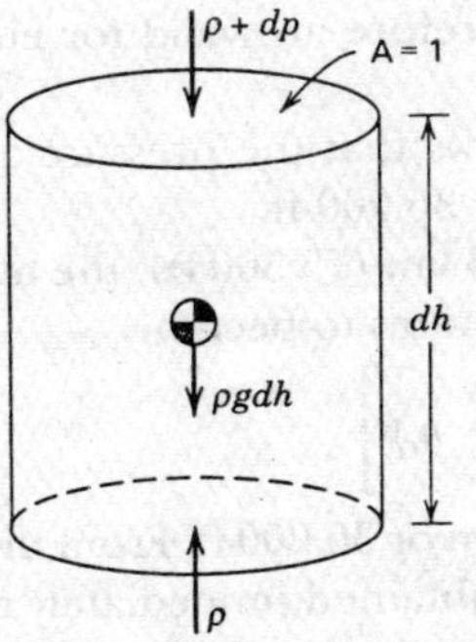

Figure 2.1 Static forces acting on a differential cylinder of air in the atmosphere.

$$\frac{dp}{dh} = -\rho g \tag{2.2}$$

As an example in the use of this equation, consider the calculation of the static pressure at some point on the bottom of the ocean floor where the depth is equal to *d*. On the surface of the water, the atmospheric pressure is equal to 2116 psf, the standard atmospheric pressure at sea level. From Equation 2.2, the pressure at any depth, *d*, is found from

$$p - \int_0^d \rho g dh = 2116$$

In water ρ is a constant and equal approximately to 2 slugs/ft^3 for salt water. It therefore follows from the above equation, for example, that the pressure 50 ft below the surface of the ocean is equal to 5336 psf.

The temperature in the atmosphere decreases at a nearly constant rate, known as the *lapse* rate, up to an altitude of approximately 11 km (36,000 ft). This region of the atmosphere is known as the *troposphere.* Above this altitude, in the lower part of the stratosphere, the temperature remains nearly constant up to an altitude of approximately 23 km (75,500 ft). If this experimentally observed behavior for the temperature is utilized together with Equations 2.1 and 2.2, a set of equations can be formulated that model closely the standard atmosphere as tabulated in Appendix A.2. Up to 11 km, the pressure and temperature are related by

$$\frac{dp}{p} = -\frac{dT}{T}\frac{1}{R(dT/dh)} \tag{2.3}$$

The standard lapse rate is 6.51° Kelvin per kilometer or 3.57° Rankine per 1000 ft. Thus, the following equation, which relates the atmospheric temperature and static pressure, is easily obtained for a constant lapse rate.

$$\frac{dp}{p} = -\frac{g}{R(dT/dh)}\frac{dT}{T} \tag{2.4}$$

Integrating the above from sea level to *h* gives the pressure as a function of temperature.

$$\delta = \theta^{5.256} \qquad h < 11 \text{ km} (36{,}000 \text{ ft}) \tag{2.5}$$

δ is the ratio of the pressure of any altitude to the pressure at sea level and θ is the corresponding ratio for the absolute temperature.

The density ratio, σ, can be obtained immediately from the equation of state and Equation 2.5.

$$\sigma = \theta^{4.256} \qquad h < 11 \text{ km} (36{,}000 \text{ ft}) \tag{2.6}$$

Equations 2.4 and 2.5 are dimensionless and therefore are valid for either the SI or English system of units.

Since θ is a function of the altitude, h, it follows that the pressure and density can also be found for any altitude up to 11 km or 36,000 ft.

Above 11 km (36,000 ft) up to approximately 23 km (75,500 ft), the temperature is nearly constant. In this case, Equation 2.2 integrates to become

$$\delta = \delta_c \exp\left[-\frac{g}{RT_c}(h - h_c)\right] \tag{2.7}$$

In this equation, a sub c denotes conditions at 11 km or 36,000 ft. From the equation of state and Equation 2.7, the density ratio, σ, is obtained immediately as

$$\sigma = \frac{\delta}{\theta_c} \tag{2.8}$$

The relationship between θ and h as well as the needed values of p, ρ, and T to model the atmosphere are tabulated in Table 2.1 for convenience for both the SI and English systems of units.

Figure 2.2 presents static pressure, mass density, and absolute temperature as a ratio to their sea level values for altitudes up to 24.4 km (80,000 ft). This graph was prepared using the preceding relationships and is not a plot of the tabulated values given in Appendix A.2. It agrees closely with the appendix so that any computer program using the relationships for the atmosphere can be used with confidence. Figure 2.3 presents the kinematic viscosity as a ratio to the sea level value as calculated from the empirical equation given in the following computer exercise.

Computer Exercise 2.1 "ATMOS"

Formulate a program that will return the properties of the atmosphere using the above relationships. The program should allow for input in either the SI or English system and return the atmospheric state variables in either system. This program will be used in later numerical calculations as a subroutine. For the sake of completeness the following empirical fit to the kinematic viscosity can also be included in the program.

$$\nu \times 10^4 = A0 + A1(h/1000) + A2(h/1000)^2 + \cdots + A7(h/1000)^7$$

Table 2.1 Approximate Relationships for the Standard Atmosphere

	SI System	English System
sea level p	101,300 N/m^2	2116 psf
sea level ρ	1.225 kg/m^3	0.002377 slugs/ft^3
sea level T	288.16 °K	518.7 °R
sea level a	340.3 m/s	1116 fps
gas constant, R	286.97 m^2/s^2/°K	1716 ft^2/s^2/°R
δ_c	0.225	0.225
θ_c	0.752	0.752

For Altitude Less than 11 km or 36,000 ft

$\theta = 1 - .0226(h/1000)$ (h in meters)

$\theta = 1 - .00688(h/1000)$ (h in feet)

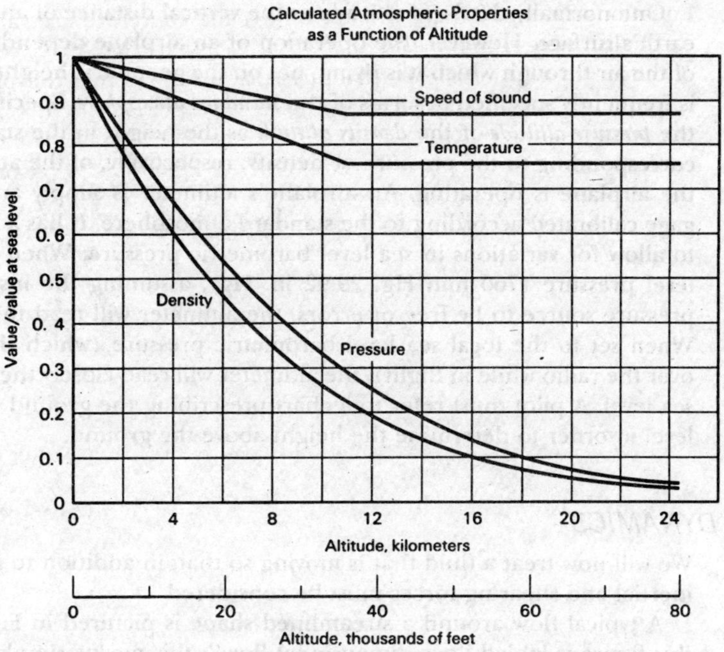

Figure 2.2 Calculated atmospheric properties as a function of altitude.

In the English system, υ has the units of ft^2/s with the coefficients being

A0 = 1.5723 A1 = 8.73065E-2 A2 = − 1.18412E-2 A3 = 1.16978E-3
A4 = − 5.27207E-5 A5 = 1.22466E-6 A6 = − 1.369780E-8 A7 = 5.94238E-11

In the SI system, the kinematic viscosity is simply multiplied by the square of 0.3048 with the units of υ becoming m^2/s. If the number of significant digits shown are used, the results are accurate to within 0.1% at an altitude of 70,000 ft.

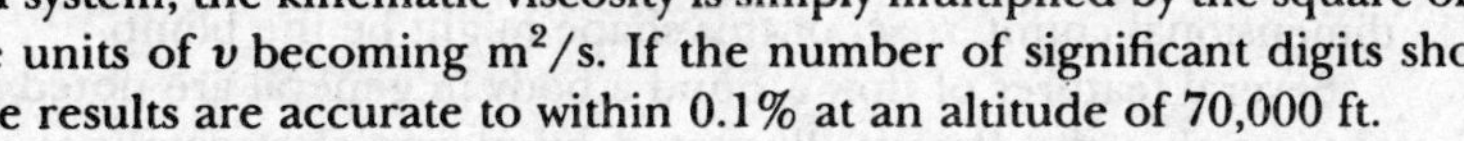

Figure 2.3 Kinematic viscosity as a function of altitude.

One normally thinks of altitude as the vertical distance of an airplane above the earth's surface. However, the operation of an airplane depends on the properties of the air through which it is flying, not on the geometric height. Thus, the altitude is frequently specified in terms of the *standard atmosphere.* Specifically, one refers to the *pressure altitude* or the *density altitude* as the height in the standard atmosphere corresponding to the pressure or density, respectively, of the atmosphere in which the airplane is operating. An airplane's altimeter is simply an absolute pressure gage calibrated according to the standard atmosphere. It has a manual adjustment to allow for variations in sea level barometric pressure. When set to standard sea level pressure (760 mm Hg, 29.92 in. Hg), assuming the instrument and static pressure source to be free of errors, the altimeter will read the pressure altitude. When set to the local sea level barometric pressure (which the pilot can obtain over the radio while in flight), the altimeter will read closely the true altitude above sea level. A pilot must refer to a chart prescribing the ground elevation above sea level in order to determine the height above the ground.

FLUID DYNAMICS

We will now treat a fluid that is moving so that, in addition to gravitational forces, inertial and shearing forces must be considered.

A typical flow around a streamlined shape is pictured in Figure 2.4. Note that this figure is labled "two-dimensional flow"; this means simply that the flow field is a function only of two coordinates (x and y, in the case of Figure 2.4) and does not depend on the third coordinate. For example, the flow of wind around a tall, cylindrical smokestack is essentially two-dimensional except near the top. Here, the wind goes over as well as around the stack, and the flow is three-dimensional. As another example, Figure 2.4. might represent the flow around a long, streamlined strut such as the one that supports the wing of a high-wing airplane. The three-dimensional counterpart of this shape might be the blimp.

Several features of flow around a body in general are noted in Figure 2.4. First, observe that the flow is illustrated by means of streamlines. A streamline is an imaginary line characterizing the flow such that, at every point along the line, the velocity vector is tangent to the line. Thus, in two-dimensional flow, if $y(x)$ defines the position of a streamline, $y(x)$ is related to the x and y components of the velocity, $u(x)$ and $v(x)$, by

$$\frac{dy}{dx} = \frac{v(x)}{u(x)} \tag{2.9}$$

Note that the body surface itself is a streamline.

In three-dimensional flow a surface swept by streamlines is known as a stream surface. If such a surface is closed, it is known as a *stream tube.*

The mass flow accelerates around the body as the result of a continuous distribution of pressure exerted on the fluid by the body. An equal and opposite reaction must occur on the body. This static pressure distribution, acting everywhere normal to the body's surface, is pictured on the lower half of the body in Figure 2.4. The small arrows represent the local static pressure, p, relative to the static pressure, p_0, in the fluid far removed from the body. Near the nose, p is greater than p_0; further aft the pressure becomes negative relative to p_0. If this static pressure distribution, acting normal to the surface, is known, forces on the body can be determined by integrating this pressure over its surface.

In addition to the local static pressure, shearing stresses resulting from the fluid's

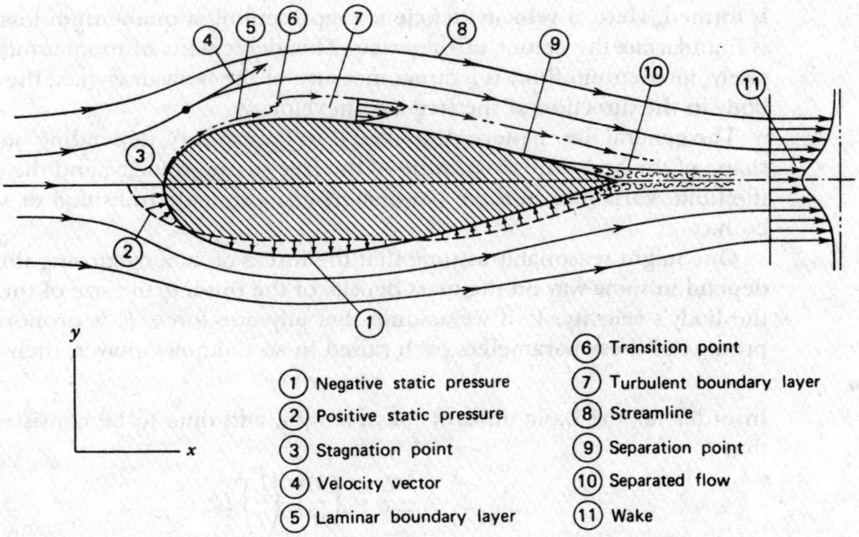

Figure 2.4 Two-dimensional flow around a streamlined shape.

viscosity also give rise to body forces. As fluid passes over a solid surface, the fluid particles immediately in contact with the surface are brought to rest. Moving away from the surface, successive layers of fluid are slowed by the shearing stresses produced by the inner layers. (The term "layers" is used only as a convenience in describing the fluid behavior. The fluid shears in a continuous manner and not in discrete layers.) The result is a thin layer of slower moving fluid, known as the *boundary layer*, adjacent to the surface. Near the front of the body, this layer is very thin, and the flow within it is smooth without any random or turbulent fluctuations. Here, the fluid particles might be described as moving along in the layer on parallel planes, or laminae; hence, the flow is referred to as *laminar.*

At some distance back from the nose of the body, disturbances to the flow (e.g., from surface roughnesses) are no longer damped out. These disturbances suddenly amplify, and the laminar boundary layer undergoes transition to a turbulent boundary layer. This layer is considerably thicker than the laminar one and is characterized by a mean velocity profile on which small, randomly fluctuating velocity components are superimposed. These flow regions are shown in Figure 2.4. The boundary layers are pictured considerably thicker than they actually are for purposes of illustration. For example, on the wing of an airplane flying at 100 m/s at low altitude, the turbulent boundary 1.0 m back from the leading edge would be only approximately 1.6 cm thick. If the layer were still laminar at this point, its thickness would be approximately 0.2 cm.

Returning to Figure 2.4, the turbulent boundary layer continues to thicken toward the rear of the body. Over this portion of the surface the fluid is moving into a region of increasing static pressure that is tending to oppose the flow. The slower moving fluid in the boundary layer may be unable to overcome this adverse pressure gradient, so that at some point the flow actually separates from the body surface. Downstream of this separation point, reverse flow will be found along the surface with the static pressure nearly constant and equal to that at the point of separation.

At some distance downstream of the body, the separated flow closes and a wake is formed. Here, a velocity deficiency representing a momentum loss by the fluid is found near the center of the wake. This decrement of momentum (more precisely, momentum flux) is a direct measure of the body drag (i.e., the force on the body in the direction of the free-stream velocity).

The general flow pattern described thus far can vary, depending on the size and shape of the body, the magnitude of the free-stream velocity, and the properties of the fluid. Variations in these parameters can eliminate transition or separation or both.

One might reasonably assume that the forces on a body moving through a fluid depend in some way on the mass density of the fluid, ρ, the size of the body, l, and the body's velocity, V. If we assume that any one force, F, is proportional to the product of these parameters each raised to an unknown power, then

$$F \propto \rho^a V^b l^c \tag{2.10}$$

In order for the basic units of mass, length, and time to be consistent, it follows that

$$\frac{ML}{T^2} = \left(\frac{M}{L^3}\right)^a \left(\frac{L}{T}\right)^b L^c \tag{2.11}$$

Considering M, L, and T in order leads to three equations for the unknown exponents of a, b, and c from which it is found that $a = 1$, $b = 2$, and $c = 2$. Hence,

$$F \propto \rho V^2 l^2 \tag{2.12}$$

For a particular force, the constant of proportionality in Equation 2.12 is referred to as a *coefficient* and is modified by the name of the force, for example, the *lift coefficient*. Thus, the lift and drag forces, L and D, can be expressed as

$$L = \frac{1}{2}\rho\, V^2 S\, C_L \tag{2.13a}$$

$$D = \frac{1}{2}\rho\, V^2 S\, C_D \tag{2.13b}$$

Note that the square of the characteristic length, l^2, has been replaced by a reference area, S. Also, a factor of 1/2 has been introduced. This can be done, since the lift and drag coefficients, C_L and C_D, are arbitrary at this point. The quantity of $\rho V^2/2$ is referred to as the *dynamic pressure*, the significance of which will be made clear shortly.

For many applications, the coefficients C_L and C_D remain constant for a given geometric shape over a wide range of operating conditions or body size. For example, a two-dimensional airfoil at a 1° angle of attack will have a lift coefficient of approximately 0.1 for velocities from a few meters per second up to 100 m/s or more. In addition, C_L will be almost independent of the size of the airfoil. However, a more rigorous application of dimensional analysis [see Buckingham's π theorem (Ref. 2.1)] will result in the constant of proportionality in Equation 2.12 possibly being dependent on a number of dimensionless parameters. Two of the most important of these are known as the *Reynolds number*, R, and the *Mach number*, M, defined by

$$R = \frac{Vl\rho}{\mu} \tag{2.14a}$$

$$M = \frac{V}{a} \tag{2.14b}$$

where l is a characteristic length, V is the free-stream velocity, μ is the coefficient of viscosity, and a is the velocity of sound. The velocity of sound is the speed at

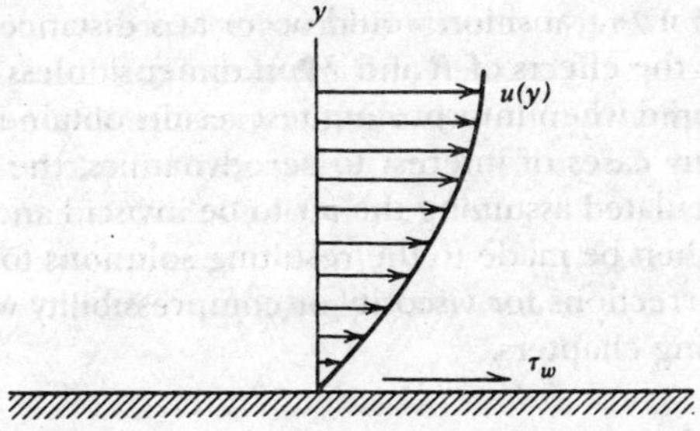

Figure 2.5 Viscous flow adjacent to a body shape.

which a small pressure disturbance is propagated through the fluid; at this point, it requires no further explanation. The coefficient of viscosity, however, is not as well known and will be elaborated on by reference to Figure 2.5. Here, the velocity profile is pictured in the boundary layer of a laminar, viscous flow over a surface. The viscous shearing produces a shearing stress of τ_w on the wall. This force per unit area is related to the gradient of the velocity $u(y)$ at the wall by

$$\tau_w = \mu \left(\frac{du}{dy}\right)_{y=0} \tag{2.15}$$

Actually, Equation 2.15 is applicable to calculating the shear stresses between fluid elements and is not restricted simply to the wall. Generally, the viscous shearing stress in the fluid in any plane parallel to the flow and away from the wall is given by the product of μ and the velocity gradient normal to the direction of flow.

The kinematic viscosity, ν, is defined as the ratio of μ to ρ.

$$\nu = \frac{\mu}{\rho}$$

ν is defined as a matter of convenience, since it is the ratio of μ to ρ that governs the Reynolds number. The kinematic viscosity for the standard atmosphere is included in Figure 2.3 as a fraction of the standard sea level value.

A physical significance can be given to the Reynolds number by multiplying numerator and denominator by V and dividing by l.

$$R = \frac{\rho V^2}{(\mu V/l)}$$

In the following material (see Eq. 2.28) the normal pressure will be shown to be proportional to ρV^2 whereas, from Equation 2.15, $\mu V/l$ is proportional to the shearing stress. Hence for a given flow the Reynolds number is proportional to the ratio of normal pressures (inertia forces) to viscous shearing stresses. Thus, relatively speaking, a flow is less viscous than another flow if its Reynolds number is higher than that of the second flow.

The Mach number determines to what extent fluid compressibility can be neglected (i.e., the variation of mass density with pressure). Current jet transports, for example, can cruise at Mach numbers up to approximately 0.8 before significant compressibility effects are encountered.

At lower Mach numbers, two flows are geometrically and dynamically similar if the Reynolds numbers are the same for both flows. Hence, for example, for a given shape, C_D for a body 10 m long at 100 m/s will be the same as C_D for a body 100 m long at 10 m/s. As another example, suppose transition occurs 2 m back from the leading edge of a flat plate aligned with a flow having a velocity of 50 m/s.

Then, a 25 m/s transition would occur at a distance of 4 m from the leading edge. Obviously, the effects of R and M on dimensionless aerodynamic coefficients must be considered when interpreting test results obtained with the use of small models.

For many cases of interest to aerodynamics, the pressure field around a shape can be calculated assuming the air to be inviscid and incompressible. Small corrections can then be made to the resulting solutions to account for these "real fluid" effects. Corrections for viscosity or compressibility will be considered as needed in the following chapters.

Conservation of Mass

Fluid passing through an area at a velocity of V has a mass flow rate equal to ρAV. This is easily seen by reference to Figure 2.6. Here, flow is pictured along a streamtube of cross-sectional area A. The fluid velocity is equal to V. At time $t = 0$, picture a small slug of fluid of length, l, about to cross a reference plane. At time l/V, this entire slug will have passed through the reference plane. The volume of the slug is Al, so that a mass of ρAl was transported across the reference plane during the time l/V. Hence, the mass rate of flow, m, is given by

$$m = \frac{\rho Al}{(l/V)}$$

$$= \rho AV \tag{2.16}$$

Along a streamtube (which may be a conduit with solid walls) the quantity ρAV must be a constant if mass is not to accumulate in the system. For incompressible flow, ρ is a constant, so that the conservation of mass leads to the continuity principle

$$AV = \text{constant}$$

AV is the volume flow rate and is sometimes referred to as the flux. Similarly, ρAV is the mass flux. The mass flux through a surface multiplied by the velocity vector at the surface is defined as the momentum flux. Generally, if the velocity vector is not normal to the surface, the mass flux will be

$$\rho \mathbf{AV} \cdot \mathbf{n}$$

with the momentum flux written as

$$(\rho \mathbf{AV} \cdot \mathbf{n})\mathbf{V}$$

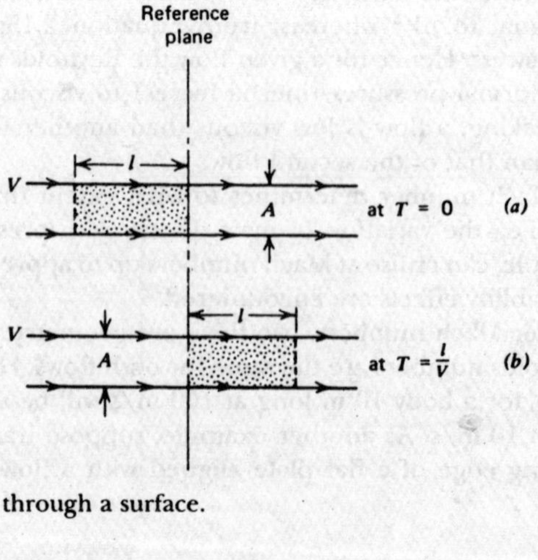

Figure 2.6 Mass flow through a surface.

Here **n** is the unit vector normal to the surface and in the direction in which the flux is defined to be positive. For example, if the surface encloses a volume and the net mass flux out of the volume is to be calculated, **n** would be directed outward from the volume, and the following integral would be evaluated over the entire surface.

$$\iint_S \rho \mathbf{V} \cdot \mathbf{n}\, d\mathbf{S}$$

Consider the conservation of mass applied to a differential control surface. For simplicity, a two-dimensional flow will be treated. A rectangular contour is shown in Figure 2.7. The flow passing through this element has velocity components of u and v in the center of the element in the x and y directions, respectively. The corresponding components on the right face of the element are found by expanding them in a Taylor series in x and y and dropping second-order and higher terms in Δx. Hence, the mass flux out through the right face will be

$$\left[\rho u + \frac{\partial(\rho u)}{\partial x}\frac{\Delta x}{2}\right]\Delta y$$

Writing similar expressions for the other three faces leads to the net mass flux out being

$$\left[\frac{\partial(\rho u)}{\partial x} + \frac{\partial(\rho v)}{\partial y}\right]\Delta x\, \Delta y$$

The net mass flux out of the differential element must equal the rate at which the mass of the fluid contained within the elements is decreasing, given by

$$-\frac{\partial}{\partial t}(\rho\, \Delta x\, \Delta y)$$

Since Δx and Δy are arbitrary, it follows that, in general,

$$\frac{\partial \rho}{\partial t} + \frac{\partial(\rho u)}{\partial x} + \frac{\partial(\rho v)}{\partial y} = 0$$

In three dimensions the preceding equation can be written in vector notation as

$$\frac{\partial \rho}{\partial t} + \nabla \cdot (\rho \mathbf{V}) = 0 \qquad (2.17)$$

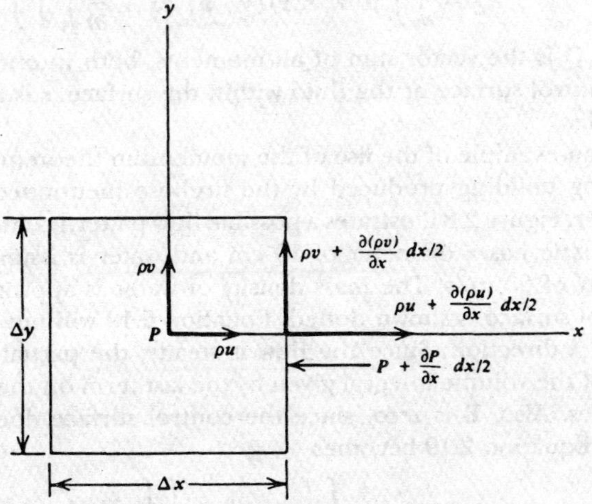

Figure 2.7 A rectangular differential control surface.

where ∇ is the vector operator, del, defined by

$$\nabla = \mathbf{i}\frac{\partial}{\partial x} + \mathbf{j}\frac{\partial}{\partial y} + \mathbf{k}\frac{\partial}{\partial z}$$

Any physically possible flow must satisfy Equation 2.17 at every point in the flow.

For an incompressible flow, the mass density is a constant, so Equation 2.17 reduces to

$$\nabla \cdot \mathbf{V} = 0 \tag{2.18}$$

The above is known as the divergence of the velocity vector, div **V**.

The Momentum Theorem

The momentum theorem in fluid mechanics is the counterpart of Newton's second law of motion in solid mechanics, which states that a force imposed on a system produces a rate of change in the momentum of the system. The theorem can be easily derived by treating the fluid as a collection of fluid particles and applying the second law. The details of the derivation can be found in several texts (e.g., Ref. 2.1) and will not be repeated here.

Defining a control surface as an imaginary closed surface through which a flow is passing, the momentum theorem states:

The sum of external forces (or moments) acting on a control surface and internal forces (or moments) acting on the fluid within the control surface produces a change in the flux of momentum (or angular momentum) through the surface and an instantaneous rate of change of momentum (or angular momentum) of the fluid particles within the control surface.

Mathematically, for linear motion of an inviscid fluid, the theorem can be expressed in vector notation by

$$-\iint_S p\mathbf{n}\, dS + \mathbf{B} = \iint_S \rho\mathbf{V}(\mathbf{V}\cdot\mathbf{n})\, dS + \frac{\partial}{\partial t}\iiint_V \rho\mathbf{V}\, d\tau \tag{2.19}$$

In Equation 2.19, **n** is the unit normal directed outward from the surface, *S*, enclosing the volume, *V*. **V** is the velocity vector, which generally depends on position and time. **B** represents the vector sum of all body forces within the control surface *acting on the fluid*. ρ is the mass density of the fluid defined as the mass per unit volume.

For the angular momentum,

$$\mathbf{Q} = \iint_S \rho(\mathbf{V}\times\mathbf{r})(\mathbf{V}\cdot\mathbf{n})\, dS + \frac{\partial}{\partial t}\iiint_V \rho(\mathbf{V}\times\mathbf{r})\, d\tau \tag{2.20}$$

Here, **Q** is the vector sum of all moments, both internal and external, acting on the control surface or the fluid within the surface. **r** is the radius vector to a fluid particle.

As an example of the use of the momentum theorem, consider the force on the burning building produced by the firehose mentioned at the beginning of this chapter. Figure 2.8 illustrates a possible flow pattern, admittedly simplified. Suppose the nozzle has a diameter of 10 cm and water is issuing from the nozzle with a velocity of 60 m/s. The mass density of water is approximately 1000 kg/m^3. The control surface is shown dotted. Equation 2.19 will now be written for this system in the *x* direction. Since the flow is steady, the partial derivative with respect to time of the volume integral given by the last term on the right side of the equation vanishes. Also, **B** is zero, since the control surface does not enclose any bodies. Thus, Equation 2.19 becomes

$$-\iint_S p\mathbf{n}\, dS = \iint_S \rho\mathbf{V}(\mathbf{V}\cdot\mathbf{n})\, dS$$

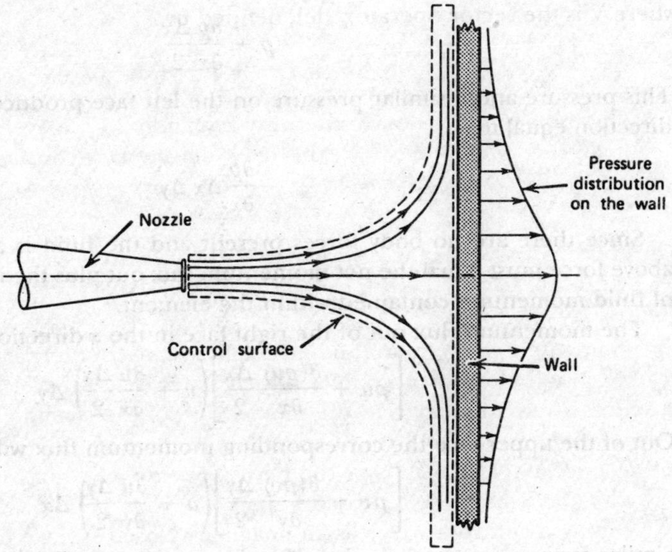

Figure 2.8 A jet of water impacting on a wall.

Measuring p relative to the atmospheric static pressure, p is zero everywhere along the control surface except at the wall. Here $\mathbf{n}$ is directed to the right so that the surface integral on the left becomes the total force exerted on the fluid by the pressure on the wall. If F represents the magnitude of the total force on the wall, then

$$-\mathbf{i}F = \int_S \int \rho \mathbf{V}(\mathbf{V} \cdot \mathbf{n}) \; dS$$

For the fluid entering the control surface on the left,

$$\mathbf{V} = 60\mathbf{i}$$
$$\mathbf{n} = -\mathbf{i}$$

For the fluid leaving the control surface, the unit normal to this cylindrical surface has no component in the x direction. Hence,

$$-\mathbf{i}F = -\int_S \int (1000)60\mathbf{i}(-60) \; dS$$

$$= -36 \times 10^5 \mathbf{i} \int_S \int dS$$

The surface integral reduces to the nozzle area of 7.85×10^{-3} m^2. Thus, without actually determining the pressure distribution on the wall, the total force on the wall is found from the momentum theorem to equal 28.3 kN.

Euler's Equation of Motion

The principle of conservation of mass applied to an elemental control surface led to Equation 2.17, which must be satisfied everywhere in the flow. Similarly, the momentum theorem applied to the same element leads to another set of equations that must hold everywhere.

Referring again to Figure 2.7, if p is the static pressure at the center of the element, then, on the center of the right face, the static pressure will be

$$p + \frac{\partial p}{\partial x}\frac{\Delta x}{2}$$

This pressure and a similar pressure on the left face produce a net force in the x direction equal to

$$-\frac{\partial p}{\partial x}\Delta x\,\Delta y$$

Since there are no body forces present and the fluid is assumed inviscid, the above force must equal the net momentum flux out plus the instantaneous change of fluid momentum contained within the element.

The momentum flux out of the right face in the x direction will be

$$\left[\rho u + \frac{\partial(\rho u)}{\partial x}\frac{\Delta x}{2}\right]\left(u + \frac{\partial u}{\partial x}\frac{\Delta x}{2}\right)\Delta y$$

Out of the upper face the corresponding momentum flux will be

$$\left[\rho v + \frac{\partial(\rho v)}{\partial y}\frac{\Delta y}{2}\right]\left(u + \frac{\partial u}{\partial y}\frac{\Delta y}{2}\right)\Delta x$$

Similar expressions can be written for the momentum flux in through the left and bottom faces.

The instantaneous change of the fluid momentum contained within the element in the x direction is simply

$$\frac{\partial}{\partial t}(\rho u\,\Delta x\,\Delta y)$$

Thus, equating the net forces in the x direction to the change in momentum and momentum flux and using Equation 2.17 leads to

$$\frac{\partial u}{\partial t} + u\frac{\partial u}{\partial x} + v\frac{\partial u}{\partial y} = -\frac{1}{\rho}\frac{\partial p}{\partial x} \tag{2.21}$$

Generalizing this to three dimensions results in a set of equations known as *Euler's equations of motion.*

$$\frac{\partial u}{\partial t} + u\frac{\partial u}{\partial x} + v\frac{\partial u}{\partial y} + w\frac{\partial u}{\partial z} = -\frac{1}{\rho}\frac{\partial p}{\partial x} \tag{2.22a}$$

$$\frac{\partial v}{\partial t} + u\frac{\partial v}{\partial x} + v\frac{\partial v}{\partial y} + w\frac{\partial v}{\partial z} = -\frac{1}{\rho}\frac{\partial p}{\partial y} \tag{2.22b}$$

$$\frac{\partial w}{\partial t} + u\frac{\partial w}{\partial x} + v\frac{\partial w}{\partial y} + w\frac{\partial w}{\partial z} = -\frac{1}{\rho}\frac{\partial p}{\partial z} \tag{2.22c}$$

Notice that if u is written as $u(x, y, z, t)$, the left side of Equation 2.22 is the total derivative of u. The operator, $\partial(\ )/\partial t$, is the local acceleration and exists only if the flow is unsteady.

In vector notation Euler's equation can be written

$$\frac{\partial \mathbf{V}}{\partial t} + (\mathbf{V}\cdot\nabla)\mathbf{V} = -\frac{1}{\rho}\nabla p \tag{2.23}$$

If the vector product of the operator ∇ is taken with each term in Equation 2.23, Equation 2.24 results.

$$\frac{\partial \boldsymbol{\omega}}{\partial t} + (\mathbf{V}\cdot\nabla)\boldsymbol{\omega} = 0 \tag{2.24}$$

$\boldsymbol{\omega}$ is the curl of the velocity vector, $\nabla \times \mathbf{V}$, and is known as the *vorticity.*

$$\nabla \times \mathbf{V} = \begin{vmatrix} \mathbf{i} & \mathbf{j} & \mathbf{k} \\ \dfrac{\partial}{\partial x} & \dfrac{\partial}{\partial y} & \dfrac{\partial}{\partial z} \\ u & v & w \end{vmatrix} \tag{2.25}$$

One can conclude from Equation 2.24 that, for an inviscid fluid, the vorticity is constant along a streamline. Since, far removed from a body, the flow is usually taken to be uniform, the vorticity at that location is zero; hence, it is zero everywhere.

Bernoulli's Equation

Bernoulli's equation is well known in fluid mechanics and relates the pressure to the velocity *along a streamline* in an inviscid, incompressible flow. It was first formulated by Euler in the middle 1700s. The derivation of this equation follows from Euler's equations using the fact that along a streamline the velocity vector is tangential to the streamline.

$$\frac{dx}{u} = \frac{dy}{v} = \frac{dz}{w} \tag{2.26}$$

First, multiply Equation 2.22*a* through by dx and then substitute Equation 2.26 for $v\,dx$ and $w\,dx$. Also, the first term of the equation will be set equal to zero; that is, at this time only steady flow will be considered.

$$u\frac{\partial u}{\partial x}dx + u\frac{\partial u}{\partial y}dy + u\frac{\partial u}{\partial z}dz = -\frac{1}{\rho}\frac{\partial p}{\partial x}dx$$

Similarly, multiply Equation 2.22*b* by dy, Equation 2.22*c* by dz, and substitute Equation 2.26 for $u\,dy$, $w\,dy$ and $u\,dz$, $v\,dz$, respectively. Adding the three equations results in perfect differentials for p and V^2, V being the magnitude of the resultant velocity along the streamline. This last term results from the fact that

$$u\frac{\partial u}{\partial x} = \frac{1}{2}\frac{\partial u^2}{\partial x}$$

and

$$V^2 = u^2 + v^2 + w^2$$

Thus, along a streamline, Euler's equations become

$$V\,dV + \frac{dp}{\rho} = 0 \tag{2.27}$$

If ρ is not a function of p (i.e., the flow is incompressible), Equation 2.27 can be integrated immediately to give

$$p + \frac{1}{2}\rho V^2 = \text{constant} \tag{2.28}$$

If the flow is uniform at infinity, Equation 2.28 becomes

$$p + \frac{1}{2}\rho V^2 = \text{constant} = p_\infty + \frac{1}{2}\rho V_\infty^{\,2} \tag{2.29}$$

Here, V is the magnitude of the local velocity and p is the local static pressure. V_∞ and p_∞ are the corresponding free-stream values. Equation 2.29 is known as *Bernoulli's equation.*

The counterpart to Equation 2.29 for compressible flow is obtained by assuming pressure and density changes to follow an isentropic process. For such a process,

$$p/\rho^\gamma = \text{constant} \tag{2.30}$$

γ is the ratio of the specific heat at constant pressure to the specific heat at constant volume and is equal approximately to 1.4 for air. Substituting Equation 2.30 into Equation 2.27 and integrating leads to an equation sometimes referred to as the *compressible Bernoulli's equation.*

$$\frac{V^2}{2} + \frac{\gamma}{\gamma - 1}\frac{p}{\rho} = \text{constant} \tag{2.31}$$

This equation can be written in terms of the acoustic velocity. First, it is necessary to derive the acoustic velocity, which can be done by the use of the momentum theorem and continuity. Figure 2.9 assumes the possibility of a stationary disturbance in a steady flow across which the pressure, density, and velocity change by small increments. In the absence of body forces and viscosity, the momentum theorem gives

$$-\,dp = (\rho + d\rho)(u + du)^2 - \rho u^2$$

But from continuity,

$$(\rho + d\rho)(u + du) = \rho u$$

or

$$u\,d\rho = -\rho du$$

Thus,

$$u^2 = \frac{dp}{d\rho} \tag{2.32}$$

If the small disturbance is stationary in the steady flow having a velocity of u, then obviously u is the velocity of the disturbance relative to the fluid. By definition, it follows that u, given by Equation 2.32, is the *acoustic velocity.*

By the use of Equation 2.30, the acoustic velocity is obtained as

$$a = \left(\frac{\gamma p}{\rho}\right)^{1/2} \tag{2.33}$$

An alternate form, using the equation of state (Equation 2.1), is

$$a = (\gamma RT)^{1/2} \tag{2.34}$$

Thus, Equation 2.31 can be written

$$\frac{V^2}{2} + \frac{a^2}{\gamma - 1} = \text{constant} \tag{2.35}$$

The acoustic velocity is also included in Figure 2.2 for the standard atmosphere.

Determination of Airspeed

A typical installation for the determination of airspeed is shown schematically in Figure 2.10. Here, a total head, or *pitot,* tube is shown mounted close to the lower surface of the wing. The use of the term "total head" is believed to be a carryover from the field of hydraulics. Henceforth, total pressure will be used to designate the pressure measured by the total head tube. The location for the total head tube shown in Figure 2.10 is favorable because the flow direction will always be nearly parallel to the tube regardless of the angle of attack. Also, the boundary layer is

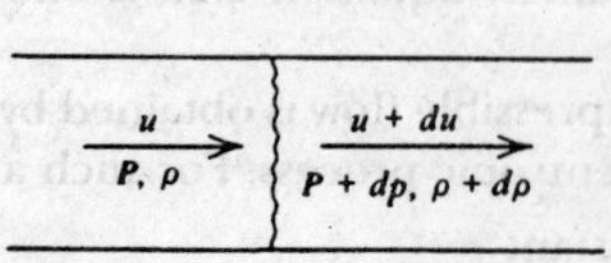

Figure 2.9 A stationary small disturbance in a steady compressible flow.

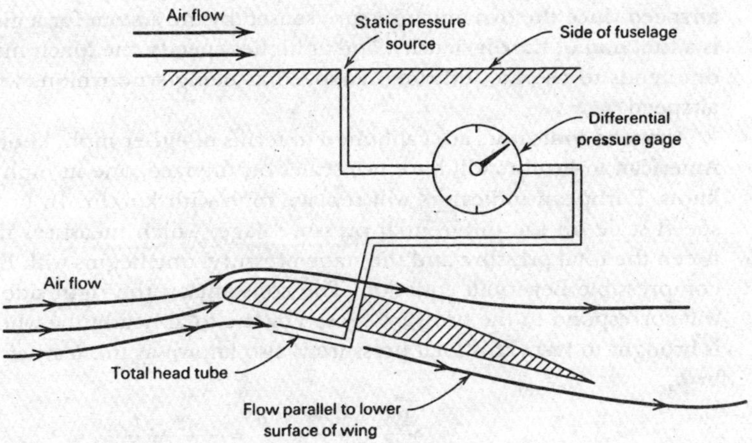

Figure 2.10 Flow parallel to lower surface of wing.

thin on the lower surface, assuring that there will be no loss in free stream total pressure at the inlet to the pitot tube. The static pressure source, in this example, is shown located on the side of the fuselage in a location where the static pressure is equal to the free stream static pressure. A satisfactory location for the static source can sometimes be difficult to find since, for most locations on an airplane, the static pressure is not equal to the free stream pressure. In addition, the static pressure at a given location on the airplane will vary with angle of attack. Sometimes the total pressure and static pressure measurements are combined in one instrument known as a *pitot-static tube.* The pitot-static tube is frequently used in wind tunnels to measure the airspeed in the test section. When used on an airplane, the tube normally extends some distance ahead of the wing's leading edge near the tip.

The total pressure and the static pressure are connected to a differential pressure gage, which indicates the difference between the two pressures. Since the total pressure is the sum of the static pressure and the dynamic pressure, the gage measures only the dynamic pressure which depends upon the velocity and air mass density. Suitably calibrated in the units of velocity, the gage then provides to the pilot a measure of the true airspeed. As to be explained, the velocity obtained from the *airspeed indicator* is called the *indicated airspeed* and must be suitably corrected to obtain the true airspeed.

When certified, an airplane must have the airspeed system calibrated to assure that the measured airspeed will result in the true airspeed when suitably corrected. Normally, two errors can exist in the indicated airspeed. The first is simply a mechanical error in the differential pressure gage, which is known as the *instrument error.* This is corrected by assuring that an accurate instrument is used. The other error is that associated with the location of the static pressure source. By choosing the proper location for the static pressure source, this error, known as the *position error,* can be made nonexistent for all but the lowest speeds where the angle of attack is high. It can, however, be a particular problem with helicopters at low speeds where the rotor wake is nearly vertical and perturbs the flow at the static pressure source.

The indicated airspeed, when corrected for instrument and position errors, becomes the *calibrated airspeed.* The calibrated airspeed is still not equal to the true airspeed since the dynamic pressure sensed by the system for a given true airspeed is a function of the air density, and at higher speeds, the Mach number. Generally, one tends to use the indicated airspeed as being synonymous with the calibrated airspeed.

Airspeed indicators are calibrated in terms of either mph, knots or km/hr. Most American indicators will have two scales on the face, one in mph and the other in knots. European indicators will replace mph with km/hr. In order to put an airspeed scale on the differential pressure gage, which measures the difference between the total pressure and the static pressure, one begins with Equation 2.31, the compressible Bernoulli equation. The constant on the right side of that equation will correspond to the pressure sensed by the total head tube when the moving air is brought to rest. This total pressure is also known as the *reservoir pressure,* denoted by p_o.

$$\frac{V^2}{2} + \frac{\gamma}{\gamma - 1}\frac{p}{\rho} = \frac{\gamma}{\gamma - 1}\frac{p_o}{\rho_o} \tag{2.36}$$

The above can be solved for the velocity as

$$V^2 = \frac{2a^2}{\gamma - 1}\left[\left(\frac{\Delta p}{p} + 1\right)^{\frac{\gamma - 1}{\gamma}} - 1\right] \tag{2.37}$$

The quantity Δp in the above is the difference between the total pressure and the static pressure as measured by the differential pressure gage, $p_o - p$. Unfortunately, this equation cannot be used to calibrate the gage in terms of airspeed since the scale will depend upon the ambient static pressure, p, and the speed of sound, a, both of which vary with altitude. Therefore, in lieu of a scale which will hold for all altitudes, the measured pressure is converted to a velocity, defined as the *calibrated airspeed,* by using the standard sea level values for these quantities.

$$V_{cal} = \sqrt{\frac{2a_{sl}^2}{\gamma - 1}\left[\left(\frac{\Delta p}{p_{sl}} + 1\right)^{\frac{\gamma - 1}{\gamma}} - 1\right]} \tag{2.39}$$

Thus an airspeed indicator shows the true airspeed only at sea level on a standard day.

During training, a pilot soon learns that the airspeed that appears on the airspeed indicator is not the true airspeed. Instead, in order to determine the true airspeed, the pilot must also read the altimeter and outside air temperature. The pilot then resorts to a small hand calculator or, in some instances, adjusts the dial on the airspeed indicator accordingly to allow for the atmospheric properties.

The true airspeed can be obtained from the calibrated airspeed by using the fact that the pressure difference can be determined from V_{cal}.

$$\Delta p = \left[\left(\frac{\gamma - 1}{2}(V_{cal}/a_{sl})^2 + 1\right)^{\frac{\gamma}{\gamma - 1}} - 1\right] p_{sl} \tag{2.39}$$

This equation can then be substituted into Equation 2.37 to obtain the true airspeed as a function of the calibrated airspeed, the ambient temperature ratio, θ, and the ambient pressure ratio, δ.

$$\frac{V}{V_{cal}} = \sqrt{\theta\,\frac{(1 + \Delta p/p)^{(\gamma-1)/\gamma} - 1}{(1 + \Delta p/p_{sl})^{(\gamma-1)/\gamma} - 1}} \tag{2.40}$$

This expression is rather unwieldy, and it is convenient to use an alternate approach at low Mach numbers.

At sea level, the true airspeed is equal to the calibrated airspeed as read directly from the airspeed indicator regardless of Mach number since the calibration of the indicator considers compressibility. At other altitudes, as the Mach approaches zero, the differential pressure, Δp, sensed by the airspeed indicator is equal to the incompressible dynamic pressure, $\rho V^2/2$. Defining this dynamic pressure as q, the ratio of q to Δp can be obtained from Equation 2.31.

$$\frac{q}{\Delta p} = \frac{\gamma M^2}{2}\left[\left(1 + \frac{\gamma - 1}{2} M^2\right)^{\gamma/\gamma - 1} - 1\right]^{-1} \tag{2.41}$$

Thus, the true airspeed can be obtained approximately by simply dividing the calibrated airspeed by the square root of the density ratio. In this case, the correction for compressibility is only an approximate one since V_{cal} is obtained using sea level values for the speed of sound. Thus, the true airspeed is given closely by

$$V = \frac{V_{cal}}{\sqrt{\sigma}} \tag{2.42}$$

The *equivalent airspeed,* V_e, is defined as the product of the true airspeed and the square root of the density ratio. This is not quite the same as Equation 2.42 since V_{cal} accounts for compressibility, whereas the V_e does not. The above equation is approximate whereas the following is exact by definition.

$$V = \frac{V_e}{\sqrt{\sigma}} \tag{2.43}$$

The error incurred by using Equation 2.42 instead of Equations 2.39 and 2.40 can be seen by reference to Figure 2.11. Here, the ratio of the true airspeed to that airspeed obtained from Equation 2.42 is graphed as a function of Mach number at standard altitudes of sea level, 6000 and 12,000 m. The correction factor is seen, for example, to be less than 3% for Mach numbers less than 0.5 at 6000 m.

Example of Airspeed Calculation

As an example in determining true airspeed, suppose a pilot reads an indicated (calibrated) airspeed of 350 kts for an outside air temperature (OAT) of 239°K and an altimeter reading of 6 km. From Appendix A.2 or Table 2.1, θ for the standard

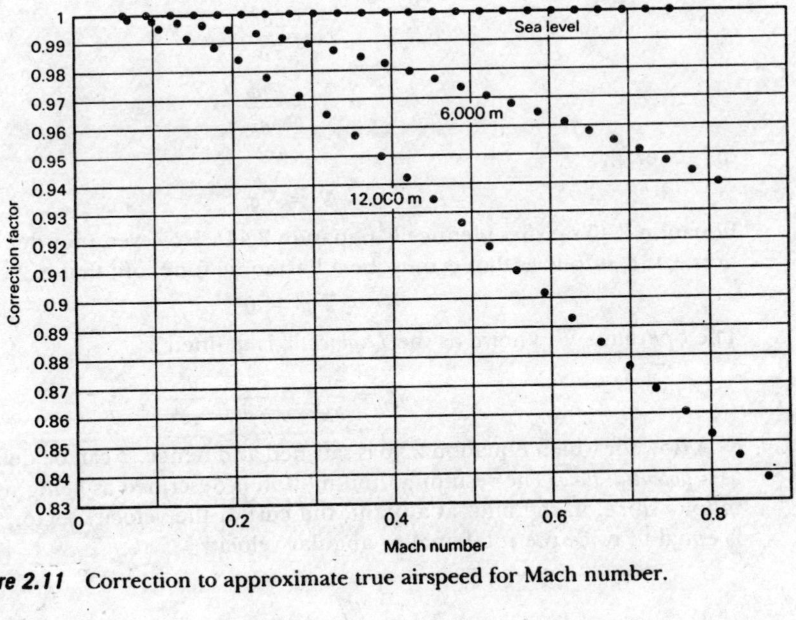

Figure 2.11 Correction to approximate true airspeed for Mach number.

atmosphere is found to equal 0.864 resulting in a δ of 0.464 from Equation 2.5. From the OAT, the actual θ is equal to 0.826 resulting in a density ratio, σ, of 0.562. The calibrated airspeed of 350 kts is equivalent to a speed of 180.3 m/s. Thus, from Equation 2.39, Δp is found to equal 21,357 Pa. Substituting this value into Equation 2.40 results in a true airspeed of 231.9 m/s or 450.1 kts. Using Equation 2.42 results in an approximate true airspeed of 466.9 kts, which is 3.7% higher than the true value. This difference does not agree exactly with Figure 2.11 because the atmosphere for the example is not standard.

POTENTIAL FLOW

For a steady, inviscid, incompressible flow, Euler's equations of fluid motion reduce to two relatively simple relationships that govern the velocity vector.

$$\text{div } \mathbf{V} = \nabla \cdot \mathbf{V} = 0 \qquad (2.44a)$$

$$\mathbf{V} = \nabla \times \mathbf{V} = 0 \qquad (2.44b)$$

The first equation satisfies conservation of mass; the second one assures that the dynamics of the flow is treated correctly. The second term, curl **V**, is called the vorticity.

In addition to satisfying Equation 2.43, one must assure that any mathematical description of the flow field around a given body shape satisfies the boundary condition that there be no velocity normal to the body at all points on its surface. If **n** is the unit vector normal to the surface, the following must hold.

$$\mathbf{V} \cdot \mathbf{n} = 0 \qquad (2.45)$$

Velocity Potential and Stream Function

To assist in the solution of Equation 2.44, two functions are introduced. The first of these is known as the *velocity potential*, ϕ, and is defined such that

$$u = \frac{\partial \phi}{\partial x}$$

$$v = \frac{\partial \phi}{\partial y}$$

$$w = \frac{\partial \phi}{\partial z}$$

or, generally,

$$\mathbf{V} = \nabla \phi \qquad (2.46)$$

Equation 2.46 satisfies identically Equation 2.44*b*. However, in order to satisfy Equation 2.43*a*, it follows that ϕ must be a harmonic function; that is,

$$\nabla^2 \phi = 0 \qquad (2.47)$$

The operator, ∇^2, known as the *Laplacian*, is defined as

$$\nabla^2 = \frac{\partial^2}{\partial x^2} + \frac{\partial^2}{\partial y^2} + \frac{\partial^2}{\partial z^2}$$

A flow for which Equation 2.46 is satisfied and hence, ϕ can be defined, is known as a *potential flow*. The resulting fluid motion is described as being *irrotational*. This follows since, in the limit at a point, the curl of the velocity vector, which is zero, is equal to twice the rotational or angular velocity.

The stream function, ψ, is related to the velocity components by

$$u = \frac{\partial \psi}{\partial y}$$

$$v = -\frac{\partial \psi}{\partial x} \tag{2.48}$$

ψ can only be defined easily for two-dimensional, or axisymmetric, flow. To obtain a particular component, the partial derivative of ψ is taken in the direction normal to the velocity and to the left as one looks in the direction of the velocity.

A line element is pictured in Figure 2.12 with flow passing through it. This element is a segment of an arbitrary line connecting two points A and B. The differential flux through this element will be

$$dQ = \mathbf{V} \cdot \mathbf{n}\, ds$$

But

$$\mathbf{n} = (\mathbf{i}\, dy - \mathbf{j}\, dx)/ds$$

$$\mathbf{V} = \mathbf{i}\, u + \mathbf{j}\, v$$

Substituting **n** and **V** into dQ and using Equation 2.47 results in

$$dQ = \frac{\partial \psi}{\partial x} dx + \frac{\partial \psi}{\partial y} dy$$

or

$$dQ = d\psi$$

Thus,

$$\psi(B) - \psi(A) = \int_A^B \mathbf{V} \cdot \mathbf{n}\, ds \tag{2.49}$$

That is, the change in the stream function between two points is equal to the flux between the points. It follows that ψ is a constant along a streamline. This can be shown by noting that along a streamline

$$\frac{v}{u} = \frac{dy}{dx}$$

and

$$d\psi = dQ = u\, dy - v\, dx$$

Combined, the two relationships give

$$d\psi = 0$$

or

$$\psi = \text{constant (along a streamline)}$$

The stream function, as a measure of the flux, satisfies identically Equation 2.44*a*. For an irrotational flow, however, in order to meet Equation 2.44*b*, it follows that ψ must also be harmonic.

$$\nabla^2 \psi = 0 \tag{2.50}$$

In a manner similar to the derivation of Equation 2.48, the change in ϕ between two points can also be easily obtained. If

$$\phi = \phi(x, y)$$

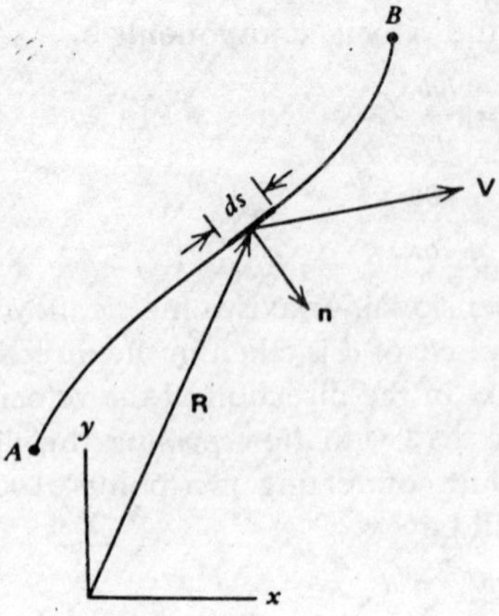

Figure 2.12 Two-dimensional flow through a line element.

then

$$d\phi = \frac{\partial \phi}{\partial x} dx + \frac{\partial \phi}{\partial y} dy$$
$$= u\, dx + v\, dy$$

or, using vector notation,

$$\phi(B) - \phi(A) = \int_A^B \mathbf{V} \cdot \mathbf{dR} \tag{2.51}$$

where **R** is the radius vector to the curve along which the integration is being performed, as shown in Figure 2.12. **dR** is then the differential vector along the curve, directed positively, with a magnitude of *ds*.

As an example in the use of ϕ and ψ, consider the uniform flow pictured in Figure 2.13. For this flow,

$$u = U = \text{constant}$$
$$v = 0$$

ψ will be taken to be zero along the x-axis. This choice is arbitrary, since the values of both ϕ and ψ can be changed by a constant amount without affecting the velocity field obtained from their derivatives. Equation 2.49 will be zero if the integral is performed along a line for which y is a constant. Thus ψ changes only in the y direction. Integrating Equation 2.49 in this direction gives

$$\psi(y) - \psi(0) = \int_0^y (\mathbf{i}\, U) \cdot \mathbf{i}\, dy$$

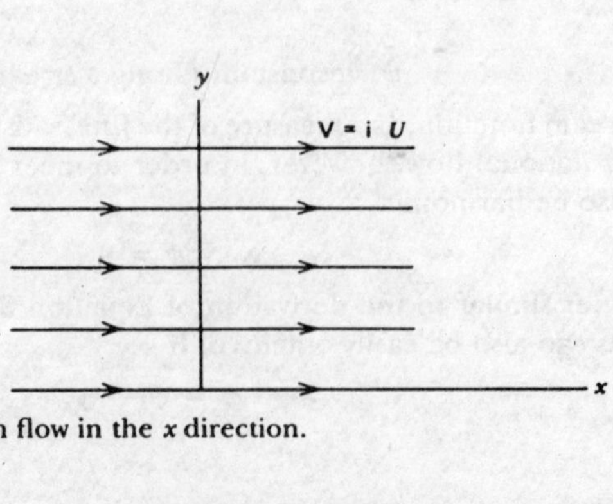

Figure 2.13 Uniform flow in the x direction.

or

$$\psi = Uy$$

If the uniform flow contains, in addition to U, a constant velocity component V in the y direction, ψ becomes

$$\psi = Uy - Vx \tag{2.52}$$

The minus sign in Equation 2.52 is in accordance with the positive direction of **n**, as shown in Figure 2.12. **n** is directed to the right as one looks in the direction of point B from point A. In this case the integration for the second term is in the positive x direction, so **n** equals $-\mathbf{j}$.

In a more formal manner, ϕ will be derived for this same flow using Equation 2.51.

$$\mathbf{V} = \mathbf{i}\,U + \mathbf{j}\,V$$

$$\mathbf{R} = \mathbf{i}\,x + \mathbf{j}\,y$$

Hence,

$$d\mathbf{R} = \mathbf{i}\,dx + \mathbf{j}\,dy$$

so that, taking $\phi = 0$ at $x, y = 0$,

$$\phi(x, y) = \int_0^{x,y} U\,dx + V\,dy$$

$$= Ux + Vy \tag{2.53}$$

ELEMENTARY FLOW FUNCTIONS

If ϕ_1 and ϕ_2 are functions satisfying Equation 2.46 then, because this equation is linear, their sum will also satisfy Equation 2.46. In general, both the velocity potential and stream function can be constructed by summing less complicated functions.

$$\phi(x, y) = \sum_{i=1}^{n} \phi_i(x, y) \tag{2.54a}$$

$$\psi(x, y) = \sum_{i=1}^{n} \psi_i(x, y) \tag{2.54b}$$

Equation 2.53 represents the real benefit to be gained in describing a flow in terms of ϕ and ψ. This statement will become obvious as the developments proceed.

The simple flows from which more complicated patterns can be developed are referred to as *elementary flow functions.* There are three of them: *uniform rectilinear flow, vortex,* and *source.* The first of these has already been covered with ϕ and ψ given by Equations 2.52 and 2.51, respectively.

Vortex

A *vortex* is pictured in Figure 2.14. This flow in two dimensions is purely circular around a point with no radial velocity component. Denoting the tangential velocity component by v_θ, the problem is to find v_θ as a function of r that will satisfy the set of Equations 2.43*a* and 2.43*b*. v_θ is to be independent of θ.

In polar coordinates,

$$\nabla \times \mathbf{V} = \frac{\partial v_\theta}{\partial r} + \frac{v_\theta}{r} - \frac{1}{r}\frac{\partial v_r}{\partial \theta} \tag{2.55a}$$

$$\nabla \cdot \mathbf{V} = \frac{1}{r}\frac{\partial v_\theta}{\partial \theta} + \frac{\partial v_r}{\partial r} + \frac{v_r}{r} \tag{2.55b}$$

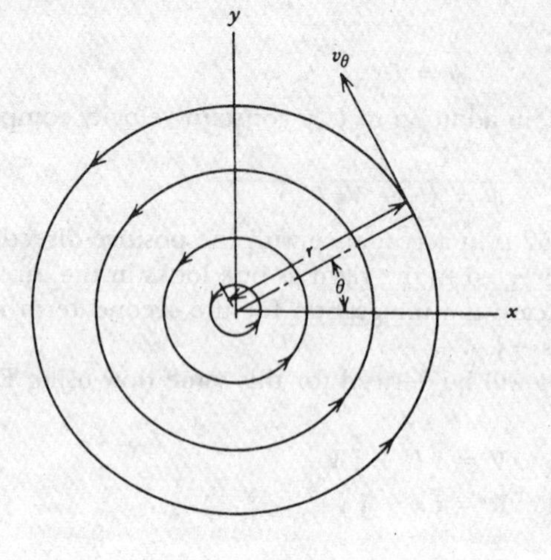

Figure 2.14 Flow field around a vortex.

where r and θ are the polar coordinates, with v_r being the radial component of velocity and v_θ the tangential component.

Since v_r is zero in Figure 2.14 and v_θ is independent of θ, Equation 2.54*b* is satisfied identically, and from Equation 2.54*a*

$$\frac{dv_\theta}{dr} + \frac{v_\theta}{r} = 0$$

or, after integrating,

$$rv_\theta = \text{constant}$$

Thus, for potential flow, the tangential velocity around a vortex must vary inversely with the radial distance from the center of the vortex.

The strength of a vortex, denoted by γ, is measured by integrating the tangential velocity completely around the vortex. The value of this integral is independent of the path providing it encloses the singular point at the center of the vortex.

$$\gamma = \oint \mathbf{V} \cdot \mathbf{dR} \tag{2.56}$$

This closed-line integral of the velocity is known as the *circulation*. Evaluating Equation 2.56 on a constant radius leads to the relationship between the tangential velocity around a vortex, the radius, and the vortex strength.

$$v_\theta = \frac{\gamma}{2\pi r} \tag{2.57}$$

Equation 2.57 is a well-known relationship and can be easily remembered from the definition of γ.

ϕ and ψ for a vortex follow immediately from Equations 2.51, 2.49, and 2.57.

$$\phi(\mathrm{B}) - \phi(A) = \int_A^B \frac{\gamma}{2\pi r}(r\,d\theta)$$

$$= \frac{\gamma}{2\pi}[\theta(B) - \theta(A)]$$

If θ is measured relative to zero and $\theta(B)$ is taken to be any value of θ,

$$\phi = \frac{\gamma\theta}{2\pi} \tag{2.58}$$

The stream function for a vortex is found from

$$\psi(B) - \psi(A) = -\int_A^B \frac{\gamma}{2\pi r}\, dr$$

$$= -\frac{\gamma}{2\pi} \ln \frac{r(B)}{r(A)}$$

Letting $\psi(A)$ be zero and $r(A)$ be an arbitrary radius, a, leads to

$$\psi = -\frac{\gamma}{2\pi} \ln \frac{r}{a} \tag{2.59}$$

The minus sign results from the choice of positive coordinate directions.

Source

The *source* is the counterpart of a vortex. Here, the flow pictured in Figure 2.15 is again symmetrical about the center, but it is entirely radial with no tangential velocity component.

The strength of a source, q, is measured by the total flux emanating from the center. From Figure 2.15, q is obviously given by

$$q = 2\pi r v_r$$

or

$$v_r = \frac{q}{2\pi r} \tag{2.60}$$

In a manner similar to that followed for the vortex, one may verify that for the source

$$\phi = \frac{q}{2\pi} \ln r \tag{2.61}$$

$$\psi = \frac{q\theta}{2\pi} \tag{2.62}$$

Equations 2.56 and 2.59, which define the velocities around vortices and sources, can be extended to three dimensions. If Q is the strength of a three-dimensional source, this flux will equal the product of the radial velocity and the surface area through which the velocity is passing. Thus, one can write v_r immediately as

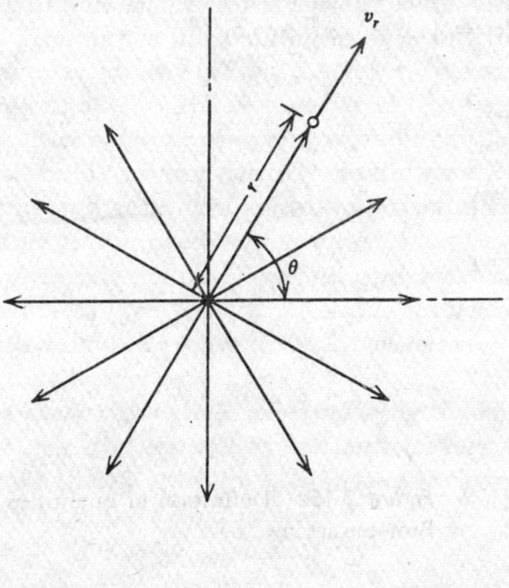

Figure 2.15 Flow from a source.

$$v_r = \frac{Q}{4\pi r^2} \tag{2.63}$$

Biot-Savart Law

The three-dimensional velocity field associated with a vortex line is considerably more complicated and is given by the *Biot-Savart law*. The derivation of this law is beyond the scope of this text. Figure 2.16*a* illustrates a portion of a vortex line about which at any point the circulation, γ, is constant. If $\mathbf{v}_i$ is the velocity vector induced at any point, *P*, in the field by the vortex line, the Biot-Savart law states

$$d\mathbf{v}_i = \frac{\gamma}{4\pi} \frac{\mathbf{r} \times \mathbf{dR}}{|\mathbf{r}|^3} \tag{2.64}$$

This is the most general form of the Biot-Savart law. **dR** is the derivative of the radius vector from the origin to the vortex line and is thus the directed differential distance along the line. **r** is the radius vector from the *P* to the line element **dR**. The positive direction of the circulatory strength, Γ, is defined according to the right-hand rule. The *x*, *y*, *z* orthogonal coordinate system is also right-handed.

A special form of the Biot-Savart law for a straight-line vortex segment found in many texts can be obtained by integrating Equation 2.63. Referring to Figure 2.16*b*, for convenience the line vortex is placed on the *x*-axis and lies between 0 and *x*. The *z*-axis will project out of the paper according to the right-hand rule. The circulation Γ is taken to be positive in the *x* direction which means it will be clockwise when viewed in that direction. For this figure,

$$\mathbf{R} = \mathbf{i}\, x$$
$$\mathbf{OP} + \mathbf{r} = \mathbf{R}$$
$$\mathbf{OP} = \mathbf{i}\, x_p + \mathbf{j}\, y_p$$

Thus,

$$\mathbf{r} = \mathbf{i}(x - x_p) - \mathbf{j}\, y_p$$
$$d\mathbf{R} = \mathbf{i}\, dx$$

so that

$$\mathbf{r} \times d\mathbf{R} = \begin{vmatrix} \mathbf{i} & \mathbf{j} & \mathbf{k} \\ (x - x_p) & -y_p & 0 \\ dx & 0 & 0 \end{vmatrix}$$
$$= \mathbf{k}\, y_p\, dx$$
$$|\mathbf{r}| = [(x - x_p)^2 + y_p^2]^{1/2}$$

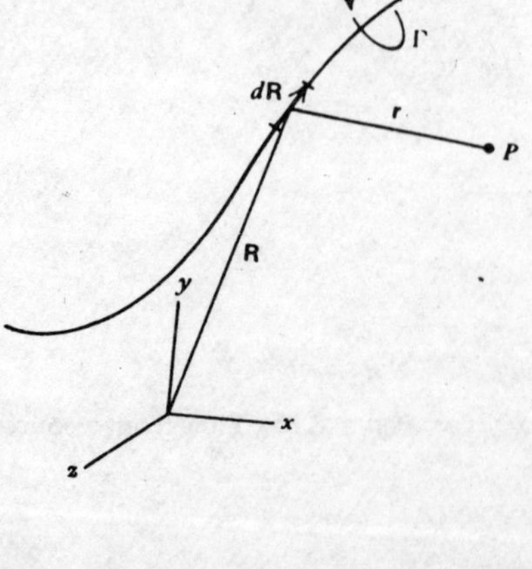

Figure 2.16a Definition of quantities used in the Biot–Savart law.

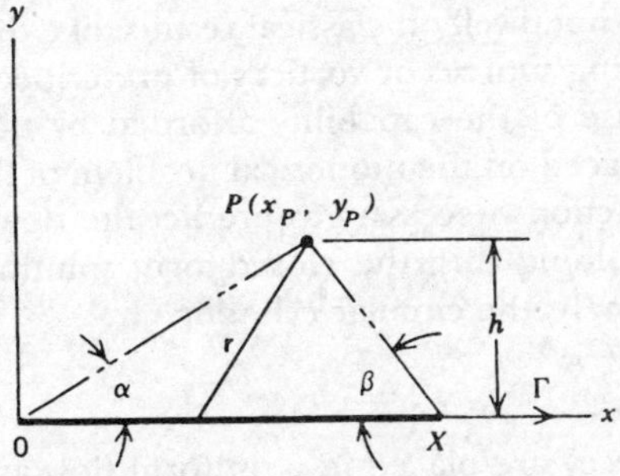

Figure 2.16b The Biot–Savart law for a straight-line vortex.

Equation 2.64 then becomes

$$\mathbf{v}_i = \frac{\mathbf{k}\gamma y_p}{4\pi}\int_0^X \frac{dx}{[(x - x_p)^2 + y_p{}^2]^{3/2}}$$

This reduces to

$$\mathbf{v}_i = \mathbf{k}\frac{\gamma}{4\pi h}(\cos\alpha + \cos\beta) \tag{2.65}$$

α, β, and h are defined in Figure 2.16*b*. Notice that the velocity has only a z component. As the line becomes infinite in length, the angles α and β approach zero, and Equation 2.64 reduces to the expression for the velocity around a two-dimensional point vortex given by Equation 2.56.

Computer Exercise 2.2 "BIOTSUB"

Formulate a subroutine to predict the velocity vector induced in three dimensions at an arbitrary point in space by a straight line vortex segment arbitrarily oriented in space and having a unit strength, γ. Equation 2.64 can be integrated to form the basis for this program, but the integration becomes involved and Equation 2.65 provides a much simpler way of doing it. Let the line begin at point 1 and extend to point 2, defining the circulation as positive according to the right hand rule in going from 1 to 2. Let point P be the point at which the velocity vector is desired. Equation 2.65 can be applied in the plane formed by the line and point P using vector analysis. Let **A** be the vector from point 1 to 2, **B** the vector from 1 to P and **C** the vector from 2 to P. The scalar product of the vector **A** with the vector **B** will give the angle α and the scalar product of the vector **A** with the vector **C** will give the other angle β. The distance h is then obtained as the magnitude of the vector **B** multiplied by the sin (α). The magnitude of the induced velocity can then be found from Equation 2.65. Its direction will be the same as the vector product **A** × **B**. Thus, multiplying the magnitude of the induced velocity by this vector product divided by the magnitude of the product gives the desired induced velocity vector.

Numerically, a problem can arise if the point P lies too close to the line. One should check on the value of cos (α), and if it is greater in magnitude than say, 0.999, then h should be determined by locating the x, y, z location along the line opposite the point by simple geometric proportioning. Then knowing xp, yp, zp at the point and x,y,z at the line, h can be found as the square root of the sum of squares of the component distances of the point from the line.

The Calculation of Flows around a Two-Dimensional Body Shape

The preceding elementary flow functions can be combined to predict the flow around a given two-dimensional body. Also, the same techniques can be extended to three-dimensional shapes. We will not dwell on classical results, like the Rankine Oval, which can be obtained by placing sources or vortices of prescribed strengths into a uniform flow. Instead, because of the capability afforded by the modern digital computer, emphasis will be placed on the numerical problem of finding the strengths of the elementary flow functions necessary to predict the flow around a prescribed shape. However, before doing this, the closed-form solution for one particular shape will be covered; namely, the circular cylinder.

Circular Cylinder

A source and a sink (a negative source) are placed in a uniform flow as shown in Figure 2.17. The source and sink are placed a small distance, ϵ, to either side of the origin along the *x*-axis. We will now move them together toward the origin, but increase their strengths in inverse proportion to the distance between them. In the limit, as the distance between the source and the sink goes to zero, a so-called source-sink doublet is obtained.

Letting 2ϵ equal the distance between the source and the sink and m the constant doublet strength, $2\epsilon Q$, the combined stream function for the source, sink, and uniform flow, from Equations 2.52 and 2.62, can be written as

$$\psi = \frac{m}{4\pi\epsilon}\left(\tan^{-1}\frac{y}{\epsilon + x} + \tan^{-1}\frac{y}{\epsilon - x} - \pi\right) + U_y$$
$$\lim \epsilon \rightarrow 0$$

In the limit, this becomes

$$\psi = -\frac{m}{2\pi}\frac{y}{x^2 + y^2} + Uy$$

For $\psi = 0$, since y is not generally zero, it follows that

$$x^2 + y^2 = \frac{m}{2\pi U}$$

This is the equation of a circle of radius

$$R = (m/2\pi U)^{1/2}$$

Thus, ψ can be written in polar coordinates as

$$\psi = Ur\sin\theta\left[1 - \left(\frac{R}{r}\right)^2\right] \tag{2.66}$$

where $r^2 = x^2 + y^2$

$$\theta = \tan^{-1}\left(\frac{y}{x}\right)$$

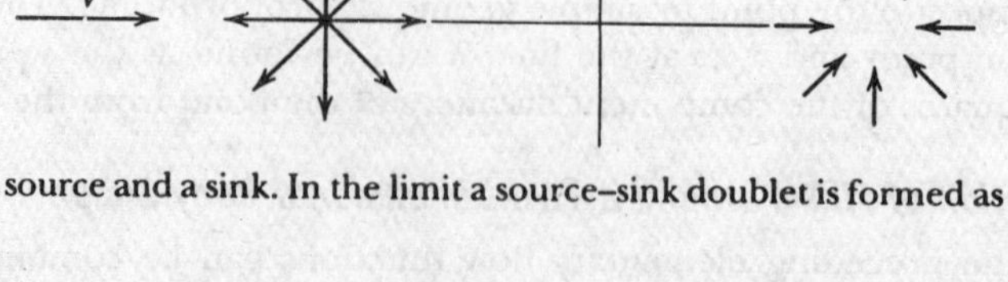

Figure 2.17 A source and a sink. In the limit a source–sink doublet is formed as ϵ approaches zero.

The tangential velocity along the surface of the cylinder is found from Equation 2.48 by differentiating ψ with respect to r and evaluating the result at $r = R$. In this way, v_θ is found to be

$$v_\theta = 2U \sin\theta \tag{2.67}$$

The pressure coefficient, C_p, is a dimensionless measure of the static pressure, which, for potential flow, depends only on the geometric shape of a body and not on the velocity and density. It is defined as

$$C_p = \frac{p - p_\infty}{\frac{1}{2} V_\infty^2} \tag{2.68}$$

where p_∞ is the free-stream static pressure and V_∞ is the freestream velocity. Using Equation 2.29, C_p is found from the ratio of local velocity to free-stream velocity.

$$C_p = 1 - \left(\frac{V}{V_\infty}\right)^2 \tag{2.69}$$

Thus, for the circular cylinder, the pressure distribution is predicted to be

$$C_p = 1 - 4\sin^2\theta \tag{2.70}$$

In Chapter 4, it will be seen that Equation 2.70 agrees fairly well with experimental results over the front half of the cylinder, but departs from actual measurements over the rear portion as the result of viscosity.

A point vortex of strength γ can be placed at the origin without altering the streamline representing the surface of the cylinder. If this is done, Equation 2.67 becomes

$$v_\theta = 2\,U\sin\theta + \frac{\gamma}{2\pi R} \tag{2.71}$$

Relative to p_0 the pressure on the surface of the cylinder will be

$$p - p_0 = \frac{1}{2}\rho U^2 - \frac{1}{2}\rho\left|2\,U\sin\theta + \frac{\gamma}{2\pi R}\right|^2 \tag{2.72}$$

Referring to Figure 2.18, the net vertical force, or lift, on the cylinder resulting from the pressure distribution will be

$$L = -\int_0^{2\pi} pR\sin\theta\, d\theta$$

or, from Equation 2.78, this reduces to

$$L = \rho U\gamma \tag{2.73}$$

This is referred to as the *Kutta-Joukowski law.* Although derived here specifically for a circular cylinder, it can be applied to other shapes where γ represents generally the circulation around the shape. This will be amplified further in Chapter 3.

The net horizontal force, or drag, on the cylinder is found from

$$D = \int_0^{2\pi} pR\cos\theta\, d\theta$$

Using Equation 2.72, the drag is found to be zero, a result that is true in general for a closed body in steady potential flow. This result is known as *D'Alembert's paradox,* after Jean le Rond D'Alembert, a French mathematician who first reached this conclusion around 1743.

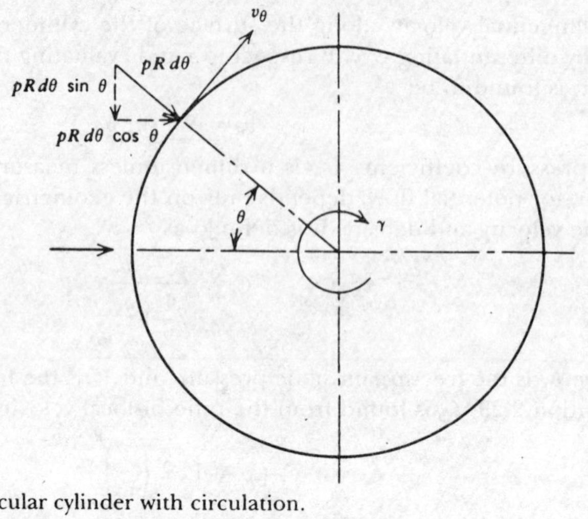

Figure 2.18 Circular cylinder with circulation.

Numerical Modeling

The foregoing analysis obtained a closed-form solution for the pressure distribution around a two-dimensional shape. However, the shape was a well-defined, relatively simple shape. Generally, such a solution cannot be obtained. Therefore, at this point, a two-dimensional body of arbitrary shape will be modeled by the use of distributed sources in the manner of Reference 2.4. In order to do this, we introduce the concept of a source panel.

The source panel is simply an extension of the point source as illustrated in Figure 2.19. In this figure, a continuous distribution of point sources are placed along a line such that the total source strength in an incremental length, Δx, is equal to $q(x)\,\Delta x$, $q(x)$ being the source strength per unit length. Now consider the dashed contour shown in Figure 2.19 having a length of Δx and a vanishingly small height. In the limit as the height vanishes, the velocity parallel to the line attributable to the sources also vanishes so that the only velocity along the contour must be directed normally outward from the line. Thus, the total flux passing out from the line element on both sides of the line must equal the total source strength inside the dashed contour. If the line represents the surface of a body and n the unit vector directed normal and outward from the surface, it follows that the velocity vector produced by the I source panel at its own surface is given by

$$\mathbf{V}_n\,(I) = \frac{q(I)}{2}\,\mathbf{n}(I) \tag{2.74}$$

Figure 2.20 depicts an arbitrary body shape at an angle of attack, α, to the freestream velocity vector, **V**. Note that the coordinate system is fixed with respect to the body so that the velocity vector, **V**, changes with the angle of attack, whereas the geometry of the body remains fixed. The body shape is approximated with $I = 1, 2, 3, \ldots, N$ source panels each having a distributed constant source strength, $q(I)$ is the length of the I panel, then the total source strength of that panel will be

$$Q(I) = q(I)\ S(I) \tag{2.75}$$

Following the lead of Reference 2.4, it is assumed that the velocity induced at a panel by any other panel can be calculated as if the distributed source on the other

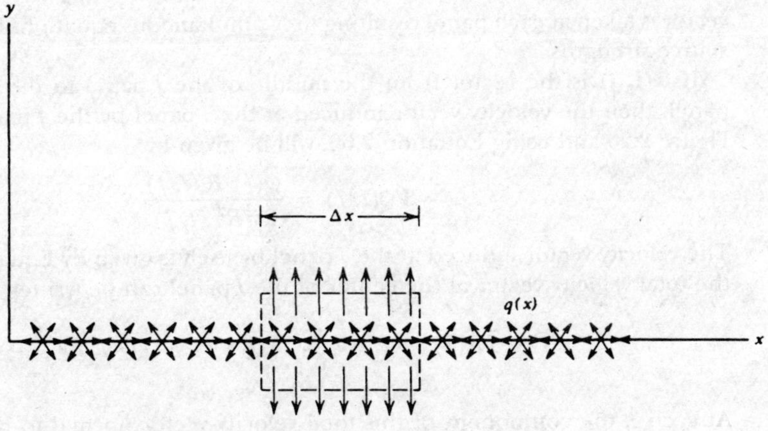

Figure 2.19 Distributed sources in two-dimensional flow.

panel is concentrated into one point source located at its middle. The velocity induced at a panel by its own distributed sources is calculated according to Equation 2.74.

The source strengths of the panels must now be found, which will give a resultant velocity vector at each panel that is tangent to the panel. In other words, at each panel the velocity vector normal to the panel must be zero. To assure this boundary condition at each panel, the resultant velocity is first written in terms of $Q(I)$ and the freestream velocity. Then the scalar product of this velocity and the unit normal

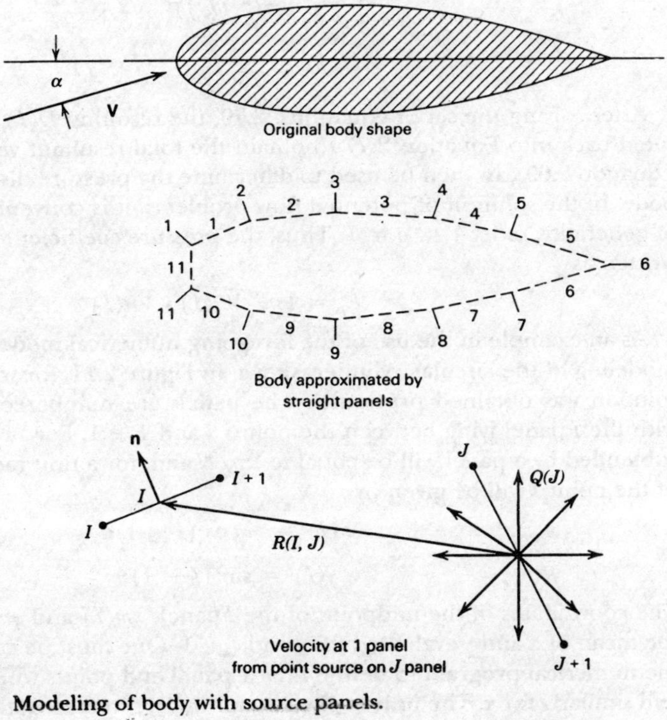

Figure 2.20 Modeling of body with source panels.

vector is taken at each panel resulting in N simultaneous equations for the unknown source strengths.

If **R(I, J)** is the vector from the middle of the J panel to the middle of the I panel, then the velocity vector induced at the I panel by the J panel, as shown in Figure 2.20 and using Equation 2.60, will be given by

$$\mathbf{VQ}(I, J) = \frac{Q(J)\, \mathbf{R}(I, J)}{2\pi R^2 (I, J)} \tag{2.76}$$

The velocity vector induced at the I panel by itself is given by Equation 2.74. Thus, the total velocity vector at the center of the I panel can be written as

$$\overline{VR}(I) = V + \sum_{\substack{J=1 \\ (J \neq I)}}^{N} \overline{VQ}(I, J) + \overline{VQ}(I, I) \tag{2.77}$$

At each I, the component of this total velocity vector normal to the surface must vanish. Thus,

$$\mathbf{n}\,(I) \cdot \mathbf{VR}\,(I) = 0 \tag{2.78}$$

Evaluated at N panels, the above leads to N simultaneous equations for the unknown source strengths, $Q(I)$, of the form

$$\sum_{J=1}^{N} A(I, J)\; Q\,(J) = B\,(I) \tag{2.79}$$

where the coefficients $A(I, J)$ and $B(I)$ are given by

$$B(I) = -\mathbf{n}(I) \cdot \mathbf{V} \tag{2.80a}$$

$$A(I, J) = \frac{\mathbf{n}(I) \cdot \mathbf{R}(I, J)}{2\pi R^2\,(I, J)} \qquad (I \neq J) \tag{2.80b}$$

$$A(I, J) = \frac{1}{2S(I)} \qquad (I = J) \tag{2.80c}$$

After solving the set of equations, 2.79, the resulting $Q(I)$ values can be substituted back into Equation 2.77 to obtain the total resultant velocity at the I panel. Equation 2.69 can then be used to determine the pressure distribution around the body. In the solution of potential flow problems, it is convenient, without any loss of generality, to let $V = \rho = 1$. Thus, the pressure coefficient at the I panel can be written as

$$C_p = 1 - \overline{VR}(I) \cdot \overline{VR}(I) \tag{2.81}$$

As an example in the use of the foregoing numerical model, consider the panel modeling of the circular cylinder, shown in Figure 2.21, for which the closed-form solution was obtained previously. The panels are numbered clockwise as shown with the I panel lying between the points I and $I + 1$. For N panels, the angle $\Delta\boldsymbol{\theta}$ subtended by a panel will be equal to $2\pi/N$ and, for a unit radius, the coordinates of the point I will be given by

$$x(I) = -\cos\,(I - 1)\theta \tag{2.82a}$$

$$y(I) = \sin\,(I - 1)\theta \tag{2.82b}$$

The coordinates of the midpoint of the I panel, $xm(I)$ and $ym(I)$, are then simply the mean of x and y evaluated for I and $I + 1$. One must be careful in formulating the numerical program to define $N + 1$ panel end points where $x(N + 1) = x(1)$ and similarly for y. The unit normal vector, $\mathbf{n}$(I), directed outward from the panel, can be written as

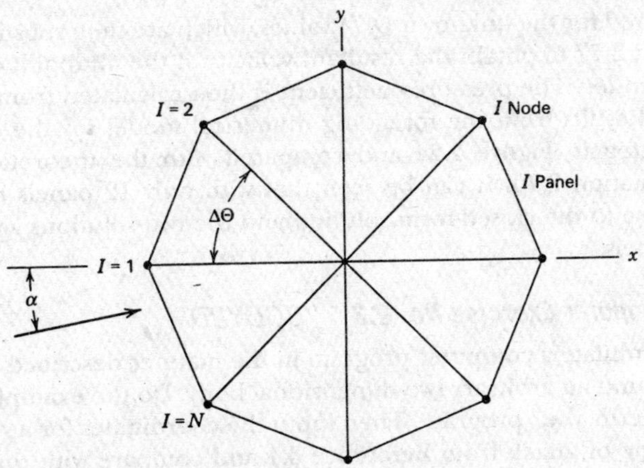

Figure 2.21 Panel model of circular cylinder.

$$n(I) = -i\frac{y(I+1) - y(I)}{S(I)} + j\frac{x(I+1) - x(I)}{S(I)} \tag{2.83}$$

where $S(I)$ is the length of the I panel. The vector, $\mathbf{R}(\mathrm{I,J})$ from the middle of the J panel to the middle of the I panel, is given by

$$R = i\,[xm(I) - xm(J)] + j\,[ym(I) - ym(J)] \tag{2.84}$$

Since this particular example is symmetrical, the angle of attack is set to zero so that the free-stream velocity is given simply by

$$\mathbf{V} = \mathrm{iV} \tag{2.85}$$

Thus, substituting Equations 2.83, 2.84, and 2.85 into the set of Equations 2.8 will result in N simultaneous equations corresponding to Equation 2.79. These are then

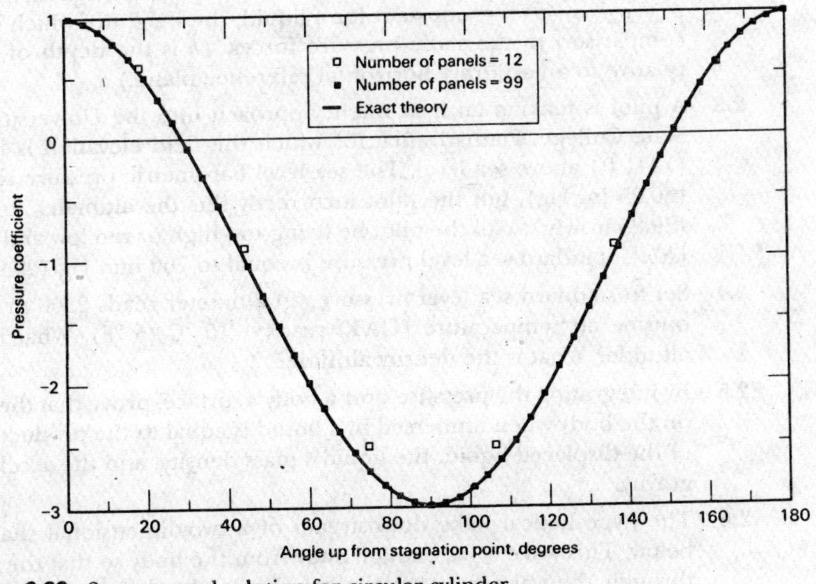

Figure 2.22 Source panel solution for circular cylinder.

solved for the unknown $Q(I)$ values, which are then substituted into Equations 2.76 and 2.77 to obtain the resultant velocity at the midpoint of each panel around the cylinder. The pressure coefficient is then calculated from Equation 2.69.

Results from the foregoing numerical model for the circular cylinder are presented in Figure 2.22 and compared with the theoretical distribution given by Equation 2.70. It can be seen that with only 12 panels the numerical solution is close to the closed-form solution and the two solutions agree almost exactly for 99 panels.

Computer Exercise No. 2.3 "BODY2D"

Formulate a computer program in the manner described above to predict the flow around an arbitrary two-dimensional body. Do the example of the circular cylinder to verify your program. Then input the coordinates for a symmetrical airfoil at zero angle of attack from Reference 3.1 and compare with the tabulated pressure and local velocity distributions given in the reference.

SUMMARY

This chapter has presented some important basic principles of fluid mechanics on which much of the material in succeeding chapters will depend. It is important to realize that potential flow methods are a valuable tool in analyzing aerodynamic systems despite the fact that viscosity is neglected. In modeling complex configurations, one should remember that elementary flow functions can be superimposed and that their velocity fields add vectorially.

PROBLEMS

2.1 How should the thickness of a dam of constant width vary with depth to ensure a constant shearing stress in the dam's material?

2.2 Show that the incompressible Bernoulli's equation (Equation 2.28) becomes $p + \rho gh + \frac{1}{2}\rho V^2 =$ constant for a liquid, the weight of which is significant in comparison to the static pressure forces. (h is the depth of the streamline relative to an arbitrary horizontal reference plane.)

2.3 A pilot is making an instrument approach into the University Park Airport, State College, Pennsylvania, for which the field elevation is listed at 378 m (1241 ft) above sea level. The sea level barometric pressure is 763.3 mm Hg (30.05 in. Hg), but the pilot incorrectly sets the altimeter to 758.2 mm Hg (29.85 in. Hg). Will the pilot be flying too high or too low and by how much? [*Note.* Standard sea level pressure is equal to 760 mm Hg (29.92 in. Hg).]

2.4 Set to standard sea level pressure, an altimeter reads 2500 m (8200 ft). The outside air temperature (OAT) reads −15 °C (5 °F). What is the pressure altitude? What is the density altitude?

2.5 By integrating the pressure over a body's surface, prove that the buoyant force on the body when immersed in a liquid is equal to the product of the volume of the displaced liquid, the liquid's mass density, and the acceleration due to gravity.

2.6 The hypothetical wake downstream of a two-dimensional shape is pictured below. This wake is far enough away from the body so that the static pressure through the wake is essentially constant and equal to the free-stream static pressure. Calculate the drag coefficient of the shape based on its projected frontal area.

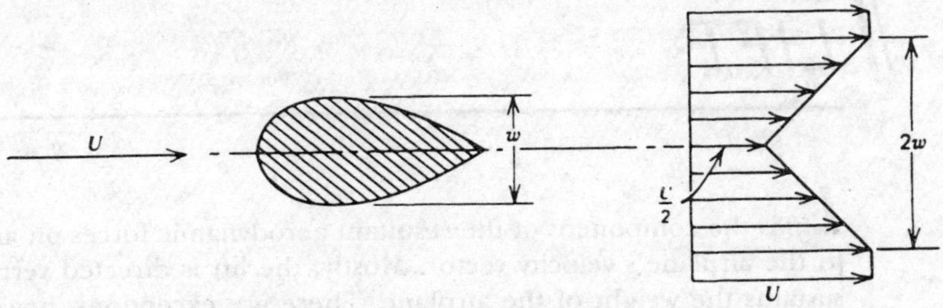

2.7 An incompressible flow has velocity components given by $u = -\omega y$ and $v = \omega x$, where ω is a constant. Is such a flow physically possible? Can a velocity potential be defined? How is ω related to the vorticity? Sketch the streamlines.

2.8 Derive Bernoulli's equation directly by applying the momentum theorem to a differential control surface formed by the walls of a small streamtube and two closely spaced parallel planes perpendicular to the velocity.

2.9 A jet of air exits from a tank having an absolute pressure of 152,000 Pa (22 psi). The tank is at standard sea level (SSL) temperature. Calculate the jet velocity if it expands isentropically to SSL pressure.

2.10 A light aircraft indicates an airspeed of 266 km/hr (165.2 mph) at a pressure altitude of 2400 m (7874 ft). If the outside air temperature is −10 °C, what is the true airspeed?

2.11 Prove that the velocity induced at the center of a ring vortex (like a smoke ring) of strength Γ and radius R is normal to the plane of the ring and has a magnitude of $\Gamma/2R$.

REFERENCES

2.1 Streeter, Victor L., and Wylie, E. Benjamin, *Fluid Mechanics*, 6th edition, McGraw-Hill, New York, 1975.

2.2 Roberson, John A., and Crowe, Clayton T., *Engineering Fluid Mechanics*, Houghton Mifflin, Boston, 1975.

2.3 Minzner, R. A., Champion, K. S. W., and Pond, H. L. "The ARDC Model Atmosphere," AF CRC-TR-59-267, 1959.

2.4 Smith, A. M. O., "Incompressible Flow About Bodies of Arbitrary Shape," IAS Paper No. 62–143, presented at the IAS National Sciences Meeting, Los Angeles, June 1962.

3 LIFT

Lift is the component of the resultant aerodynamic forces on an airplane normal to the airplane's velocity vector. Mostly, the lift is directed vertically upward and sustains the weight of the airplane. There are exceptions, however. A jet fighter with a thrust-to-weight ratio close to unity in a steep climb may be generating very little lift with its weight being opposed mainly by the engine thrust.

The component that is the major lift producer on an airplane and on which this chapter will concentrate is the wing. Depending on the airplane's geometry, other components can contribute to or significantly affect the lift, including the fuselage, engine nacelles, and horizontal tail. These latter components will be considered, but to a lesser extent than the wing.

WING GEOMETRY

Wing geometry, discussed briefly in Chapter 1, will now be considered in some detail. Figure 3.1 is a top view, or planform, of an isolated wing, which illustrates the parameters that define the planform shape. As shown, the distance from one wing tip to the other is defined as the span, b. Note that b is measured parallel to the y-axis (Figure 1.10) and not along any swept-back line such as the leading edge. Similarly, the chord, c, at any spanwise station, y, is the distance from the leading edge to the trailing edge measured parallel to the x-axis.

The wing shown in Figure 3.1 is linearly tapered where the taper ratio, λ, is defined as the ratio of the tip chord, c_t, to the midspan chord, c_o.

$$\lambda = \frac{c_t}{c_o} \tag{3.1}$$

For the linearly tapered wing, the chord will vary directly with the absolute value of the distance from the center of the wing as a fraction of the semispan.

$$c = c_o\left[1 - (1 - \lambda)\left|\frac{2y}{b}\right|\right] \tag{3.2}$$

Since the planform area, S, is given by

$$S = \frac{c_o\,(1 + \lambda)\,b}{2} \tag{3.3}$$

it follows from Equation 1.2 that the aspect ratio, taper ratio, midspan chord, and span are related by

$$A = \frac{2b}{c_o(1 + \lambda)} \tag{3.4}$$

The quarter-chord line of a wing is the locus of points one-quarter of the chord back from the leading edge. For a linearly tapered wing this is a straight line. The sweep angle of this line, $\Lambda_{1/4}$, and that of the leading edge, Λ, are related by

$$\tan\Lambda_{1/4} = \tan\Lambda - \frac{4\,(1 - \lambda)}{A\,(1 + \lambda)} \tag{3.5}$$

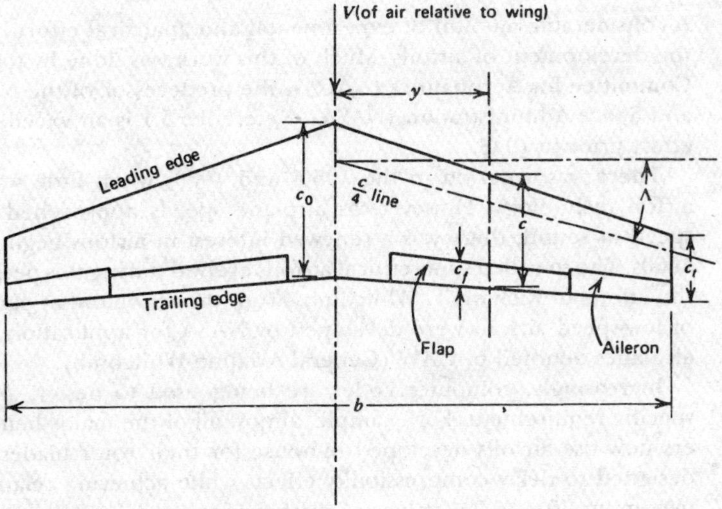

Figure 3.1 Top view of a wing planform.

Usually the center portion of a wing is enclosed by the fuselage. In such an instance the wing's aspect ratio and taper ratio are determined by ignoring the fuselage and extrapolating the planform shape into the centerline. The midspan chord in this instance is thus somewhat fictitious. The wing root is defined as the wing section at the juncture of the wing and fuselage. Occasionally, in the literature, one will find wing geometry characterized in terms of the wing root chord instead of the midspan chord.

Approximately the aft 20 to 30% of a wing's trailing edge is movable. On the outer one-third or so of the span, the trailing edge on one side of the wing deflects opposite to that on the other. These oppositely moving surfaces are called *ailerons*; ailerons provide a rolling moment about the airplane's longitudinal axis. For example, when the aileron on the left wing moves down and the one on the right moves up, a moment is produced that tends to lift the left wing and lower the right one; this is a maneuver necessary in making a coordinated turn to the right.

The inner movable portions of the wing's trailing edge on both sides of the wing are known as the *flaps*. For takeoff and landing the flaps are lowered the same on both sides. There is no differential movement of the flaps on the left and right sides of the wing. The purpose of the flaps is to allow the wing to develop a higher lift coefficient than it would otherwise. Thus, for a given weight, the airplane can fly slower with the flaps down than with them up. Flaps, including leading edge flaps and the many different types of trailing edge flaps, will be discussed in more detail later.

For some applications both ailerons are lowered to serve as an extension to the flaps. In such a case they are referred to as *drooped ailerons*, or *flaperons*. When flaperons are employed, additional roll control is usually provided by spoilers. These are panels that project into the flow near the trailing edge to cause separation with an attendant loss of lift.

In order to understand and predict the aerodynamic behavior of a wing, it is expedient to consider first the behavior of two-dimensional airfoils. An airfoil can be thought of as a constant chord wing of infinite aspect ratio.

AIRFOILS

A considerable amount of experimental and analytical effort has been devoted to the development of airfoils. Much of this work was done by the National Advisory Committee for Aeronautics (NACA), the predecessor of the National Aeronautics and Space Administration (NASA). Reference 3.1 is an excellent summary of this effort prior to 1948.

There was a period in the 1950s and 1960s when little was done to advance airfoil technology. However, as airplane speeds approached and exceeded the speed of sound, there was a renewed interest in airfoils beginning in the middle 1960s. The so-called supercritical airfoils evolved during this period, mainly through the efforts of Richard T. Whitcomb. From these, somewhat paradoxically, a family of low-speed airfoils were developed by NASA for application to general aviation airplanes denoted by GAW (General Aviation-Whitcomb).

Increasingly, computer codes are being used to design airfoils tailored to a specific requirement. For example, almost all of the major helicopter manufacturers now use airfoils developed in-house for their rotor blades. These airfoils are designed to delay compressibility effects while achieving relatively high values of maximum lift and low values of pitching moment. A more detailed discussion of these codes will follow later in this chapter.

The development of a standard NACA airfoil is illustrated in Figure 3.2. First, in Figure 3.2*a*, the chord line, *c*, is drawn. Next, in Figure 3.2*b*, the camber line is plotted up from the chord a small distance *z*, which is a function of the distance from the leading edge. Next, as shown in Figure 3.2*c*, the semithickness is added to either side of the camber line. Also, the nose circle is centered on a tangent to the camber line at the leading edge and passes through the leading edge. Finally, an outer contour is faired around the skeleton to form the airfoil shape. Observe that the chord line is the line joining the ends of the mean camber line.

The early NACA families of airfoils were described in this way, with the camber and thickness distributions given as algebraic functions of the chordwise position.

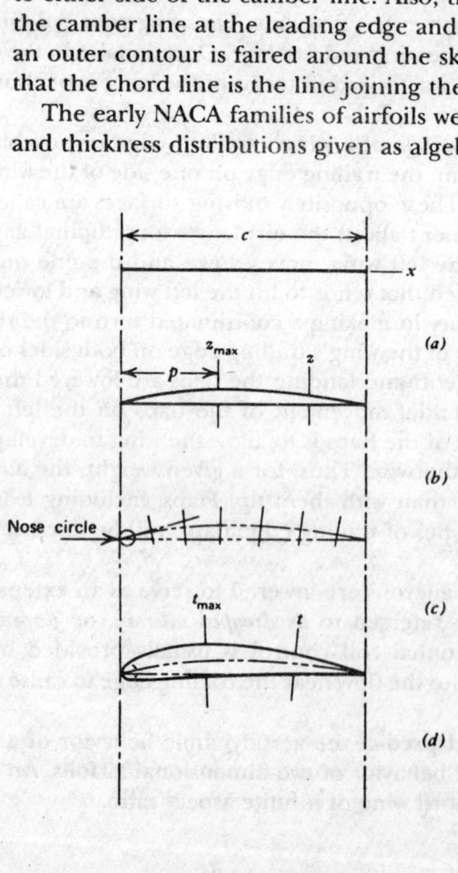

Figure 3.2 The construction of an airfoil contour.

However, for certain combinations of maximum thickness-to-chord ratios, maximum camber-to-chord ratios, and chordwise position of maximum camber, tabulated ordinates for the upper and lower surfaces are available (Ref. 3.1).

Before discussing the various families of airfoils in detail, we will generally consider the aerodynamic characteristics for airfoils, all of which can be influenced by airfoil geometry.

To begin, an airfoil derives its lift from the pressure being higher on the lower surface of the airfoil than on the upper surface. If a subscript l denotes lower surface and "u" denotes upper surface, then the total lift (per unit span) on the airfoil will be

$$L = \int_0^c (p_l - p_u)\, dx \qquad (3.6)$$

The moment about the leading edge, defined positive nose up, will be

$$M_{\text{LE}} = -\int_0^c x(p_l - p_u)\, dx \qquad (3.7)$$

The lift and moment can be expressed in terms of dimensionless coefficients.

$$C_l = \frac{L}{(1/2)\rho V^2 c} \qquad (3.8)$$

$$C_{m_{\text{LE}}} = \frac{M_{\text{LE}}}{(1/2)\rho V^2 c^2} \qquad (3.9)$$

Note that lowercase subscripts are used to denote coefficients for a two-dimensional airfoil, whereas uppercase subscripts are used for the three-dimensional wing. Writing

$$\frac{p_l - p_u}{(1/2)\rho V^2} = \frac{p_l - p_\infty}{(1/2)\rho V^2} - \frac{p_u - p_\infty}{(1/2)\rho V^2}$$

and redefining x as the distance in chord lengths from the leading edge, Equations 3.5 and 3.6 become

$$C_l = \int_0^1 (C_{p_l} - C_{p_u})\, dx \qquad (3.10)$$

and

$$C_{m_{\text{LE}}} = -\int_0^1 x(C_{p_l} - C_{p_u})\, dx \qquad (3.11)$$

where the upper and lower pressure coefficients are defined according to Equation 2.73.

The moment calculated from Equation 3.10 can be visualized as being produced by the resultant lift acting at a particular distance back from the leading edge. As a fraction of the chord, the distance x_{cp} to this point, known as the *center of pressure*, can be calculated from

$$-x_{\text{cp}} C_l = C_{m_{\text{LE}}} \qquad (3.12)$$

Knowing x_{cp}, the moment coefficient about any other point, x, along the airfoil can be written, referring to Figure 3.3, as

$$C_m = -(x_{\text{cp}} - x)\, C_l \qquad (3.13)$$

As an example in the use of the foregoing, consider the idealized chordwise pressure distribution along an airfoil shown in Figure 3.4. Here, the difference between the pressure coefficients on the lower and upper surfaces is given by

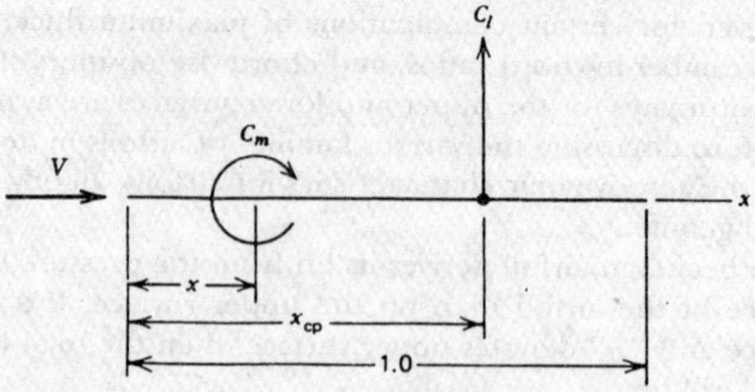

Figure 3.3 Dimensionless moment at x produced by dimensionless lift acting at x_{cp}.

$$C_{p_l} - C_{p_u} = C_{p_o}(1 - x) \tag{3.14}$$

Thus, using Equation 3.10, the airfoil section C_l and C_{p_o} are related by

$$C_l = \frac{C_{p_o}}{2}$$

The moment coefficient at the leading edge becomes

$$C_{m_{LE}} = -\frac{C_{p_o}}{6}$$

Stating that the moment is equal to some value at a particular location implies that the lift is taken as acting at that particular location. Thus, visualizing the above lift and moment acting at the leading edge of the airfoil in Figure 3.4, the dimensionless moment about any dimensionless chordwise position, x, can be written as the sum of the moment coefficient about the leading edge and the product of the dimensionless lift, C_l, and the dimensionless distance, x. Thus,

$$C_{m_x} = C_{m_{LE}} + xC_l \tag{3.15}$$

Equating the above to zero, which is equivalent to Equation 3.11, leads to the center of pressure for this particular distribution.

$$x_{cp} = \frac{1}{3}$$

It will be shown later that a point exists on an airfoil called the *aerodynamic center* about which the moment coefficient is constant and does not depend on C_l. De-

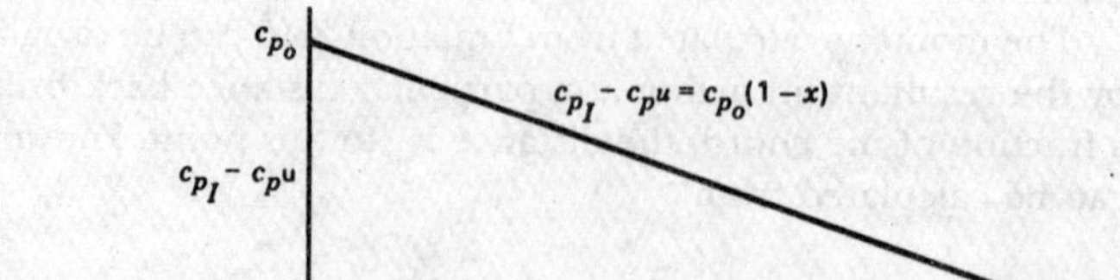

Figure 3.4 Airfoil with a triangular chordwise pressure coefficient distribution.

noting the location of the aerodynamic center by x_{ac}, Equation 3.13 can be solved for the location of the center of pressure.

$$x_{cp} = x_{ac} - \frac{C_{m_{ac}}}{C_l} \tag{3.16}$$

Do not confuse the aerodynamic center with the center of pressure. Again, the aerodynamic center is the location about which the moment is constant, and the center of pressure is the point at which the resultant lift acts.

The progressive development of an airfoil shape is illustrated by reference to Figure 3.5. Historically, airfoils developed approximately in this manner. Consider first the simple shape of a thin, flat plate.

Beginning with Figure 3.5*a*, if the angle of attack of a thin, flat plate is suddenly increased from zero, the flow will appear for a moment as shown. Because of near-symmetry, there is practically no lift produced on the plate. However, because of viscosity, the flow at the trailing edge cannot continue to turn the sharp edge to flow upstream. Instead, it quickly adjusts to the pattern shown in Figure 3.5*b*. Here, the flow leaves nearly tangent to the trailing edge. This condition is known as the *Kutta condition* after the German scientist, W. M. Kutta, who in 1902 first imposed the trailing edge condition in order to predict the lift of an airfoil theoretically. In Figure 3.5*b*, observe that there is one streamline that divides the flow that passes over the plate from that below. Along this "dividing streamline," the flow comes to rest at the stagnation point, where it joins perpendicular to the lower surface of the plate near the leading edge. As the flow progresses forward along this line, it is unable to adhere to the surface around the sharp leading edge and separates from the plate. However, it is turned backward by the main flow and reattaches to the upper surface a short distance from the leading edge. The resulting nonsymmetrical flow pattern causes the fluid particles to accelerate over the upper surface and decelerate over the lower surface. Hence, from Bernoulli's equation, there is a decrease in air pressure above the plate and an increase below it. This pressure difference acting on the airfoil produces a lift.

If the angle of attack of the plate is too great, the separated flow at the leading edge will not reattach to the upper surface, as shown in Figure 3.5*c*. When this occurs, the large separated region of unordered flow on the upper surface produces an increase in pressure on that surface and hence, a loss in lift. This behavior of the airfoil is known as *stall*. Thus, the limit in C_l, that is, $C_{l_{max}}$, is the result of flow separation on the upper surface of the airfoil.

To improve this condition, one can curve the leading edge portion of the flat plate, as shown in Figure 3.5*d*, to be more nearly aligned with the flow in that region. Such a shape is similar to that used by the Wright Brothers. This solution to the separation problem, as one might expect, is sensitive to angle of attack and only holds near a particular design angle. However, by adding thickness to the thin, cambered plate and providing a rounded leading edge, the performance of the airfoil is improved over a range of angles, with the leading edge separation being avoided altogether. Thus, in a qualitative sense, we have defined a typical airfoil shape. Camber and thickness are not needed to produce lift (lift can be produced with a flat plate), but instead, to increase the maximum lift that a given wing area can deliver.

Even a cambered airfoil of finite thickness has its limitations, as shown in Figure 3.5*f*. As the angle of attack is increased, the flow can separate initially near the trailing edge, with the *separation point* progressively moving forward as the angle of attack continues to increase.

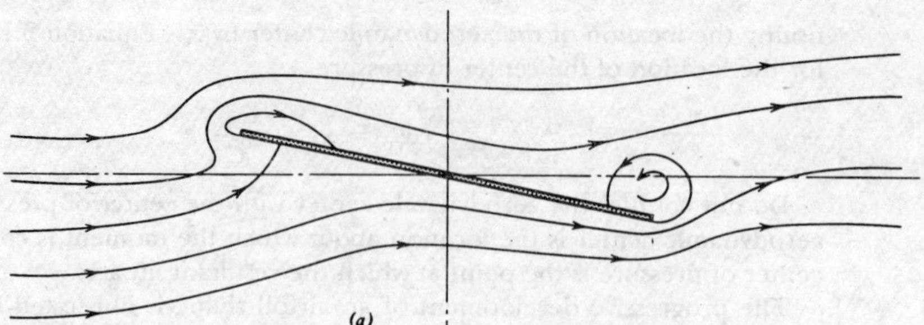

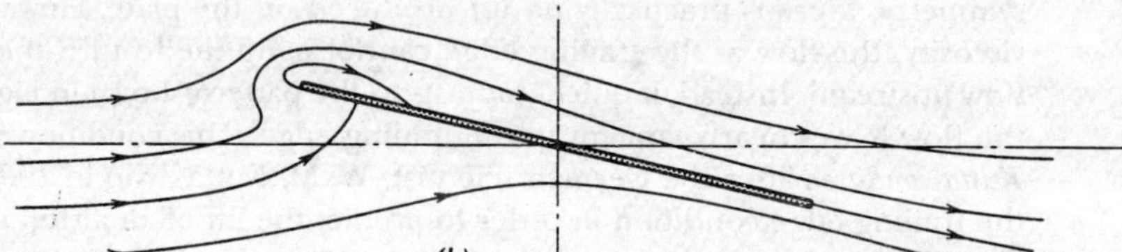

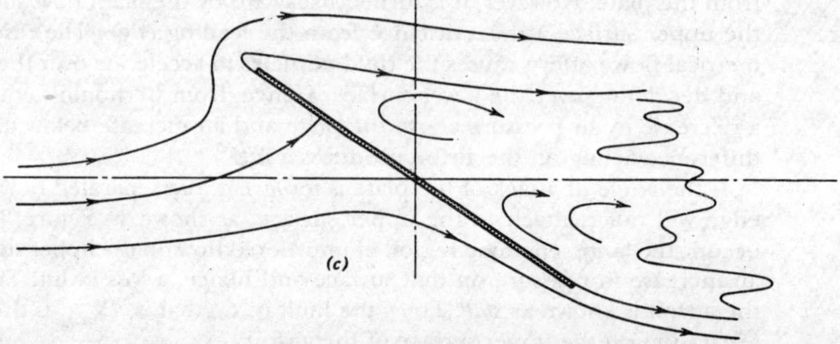

Figure 3.5 Progressive development of airfoil shapes (*a*) Flat plate at sudden angle of attack—no lift. (*b*) Flat plate at angle of attack and generating lift. (*c*) Flat plate experiencing leading edge separation and loss of lift (stall). (*d*) Flat plate with curved leading edge to prevent leading separation. (*e*) Airfoil with thickness and camber to delay stall. (*f*) Airfoil with trailing edge separation.

The degree to which the flow separates from the leading or trailing edge depends on the Reynolds number and the airfoil geometry. Thicker airfoils with more rounded leading edges tend to delay leading edge separation. This separation also improves with increasing values of the Reynolds number.

Leading edge separation results in flow separation over the entire airfoil and a sudden loss in lift. On the other hand, trailing edge separation is progressive with angle of attack and results in a more gradual stalling. The situation is illustrated in Figures 3.6 and 3.7 (taken from Ref. 3.1). In Figure 3.6, note the sharp drop in C_l at an α of 12° for $R = 3 \times 10^6$, whereas for $R = 9 \times 10^6$, the lift curve is more

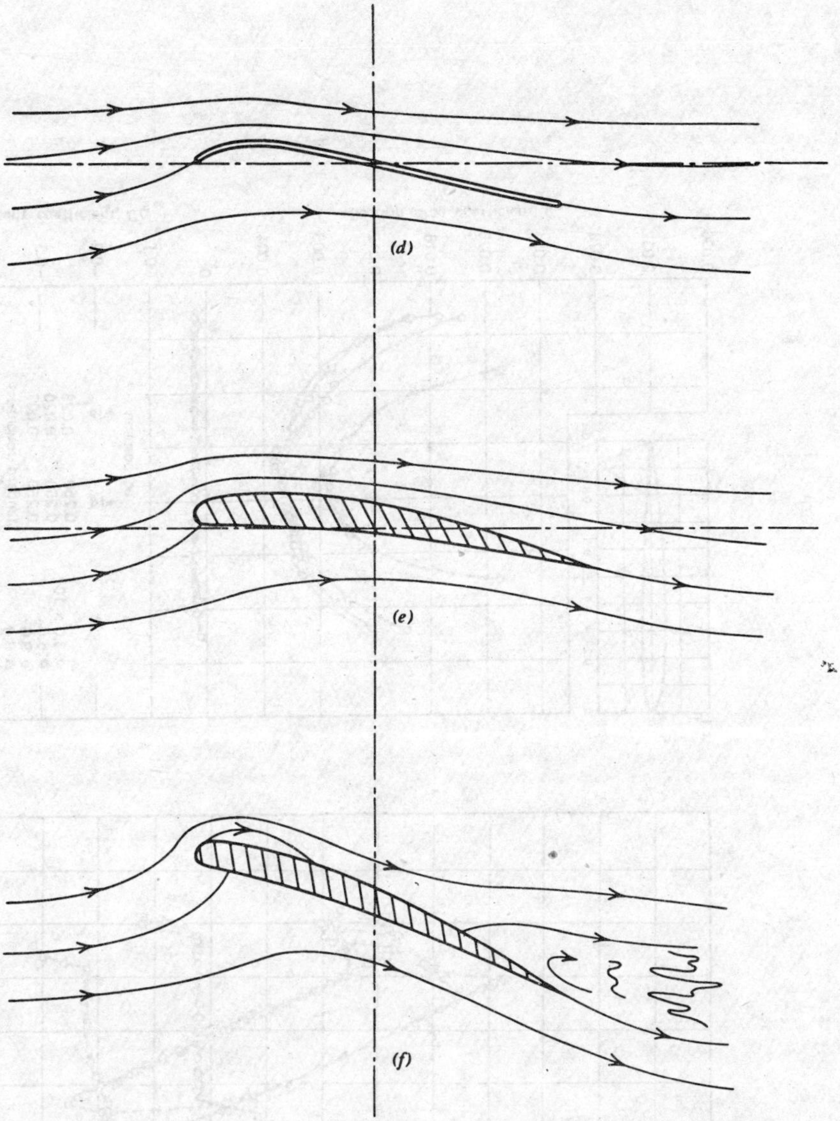

rounded, with a gradual decrease in C_l beyond an α of 14°. In Figure 3.7, for a thicker airfoil with the same camber, the lift increases up to an angle of approximately 16° for all R values tested. At this higher angle, even for $R = 9 \times 10^6$, it appears that leading edge separation occurs because of the sharp drop in C_l for α values greater than 16°. From a flying qualities standpoint, an airfoil with a well-rounded lift curve is desirable in order to avoid a sudden loss in lift as a pilot slows down the airplane. However, other factors such as drag and Mach number effects must also be considered in selecting an airfoil. Hence, as is true with most design decisions, the aerodynamicist chooses an airfoil that represents the best compromise to conflicting requirements, including nonaerodynamic considerations such as structural efficiency.

Figures 3.6 and 3.7 illustrate other characteristics of airfoil behavior that will be

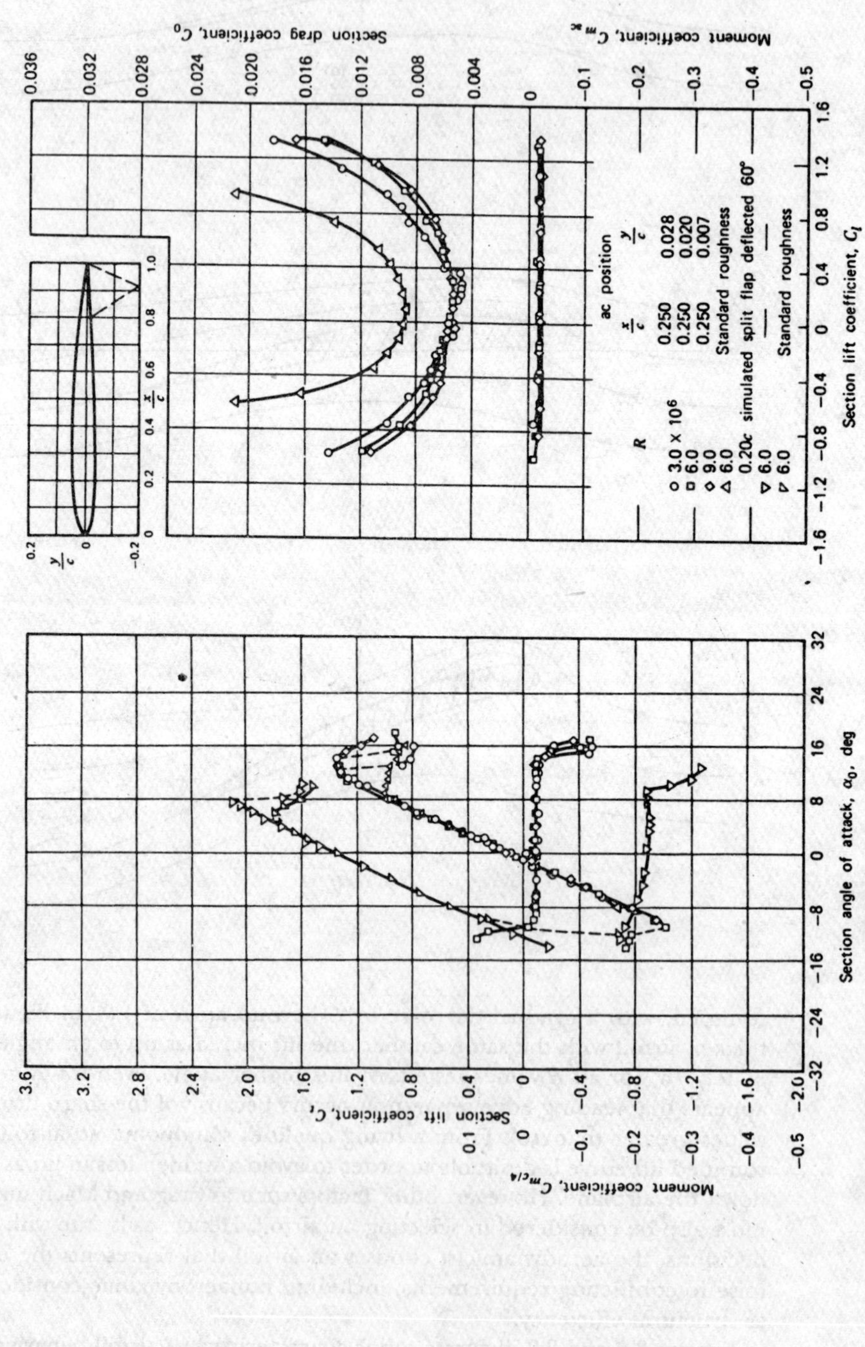

Figure 3.6 Aerodynamic characteristics of the NACA 1408 airfoil.

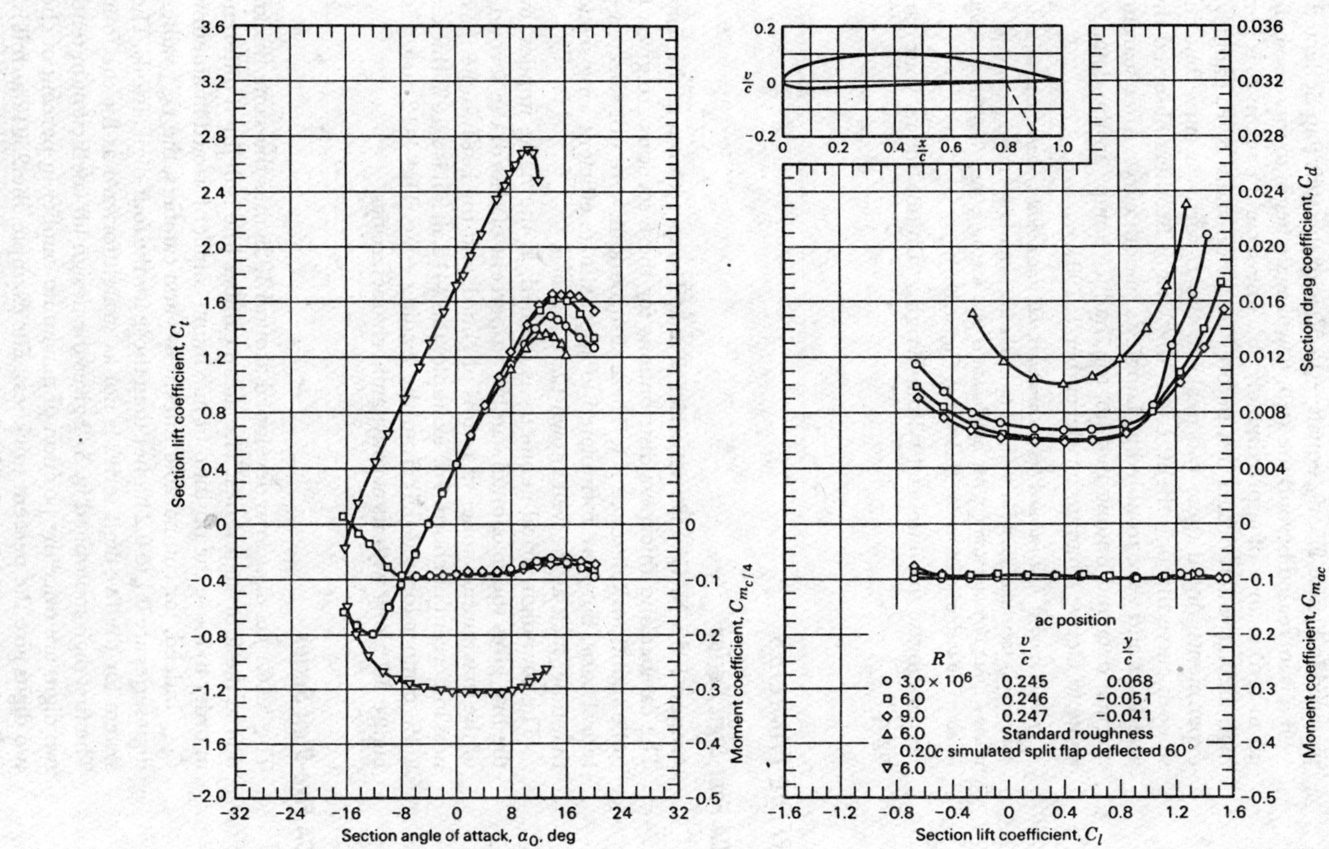

Figure 3.7 Aerodynamic characteristics of the NACA 4412 airfoil.

considered in more detail later. Observe that the lift curve, C_l versus α, is nearly linear over a range of angles of attack. Notice also that the slope, $dC_l/d\alpha$, of the lift curve over the linear portion is unchanged by deflecting the split flap. The effect of lowering the flap or, generally, of increasing camber is to increase C_l by a constant increment for each α in the linear range. Thus, the angle of attack for zero lift, α_{0l}, is negative for a cambered airfoil. In the case of the 1408 airfoil pictured in Figure 3.5, α_{0l} equals $-12.5°$, with the split flap deflected 60°.

If α is increased beyond the stall, C_l will again begin to increase before dropping off to zero at an α of approximately 90°. The second peak in C_l is generally not as high as that which occurs just before the airfoil stalls. S. P. Langley, in his early experiments, noted these two peaks in the C_l versus α curve but chose to fair a smooth curve through them. Later, the Wright Brothers observed the same characteristics and were troubled by Langley's smooth curve. After searching Langley's original data and finding that he, too, had a "bump" in the data, Wilbur Wright wrote to Octave Chanute on December 1, 1901.

> *If he (Langley) had followed his observations, his line would probably have been nearer the truth. I have myself sometimes found it difficult to let the lines run where they will, instead of running them where I think they ought to go. My conclusion is that it is safest to follow the observations exactly and let others do their own correcting if they wish (Ref. 1.1).*

To paraphrase the immortal Wilbur Wright, "Do not 'fudge' your data—it may be right."

AIRFOIL FAMILIES

NACA Four-Digit Series

Around 1932, NACA tested a series of airfoil shapes known as the *four-digit sections.* The camber and thickness distributions for these sections are given by equations to be found in Reference 3.1. These distributions were not selected on any theoretical basis, but were formulated to approximate efficient wing sections in use at that time, such as the well-known *Clark-Y section.*

The four-digit airfoil geometry is defined, as the name implies, by four digits; the first gives the maximum camber in percent of chord; the second the location of the maximum camber in tenths of chord; and the last two the maximum thickness in percent of chord. For example, the 4412 airfoil is a 12% thick airfoil having a 4% camber located $0.4c$ from the leading edge. The 4412 airfoil is pictured in Figure 3.7 along with its aerodynamic characteristics.

NACA Five-Digit Series

The *NACA five-digit series* developed around 1935 uses the same thickness distribution as the four-digit series. The mean camber line is defined differently, however, in order to move the position of maximum camber forward in an effort to increase $C_{l_{max}}$. Indeed, for comparable thicknesses and cambers, the $C_{l_{max}}$ values for the five-digit series are 0.1 to 0.2 higher than those for the four-digit airfoils. The numbering system for the five-digit series is not as straightforward as for the four-digit series. The first digit multiplied by 3/2 gives the design lift coefficient in tenths. The next two digits are twice the position of maximum camber in percent of chord. The last two digits give the percent thickness. For example, the 23012 airfoil is a 12% thick airfoil having a design C_l of 0.3 and a maximum camber located 15% of c back from the leading edge. This airfoil is also pictured in Figure 3.8.

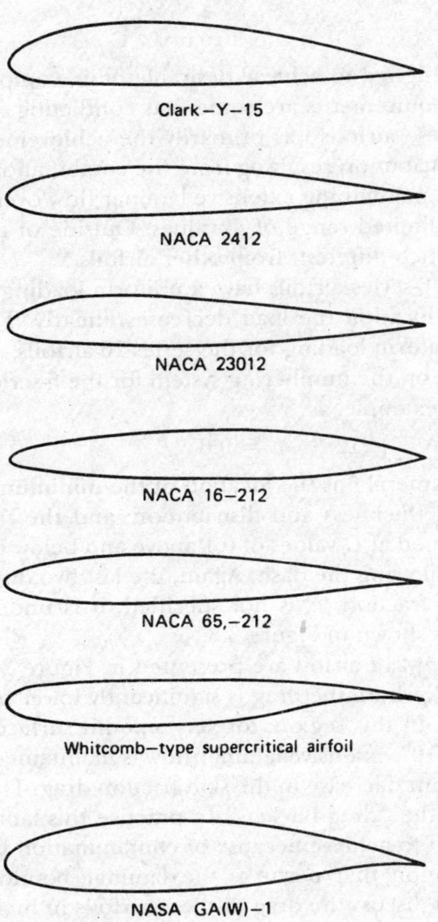

Figure 3.8 Comparison of various airfoil shapes.

NACA 1-Series (Series 16)

The *NACA 1-series* of wing sections developed around 1939 was the first series based on theoretical considerations. The most commonly used 1-series airfoils have the minimum pressure located at the 0.6c point and are referred to as *series-16 airfoils.* The camber line for these airfoils is designed to produce a uniform chordwise pressure difference across it. In the thin airfoil theory to follow, this corresponds to a constant chordwise distribution of vorticity.

Operated at its design C_l, the series-16 airfoil produces its lift while avoiding low-pressure peaks corresponding to regions of high local velocities. Thus, the airfoil has been applied extensively to both marine and aircraft propellers. In the former application, low-pressure regions are undesirable from the standpoint of cavitation (the formation of vaporous cavities in a flowing liquid). In the latter, the use of series-16 airfoils delays the onset of deleterious effects resulting from shock waves being formed locally in regions of high velocities.

Series-1 airfoils are also identified by five digits as, for example, the NACA 16–212 section. The first digit designates the series; the second digit designates the location of the minimum pressure in tenths of chord. Following the dash, the first number gives the design C_l in tenths. As for the other airfoils, the last two digits designate the maximum thickness in percent of chord. The 16–212 airfoil is shown in Figure 3.8.

NACA 6-Series

The *NACA 6-series airfoils* were designed to achieve desirable drag, compressibility, and $C_{l_{max}}$ performance. These requirements are somewhat conflicting, and it appears that the motivation for these airfoils was primarily the achievement of low drag. The chordwise pressure distribution resulting from the combination of thickness and camber is conducive to maintaining extensive laminar flow over the leading portion of the airfoil over a limited range of C_l values. Outside of this range, C_d and $C_{l_{max}}$ values are not too much different from other airfoils.

The mean lines used with the 6-series airfoils have a uniform loading back to a distance of $x/c = a$. Aft of this location the load decreases linearly. The $a = 1$ mean line corresponds to the uniform loading for the series-16 airfoils.

There are many perturbations on the numbering system for the 6-series airfoils. The later series is identified, for example, as

$$\text{NACA } 65_1\text{-}212 \qquad a = 0.6$$

Here 6 denotes the series; the numeral 5 is the location of the minimum pressure in tenths of chord for the basic thickness and distribution; and the subscript 1 indicates that low drag is maintained at C_l values of 0.1 above and below the design C_l of the 0.2, denoted by the 2 following the dash. Again, the last two digits specify the percentage thickness. If the fraction, a, is not specified, it is understood to equal unity. The 65_1-212 airfoil is shown in Figure 3.8.

Lift and drag curves for the 65_1-212 airfoil are presented in Figure 3.9. Notice the unusual shape of C_d versus C_l, where the drag is significantly lower between C_l values of approximately 0 to 0.3. In this region, for very smooth surfaces and for Reynolds numbers less than 9×10^6, extensive laminar flow is maintained over the surface of the foil with an attendant decrease in the skin friction drag. This region, for obvious reasons, is known as the "drag bucket." In practice this laminar flow, and resulting low drag, is difficult to achieve because of contamination by bugs or by structurally transmitted vibration that perturbs the laminar boundary layer, causing transition. Chapter 4 will discuss the drag of these airfoils in more detail.

MODERN AIRFOIL DEVELOPMENTS

Systematic series of airfoils have given way, at least in part, to specialized airfoils designed to satisfy particular requirements. These airfoils are synthesized with the use of sophisticated computer programs such as the ones described in References 3,2, 3.3, and 3.4, which will be discussed in more detail later. One such special purpose airfoil is the so-called *supercritical airfoil* reported on in References 3.5, 3.6, 3.7, and 3.8. This airfoil has a well-rounded leading edge and is relatively flat on top with a drooped trailing edge. For a constant thickness of 12%, wind tunnel studies indicate a possible increase of approximately 15% in the drag-divergence Mach number for a supercritical airfoil as compared to a more conventional 6-series airfoil. In addition, the well-rounded leading edge provides an improvement in $C_{l_{max}}$ at low speeds over the 6-series which has sharper leading edges.

A qualitative explanation for the superior performance of the supercritical airfoil is found by reference to Figure 3.10. At a free-stream Mach number as low as 0.7 or so depending on the shape and C_l, a conventional airfoil will accelerate the flow to velocities that are locally supersonic over the forward or middle portion of its upper surface. The flow then decelerates rapidly through a relatively strong shock wave to subsonic conditions. This compression wave with its steep positive pressure gradient causes the boundary layer to thicken and, depending on the strength of

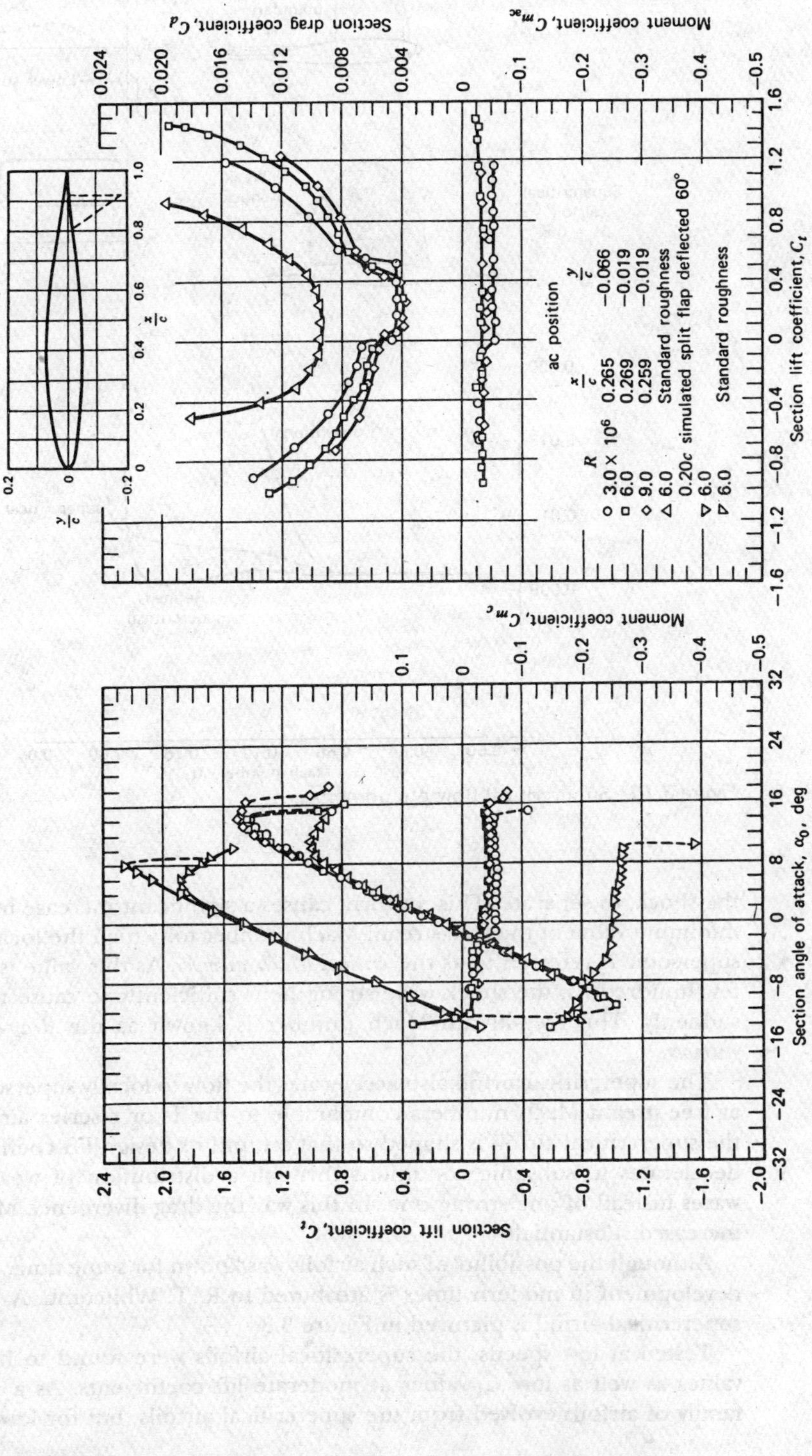

Figure 3.9 Aerodynamic characteristics of the NACA 65_1-212 a = 0.6 airfoil.

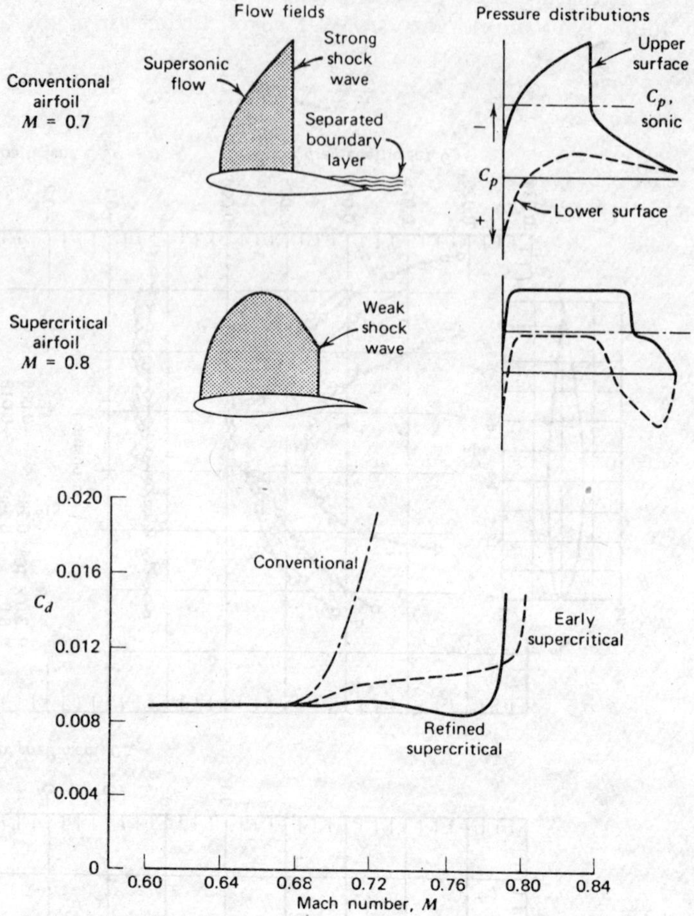

Figure 3.10 Supercritical flow phenomena.

the shock, to separate. This, in turn, causes a significant increase in the drag. The minimum value of the free-stream Mach number for which the local flow becomes supersonic is referred to as the *critical Mach number.* As this value is exceeded by a few hundredths, the shock wave strengthens sufficiently to cause the drag to rise suddenly. This free-stream Mach number is known as the *drag-divergence Mach number.*

The supercritical airfoil also accelerates the flow to locally supersonic conditions at free-stream Mach numbers comparable to the 1- or 6-series airfoils. However, the supercritical airfoil is shaped so that around its design lift coefficient, the flow decelerates to subsonic conditions through a distribution of weak compression waves instead of one strong one. In this way the drag-divergence Mach number is increased substantially.

Although the possibility of such airfoils was known for some time, their successful development in modern times is attributed to R. T. Whitcomb. A Whitcomb-type supercritical airfoil is pictured in Figure 3.8.

Tested at low speeds, the supercritical airfoils were found to have good $C_{l_{max}}$ values as well as low C_d values at moderate lift coefficients. As a result, another family of airfoils evolved from the supercritical airfoils, but for low-speed applications. These are the "general aviation" airfoils, designated GA(W) for general aviation (Whitcomb). The GA(W)-1 airfoil is the last of the airfoils pictured in Figure 3.10. Test results for this airfoil are reported in Reference 3.9, where its $C_{l_{max}}$ values are shown to be about 30% higher than those for the older NACA 65-series airfoils. In addition, above C_l values of around 0.6, its drag is lower than the older laminar flow series with standard roughness. These data are presented in Figure 3.11 for the GA(W)-1 airfoil. Comparisons of $C_{l_{max}}$ and C_d for this airfoil with similar coefficients for other airfoils are presented in Figures 3.12 and 3.13.

Observe that the performance of the GA(W)-1 airfoil is very Reynolds number-dependent, particularly $C_{l_{max}}$, which increases rapidly with Reynolds numbers from 2 to 6 million. NASA has adopted an alternate notation for the GA(W). Currently, they are designated by LS (low speed) or MS (medium speed) followed by four digits. For example, the GA(W)-1 airfoil is now designated as the LS(1)-0417. The (1) refers to a family. The 04 defines a design lift coefficient of 0.4, and 17 is the maximum thickness ratio in percent.

The GA(W) airfoils have found limited application to general aviation airplanes. However, supercritical airfoils have found extensive application to high-subsonic transports.

The Eppler Code, Reference 3.3, is one that finds application to the inverse design problem. In the inverse problem, instead of predicting the velocity field around a given airfoil shape, the velocity field is prescribed and the airfoil shape then determined, which will give the desired velocity profile. The Eppler code will then find the airfoil shape using a *conformal-mapping* technique (a potential flow method beyond the level of this text). The conformal mapping is connected within the code to a boundary layer prediction method that allows the designer to examine the stability of the layer during the airfoil design process. This can allow airfoils to be optimized for multipoint applications by assuring favorable *pressure gradients* on the upper or lower surfaces.

In 1977, the Eppler code was used by Somers to design an airfoil designated NASA NLF(1)-0416 (Ref. 3.10) for application to an advanced, light, single-engine, general-aviation aircraft. Several objectives were defined for the design. It was to have extensive laminar flow with a maximum C_l at least as high as the best NASA airfoils without laminar flow. Concurrently with this requirement, the $C_{l_{max}}$ was not to be affected if the laminar flow became turbulent as the result of roughness (bugs, heavy rain, etc.). Second, the drag coefficient corresponding to the C_l for cruise was to be similar to the values obtained by the series-6 airfoils. Finally, the drag coefficient at a C_l of 1.0 was to be lower than the values obtained by existing airfoils.

With these objectives in mind, four constraints were placed on the design. First, the extent of the favorable pressure gradient on the upper surface at the cruise C_l was set at 30% of the chord. This was a conservative condition premised by existing experimental results. Second, the minimum thickness of the airfoil was set at 12% of the chord from structural considerations. Third, the pitching moment coefficient about the quarter-chord point was specified to be no more negative than -0.1. Finally, the airfoil was not to employ a flap.

The resulting shape of this airfoil is given in Figure 3.14 with some experimental results. Suffice it to say that the measured and predicted chordwise pressure distributions for C_l values of 0.45 and 1.0 are in close agreement. It can be seen from Figure 3.14 that the maximum C_l for this airfoil is unaffected by roughness. This is attributed to the fact that the transition point on the upper surface moves steadily toward the leading edge as C_l increases. Thus at $C_{l_{max}}$ the transition point for the smooth airfoil is the same as it would be if the laminar boundary was tripped by roughness.

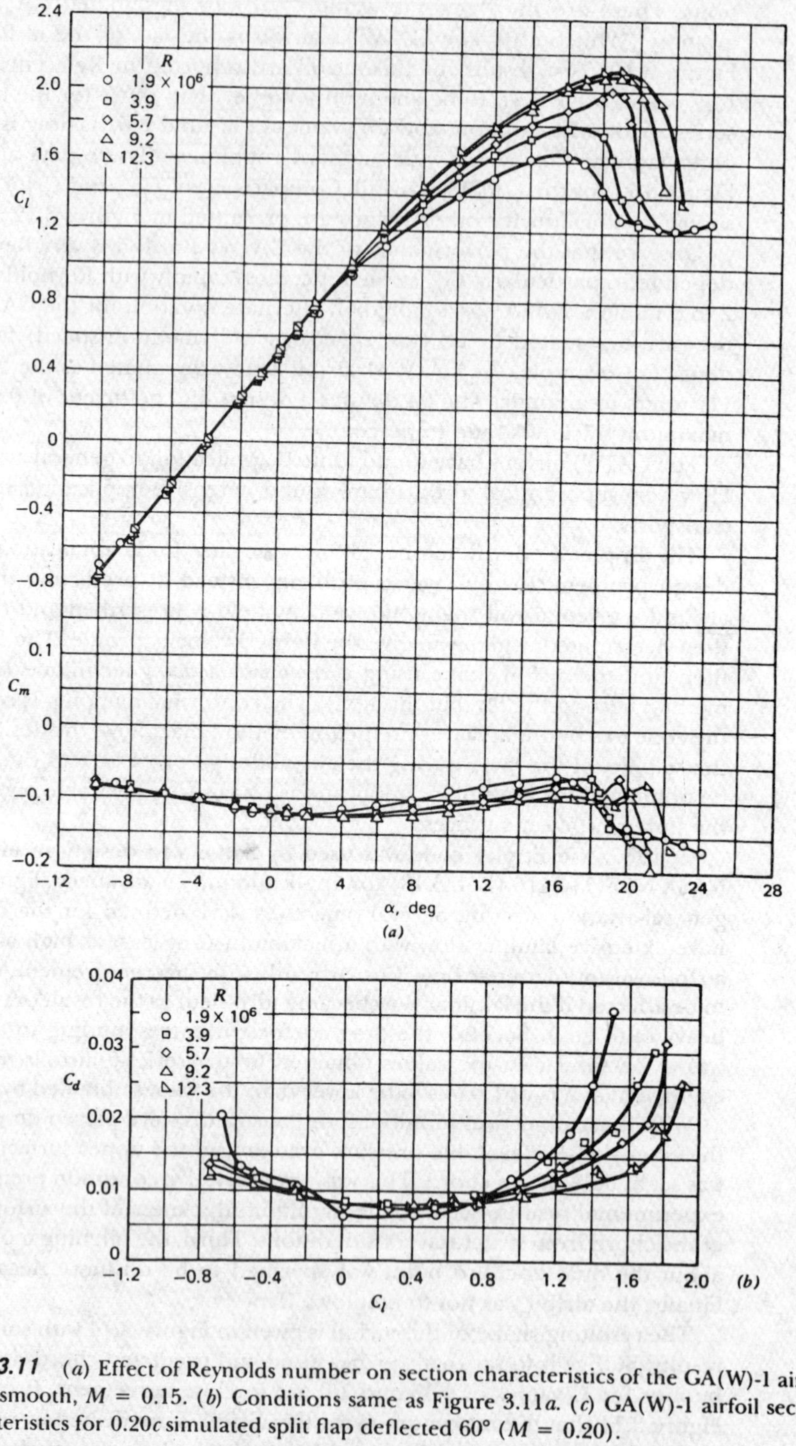

Figure 3.11 (*a*) Effect of Reynolds number on section characteristics of the GA(W)-1 airfoil model-smooth, $M = 0.15$. (*b*) Conditions same as Figure 3.11*a*. (*c*) GA(W)-1 airfoil section characteristics for $0.20c$ simulated split flap deflected 60° ($M = 0.20$).

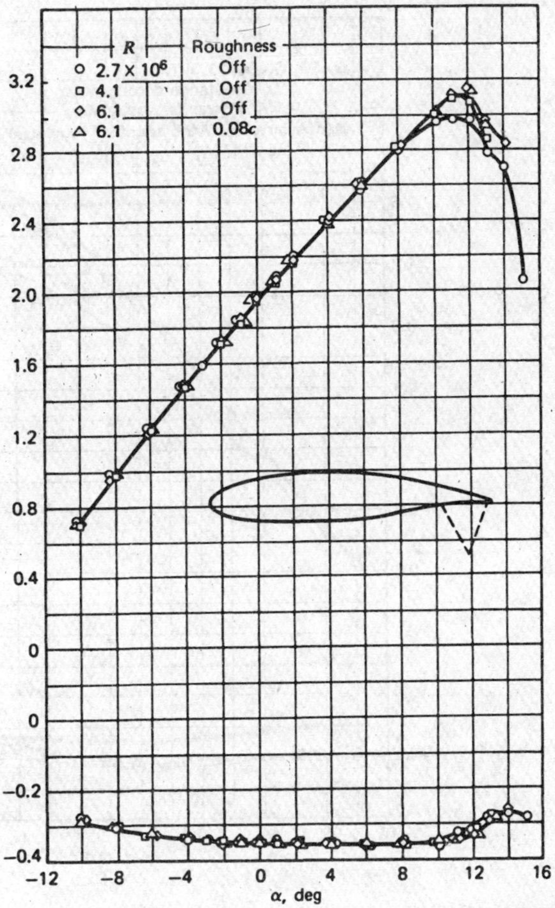

(c)

THIN AIRFOIL THEORY

In Chapter 2 it was noted that the concepts of a point vortex and a point source could be extended to a continuous distribution of the elementary flow functions. In that chapter, a distribution of sources in a uniform flow was found to produce a nonlifting body of finite thickness. In the case of the circular cylinder, the addition of a vortex also produced lift.

Comparable to the continuous distribution of sources pictured in Figure 2.20, consider a similar distribution of vortices as illustrated in Figure 3.15. Such a distribution is referred to as a *vortex sheet*. If γ is the strength per unit length of the sheet, $\gamma\,\Delta x$ will be the total strength enclosed by the dashed contour shown in the figure. The contour is taken to lie just above and below the sheet. Δx is sufficiently small so that the velocity tangent to the sheet, v, can be assumed to be constant. Because of the symmetry to the flow provided by any one segment of the sheet, the tangential velocity just below the sheet is equal in magnitude but opposite in direction to that just above the sheet. From Equation 2.55, relating circulation to the strength of a vortex, it follows that

$$\gamma\,\Delta x = 2v\,\Delta x$$

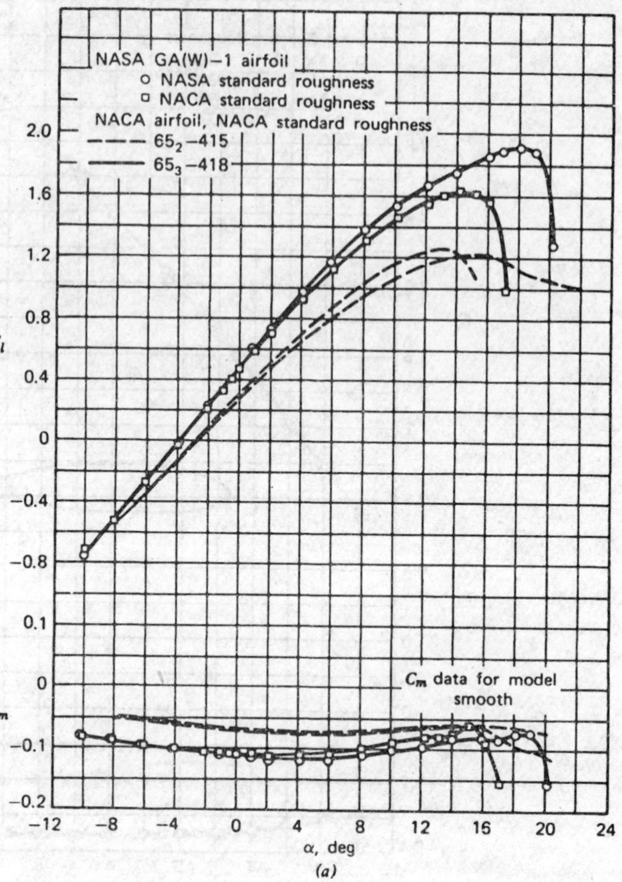

(a)

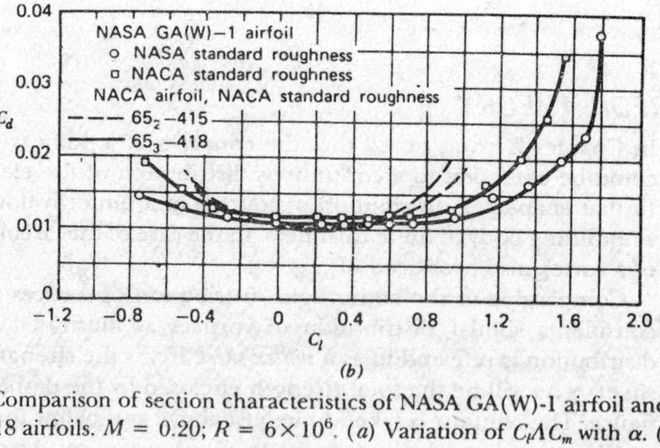

(b)

Figure 3.12 Comparison of section characteristics of NASA GA(W)-1 airfoil and NACA 65_2-415 and 65_3-418 airfoils. $M = 0.20$; $R = 6 \times 10^6$. (*a*) Variation of C_l/C_m with α. (*b*) Variation of C_d with C_l.

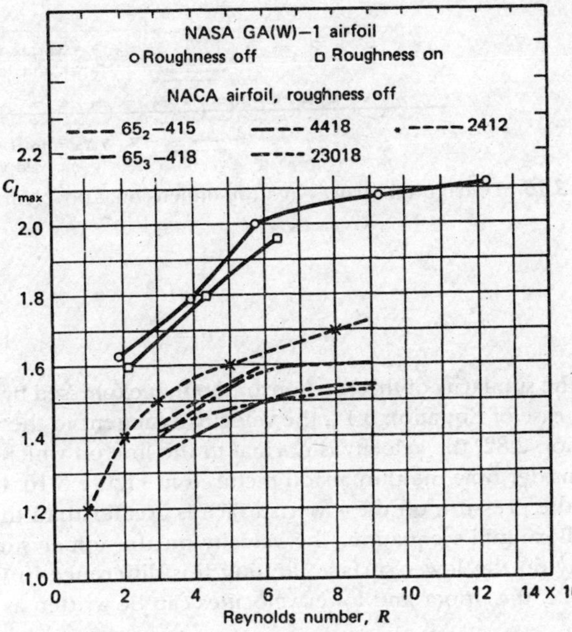

Figure 3.13 Comparison of maximum lift coefficient of the GA(W)-1 airfoil with other NACA airfoils $M = 0.15$.

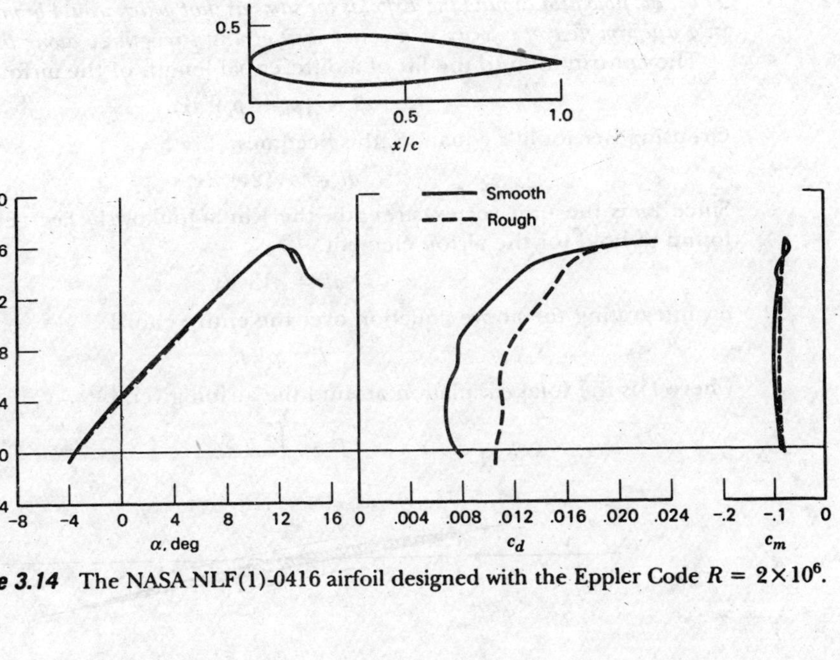

Figure 3.14 The NASA NLF(1)-0416 airfoil designed with the Eppler Code $R = 2\times10^6$.

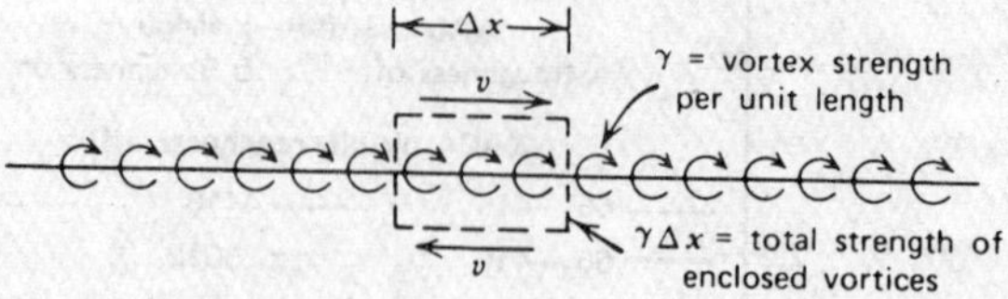

Figure 3.15 Distributed vortices in a two-dimensional flow (vortex sheet).

or

$$v = \frac{\gamma}{2} \tag{3.17}$$

Note the similarity of this relationship to that expressed by Equation 2.82. However, in the case of Equation 3.14, the velocity is tangent to the vortex sheet, whereas for Equation 2.82, the velocity is normal to the line on which the sources lie.

Consider now the thin airfoil pictured in Figure 3.16. If the airfoil is producing a lift, the pressure on the lower surface is greater than that on the upper. Hence, from Bernoulli's equation, the velocity on the upper surface is greater than the velocity on the lower surface. Letting this difference in velocity across the airfoil equal $2v$, the upper and lower velocities can be written as

$$V_u = V + v$$

and

$$V_l = V - v$$

Thus, the flow field around the airfoil is the same as that which would be produced by placing, in a uniform flow of velocity V, a vortex sheet of unit strength 2v along the airfoil.

The contribution to the lift of a differential length of the airfoil will be

$$dl = (p_l - p_u)\ dx$$

Or, using Bernoulli's equation, this becomes

$$dl = \rho V(2v)\ dx$$

Since $2v$ is the unit vortex strength, the Kutta–Joukowski law (Equation 2.81) is found to hold for the airfoil element.

$$dl = \rho V\gamma\ dx$$

or, integrating the above equation over the entire chord,

$$l = \rho V\Gamma \tag{3.18}$$

where Γ is the total circulation around the airfoil given by

$$\Gamma = \int_0^c \gamma\ dx \tag{3.19}$$

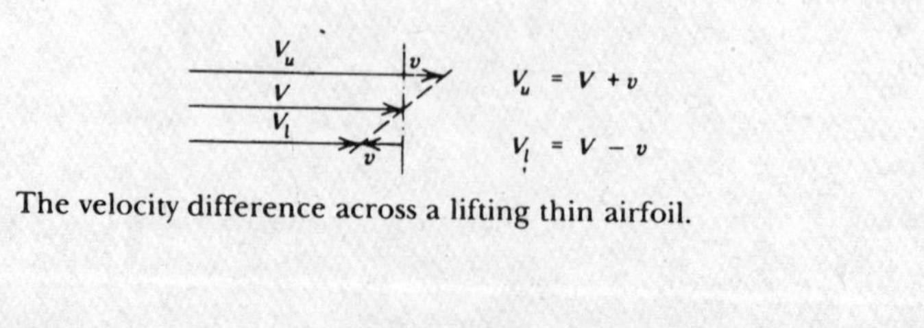

Figure 3.16 The velocity difference across a lifting thin airfoil.

In order to predict the lift and moment on the airfoil, one must find the chordwise distribution of $\gamma(x)$ that will produce a resultant flow everywhere tangent to the mean camber line (thin airfoil approximation). In addition, the Kutta condition is applied at the trailing edge to assure that the flow *leaves* the trailing edge tangent to the mean camber line at that point. This is a necessary condition; otherwise, the resulting flow will appear similar to Figure 3.4*a* with the lift being equal to zero.

Analytical Solution

Analytical solutions to the thin airfoil can be found in several texts (e.g., Refs. 3.11 and 3.12). Here, the airfoil is replaced by a continuous distribution of vortices instead of discrete point vortices as used with the numerical solution.

Referring to Figure 3.17, without any loss of generality, the airfoil is taken to have a unit chord lying along the x-axis with the origin at the leading edge. The shape of the camber line is given by $z(x)$, and it is assumed that

$$z(x) \ll 1$$

With this assumption, the problem is linearized and made tractable by replacing the airfoil with a vortex sheet of unit strength $\gamma(x)$ lying along the chord line instead of along the camber line.

At the point x_0, the downward velocity induced by an elemental vortex of strength $\gamma(x)\ dx$ located at x, according to Equation 2.56, will be given by

$$dw(x_0) = \frac{\gamma(x)\ dx}{2\pi(x_0 - x)}$$

or, integrating over the chord,

$$w(x_0) = \frac{1}{2\pi}\int_0^1 \frac{\gamma(x)\ dx}{x_0 - x} \tag{3.20}$$

In order to satisfy the boundary condition that the flow be tangent everywhere to the mean camber line, it follows that, to a small angle approximation

$$\frac{w(x_0)}{V} = \alpha - \left(\frac{dz}{dx}\right)_{x_0} \tag{3.21}$$

Thus, given α and $z(x)$, the following integral equation must be solved for $\gamma(x)$.

$$\frac{1}{2\pi V}\int_0^1 \frac{\gamma(x)\ dx}{x_0 - x} = \alpha - \left(\frac{dz}{dx}\right)_{x_0} \tag{3.22}$$

In addition, $\gamma(x)$ must vanish at the trailing edge in order to satisfy the Kutta condition. Otherwise, the induced velocity will be infinite just downstream of this point.

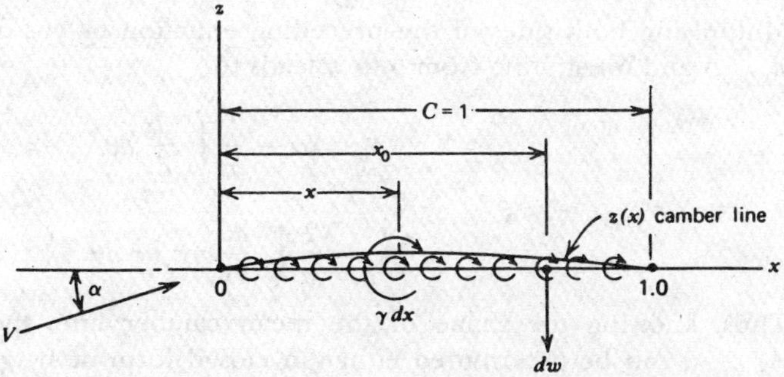

Figure 3.17 The modeling of a thin airfoil by a vortex sheet.

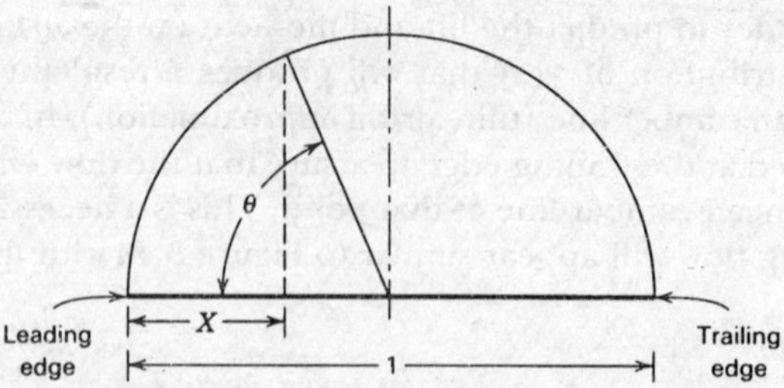

Figure 3.18 Transformation from rectangular to polar coordinates for an airfoil.

In order to solve Equation 3.22, a transformation to polar coordinates is made as shown in Figure 3.18. From this figure, the dimensionless distance x can be expressed in terms of the angle, θ.

$$x = \frac{1}{2}(1 - \cos\theta) \tag{3.23}$$

Equation 3.22 becomes

$$\frac{1}{2\pi V}\int_0^\pi \frac{\gamma(\theta)\sin\theta\, d\theta}{\cos\theta - \cos\theta_0} = \alpha - \left(\frac{dz}{dx}\right)_{x_0} \tag{3.24}$$

On the basis of the more sophisticated method of conformal mapping (e.g., see Ref. 3.13), it is known that $\gamma(x)$ is generally singular at the leading edge approaching infinity as $1/x$. Thus, we will assume a priori that Equation 3.24 can be satisfied by a $\gamma(\theta)$ distribution of the form

$$\gamma = 2V\left[A_0\frac{(1+\cos\theta)}{\sin\theta} + \sum_{n=1}^{\infty} A_n \sin n\theta\right] \tag{3.25}$$

Using the relationships

$$\frac{1}{2}\left[\cos(n-1)\theta - \cos(n+1)\theta\right] = \sin n\theta \sin\theta$$

and

$$\int_0^\pi \frac{\cos n\theta\, d\theta}{\cos\theta - \cos\theta_0} = \pi\frac{\sin n\theta_0}{\sin\theta_0} \tag{3.26}$$

Equation 3.24 becomes

$$A_0 - \sum_{n=1}^{\infty} A_n \cos n\theta = \alpha - \frac{dz}{dx}$$

Multiplying both sides of the preceding equation by $\cos m\theta$ ($m = 0, 1, 2, \ldots, n, \ldots$) and integrating from 0 to π leads to

$$A_0 = \alpha - \frac{1}{\pi}\int_0^\pi \frac{dz}{dx}\, d\theta \tag{3.25a}$$

$$A_n = \frac{2}{\pi}\int_0^\pi \frac{dz}{dx}\cos n\theta\, d\theta \tag{3.25b}$$

Thus, knowing the shape of the mean camber line, the coefficients, A_0, A_1, A_2, . . ., can be determined either in closed form or by graphical or numerical means (see Ref. 3.1). Having these coefficients, C_l and C_m can then be easily determined from the Kutta–Joukowski relationship.

The lift and moment about the leading edge are given by

$$L = \int_0^1 \rho V \gamma(x)\, dx$$

$$M_{\mathrm{LE}} = -\int_0^1 \rho V \gamma(x)\, x\, dx$$

From these and using Equation 3.25,

$$C_l = 2\pi A_0 + \pi A_1 \tag{3.26}$$

$$C_{m_{\mathrm{LE}}} = -\frac{\pi}{2}\left(A_0 + A_1 - \frac{A_2}{2}\right) \tag{3.27}$$

It follows that C_m about the quarter-chord point is independent of α, so that this point is the aerodynamic center, with the moment coefficient being given by

$$C_{m_{ac}} = -\frac{\pi}{4}(A_1 - A_2) \tag{3.28}$$

Since α is contained only in the A_0 coefficient, it can be concluded immediately without considering the actual form of $z(x)$ that C_l is given by a linear combination of α and a function of z. Thus, camber changes can be expected to affect the angle of zero lift but not the slope of the lift curve.

Reference to airfoil data, such as that presented in Figures 3.6 and 3.7, will show that the predictions of thin airfoil theory are essentially correct. There is a range of angles of attack over which the lift coefficient varies linearly with α. The slope of this lift curve is usually not as high as the theory predicts, being approximately 4 to 8% less than the theoretical value. For many purposes an assumed value of 0.1 C_l/deg is sufficiently accurate and is a useful number to remember. Experimental data also show the aerodynamic center to be close to the quarter-chord point. The effects of camber of C_l, $dC_l/d\alpha$, and $C_{m_{ac}}$ are also predicted well.

Leading Edge Suction Force

The form of Equation 3.25 shows that generally a thin airfoil at an angle of attack will experience a leading edge singularity. As the flow attempts to curve around the leading edge without separating, very low pressures are produced at the nose. In the limit of zero thickness, the pressure at the nose becomes infinitely negative as the area over which it acts approaches zero. It can be shown that the product of the pressure and the area over which it acts approaches a limiting force known as the leading edge suction force.

The leading edge force can be determined easily from D'Alembert's paradox, which says there can be no drag in a potential flow. For a flat plate airfoil, the pressure can only be normal to the surface. Thus, the pressure difference across the surface producing a lift coefficient of C_l must produce a dimensionless coefficient in the drag direction, $C_l\alpha$. Since the drag must be zero, $C_l\alpha$ must also be the value of the suction force in the forward direction. In the general case, the magnitude of the suction force can be obtained by integrating the downstream component of the pressure over the chord.

The Circular Arc Airfoil

As an example in the use of the foregoing relationships, consider a thin airfoil whose mean camber line is a circular arc having a radius of R and a chord of one. Such an airfoil is shown in Figure 3.19. The maximum value of the camber ratio can be written in terms of the angle Φ and R as,

$$z_{\max} = R(1 - \cos\Phi) \tag{3.29}$$

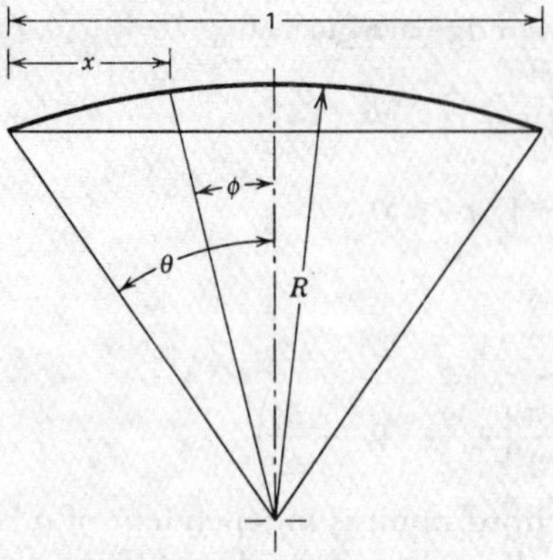

Figure 3.19 Geometry of a circular arc airfoil.

For $z_{max} \ll 1$ the angle Φ is small so that $\cos(\Phi) \simeq 1 - \frac{\Phi^2}{2}$. Thus it follows that

$$R = \frac{1}{8z_{max}} \tag{3.30}$$

The camber ratio, z, will be simply

$$z = R[\cos(\theta) - \cos(\Phi)] \tag{3.31}$$

or for small angles,

$$z = z_{max}\left[1 - \left(\frac{\Phi}{\theta}\right)^2\right] \tag{3.32}$$

In terms of the distance from the leading edge, x, the camber becomes

$$\frac{z}{z_{max}} = 1 - (1 - 2x)^2 \tag{3.33}$$

If the coordinate transformation given by Equation 3.23 is substituted into the above, it is found that the circular arc airfoil is described simply by

$$\frac{z}{z_{max}} = \sin^2(\theta) \tag{3.34}$$

The slope, dz/dx, is obtained immediately from Figure 3.19 as equal to the angle Φ.

$$\frac{dz}{dx} = \Phi = 4z_{max}\cos(\theta) \tag{3.35}$$

If these results are substituted into Equations 3.25 (*a*) and (*b*), then the following is obtained.

$$A_0 = \alpha \tag{3.36}$$

$$A_1 = 4z_{max} \tag{3.37}$$

$$A_2 = 0 \tag{3.38}$$

Thus, for a thin airfoil with a circular arc camber line,

$$C_l = 2\pi\alpha + 4\pi z_{max} \tag{3.39}$$

$$C_{m1/4} = -\pi z_{max} \tag{3.40}$$

$$C_{m_{l_e}} = -\frac{\pi}{2}(\alpha + 4z_{max}) \tag{3.41}$$

Observe, from Equations 3.11, 3.39, and 3.41, that as C_l decreases, the center of pressure moves aft, approaching infinity as C_l goes to zero. This movement of the center of pressure is opposite to what was believed to be true by the early pioneers in aviation. The Wright Brothers were probably the first to recognize the true nature of the center of pressure movement as a result of their meticulous wind tunnel tests.

Numerical Solution

A numerical solution to the thin airfoil can be obtained by replacing the continuous distribution of vorticity, $\gamma(x)$, by a row of discrete vortices. The airfoil chord is divided into N equal segments with a vortex of unknown strength placed at the *quarter-chord point of each segment.* The unknown strengths of the vortex elements are determined by assuring that the normal component of the resultant velocity to the chord vanishes at the three-quarter chord point of each segment. With the last control point downstream of the last vortex singularity, the Kutta condition is assured. Such a model is illustrated in Figure 3.20. Consistent with the approximation used in the analytical solution for a thin airfoil, the vortex elements and control points are placed along a straight line.

Referring to Figure 3.20, the chord in this instance is divided into eight elements. One of the elements, the Jth element, is shown toward the leading edge with a vortex of unknown strength, $\gamma(J)$, placed a quarter of the element's length back from the leading edge of the element. Another element, the Ith element, is shown downstream of the Jth element with a control point at three-quarters of the element's length back from the leading edge of that element. These elements are shown expanded in the lower part of the figure together with the velocity components at the Ith control point.

The downward velocity induced at the Ith control point by the vortex at the Jth element is denoted by $w(I, J)$ and can be calculated by

$$w(I, J) = \frac{\gamma(J)}{2\pi\,[xc(I) - xv(J)]} \tag{3.42}$$

where $xc(I)$ is the distance of the control point from the leading edge of the airfoil and $xv(J)$ is the similar distance for the vortex element.

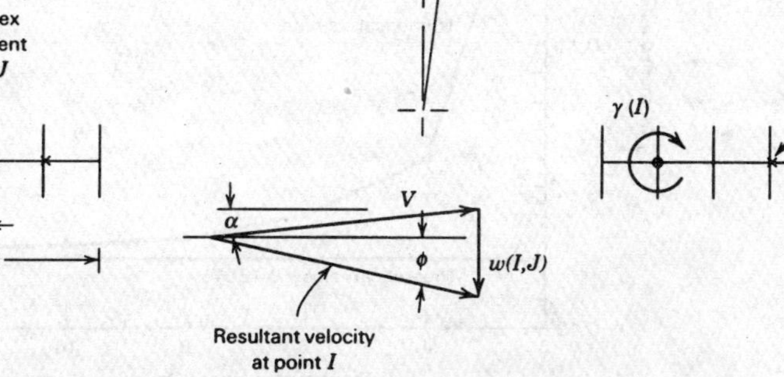

Figure 3.20 Numerical model for a thin, cambered airfoil at an angle of attack.

The total velocity induced downward at the Ith control point by the distribution of vortex elements over the chord is obtained by summing Equation 3.42 over the number of elements. The resultant of this velocity, $w(I)$, and the free-stream velocity, V, shown in Figure 3.19, must be tangent to the mean camber line at the Ith point. Since the angle Φ in this figure is the negative of that shown in Figure 3.18, it follows that the vortex strengths, $\gamma(J)$, must be such as to assure that the following holds.

$$\sum_{J=1}^{N} \frac{\gamma(J)}{2\pi\,[xc(I) - xv(J)]} = V\left[\alpha - \frac{dz}{dx}(I)\right] \qquad (3.43)$$

Satisfying this relationship at $I = 1, 2, \ldots, N$, control points results in N simultaneous, linear, algebraic equations, which can be solved for the vortex strengths, $\gamma(I)$. In performing these calculations, one can let ρ and V equal unity without any loss of generality in solving for the dimensionless coefficients.

Knowing the unknown vortex strengths, $\gamma(I)$, the section lift coefficient and moment coefficient about the leading edge can immediately be obtained as

$$C_l = 2\sum_{I=1}^{N} \gamma(I) \qquad (3.44)$$

$$C_{m_{le}} = -2\sum_{i=1}^{N} xv(I)\,\gamma(I) \qquad (3.45)$$

The above equations assume ρ and V equal to 1. The moment coefficient about the aerodynamic center is then determined by transferring the above force and moment to the quarter-chord point.

$$C_{m_{1/4}} = C_{m_{le}} + \frac{C_l}{4}$$

This numerical model has been run for the circular arc airfoil discussed earlier with the results shown in Figures 3.20 and 3.21. Figure 3.21 illustrates how rapidly the numerical model converges to the closed-form predictions as the number, N, of elements is increased.

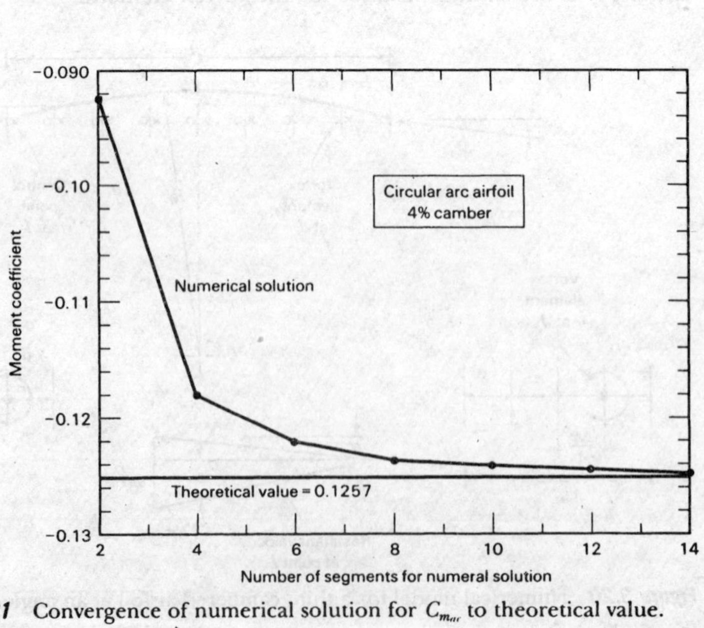

Figure 3.21 Convergence of numerical solution for $C_{m_{ac}}$ to theoretical value.

The predicted section lift coefficient of 0.503 agrees exactly with the theory for all values of N. In other words, a single vortex placed at the airfoil's quarter-chord point and set to a strength, which satisfies the boundary condition at only one control point, namely, the three-quarter chord point, results in the correct lift coefficient. This very simple two-point model will be applied later to a wing. Known as *Weissinger's approximation,* the model works well if there are no discontinuities in the shape of the mean camber near the three-quarter chord point such as a deflected flap.

The predicted value of C_l is somewhat higher than that for the 4412 airfoil shown in Figure 3.7. However, the maximum camber for this airfoil is at 40% of the chord instead of 50%. This would reduce the slope at the three-quarter chord point resulting in a predicted C_l approximately 20% lower than for the circular arc airfoil.

The moment coefficient is a measure of how well the numerical model predicts the chordwise loading distribution. As shown in Figure 3.21, this coefficient converges rapidly to the theoretical value of $C_{m_{1/4}} = 0.1257$ as N is increased above 2. For N equal to 16, as in this figure, $C_{m_{le}}$ from the numerical model agrees within 0.2% of the theoretical value given by Equation 3.41. This close agreement is also obvious from Figure 3.22, which presents the chordwise distribution of the vortex strength over an element as calculated numerically and from Equation 3.25.

Figure 3.23 is a comparison of the chordwise loading distribution for two airfoils operating at the same lift coefficient. In the one case, the lift is obtained purely by camber, whereas the other airfoil has no camber and generates its lift by angle of attack. Generally, one will favor the type of loading obtained through camber since it avoids the high loading at the leading edge resulting from angle of attack. By distributing the loading more uniformly over the airfoil, one avoids high local velocities, which can result in unfavorable compressibility effects. Further, the unfavorable pressure gradient on the upper surface downstream of the leading edge can result in premature flow separation.

Computer Exercise 3.1 "AIRFOIL"

Formulate a numerical model using point vortices as described above to predict the chordwise loading and aerodynamic coefficients for a cambered airfoil with a

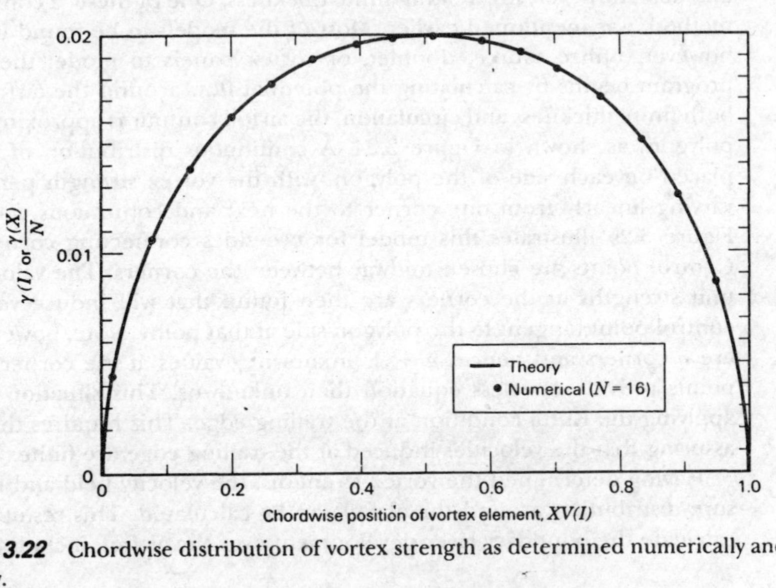

Figure 3.22 Chordwise distribution of vortex strength as determined numerically and from theory.

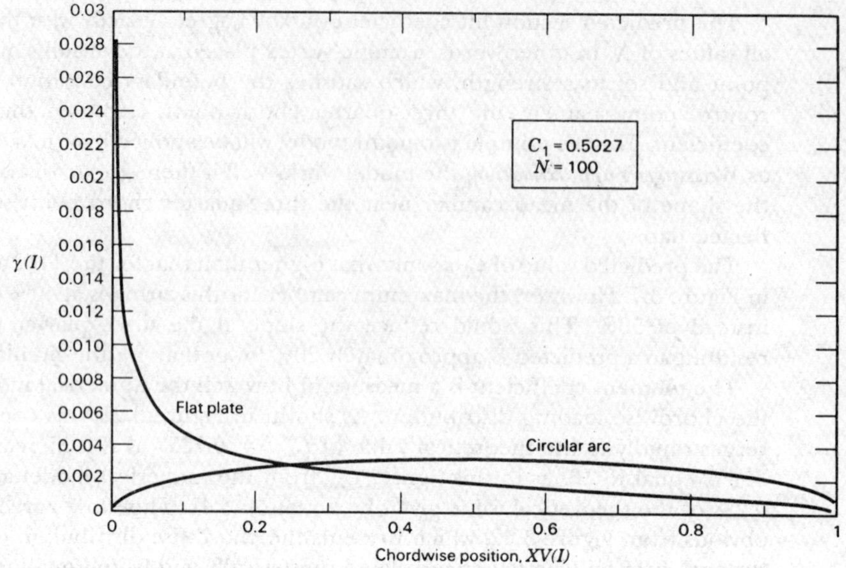

Figure 3.23 Chordwise loading for a cambered airfoil compared to the loading for a flat plate airfoil at the same lift coefficient.

flap at an angle of attack. Use the model to compare with Figure 3.23 and with Equations 3.39 and 3.40.

Numerical Model Including Thickness

The thin airfoil model is useful in predicting overall characteristics such as lift and moment coefficients but is not sufficient to predict accurately the local static pressure on the upper and lower surfaces. In order to predict these distributions, it is necessary to consider the finite thickness of the airfoil. This is difficult to do in a closed form except for a few special cases. There are several numerical models available, however, for treating finite thickness. One of these, a conformal mapping method, was mentioned earlier. Most of the models to be found in the literature, however, utilize source, doublet, or vortex panels to model the airfoil. Such a program begins by calculating the potential flow around the airfoil. To allow for both finite thickness and circulation, the airfoil contour is approximated by a closed polygon, as shown in Figure 3.24. A continuous distribution of vortices is then placed on each side of the polygon, with the vortex strength per unit length, γ, varying linearly from one corner to the next and continuous across the corner. Figure 3.25 illustrates this model for two sides connecting corners 3, 4, and 5. Control points are chosen midway between the corners. The values of the vortex unit strengths at the corners are then found that will induce velocities at each control point tangent to the polygon side at that point. Note, however, that if there are n corners and hence, $n + 1$ unknown γ values at the corners, the n control points provide one less equation than unknowns. This situation is remedied by applying the Kutta condition at the trailing edge. This requires that $\gamma_{n+1} = -\gamma_1$, assuring that the velocities induced at the trailing edge are finite.

Having determined the vortex strengths, the velocity field and hence, the pressure distribution around the airfoil can be calculated. This result is then used to calculate the boundary layer development over the airfoil, including the growth of

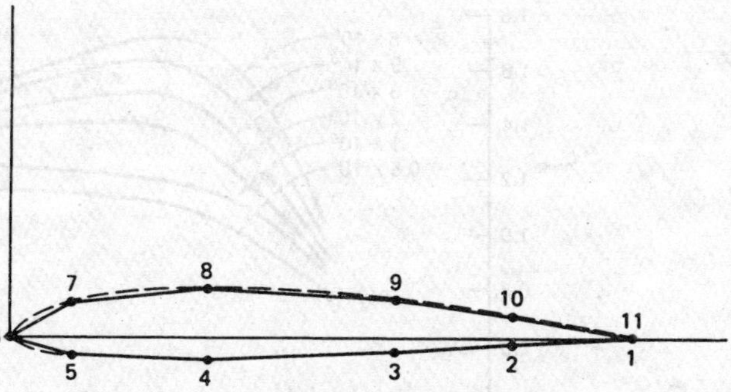

Figure 3.24 Approximation of airfoil contour by closed polygon.

the laminar layer, transition, growth of the turbulent layer, and possible boundary layer separation. The airfoil shape is then enlarged slightly to allow for the boundary layer thickness and the potential flow solutions are repeated. The details of this iterative procedure are beyond the scope of this text.

MAXIMUM LIFT

Airfoil theory based on potential flow methods predicts the lift of an airfoil in its linear range but does not provide any information concerning maximum lift capability. As discussed previously, $C_{l_{max}}$ is determined by flow separation, which is a "real fluid" effect. Separation is difficult to predict analytically, so the following material on $C_{l_{max}}$ is mainly empirical.

Typically, conventional airfoils without any special high-lift devices will deliver $C_{l_{max}}$ values of approximately 1.3 to 1.7, depending on Reynolds number, camber, and thickness distribution. The appreciable dependence of $C_{l_{max}}$ on R shown in Figure 3.13 for the GA(W)-1 airfoil is typical of other airfoils. Figure 3.26 presents $C_{l_{max}}$ as a function of R and thickness ratio for NACA four-digit airfoils having a maximum camber of 2%, located 40% of the chord back from the leading edge. At intermediate thickness ratios of around 0.12, the variation of $C_{l_{max}}$ with R parallels that of the 17% thick GA(W)-1 airfoil. Note at least for this camber function that a thickness ratio of 12% is about optimum. This figure is taken from Reference 3.16. This same reference presents the following empirical formula for $C_{l_{max}}$ for NACA four-digit airfoils at an R of 8×10^6.

$$C_{l_{max}} = 1.67 + 7.8\,pz - 2.6\frac{(0.123 + 0.022p - 0.5z - t)^2}{t^{3/2}} \tag{3.46}$$

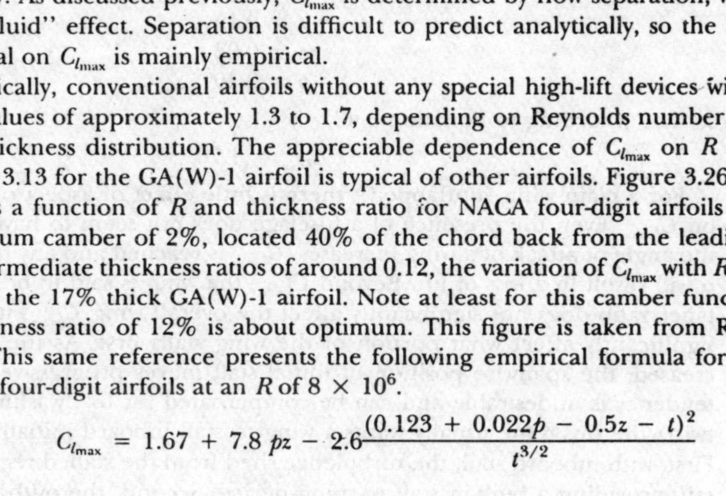

Figure 3.25 Vortex distributions representing airfoil contour.

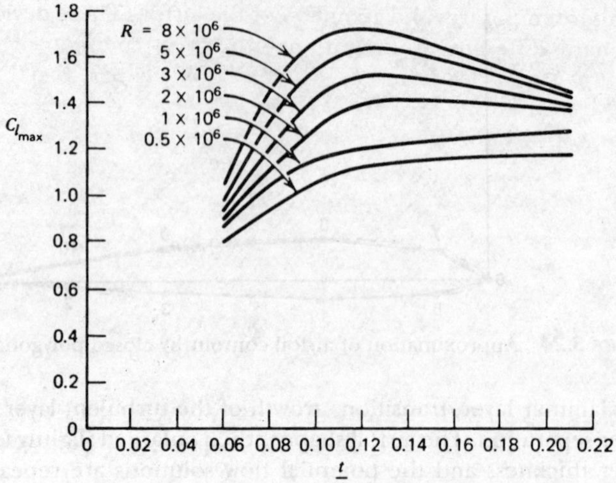

Figure 3.26 Variation of $C_{l_{max}}$ with thickness ratio of NACA 24xx airfoils for various Reynolds numbers. (B. W. McCormick, Aerodynamics of V/STOL Flight. Academic Press, Inc. 1967. Reprinted by permission of Academic Press, Inc.)

t, z, and p are thickness, maximum camber, and position of maximum camber, respectively, expressed as a fraction of the chord. For example, for a 2415 airfoil,

$$t = 0.15$$
$$z = 0.02$$
$$p = 0.40$$

so that according to Equation 3.42,

$$C_{l_{max}} = 1.70$$

For a plain wing (unflapped), there is little effect of aspect ratio or taper ratio on $C_{L_{max}}$. Even the presence of a fuselage does not seem to have much effect. As the angle of attack of a wing increases, $C_{L_{max}}$ is reached and any further increase in α will result in a loss of lift. Beyond $C_{L_{max}}$, the wing is said to be stalled. Although taper ratio does not significantly affect the overall wing $C_{L_{max}}$, it (and wing twist) significantly affect what portion of the wing stalls first. As the taper ratio is decreased, the spanwise position of initial stall moves progressively outboard. This tendency is undesirable and can be compensated for by "washing out" (negative twist) the tips. One usually wants a wing to stall inboard initially for two reasons. First, with inboard stall, the turbulence shed from the stalled region can shake the tail, providing a built-in stall warning device. Second, the outboard region being unstalled will still provide aileron roll control even though the wing has begun to stall. The lift characteristics of three-dimensional wings will be treated in more detail later.

Flaps

An examination of all of the airfoil data presented in Reference 3.1 discloses that the greatest value of $C_{l_{max}}$ one can expect at a high Reynolds number from an ordinary airfoil is around 1.8. This maximum value is achieved by the NACA 23012 airfoil. Another 12% thick airfoil, the 2412, delivers the second highest value, 1.7.

To achieve higher $C_{l_{max}}$ values for takeoff and landing without unduly penalizing an airplane's cruising performance, one resorts to the use of mechanical devices to alter temporarily the geometry of the airfoil. These devices, known as *flaps*, exist in many different configurations—the most common of which are illustrated in Figure 3.27. In addition to the purely mechanical flaps, this figure depicts flaps that can be formed by sheets of air exiting at the trailing edge. These "jet flaps" can produce $C_{l_{max}}$ values in excess of those from mechanical flaps, provided sufficient energy and momentum are contained in the jet. Frequently, one uses the terms "powered" and "unpowered" to distinguish between jet and mechanical flaps.

The effect of a mechanical flap can be seen by referring once again to Figure 3.6. Deflecting the flap, in this case a split flap, is seen to shift the lift curve upward without changing the slope. This is as one might expect from Equation 3.24 or 3.39, since deflecting the flap is comparable to adding camber to the airfoil.

Some flap configurations appear to be significantly better than others simply because, when deflected, they extend the original chord on which the lift coefficient is based. One can determine if the flap is extensible, such as the Fowler or Zap flaps in Figure 3.27, by noting whether or not the slope of the lift curve is increased with the flap deflected. Consider a flap that, when deflected, extends the original chord by the fraction x. The physical lift curve would have a slope given by

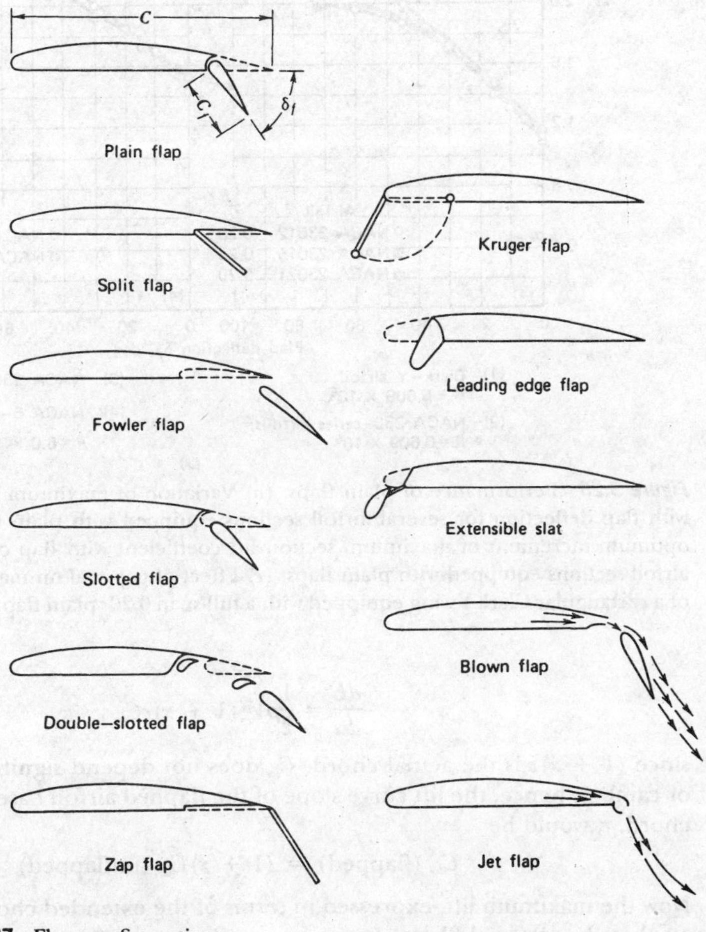

Figure 3.27 Flap configurations.

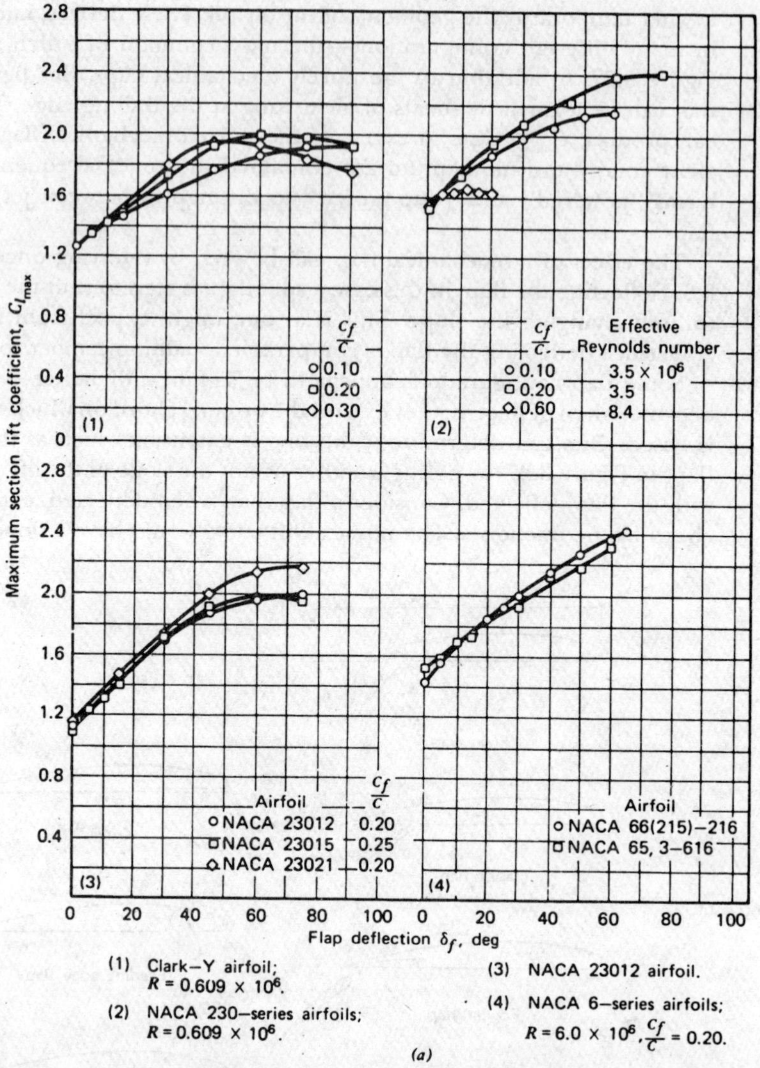

Figure 3.28 Performance of plain flaps. (*a*) Variation of maximum section lift coefficient with flap deflection for several airfoil sections equipped with plain flaps. (*b*) Variation of optimum increment of maximum, section lift coefficient with flap chord ratio for several airfoil sections equipped with plain flaps. (*c*) Effect of gap seal on maximum lift coefficient of a rectangular Clark-Y wing equipped with a full-span 0.20*c* plain flap. $A = 6$, $R = 0.6 \times 10^4$.

$$\frac{dL}{d\alpha} = \frac{1}{2}\rho V^2 (1 + x)\, c C_{l\alpha} \tag{3.47}$$

since $(1 + x)c$ is the actual chord. $C_{l\alpha}$ does not depend significantly on thickness or camber; hence, the lift curve slope of the flapped airfoil based on the unflapped chord, c, would be

$$C_{l\alpha}\ (\text{flapped}) = (1 + x)\, C_{l\alpha}\ (\text{unflapped}) \tag{3.48}$$

Now the maximum lift, expressed in terms of the extended chord and $C_{l_{max}}$, (based on that chord) would be

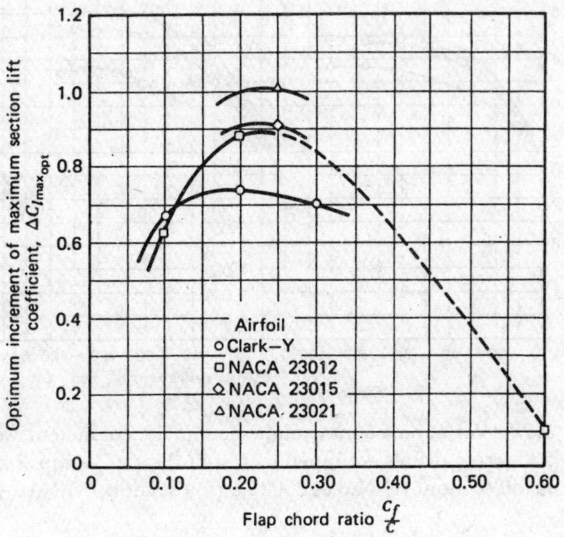

(b)

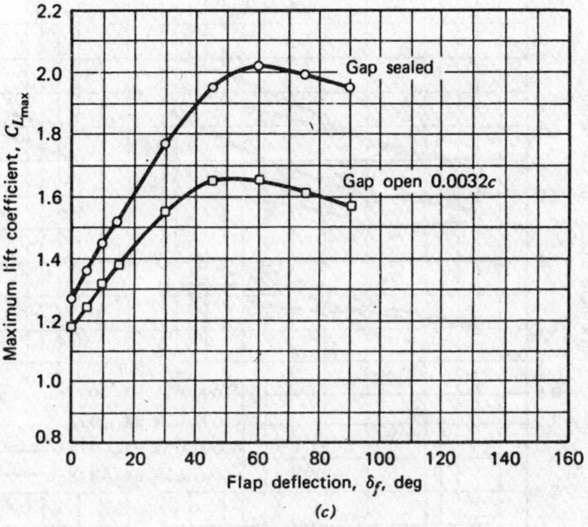

(c)

$$L_{max} = \frac{1}{2}\rho V^2 (1 + x) c C_{l_{max_f}}$$

Thus, $C_{l_{max}}$ based on the original chord becomes

$$C_{l_{max}} = (1 + x) C_{l_{max_f}} \tag{3.49}$$

Figures 3.28 to 3.33 and Tables 3.1 and 3.2 present section data on plain, split, and slotted flaps as reported in Reference 3.17. With these data one should be able to estimate reasonably accurately the characteristics of an airfoil section equipped with flaps.

A study of this data suggests the following:

Plain Flaps

1. The optimum flap chord ratio is approximately 0.25.
2. The optimum flap angle is approximately 60°.

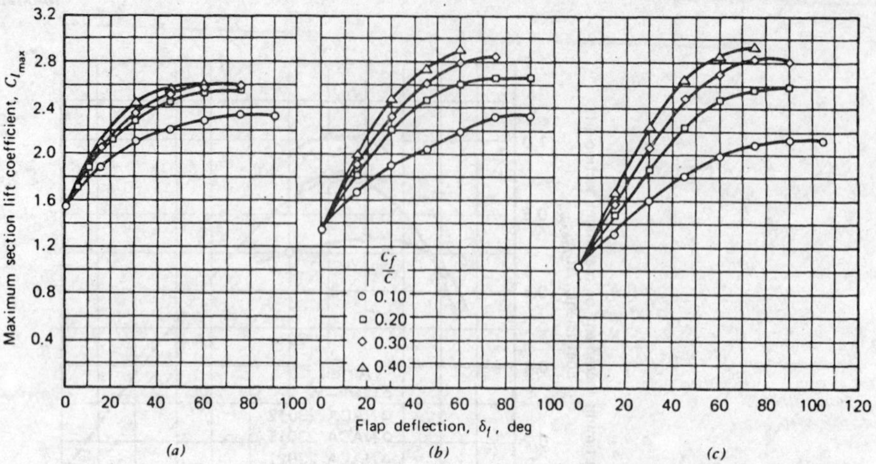

Figure 3.29 Variation of maximum section lift coefficient with flap deflection for three NACA 230-series airfoils equipped with split flaps of various sizes, $R = 3.5 \times 10^6$. (*a*) NACA 23012 airfoil section. (*b*) NACA 23021 airfoil section. (*c*) NACA 23030 airfoil section.

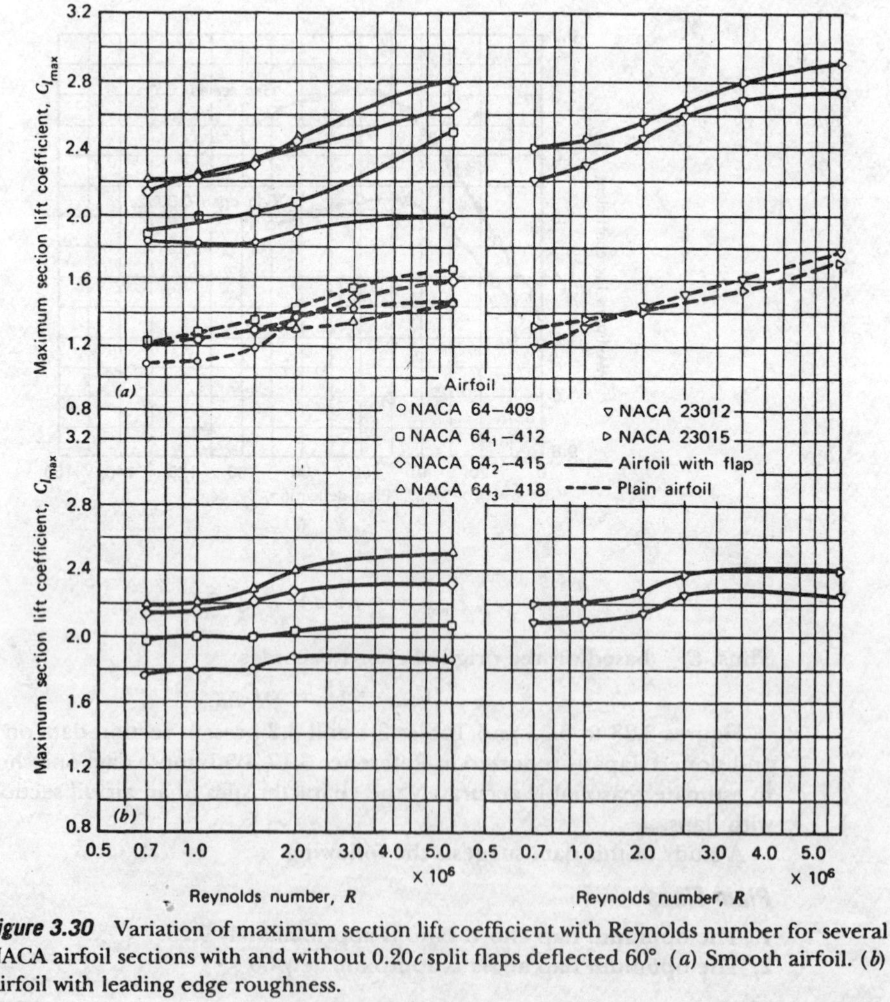

Figure 3.30 Variation of maximum section lift coefficient with Reynolds number for several NACA airfoil sections with and without $0.20c$ split flaps deflected 60°. (*a*) Smooth airfoil. (*b*) Airfoil with leading edge roughness.

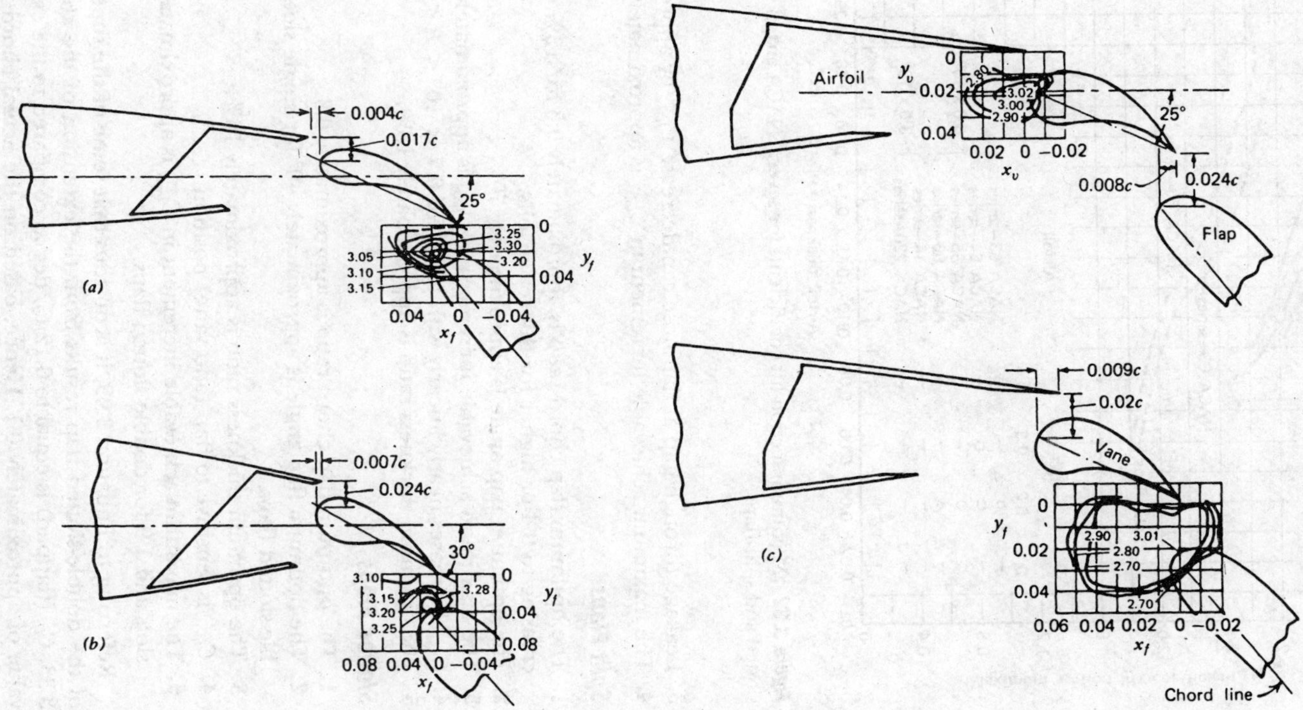

Figure 3.31 Contours of flap and vane positions for maximum section lift coefficient for several airfoil sections equipped with double-slotted flaps. (*a*) NACA 23012 airfoil section; $\delta_f = 60°$. (*b*) NACA 23021 airfoil section; $\delta_f = 60°$. (*c*) NACA 61_1-212 airfoil section, $\delta_f = 60°$.

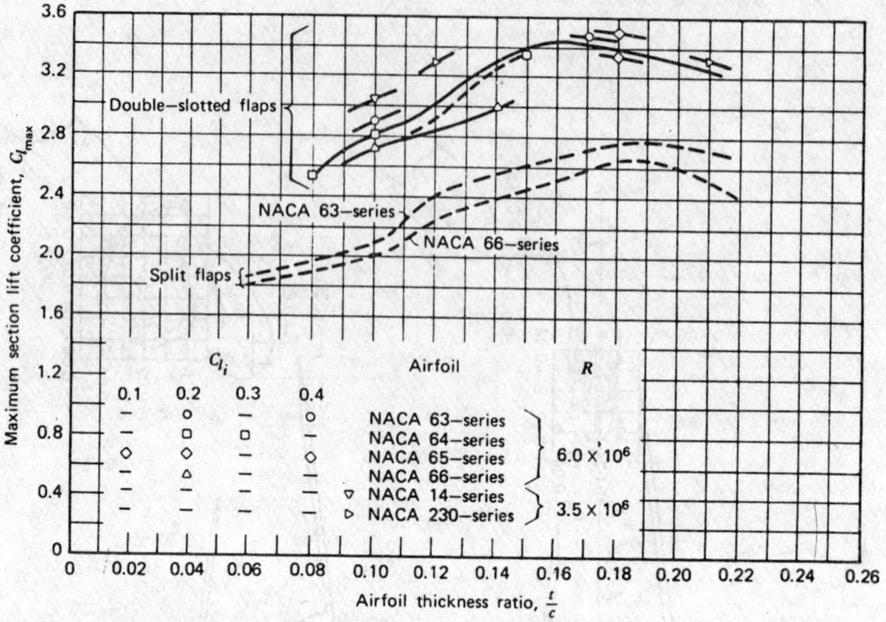

Figure 3.32 Maximum section lift coefficients for several NACA airfoil sections with double-slotted and split flaps.

3. Leakage through gap at flap nose can decrease $C_{l_{max}}$ by approximately 0.4.
4. The maximum achievable increment in $C_{l_{max}}$ is approximately 0.9.

Split Flaps

1. The optimum flap chord ratio is approximately 0.3 for 12% thick airfoils, increasing to 0.4 or higher for thicker airfoils.
2. The optimum flap angle is approximately 70°.
3. The maximum achievable increment in $C_{l_{max}}$ is approximately 0.9.
4. $C_{l_{max}}$ increases nearly linearly with log R for $0.7 \times 10^6 < R < 6 \times 10^6$.
5. The optimum thickness ratio is approximately 18%.

Slotted Flaps

1. The optimum flap chord ratio is approximately 0.3.
2. The optimum flap angle is approximately 40° for single slots and 70° for double-slotted flaps.
3. The optimum thickness ratio is approximately 16%.
4. $C_{l_{max}}$ is sensitive to flap (and vane) position.
5. The maximum achievable increment in $C_{l_{max}}$ is approximately 1.5 for single slots and 1.9 for double-slotted flaps.

Referring to Equation 3.49, it is obvious that some of the superior performance of the double-slotted flap results from the extension of the chord. From Figure 3.33, $C_{l\alpha}$ (flapped) is equal to 0.12 C_l/deg as compared to the expected unflapped value of approximately 0.1. Hence, based on the actual chord, the increment in $C_{l_{max}}$ for the double-slotted flap is only 1.6. However, this is still almost twice that afforded by plain or split flaps and points to the beneficial effect of the slot in delaying separation.

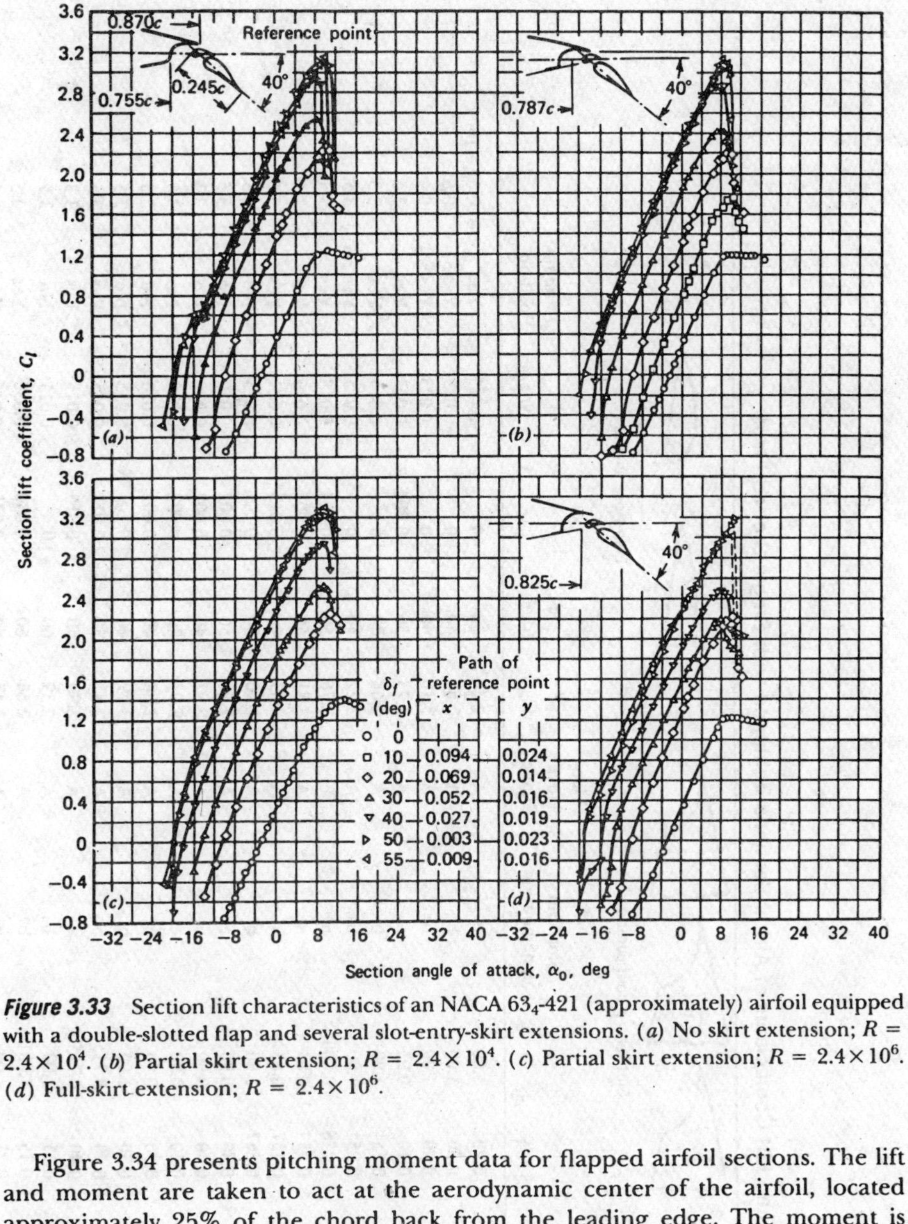

Figure 3.33 Section lift characteristics of an NACA 63_4-421 (approximately) airfoil equipped with a double-slotted flap and several slot-entry-skirt extensions. (*a*) No skirt extension; $R = 2.4 \times 10^4$. (*b*) Partial skirt extension; $R = 2.4 \times 10^4$. (*c*) Partial skirt extension; $R = 2.4 \times 10^6$. (*d*) Full-skirt extension; $R = 2.4 \times 10^6$.

Figure 3.34 presents pitching moment data for flapped airfoil sections. The lift and moment are taken to act at the aerodynamic center of the airfoil, located approximately 25% of the chord back from the leading edge. The moment is positive if it tends to increase the angle of attack.

From Figure 3.34, the lowering of a flap results in an incremental pitching moment. In order to trim the airplane, a download must be produced on the horizontal tail. The wing must now support this download in addition to the aircraft's weight. Hence, the effective increment in lift due to the flap is less than that which the wing-flap combination produces alone. This correction can typically reduce $\Delta C_{L_{max}}$ by 0.1 to 0.3.

In a high-wing airplane, lowering the flaps can cause the nose to pitch up. This is due to the moment produced about the center of gravity from the increase in

(Text continued on p. 98.)

Table 3.1 Maximum Lift Coefficients of Airfoil Sections Equipped with Single-Slotted Flaps

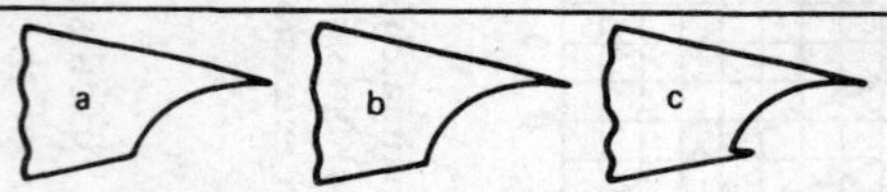

Slot–entry configuration

Flap nose shapes

Airfoil Section	c_f/c	c_a	Slot-Entry configuration	Flap Nose Shape	$C_{l_{max}}$	δf (deg)	x_f	y_f	Optimum Position	R
Clark Y	0.20	1.00	b	A	2.44	30	0	−0.025	Yes	0.61 × 10
Clark Y	0.30	1.00	b	A	2.83	40	0	−0.025	Yes	0.61
Clark Y	0.40	1.00	b	A	3.10	40	0	−0.025	Yes	0.61
23012	0.10	0.93	a	A	2.25	50	0.004	0.005	Yes	3.5
23012	0.15	1.00	b	A	2.68	30	0	0.015	Yes	3.5
23012	0.25	1.00	b	A	3.22	40	0	0.015	Yes	3.5
23012	0.256	0.800	a	B	2.76	50	0.005	0.018	Yes	3.5
23012	0.257	0.83	a	A	2.81	50	0.005	0.016	Yes	3.5
23012	0.257	0.83	a	A	2.83	40	0.013	0.024	Yes	3.5
23012	0.267	1.00	b	A	2.90	30	0	0.025	No	3.5
23012	0.30	0.90	c	A	2.92	50	0.002	0.010	No	3.5
23012	0.30	0.90	c	A	2.92	40	0.002	0.020	No	3.5
23012	0.30	0.90	c	A	2.93	30	0.002	0.030	No	3.5
23012	0.30	0.90	b	A	2.88	40	0.002	0.020	No	3.5
23012	0.30	1.00	b	A	3.29	40	0	0.015	No	3.5
23012	0.40	0.715	b	A	2.87	50	0.015	0.015	Yes	3.5
23012	0.40	0.715	a	A	2.90	50	0.015	0.015	Yes	3.5
23012	0.15	1.00	b	A	2.59	60	0	0.015	No	3.5
23021	0.15	1.00	b	A	2.66	60	0.050	0.030	Yes	3.5
23021	0.25	1.00	b	A	3.17	40	0.025	0.015	Yes	3.5
23021	0.257	0.827	b	B	2.69	60	0	0.015	Yes	3.5
23021	0.257	0.827	a	B	2.74	60	0	0.015	Yes	3.5
23021	0.257	0.827	b	A	2.71	60	0.005	0.020	Yes	3.5
23021	0.257	0.827	a	A	2.82	50	0	0.025	Yes	3.5

23021	0.40	0.715	b	A	2.79	50	0.015	0.025	Yes	3.5
23021	0.40	0.715	a	A	2.88	50	0.015	0.045	Yes	3.5
23030	0.257	0.860	b	B	2.59	60	0.025	0.040	Yes	3.5
23030	0.257	0.860	a	B	2.68	60	−0.005	0.040	Yes	3.5
23030	0.40	0.775	b	B	2.82	50	0.025	0.060	Yes	3.5
23030	0.40	0.775	a	B	2.90	50	0.025	0.060	Yes	3.5
63,4–420	0.25	0.88	b	B	3.00	35	0.018	0.045	No	6.0
3,4–421 (approximately)	0.243	0.835	a	A	3.21	40	0	0.027	Yes	9.0
65–210	0.25	0.84	c	A	2.47	45	0.009	0.010	Yes	6.0
65–210	0.25	0.90	c	A	2.48	41.3	0.014	0.009	Yes	6.0
65–210	0.25	0.975	c	A	2.45	35	0.004	0.020	Yes	6.0
$5_{(112)}$A111 (approximately)	0.35	0.839	c	A	2.69	35	−0.020	0.032	Yes	9.0
5_1-213 (approximately)	0.336	0.889	c	A	2.63	40	0.019	0.046	No	9.0
$5_{(215)}$-114	0.259	0.915	c	A	2.80	40	0.019	0.038	No	9.0
5,2–221 (approximately)	0.263	0.832	a	B	2.83	30	0.025	0.046	Yes	9.95
6(215)-116, $a = 0.6$	0.25	0.824	c	B	2.70	55	0	0.028	No	6.0
6,2–116, $a = 0.6$	0.25	0.827	a	A	2.69	45	0.017	0.038	No	6.0
6,2–216, $a = 0.6$	0.30	0.90	c	A	2.92	37	0	0.016	No	6.0
6,2–216, $a = 0.6$	0.25	0.824	a	A	2.89	40	0.023	0.040	Yes	5.1
6,2–216, $a = 0.6$	0.25	0.834	c	A	2.88	45	0.011	0.031	Yes	5.1
6,2–118	0.25	0.90			2.68	32.5	--------	--------	No	6.0

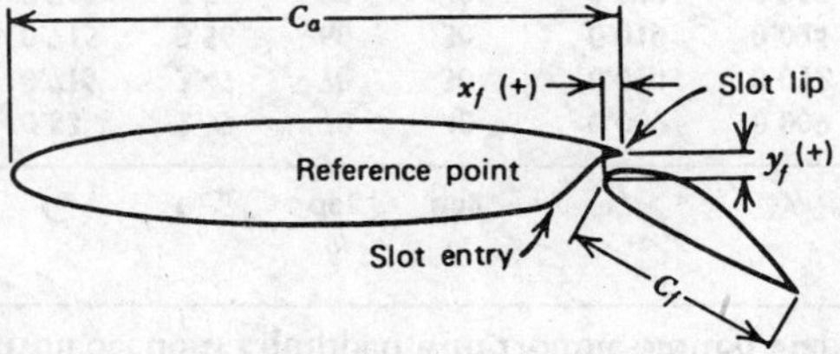

Typical single–slotted flap configuration.
(All dimensions are given in fractions of airfoil chord.)

Table 3.2 Maximum Lift Coefficients of Airfoil Sections Equipped with Double-Slotted Flaps

Airfoil Section	c_f/c	c_v/c	C_a	$C_{l_{max}}$	δ_f deg	δ_s deg	x_f	y_f	x_v	y_v	Optimum Position	R
23012	0.10	0.189	0.83	2.99	70	40	0.009	0.009	0.014	0.024	Yes	3.5×10^6
23012	0.257	0.227	0.715	3.47	70	30	0.014	0.012	0.015	0.035	No	3.5
23021	0.257	0.227	0.715	3.56	60	30	0.019	0.024	0.025	0.065	No	3.5
23030	0.257	0.260	0.715	3.71	80	40	0.049	0.050	0.045	0.040	No	3.5
23012	0.257	0.117	0.826	3.30	60	25	−0.016	0.010	−0.004	0.017	Yes	3.5
23021	0.257	0.147	0.827	3.32	70	30	0.017	0.027	0.007	0.024	Yes	3.5
63–210	0.25	0.075	0.84	2.91	50	25	0.022	0.024	0.024	0.018	Yes	6.0
63,4–421 (approximately)	0.195	0.083	0.87	3.30	55	14	0.038	0.012	0.009	0.016	No	6.0
64–208	0.25	0.075	0.84	2.51	45	30	0.015	0.015	0.015	0.019	Yes	6.0
64–208	0.25	0.056	0.84	2.40	50	25	0.018	0.014	0.015	0.024	Yes	6.0
64–210	0.25	0.075	0.84	2.82	55	30	0.023	0.006	0.012	0.018	Yes	6.0
64_1–212	0.25	0.075	0.84	3.03	50	30	0.021	0.020	0.010	0.019	Yes	6.0
64_1A212	0.229	0.083	0.833	2.83	55	26	0.044	0.005	0.004	0.014	Yes	6.0
65–210	0.25	0.075	0.84	2.72	50	25	0.025	0.011	0.009	0.024	Yes	6.0
65(216)–215, $a = 0.8$	0.248	0.096	0.82	3.38	70	12	0.024	0.010	0.025	0.032	No	6.3
65_3–118	0.244	0.10	0.864	3.35	65	23	0.038	0.007	0.009	0.025	Yes	6.0
65_3–418	0.236	0.106	0.851	3.50	65	21	0.027	0.007	0.012	0.028	Yes	6.0
65_4–421	0.236	0.109	0.85	3.08	51	20	0.029	0.017	0.012	0.024	Yes	2.2
66–210	0.25	0.075	0.84	2.64	55	25	0.029	0.023	0.012	0.022	Yes	6.0
66–210	0.25	0.100	0.84	2.72	60	25	0.027	0.039	0.024	0.021	Yes	6.0
66_2–214 (approximately)	0.227	0.085	0.854	3.00	55	20	0.044	0.009	0.004	0.025	Yes	9.0
1410	0.25	0.075	0.84	3.06	50	25	0.026	0.016	0.012	0.019	Yes	6.0

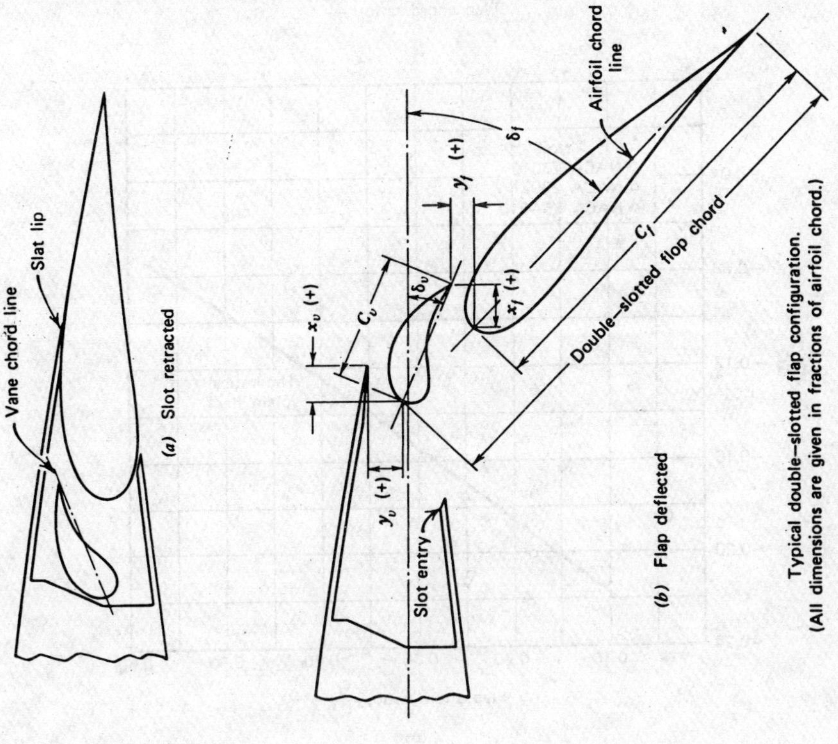

Definition of Flap Geometry given in Table 3.2

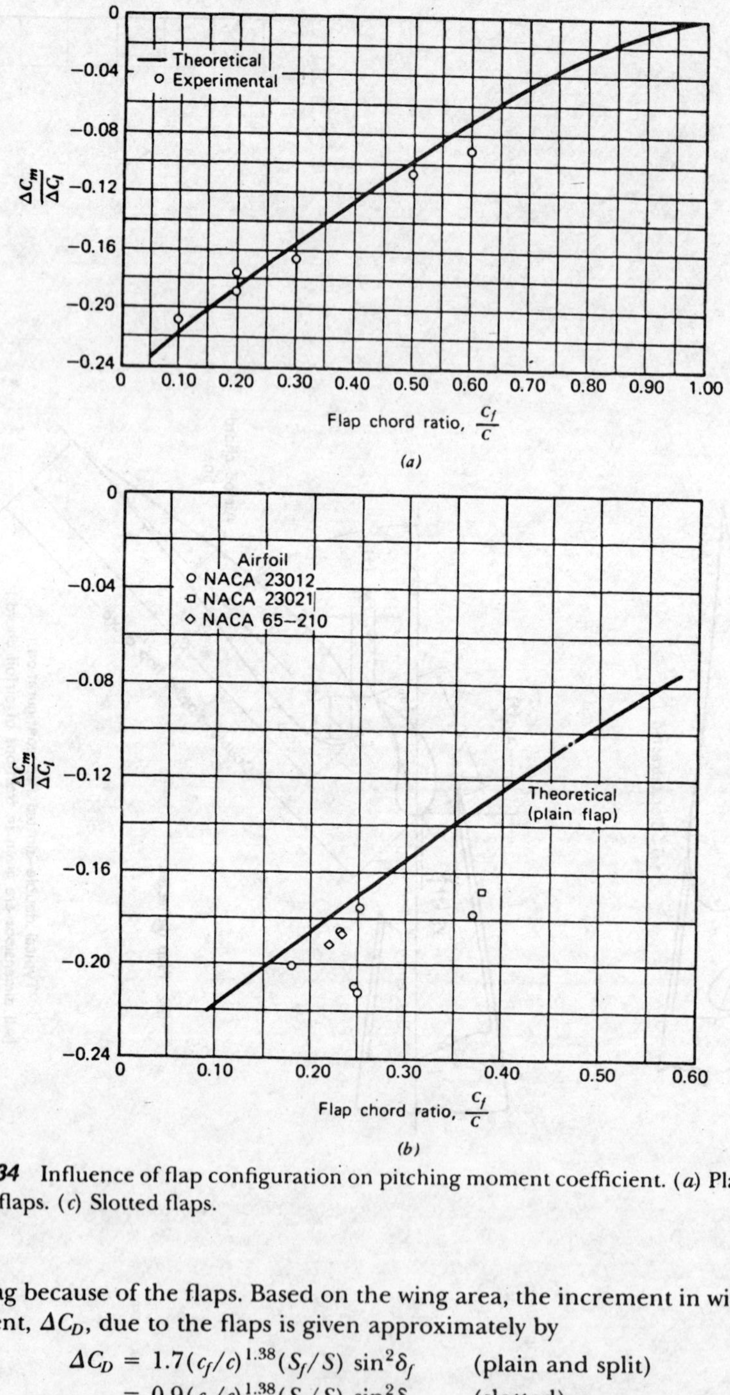

Figure 3.34 Influence of flap configuration on pitching moment coefficient. (*a*) Plain flaps. (*b*) Split flaps. (*c*) Slotted flaps.

wing drag because of the flaps. Based on the wing area, the increment in wing drag coefficient, ΔC_D, due to the flaps is given approximately by

$$\Delta C_D = 1.7(c_f/c)^{1.38}(S_f/S)\sin^2\delta_f \qquad \text{(plain and split)} \qquad (3.50)$$

$$= 0.9(c_f/c)^{1.38}(S_f/S)\sin^2\delta_f \qquad \text{(slotted)} \qquad (3.51)$$

If the wing is located a height of h above the center of gravity, a balancing upload is required on the tail. The effect of trim on $C_{L_{\max}}$ for a complete airplane will be discussed in more detail later.

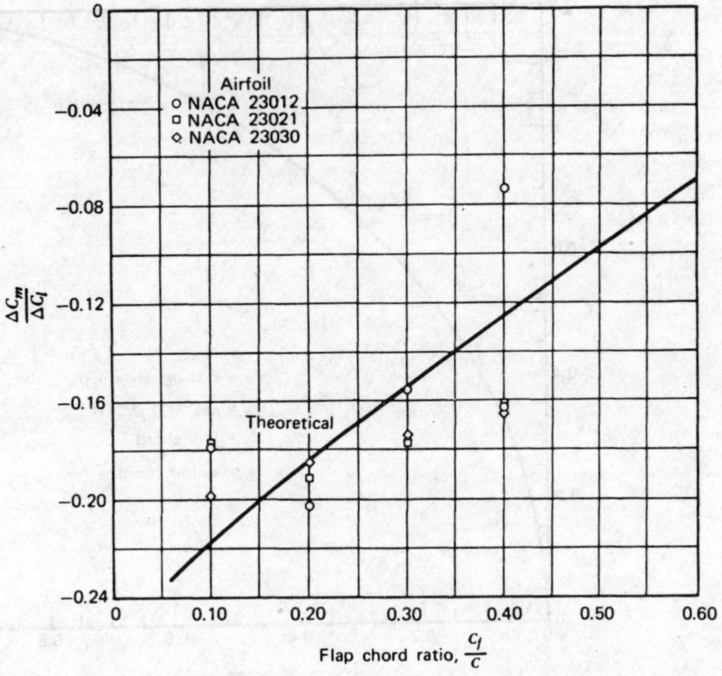

(c)

Flap Effectiveness in the Linear Range

Frequently, one needs to estimate the increment in C_l below stall, ΔC_l, produced by a flap deflection. Not only is this needed in connection with the wing lift, but ΔC_l is required also in analyzing the effectiveness of movable control surfaces, which frequently resemble plain flaps.

If an airfoil section has a lift curve slope of $C_{l\alpha}$ and lowering its flap produces an increment of ΔC_l, the angle of zero lift, α_{0l}, is decreased by

$$\Delta\alpha_{0l} = \frac{\Delta C_l}{C_{l\alpha}} \tag{3.52}$$

The rate of decrease of α_{0l} per unit increase in the flap angle δ_f is referred to as the *flap effectiveness factor*, τ. Thus, for a flapped airfoil, the lift coefficient can be written as

$$C_l = C_{l\alpha}(\alpha + \tau\,\delta_f) \tag{3.53}$$

where α is the angle of attack of the airfoil's zero lift line with the flap undeflected.

Theoretically, τ is a constant for a given flap geometry but, unfortunately, flap behavior with δ_f is rather nonlinear and hence, τ must be empirically corrected by a factor η to account for the effects of viscosity. Including η, Equation 3.53 becomes

$$C_l = C_{l\alpha}(\alpha + \tau\eta\,\delta_f) \tag{3.54}$$

The functions τ and η can be obtained from Figures 3.35 and 3.36. Figure 3.36 is empirical and is based on data from References 3.17-3.23. Although there is some scatter in the data, as faired, the comparisons among the various types of flaps are consistent. The double-slotted flap delays separation on the upper surface, so that

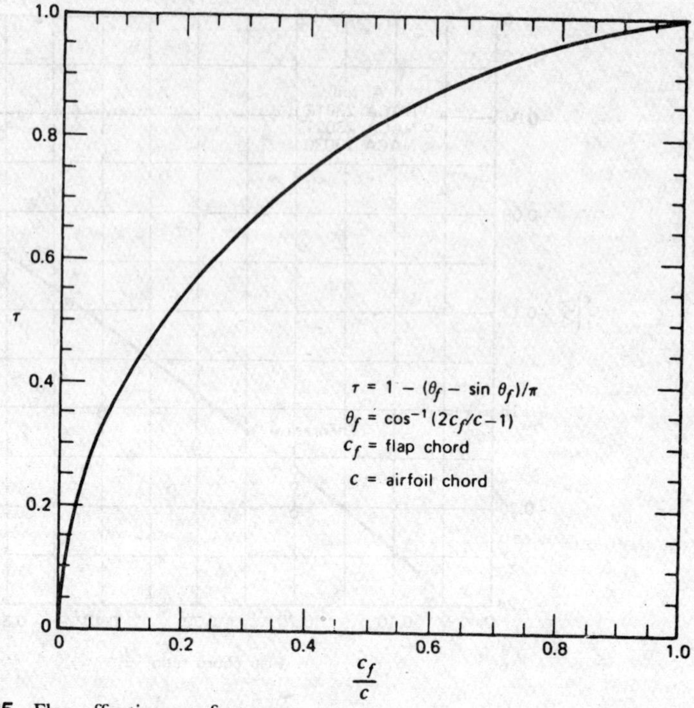

Figure 3.35 Flap effectiveness factor.

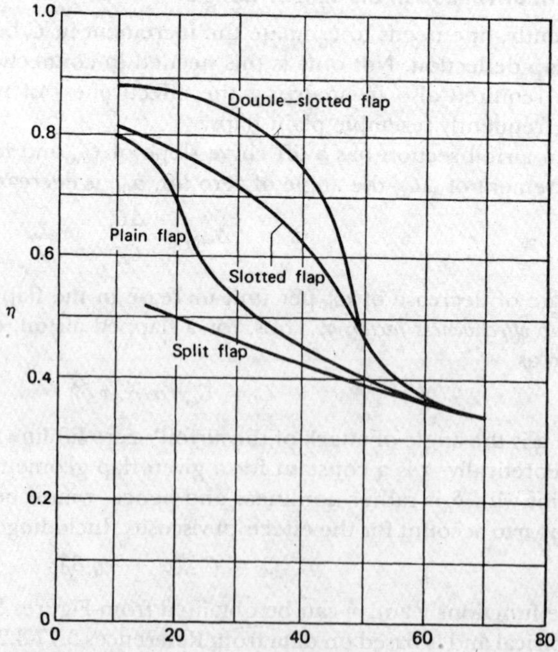

Figure 3.36 Correction factor to flap effectiveness factor τ. Note that curves apply for thickness ratios of approximately 12% and flap-chord fractions of 40% or less.

the decrease in flap effectiveness occurs at higher flap angles than for the other flap types. The same can be said of the slotted flap relative to the plain and split flaps. The plain flap is fairly good out to about 20° and then apparently the flow separates from the upper surface and the effectiveness drops rapidly, approaching the curve for split flaps at the higher flap angles. In a sense, the flow is always separated on the upper surface of a split flap. Thus, even for small flap angles, the effective angular movement of the mean camber line at the trailing edge of an airfoil with a split flap would only be about half of the flap displacement.

In the case of the double-slotted flap it should be emphasized that this curve in Figure 3.36 is for an optimum flap geometry. The trailing segment of the flap is referred to as the *main flap* and the leading segment is called the *vane*. In applying Equation 3.54 and Figures 3.35 and 3.36 to the double-slotted flap, the total flap chord should be used together with the flap angle of the main flap. Usually, the deflection angle of the vane is less than that for the main flap for maximum lift performance.

The angle of attack at which the flapped airfoil stalls is generally less than that for the plain airfoil. Hence, the increment in $C_{l_{max}}$ because the flap is not as great as the increment in C_l at an angle below the stall. Denoting these increments by $\Delta C_{l_{max}}$ and ΔC_l, respectively, it is obvious that the ratio $\Delta C_{l_{max}}/\Delta C_l$ must depend on c_f/c. If c_f/c, for example, is equal to 1.0, in a sense, the entire airfoil is the flap and $\Delta C_{l_{max}}$ must be zero. Systematic data on $\Delta C_{l_{max}}/\Delta C_l$ are sparse. Figure 3.37 has been drawn based on a limited number of data points and should be used with discretion.

The theoretical values for the flap effectiveness factor, τ, and the ratio of $\Delta C_{m1/4}$ to ΔC_l can be obtained from thin airfoil theory by treating a flat plate airfoil with a deflected flap as an equivalent cambered airfoil. This equivalency is illustrated in Figure 3.38. Assuming a small flap angle, it is obvious from the figure that a flat plate airfoil at zero angle of attack with a deflected flap is equivalent to a cambered airfoil at an angle of attack of

$$\alpha = \frac{c_f}{c}\delta_f \tag{3.55}$$

The slope of the camber line is discontinuous having a value of α back to the flap and then decreasing by δ_f at that point. Using this discontinuous function for dz/dx and the above value for α leads to the following:

$$\tau = 1 - \frac{(\theta_f - \sin\theta_f)}{\pi} \tag{3.56}$$

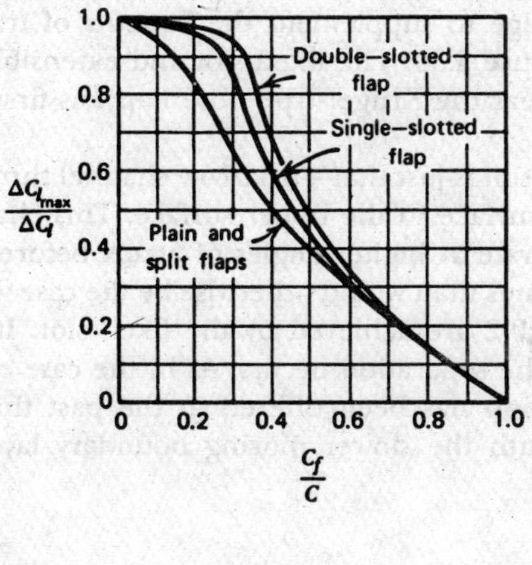

Figure 3.37 $C_{l_{max}}$ increment ratio as a function of flap chord ratio.

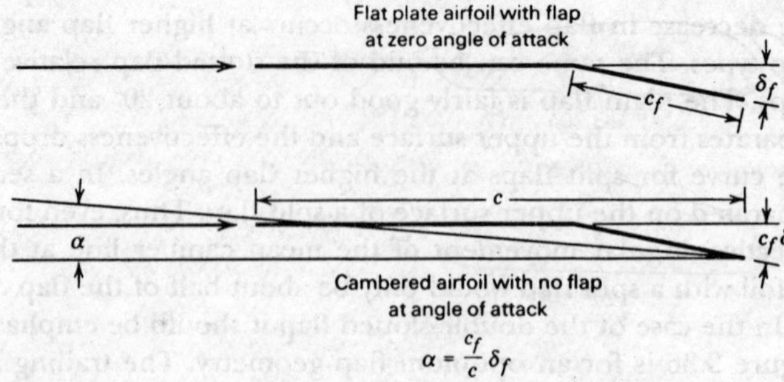

Figure 3.38 An airfoil with a flap represented as a cambered airfoil at an angle of attack.

$$\frac{\Delta C_{m_{1/4}}}{\Delta C_l} = -\frac{2\sin\theta_f - \sin 2\theta_f}{8\,(\pi - \theta_f + \sin\theta_f)} \tag{3.57}$$

As an example, in using Figures 3.35, 3.36, and 3.37, consider the prediction of $C_{l_{\max}}$ for a 23012 airfoil equipped with a 30% chord split flap deflected 60° and operating at a Reynolds number of 3.5 × 10⁶. From Figure 3.35, $\tau = 0.66$ for $c_f/c = 0.3$ and from Figure 3.36, $\eta = 0.35$ for a split flap deflected 60°. Hence, from Equation 3.54, ΔC_l is equal to

$$\begin{aligned}\Delta C_l &= C_{l_\alpha}\tau\eta\,\delta_f \\ &= (0.105)(0.66)(0.35)(60) \\ &= 1.46\end{aligned} \tag{3.58}$$

In Equation 3.58, C_{l_α} of 0.105 is obtained from Reference 3.1. Using Figure 3.37, the ratio of $\Delta C_{l_{\max}}$ to ΔC_l is obtained as 0.66. Hence,

$$\Delta C_{l_{\max}} = 0.96$$

From Figure 3.30, $C_{l_{\max}}$ for a plain 23012 airfoil equals 1.65 at $R = 3.5 \times 10^6$. Thus, for the flapped airfoil, $C_{l_{\max}}$ is predicted to be 1.65 + 0.96, or 2.61. This result compares closely with Figure 3.28*a*. If the procedure is repeated for other flap angles, close agreement is also obtained with the figure. However, for a flap chord ratio of 0.1, the predicted values of $C_{l_{\max}}$ based on Figures 3.35 to 3.37 are higher than those shown in Figure 3.28*a*.

Leading Edge Devices

To avoid leading edge separation, particularly at low Reynolds numbers or for airfoils with relatively sharp leading edges, special high-lift devices can also be incorporated into the leading edge to supplement the benefits of trailing edge flaps. These are illustrated in Figure 3.39. The fixed slot and extensible slat have been in use for some time, whereas the Kruger-type nose flap was first employed on the turbojet transport.

As the name implies, the fixed slot is just that—a narrow channel through which the air can flow from the lower surface to the upper surface. This channeling of the flow allows the airfoil to operate at higher angles of attack before the upper surface of the leading edge separates than would otherwise be the case. Increments in $C_{l_{\max}}$ of approximately 0.1 or 0.2 are achieved by the fixed slot. It is a moot question as to why this delay in the separation occurs. As in the case of slots with trailing edge flaps, the explanation has been offered in the past that the flow through the slot feeds energy into the slower moving boundary layer, thereby

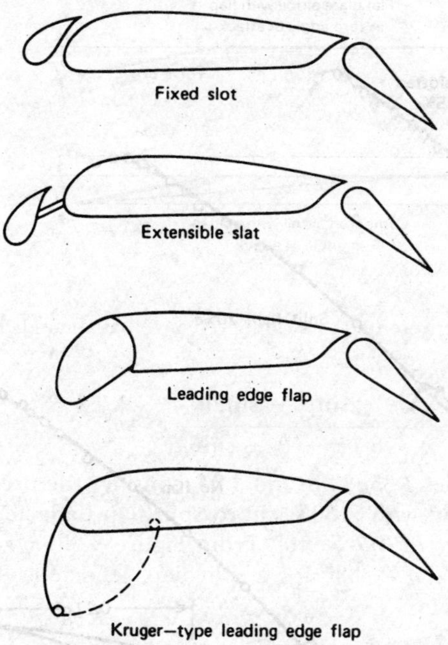

Figure 3.39 Various methods for delaying leading edge separation.

decreasing its tendency to separate. More recently, however, in a Wright Brothers' Lecture (Ref. 3.16), Smith, in examining numerical results on multielement airfoils, concluded that improved stall performance from slots is most likely the result of more favorable pressure gradients being produced on one airfoil element by the other.

The extensible slat is similar in its performance to the slot, but it is considerably more efficient because it can be positioned to optimize its contribution to $C_{l_{max}}$. The mechanically extended slat is finding increased application, particularly with the use of thinner airfoil sections for high-speed applications. Figure 3.40 presents some data on slats taken from Reference 3.17. Here, a NACA 64A010 airfoil was tested using a slat in combination with split and double-slotted trailing edge flaps. The slat is seen to improve $C_{l_{max}}$ significantly, producing increments in $C_{l_{max}}$ of approximately 0.9, 0.8, and 0.6 for the no-flap, split-flap, and double-slotted flap configurations, respectively. Unlike the trailing edge flap, the primary effect of the slat is seen to be an extension of the lift curve without the slat; that is, opening the slat does not change C_l by a large increment at a fixed angle of attack. The same is true of leading edge flaps and is not unexpected in view of Figure 3.35.

The performance of a leading edge flap is presented in Figure 3.41 for the same airfoil as for Figure 3.40. Comparing the two figures, it is obvious that the two leading edge devices are nearly comparable in performance.

Figure 3.42 shows a section of a sophisticated Kruger-type flap. As this flap swings down and forward, it assumes the curved shape that is shown. With this optimum shaping, its performance probably exceeds to some extent the data presented in Figures 3.43 and 3.44. Figure 3.43 (taken from Ref. 3.20) shows that without a trailing edge flap, the Kruger flap gives an increment in $C_{l_{max}}$ to the 64-012 airfoil of only 0.4. However, the plain airfoil has a higher $C_{l_{max}}$ to begin with than that of Figures 3.40 and 3.41. Hence, the total $C_{l_{max}}$ for the Kruger-flapped airfoil without a trailing edge flap is about the same as for the other two leading edge devices.

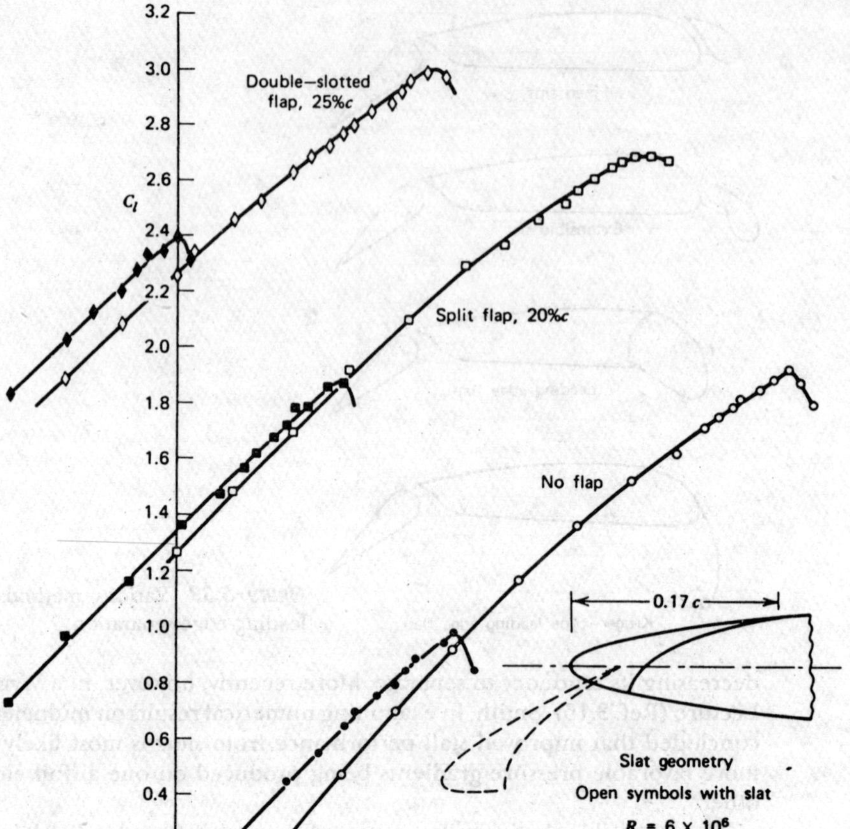

Figure 3.40 Effect of leading edge slat on NACA 64A010 airfoil with and without flaps.

However, with the split flap, the Kruger flap produces a combined $C_{l_{max}}$ equal to 3.0, which is 0.3 to 0.4 higher than the corresponding data of Figures 3.40 and 3.41.

The data of Figure 3.44 (taken from Ref. 3.21) are based on Kruger's original work.

The Optimum Airfoil for High Lift

Stratford, in References 3.25 and 3.26, examined both theoretically and experimentally the possibility of diffusing a turbulent boundary layer in such a way as to produce zero wall shear. Known as "imminent separation pressure recovery," it was found by Stratford that it is indeed possible, with the proper pressure gradient,

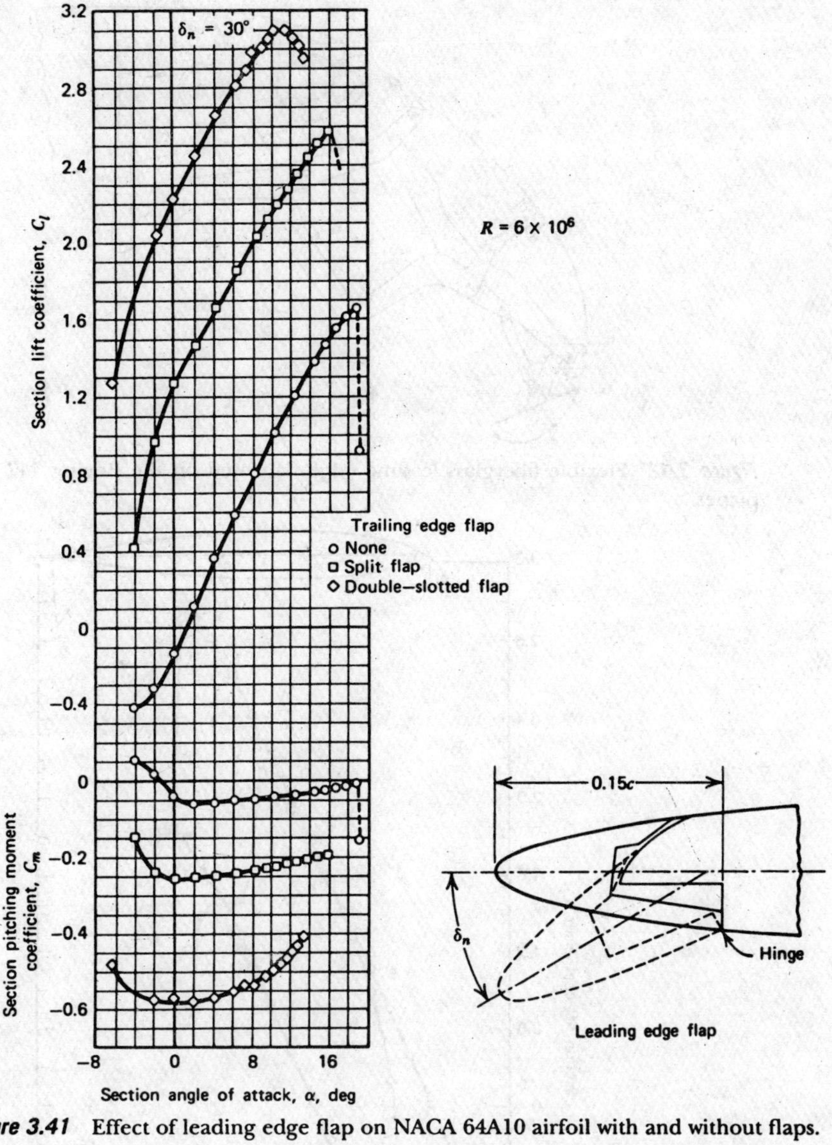

Figure 3.41 Effect of leading edge flap on NACA 64A10 airfoil with and without flaps.

to maintain a velocity profile along a diffuser such that $\partial u(y)/\partial y$ is equal to zero at the wall. $u(y)$ is the velocity in the boundary layer parallel to the wall and is a function of the distance, y, from the wall. With the velocity gradient at the wall equal to zero, the boundary layer is just on the verge of separating, since a negative value of this gradient will result in reverse flow, as illustrated in Figure 3.45.

In the abstract to Reference 3.24, Stratford states:

No fundamental difficulty was encountered in establishing the flow and it had, moreover, a good margin of stability. The dynamic head in the zero skin friction boundary layer was found to be linear at the wall (i.e., $u \infty y^{1/2}$), as predicted theoretically in the previous paper. (Author's note, Stratford is referring to Ref. 3.25.)

The flow appears to achieve any specified pressure rise in the shortest possible distance and with

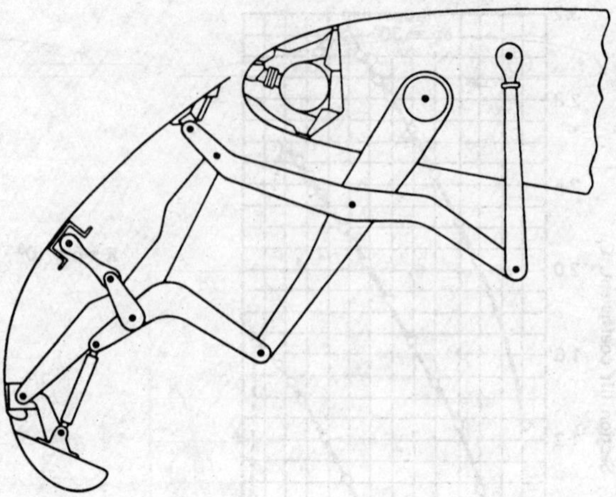

Figure 3.42 Flexible fiberglass leading edge flap used on the Boeing 747 and YC-14 airplanes.

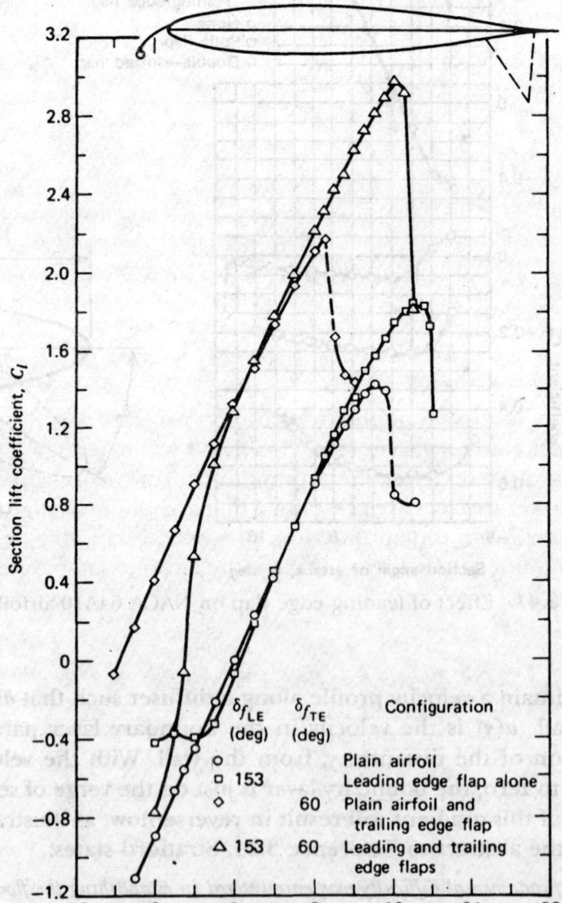

Figure 3.43 Section lift characteristics for the NACA 64_1-012 airfoil section equipped with a 0.10c upper-surface leading edge flap alone, and in combination with a 0.20c trailing edge split flap $R = 6.0 \times 10^6$.

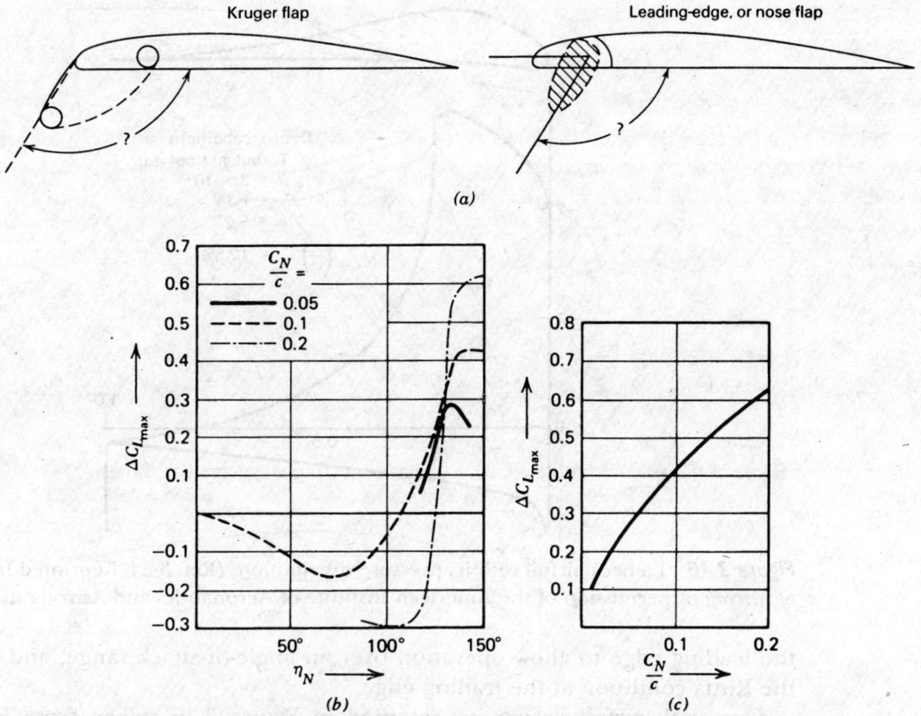

Figure 3.44 Characteristics of Kruger flaps. (*a*) Illustration of Kruger nose flap and simple, hinged leading edge flap. (*b*) Effect of flap angle on maximum lift coefficient. (*c*) Effect of flap chord on maximum lift coefficient.

probably the least possible dissipation of energy for a given initial boundary layer. Thus, an airfoil which could utilize it immediately after transition from laminar flow would be expected to have a very low drag.

The Stratford imminent separation pressure recovery was adopted for airfoils by Liebeck and Ormsbee (Ref. 3.27) and was extended later by Liebeck (Ref. 3.24). Using variational calculus, optimum chordwise pressure distributions for the upper and lower surfaces are prescribed that are modified slightly by additional constraints not present in the optimization process. Specifically, the optimum C_p distributions are modified in order to (1) close the airfoil contour at the trailing edge, (2) round

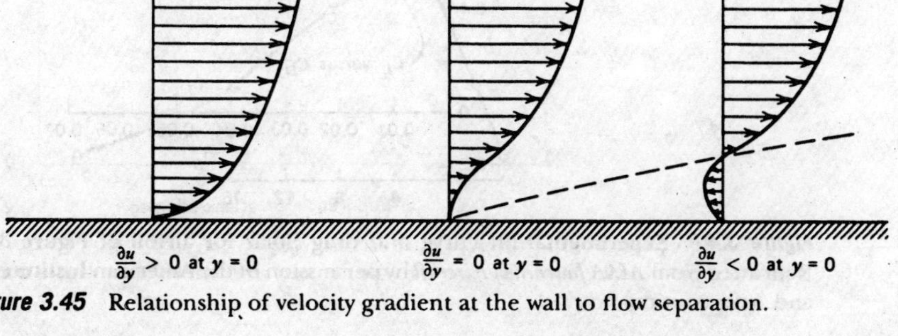

Figure 3.45 Relationship of velocity gradient at the wall to flow separation.

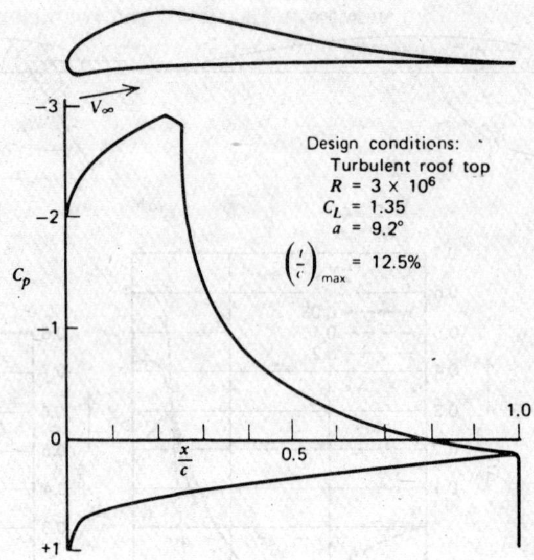

Figure 3.46 Liebeck airfoil with its pressure distribution. (Ref. 3.24. Reprinted from *Journal of Aircraft* by permission of the American Institute of Aeronautics and Astronautics)

the leading edge to allow operation over an angle-of-attack range, and (3) satisfy the Kutta condition at the trailing edge.

One such airfoil design is presented in Figure 3.46 (taken from Ref. 3.24). Included on the figure is the pressure distribution to which the airfoil was designed. Test data on this airfoil obtained at a Reynolds number of 3×10^6 are presented in Figure 3.47. Although this configuration is referred to by the reference as a "turbulent rooftop" case, transition does not occur until the start of the Stratford pressure recovery. In this case, the performance of the airfoil is seen to be good

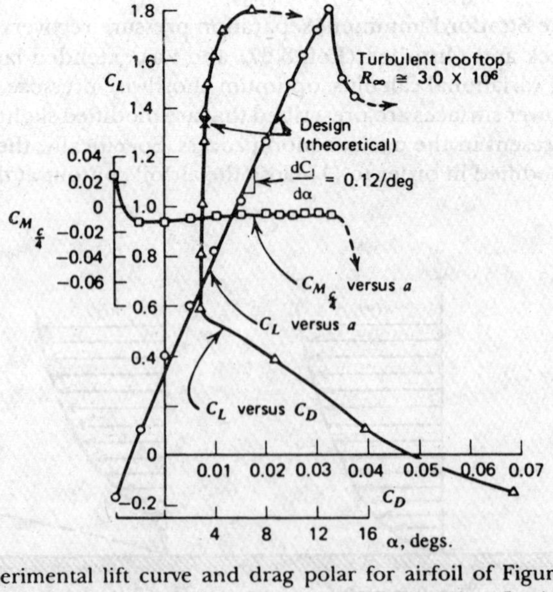

Figure 3.47 Experimental lift curve and drag polar for airfoil of Figure 3.46 (Ref. 3.24. Reprinted from *AIAA Journal of Aircraft* by permission of the American Institute of Aeronautics and Astronautics)

from the standpoint of $C_{l_{max}}$ and C_d. The drag coefficient remains below the value of 0.01 over a wide range of C_l values from 0.6 to 1.6.

Powered-Lift Systems

Figure 3.48 (taken from Ref. 3.26) presents the growth of $C_{L_{max}}$ over the years since the Wright Brothers' success. The two points labeled *K* and *L* are somewhat misleading, since these two aircraft were experimental in nature and used distributed suction over the wing to delay separation. From this figure and the preceding information on flapped and plain airfoils, $C_{L_{max}}$ of slightly over 3 is probably the best that can be achieved without the addition of power. Although two-dimensional airfoils with double-slotted flaps can do better than this, as will be seen later, their full potential cannot be achieved when applied to an airplane. Generally, the flaps cannot be applied over the entire span of the wing. In addition to this loss in $C_{L_{max}}$, an added penalty results from the fuselage and tail download required for trim.

$C_{L_{max}}$ values considerably in excess of those presented thus far can be achieved by the use of power. The power can be applied directly to the wing, such as blowing a jet of air over the flap, or the propeller slipstream or a jet exhaust can be deflected by the flap system to produce high lift. These high lift systems are important to achieving vertical or short takeoff and landing (V/STOL) performance. Therefore, coverage of powered lift will be deferred until Chapter 8.

THE LIFTING CHARACTERISTICS OF A FINITE WING

A two-dimensional airfoil with its zero lift line at an angle of attack of 10° will deliver a lift coefficient, C_l, of approximately 1.0. When incorporated into a wing of finite aspect ratio, however, this same airfoil at the same angle of attack will produce a

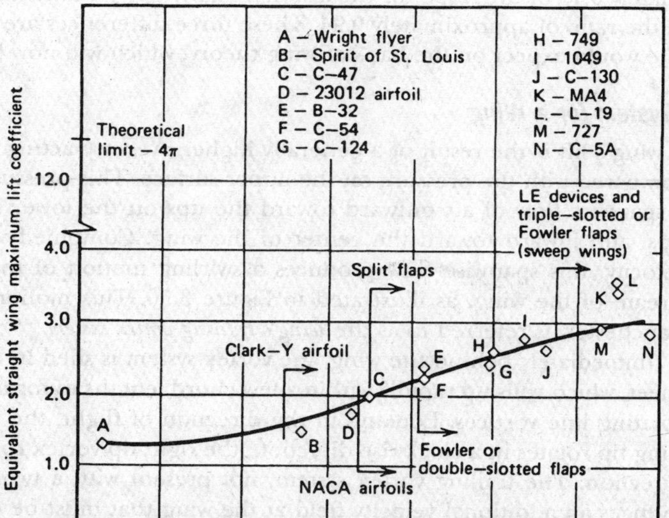

Figure 3.48 History of maximum lift coefficients for mechanical lift systems. (Ref. 3.28. Reprinted from the *AIAA Journal of Aircraft* by permission of the American Institute of Aeronautics and Astronautics)

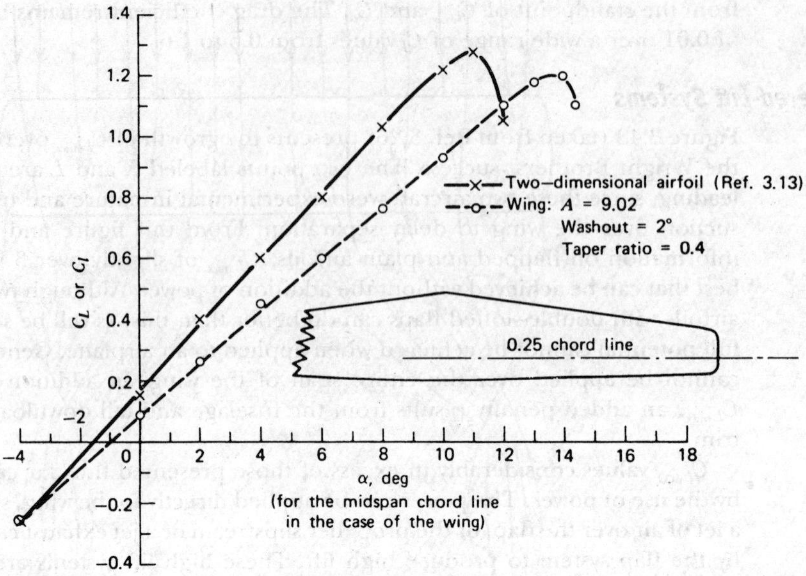

Figure 3.49 Comparison of NACA 65-210 airfoil lift curve with that of a wing using the same airfoil.

wing lift coefficient, C_L, significantly less than 1.0. The effect of aspect ratio is to decrease the slope of the lift curve $C_{L\alpha}$ as the aspect ratio decreases. Figure 3.49 illustrates the principal differences in the lift behavior of a wing and an airfoil. First, because the wing is twisted so that the tip is at a lower angle of attack than the root (washout), the angle for zero lift, measured at the root, is higher for the wing by approximately 0.6°. Next, the slope of the wing's lift curve $C_{L\alpha}$, is approximately 0.79 of the slope for the airfoil. Finally, $C_{L_{\max}}$ is only slightly less than $C_{l_{\max}}$, in the ratio of approximately 0.94. These three differences are almost exactly what one would expect on the basis of wing theory, which will now be developed.

The Vortex System for a Wing

A wing's lift is the result of a generally higher pressure acting on its lower surface compared with the pressure on the upper surface. This pressure difference causes a spanwise flow of air outward toward the tips on the lower surface, around the tips, and inward toward the center of the wing. Combined with the free-stream velocity, this spanwise flow produces a swirling motion of the air trailing downstream of the wing, as illustrated in Figure 3.50. This motion, first perceived by Lanchester, is referred to as the *wing's trailing vortex system.*

Immediately behind the wing, the vortex system is shed in the form of a vortex sheet, which rolls up rapidly within a few chord lengths to form a pair of oppositely rotating line vortices. Looking in the direction of flight, the vortex from the left wing tip rotates in a clockwise direction; the right tip vortex rotates in the opposite direction. The trailing vortex system, not present with a two-dimensional airfoil, induces an additional velocity field at the wing that must be considered in calculating the behavior of each section of the wing.

At any airfoil section along the wing, except near the tips, the flow field will appear qualitatively as shown in the sketch of Figure 3.51. Far ahead of the wing the velocity induced by the trailing vortex system vanishes, whereas far behind the

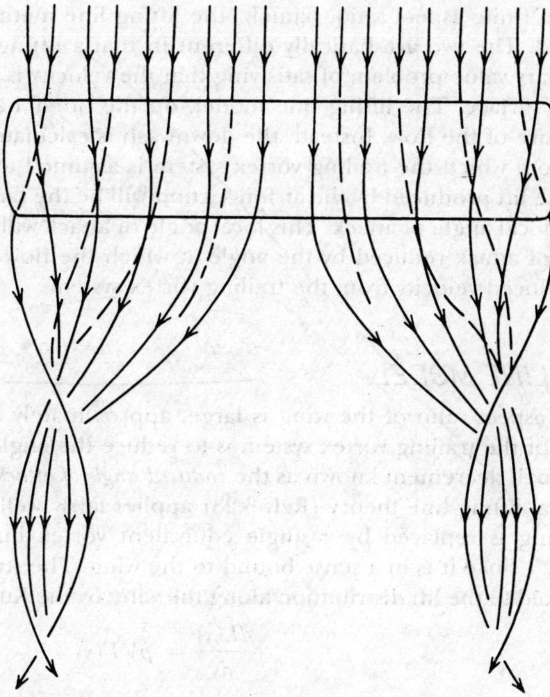

Figure 3.50 Generation of vortex system by finite aspect ratio wing.

wing, the induced downwash approaches a value equal to approximately twice that induced at the wing. Two effects, shown on the figure, are caused at the wing by this induced flow, which is not present for the two-dimensional airfoil. First, the flow is curved because of the fact that the downwash increases from the leading edge of the chord to the trailing edge. This curvature of the flow effectively decreases the camber of the airfoil section. Second, the downward induced flow added vectorially to the free-stream velocity effectively decreases the angle of attack. Both effects decrease the section lift compared to the lift, which the section would produce as a 2-D airfoil at the angle of attack, α.

Two methods are generally used to predict the aerodynamic characteristics of a

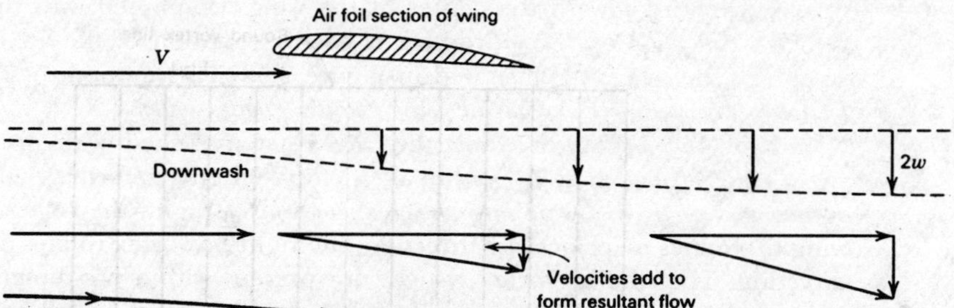

Figure 3.51 Flowfield induced downward by vortex system trailing from a wing.

wing of finite aspect ratio, namely, the lifting line method and the lifting surface method. The two are basically different in that a lifting surface model solves the boundary value problem of satisfying that the velocity is everywhere tangent to the wing's surface. The lifting line model, on the other hand, neglects the induced curvature of the flow. Instead, the downwash is calculated along a single spanwise line from which the trailing vortex system is assumed to leave. It is then assumed that the lift produced by the airfoil section will be the same as the 2-D airfoil acting at the local angle of attack. This local angle of attack will be equal to the geometric angle of attack reduced by the angle at which the flow is deflected downward by the induced velocity from the trailing vortex system.

THE LIFTING LINE MODEL

If the aspect ratio of the wing is large, approximately 5 or higher, the principal effect of the trailing vortex system is to reduce the angle of attack of each section by a small decrement known as the *induced angle of attack*, α_i. In this case, Prandtl's classical lifting line theory (Ref. 3.28) applies fairly well. As shown in Figure 3.52, the wing is replaced by a single equivalent vortex line, known as the "bound vortex," since it is in a sense bound to the wing. The strength of this vortex, $\Gamma(y)$, is related to the lift distribution along the wing by the Kutta-Joukowski relationship.

$$\frac{dL(y)}{dy} = \rho V \Gamma(y) \tag{3.59}$$

Expressing the lift per unit length of span in terms of the section chord length, $c(y)$, and section lift coefficient, $C_l(y)$, leads to

$$\Gamma(y) = \frac{1}{2} c(y)\, C_l(y)\, V \tag{3.60}$$

With no physical surface outboard of the wing tips to sustain a pressure difference, the lift, and hence Γ, must vanish at the tips.

According to the Helmholz theorem regarding vortex continuity (Ref. 1.3, p. 120), a vortex line or filament can neither begin nor end in a fluid; hence, it appears as a closed loop, ends on a boundary, or extends to infinity. Thus, it follows that if in going from y to $y + dy$, the bound circulation around the wing increases from Γ to $\Gamma + d\Gamma$, a free vortex filament of strength $d\Gamma$, lying in the direction of

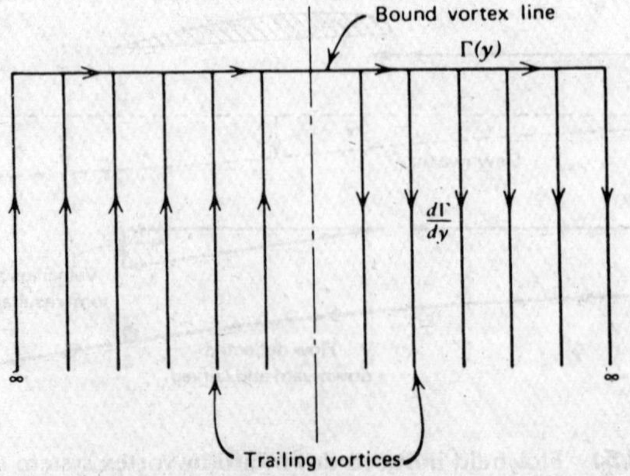

Figure 3.52 Lifting line model of a wing and trailing vortex system.

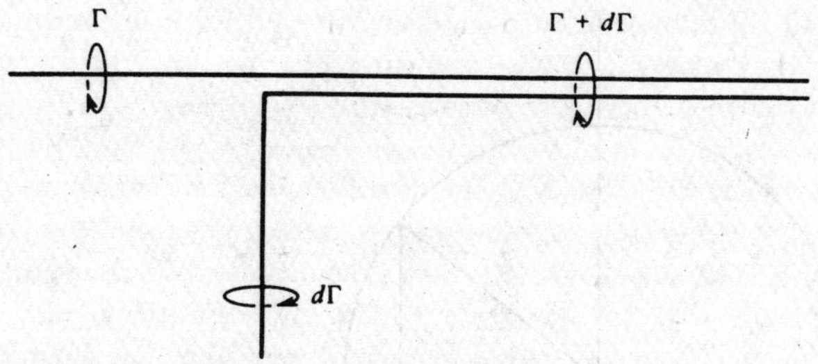

Figure 3.53 Illustration of vortex continuity.

the free-stream velocity, must be feeding into Γ in order to satisfy vortex continuity. This statement may be clarified by reference to Figure 3.53.

The entire vortex system shown in Figure 3.52 can be visualized as being closed infinitely far downstream by a "starting" vortex. This vortex, opposite in direction to the bound vortex, would be shed from the trailing edge of the wing as its angle of attack is increased from zero.

The trailing vortex system of strength $d\Gamma$ induces a downwash, $w(y)$, at the lifting line, as mentioned earlier. As pictured in Figure 3.54, this reduces the angle of attack by the small angle α_i. Thus, the section lift coefficient will be given by

$$C_l = C_{l_\alpha} (\alpha - \alpha_i) \tag{3.61}$$

α being measured relative to the section zero lift line.

To a small angle approximation, the induced angle of attack, α_i, is given by w/V. The downwash, w, can be determined by integrating the contributions of the elemental trailing vortices of strength $d\Gamma$. If the vortex strength $d\Gamma$ trails from the wing at a location of y, its contribution to the downwash at another location y_0 can be found by Equation 2.65 to be

$$dw(y_0) = \frac{d\Gamma(y)}{4\pi(y_0 - y)} \tag{3.62}$$

Thus, α_i becomes

$$\alpha_i(y_0) = \frac{1}{4\pi V}\int_{-b/2}^{b/2} \frac{d\Gamma(y)}{y_0 - y} \tag{3.63}$$

Equations 3.60, 3.61, and 3.63 together relate $\Gamma(y)$ to $c(y)$ and $\alpha(y)$ so that, given the wing geometry and angle of attack, one should theoretically be able to solve

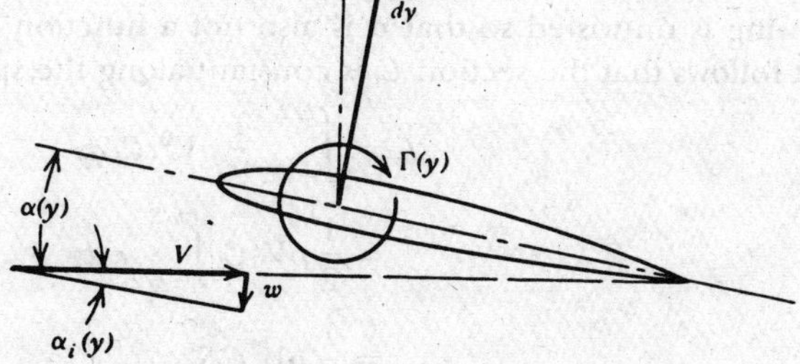

Figure 3.54 A wing section under the influence of the free-stream velocity and the downwash.

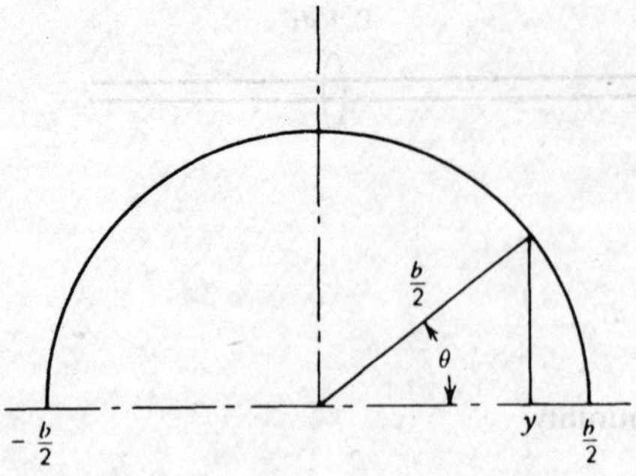

Figure 3.55 Coordinate transformation used in the reduction of the lifting line equations.

for Γ and hence the wing lift. In order to accomplish the solution, it is expedient to make the coordinate transformation pictured in Figure 3.55.

$$y = \frac{b}{2}\cos\theta$$

Hence, Equation 3.62 becomes

$$\alpha_i(\theta_0) = \frac{1}{2\pi bV}\int_0^{\pi}\frac{d\Gamma(\theta)}{\cos\theta - \cos\theta_0} \tag{3.64}$$

Since more elaborate and comprehensive treatments of wing theory can be found in texts devoted specifically to the subject (e.g., see Ref. 3.32), only the classical solution for the elliptic Γ distribution will be covered here. This particular case is easily handled and results in the essence of the general problem.

Assume that Γ is of the form

$$\Gamma = \Gamma_0\sqrt{1 - \left(\frac{2y}{b}\right)^2} \tag{3.65}$$

This transforms to

$$\Gamma = \Gamma_0 \sin\theta$$

Here, Γ_0 is obviously the midspan value of the bound circulation. Thus, Equation 3.64 becomes

$$\alpha_i(\theta_0) = \frac{\Gamma_0}{2\pi bV}\int_0^{\pi}\frac{\cos\theta\, d\theta}{\cos\theta - \cos\theta_0}$$

The preceding integral was encountered previously in thin airfoil theory and has a value of π. Thus, for an elliptic Γ distribution, α_i and hence the downwash is found to be a constant independent of y.

$$\alpha_i = \frac{\Gamma_0}{2bV} \tag{3.66}$$

If the wing is untwisted so that α is also not a function of y then, from Equation 3.61, it follows that the section C_l is constant along the span. Thus,

$$\begin{aligned} L &= \int_{-b/2}^{b/2}\frac{1}{2}\rho V^2 c C_l\, dy \\ &= \frac{1}{2}\rho V^2 C_l \int_{-b/2}^{b/2} c\, dy \\ &= \frac{1}{2}\rho V^2 C_l S \end{aligned}$$

But

$$C_L = \frac{L}{qS}$$

Hence,

$$C_l = C_L$$

Equation 3.60 then gives

$$\Gamma(y) = \frac{1}{2}c(y)\,C_L V$$

or

$$c(y) = \frac{2\Gamma_0}{C_L V}\sqrt{1 - \left(\frac{2y}{b}\right)^2} \tag{3.67}$$

$$= c_0\sqrt{1 - \left(\frac{2y}{b}\right)^2}$$

Thus, it is found that, according to lifting line theory, an untwisted wing with an elliptical planform will produce an elliptic Γ distribution. Such a wing will have a constant downwash and section C_l.

Since C_l is constant and equal to C_L, Equations 3.60, 3.61, and 3.66 can be applied at the midspan position in order to determine the slope of the wing lift curve. First, from Equations 3.60 and 3.66,

$$\alpha_i = \frac{c_0 C_L}{4b}$$

But, for the planform given by Equation 3.67, c_0 and b are related to the aspect ratio by

$$A = \frac{4b}{\pi c_0}$$

Thus, α_i becomes

$$\alpha_i = \frac{C_L}{\pi A} \tag{3.68}$$

Inserted into Equation 3.66, the preceding results in

$$C_L = C_{l_\alpha}\left(\alpha - \frac{C_L}{\pi A}\right)$$

or

$$C_L = C_{l_\alpha}\left[\frac{1}{1 + (C_{l_\alpha}/\pi A)}\right]\alpha$$

Using the theoretical value of 2π C_l/rad derived earlier, the preceding becomes

$$C_L = C_{l_\alpha}\left(\frac{A}{A + 2}\right)\alpha \tag{3.69}$$

Equations 3.68 and 3.69 are important results. The induced angle of attack is seen to increase with decreasing aspect ratio of which reduces the slope of the lift curve, $C_{L\alpha}$. A wing having a low aspect ratio will require a higher angle of attack than a wing with a greater aspect ratio in order to produce the same C_L.

It was stated previously that the comparative performance between the wing and airfoil shown in Figure 3.49 could be explained theoretically. In this case, $A = 9.02$ so that, on the basis of Equation 3.69,

$$C_{L_\alpha} = 0.819\ C_{l_\alpha}$$

This result is within about 2% of the experimental results presented in Figure 3.49.

Note from Figure 3.54 that, because of the induced angle of attack, the lift vector is tilted backward, resulting in a component of force in the drag direction. This drag is known as induced drag and is covered more thoroughly in Chapter 4. At this time it will be noted simply that

$$D_i = L\alpha_i$$

which leads to

$$C_{D_i} = \frac{C_L^2}{\pi A}$$

As the aspect ratio decreases, the lifting line becomes progressively less accurate. For example, for an aspect ratio of 4.0, Equation 3.69 is approximately 11% higher than that predicted by more exact methods.

As described in Reference 3.12, a more accurate estimate of C_{L_α} is obtained from

$$C_{L_\alpha} = C_{l_\alpha}\frac{A}{A + [2(A + 4)/(A + 2)]} \tag{3.70}$$

An alternate to Equation 3.70 is offered by Reference 3.32 and is referred to as the *Helmbold equation*, after the original source noted in the reference. The Helmbold equation reads

$$C_{L_\alpha} = C_{l_\alpha}\frac{A}{(C_{l_\alpha}/\pi) + \sqrt{(C_{l_\alpha}/\pi)^2 + A^2}}$$

Replacing C_{l_α} by 2π in the denominator,

$$C_{L_\alpha} = C_{l_\alpha}\frac{A}{2 + \sqrt{4 + A^2}} \tag{3.71}$$

Equations 3.70 and 3.71 agree within a couple of percent over the range of practical aspect ratios and approach each other in the limits of $A = 0$ or $A = \infty$. This holds for high or low aspect ratios and is based on an approximate lifting surface theory, which accounts for the chordwise distribution of bound circulation as well as the spanwise distribution.

Let us visualize an aerodynamically untwisted wing; that is, one for which the zero lift lines all lie in a plane. Imagine this wing to be at a zero angle of attack and hence, operating at zero C_L. Holding the midspan section fixed, let us now twist the tip *up* through an angle ϵ_T. We will assume that the twist along the wing is linear, so that at any spanwise location y, the twist relative to the midspan section is given by

$$\epsilon = \frac{2\epsilon_T}{b}|y|$$

Obviously, the wing will now develop a positive C_L, since every section except the midspan is at a positive angle of attack. If we define the angle of attack of the wing to be that of the zero lift line at the root, the angle of attack of the wing for zero lift will be negative; that is, we must rotate the entire wing nose downward in order to return to the zero lift condition. For a C_L of zero, Equation 3.68 shows that on the average, α_i equals zero; thus, at any section,

$$C_l = C_{l_\alpha}\left(\epsilon_T \frac{|y|}{b/2} + \alpha_{w_0}\right)$$

where α_{w_0} is the angle of attack of the wing for zero lift. To find this angle, an expression is written for the total wing lift and is equated to zero.

$$\begin{aligned} L &= \int_{-b/2}^{b/2} qcC_l\, dy \\ &= \int_{-b/2}^{b/2} qcC_{l_\alpha}\left(\epsilon_T \frac{|y|}{b/2} + \alpha_{w_0}\right) dy \end{aligned}$$

Equating this to zero and taking q and C_{l_α} to be constant (this is not quite true for C_{l_α}, but close) leads to

$$\alpha_{w_0} = -\frac{2\epsilon_T}{bS}\int_{-b/2}^{b/2} c|y|dy \tag{3.72}$$

If ϵ is not linear,

$$\alpha_{w_0} = \frac{-1}{S}\int_{-b/2}^{b/2} c\epsilon\, dy \tag{3.73}$$

Now consider a linearly tapered wing for which the chord distribution is given by

$$c = c_0 - (c_0 - c_T)\frac{2|y|}{b}$$

Defining the taper ratio λ as the ratio of the tip chord, c_T, to the midspan chord, c_0, the preceding equation can be written as

$$c = c_0\left[1 - (1 - \lambda)\frac{2|y|}{b}\right]$$

Substituting this into Equation 3.75 and integrating results in the angle of attack of the wing for zero lift as a function of twist and taper ratio.

$$\alpha_{w_0} = -\frac{\epsilon_T}{3}\frac{(1 + 2\lambda)}{(1 + \lambda)} \tag{3.74}$$

Most wings employ a negative twist referred to as "washout." Generally, ϵ_T is of the order of -3 or 4° to assure that the inboard sections of the wing stall before the tip sections. Thus, as the wing begins to stall, the turbulent separated flow from the inboard portion of the wing flows aft over the horizontal tail, providing a warning of impending stall as the pilot feels the resulting buffeting. In addition, with the wing tips still unstalled, the pilot has aileron control available to keep the wings level in order to prevent the airplane from dropping into a spin. At the present time, the stall-spin is one of the major causes of light airplane accidents.

The wing of the airplane in Figure 3.49 has a 2° washout and a taper ratio of 0.4. According to Equation 3.74, α_{w_0} for this wing will be $+0.8°$. This is close to the results presented in Figure 3.49, where the angle of attack of the wing for zero lift is seen to be approximately 0.6° greater than the corresponding angle for the airfoil. Thus, knowing an airfoil lift curve, one can estimate with reasonable accuracy the lift curve of a wing incorporating that airfoil by calculating the slope and angle for zero lift with the use of Equations 3.70 and 3.74, respectively.

As a further example, in the use of these equations, consider a wing having an NACA 63,4-421 airfoil (Figure 3.33) at its midspan that fairs linearly into an NACA 0012 airfoil at the tip. The wing has no geometric twist; that is, the section chord lines all lie in the same plane. The wing's aspect ratio is equal to 6.0, and it has a taper ratio of 0.5. The problem is to find the angle of attack of the wing measured relative to the midspan chord, which will result in a C_L value of 0.8.

To begin, we note from Figure 3.33 that the 63,4-421 airfoil has an angle of zero lift of $-3°$; thus, the zero lift line at the midspan is up 3° from the chord line. It follows that the aerodynamic twist ϵ_T is $-3°$, since the tip airfoil is symmetrical. Inserting this and the taper ratio into Equation 3.74 results in $\alpha_{wo} = 1.3°$. From Equation 3.70, for an aspect ratio of 6.0,

$$C_{L\alpha} = 0.706\ C_{l\alpha}$$

From Figure 3.33 or Reference 3.1, $C_{l\alpha}$ is nearly the same for the midspan and tip sections and is equal approximately to 0.107 C_l/deg. Hence, $C_{L\alpha} = 0.076$ C_L/deg. Therefore, over the linear portion of the lift curve, the equation for the wing C_L relative to the midspan zero lift line becomes

$$C_L = 0.076(\alpha - 1.3)$$

For C_L of 0.8, α is found to equal 11.8°. Thus, to answer the original problem, the angle of attack of the midspan chord, α_w, will equal $11.8 - 3$, or 8.8°.

One should remember that, in applying the lifting line model to predicting the characteristics of a wing, the aerodynamic twist is measured relative to the zero lift lines of the wing airfoil sections. Thus, the lifting line model can be applied to a wing with deflected flaps by determining the increase in the angle of attack of the zero lift lines for the sections over the flap when it is deflected.

A numerical solution of a wing paralleling the closed-form solution for the lifting line model can be written without too much difficulty. To avoid complexities associated with satisfying Helmholz's law of vortex continuity, the wing can be modeled by a series of horseshoe vortices of unknown strengths as shown in Figure 3.56. The conditions set by Equations 3.60 through 3.63 are then satisfied at points along the lifting line at the middle of each horseshoe vortex. This will lead to a set of simultaneous equations, equal in number to the number of horseshoe vortices, for the unknown horseshoe vortex strengths.

This model, for high aspect ratio wings, may be preferable to the more exact lifting surface theory to be discussed next since experimental section lift curve

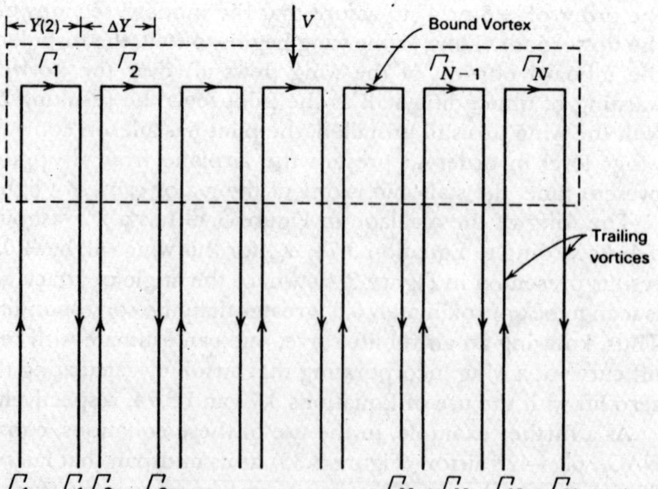

Figure 3.56 Lifting line model of a wing using horseshoe vortex elements.

slopes can be incorporated into the lifting line model. In formulating the expressions for the downwash at N points in terms of the unknown $\Gamma(I)$ values, the circulation along the bound vortex line itself should not be included. First, this will lead to a singularity in the application of the Biot–Savart law and, second, the effect of the distributed bound circulation is included in the use of the section $C_{l\alpha}$.

Computer Exercise 3.2 "LIFTLINE"

Formulate a computer program to model a linearly tapered wing or an elliptic wing by means of a lifting line. Run the program for a flat, untwisted elliptic wing having an aspect ratio of 8 and compare the C_l and α_i distributions with predictions from the closed form solution. Also compare the total wing C_L with Equation 3.69.

LIFTING SURFACE MODEL

A theoretical treatment of a finite lifting surface, more exact than the lifting line model, is known as a *lifting surface* model. Here, instead of concentrating the bound circulation along a lifting line and correcting the section angle of attack for induced effects, the boundary value problem is solved for a spanwise and chordwise distribution of bound circulation. One particular model will be presented here, which again makes use of horseshoe vortices to automatically satisfy vortex continuity.

One possible scheme for paneling the wing is shown in Figure 3.57. To begin, note that the span is given a value of 2 so that any spanwise location represents a fraction of the semispan. Next, M spanwise stations and N chordwise stations are defined. Thus, the wing is divided into $(N - 1)$ $(M - 1)$ or NP panels. For the example shown in Figure 3.57, $M = 9$ and $N = 5$ resulting in NP equal to 32. A right-handed coordinate system is chosen with the origin at the midspan point of

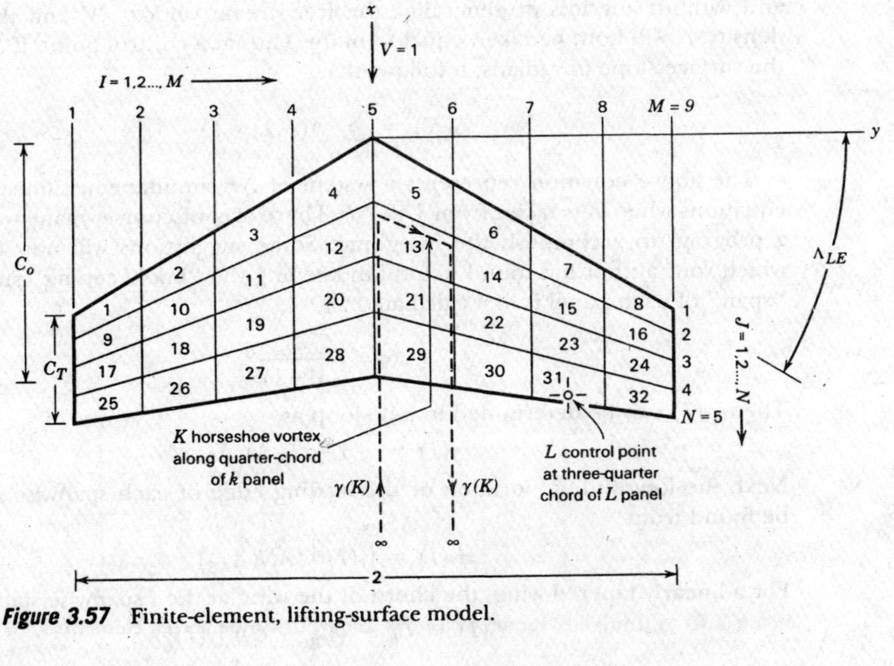

Figure 3.57 Finite-element, lifting-surface model.

the leading edge. The x coordinate is defined positive forward, y positive to the right and z downward. This assures that the calculated induced velocity components will have the correct sign when applying the Biot–Savart law to the vortex elements.

Horseshoe vortex elements are placed on each panel. As shown, the left leg of the horseshoe begins infinitely far downstream and extends up the left side of a panel to a point located a quarter of the panel's chord on the left side behind the panel's leading edge. The vortex line then runs along the panel's "quarter-chord" line to the right side of the panel. From this point, it trails infinitely far downstream. The program is written to define the four corners of each horseshoe vortex in terms of $N,M,$ and the wing geometry. The direction of the circulation is taken according to the right-hand rule with the thumb pointing in the direction of the arrows shown in the figure along the horseshoe vortex.

Control points are located at the middle of each panel three-quarters of the panel's chord back from its leading edge. The problem then becomes that of finding the strengths of the NP horseshoe vortices, which will induce a resultant flow tangent to the panel surfaces at all of the NP control points. The Kutta–Joukowski relationship can then be applied to the spanwise component of each horseshoe vortex to determine the lift on that particular panel. In specifying the surface slope at each control point, one can include angle of attack, camber, twist, flaps, and ailerons. Note that the last row of control points are downstream of the last row of horseshoe vortices, assuring that the flow leaves tangent to the trailing edge in accordance with the Kutta condition.

In formulating the numerical solution it is expedient to determine influence coefficients, $A(K,L)$. $A(K,L)$ is the downwash induced at control point K by a horseshoe vortex placed on the panel L and having a unit circulation strength. Knowing $A(K,L)$, the downwash at point K can then be written as

$$w(K) = \sum_{L=1}^{NP} A(K,L)\,\gamma(L) \tag{3.75}$$

As was done earlier in the numerical model for the thin airfoil, for convenience and without any loss of generality, the free-stream velocity, V, and the air mass density, ρ, will both be taken equal to unity. Thus at a control point, if $S(K)$ equals the surface slope in radians, it follows that

$$S(K) = \sum_{L=1}^{NP} A(K,L)\,\gamma(L) \tag{3.76}$$

The above equation represents a system of NP simultaneous, linear algebraic equations when K is taken from 1 to NP. There are, of course, many ways to write a program to accomplish the foregoing. Some suggestions will now be offered, which your author has found helpful in keeping the "bookkeeping" straight. The "span" of each panel is first calculated as

$$\Delta y = \frac{2}{(M-1)}$$

Then, $y(I)$ can be determined in a do-loop as

$$y(I) = -1 + (I-1)\,\Delta y$$

Next, the longitudinal location or the leading edge of each spanwise station can be found from

$$xle(I) = |y(I)|\,TAN(\Lambda_{LE})$$

For a linearly tapered wing, the chord of the wing at the I spanwise station is

$$C(I) = C_0 - (C_0 - C_T)\,|y(I)|$$

A transformation will be required in the numerical program to define the panel number, or control point, for a particular spanwise and chordwise location. In other words, the problem is formulated in terms of two indices, I and J, but must then be transformed to a single index, K. This is done by realizing that, for a given I and J, the following holds:

$$K = I + (M - 1)(J - 1)$$

In order to determine the influence coefficients $A(K,L)$ the procedure indicated by the flow chart of Figure 3.58 is followed. Here, the indices I and J have a C or V added to indicate that either the control point or vortex is being located. Looping over IC and JC, one calculates the index K and the location for the control point. Then for each point, the program loops over IV and JV to determine the index L and the velocity induced by the L vortex at the point K.

Computer Exercise 3.3 "LATTICE"

Write a numerical model of a lifting surface in the manner just described for a linearly tapered wing. Include wing twist and flaps. Run this program and the lifting line model for an untwisted, linearly tapered wing having an aspect ratio of 5 and a taper ratio of 0.5 and compare results. Also, use the model to compare with the prediction of $C_{L_{max}}$ for a finite wing, which follows.

THE MAXIMUM LIFT OF A FINITE WING

The maximum lift coefficient of a finite wing is influenced by several factors. Obviously, $C_{L_{max}}$ is strongly dependent on $C_{l_{max}}$; that is, the wing's performance depends on its airfoil performance. Second, the spanwise extent to which the wing is flapped has a significant influence on the maximum lift. Also, in estimating $C_{L_{max}}$, one must account for the presence of the fuselage, the tail download required to trim the aerodynamic pitching moment, and the spanwise distribution of loading over the wing.

The effect of aspect ratio on $C_{L_{max}}$ is slight, as one might expect from preceding considerations on the elliptic wing. The wing lift coefficient and section lift coefficients are nearly equal. The detailed estimation of a wing's $C_{L_{max}}$ begins with a calculation of its spanwise section C_l distribution. In the general case of a cambered, flapped wing with an arbitrary planform, this calculation can prove difficult. Before

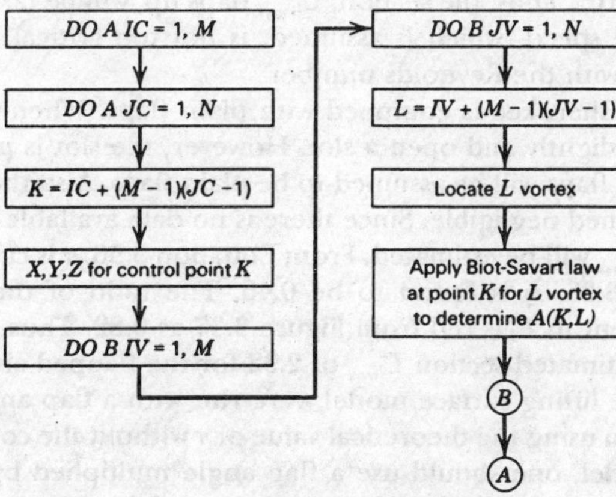

Figure 3.58 Flow chart to determine influence coefficients.

the availability of computers, aeronautical engineers used an approximate method, known as *Schrenk's Approximation,* in order to estimate the section C_l distribution for preliminary design purposes. This method was presented in the first edition of this book but is deleted from this edition since a numerical model, once developed, is considerably more accurate and just as convenient to apply.

The method of predicting $C_{L_{max}}$ for an isolated wing is best explained by means of an example. Consider the following wing geometry, which corresponds to the Piper PA-28 Cherokee shown in Figure 3.59.

Rectangular planform
Wing span = 30 ft.
Aspect ratio = 5.625
2° of washout
Plain flaps extend from 13 to 60% of the semispan
Flap chord is 20% of the airfoil chord
Two cases: flaps up and flaps down 40°
Airfoil section is NACA 65_2-415

From Reference 3.15, $C_{l_{max}}$ values for this airfoil with no flaps are as follows:

$$R = 8.9 \times 10^6 \qquad C_{l_{max}} = 1.6$$
$$R = 6.0 \times 10^6 \qquad C_{l_{max}} = 1.58$$
$$R = 3.0 \times 10^6 \qquad C_{l_{max}} = 1.45$$

The lifting line or lifting surface numerical models described previously can be used to obtain the section spanwise C_l at different angles of attack. However, a somewhat different approach must be taken in applying the one method or the other and some judgment exercised.

Basically, in predicting $C_{L_{max}}$ for a wing, one compares the distribution of the section C_l for a given wing C_L to the spanwise distribution of the section $C_{l_{max}}$, including the effects of flaps. If the section C_l exceeds $C_{l_{max}}$ at any point along the span, then the wing is predicted to be stalled in that area. Thus, one must estimate the spanwise distribution of both the section C_l and the section $C_{l_{max}}$ at a given wing C_L in order to determine $C_{L_{max}}$.

In order to estimate the Reynolds number, it will be necessary to select by some means, possibly iteratively if the stalling speed is being determined, at what speed the wing is operating when it stalls. For this example, a calibrated airspeed of 60 mph at sea level will be assumed. The constant chord is found from the span and aspect ratio to equal 5.33 ft. These numbers result in a Reynolds number of 2.98×10^6. Thus, the section, $C_{l_{max}}$, flaps up, will be taken as equal to 1.45. Observe that the speed, which is assumed, is not too critical because $C_{l_{max}}$ does not vary rapidly with the Reynolds number.

The Cherokee is equipped with plain flaps. When the flaps are deflected they extend slightly and open a slot. However, the slot is probably not too effective so that the flaps will be assumed to be plain flaps. Also the slight chord extension will be assumed negligible. Since there is no data available on the airfoil with the Piper flap, $C_{l_{max}}$ will be estimated. From Equation 3.56, τ is calculated to be 0.55 and from Figure 3.36, η is found to be 0.46. The ratio of the increment in $C_{l_{max}}$ to the increment in C_l is red from Figure 3.37 as 0.82. Thus, $\Delta C_{l_{max}}$ equals 0.87 resulting in an estimated section $C_{l_{max}}$ of 2.32 for the flapped airfoil.

If the lifting surface model were run with a flap angle of 40°, this would correspond to using the theoretical value of τ without the correction η. Thus, in running the model, one should use a flap angle multiplied by η. Thus, the effective flap angle to be used in the lifting surface model is 18°.

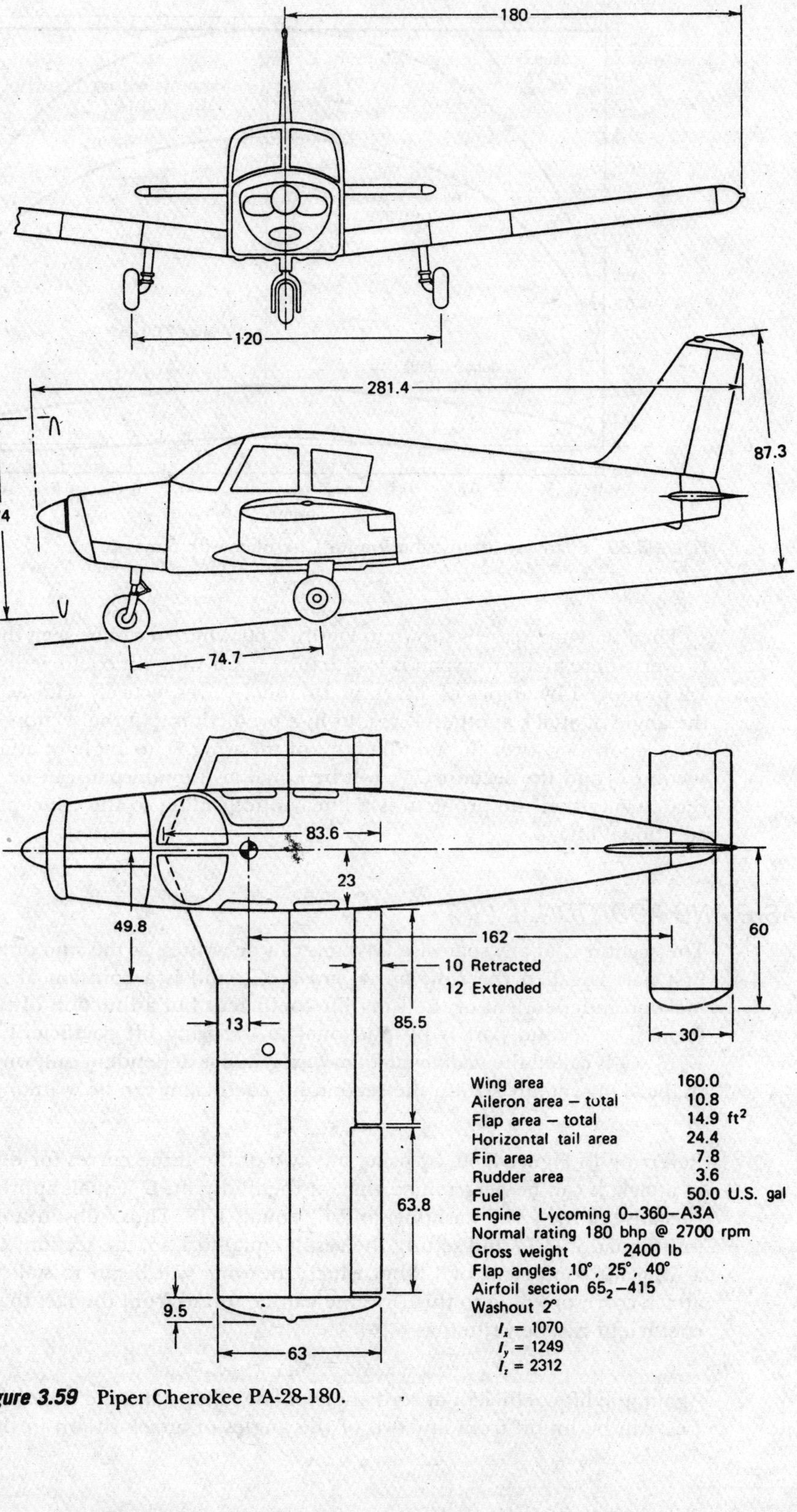

Figure 3.59 Piper Cherokee PA-28-180.

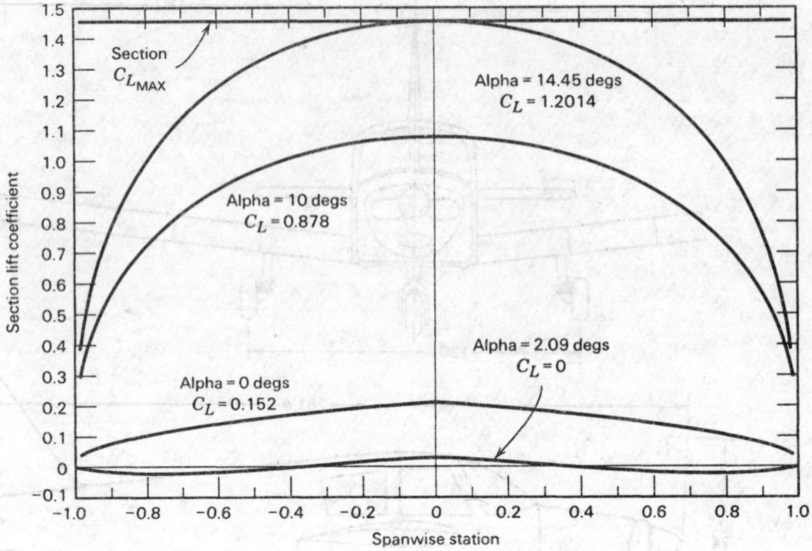

Figure 3.60 Spanwise lift distribution for Cherokee with flaps up.

The "flaps up" case is shown in Figure 3.60, where it can be seen that the section C_l everywhere along the span is less than $C_{l_{max}}$ for wing lift coefficients of 0.878 and 1.2 produced by angles of attack of 10 and 14°, respectively. However, increasing the angle of attack another 4° results in a prediction that the section C_l will exceed the section $C_{l_{max}}$ over the middle 50% of the wing. The angle of attack where the section C_l and the section $C_{l_{max}}$ will be equal at the midspan can be calculated by recognizing that the problem is a linear one leading to the concept of basic and additional lift.

BASIC AND ADDITIONAL LIFT

The section C_l at any spanwise location can be written as the sum of two parts. The first part is called the *basic lift coefficient*, C_{l_b}, and is a constant at any particular location independent of the wing lift coefficient but a function of twist and planform. The second part is proportional to the wing lift coefficient and given by $C_{l_a} C_L$. C_{l_a} is called the *additional lift coefficient* and is dependent only on the planform shape. Thus, at any station the section lift coefficient can be written as

$$C_l = C_{l_b} + C_{l_a} C_L \tag{3.77}$$

Referring to Figure 3.60, by using any two of the three curves for different angles of attack, it can be determined that, at the midspan, C_{l_b} equals approximately 0.03 (because of the small washout) and C_{l_a} equals 1.18. Thus, substituting these values into Equation 3.77 and setting the result equal to 1.45, the section, $C_{l_{max}}$, results in a wing lift coefficient of 1.20 at which the wing will begin to stall. The angle of attack corresponding to this C_L value can be found from the fact that the wing lift coefficient can be written as

$$C_L = C_{L_0} + C_{L_\alpha} \alpha \tag{3.78}$$

Again, the lift coefficient at zero angle of attack, C_{L_0}, and the slope of the lift curve, C_{L_α}, can be found from any two of the angles of attack shown in the figure. The

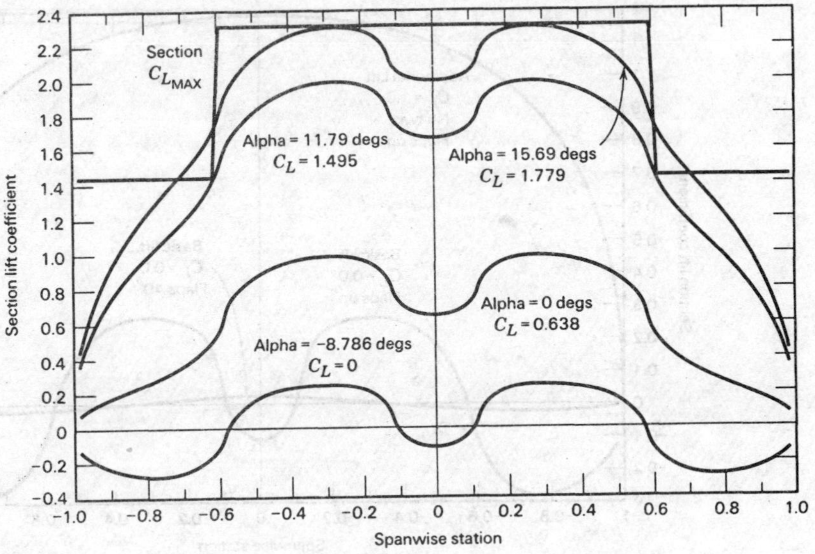

Figure 3.61 Spanwise lift distribution for Cherokee with flaps down 40°.

value of 1.20 for $C_{L_{max}}$ agrees almost exactly with the airplane's manual. In view of the uncertainties to be discussed later, this agreement may be somewhat fortuitous.

In the case of Figure 3.61, the wing is predicted to stall first just outboard of the flaps at a wing lift coefficient of 1.50. However, the inboard portion of the flaps is well below stalling at that C_L, so that the lift of the wing will probably continue to increase as the angle of attack increases beyond this point. Indeed, flight test results obtained by the author using a Cherokee showed a $C_{L_{max}}$ close to the value of 1.80 shown on the figure as that needed to stall the center of the wing. The manual shows a number slightly lower at approximately 1.75. Tuft studies, which were done in conjunction with the flight testing, do not show the initial stall outboard of the flaps, but instead indicate that the stall occurs first at the wing root. This may be the result of fuselage interference affecting the lift distribution.

A graph of C_{l_a} and C_{l_b} for the Cherokee is presented in Figure 3.62. C_{l_a} was obtained by running the lifting surface program with no twist or flaps.

UNCERTAINTIES IN THE CALCULATION OF $C_{L_{max}}$

The fuselage affects the spanwise loading distribution of a wing and develops a lift of its own. In the preceding developments the fuselage has been neglected except for terminating the flaps at the sides of the fuselage. The rest of the wing is assumed to extend through the fuselage, an assumption that is compensated for, in part, by the fuselage lift. The drop in the middle of the lift distributions of Figure 3.61 is due entirely to the lack of flaps in that region. This decrease in C_l should probably be even greater, and a similar but smaller drop should appear in Figure 3.60 because of the effect of the fuselage.

In order to estimate the effect of the fuselage on $C_{L_{max}}$ it will be assumed that the fuselage effects a drop in the loading per unit span over the fuselage width, which is proportional to the loading per unit span at the midspan. Thus, the lift decrement from the fuselage will be

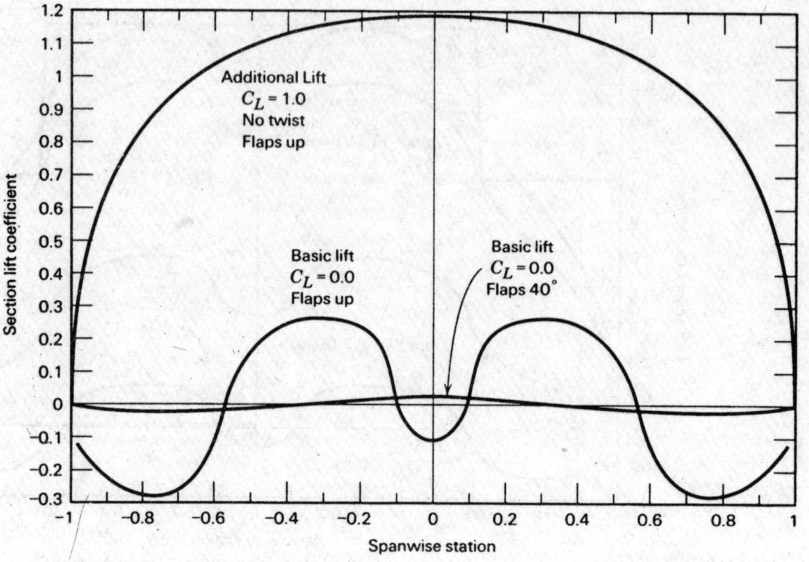

Figure 3.62 Basic and additional lift distributions for 0 and 40° of flaps.

$$\Delta L = -kqC_{l_0}S_{\text{fuse}}$$

where S_{fuse} is the wing planform submerged in the fuselage, C_{l_0} is the midspan section lift coefficient, and k is a constant of proportionality. Thus, the total C_L with the fuselage, $C_{L_{\text{fuse}}}$, can be written in terms of C_L without the fuselage as

$$C_{L_{\text{fuse}}} = C_L\left(1 - k\frac{C_{L0}}{C_L}\frac{S_{\text{fuse}}}{S}\right) \tag{3.78}$$

In Reference 3.29, two wings equipped with partial and full-span, split, single-slotted, and double-slotted flaps were tested with and without a fuselage. The fuselage was circular in cross-section, and the wing was mounted slightly above the middle of the fuselage. The ratio S_{fuse}/S was equal to 0.083. The results of these tests are plotted in Figure 3.63 and compared with Equation 3.78 using kC_{l_0}/C_L = 1.0. Also plotted on Figure 3.63 are test results from References 3.36 and 3.37. The ratio S_{fuse}/S was nearly the same for these two references as for Reference 3.29. These data support the form of Equation 3.78, at least to the extent that the correction to $C_{L_{\max}}$ of the fuselage appears to increase linearly with $C_{L_{\max}}$ of the wing alone. The correction depends on the cross-sectional shape of the fuselage and seems to vanish or even be slightly favorable for a rectangularly shaped section. Reference 3.37 also shows the correction to be slight for elliptical shapes, where the height is greater than the width. The decrement in $C_{L_{\max}}$ also depends on wing position and appears to be a maximum for the midwing configuration.

Aerodynamic Center Location for a Finite Wing, $C_{M_{ac}}$, and Mean Aerodynamic Chord (MAC)

As in the case of an airfoil, a finite wing has a chordwise location about which the aerodynamic moment is constant, independent of angle of attack. This moment is expressed as a coefficient, $C_{M_{ac}}$, in three dimensions by referencing it to the product of the wing planform and a fictitious chord length called the *mean aerodynamic chord* (MAC). The MAC, $\bar{c}$, is defined as the chord length that, when multiplied by the

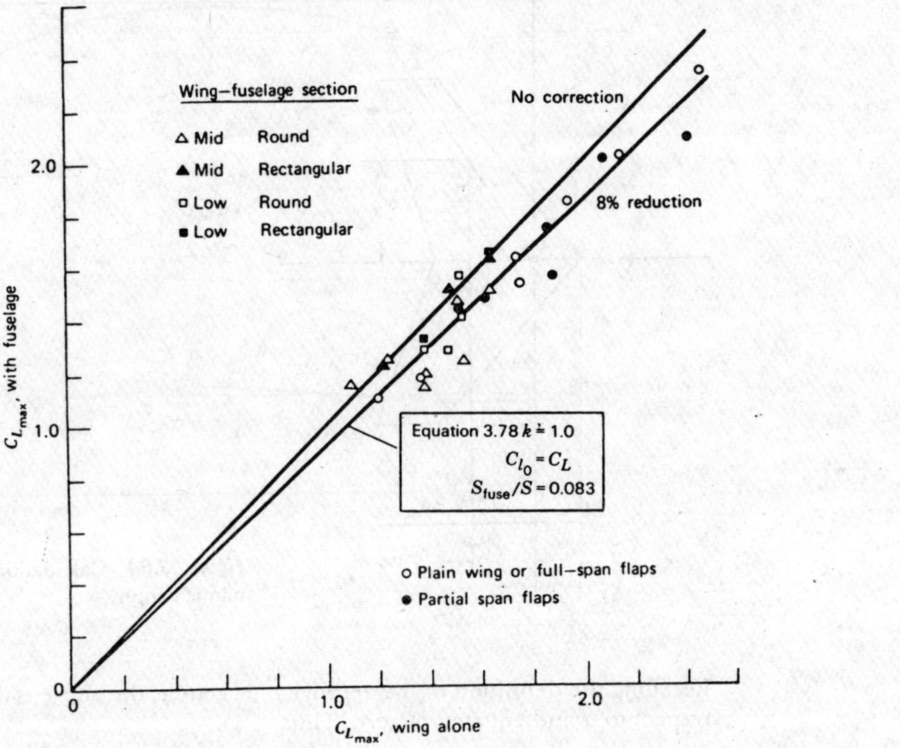

Figure 3.63 Effect of fuselage on $C_{L_{max}}$.

wing area, the dynamic pressure and an average $C_{m_{ac}}$, gives the total moment about the wing's aerodynamic center. If the $C_{m_{ac}}$ is not constant along the span, one has a problem in defining "average." Since the definition of $\bar{c}$ is somewhat arbitrary, it is best to define it always as if the section $C_{m_{ac}}$ is constant. Thus,

$$\bar{c} = \frac{1}{S} \int_{-b/2}^{b/2} c^2 \, dy \tag{3.79a}$$

For a linearly tapered wing with a midspan chord of c_0 and a taper ratio of λ, the above can be integrated to give

$$\bar{c} = \frac{2c_0}{3} \frac{1 + \lambda + \lambda^2}{1 + \lambda} \tag{3.79b}$$

The chordwise location of the aerodynamic center and the moment about this spanwise line for a finite wing can be determined by applying the lifting surface methods presented earlier to determine section properties.

With reference to Figure 3.64, the line through point A swept back through the angle Λ is the locus of the section aerodynamic centers. Therefore, the pitching moment about a spanwise line through A will be

$$M_A = q \int_{-b/2}^{b/2} c^2 \, C_{m_{ac}} \, dy \text{-} q \int_{-b/2}^{b/2} cC_l y \tan\Lambda \, dy$$

If X_A is the distance of the aerodynamic center behind the point A, then

$$M_{ac} = M_A + LX_A$$

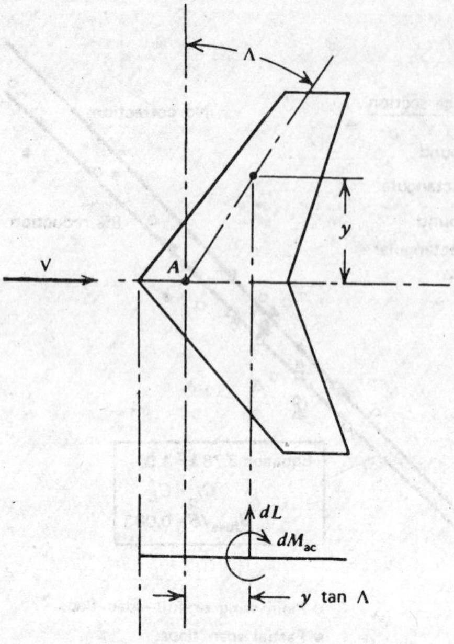

Figure 3.64 Calculation of wing aerodynamic center.

Recalling the definition of the aerodynamic center, the above is differentiated with respect to α and equated to zero. Thus,

$$X_A = -\frac{M_{A\alpha}}{L_\alpha} \tag{3.80}$$

The above can then be evaluated numerically to determine the location of the aerodynamic center.

An approximate closed form solution for X_A can be obtained by remembering that the section C_l is given by the product of the section lift curve slope and $\alpha - \alpha_i$. For the elliptic spanwise load distribution, α_i is constant and, assuming that $C_{l\alpha}$ is constant, the expression for X_A becomes

$$X_A = \tan \lambda \frac{\int_{-b/2}^{b/2} cy dy}{\int_{-b/2}^{b/2} c dy}$$

or

$$X_A = \bar{y} \tan \Lambda \tag{3.81}$$

where $\bar{y}$ is the spanwise distance out from the centerline to the centroid of the half-wing area. For a linearly tapered wing, this expression becomes

$$X_A = \frac{b}{6}\left(\frac{1 + 2\lambda}{1 + \lambda}\right) \tan \Lambda$$

Effect of Trim on $C_{L_{max}}$

In order to calculate the stalling speed of an airplane in steady flight, one must consider that the tail, in general, provides either lift or a download depending upon the moment balance around the center of gravity (cg). To determine this

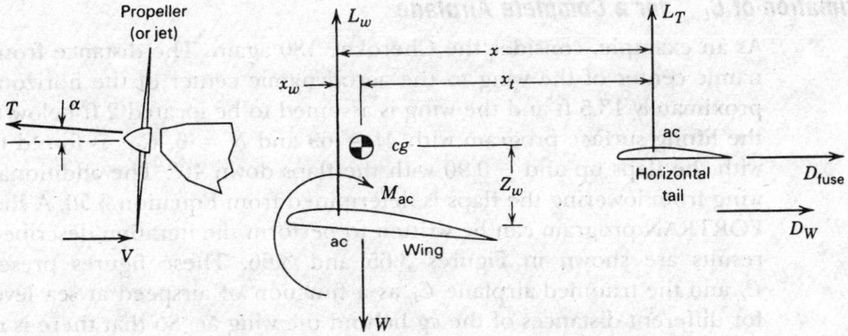

Figure 3.65 Forces and moment on an airplane in trimmed level flight.

additional trim load, refer to Figure 3.65 which shows the major forces and moments acting on an airplane. Here, the wing lift, the tail lift, the pitching moment about the wing's aerodynamic center, and the weight are all shown in a positive sense. An aerodynamic pitching moment for the tail is not included because the tail is usually symmetrical and, in any case, is small compared to the other moments. The moments about the cg because of the thrust and fuselage drag will also be neglected although they might be included in a more exact calculation if the points at which they act are known.

Summing the forces in the x and z directions and the moments about the cg results in

$$\sum F_X = 0 = T\cos\alpha - D_{\text{fuse}} - D_w \tag{3.82}$$

$$\sum f_z = 0 = W - L_w - L_t - T\sin\alpha \tag{3.83}$$

$$\sum M = 0 = M_{ac} + L_w X_w - L_t\,(X - X_w) - D_w Z_w \tag{3.84}$$

The above system can easily be solved for the wing lift to give the following:

$$L_w = W\left(1 - \frac{X_w}{X}\right)\left(1 - \frac{T}{W}\sin\alpha\right) - \frac{M_{ac}}{x} + D_w\frac{Z_w}{X}$$

The above can be put into a dimensionless form by dividing by the wing area, the mean aerodynamic chord, and the dynamic pressure.

$$C_{L_w} = C_L\left(1 - \frac{X_w}{X}\right)\left(1 - \frac{C_D}{C_L}\tan\alpha\right) - \frac{C_{M_{ac}}}{x} + C_{D_w}\frac{Z_w}{X} \tag{3.85}$$

Here, C_L is the lift coefficient based on the weight, C_{L_w} is the lift coefficient of the wing, and C_{D_w} is the drag coefficient for the wing. C_D is the drag coefficient for the whole airplane, $C_{M_{ac}}$ is the wing's moment coefficient about the ac, and all distances are relative to the mean aerodynamic chord.

Generally, this system of equations must be solved iteratively. This can be done by estimating a stalling speed and then assuming that the quantity $1 - (T/W)\sin\alpha$ is equal approximately to one and that M_{ac}/X and $C_{D_w}Z_w/X$ are small compared to the weight. Then, given the location of the cg and the weight, and having estimated a velocity, a first estimate of the wing lift coefficient is obtained. This leads to an angle of attack and drag coefficient. These can then be substituted into Equation 3.84 and a second value then obtained for the wing lift coefficient. The process is repeated until it converges to a trimmed wing lift coefficient. If this coefficient is less than $C_{L_{max}}$ for the wing, then a lower value of the speed is chosen and the process is repeated.

Estimation of $C_{L_{max}}$ for a Complete Airplane

As an example, consider the Cherokee 180 again. The distance from the aerodynamic center of the wing to the aerodynamic center of the horizontal tail is approximately 13.5 ft and the wing is assumed to be located 2 ft below the cg. Using the lifting surface program with $M = 69$ and $N = 6$, $C_{M_{ac}}$ is found to be -0.028 with the flaps up and -0.90 with the flaps down 40°. The additional drag at the wing from lowering the flaps is determined from Equation 3.50. A short (50 lines) FORTRAN program can be written to perform the iteration described above. The results are shown in Figures 3.65 and 3.66. These figures present the wing C_L and the trimmed airplane C_L as a function of airspeed at sea level conditions for different distances of the cg behind the wing ac. So that there is no misunderstanding, the trim C_L is simply the C_L based on the gross weight, wing area, and airspeed.

Observe that as the cg moves back, the tail must support more of the gross weight so that the wing C_L decreases. When the cg is on the wing ac, the negative, nose-down aerodynamic moment about the wing ac must be balanced by a tail download so one might expect the wing C_L to be greater than the trim C_L. However, the vertical component of the thrust is sufficient to more than compensate for the tail download so that the wing C_L is still less than the trim value.

Computer Exercise 3.4 "TRIMCL"

Write a computer program to solve iteratively for the wing lift coefficient given the airplane trimmed lift coefficient. Compare your results with Figures 3.67 and 3.68.

By comparing Figures 3.66 and 3.67 let us examine the effect of the flap on the stalling speed for the loading condition where the cg is on the wing's ac. With the flaps up, the wing was predicted earlier to have a $C_{L_{max}}$ of 1.20. From Figure 3.66, this corresponds to a trim C_L of 1.32. With the flaps down 40°, the wing $C_{L_{max}}$ is 1.78 corresponding to a trim C_L of 2.0. Thus, lowering the flaps has resulted in an

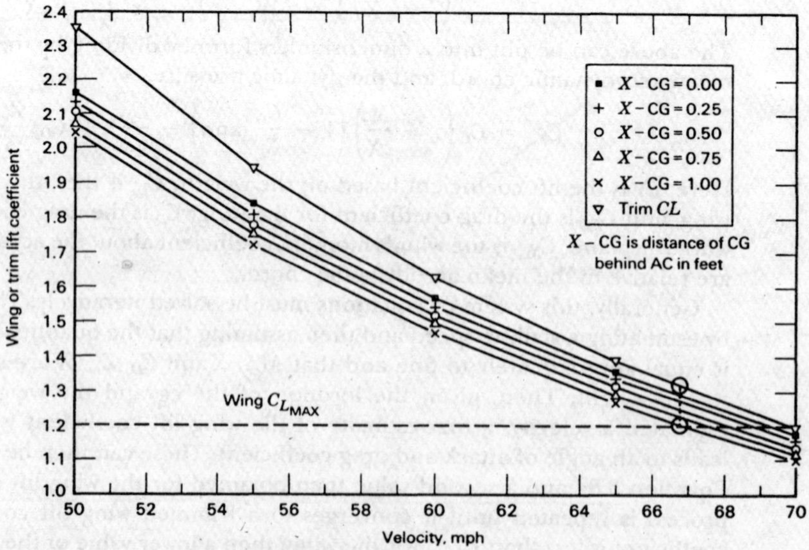

Figure 3.66 Wing C_L and trim C_L for the Cherokee with flaps up.

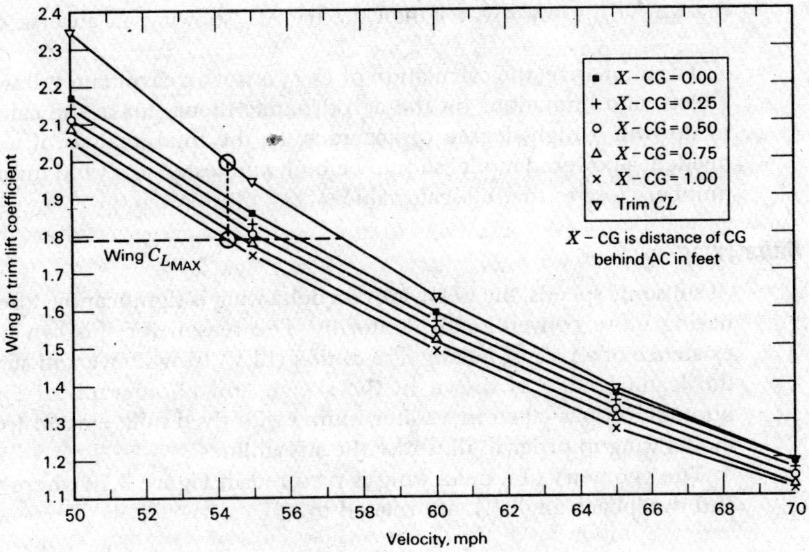

Figure 3.67 Wing C_L and trim C_L for the Cherokee with flaps down.

increase in the trim $C_{L_{max}}$ of 0.68. However, for the wing alone, the effect of the flaps was to increase $C_{L_{max}}$ from 1.20 to 1.78, or an increase of 0.58. Thus, the increase in the trim $C_{L_{max}}$ in this instance is greater than that for the wing alone despite the added nose-down pitching moment. The reason for this is increased angle of attack of the thrust vector with the flaps down.

To emphasize this point further, Figure 3.68 was prepared with the thrust set to zero. Here, the lift coefficient of the wing has increased by the order of 0.2 resulting in a corresponding decrease in the trim $C_{L_{max}}$. Also note that with the cg forward,

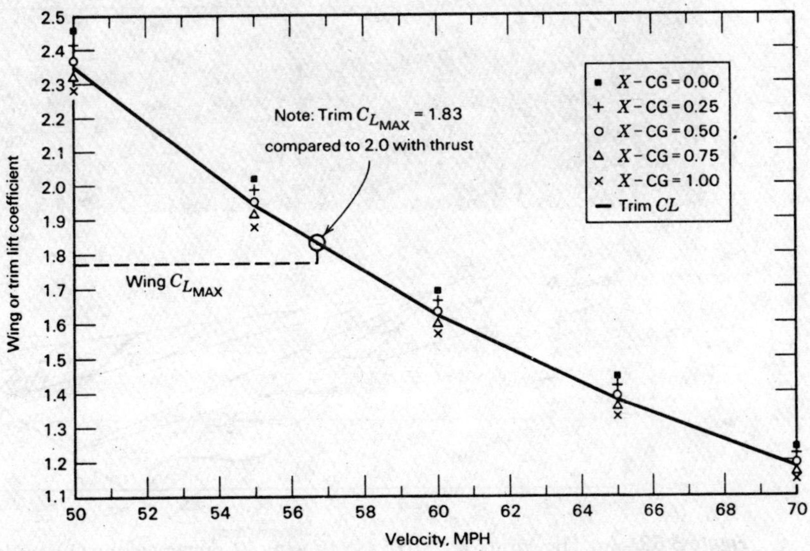

Figure 3.68 Wing C_L and trim C_L for the Cherokee with flaps down but with no thrust.

the wing C_L is now greater than the trim C_L showing an adverse effect of trim on $C_{L_{max}}$.

To summarize, the calculation of $C_{L_{max}}$ is not an easy matter. Power, interference effects, and limitations on the aerodynamic theory make the calculation difficult to do with a high degree of accuracy. In the final analysis, if a precise value is needed, a powered model should be built and tested in a wind tunnel at a Reynolds number close to the full-scale value.

The Delta Wing

At subsonic speeds, the behavior of a delta wing is significantly different than wings having more conventional planforms. The reason for this can be traced to the existence of a pair of *leading edge vortices* (LEV) lying above and slightly inboard of the leading edge as shown in the sketch and photograph of Figure 3.69. This photograph was taken in a water tunnel with dyed milk ejected from holes in the model wing in order to illustrate the streamlines.

The geometry of a delta wing is pictured in Figure 3.70 where the aspect ratio and sweepback angle, Λ, are related by

$$A = \frac{4}{\tan \Lambda}$$

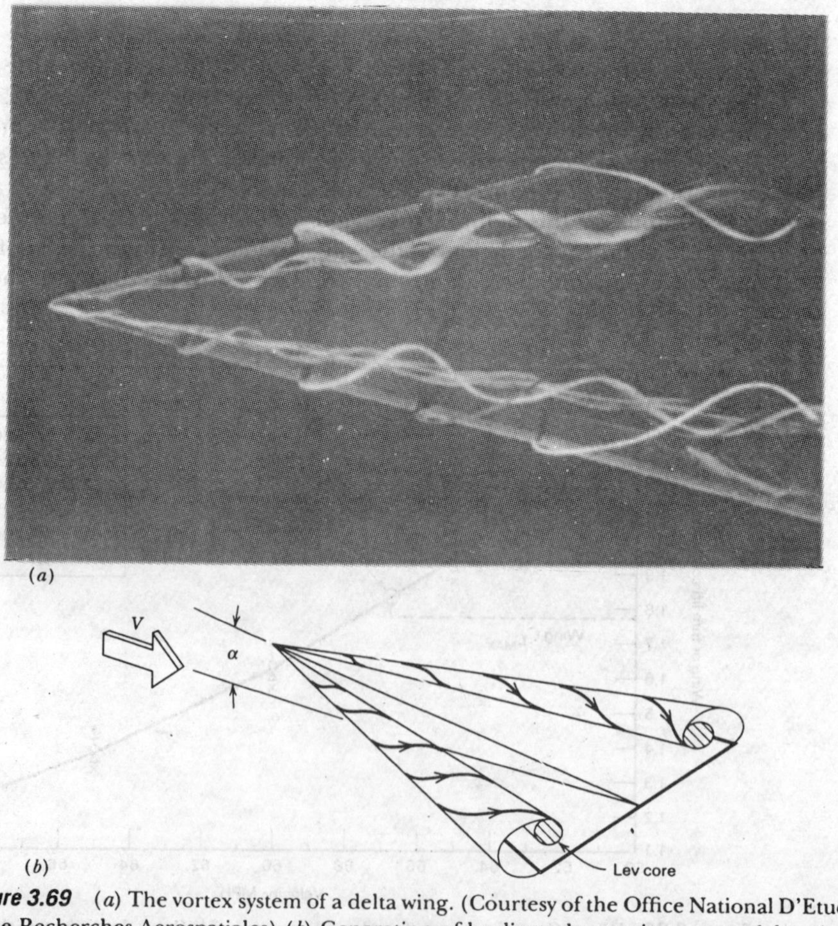

Figure 3.69 (*a*) The vortex system of a delta wing. (Courtesy of the Office National D'Etudes et de Recherches Aerospatiales) (*b*) Generation of leading edge vortices over a delta wing.

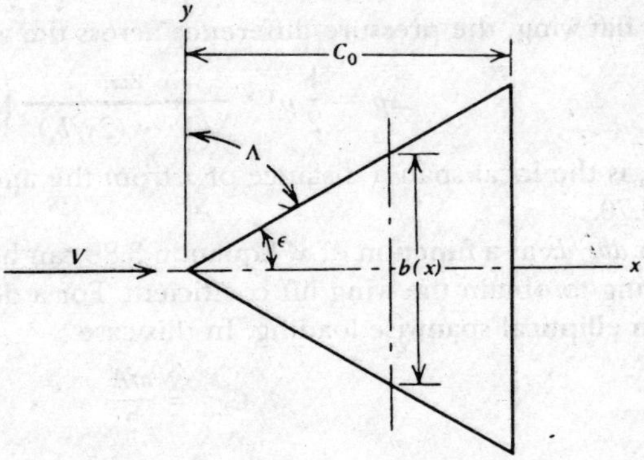

Figure 3.70 The planform of a delta wing.

At small angles of attack and low aspect ratios one can apply a linearized approach referred to as slender wing (or wing-body) theory. The details of this theory will not be developed here; instead, refer to two of the original sources (Refs. 3.38 and 3.39) for the treatment of this theory. A brief treatment for the wing-alone case can be found in Reference 3.12. Basically, the theory assumes that the flow in any transverse plane is essentially two dimensional. This assumption eliminates the Mach number-dependent term developed later in Equation 5.56 that governs the perturbation velocity potential, leaving only

$$\frac{\partial^2 \phi}{\partial y^2} + \frac{\partial^2 \phi}{\partial z^2} = 0 \tag{3.87}$$

Since $V_\infty \alpha$ is the velocity component normal to the plane of the wing, the problem is reduced to finding the two-dimensional flow in a transverse plane, as illustrated in Figure 3.71.

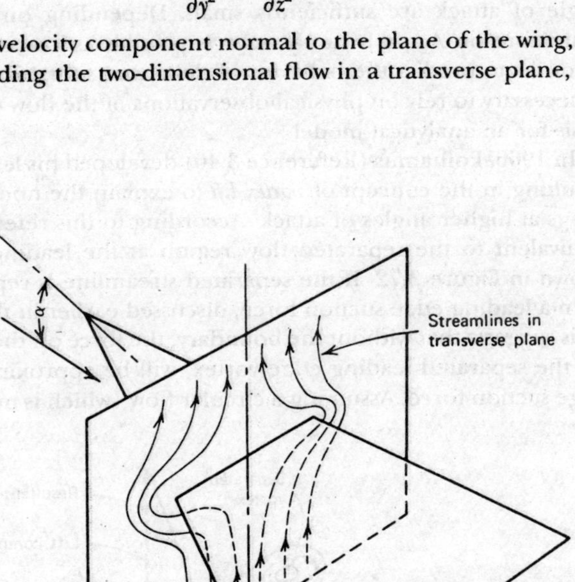

Figure 3.71 Two-dimensional flow approximation for a slender wing.

For a flat wing, the pressure difference across the wing predicted to

$$\Delta p = \frac{1}{2}\rho V^2 \frac{2\alpha}{\sqrt{1-(2y/b_x)^2}}\left(\frac{db_x}{dx}\right) \tag{3.88}$$

where b_x is the local span a distance of x from the apex. y, b_x, and x are shown in Figure 3.70.

Given db_x/dx as a function of x, Equation 3.88 can be integrated over the surface of the wing to obtain the wing lift coefficient. For a delta wing, db_x/dx is constant, giving an elliptical spanwise loading. In this case,

$$C_{L_\alpha} = \frac{\pi A}{2} \tag{3.89}$$

$$C_{D_i} = \frac{C_L{}^2}{\pi A} \tag{3.90}$$

Notice that because of the assumption of slenderness, these results do not depend on Mach number.

Since the shape of the local spanwise pressure distribution is the same at all chordwise positions for a delta wing (db_x/dx is constant), it follows that the center of pressure is predicted to be at the centroid of the planform area, a distance of $2c_0/3$ back from the apex. Based on the wing area and geometric mean chord, the pitching moment coefficient about the apex is thus,

$$C_{M_{LE}} = -\frac{2}{3}\pi A\alpha \tag{3.91}$$

Again, these results hold for any Mach number, providing the aspect ratio and angle of attack are sufficiently small. Depending on the desired accuracy, the limitations on A and α can be severe, possibly as low as 0.5 for A and 2 or 3° on α. In order to predict the behavior of delta wings for higher α's and aspect ratios, it is necessary to rely on physical observations of the flow over such wings to form the basis for an analytical model.

In 1966, Polhamus (Reference 3.40) developed his leading edge suction analogy, resulting in the concept of *vortex lift* to explain the nonlinear lift behavior of delta wings at higher angles of attack. According to this reference, the separated LEV is equivalent to the separated flow region at the leading edge of a thin airfoil, as shown in Figure 3.72. If the separated streamline is replaced by a solid boundary, then a leading edge suction force, discussed earlier in this chapter, will be present. This suggests that without the boundary, the force on the separated flow, and hence on the separated leading edge vortex, will be approximately equal to the leading edge suction force. Assuming a circular flow (which is not quite true) for the vortex

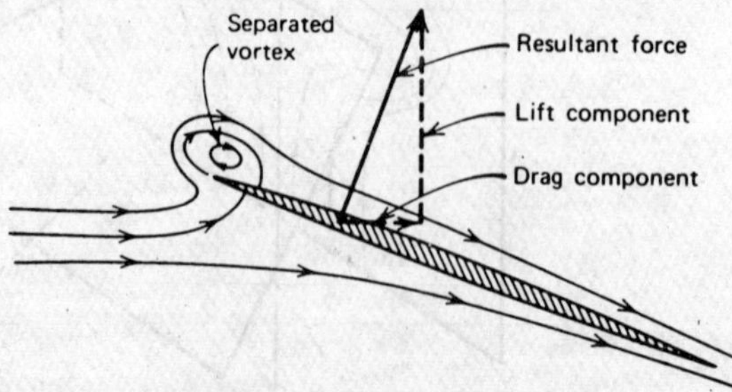

Figure 3.72 Separated leading edge flow with standing vortex on a thin airfoil.

leads to a reaction in the upward direction normal to the chord and near the leading edge that is nearly equal in magnitude to the leading edge suction force for the fully wetted, potential flow case. This is the basis for the subsonic Polhamus's delta wing theory presented in References 3.40 and 3.41.

Without delving further into the details of Polhamus's theory, the following results are obtained:

$$C_L = K_p \sin\alpha \cos^2\alpha + K_v \cos\alpha \sin^2\alpha \tag{3.92}$$

The term $K_p \sin\alpha \cos^2\alpha$ represents the potential lift on the wing for the fully wetted case minus a small vertical component of the leading-edge, suction-force, which is lost when the flow separates. The second term represents the added lift caused by the presence of the leading edge vortex. Not surprisingly, this is referred to as "vortex lift."

As α approaches zero, Equation 5.97 becomes

$$C_L = K_p\alpha \qquad (\alpha \rightarrow 0)$$

Thus, it follows that the constant, K_p, is simply the slope of the wing lift curve for the fully wetted case according to the usual lifting surface theories. The constant, K_v, is obtained from

$$K_v = \frac{K_p - K_p{}^2 K_i}{\cos\Lambda} \tag{3.93}$$

where $K_i = dC_{D_i}/dC_L{}^2$ and Λ is the sweepback angle of the leading edge.

Again, K_i can be evaluated from the usual lifting surface methods. However, K_p and K_v have already been evaluated in the references and are presented here in Figure 3.73 as a function of aspect ratio.

Having C_L, the drag coefficient is easily obtained. Since there is no leading edge suction force parallel to the chord for the separated leading edge, it follows simply that

$$C_D = C_L \tan\alpha \tag{3.94}$$

Predictions of the lift of delta wings are improved considerably by the added vortex lift obtained from the leading-edge, suction-force analogy. Comparisons between theory and experiment for a delta wing with 70° of sweep are presented

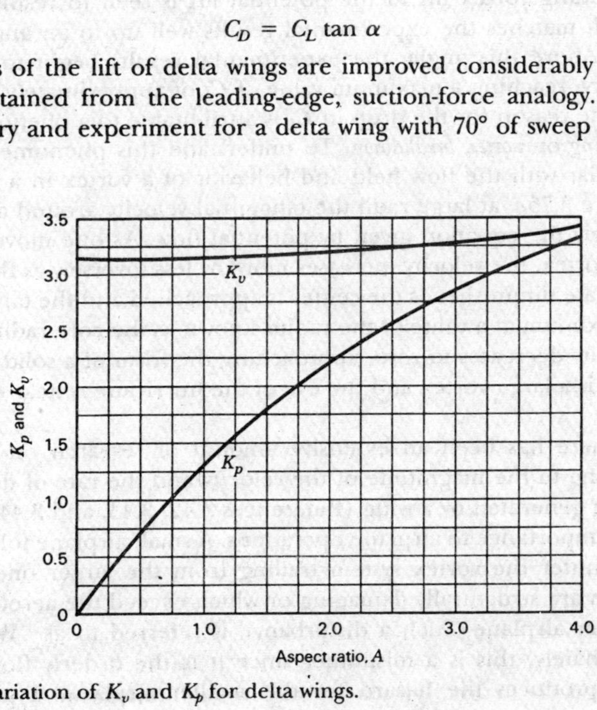

Figure 3.73 Variation of K_v and K_p for delta wings.

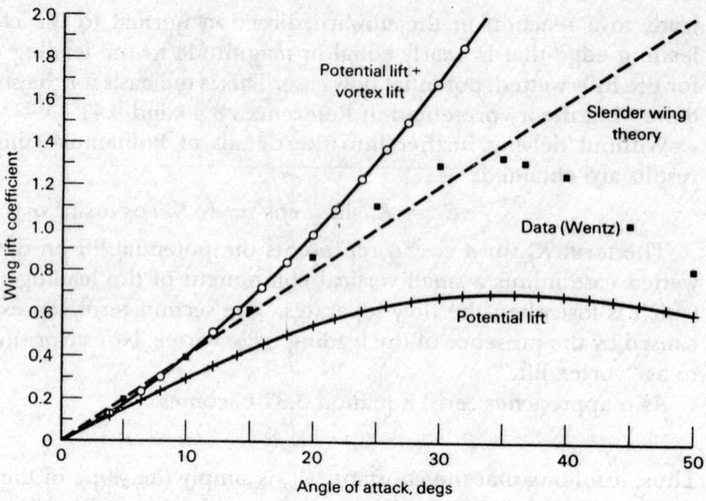

Figure 3.74 Components of lift on a delta wing having 70° of sweep.

in Figure 3.74. From this figure it is seen that predictions of C_L for a delta wing based on the usual potential flow methods are appreciably lower than the experimental results. It should be noted that the curve for the potential, or circulatory, lift is not linear with the angle of attack because α is not assumed to be small. The slender wing theory agrees fairly well with the experimental results up to an angle of attack of approximately 35°. However, the agreement is fortuitous as a result of the linearizing assumptions applied to potential flow. One would expect the theory to be applicable only as the angle of attack approaches zero.

Adding vortex lift to the potential lift is seen to result in a nonlinear curve, which matches the experimental results well up to an angle of attack of around 20°. Above this angle, the experimental results begin to fall below Polhamus's theory, reaching a maximum value of C_L of approximately 1.35 at an α of 35°.

The reason for the drop in C_L is attributable to a phenomenon known as *vortex bursting* or *vortex breakdown*. To understand this phenomenon, one must first be familiar with the flow field and behavior of a vortex in a real fluid. Referring to Figure 3.75*a*, at large radii the tangential velocity around a vortex follows approximately the equation given by potential flow. As one moves toward the center of the vortex, the velocity increases more or less inversely as the radius. However, this increase diminishes as the center is approached and the tangential velocity reaches a maximum at a value of the radius known as the core radius. Within the core, the velocity decreases to zero, approaching the form of a solid body rotation. A hurricane is a large vortex and the eye of the hurricane is near the center of the vortex core.

There has been an extensive amount of research on trailing vortex systems relating to the magnitude of the velocity and the rate of decay of the vortex after being generated by a wing (References 3.42, 3.43, and 3.44). The topic has particular importance to airplane operations. A small airplane following a larger one can encounter the vortex system trailing from the larger one resulting in motions, which are structurally damaging or which exceed the aerodynamic controls of the smaller airplane. Such a disturbance is referred to as "Wake Turbulence." Unfortunately, this is a misnomer since it is the orderly flow and not turbulence, that produces the hazard for the smaller airplane. Typically, a light airplane

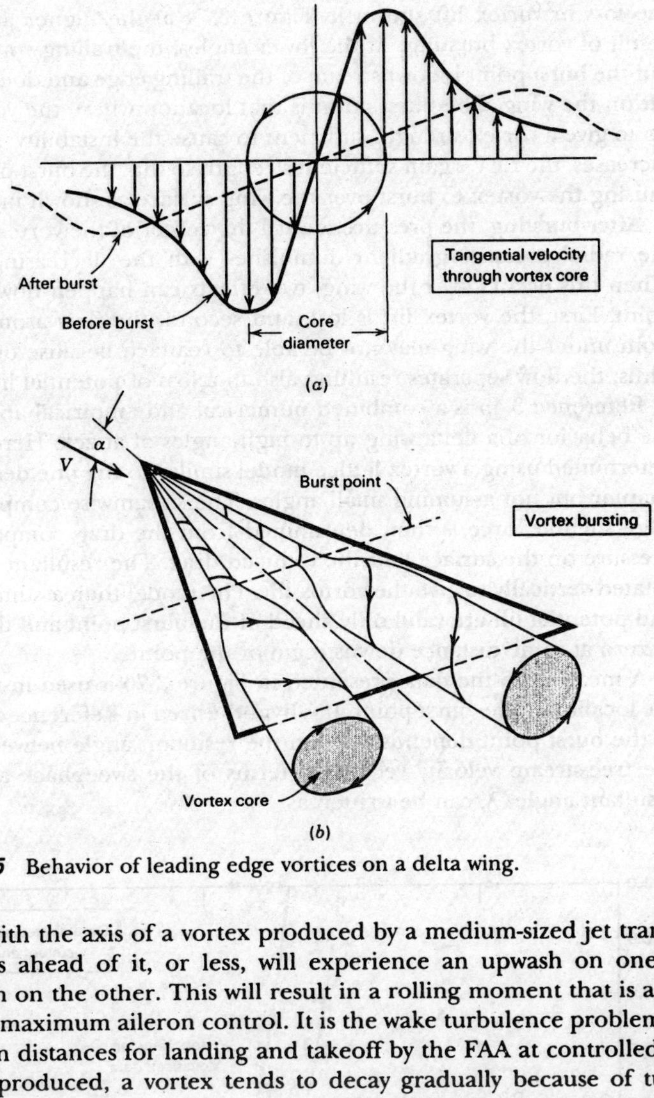

Figure 3.75 Behavior of leading edge vortices on a delta wing.

aligned with the axis of a vortex produced by a medium-sized jet transport, which is 5 miles ahead of it, or less, will experience an upwash on one wing and a downwash on the other. This will result in a rolling moment that is approximately twice the maximum aileron control. It is the wake turbulence problem that sets the separation distances for landing and takeoff by the FAA at controlled airports.

Once produced, a vortex tends to decay gradually because of turbulent and viscous diffusion. The core size increases and the maximum tangential velocity decreases while maintaining approximately a constant circulation at the core radius; that is,

$$\Gamma(a) = 2\pi a V_\theta(a) = \text{constant} \tag{3.95}$$

where $\Gamma(a)$ is the circulation at the core radius, a, and $V_\theta(a)$ is the tangential velocity at that radius.

If the strength of the vortex is sufficiently high so that the swirl angle given by $tan^{-1}(V_\theta(a)/V)$ is high, the vortex can exhibit an instability referred to earlier as vortex burst. When bursting occurs, the core suddenly enlarges significantly in size with a reduction in V_θ compatible with the above equation. The burst point probably represents a standing wave through which momentum is conserved while the energy state drops. It is a three-dimensional analogy to the two-dimensional hydraulic jump seen when a sheet of water runs down a gradually sloping surface.

Returning to the delta wing, it is obvious from experimental observations that the loss in vortex lift shown in Figure 3.74 at the higher angles of attack is the result of vortex bursting. At the lower angles, the trailing vortices may be bursting but the burst point is downstream of the trailing edge and does not affect the vortex lift on the wing. The burst point is that location where the vortex sheet has rolled up to give a vortex strength sufficient to cause the instability. As the angle of attack increases, the LEV's gain sufficient strength so that the burst point moves upstream, causing the vortex to burst over the wing surface as shown in Figure 3.75*b*.

After bursting, the pressure toward the center of the vortex rises considerably as the radial pressure gradient diminishes with the decreasing tangential velocity. When this occurs over the wing, two effects can happen downstream of the burst point. First, the vortex lift is lost, and second, the flow around the leading edge from under the wing may not be able to reattach because of the weakened swirl. Thus, the flow separates resulting also in a loss of potential lift.

Reference 3.45 is a combined numerical and empirical approach to predicting the behavior of a delta wing up to high angles of attack. Here, the potential lift is determined using a vortex lattice model similar to the one described earlier in this chapter but not assuming small angles. The streamwise component of the leading edge suction force is then determined from the drag component of the normal pressure on the surface and the induced drag. The resultant suction force is then rotated vertically to give the vortex lift. The model then assumes that the vortex lift and potential lift are valid only ahead of the burst point and then decrease linearly to zero at some distance downstream of the point.

A mean fit to the data presented in Figure 3.76 is used in the model to predict the location of the burst point. It is hypothesized in Reference 3.47 that the location of the burst point depends only on the resultant angle between leading edge and the free-stream velocity vector. In terms of the sweepback angle, Λ, and α, this resultant angle, λ, can be written as

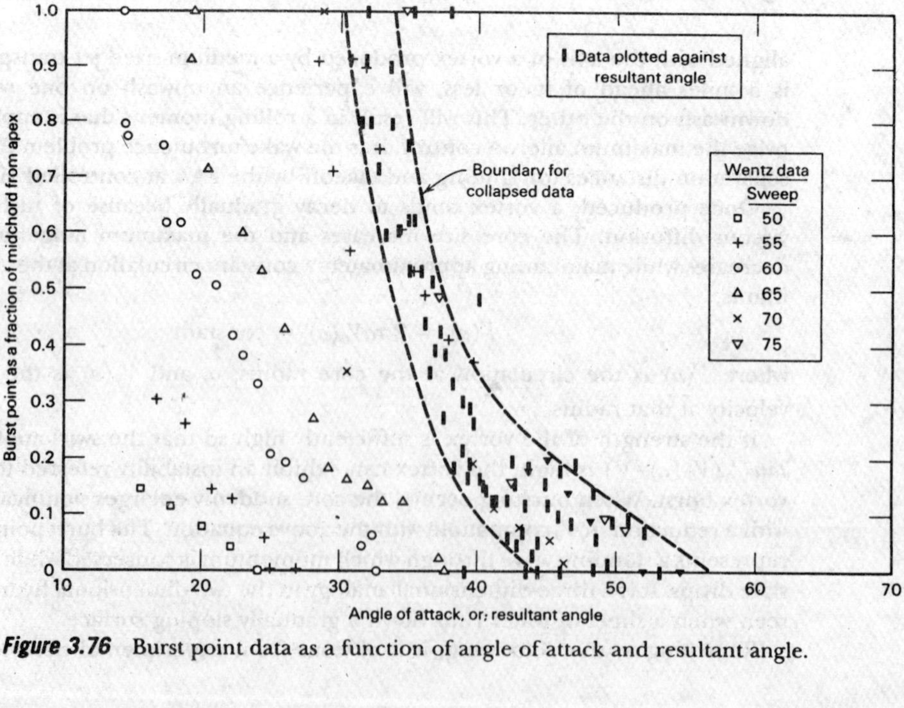

Figure 3.76 Burst point data as a function of angle of attack and resultant angle.

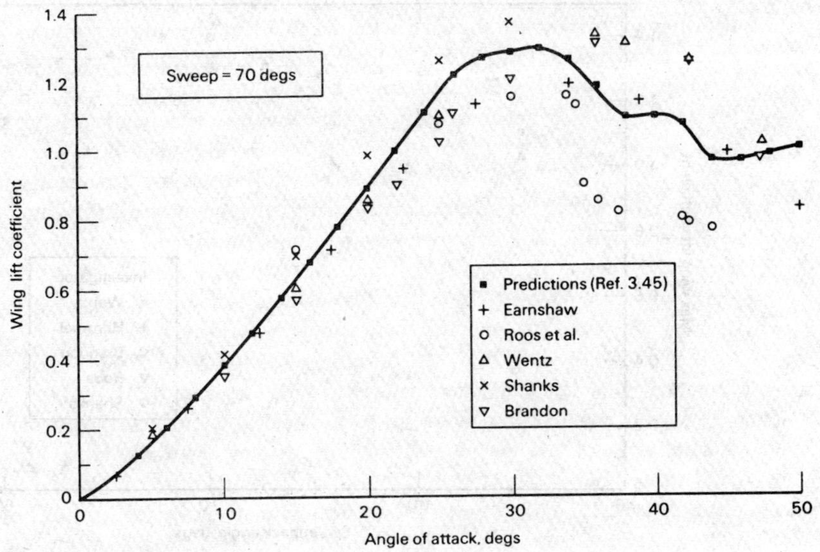

Figure 3.77 Comparison of tests with predictions for delta wings.

$$\lambda = \arccos\,(\sin\Lambda \cos\alpha) \tag{3.96}$$

This figure certainly tends to confirm the above hypothesis within the scatter of the data. Plotted against angle of attack, the burst point locations lie along curves for the different sweep angles, which are widely separated. However, when the data is plotted against the resultant angle, almost all of the data falls within the two dashed boundaries. Although the scatter shown here may seem to be excessive, your author has participated in wind tunnel smoke tests to determine the burst point and can say, firsthand, that it is difficult to measure the same point repeatedly since it is not a sharply defined location.

A comparison of the semiempirical theory of Reference 3.45 with data from a number of investigators is presented in Figure 3.77. Observe the wide scatter in the data above the angle of attack for maximum C_L. This may be the result of different edge geometries used by the different investigators or by the wind tunnel boundary conditions that were used. Much of the wind tunnel testing of delta wings reported in the literature was done by Hummel (Reference 3.50). His testing was performed in an open-jet tunnel, and his results show $C_{L_{\max}}$ values, which are 0.2–0.4 lower than those obtained by others using closed-jet tunnels. It is difficult to say which set of data more accurately represents flight in an unrestricted atmosphere. Both the free-streamline along the open jet and the solid walls of the closed-jet introduce boundary conditions that are not present in flight.

Figure 3.78 presents $C_{L_{\max}}$ for delta wings as a function of sweepback angle as obtained by several investigators. Again the data scatters, but (except for Hummel's data) lies within a band of approximately ± 0.1 wide.

AIRFOIL CHARACTERISTICS AT LOW REYNOLDS NUMBERS

For many applications such as model airplanes or remote-piloted vehicles (RPV), the need arises for airfoil characteristics at Reynolds number values much lower than those for which most of the NACA and NASA data were obtained. These data

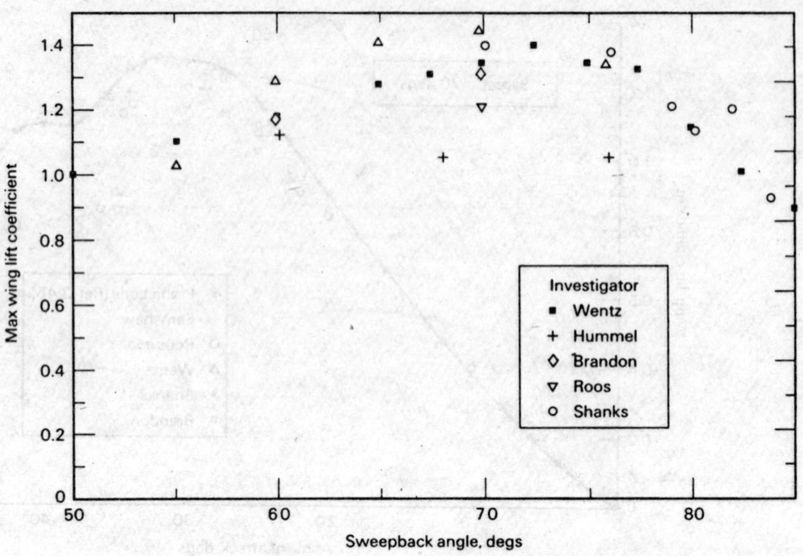

Figure 3.78 Experimental variation of maximum wing lift coefficients for delta wings as a function of sweepback angle.

were obtained at R values of 3×10^6 or higher. Prior to the development of modern airfoil codes discussed earlier, there was little data to be found in the literature. The most reliable of this data appear to be those given in Reference 3.52 where tests of five different airfoil shapes are reported for R values as low as 42,000. These tests were conducted in a low-turbulence tunnel.

The five airfoil shapes that were tested in Reference 3.37 are shown in Figure 3.79. These are seen to comprise a thin, flat plate, a thin, cambered plate, two 12% thick airfoils with 3 and 4% camber, and one 20% thick airfoil with 6% camber. The airfoil shapes are similar in appearance to the NACA four-digit series.

The lift curves for these airfoils are presented in Figure 3.80 for four different

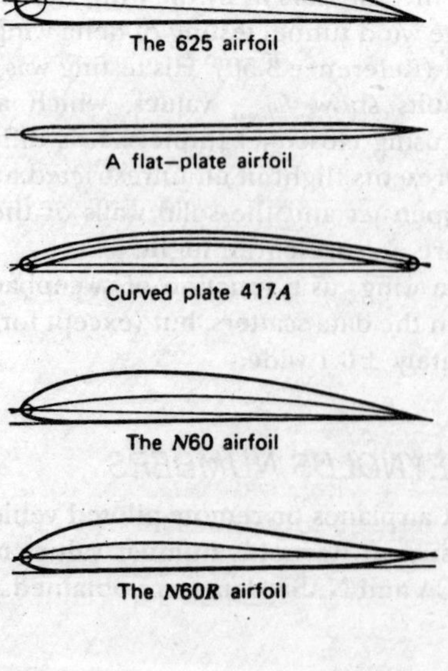

Figure 3.79 Airfoil shapes tested at low Reynolds numbers.

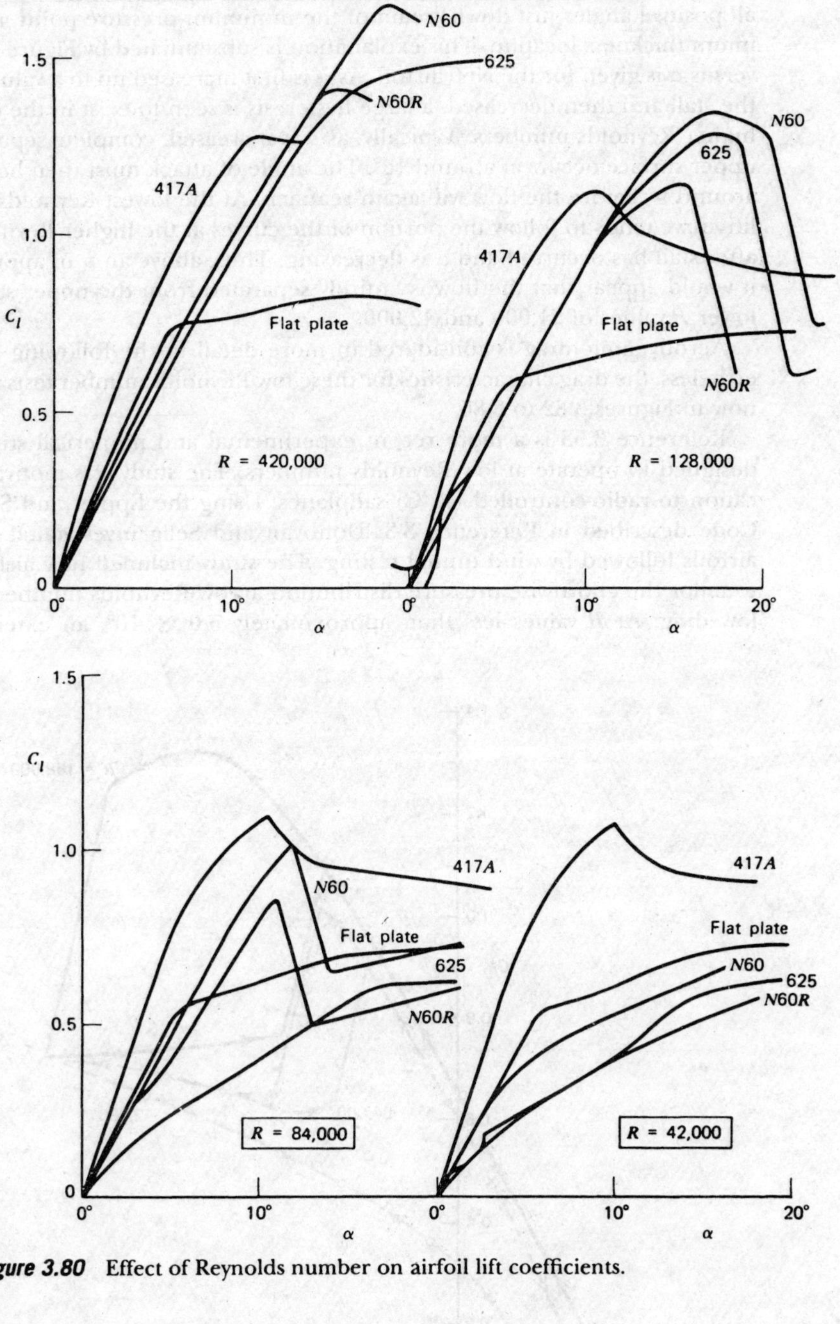

Figure 3.80 Effect of Reynolds number on airfoil lift coefficients.

Reynolds numbers. As one might expect, the flat-plate results are nearly independent of *R* since the separation point at the leading edge is well defined. To a slightly lesser degree, the same can be said for the cambered plate. The form of the lift curves for the three airfoils is seen to change substantially, however, over the *R* range from 4.2×10^5 down to 0.42×10^5. Particularly at the very lowest Reynolds number, the C_l versus α curve is no longer linear. The flow apparently separates at all positive angles just downstream of the minimum pressure point, near the maximum thickness location. This explanation is substantiated by Figure 3.81. Here, C_l versus α is given for the N60 airfoil. As α is first increased up to a value well beyond the stall and then decreased, a large hysteresis is seen to exist in the curves for the higher Reynolds numbers. Typically, as α is increased, complete separation on the upper surface occurs at around 12°. The angle of attack must then be decreased to around 5° before the flow will again reattach. At the lowest Reynolds number, the lift curve tends to follow the portion of the curves at the higher Reynolds numbers after stall has occurred and α is decreasing. Thus, above an α of approximately 0°, it would appear that the flow is entirely separated from the upper surface for the lower R values of 21,000 and 42,000.

Aerodynamic drag is considered in more detail in the following chapter. Nevertheless, the drag characteristics for these low Reynolds number tests are presented now in Figures 3.82 to 3.86.

Reference 3.53 is a more recent experimental and numerical study of airfoils designed to operate at low Reynolds numbers. The study was motivated by application to radio-controlled (R/C) sailplanes. Using the Eppler and Somers Airfoil Code described in Reference 3.3, Donovan and Selig investigated a number of airfoils followed by wind tunnel testing. The study included new airfoils designed to tailor the chordwise pressure distribution at low Reynolds numbers to promote low drag. At R values less than approximately 5.0×10^5, an extensive laminar

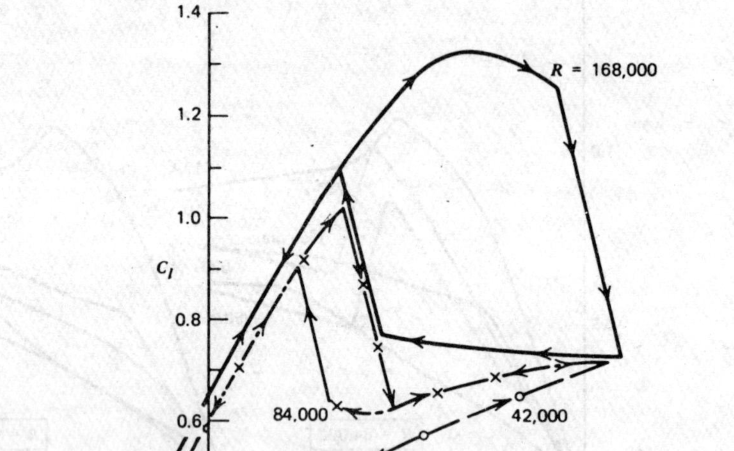

Figure 3.81 Lift curve for the N60 airfoil.

(Text continued on p. 146.)

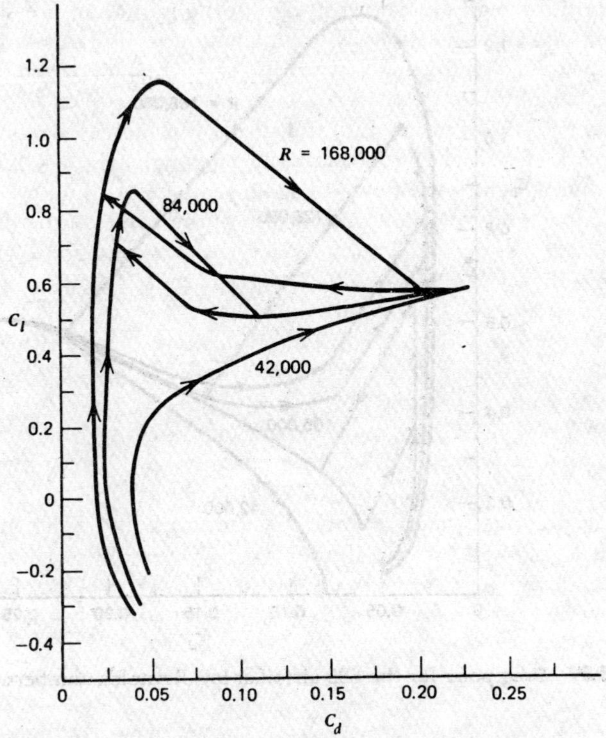

Figure 3.82 Drag polar for the N60R airfoil at low Reynolds numbers.

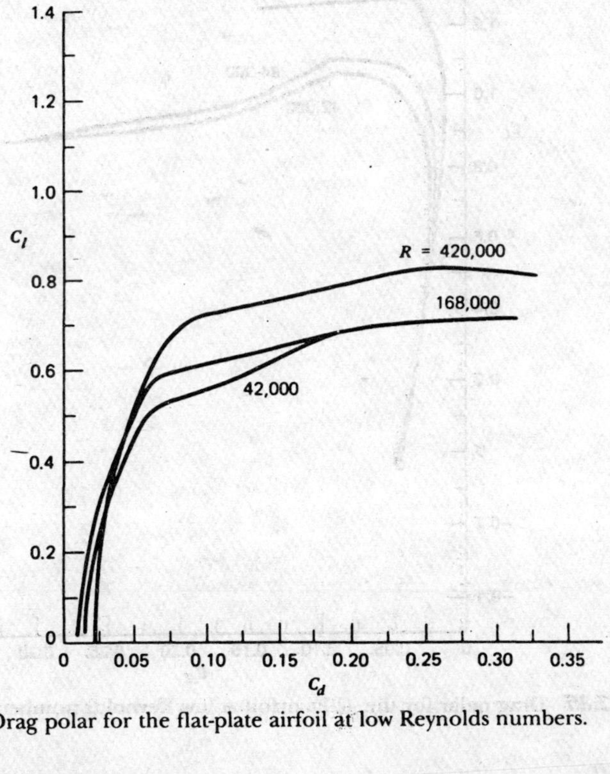

Figure 3.83 Drag polar for the flat-plate airfoil at low Reynolds numbers.

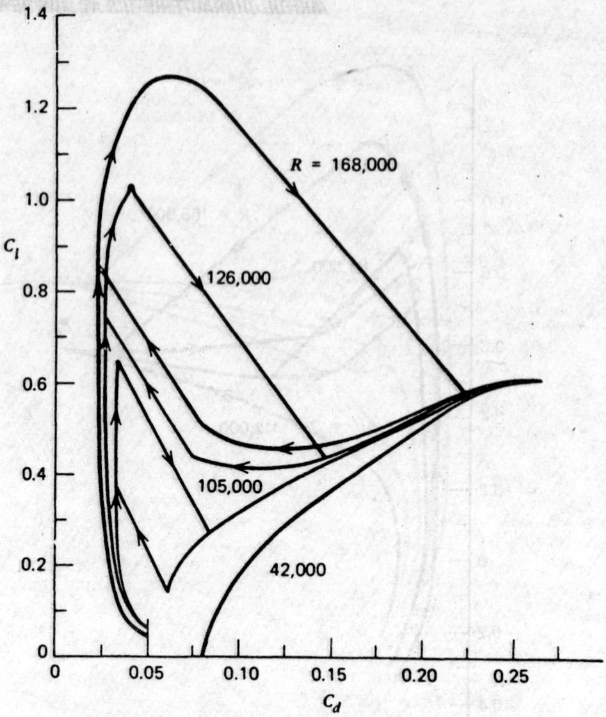

Figure 3.84 Drag polar for the 625 airfoil at low Reynolds numbers.

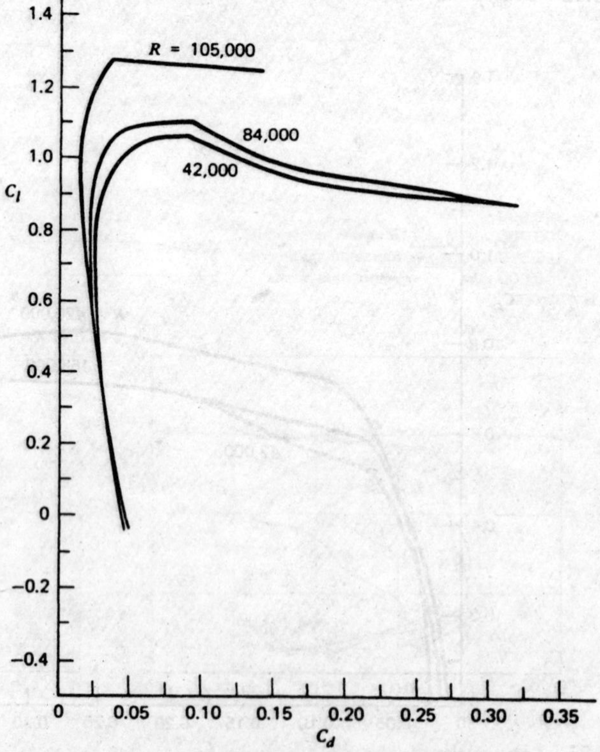

Figure 3.85 Drag polar for the 417a airfoil at low Reynolds numbers.

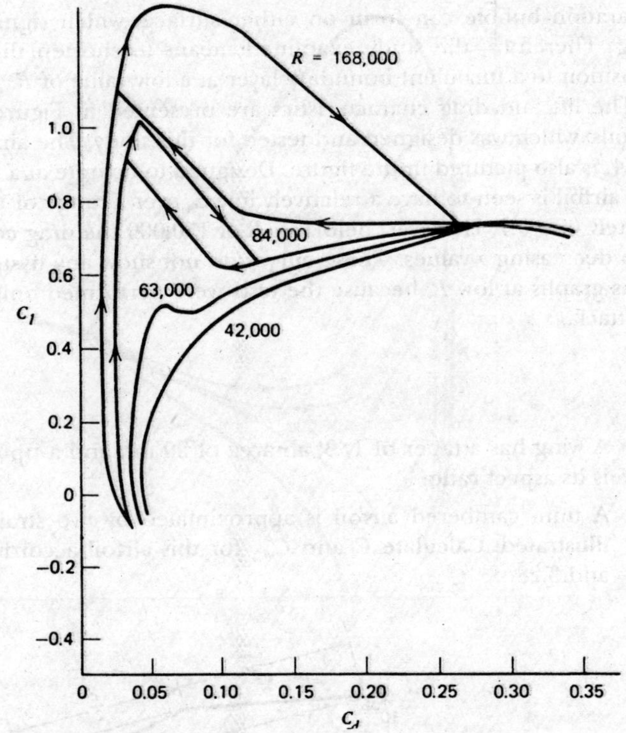

Figure 3.86 Drag polar for the N60 airfoil at low Reynolds numbers.

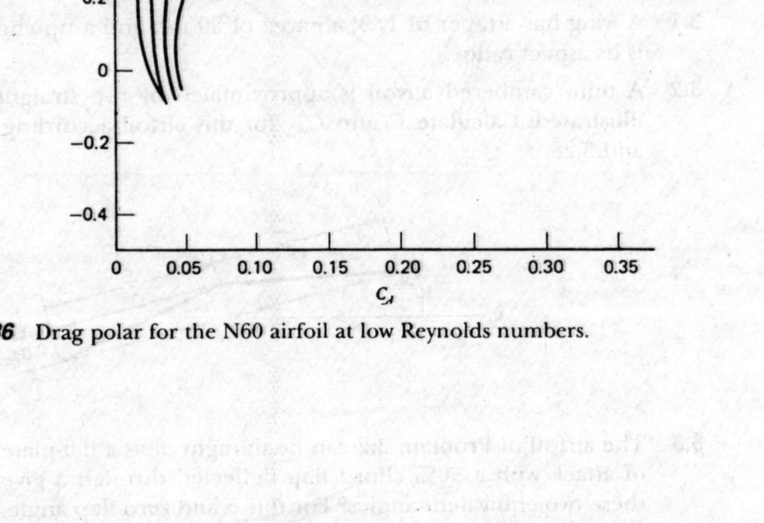

Figure 3.87 Test results for the E374 airfoil.

separation bubble can form on either surface, which significantly increases the drag. Therefore, the study examined means to shorten the bubble or promote transition to a turbulent boundary layer at a low value of R.

The lift and drag characteristics are presented in Figure 3.87 for one of the airfoils, which was designed and tested for this study. The airfoil shape, designated E374, is also pictured in the figure. Designed to operate at a lift coefficient of 0.55, this airfoil is seen to have a relatively low C_d over a range of C_l values from approximately 0 to 0.7. However, below an R of 150,000 the drag coefficient rises rapidly with decreasing r values. These curves do not show any hysteresis, unlike the previous graphs at low R, because the tests were performed only for increasing angle of attack.

PROBLEMS

3.1 A wing has a taper of 1/3, an area of 20 m^2, and a tip chord of 1.5 m. What is its aspect ratio?

3.2 A thin, cambered airfoil is approximated by two straight-line segments, as illustrated. Calculate C_l and $C_{m_{ac}}$ for this airfoil according to Equations 3.26 and 3.28.

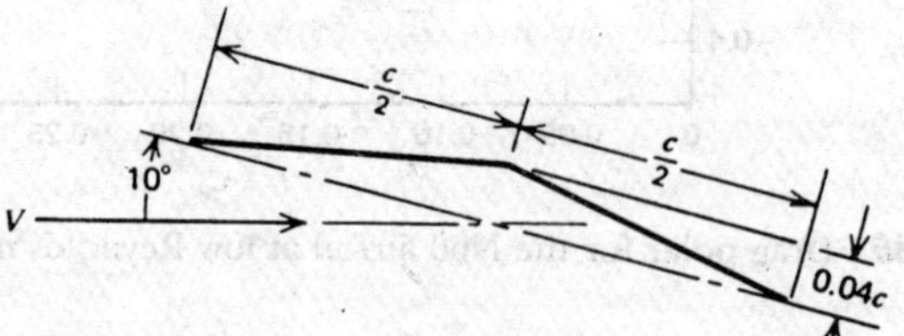

3.3 The airfoil of Problem 3.2 can be thought of as a flat-plate airfoil at an angle of attack with a 50% chord flap deflected through a given angle. What are these two equivalent angles? For this α and zero flap angle, what would C_l be? Comparing this C_l to the value from Problem 3.2, calculate the flap effectiveness factor τ and compare it with Figure 3.35.

3.4 A 23015 airfoil is equipped with a 25% fully extensible, double-slotted flap deflected at an optimum angle. It has a 6 ft chord and is operating at 100 mph at standard sea level (SSL) conditions. Estimate $C_{l_{max}}$ from: (*a*) the summary observations listed at the beginning of the section on flaps, (*b*) the numerous tables and graphs of data, and (*c*) Figures 3.32, 3.33, and 3.34.

3.5 Estimate C_l for a thin flat-plate airfoil at a 5° angle of attack having a 33% c plain flap deflected 15°. Divide the chord into three equal segments and model the airfoil with three suitably placed point vortices.

3.6 This is an open-ended problem. For the same wing area, a low-aspect wing will give a smoother ride than one with a higher aspect ratio. For a given wing loading, W/S, examine the relationship between aspect ratio, vertical gust velocity, airspeed, $C_{L_{max}}$, and vertical acceleration. For an airplane weighing 5000 lb, select the wing area and aspect ratio that will limit the vertical acceleration to 0.1 g when encountering a sharp-edged gust having a vertical velocity, V_g, of 30 fps. The wing is unflapped and is flying at a speed equal to three times the stalling speed. (Note: When an airplane encounters a gust, its angle of attack is suddenly increased by V_g/V.)

3.7 Two 2-D airfoils are in tandem, each at an angle of attack of 10°. The distance from one airfoil to the corresponding point on the other airfoil is 3 chord lengths. Calculate C_l for each airfoil using Weissinger's approximation.

3.8 An untwisted, elliptic wing with a span of 40 ft produces a lift of 10,000 lb at a speed of 200 kts at standard sea level (SSL) conditions. It has an aspect ratio of 6.0 and the 2-D slope of the lift curve is 0.1 C_l per degree. The trailing vortex system eventually rolls up into two trailing vortices spaced a spanwise distance $\pi b/4$ apart. What is the downwash induced midway between the trailing vortices far behind the wing?

3.9 A vortex line has a strength of $10 m^2/s$ and is shaped like a parabola, $y = x^2$. Calculate the velocity induced at $x = 0$ and $y = 10$ m.

3.10 A vortex line extends from the origin to the point $x, y, z = 10, 5, 5$ ft. The circulation, Γ, around the line equals 20 ft^2/s. Calculate the magnitude of the velocity induced at the point 3,0,0.

3.11 A thin, flat-plate airfoil has a 10% leading edge flap and a 25% trailing edge flap. The airfoil is placed at a zero angle of attack. Then, the leading edge flap is lowered 10° and the trailing edge flap is lowered 15°. Calculate the lift coefficient for this configuration.

3.12 An untwisted, rectangular wing with an aspect ratio of 6 is modeled approximately with a lifting line and four trailing vortices symmetrically placed as shown. The angle of attack is 10°. Let $V = 1$ and $b = 2$. Choose control points at the midspan and midway between Γ_1 and Γ_2. Find Γ_1, Γ_2 and the wing lift coefficient.

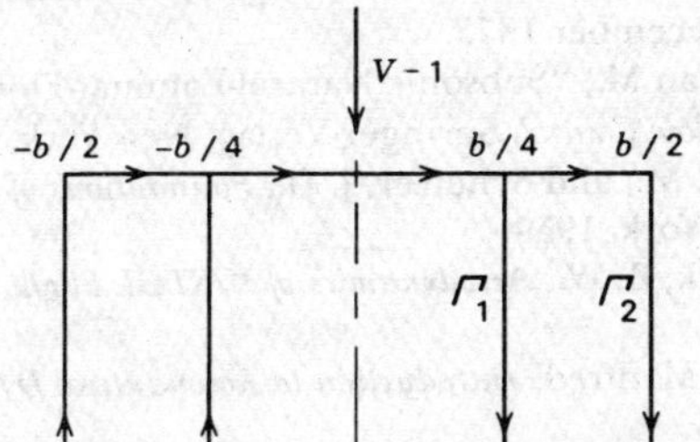

3.13 A rectangular wing has its maximum section C_l at midspan. At a wing C_L of 0, the midspan $C_l = 0.1$ and at a C_L of 1.0, $C_l = 1.3$. The wing is equipped with a full-span, double-slotted flap deflected 30°. The airfoil section is GA(W) −1 shown in Figure 3.11*a*. If the chord of the wing is 3 m. and the airspeed is 70 m/s at SSL, at what C_L would the wing first begin to stall?

3.14 A delta wing has an aspect ratio of 2.0. By means of helium bubbles in the flow, the vortices are observed to be bursting 2/3 of the midspan chord back from the apex. What would you estimate the angle of attack of the wing to be?

3.15 The landing weight of the Concorde, a supersonic transport, is approximately 245,000 lb. The planform has a curved, swept leading edge and is referred to as an *ogee;* however, assume that it can be approximated as a delta wing having an aspect ratio of 1.7. The wing span is 25.5 m. If the airplane lands at a speed that is 20% above the stalling speed, what is the landing speed and angle of attack of the wing when landing at SSL conditions? (Neglect the effects of vortex bursting in determining α but not in finding the landing speed.)

REFERENCES

3.1 Abbott, Ira H., and von Doenhoff, Albert E., *Theory of Wing Sections (including a summary of airfoil data),* Dover Publications, New York, 1958.

3.2 Stevens, W. A., Goradia, S. H., and Braden, J. A., *Mathematical Model for Two-Dimensional Multi-Component Airfoils in Viscous Flow,* MASA CR-1843, 1071.

3.3 Eppler, R., and Somers, D. M., "A Computer Program for the Design and Analysis of Low-Speed Airfoils, Including Transition," *NASA TM-80210,* August 1980 (Supplement: NASA TM-81862, 1980.

3.4 Drela, M., and Giles, M. B., "ISES: A Two-Dimensional Viscous Aerodynamic Design and Analysis Code," *AIAA Paper* 87-04224, January 1987.

3.5 Whitcomb, R. T., and Clark, L. R., "An Airfoil Shape for Efficient Flight at Supercritical Mach Numbers," *NASA TM X-1109,* NASA Langley Research Center, July 1965.

3.6 Ayers, T. G., "Supercritical Aerodynamics Worthwhile over a Range of Speeds," *Astronautics and Aeronautics,* 10(8), August 1972.

3.7 Carlson, F. A., "Transonic Airfoil Analysis and Design Using Cartesian Coordinates." *AIAA J. of Aircraft,* 13(5), May 1976 (see also NASA CR-2578, 1976).

3.8 Hurley, F. X., Spaid, F. W., Roos, F. W., Stivers, L. S., and Bandettini, A., "Supercritical Airfoil Flowfield Measurements," *AIAA J. of Aircraft,* 12(9), September 1975.

3.9 McGhee, R. J., and Beasley, W. D., "Low-Speed Aerodynamic Characteristics of a 17-Percent Thick Airfoil Section Designed for General Aviation Applications," *NASA TN D-7428,* December 1973.

3.10 Somers, Dan M., "Subsonic Natural-Laminar-Flow Airfoils," *Natural Laminar Flow and Laminar Flow Control,* Springer-Verlag, New York, 1991.

3.11 Kuethe, A. M., and Schetzer, J. D., *Foundations of Aerodynamics,* John Wiley & Sons, Inc., New York, 1959.

3.12 McCormick, B. W., *Aerodynamics of V/STOL Flight,* Academic Press, New York, London, 1967.

3.13 Rauscher, Manfred, *Introduction to Aeronautical Dynamics,* John Wiley & Sons, Inc., New York, 1953.

3.14 Lindsey, W. F., Stevenson, D. B., and Daley, B. N., "Aerodynamic Characteristics of 24 NACA 16-Series Airfoils at Mach Numbers between 0.3 and 0.8," *NACA TN 1546,* September 1948.

3.15 Abbott, I. H., von Doenhoff, A. E., and Stivers, Louis S., "Summary of Airfoil Data," *NACA R 824,* 1945.

3.16 Anonymous, "Airfoil Information for Propeller Design," Ordnance Research Laboratory, The Pennsylvania State University, Report No. NOrd 7958-71, November 1947.

3.17 Cahill, J. F., *Summary of Section Data on Trailing-Edge Devices,* NACA R 938, 1949.

3.18 Smith, A. M. O., "High-Lift Aerodynamics," *AIAA J. of Aircraft,* 12(6), June 1975.

3.19 Kelly, J. A., and Hayter, N. F., "Lift and Pitching Moment at Low Speeds of the NACA 64A010 Airfoil Section Equipped with Various Combinations of a Leading-Edge Slat, Leading-edge Flap, Split Flap, and Double-Slotted Flap," *NACA TN 3007,* September 1953.

3.20 Fullmer, F. F., "Two-dimensional Wind Tunnel Investigation of the NACA 64,*012 Airfoil Equipped with Two Types of Leading-Edge Flap,"* *NACA TN 1277,* May 1947.

3.21 Harris, T. A., and Recant, I. G., "Wind Tunnel Investigation of NACA 23012, 23021, and 23030 Airfoils Equipped with 40-Percent Chord Double-Slotted Flaps," *NACA R 723,* 1941.

3.22 Wenzinger, C. J., and Rogallo, F. M., "Resume of Air-Load Data on Slats and Flaps," *NACA TN 690,* March 1939.

3.23 Young, A. D., "The Aerodynamic Characteristics of Flaps," *ARC R&M 2622,* 1953.

3.24 Liebeck, R. H., "A Class of Airfoils Designed for High Lift in Incompressible Flow," *AIAA J. of Aircraft* 10(10), October 1973.

3.25 Stratford, B. S., "The Prediction of the Separation of the Turbulent Boundary Layer," *J. of Fluid Mechanics,* 5, 1959.

3.26 Stratford, B. S., "An Experimental Flow with Zero Skin Friction Throughout its Region of Pressure Rise," *J. of Fluid Mechanics,* 5, 1959.

3.27 Liebeck, R. H., and Ormsbee, A. I., "Optimization of Airfoils for Maximum Lift," *AIAA J. of Aircraft,* 7(5), September–October 1970.

3.28 Cleveland, F. A., "Size Effects in Conventional Aircraft Design," *AIAA J. of Aircraft,* 7(6), November–December 1970.

3.29 Sivells, J. C., and Spooner, S. H., "Investigation in the Langley 19-Foot Pressure Tunnel of Two Wings of NACA 65-210 Airfoil Sections with Various Type Flaps," *NACA R 941,* 1949.

3.30 Lowry, J. G., and Polhamus, E. C., *A Method for Predicting Lift Increments Due to Flap Deflection at Low Angles of Attack in Incompressible Flow,* NACA TN 3911, January 1957.

3.31 Prandtl, L., and Betz, A., "Vier Abhandlungen Zur Hydrodynamik und Aerodynamic," Göttingen, 1927 (reprint Edward Bros. 1943, Ann Arbor, Mich.).

3.32 Ashley, H., and Landahl, M., *Aerodynamics of Wings and Bodies,* Addison-Wesley, Reading, Mass. 1965.

3.33 Faulkner, V. M., "The Calculation of Aerodynamic Loading on Surfaces of Any Shape," *ARC R&M 1910,* 1943.

3.34 Schrenk, O., "A Simple Approximation Method for Obtaining the Spanwise Lift Distribution," *NACA TM 1919,* 1940.

3.35 Jacobs, Eastman M., and Ward, Kenneth E., *Interference of Wing and Fuselage from Tests of 209 Combinations in the NACA Variable-Density Tunnel,* NACA R 540, 1936.

3.36 Sherman, Albert, "Interference of Wing and Fuselage from Tests of 28 Combinations in the NACA Variable-Density Tunnel," *NACA R 575,* 1936.

3.37 Sherman, Albert, "Interference of Wing and Fuselage from Tests of 30 Combinations with Triangular and Elliptical Fuselages in the NACA Variable-Density Tunnel," *NACA TN 1272,* 1947.

3.38 Jones, R. T., "Properties of Low-Aspect-Ratio Pointed Wings at Speeds Below and Above the Speed of Sound," *NACA Report 835,* 1946.

3.39 Spreiter, J. R., "The Aerodynamic Forces on Slender Plane- and Cruciform Wing and Body Combinations," *NACA Report 962,* 1950.

3.40 Polhamus, E. C., "A Concept of the Vortex Lift of Sharp Edge Delta Wings Based on a Leading-Edge Suction Analogy," *NASA TN D-3767,* December 1966.

3.41 Polhamus, E. C., "Prediction of Vortex-Lift Characteristics by a Leading-Edge Suction Analogy," *AIAA J. of Aircraft,* 8(4), April 1971.

3.42 McCormick, B. W., Tangler, J. L., and Sherrieb, H. E., "The Structure of Trailing Vortices," *AIAA J. of Aircraft,* 5(3), May 1968.

3.43 ———, "Vortex Flow Aerodynamics," *AGARD Conference Proceedings, AGARD-CP-494,* October 1990.

3.44 Lambourne, N. C., and Bryer, D. W., "The Bursting of Leading-Edge Vortices—Some Observations and Discussion of the Phenomenon," *J. of Fluid Mechanics,* 14(4), 1962.

3.45 McCormick, B. W., "Aerodynamics of Delta Wings with Application to High-Alpha Flight Mechanics," Third Annual Aerospace Symposium, TU Braunschweig, Germany, August 26–28, 1991. (Proceedings published by Springer-Verlag Berlin)

3.46 Wentz, William H., and Kohlman, David L., *Wind Tunnel Investigation of Vortex Breakdown on Slender Sharp-Edge Wings,* Ph.D. Thesis, University of Kansas, 1969 (NASA CR 98737, 1969).

3.47 Roos, F. W., and Kegelman, J. T., *An Investigation of Sweep-Angle Influence on Delta-Wing Flows,* AIAA Paper 90-0383, January 8–11, 1990.

3.48 Earnshaw, P. B., and Lawford, J. A., *Low-Speed Wind Tunnel Experiments on a Series of Sharp-Edged Delta Wings,* ARC R&M No. 3424, 1966.

3.49 Shanks, R. E., *Low Subsonic Measurements of Static and Dynamic Stability Derivatives of Six Flat-Plate Wings Having Leading Edge Sweep Angles of 70 Deg. to 84 Deg.,* NASA TN D-1822, July 1963.

3.50 Hummel, D., and Redeker, G. *Über den Einfluß des Aufplatzens der Wirbel auf die aerodynamischen Beiwerte von Deltaflügeln mit kleinem Seitenverhältnis beim Schiebeflug.* Jahrbuch der WGLR, 1967.

3.51 Brandon, J. M., and Shah, G. H., *Effect of Large Amplitude Pitching Motions on the Unsteady Aerodynamic Characteristics of Flat-Plate Wings,* AIAA Paper No. 88-4331, August 1988.

3.52 Schmitz, F. W., *Aerodynamics of Model Aircraft Wing Measurements I,* R. T. P. Translation No. 2460. Issued by Ministry of Aircraft Production.

3.53 Donovan, J. F., and Selig, M. S., *Low Reynolds Number Airfoil Design and Wind Tunnel Testing at Princeton University,* Low Reynolds Conference, Notre Dame University, June 5–7, 1989. Proceedings (T. J. Meuller, editor) published by Springer-Verlag, New York.

3.54 Eppler, R., and Somers, D. M., *Airfoil Design for Reynolds Numbers between 50,000 and 500,000,* Proceedings of the Conference on Low Reynolds Number Airfoil Aerodynamics, Notre Dame University, South Bend, Ind., June 1986.

4 DRAG

As a child, it was fun to stick your hand out of the car window and feel the force of the moving, invisible air. To the aeronautical engineer, however, there is nothing very funny about aerodynamic drag. A continuing struggle for the practicing aerodynamicist is that of minimizing drag whether it is for an airplane, missile, or ground-based vehicle such as an automobile or train. It takes power to move a vehicle through the air. This power is required to overcome the aerodynamic force on the vehicle opposite to its velocity vector. Any reduction of this force, known as the *drag*, represents either a direct saving in fuel or an increase in performance.

The estimation of the drag of a complete airplane is a difficult and challenging task, even for the simplest configurations. A list of definitions of various types of drag partly reveals why this is so.

Induced Drag The drag that results from the generation of a trailing vortex system downstream of a lifting surface of finite aspect ratio.

Parasite Drag The total drag of an airplane minus the induced drag. Thus, it is the drag not directly associated with the production of lift. The parasite drag is composed of many drag components, the definitions of which follow.

Skin Friction Drag The drag on a body resulting from viscous shearing stresses over its wetted surface.

Form Drag (Sometimes Called Pressure Drag) The drag on a body resulting from the integrated effect of the static pressure acting normal to its surface resolved in the drag direction.

Interference Drag The increment in drag resulting from bringing two bodies in proximity to each other. For example, the total drag of a wing-fuselage combination will usually be greater than the sum of the wing drag and fuselage drag independent of each other.

Trim Drag The increment in drag resulting from the aerodynamic forces required to trim the airplane about its center of gravity. Usually this takes the form of added induced and form drag on the horizontal tail.

Profile Drag Usually taken to mean the total of the skin friction drag and form drag for a two-dimensional airfoil section.

Cooling Drag The drag resulting from the momentum lost by the air that passes through the power plant installation for purposes of cooling the engine, oil, and accessories.

Base Drag The specific contribution to the pressure drag attributed to the blunt after-end of a body.

Wave Drag Limited to supersonic flow, this drag is a pressure drag resulting from noncanceling static pressure components to either side of a shock wave acting on the surface of the body from which the wave is emanating.

With the exception of wave drag, the material to follow will consider these various types of drag in detail and will present methods of reasonably estimating their magnitudes. Wave drag will be discussed in Chapter 5.

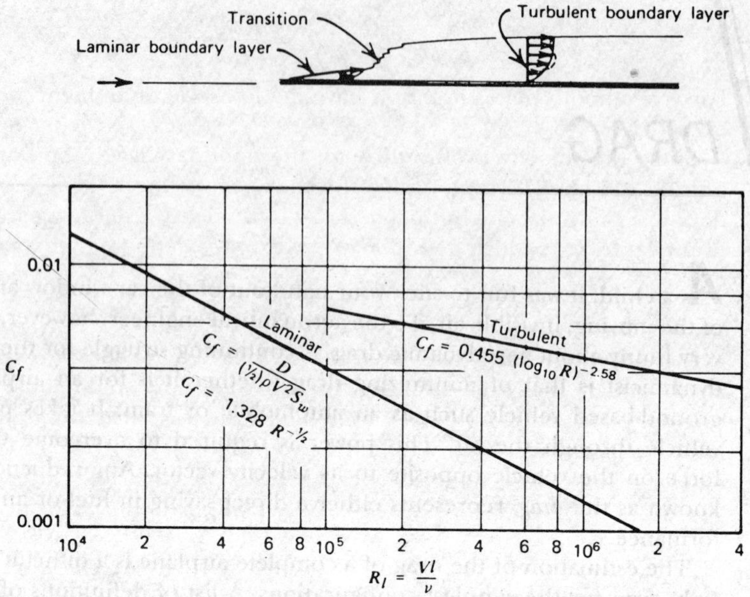

Figure 4.1 Drag of a thin, flat plate.

SKIN FRICTION DRAG

Figure 4.1 depicts a thin, flat plate aligned with the free-stream velocity. Frequently, the drag of a very streamlined shape such as this is expressed in terms of a skin friction drag coefficient, C_f, defined by

$$C_f = \frac{D_f}{qS_w} \tag{4.1}$$

where S_w is the wetted surface area that is exposed to the flow. This coefficient is presented in Figure 4.1 as a function of Reynolds number for the two cases where the flow in the boundary layer is entirely laminar or entirely turbulent over the plate. Here, the Reynolds number is based on the total length of the plate in the direction of the velocity. In a usual application, the boundary layer is normally laminar near the leading edge of the plate undergoing transition to a turbulent layer at some distance back along the surface, as described in Chapter 2. The situation is pictured in Figure 4.1, where the velocity profile through the layer is shown. To illustrate it, the thickness of the layer is shown much greater than it actually is.

As shown in this figure, a laminar boundary layer begins to develop at the leading edge and grows in thickness downstream. At some distance from the leading edge, the laminar boundary becomes unstable and is unable to suppress disturbances imposed on it by surface roughness or fluctuations in the free stream. In a short distance the boundary layer undergoes transition to a turbulent boundary layer. Here, the layer suddenly increases in thickness and is characterized by a mean velocity profile on which a random fluctuating velocity component is superimposed. The distance, x, from the leading edge of the plate to the transition point can be calculated from the transition Reynolds number, R_x. R_x is typically, for a flat plate, of the order of 3×10^5, R_x being defined by

$$R_x = \frac{\rho V x}{\mu} \tag{4.2}$$

For very smooth plates in a flow having a low level of ambient turbulence, R_x can exceed 1×10^6.

Since the velocity profile through the boundary layer approaches the velocity outside the layer asymptotically, the thickness of the layer is vague. To be more definitive, a displacement thickness, δ^*, is frequently used to measure the thickness of the layer. δ^* is illustrated in Figure 4.2 and is defined mathematically by

$$\delta^* = \int_0^\infty \left(1 - \frac{u}{V}\right) dy \tag{4.3}$$

where y is the normal distance from the plate at any location such that, without any boundary layer, the total flow past that location would equal the flow for the original plate with a boundary layer. To clarify this further, let δ be the boundary layer thickness where, for all intents and purposes, $u = V$. Then

$$V(\delta - \delta^*) = \int_0^\infty u \, dy \tag{4.4}$$

Allowing δ to become infinite leads to Equation 4.3.

If we arbitrarily define δ as the value of y at which $u = 0.99V$ then, for a laminar layer,

$$\delta \simeq 5.2 \left(\frac{vx}{V}\right)^{1/2} = 5.2x \left(\frac{Vx}{v}\right)^{-1/2} \tag{4.5}$$

$$\delta^* = \frac{\delta}{3} \tag{4.6}$$

For a turbulent layer,

$$\delta \simeq 0.37x \left(\frac{Vx}{v}\right)^{-1/5} \tag{4.7}$$

$$\delta^* = \frac{\delta}{8} \tag{4.8}$$

Observe that relatively speaking, the turbulent boundary layer is more uniform, with δ^* being only one-eighth of δ as compared to one-third for the laminar layer.

To clarify the use of Figure 4.1 and Equations 4.5 to 4.8, let us consider the horizontal tail of the Cherokee pictured in Figure 3.62 at a velocity of 60.4 m/s (135 mph) at a 1524 m (5000 ft) standard altitude. We will assume that the tail can be approximately treated as a flat plate at zero angle of attack.

From Figure 3.59, the length of the plate is 30 in. or 0.762 m. The total wetted area, taking both sides and neglecting the fuselage, is 4.65 m² (50 ft²). At an altitude

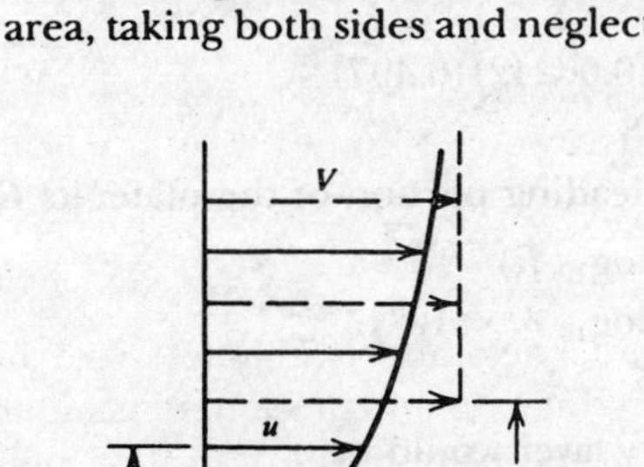

Figure 4.2 Displacement thickness.

of 1.52 km, $\rho = 1.054\ \text{kg/m}^3$ and $\upsilon = 1.639 \times 10^{-5}\ \text{m}^2/\text{s}$. We will assume that the transition Reynolds number is equal to 3×10^5.

The distance from the leading edge to the transition point is found from Equation 4.2.

$$\begin{aligned} x &= \frac{\upsilon R_x}{V} \\ &= 1.639 \times 10^{-5} \times 3 \times \frac{10^5}{60.4} \\ &= 0.0814\ \text{m (3.2 in.)} \end{aligned}$$

The Reynolds number based on the total length will be equal to

$$\begin{aligned} R_l &= \frac{Vl}{\upsilon} \\ &= \frac{60.4\ (0.762)}{1.639 \times 10^{-5}} \\ &= 2.81 \times 10^6 \end{aligned}$$

If the flow over the tail were entirely turbulent then, from Figure 4.1,

$$\begin{aligned} C_f &= 0.455\ (\log^{10} R_l)^{-2.58} \\ &= 0.00371 \end{aligned} \tag{4.9}$$

The dynamic pressure q for this case is

$$\begin{aligned} q &= \rho V^2/2 \\ &= \frac{1.054\ (60.4)^2}{2} \\ &= 1923\ \text{N/m}^2 \end{aligned}$$

Hence the total skin friction drag would be

$$\begin{aligned} D &= qS_w C_f \\ &= 1923\ (4.65)(0.00371) \\ &= 33.17\ \text{N} \end{aligned}$$

However, the leading portion of the plate is laminar. The wetted area of this portion is equal to $0.497\ \text{m}^2$. For laminar flow over this portion.

$$\begin{aligned} C_f &= 1.328 R^{-1/2} \\ &= 1.328\ (3 \times 10^5)^{-1/2} \\ &= 0.00242 \end{aligned} \tag{4.10}$$

Hence, the drag of this portion of the plate is equal to

$$\begin{aligned} D &= qC_f S_w \\ &= 1923(0.00242)(0.497) \\ &= 2.31\ \text{N} \end{aligned}$$

If the flow were turbulent over the leading portion of the plate, its C_f would be

$$\begin{aligned} C_f &= 0.455\ (\log_{10} R)^{-2.58} \\ &= 0.455\ (\log_{10} 3 \times 10^5)^{-2.58} \\ &= 0.00566 \end{aligned}$$

Thus, its drag for a turbulent boundary layer would be

$$\begin{aligned} D &= qC_f S_w \\ &= (1923)(0.00566)(0.497) \\ &= 5.35\ \text{N} \end{aligned}$$

The above is 5.35 − 2.31, or 3.04 N higher than the actual drag for laminar flow. This difference must be subtracted from the total drag of 33.17 N previously calculated assuming the boundary layer to be turbulent over the entire plate. Hence, the final drag of the total horizontal tail is estimated to be

$$\begin{aligned} D &= 33.17 - 3.04 \\ &= 30.13 \text{ N} \\ &= 6.77 \text{ lb} \end{aligned}$$

The thickness, δ, of the laminar boundary layer at the beginning of transition can be calculated from Equation 4.5.

$$\begin{aligned} \delta &= 5.2\,(0.0814)(3 \times 10^5)^{-1/2} \\ &= 7.728 \times 10^{-4} \text{ m} \\ &= 0.0304 \text{ in.} \end{aligned}$$

The thickness of the turbulent layer right after transition is found from Equation 4.7 assuming the layer to have started at the leading edge.

$$\begin{aligned} \delta &= 0.37(0.0814)(3 \times 10^5)^{-1/5} \\ &= 2.418 \times 10^{-3} \text{ m} \\ &= 0.0952 \text{ in.} \end{aligned}$$

At the trailing edge, the thickness of the turbulent layer will be

$$\begin{aligned} \delta &= 0.37(0.762)(2.81 \times 10^6)^{-1/5} \\ &= 0.0145 \text{ m} \\ &= 0.5696 \text{ in.} \end{aligned}$$

The displacement thickness at the trailing edge is thus only 0.0018 m (0.071 in.).

Before leaving the topic of skin friction drag, the importance of surface roughness should be discussed. Surface roughness can have either a beneficial or adverse effect on drag. If it causes premature transition, it can result in a reduced form drag by delaying separation. This is explained more fully in the next section. Adversely, surface roughness increases the skin friction coefficient. First, by causing premature transition, the resulting turbulent C_f is higher than C_f for laminar flow, in accordance with Figure 4.1. Second, for a given type of flow laminar or turbulent, C_f increases as the surface is roughened.

It is difficult to quantify the increment in C_f as a function of roughness, since roughness comes in many forms. For some information on this, refer to the outstanding collection of drag data noted previously (e.g., Ref. 4.4). Generally, if a roughness lies well within the boundary layer thickness, say of the order of the displacement thickness, then its effect on C_f will be minimal. Thus, for the preceding example of the horizontal tail for the Cherokee, the use of flush riveting near the trailing edge is probably not justified.

An approximate estimate of the effect of roughness, at least on streamlined bodies, can be obtained by examining the airfoil data of Reference 3.1. Results are presented for airfoils having both smooth and rough surfaces. The NACA "standard" roughness for 0.61-m (2-ft) chords consisted of 0.028-cm (0.011-in.) carborundum grains applied to the model surface starting at the leading edge and extending 8% of the chord back on both the upper and lower surfaces. The grains were spread thinly to cover 5 to 10% of the area.

An examination of the drag data with and without the standard roughness discloses a 50 to 60% increase in airfoil drag resulting from the roughness. It is difficult to say how applicable these results are to production aircraft. Probably the NACA standard roughness is too severe for high-speed aircraft employing extensive flush riveting with particular attention to the surface finish. In the case of a pro-

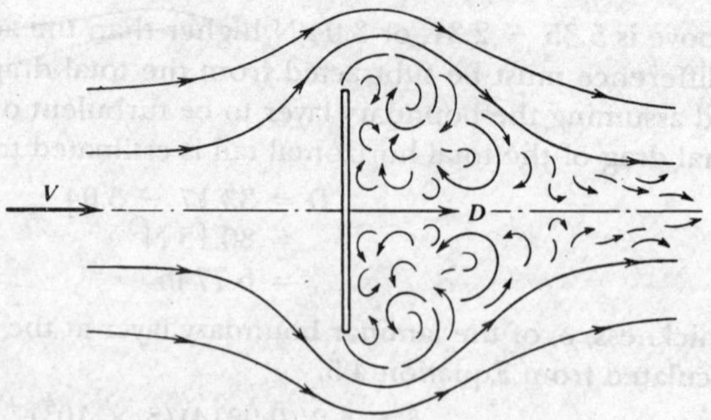

Figure 4.3 Flat plate normal to flow.

duction light aircraft for general aviation usage, the standard roughness could be quite appropriate.

FORM DRAG

In addition to skin friction drag, a body generally experiences some form drag. Unlike the skin friction drag that results from viscous shearing forces tangential to a body's surface, form drag results from the distribution of pressure normal to the body's surface. The extreme case of a flat plate normal to the flow is pictured in Figure 4.3. Here, the drag is totally the result of an imbalance in the normal pressure distribution. There is no skin friction drag present in this case.

Generally, form drag is difficult to predict. For that matter, so is skin friction drag except for the simplest cases. Thus, in general cases, such as that pictured in Figure 4.4, where the total drag results from both normal and tangential stresses (or pressures) one must usually resort to experimental data to estimate the drag.

As with skin friction drag, form drag is generally dependent on Reynolds number. To see why, consider the flow around the circular cylinder pictured in Figure 4.5. In Figure 4.5*a*, flow is pictured at a low Reynolds number. Here, beginning at the stagnation point, a laminar boundary layer develops. On the surface of the cylinder, the static pressure (normal) is highest at the stagnation point and decreases to a minimum at the top and bottom. Moving around toward the rear, beyond these points, the static pressure increases, tending toward the stagnation pressure at the very rear. In the absence of viscosity, the normal pressure distribution would be symmetrical (Equation 2.78) and there would be no drag. This is a clear example of D'Alembert's paradox, which states that a body in an inviscid fluid will experience no drag. As the slower moving fluid in the laminar boundary layer moves beyond the minimum pressure point on the cylinder, its momentum is

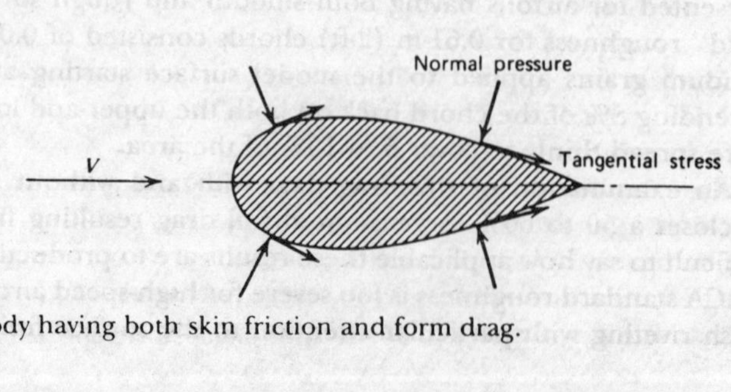

Figure 4.4 Body having both skin friction and form drag.

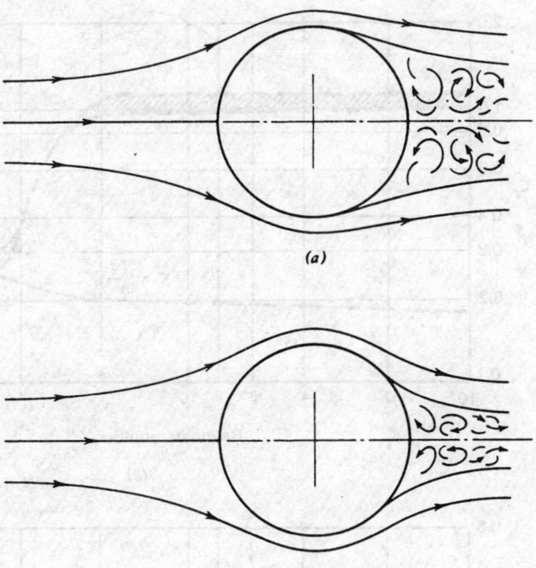

Figure 4.5 Flow over a circular cylinder. (*a*) Low Reynolds number. Separation occurs before transition. Large wake. (*b*) High Reynolds number. Transition occurs before separation. Small wake.

insufficient to move against the positive pressure gradient, referred to as an *adverse gradient* and thus, the flow separates just past the top and bottom locations on the cylinder. In the separated region over most of the rear portion of the cylinder, the static pressure is constant and equal to the low pressure at the top and bottom. Thus, the high pressure acting over the front and the low pressure over the rear result in a high form drag.

The high-Reynolds number case is shown in Figure 4.5*b*. Here, the laminar boundary layer undergoes transition to a turbulent boundary layer before separating. The subsequent turbulent mixing increases the momentum and energy of the boundary layer so that it remains attached toward the rear of the cylinder, well beyond the separation point of the laminar layer. Thus, in this case, the separation region is much smaller and the static pressure is higher on the rear of the cylinder than for the laminar case. Therefore, because of reduced form drag, the drag coefficient of a cylinder is lower at higher Reynolds numbers.

C_d as a function of Reynolds number is presented in Figure 4.6 for both spheres and two-dimensional circular cylinders. Here, C_d is based on the projected frontal area. Note the rapid drop in C_d above an R value of approximately 2×10^5. This is the so-called critical Reynolds number, where the transition point is nearly coincident with the separation point. "Subcritical" refers to flow at Reynolds numbers that are less than critical; "supercritical" denotes R values that are higher than critical. A body shape having a well-defined separation point will not exhibit a critical Reynolds number; neither will streamlined shapes.

Although not concerned with drag per se, Figure 4.6*a* also includes the quantity fd/V, known as the *Strouhal number, S. S* characterizes an interesting behavior of bluff bodies with rounded trailing edges. As such a body first begins to move through a fluid, the vorticity in the boundary layer is shed symmetrically from the upper and lower surfaces to form two vortices of opposite rotation. However, the

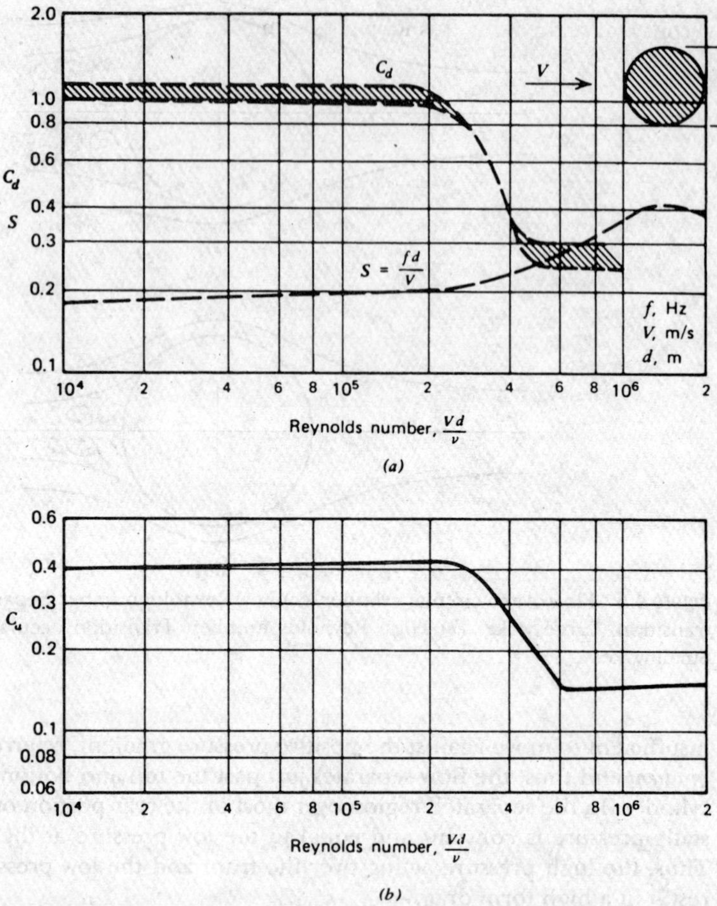

Figure 4.6 Drag coefficients of cylinders and spheres versus Reynolds number. (*a*) Two-dimensional circular cylinders. (*b*) Spheres.

symmetrical placement of the vortex pair is unstable, so that succeeding vortices are then shed alternately from the upper and lower surfaces. The resulting flow pattern of periodically spaced vortices downstream of the body is known as a *Karman vortex street.*

In the definition of Strouhal number, f is the frequency at which the vortices are shed. As a vortex is shed from one surface of the cylinder, it produces a momentary circulation around the cylinder opposite in direction to the vortex. From the Kutta-Joukowski law, a force on the cylinder normal to V results. As the next vortex is shed, the force reverses its direction, resulting in an alternating force on the cylinder. This particular phenomenon is the cause for the "singing" of telephone wires in the wind.

As an example of the use of Figure 4.6*a*, consider a wire 2 cm in diameter in a wind blowing at a speed of 8 m/s. Assuming standard sea level conditions,

$$R = \frac{Vd}{\nu}$$

$$= \frac{8(0.02)}{1.456 \times 10^{-5}}$$

$$= 1.099 \times 10^4$$

From Figure 4.6*a*, for this Reynolds number,

$$C_d = 1.1$$
$$S = 0.18$$

Thus, the drag per unit length on the wire will be

$$\begin{aligned} D &= qC_d\, d \\ &= \frac{1}{2}(1.226)(8)^2(1.1)(0.02) \\ &= 0.863 \text{ N/m} \end{aligned}$$

The frequency of the alternating lift force on the wire will be

$$\begin{aligned} f &= \frac{VS}{d} \\ &= \frac{8(0.18)}{0.02} \\ &= 72 \text{ Hz} \end{aligned}$$

Let us now consider the extreme case of form drag illustrated in Figure 4.3, where the point of flow separation is well defined and not dependent on Reynolds number. It is not too surprising to find that drag coefficients for such shapes are nearly constant over a wide range of Reynolds number values. A number of such shapes are pictured in Figure 4.7*a*.

This figure presents values for both two-dimensional and three-dimensional shapes. Three-dimensional shapes are all bodies of revolution. Observe that for the same profile shape,

$$\frac{C_d(\text{2-D})}{C_d(\text{3-D})} \simeq 1.8$$

If the ratio of the span to the height (or diameter) of a flat plate (or cylinder) normal to the flow is approximately 5 or less, C_d is nearly constant and equal to the 3-D value. For aspect ratios greater than 5, C_d varies approximately in the manner given by the normalized curve of Figure 4.7*b*. This curve is based on data from several sources, including Reference 4.4.

A qualitative evaluation of the drag coefficient for a given shape can be made using some "educated intuition." Referring to Figure 4.8, the drag coefficient of a bluff shape depends on the width of the wake behind the body (before viscosity dissipates it). Beginning with the top figure and working downward, one would expect, for the same projected frontal area, that the widths of the wakes would diminish progressively. Intuitively, such a progression is visualized by picturing the flow as separating tangent to the surface and then being turned gradually in the direction of the main flow.

With regard to drag, the trailing edge shape of a body is usually more important than the leading edge shape. For example, the drag of the top shape in Figure 4.8 can be reduced significantly by providing a body behind it to which the flow can reattach. This is illustrated in Figure 4.9. As opposed to Figure 4.8, in this case, the low pressure in the separated region between the front and the afterbody reacts on both parts, contributing little or nothing to the drag.

To provide an additional basis for estimating the drag of two-dimensional sections, the data in Figure 4.10 are provided (Ref. 4.5). This figure shows that for a shape with sharp corners, a rounding of the corners will reduce the drag coefficient as well as the critical Reynolds number.

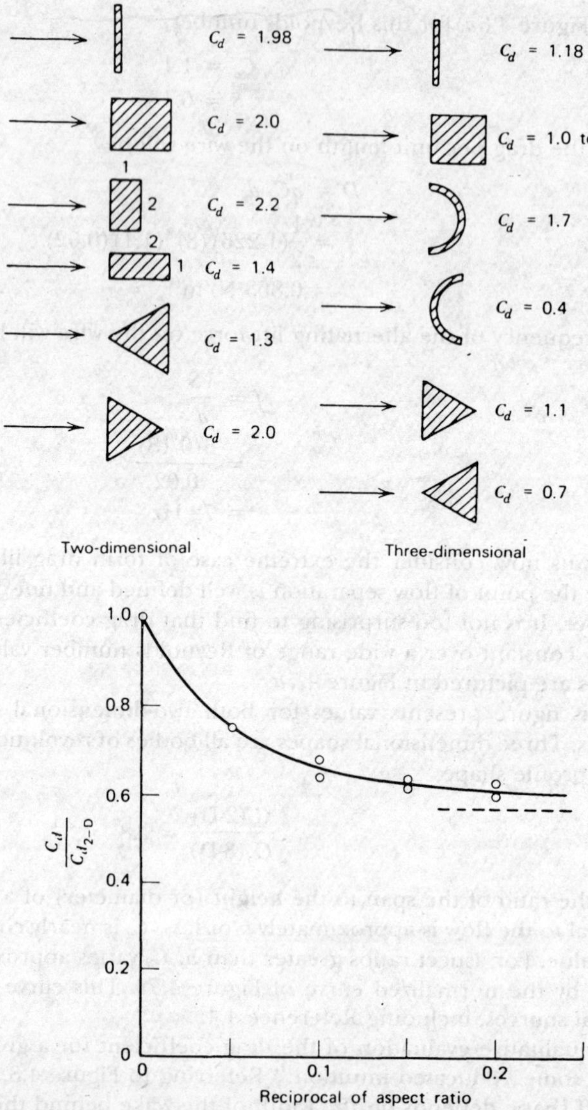

Figure 4.7 (*a*) Examples of shapes having C_d values nearly independent of Reynolds number. (*b*) Transition from three-dimensional to two-dimensional drag for cylinders at supercritical Reynolds numbers.

DRAG OF STREAMLINED SHAPES

The drag of shapes such as airfoils, fuselages, nacelles, torpedoes, submarines, and airships is composed of both form drag and skin friction drag. As the fineness ratio (length/maximum thickness) of a streamlined shape increases, more and more of its drag is attributable to skin friction. Conversely, at low fineness ratios, the drag is principally form drag.

Data on the drag of two-dimensional and three-dimensional streamlined shapes are presented in Figures 4.11, 4.12, and 4.13. Based on the projected frontal area, C_D is given as a function of fineness ratio in Figure 4.11 at a high Reynolds number.

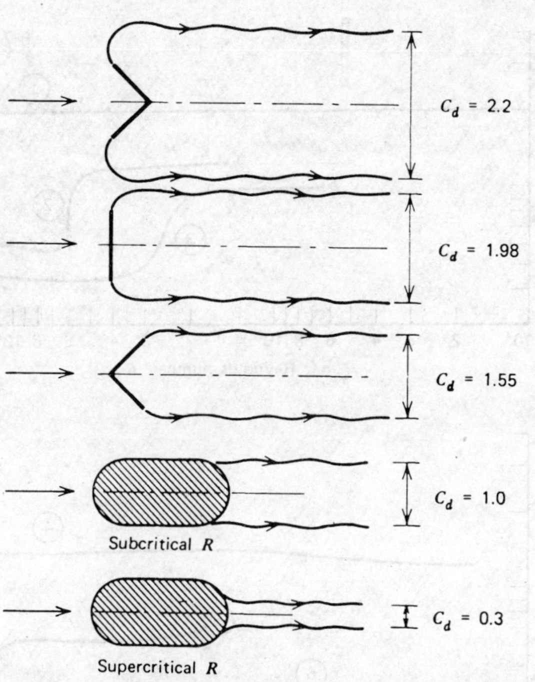

Figure 4.8 Qualitative estimate of drag for two-dimensional shapes.

In this figure, a fineness ratio of 1.0 corresponds to a circular cylinder and sphere for two- and three-dimensional shapes, respectively. Notice that the minimum drag occurs at a fineness ratio of approximately 2 for a three-dimensional shape, and at a value of approximately 3 for a two-dimensional shape. However, in view of the sharp rise in both curves at the lower fineness ratios, it might be well in either case to use fineness ratios higher than these, say around 4, if one wishes to fair a blunt shape of a given frontal area.

The crossover of the two curves in Figure 4.11 is to be expected. At low values of fineness ratio, C_d for the two-dimensional shapes is higher than that for the three-dimensional bodies, based on the data of Figures 4.6 and 4.7. At the other extreme, as the fineness ratio becomes large, the skin friction drag predominates. If C_f is assumed to be the same for either the two-dimensional or three-dimensional shapes, the ratio of the C_d values, based on the projected frontal areas, becomes

$$\frac{C_D(\text{3-D})}{C_d(\text{2-D})} = \frac{S_w(\text{3-D})}{S_w(\text{2-D})}\frac{4}{\pi D}$$

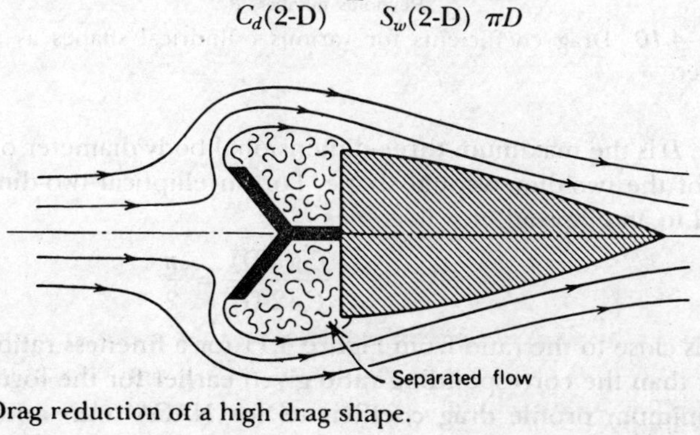

Figure 4.9 Drag reduction of a high drag shape.

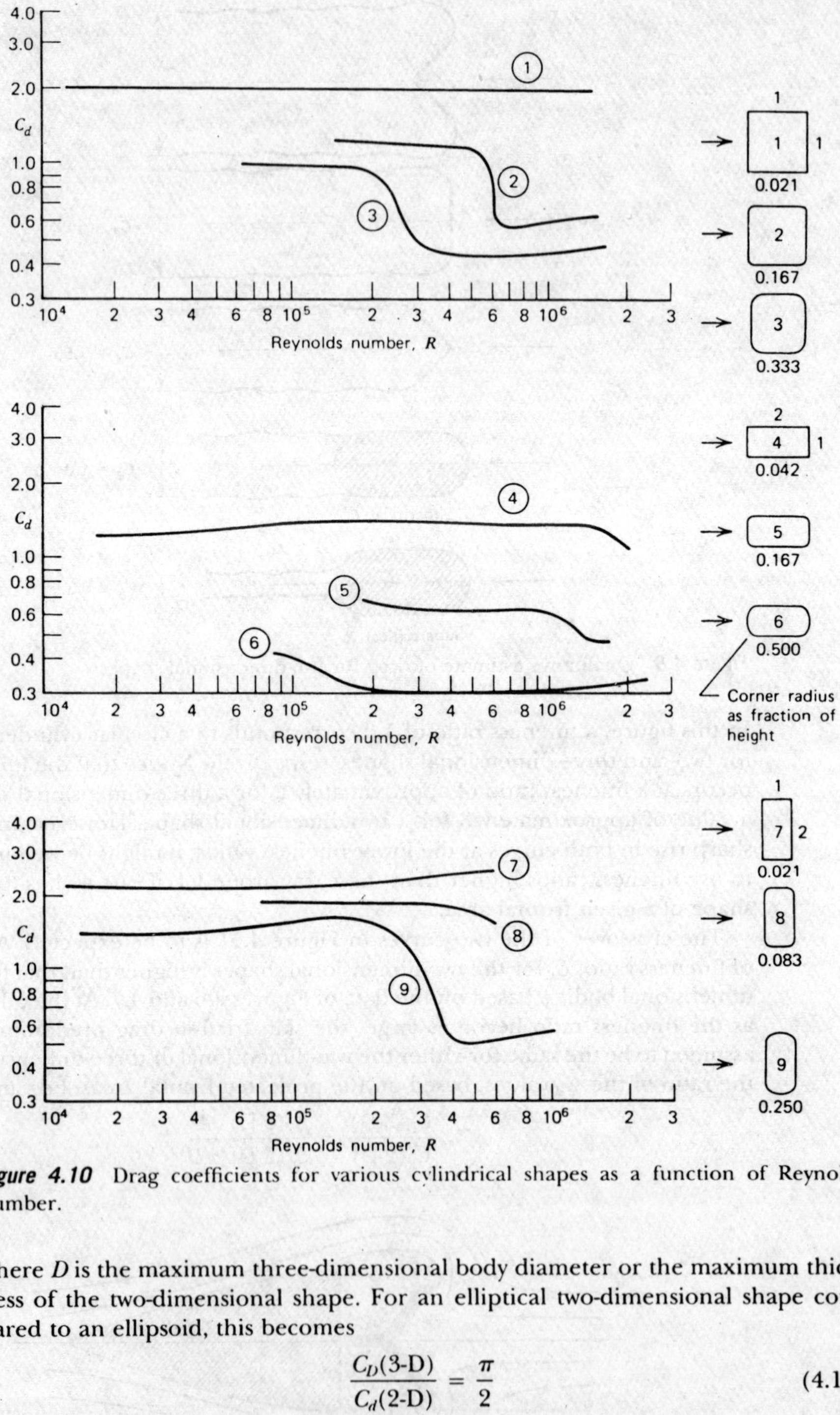

Figure 4.10 Drag coefficients for various cylindrical shapes as a function of Reynolds number.

where D is the maximum three-dimensional body diameter or the maximum thickness of the two-dimensional shape. For an elliptical two-dimensional shape compared to an ellipsoid, this becomes

$$\frac{C_D(\text{3-D})}{C_d(\text{2-D})} = \frac{\pi}{2} \tag{4.11}$$

This is close to the ratio from Figure 4.11 for a fineness ratio of 8 and only slightly lower than the corresponding ratio given earlier for the form drag.

Minimum profile drag coefficients for NACA four- and five-digit airfoils are presented in Figure 4.12 as a function of thickness ratio at a Reynolds number of

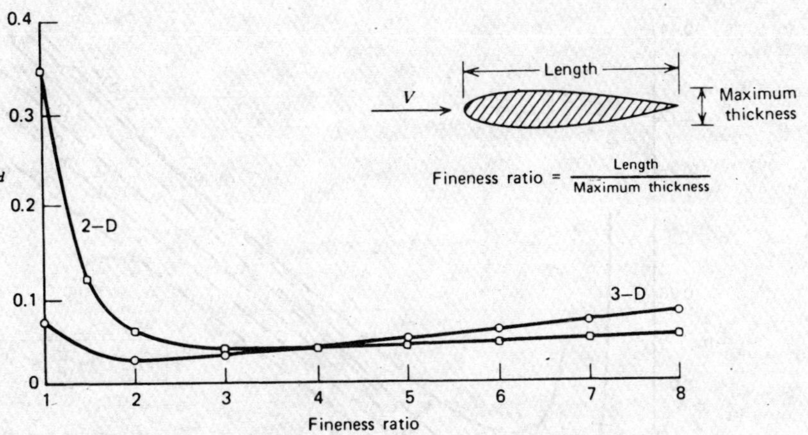

Figure 4.11 Drag coefficients for streamlined shapes as a function of fineness ratio. C_d based on frontal area; $R \simeq 10^7$ based on length.

6×10^6. Here, as is usual for airfoils, C_d is based on the chord length. The several data points at each thickness ratio result from airfoils of different camber ratios. Note that $C_{d_{min}}$ does not vary significantly with camber. $C_{d_{min}}$ appears to vary almost linearly with t/c and extrapolates to a value of 0.004 for a t/c of zero. This corresponds to a C_f value of 0.002. According to Figure 4.1, this would require laminar flow over these sections more extensive than one would expect. Probably, transition is delayed until approximately the 25% chord point, the location of maximum thickness. One would then expect a $C_{d_{min}}$ value of about 0.005.

Figure 4.13 presents three-dimensional drag data directly comparable to Figure 4.11, but with more detail. Data representing practical fuselage and nacelle construction are included in Figure 4.13 together with C_D results from torpedo-shaped bodies. Assuming a reasonable relationship between the frontal and wetted areas of such bodies, expected C_D values for various values of C_f are also included on the

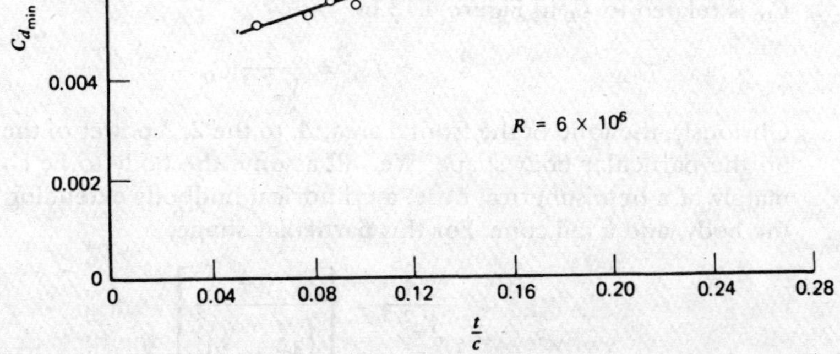

Figure 4.12 Minimum C_d for four- and five-digit airfoils of varying camber ratios as a function of thickness ratio. C_d based on chord; smooth surface.

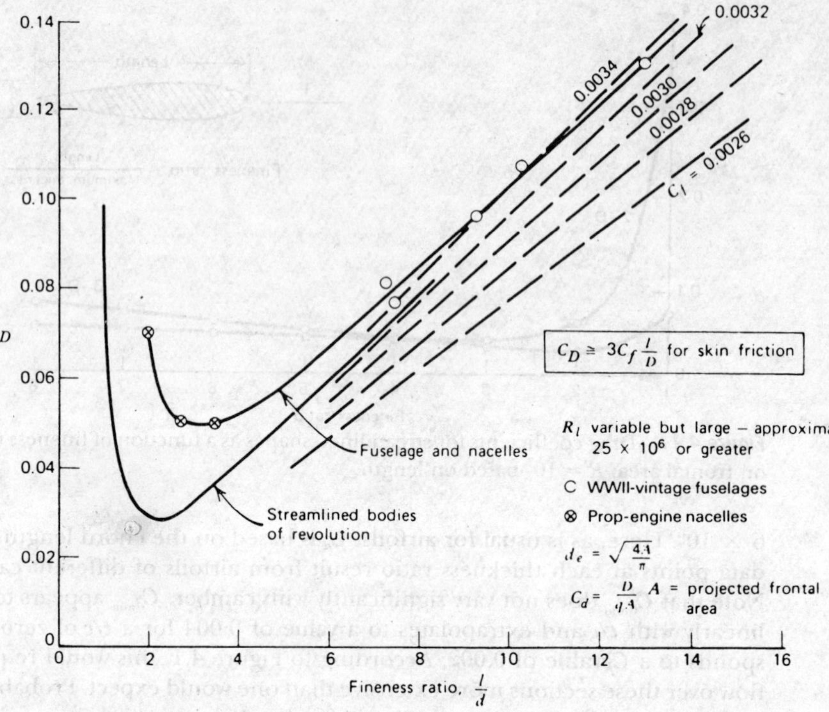

Figure 4.13 Drag of fuselages and similar shapes.

figure. For a given C_f value, the experimental results should approach one of these lines as the fineness ratio increases.

For fully turbulent flow at an R of 25×10^6, C_f for a flat plate would be 0.0026, whereas the data appears to be approaching a C_f of 0.0032 to 0.0034. The higher skin friction drag on the bodies is probably the result of surface roughness.

It is interesting to examine the data of Figure 4.13 in terms of minimum drag for a given body volume. This is particularly important for airship and underwater applications. It is also of interest to the design of tip tanks, where minimum drag for a given volume of fuel is desirable. Denoting the volume by V_m, we will define another drag coefficient.

$$C_{D_v} = \frac{D}{qV_m^{2/3}} \tag{4.12}$$

C_{D_v} is related to C_D in Figure 4.13 by

$$C_{D_v} = \frac{A}{V_m^{2/3}} C_D$$

Obviously, the ratio of the frontal area, A, to the 2/3 power of the volume depends on the particular body shape. We will assume the body to be composed approximately of a hemispherical nose, a cylindrical midbody extending to the middle of the body, and a tail cone. For this particular shape,

$$\frac{A}{V_m^{2/3}} = \left[\frac{9\pi}{\left(4\frac{l}{d} - 1\right)^2}\right]^{1/3} \tag{4.13}$$

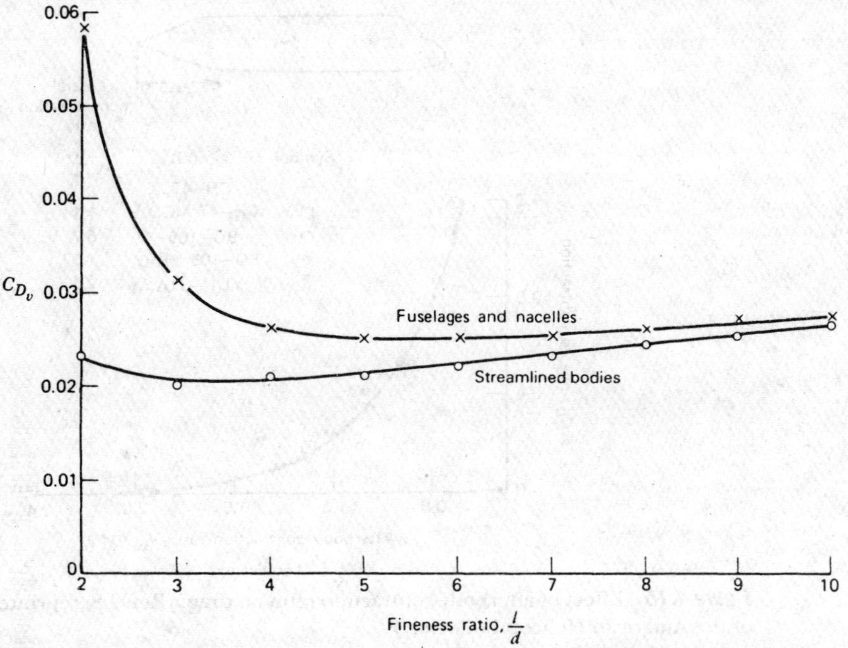

Figure 4.14 Drag coefficients based on volume for bodies as a function of fineness ratio.

Using this relationship and Figure 4.13, the graphs presented in Figure 4.14 were obtained. From this figure it can be seen that to enclose a given volume with a minimum drag body, its fineness ratio should be higher than the optimum values from Figure 4.13. Indeed, for fuselages, the drag for a given volume is nearly constant for l/d values from 4 to 10.

For certain applications, it is desirable to keep the rear portion of a fuselage as wide and bluff as possible without paying too much of a drag penalty. If the afterbody is tapered too abruptly, flow separation will occur over the rear, resulting in an unduly high form drag. Some guidance in this regard is provided by Figure 4.15 (taken from Ref. 4.8). Here, the increment in C_D (based on frontal area) resulting from afterbody contraction is presented as a function of afterbody geometry. From this figure it appears that the ratio of the afterbody length to the equivalent diameter should be no less than approximately 2.0.

The importance of streamlining is grapically illustrated in Figure 4.16, which is drawn to scale. Conservatively (supercritical flow), the ratio of C_D for a circular cylinder to a two-dimensional streamlined shape having a fineness ratio of 4 is approximately 7.5. Thus, as shown in Figure 4.16, the height of the streamlined shape can be 7.5 times greater than the circular cylinder for the same drag. For subcritical flow the comparison becomes even more impressive, with the ratio increasing to approximately 25.

INTERFERENCE DRAG

When two shapes intersect or are placed in proximity, their pressure distributions and boundary layers can interact with each other, resulting in a net drag of the combination that is higher than the sum of the separate drags. This increment in the drag is known as interference drag. Except for specific cases where data are

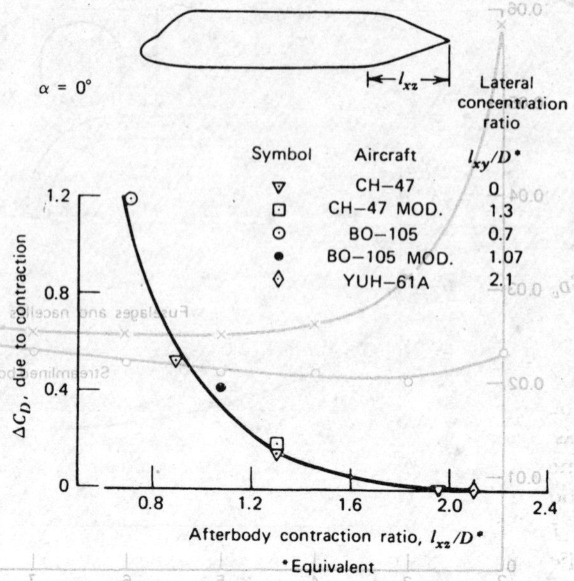

Figure 4.15 Effect of afterbody contraction ratio on drag. (Ref. 4.8, reprinted by permission of the American Helicopter Society)

available, interference drag is difficult to estimate accurately. Some examples of interference drag are presented in Figures 4.17, 4.18, and 4.19.

Figure 4.17 illustrates the drag penalty that is paid for placing an engine nacelle in proximity to a rear pylon on a tandem helicopter (like a CH-47). In this particular instance, the interference drag is nearly equal to the drag of the nacelle alone, because the nacelle is mounted very close to the pylon. For spacings greater than approximately one-half of a nacelle diameter, the interference drag vanishes.

Figure 4.18 presents the interference drag between the rotor hub and pylon for a helicopter. The trends shown in this figure are similar to those in the previous figure. In both instances the added interference drag is not necessarily on the appended member; probably, it is on the pylon.

Figure 4.19 shows a wing abutting the side of a fuselage. At the fuselage-wing juncture a drag increment results as the boundary layers from the two airplane components interact and thicken locally at the junction. This type of drag penalty will become more severe if surfaces meet at an angle other than 90°. In particular, acute angles between intersecting surfaces should be avoided. Reference 4.4, for example, shows that the interference drag of a 45% thick strut abutting a plane wall doubles as the angle decreases from 90° to approximately 60°. If acute angles cannot be avoided, filleting should be used at the juncture.

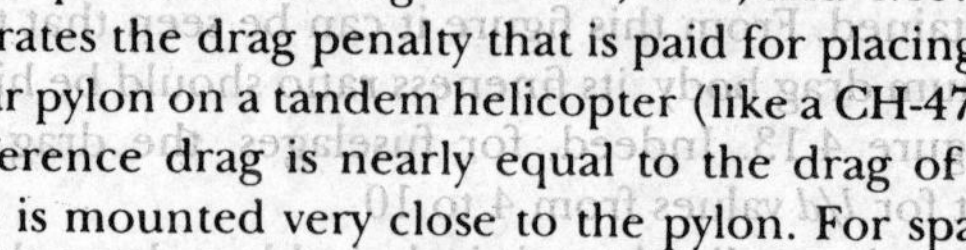

Figure 4.16 Two bodies having the same drag (supercritical flow).

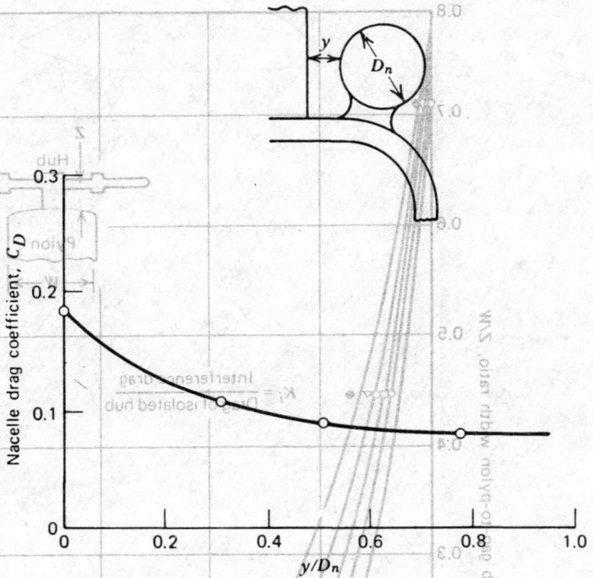

Figure 4.17 Effect of nacelle location on interference drag. (Ref. 4.8, reprinted by permission of the American Helicopter Society)

In the case of a high-wing configuration, interference drag results principally from the interaction of the fuselage boundary layer with that from the wing's lower surface. This latter layer is relatively thin at positive angles of attack. On the other hand, it is the boundary layer on the upper surface of a low wing that interferes with the fuselage boundary layer. This upper surface layer is appreciably thicker than the lower surface layer. Thus, the wing-fuselage interference drag for a low-wing configuration is usually greater than for a high-wing configuration.

The available data on wing-fuselage interference drag are sparse. Reference 4.4 presents a limited amount but, even here, there is no correlation with wing position or lift coefficient. Based on this reference, an approximate drag increment caused by wing-fuselage interference is estimated to equal 4% of the wing's profile drag for a typical aspect ratio and wing thickness.

Although data such as those in Reference 4.4 may be helpful in estimating interference drag, an accurate estimate of this quantity is nearly impossible. For example, a wing protruding from a fuselage just forward of the station where the fuselage begins to taper may trigger separation over the rear portion of the fuselage.

Sometimes interference drag can be favorable as, for example, when one body operates in the wake of another. Race car drivers frequently use this to their advantage in the practice of "drafting." Some indication of this favorable interference is provided by Figure 4.20, based on data obtained in Pennsylvania State University's subsonic wind tunnel. Here, the drag on one rectangular cylinder in tandem with another is presented as a function of the distance between the cylinders. The cylinders have a 2:1 fineness ratio. Tests were performed with the long side oriented both with and normal to the free-stream velocity. The drag is referenced with respect to D_∞, the drag on the one cylinder alone. The spacing is made dimensionless with respect to the dimension of the cylinder normal to the flow. The spacing, x, is positive when the cylinder on which the drag is being measured is downstream of the other. Notice that the cylinder's drag is reduced significantly

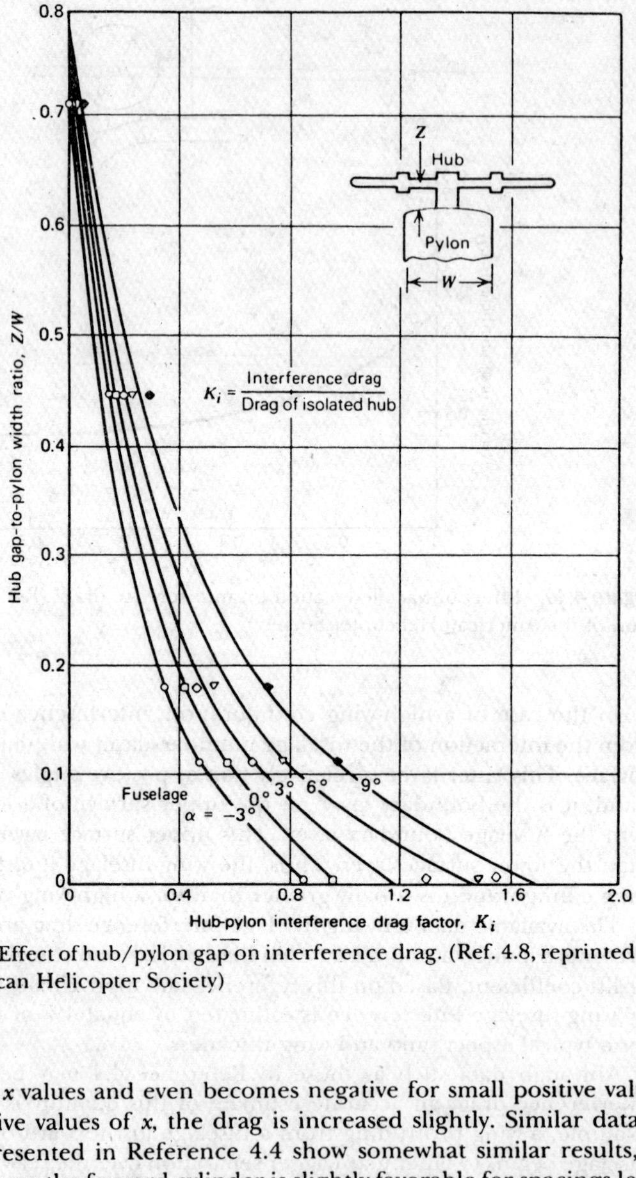

Figure 4.18 Effect of hub/pylon gap on interference drag. (Ref. 4.8, reprinted by permission of the American Helicopter Society)

for positive x values and even becomes negative for small positive values of x. For small negative values of x, the drag is increased slightly. Similar data for circular cylinders presented in Reference 4.4 show somewhat similar results, except that interference on the forward cylinder is slightly favorable for spacings less than three diameters. For the downstream cylinder, the drag is reduced by a factor of 0.3 for spacings between three and nine diameters. For less than three diameters, the downstream drag is even less and becomes negative for spacings less than approximately two diameters.

INDUCED DRAG

The two major components of the total drag of an airplane are the *induced drag* and the *parasite drag*. The parasite drag is the drag not directly associated with the production of lift. This drag, expressed as a coefficient, is nearly constant and approximately equal to the drag for an airplane lift coefficient of zero. As the lift coefficient takes on a value different than zero, the drag coefficient will increase.

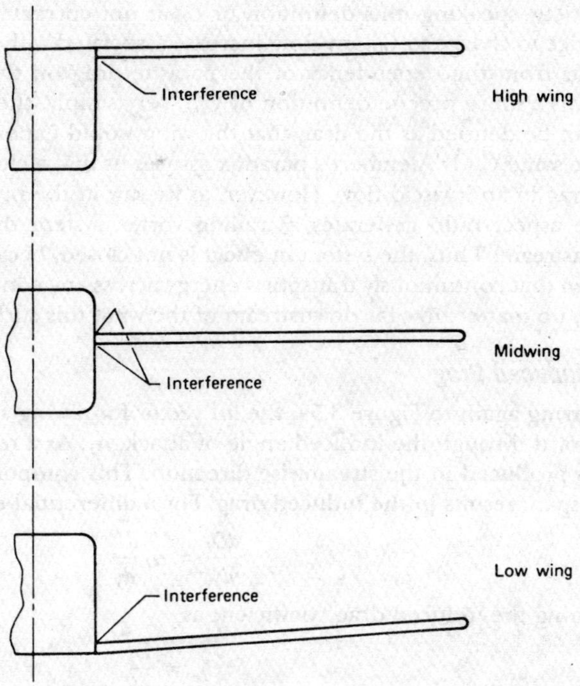

Figure 4.19 Wing-fuselage interference drag.

This increment in C_d is defined as the induced drag coefficient, C_{D_i}. Thus, for an airplane,

$$C_D = C_{D_0} + C_{D_i} \tag{4.14}$$

Here, C_{D_0} is the parasite drag coefficient and is not a function of C_L. On the other hand, the induced drag coefficient, C_{D_i}, varies approximately as the square of C_L. This dependence will be derived later.

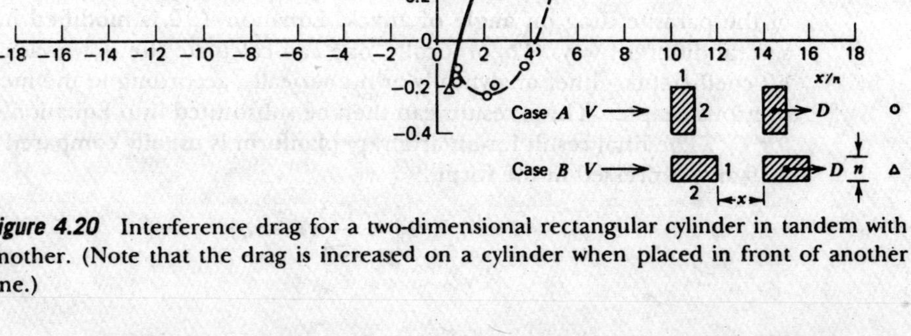

Figure 4.20 Interference drag for a two-dimensional rectangular cylinder in tandem with another. (Note that the drag is increased on a cylinder when placed in front of another one.)

Strictly speaking, this definition of C_{D_i} is not correct. Although it has become practice to charge to C_{D_i} any drag increase associated with C_L, some of this increase results from the dependency of the parasite drag on the angle of attack. What, then, is a more precise definition of C_{D_i}? Very simply, the induced drag at a given C_L can be defined as the drag that the wing would experience in an inviscid flow at the same C_L. D'Alembert's paradox assures us that a closed body can experience no drag in an inviscid flow. However, as we saw in the previous chapter, a wing of finite aspect ratio generates a trailing vortex system that extends infinitely far downstream. Thus, the system in effect is not closed, because of the trailing vortex system that continuously transports energy across any control surface enclosing the wing, no matter how far downstream of the wing this surface is chosen.

Calculation of Induced Drag

Referring again to Figure 3.54, the lift vector for a wing section is seen to be tilted rearward through the induced angle of attack, α_i. As a result, a component of the lift is produced in the streamwise direction. This component, integrated over the wingspan, results in the induced drag. For a differential element,

$$\frac{dD_i}{dy} = \alpha_i \frac{dL}{dy} \tag{4.15}$$

Defining the induced drag coefficient as

$$C_{D_i} = \frac{D_i}{qS} \tag{4.16}$$

it follows that

$$C_{D_i} = \frac{\int_{-b/2}^{b/2} \alpha_i C_l c\,dy}{S} \tag{4.17}$$

For the special case of an untwisted elliptic wing, α_i and C_l are constant over the span, so that Equation 4.17 becomes

$$C_{D_i} = \alpha_i C_L \tag{4.18}$$

The induced angle of attack for this case was given previously by Equation 3.68. Thus,

$$C_{D_i} = \frac{C_L{}^2}{\pi A} \tag{4.19}$$

This is a well-known and often-used relationship that applies fairly well to other than elliptic planforms. For a given aspect ratio and wing lift coefficient, it can be shown (Ref. 4.1) that Equation 4.19 represents the minimum achievable induced drag for a wing. In other words, the elliptic lift distribution is optimum from the viewpoint of induced drag.

To account for departures from the elliptic lift distribution and the dependence of the parasite drag on angle of attack, Equation 4.19 is modified in practice in several different ways. Theoretically, one can calculate the downwash and section lift coefficients, either analytically or numerically, according to the methods of the previous chapter. These results can then be substituted into Equation 4.17 to solve for C_{D_i}. The final result for an arbitrary planform is usually compared to Equation 4.19 and expressed in the form

$$C_{D_i} = \frac{C_L{}^2}{\pi A}(1 + \delta) \tag{4.20}$$

For a given planform shape, δ is a constant that is normally small in comparison to unity. It therefore represents, for a given wing, the fractional increase in the induced drag over the optimum elliptic case.

The numerical determination of δ will now be outlined for the simplified lifting line model and some typical results will be presented.

Numerical Model of a Lifting Line to Determine Induced Drag

In order to avoid uncertainties with the leading edge suction force, the lifting line model will be used to determine the induced drag instead of a lifting surface model. To evaluate the preceding equations numerically, a wing is replaced by the system of horseshoe vortices as was shown earlier in Figure 3.56. Referring to this figure, the velocity induced downward along the lifting line at point I by the trailing vortices from the horseshoe vortex at point J can be written as

$$w(I,J) = \frac{\Gamma(J)}{4\pi} \frac{\Delta y}{[\Delta y/2]^2 - [y(J) - y(I)]^2} \tag{4.21}$$

or

$$w(I) = W(I,J)\,\Gamma(J) \tag{4.22}$$

Now consider again Equations 3.60 and 3.61. If these two equations are solved for the section lift coefficient and equated, the following result is obtained:

$$C_l = C_{l_\alpha}(\alpha - \alpha_i) = \frac{2\Gamma}{cV} \tag{4.23}$$

Substituting Equation 4.22 into the above and setting V equal to 1 leads to

$$\frac{2\Gamma(I)}{c(I)} = C_{l_\alpha}\left[\alpha(I) - \sum_{J=1}^{N} W(I,J)\,\Gamma(J)\right] \tag{4.24}$$

Writing the above for $I = 1, 2, 3, \ldots, N$ leads to a set of N simultaneous equations of the form

$$A(I,J)\Gamma(I) = B(I) \tag{4.25}$$

Observe that $\Gamma(I)$ is on both sides of Equation 4.24 since it is contained within the summation. Thus,

$$A(I,J) = \frac{C_{l_\alpha} c(I)\, W(I,J)}{2} \quad I \neq J \tag{4.26}$$

and

$$A(I,I) = 1 + \frac{C_{l_\alpha} c(I)\, W(I,I)}{2} \tag{4.27}$$

The right-hand side of Equation 4.25 becomes

$$B(I) = \frac{C_{l_\alpha} c(I)\, \alpha(I)}{2} \tag{4.28}$$

Given the wing geometry including angle of attack, twist, and planform shape, this set of equations can be solved for $\Gamma(I)$. The section lift coefficients can then be obtained immediately from Equation 4.23. The downwash induced at each point is then obtained by summing over J at each I the influence coefficients, $W(I,J)$, multiplied by $\Gamma(I)$. The section induced drag coefficients are then determined simply as

$$C_{d_i}(I) = C_l(I)\, w(I) \tag{4.29}$$

Finally, the induced drag coefficient for the wing is found from

$$C_{D_i} = \frac{\sum_{I=1}^{N} C_{d_i}(I)\, c(I)\, \Delta y}{S} \tag{4.30}$$

The numerical model of the lifting line has been applied to a family of flat, linearly tapered, unswept wings and also to the elliptic planform. The planform for the family of linearly tapered wings is defined by

$$c = c_0\,[1 - (1 - \lambda)|(2y/b)|] \tag{4.31}$$

The calculated results are presented in Figure 4.21 where it can be observed that the induced drag for a linearly tapered wing with a taper ratio of 0.35 is less than 2% greater than the value given by Equation 4.19 for aspect ratios of 10 or less. However, it should be noted, as shown on the figure that the numerical model results in a prediction of C_{D_i} for the elliptic planform which is 0.4% less than Equation 4.19.

The rectangular wing is represented in Figure 4.21 by a taper ratio of 1.0. For this planform, used on many light, single-engine aircraft, the induced drag is seen to be 4% or higher than that for the elliptic wing for aspect ratios of 6 or higher.

The results of Figure 4.21 can be explained by reference to Figure 4.22, which presents spanwise distributions of Γ for the elliptic, rectangular, and 0.35 taper ratio wings. Observe that the distribution for $\lambda = 0.35$ is close to the elliptic distribution. The kinetic energies of the trailing vortex systems shed from these two Γ distributions are about the same. On the other hand, the Γ distribution for the rectangular planform is nearly constant inboard out to about 70% of the semispan and then drops off more rapidly than the elliptic distribution toward the tip. Thus, the kinetic energy per unit length of the trailing vortex system shed from the rectangular wing is approximately 6% higher than the energy left in the wake by the tapered or elliptic wing.

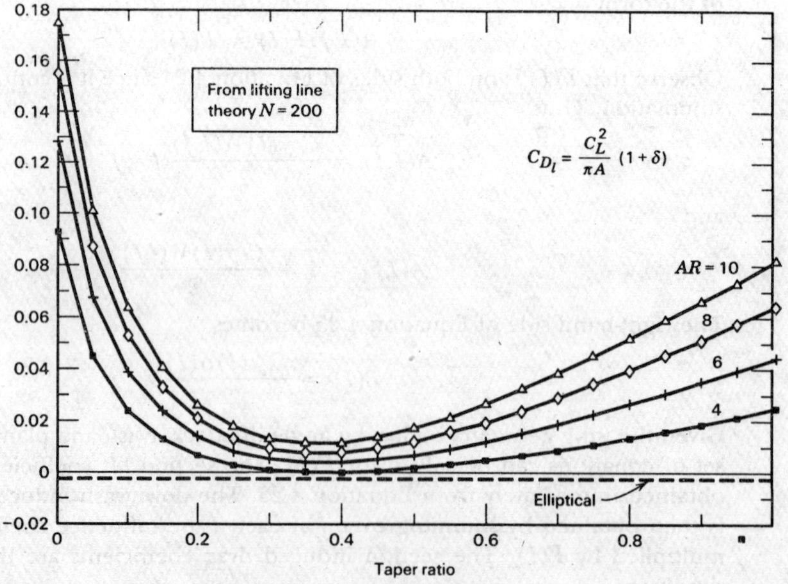

Figure 4.21 Numerical calculation of induced drag factor for linearly tapered and elliptic wings.

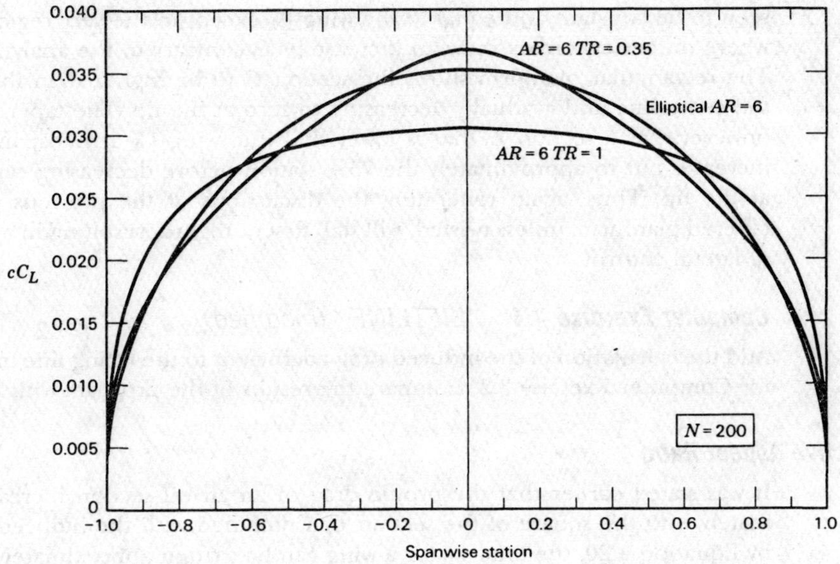

Figure 4.22 Comparison of elliptic lift distribution with those of rectangular wings.

In view of the preceding, one might ask why rectangular planforms are used in many general aviation airplanes instead of tapered planforms. Part of the answer lies with the relative cost of manufacture. Obviously, the rectangular planform with an untapered spar and constant rib sections is less costly to fabricate. Figure 4.23 discloses a second advantage to the rectangular planform. Here, the section lift coefficient is presented as a ratio to the wing lift coefficient for untwisted elliptic, rectangular, and linearly tapered planforms. For the elliptic wing, the section C_l is

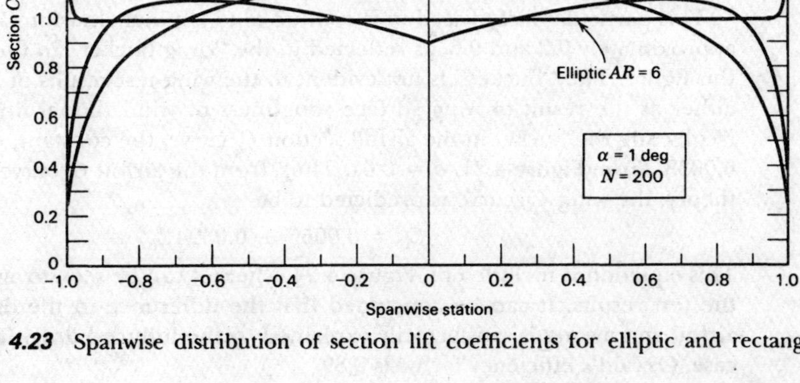

Figure 4.23 Spanwise distribution of section lift coefficients for elliptic and rectangular wings.

seen to be constant and equal to the wing C_L except in the very region of the tip, where numerical errors show an increase in C_l contrary to the analytical solution. The rectangular planform shows the section C_l to be higher than the wing C_L at the centerline and gradually decreasing to zero at the tip. The tapered planform, however, has a section C_l that is lower than the wing C_L at midspan. Its C_l then increases out to approximately the 75% station before decreasing rapidly to zero at the tip. Thus, again reiterating the discussions of the previous chapter, the tapered planform, unless twisted, will stall first outboard, resulting in a possible loss of lateral control.

Computer Exercise 4.1 *"LIFTLINE" (modified)*

Add the calculation of the induced drag coefficient to the lifting line model written for Computer Exercise 3.2. Compare the results of the program with Figure 4.21.

Effective Aspect Ratio

It was stated earlier that the profile drag of an airfoil section increases approximately with the square of the section C_l. Combined with the induced drag, given by Equation 4.20, the total C_D for a wing can be written approximately as

$$C_D = C_{D_{\min}} + kC_L^2 + \frac{C_L^2}{\pi A}(1 + \delta)$$

or

$$C_D = C_{D_{\min}} + \frac{C_L^2}{\pi A}(k\pi A + 1 + \delta) \tag{4.32}$$

where k is the constant of proportionality giving the rate of increase of C_d with C_l^2. Equation 4.32 can be rewritten as

$$C_D = C_{D_{\min}} + \frac{C_L^2}{\pi A e} \tag{4.33}$$

where

$$e = \frac{1}{1 + \delta + k\pi A}$$

The factor e is known as *Oswald's efficiency factor* (see Ref. 4.2). The product Ae is referred to as the "effective aspect ratio" and is sometimes written as A_e.

Consider data from References 3.1 and 3.27 in light of Equation 4.32. Figure 4.24 presents C_D as a function of C_L for the finite wing tested in Reference 3.27 and C_d versus C_l from Reference 3.1 for the 65-210 airfoil.

This particular airfoil section is conducive to laminar flow for C_l values between approximately 0.2 and 0.6, as reflected in the "drag bucket" in the lower curve of this figure. The "bucket" is not evident in the wing test results of Reference 3.27, either as the result of wing surface roughness or wind tunnel flow disturbances. Neglecting the bucket in the airfoil section C_d curve, the constant, k, is found to be 0.0038. From Figure 4.21, δ = 0.01. Thus, from the airfoil C_d curve and lifting line theory, the wing C_D curve is predicted to be

$$C_D = 0.0055 + 0.0394C_L^2 \tag{4.34}$$

This equation is included on Figure 4.24, where it can be seen to agree closely with the test results. It can be concluded that the difference in the drag between an airfoil and a wing is satisfactorily explained by the induced drag. In this particular case, Oswald's efficiency factor is 0.89.

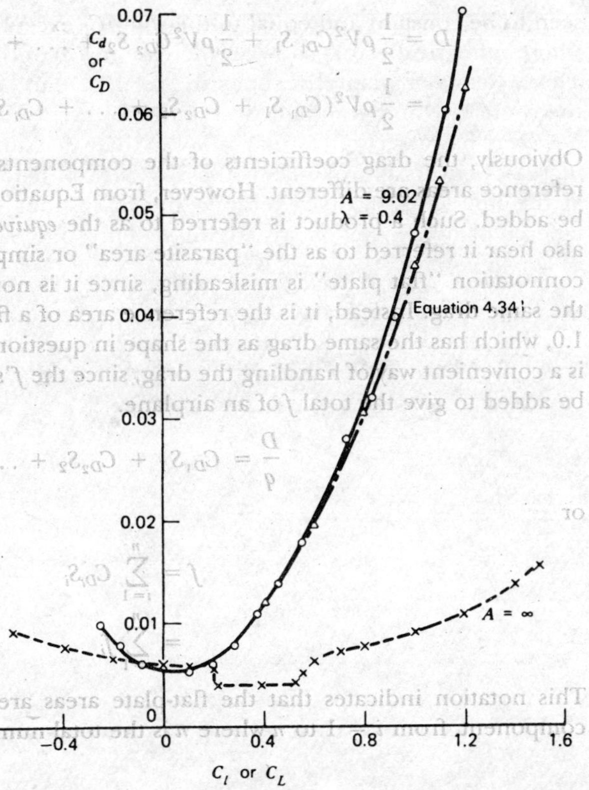

Figure 4.24 Comparison between predicted and measured drag polar for a wing having a finite aspect ratio.

Generally, for a complete airplane configuration, e is not this high because of wing-fuselage interference and contributions from the tail and other components. High-wing and low-wing airplanes show a measurable difference in Oswald's efficiency factor. Most likely as the result of interference between the boundary layer on the wing's upper surface with that on the fuselage, e values for low-wing airplanes are lower than those for high-wing airplanes. The boundary layer on the upper surface of a wing is considerably thicker than the one on the lower surface. Combining with the boundary layer over the sides of the fuselage, the wing's upper surface boundary layer, for the low-wing airplane, can cause a rapid increase in the wing and fuselage parasite drag as the angle of attack increases. For a high-wing airplane, the relatively thin boundary layer on the lower surface of the wing interferes only slightly with the fuselage boundary layer. Typically, e is equal approximately to 0.6 for low-wing airplanes and 0.8 for high-wing airplanes. These values are confirmed by the flight tests reported in Reference 4.3 and in other data from isolated sources.

DRAG BREAKDOWN AND EQUIVALENT FLAT-PLATE AREA

The parasite drag of an airplane can be estimated by estimating the drag of each component and then totaling the component drag while accounting for some interference drag. If C_{D_i} and S_i are the drag coefficient and reference area, respectively, for the ith component, then the total drag will be

$$D = \frac{1}{2}\rho V^2 C_{D_1} S_1 + \frac{1}{2}\rho V^2 C_{D_2} S_2 + \ldots + \frac{1}{2}\rho V^2 C_{D_i} S_i + \ldots$$
$$= \frac{1}{2}\rho V^2 (C_{D_1} S_1 + C_{D_2} S_2 + \ldots + C_{D_i} S_i + \ldots) \quad (4.35)$$

Obviously, the drag coefficients of the components cannot be added since the reference areas are different. However, from Equation 4.35, the products $C_{D_i}S_i$ can be added. Such a product is referred to as the *equivalent flat-plate area, f*. One will also hear it referred to as the "parasite area" or simply, the "flat-plate area." The connotation "flat plate" is misleading, since it is not the area of a flat plate with the same drag. Instead, it is the reference area of a fictitious shape having a C_D of 1.0, which has the same drag as the shape in question. f is therefore simply D/q. It is a convenient way of handling the drag, since the f's of the drag components can be added to give the total f of an airplane.

$$\frac{D}{q} = C_{D_1}S_1 + C_{D_2}S_2 + \ldots$$

or

$$f = \sum_{i=1}^{n} C_{D_i}S_i$$
$$= \sum_{i=1}^{n} f_i$$

This notation indicates that the flat-plate areas are to be summed for the ith component, from $i = 1$ to n where n is the total number of components.

DRAG COUNTS

As a measure of an airplane's drag, in practice one will frequently hear the term "drag count" used. Usually, it is used in an incremental or decremental sense, such as "fairing the landing gear reduced the drag by 20 counts." One drag count is defined simply as a change in the total airplane C_D, based on the wing planform area of 0.0001. Hence, a reduction in drag of 20 counts could mean a reduction in the C_D from, say, 0.0065 down to 0.0045.

AVERAGE SKIN FRICTION COEFFICIENTS

In examples to follow, one will see that several uncertainties arise in attempting to estimate the absolute parasite drag coefficient (as opposed to incremental effects) of an airplane. These generally involve questions of interference drag and surface irregularities. In view of these difficulties, it is sometimes better to estimate the total drag of a new airplane on the basis of the known drag of existing airplanes having a similar appearance; that is, the same degree of streamlining and surface finish.

The most rational basis for such a comparison is the total wetted area and not the wing area, since C_F depends only on the degree of streamlining and surface finish, whereas C_D depends on the size of the wing in relation to the rest of the airplane. In terms of an average C_F, the parasite drag at zero C_L for the total airplane can be written as

$$D = qC_F S_w \quad (4.36)$$

where S_w is the total wetted area of the airplane. Since

$$f = \frac{D}{q},$$

it follows that the ratio of the equivalent flat-plate area to the wetted area is

$$\frac{f}{S_w} = C_F \quad (4.37)$$

To provide a basis for estimating C_F, Table 4.1 presents a tabulation of this quantity for 23 different airplanes having widely varying configurations. These range all the way from Piper's popular light plane, the Cherokee, to Lockheed's jumbo jet, the C-5A.

The data in this table were obtained from several sources and include results obtained by students taking a course in techniques of flight testing. Thus, the absolute value of C_F for a given airplane may be in error by a few percent. For purposes of preliminary design, the C_F ranges given in Table 4.2 are suggested for various types of airplanes. Where a particular airplane falls in the range of C_F values for its type will depend on the attention given to surface finish, sealing (around cabin doors, wheel wells, etc.), external protuberances, and other drag-producing items.

Table 4.1 Typical Overall Skin Friction Coefficients for a Number of Airplanes Built from Approximately 1940 to 1976.

C_F	Airplane designation	Description
0.0100	Cessna 150	Single prop, high wing, fixed gear
0.0095	PA-28	Single prop, low wing, fixed gear
0.0070	B-17	Four props, World War II bomber
0.0067	PA-28R	Single prop, low wing, retractable gear
0.0066	C-47	Twin props, low wing, retractable gear
0.0060	P-40	Single prop, World War II fighter
0.0060	F-4C	Jet fighter, engines internal
0.0059	B-29	Four props, World War II bomber
0.0054	P-38	Twin props, twin-tail booms, World War II fighter
0.0050	Cessna 310	Twin props, low wing, rectractable gear
0.0049	Beech V35	Single prop, low wing, retractable gear
0.0046	C-46	Twin props, low wing, retractable gear
0.0046	C-54	Four props, low wing, retractable gear
0.0042	Learjet 25	Twin jets, pod-mounted on fuselage, tip tanks
0.0044	CV 880	Four jets, pod-mounted under wing
0.0041	NT-33A	Training version of P-80 (see below)
0.0038	P-51F	Single prop, World War II fighter
0.0038	C-5A	Four jets, pod-mounted under wing, jumbo jet
0.0037	Jetstar	Four jets, pod-mounted on fuselage
0.0036	747	Four jets, pod-mounted under wing, jumbo jet
0.0033	P-80	Jet fighter, engines internal, tip tanks, low wing
0.0032	F-104	Jet fighter, engines internal, midwing
0.0031	A-7A	Jet fighter, engines internal, high wing

Table 4.2 Typical Total Skin Friction Coefficient Values for Different Airplane Configurations

Airplane configuration	C_F Range at low Mach numbers
Propeller driven, fixed gear	0.008–0.010
Propeller driven, retractable gear	0.0045–0.007
Jet propelled, engines pod-mounted	0.0035–0.0045
Jet propelled, engines internal	0.0030–0.0035

Additional drag data on a number of airplanes, including supersonic airplanes, are presented in Appendix A.3 as a function of Mach number and altitude.

Finally, with regard to average C_F values, Figure 4.25 (taken from Ref. 4.11) is presented. Although only a few individual points are identified on this figure, its results agree generally with Table 4.2. This figure graphically depicts the dramatic improvement in aerodynamic cleanliness of airplanes that has been accomplished since the first flight of the Wright Brothers.

EXAMPLE ESTIMATES OF DRAG BREAKDOWN

The use of Equation 4.35 is illustrated in Table 4.3, where the author has performed an estimate of the drag breakdown for the Cherokee (Figure 3.59). Armed with a tape measure, a visual inspection of the airplane was made, noting the dimensions of all drag-producing appendages. What you see here is a first estimate without any iteration. The total value for f of 0.36 m^2 (3.9 ft^2) is obviously too low and should be approximately 50% higher. This airplane has a total wetted area of approximately 58.06 m^2 (625 ft^2).

Undoubtedly, an aerodynamicist working continuously with this type of aircraft

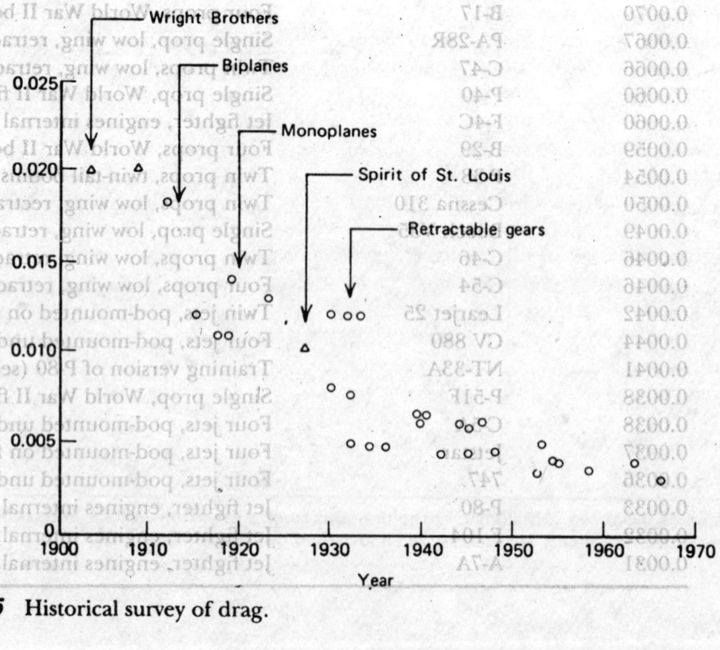

Figure 4.25 Historical survey of drag.

Table 4.3 Drag Breakdown for Piper Cherokee 180

Item	Description	Reference area	C_d	Basis for C_d	$f = C_dA$
Wing	See Figure 3.62	160 (plan)	0.0093	Figure 4.12 +50% for roughness	1.49
Fuselage	See Figure 3.62	15.2 (front)	0.058	Figure 4.13 $l/d_e = 5$	0.88
Horizontal tail	See Figure 3.62	25 (plan)	0.0084	Figure 4.12 +50% for roughness	0.21
Vertical tail	See Figure 3.62	11.5 (plan)	0.0084	Figure 4.12 +50% for roughness	0.10
Wheel struts	20 in. long 1.5 in. D half faired	0.63 (front)	0.3	Figure 4.6 supercritical	0.19
Wheel pants	12 in. high 7 in. wide streamlined	1.75 (front)	0.04	Figure 4.11	0.07
Wheels	6 in. below pants 5 in. wide	0.63 (front)	0.70	Figure 4.10 corrected to three-dimensional	0.44
Pitot-static tube	$\frac{1}{2}$ in. × 5 in. blunt/ rounded edge	0.02 (front)	1.0	Figure 4.7	0.02
Flap control horns	3 in. × $\frac{3}{4}$ in. blunt (6 total)	0.09 (front)	1.0	Figure 4.7*a*	0.09
Gas drain cocks	1 in. × $\frac{1}{2}$ in. blunt (2 total)	0.01 (front)	1.0	Figure 4.7	0.01
Rotating beacon	4 in. D × 5 in. semispherical	0.14 (front)	0.15	Figure 4.6	0.02
Tail tie-down	$3\frac{1}{2}$ in. × $\frac{3}{8}$ in. blunt	0.01 (front)	1.0	Figure 4.7	0.01
Wing tie-downs	1 in. × $\frac{3}{8}$ in. blunt (2)	0.01 (front)	1.0	Figure 4.7	0.01
Five whip antennas	$\frac{1}{4}$ in. D × 20 in. each	0.17 (front)	1.0	Figure 4.7 subcritical	0.17
Two pipes from engine	3 in. D × 2 in. each	0.08 (front)	0.2	Figure 4.6 and 4.7*b*	0.02
step	12 in. long, $\frac{3}{4}$ in. thick, 2 in. chord	0.06 (front)	0.06	Figure 4.11 +50% for roughness	0.04
OAT gage	3 in. × $\frac{3}{4}$ in. D blunt	0.02	1.0	Figure 4.7	0.02
Antenna fairing	3 in. deep × 6 in. width streamlined	0.13	0.06	Figure 4.11 +50% for roughness	0.01
Antenna supports	5 in. × $\frac{1}{2}$ in. D (two)	0.03	1.0	Figure 4.7	0.03
Interference					
Fuse vertical tail		0.11 (t^2)	0.05	Ref. 4.4	
Fuse horizontal tail		0.06 (t^2)	0.05		0.07
Fuse wing		0.6^2 (t^2)	0.1		
				Total	3.9 ft²
Leakage?*					
Cooling?*					

* In addition to the above, an estimate of the drag increments caused by leakage and cooling should be made from experiment, these drag items will increase the above total by approximately 20%.

would be able to make a drag breakdown more accurate than the one shown in Table 4.3. Based on one's experience with his or her company's aircraft, the aerodynamicist can make more certain allowances for surface roughness, interferences, and leakage. For example, in the case of the Cherokee's landing gear, the oleo struts have a bar linkage immediately behind them. The linkage, struts, wheel pants, and brake fittings produce a total drag for the entire landing gear that is probably significantly higher than the total f of 0.065 m^2 (0.7 ft^2) estimated for the wheels, wheel pants, and struts. The cylindrical oleo struts, in particular, being close to the wheel pants probably produce separation over the pants so that C_d for this item could be more like 0.4 or even higher instead of 0.04, as listed in Table 4.3. This would add another 0.06 m^2 (0.63 ft^2) to f.

Another example of a drag breakdown is provided by Reference 4.10. In this case, the airplane is the Gates Learjet Model 25 pictured in Figure 4.26. Table 4.4

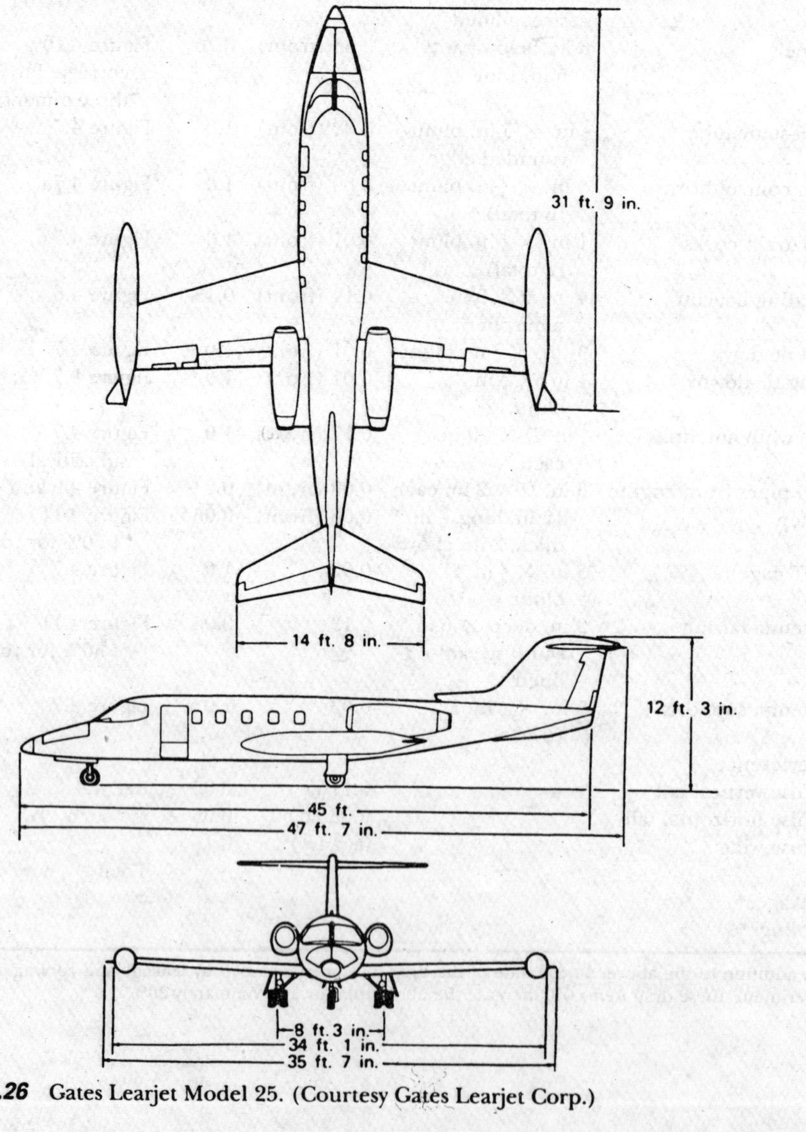

Figure 4.26 Gates Learjet Model 25. (Courtesy Gates Learjet Corp.)

Table 4.4 Parasite Drag Breakdown for Gates Learjet Model 25 (from Reference 4.10)

Item	C_d (based on wing) planform area)	Percent of total
Wing	0.0053	23.45
Fuselage	0.0063	27.88
Tip tanks	0.0021	9.29
Tip tank fins	0.0001	0.44
Nacelles	0.0012	5.31
Pylons	0.0003	1.33
Horizontal tail	0.0016	7.08
Vertical tail	0.0011	4.86
Interference	0.0031	13.72
Roughness and gap	0.0015	6.64
Total	0.0226	100.00

was prepared on the basis of Reference 4.10. The authors of the reference chose to base C_d for each item on the wing area. This is therefore the case for Table 4.4, since dimensions and areas for each item were not available. Also, the reference did not include interference or roughness and gap drag in the parasite drag. Why this was done is not clear, and these two items are included in Table 4.4. These two somewhat elusive drag items are estimated to account for 20% of the parasite drag. Although not related to its parasite drag, according to Reference 4.10, this airplane has an Oswald's efficiency factor of 0.66.

TRIM DRAG

Basically, *trim drag* is not any different from the types of drag already discussed. It arises mainly as the result of having to produce a horizontal tail load in order to balance the airplane around its pitching axis. Any drag increment that can be attributed to a finite lift on the horizontal tail contributes to the trim drag. Such increments mainly represent changes in the induced drag of the tail. To examine this further, we again write that the sum of the lifts developed by the wing and tail must equal the aircraft's weight.

$$L + L_t = W$$

Solving for the wing lift and dividing by qS leads to

$$C_{L_w} = C_L - C_{L_t}\frac{S_t}{S}$$

Here, C_{L_w} is the wing lift coefficient, C_L is the lift coefficient based on the weight and wing area, and C_{L_t} is the horizontal tail lift coefficient. Thus, the C_{D_i} of the wing, accounting for the tail lift, becomes

$$\begin{aligned} C_{D_{i_w}} &= \frac{C_{L_w}^2}{\pi A e_w} \\ &\simeq \frac{C_L^2}{\pi A e} - 2\frac{C_L^2}{\pi A e}\frac{C_{L_t}}{C_L}\frac{S_t}{S} \end{aligned} \tag{4.38}$$

The term $[C_{L_t}(S_t/S)]^2$ has been dropped as being of higher order. Since $C_L^2/\pi Ae$ is the term normally defined as C_{D_i}, it follows from Equation 4.38 that the increment in the induced drag coefficient contributed by the wing because of trim is

$$\Delta C_{D_i} = -2C_{D_i}\frac{C_{L_t}}{C_L}\frac{S_t}{S} \tag{4.39}$$

The tail itself will have an induced drag given by

$$D_{i_t} = qS_t\frac{C_{L_t}^2}{\pi A_t e_t} \tag{4.40}$$

Expressed in terms of the wing area, this becomes

$$C_{D_{i_t}} = \frac{S_t}{S}\frac{C_L{}^2}{\pi Ae}\left(\frac{C_{L_t}}{C_L}\right)^2\frac{Ae}{A_t e_t} \tag{4.41}$$

Added to Equation 4.39 the total increment in the induced drag coefficient becomes

$$\Delta C_{D_{\text{trim}}} = \frac{S_t}{S}\frac{C_{L_t}}{C_L}\left(\frac{C_{L_t}Ae}{C_L A_t e_t} - 2\right)C_{D_i} \tag{4.42}$$

To gain further insight into the trim drag, consider the simplified configuration shown in Figure 4.27. For the airplane to be in equilibrium, it follows that

$$L_w + L_t = W$$

$$xL_w = (l - x)L_t$$

where l is the distance from the aerodynamic center of the wing to the aerodynamic center of the tail. x is the distance of the center of gravity aft of the wing's aerodynamic center. Solving for l_t gives

$$L_t = \frac{x}{l}W$$

Thus,

$$\frac{S_t C_{L_t}}{SC_L} = \frac{x}{l} \tag{4.43}$$

Substituting Equation 4.43 into Equation 4.42 lead to

$$\frac{\Delta C_{D_{\text{trim}}}}{C_{D_i}} = \frac{x}{l}\left[\frac{x}{l}\left(\frac{b}{b_t}\right)^2\frac{e}{e_t} - 2\right] \tag{4.44}$$

The ratio of the wingspan to the tailspan is of the order of 3, while e is equal approximately to e_t. With these magnitudes in mind, Figure 4.28 was prepared; it presents the trim drag as a fraction of the original induced drag as influenced by x/l.

Notice that the possibility of a negative trim drag exists for small, positive, center-of-gravity positions. This results from the slight reduction in wing lift and hence, its induced drag for aft center-of-gravity positions. However, as the center of gravity moves farther aft, the induced drag from the tail overrides the saving from the wing so that the net trim drag becomes more positive.

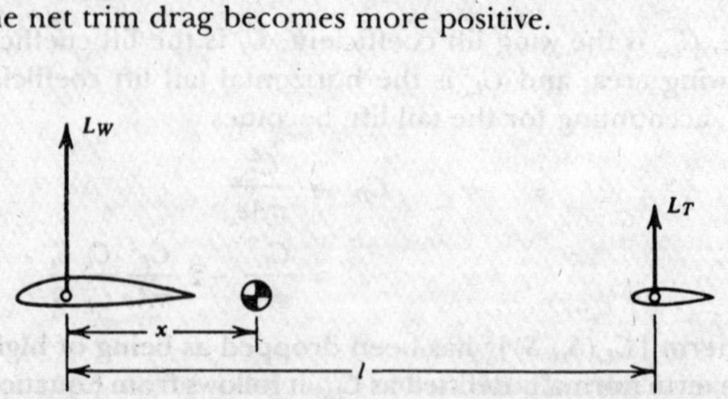

Figure 4.27 Equilibrium of wing and tail lift.

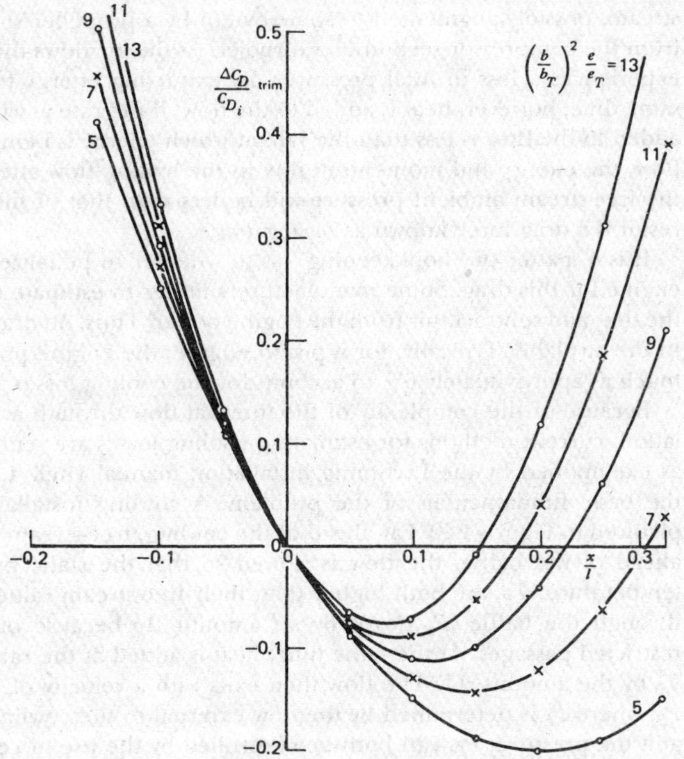

Figure 4.28 Effect of center-of-gravity position on trim drag.

The aerodynamic moment about the airplane's aerodynamic center was neglected in this analysis. By comparison to the moment contributed by the tail, M_{ac} should be small. Qualitatively, the results of Figure 4.28 should be relatively unaffected by the inclusion of M_{ac}.

The trim drag is usually small, amounting to only 1 or 2% of the total drag of an airplane for the cruise condition. Reference 4.10, for example, lists the trim drag for the Learjet Model 25 as being only 1.5% of the total drag for the cruise condition. As another example, consider the Cherokee once again at an indicated airspeed of 135 mph. At its gross weight of 2400 lb, this corresponds to a C_L of 0.322. For this weight, the most forward center of gravity allowed by the flight manual is 3% of the chord ahead of the quarter-chord point. With a chord of 63 in., and the distance between the wing and tail aerodynamic center of approximately 13 ft, x/l has a value of -0.012. Since $b \simeq 3b_t$, Figure 4.28 gives

$$\frac{\Delta C_{\text{trim}}}{C_{D_i}} = 0.028$$

With an effective aspect ratio of approximately 3.38, C_{D_i} will be equal to 0.0098, so that $C_{D_{\text{trim}}} = 0.0003$. The parasite drag coefficient is approximately 0.0037, so that the total C_D equals 0.0138. Thus, for the Cherokee in cruise, the trim drag amounts to only 2% of the total drag.

COOLING DRAG

Cylinder heads, oil coolers, and other heat exchangers require a flow of air through them for purposes of cooling. Usually, the source of this cooling air is the free stream, possibly augmented to some extent by a propeller slipstream or bleed air from the compressor section of a turbojet. As the air flows through the baffling, it experiences a loss in total pressure, Δp, extracting energy from the flow. At the same time, however, heat is added to the flow. If the rate at which the heat is being added to the flow is less than the rate at which energy is being extracted from the flow, the energy and momentum flux in the exiting flow after it has expanded to the free-stream ambient pressure will be less than that of the entering flow. The result is a drag force known as *cooling drag.*

It is a matter of "bookkeeping" as to whether to penalize the airframe or the engine for this drag. Some manufacturers prefer to estimate the net power lost to the flow and subtract this from the engine power. Thus, no drag increment is added to the airplane. Typically, for a piston engine, the engine power is reduced by as much as approximately 6% to account for the cooling losses.

Because of the complexity of the internal flow through a typical engine installation, current methods for estimating cooling losses are semiempirical in nature, as exemplified by the Lycoming installation manual (Ref. 4.12). Let us examine the basic fundamentals of the problem. A cowling installation is schematically pictured in Figure 4.29. Far ahead of the cowling, free-stream conditions exist. Just ahead of the baffle, the flow is slowed so that the static pressure, P_B, and the temperature, T_B, are both higher than their free-stream values. As the flow passes through the baffle, P_B drops by an amount Δp because of the friction in the restricted passages. At the same time, heat is added at the rate Q, which increases T_B by the amount ΔT. The flow then exits with a velocity of V_E and a pressure of P_E, where P_E is determined by the flow external to the cowling. The exit area, A_E, and the pressure, P_E, can both be controlled by the use of cowl flaps, as pictured in Figure 4.30. As the cowl flaps are opened, the amount of cooling flow increases rapidly because of the decreased pressure and the increased area. Downstream of the exit, the flow continues to accelerate (or decelerate) until the free-stream static pressure is reached. Corresponding to this state, the cooling air attains an ultimate velocity denoted by V_∞.

If m represents the mass flow rate through the system, the cooling drag, D_c, will be given by the momentum theorem as

$$D_c = m(V_0 - V_\infty) \tag{4.45}$$

The rate, ΔW, at which work is extracted from the flow is

$$\Delta W = \tfrac{1}{2}m(V_0^2 - V_\infty^2) \tag{4.46}$$

Observe that the cooling drag is obtained by dividing the increment in the energy by a velocity that is the average between V_0 and V_∞; that is,

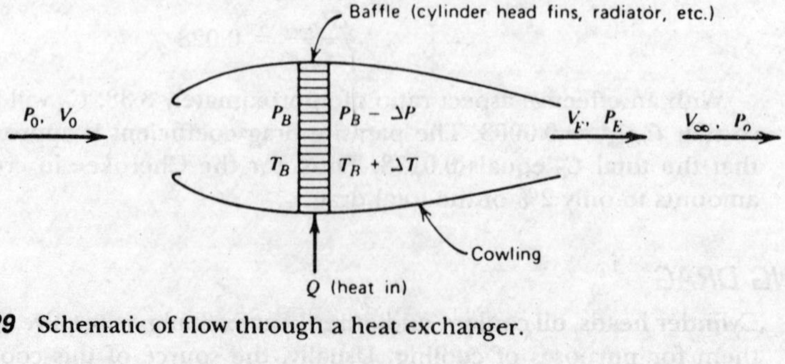

Figure 4.29 Schematic of flow through a heat exchanger.

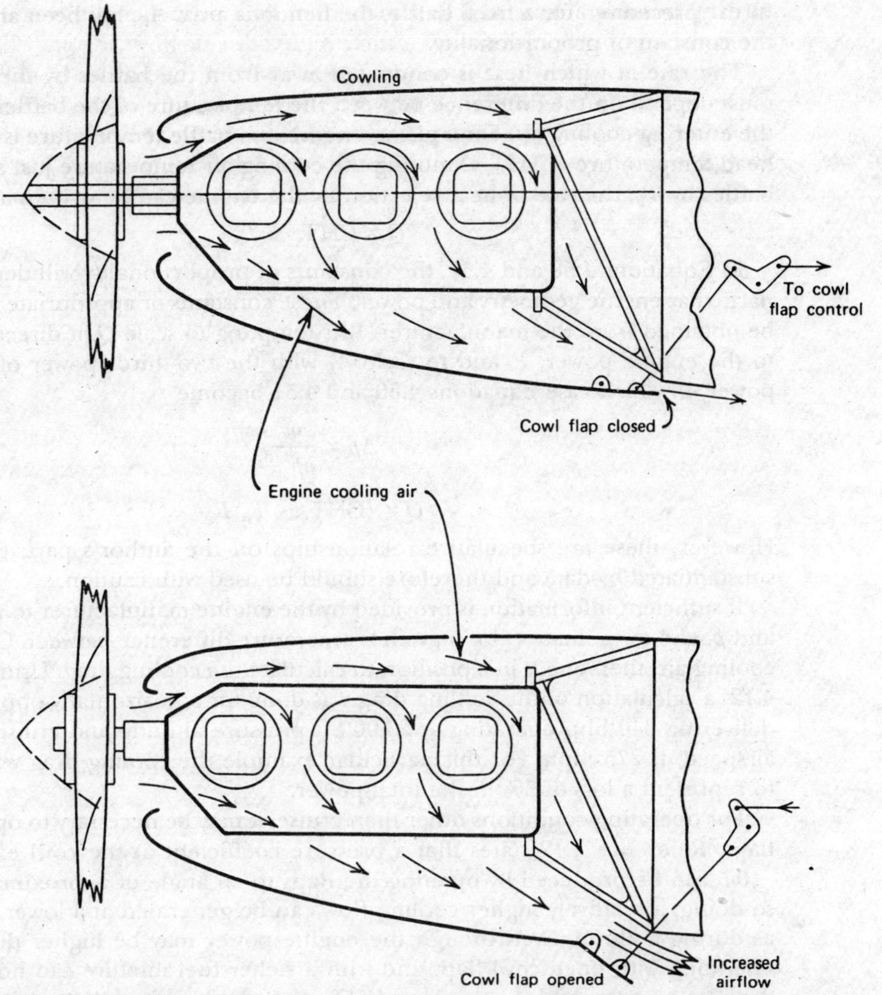

Figure 4.30 The use of cowl flaps to control engine cooling air. (*a*) Flow of cooling air around an air-cooled engine. (*b*) Typical cowl flaps on horizontally opposed engine.

$$\Delta W = D_c\left(\frac{V_0 + V_\infty}{2}\right) \tag{4.47}$$

For a fixed-baffle geometry, the pressure drop through the baffle will be proportional to the dynamic pressure ahead of the baffle.

$$\Delta p \propto \tfrac{1}{2}\rho\, V_B^2 \tag{4.48}$$

V_B can be related to the mass flow, m, through the baffle and an average flow area through the baffle.

$$m = \rho A_B V_B \tag{4.49}$$

Substitution of Equation 4.49 into Equation 4.48 results in

$$\Delta p \propto \frac{m^2}{\rho} \tag{4.50}$$

In the preceding, for a fixed baffle, the fictitious area, A_B, has been absorbed into the constant of proportionality.

The rate at which heat is conducted away from the baffles by the cooling air must depend on the difference between the temperature of the baffles and that of the entering cooling air. For a piston engine, the baffle temperature is the cylinder head temperature (CHT). Denoting the cooling air temperature just ahead of the baffles by T_B, the rate of heat rejection by the engine can be written as

$$Q \propto \text{CHT} - T_B \tag{4.51}$$

In Equations 4.50 and 4.51, the constants of proportionality will depend on the particular engine geometry and power. These constants or appropriate graphs must be obtained from the manufacturer. It is tempting to scale Q in direct proportion to the engine power, P, and to scale A_B with the two-thirds power of the engine power, in which case Equations 4.50 and 4.51 become

$$\Delta p \propto \frac{m^2}{\rho p^{4/3}} \tag{4.52}$$

$$Q \propto (\text{CHT} - T_B)P \tag{4.53}$$

However, these are speculative relationships on the author's part; they are not substantiated by data and therefore should be used with caution.

If sufficient information is provided by the engine manufacturer to relate Δp, m, and ρ, and to estimate Q for a given temperature difference between CHT and the cooling air, then one is in a position to calculate the cooling drag. Using Reference 4.12, a calculation of the cooling drag was done for a horizontally opposed engine delivering 340 bhp operating at 25,000 ft pressure altitude and cruising at a true airspeed of 275 mph. For this particular example the cooling drag was estimated to represent a loss of 3% in the total power.

For operating conditions other than cruise, it may be necessary to open the cowl flaps. Reference 4.12 states that a pressure coefficient at the cowl exit as low as −0.5 can be produced by opening the flaps to an angle of approximately 15°. By so doing, a relatively higher cooling flow can be generated at a lower speed, such as during a climb. Even though the engine power may be higher during climb, operating with open cowl flaps and with a richer fuel mixture can hold the CHT down to an acceptable value. Also, CHT values higher than the maximum continuous rating are allowed by the manufacturer for a limited period of time.

A calculation similar to the one for cruise was performed for climbing at 130 mph with cowling flaps open (Figure 4.30) at a power of 450 bhp at 19,000 ft. The result was an estimated cooling drag, which represented only 0.5% of the total engine power.

The calculation of the cooling drag is highly configuration-dependent so that it is unfeasible to present here a general method which will apply to most engine installations. Instead, suffice it to say that one should consider cooling drag in predicting the performance of an airplane and this is best done in consultation with the engine manufacturer. Large manufacturers of turbine engines will provide computer programs for estimating installation losses for their engines.

DRAG REDUCTION

Skin friction drag and induced drag are the major contributors to the total drag of an airplane, at least for a modern jet transport. For lower-speed, general aviation aircraft, form drag assumes more relative importance. A typical drag buildup (or

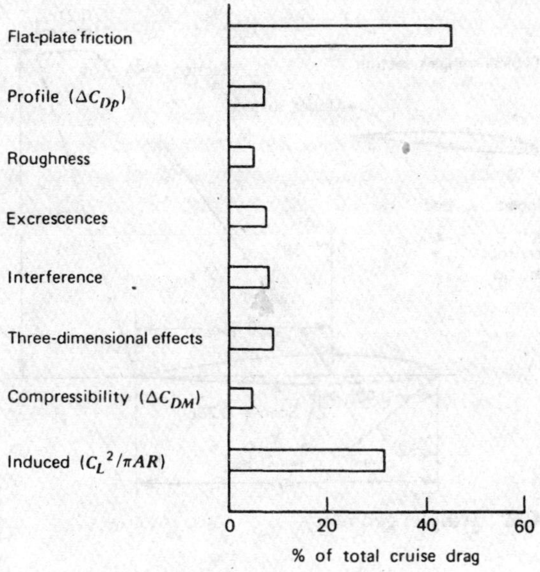

Figure 4.31 Typical drag build-up for jet transport.

breakdown, depending on your outlook) is presented in the bar graph of Figure 4.31 for a jet transport, as reproduced from Reference 4.11. It is seen that skin friction drag and induced drag account for approximately 75% of the total drag. Although the remaining 25% is not to be taken lightly, the potential for real savings in power or fuel rests with reducing the skin friction drag and induced drag.

Winglets

A recent development that holds some promise for reducing induced drag, short of increasing the aspect ratio, is the so-called *winglet*. The details of a winglet (studied in Ref. 4.15) are shown in Figure 4.32; Figure 4.33 pictures the winglet mounted on the wing tip of a first-generation jet transport (such as a Boeing 707).

The winglet is reminiscent of the tip plate, which has been tried over the years for the same purpose. These plates have never proven very successful for reasons that will become clear as the details of the winglet design are discussed.

Placing the winglet on an existing wing will alter the spanwise distribution of circulation along the wingspan and hence the structure of its trailing vortex system far downstream. One can calculate the reduction in the induced drag afforded by the winglet solely by reference to the ultimate wake, or so-called *Treffetz plane*. This is the method used by Reference 4.17 together with a numerical vortex-lattice, lifting-surface theory. Examining only the ultimate wake is not very satisfying, however, from a physical standpoint. Instead, consider the flow field into which the winglet is inserted. Figure 4.34 qualitatively illustrates the situation. Outboard of the tip, the flow is nearly circular as air from beneath the wing flows outward along the span, around the tip, and inward on the upper surface. The velocities induced by the wing are shown. To these the free-stream velocity is vectorially added. The magnitudes of the induced velocities generally increase toward the tip. At a given spanwise location, the induced velocities are highest close to the surface of the wing, just outside of the boundary layer.

Consider a section of the winglet as shown in Figure 4.34*c*. The induced velocity

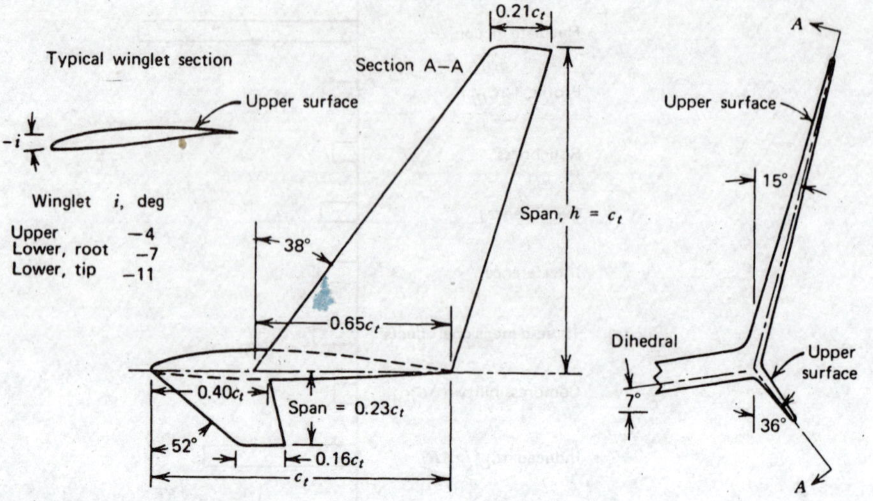

Figure 4.32 Winglet geometry.

V_i produced by the main wing combines with the free-stream velocity, V, to produce an angle of attack, α. Assuming α to be a small angle, a net forward component of force, $-dD$, results from the differential section lift and drag on the winglet. Denoting winglet quantities by a subscript w,

$$-dD = dL_w\alpha_w - dD_w$$

Observe that the same result is obtained if the winglet is mounted below the wing, where the induced velocity is outward.

Since we do not know the induced flow in sufficient detail to integrate along

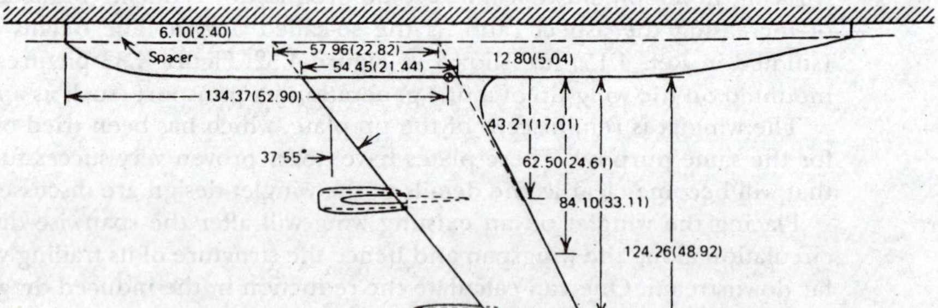

Figure 4.33 NACA model of first-generation jet transport with tip-mounted winglets.

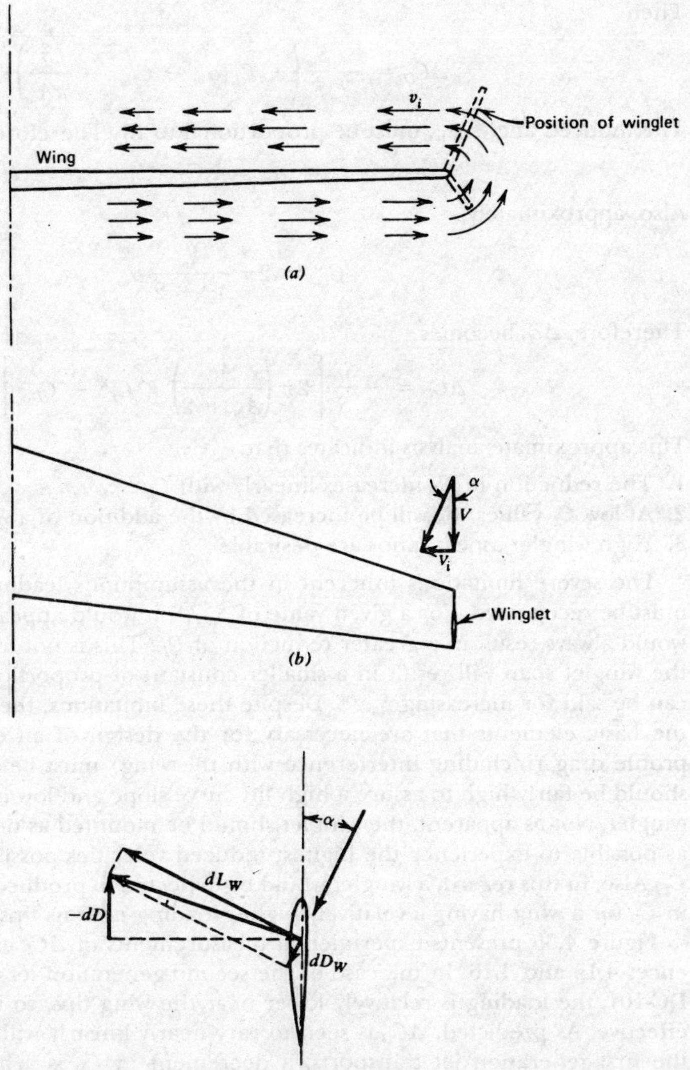

Figure 4.34 Generation of negative drag by winglet section. (*a*) Looking in direction of flight. (*b*) Planview. (*c*) Forces acting on winglet.

the span of the winglet, let us assume an average V_i acting over the winglet for the purpose of disclosing trends. Thus,

$$\Delta D = -L_w\alpha_w + D_w$$

or

$$\Delta C_D = -\frac{S_w}{S}\left(C_{L_w}\alpha_w + C_{D_w}\right)$$

Let

$$C_{D_w} = C_{D0_w} + \frac{C_{L_w}^2}{\pi A_w}$$

Then

$$\Delta C_D = -\frac{S_w}{S}\left(+ C_{L_w}\alpha_w - C_{D0_w} - \frac{C_{L2}^2}{\pi A_w}\right)$$

The induced angle, α_w, must be proportional to C_L. Therefore, let

$$\alpha_w = KC_L$$

Also, approximately,

$$C_{L_w} = 2\pi \frac{A_w}{A_w + 2}\alpha_w$$

Therefore, Δc_D becomes

$$\Delta C_D = -\frac{S_w}{S}\left[2\pi\left(\frac{A_w}{A_w + 2}\right)^2 K^2 C_L{}^2 - C_{D0_w}\right] \tag{4.54}$$

This approximate analysis indicates that:

1. The reduction in C_D increases linearly with $C_L{}^2$.
2. At low C_L values, C_D will be increased by the addition of a winglet.
3. High winglet aspect ratios are desirable.

The severe limitations inherent in the assumptions leading to Equation 4.54 must be recognized. For a given value of S_w/S, it would appear that increasing A_w would always result in a greater reduction of C_D. This is not true, since increasing the winglet span will result in a smaller constant of proportionality, K. The same can be said for increasing S_w/S. Despite these limitations, the foregoing discloses the basic elements that are necessary for the design of an effective winglet. Its profile drag (including interference with the wing) must be low. Its aspect ratio should be fairly high to assure a high lift curve slope and low induced drag for the winglet. Not as apparent, the winglet should be mounted as near the trailing edge as possible to experience the highest induced velocities possible for a given wing C_L. Also, in this regard, a winglet would be expected to produce a larger decrement in C_D for a wing having a relatively higher loading near its tips.

Figure 4.35 presents experimental measurements of ΔC_D as reported in References 4.15 and 4.16. In the case of the second-generation jet transport (such as a DC-10), the loading is relatively lower near the wing tips, so the winglets are less effective. As predicted, ΔC_D is seen to vary nearly linearly with $C_L{}^2$. In the case of the first-generation jet transports, a decrement in C_D is achieved for C_L values greater than 0.22. This number increases to 0.30 for the second-generation jet transports.

The induced drag of a wing can also be reduced simply by extending its tip and thereby increasing its aspect ratio. Reference 4.17 considers this possibility and compares the savings in drag to be gained from extending the tips with those obtained by the use of winglets. Since either method will result in greater root bending moments and hence, increased wing structure and weight, both the induced drag and wing root bending moments are treated by the reference. Typical results from this study are presented in Figure 4.36. It is emphasized that these results are from potential flow calculations and thus, do not include the profile drag of the winglet or any interference drag. The trends determined by the reference are probably valid but somewhat optimistic with regard to the winglets. For identical increases in bending moment, the winglet can provide a greater reduction in induced drag than can be achieved with a tip extension. Referring to Figure 4.36, the ratio $e_{\text{with}}/e_{\text{without}}$ is simply the ratio of the induced drag coefficient of the

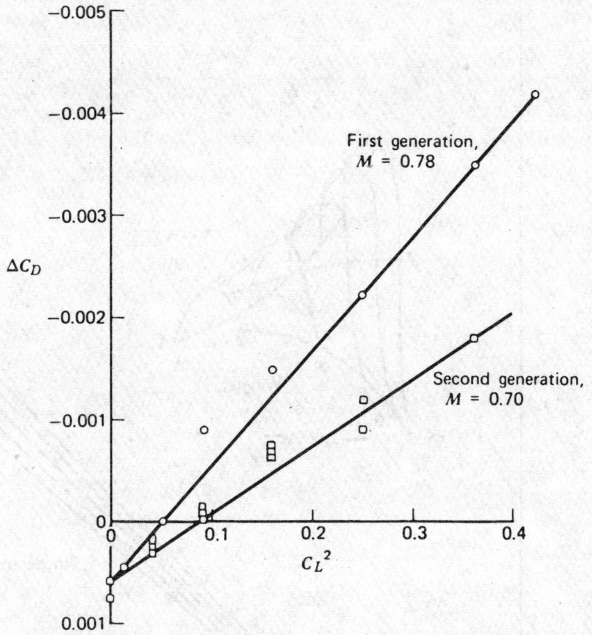

Figure 4.35 Effect of winglets on drag of first- and second-generation jet transports.

original wing to the coefficient with a winglet or tip extension. Consider, for example, an untwisted wing with an aspect ratio of 8 and a taper ratio of 0.5. The leading edge, as with all the wing studies in Reference 4.17, is swept back 30°. With a winglet, the induced drag can be reduced by 24% with only a 2.6% increase in bending moment. For the same bending moment increase, extending the tip would save only 6% in the induced drag. To achieve the same reduction in the induced drag with a tip extension as with the winglet would require a 13% increase in the bending moment.

Recently, there has been renewed interest in induced drag. Studies have been conducted using computational methods and wind tunnel testing to study the effect of planform shape on induced drag. In particular, the crescent wing and sheared tips, shown in Figure 4.37, have been investigated.

van Dam (Ref. 4.18) concludes from numerical computations using a low-order panel method that the crescent planform shape shown in Figure 4.37 has a smaller induced drag than the unswept elliptic planform For an aspect ratio of 7.0, the reduction using the nonplanar panel method amounts to approximately 8%. However, Smith and Kroo (Ref. 4.19) question these computational results stating that the calculation of the drag from pressure over the wing is highly dependent upon the panel geometry. Their calculations, using the Treffetz plane, show possibly a slight reduction in C_{D_i} of only 0.25%.

Wind tunnel tests (Ref. 4.20) profess to support the predictions of reference 4.18 regarding the reduction in the induced drag. However, the error analysis contained in the reference indicates that the measured drag reduction lies within the error band of the experiment. Therefore, at this time, it has yet to be proven conclusively that the planform shapes shown in Figure 4.37 offer any reduction in the induced drag compared to the unswept elliptic planform. Of course, one should remember that it is the elliptic loading distribution that results in the minimum

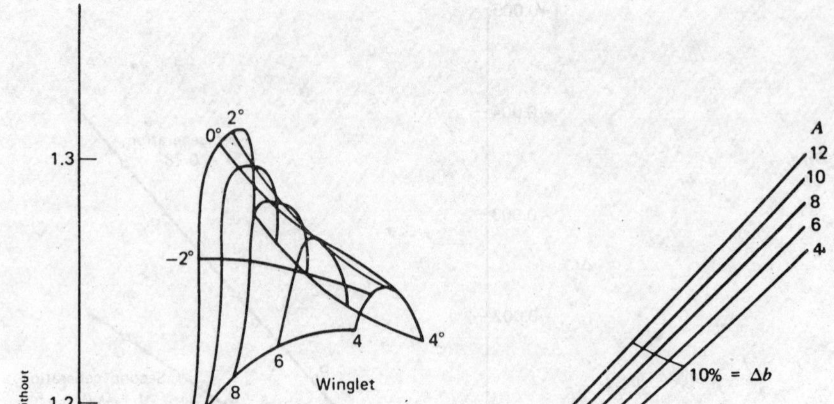

Figure 4.36 Comparison of tip extension and winglet when added to an untwisted wing.

induced drag. This distribution and the elliptic planform are equivalent only within the limitations of lifting line theory. Strictly speaking, an untwisted, elliptic planform will result in a spanwise load distribution, which is close to an ellipse but differs slightly from this shape near the tips.

Reduction of Skin Friction Drag

The contribution to the drag of a streamlined shape from skin friction can be reduced appreciably if transition from laminar to turbulent flow can be delayed. An estimate of the gains to be realized can be seen from Figure 4.38. This figure has been prepared based on Figure 4.1 and the methods outlined earlier for calculating the drag of a flat plate over which the boundary layer is partly laminar and partly turbulent. In this case, the total C_F is calculated from

$$C_F = C_{F_T}(R) - x_t[C_{F_T}(R_t) - C_{F_L}(R_t)] \tag{4.55}$$

In this equation, a subscript T refers to turbulent flow and a subscript L to a laminar flow. R_t is the transition Reynolds number based on the transition length, l_t, shown in Figure 4.38. x_t is the relative distance from the leading edge to the

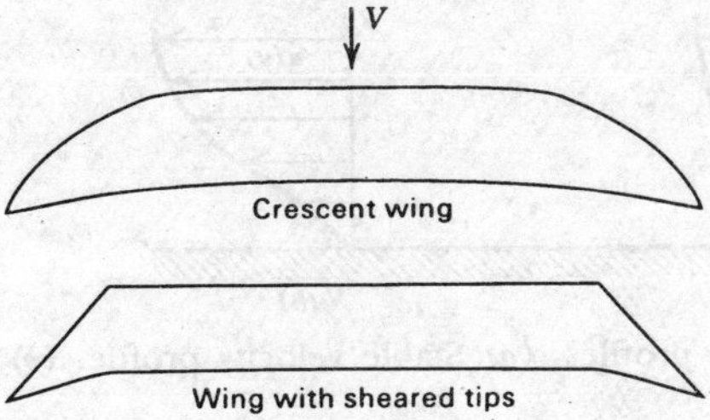

Figure 4.37 Planform shapes, which may reduce induced drag. (*a*) The crescent planform. (*b*) Sheared wing tips.

transition point expressed as a fraction of the total length. The notation $C_{FT}(R_t)$, for example, does not indicate a product, but instead shows that C_{FT} is to be evaluated at the Reynolds number R_t.

From Figure 4.38, it is obvious that the skin friction can be reduced significantly if some means can be found to stabilize the laminar layer so as to prevent or delay transition to a turbulent layer. This delay can be produced by suitably shaping the body under consideration so as to produce streamwise pressure gradients conducive to stabilizing the boundary. This method has come to be known as *natural laminar flow* (NLF). Transition can also be delayed by the use of power in some manner. This will be described later in more detail and is known as *laminar flow control* (LFC).

It is not the purpose of this textbook to consider in detail the fluid mechanics involved in stabilizing the laminar layer. Generally, the problem is that of maintaining a boundary layer that is thin with a full velocity profile. This latter statement is clarified in Figure 4.39.

A good example of NLF is the family of airfoils, the NACA 6-series airfoils discussed briefly in Chapter 3. One of these, the NACA 66_2-015 airfoil, is pictured in Figure 4.40 along with its chordwise pressure distribution. Note that because of the airfoil shape, the pressure decreases with distance all the way back to the 65% chord position. This favorable pressure gradient is conducive to maintaining a thin boundary layer with a stable velocity profile. One might assume, as a first estimate, that its transition point is close to the 65% chord position.

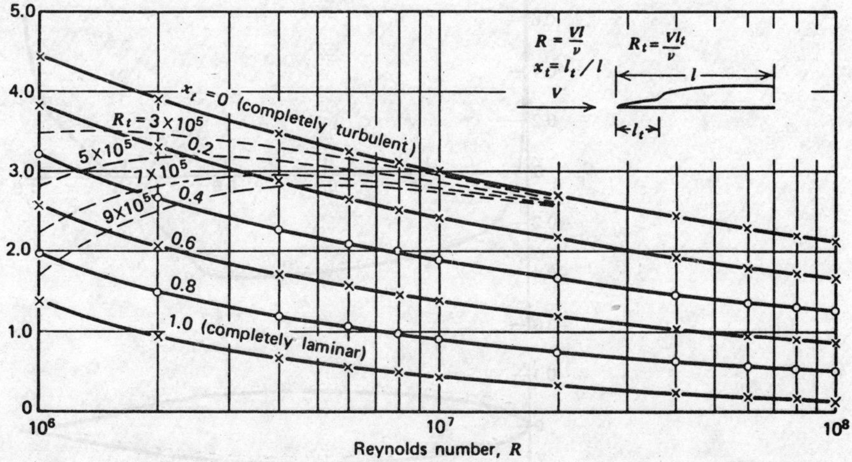

Figure 4.38 Skin friction coefficient for a flat plate as a function of Reynolds number for constant transition lengths.

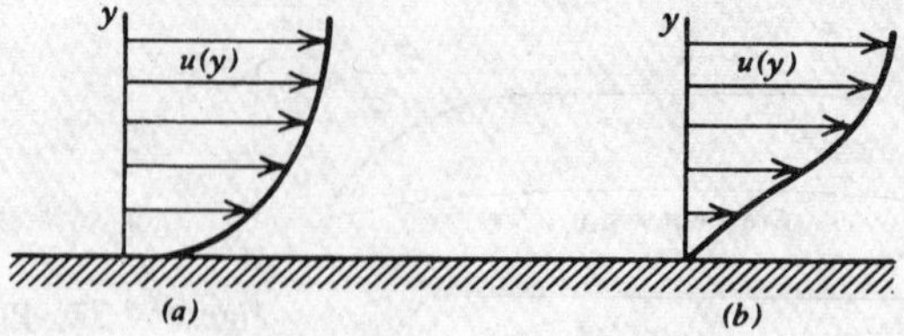

Figure 4.39 Boundary layer velocity profiles. (*a*) Stable velocity profile. (*b*) Less stable velocity profile.

For comparison, consider the NACA 0015 airfoil having the same thickness ratio but with its maximum thickness located farther forward than the 66_2-015 airfoil. For this airfoil, one might expect transition to occur at around the 20% chord, where the flow first encounters an adverse pressure gradient.

Both the 66_2-015 and 0015 airfoils lie within the families of airfoils considered in Figure 4.41. For rough surfaces, C_d is approximately the same for both airfoil families. The roughness causes transition in both cases to occur near the leading edge. The picture is different in the case of smooth surfaces. Here, C_d equals 0.0064 for the 0015 airfoil but only 0.0036 for the laminar flow airfoil. These correspond to C_f values of approximately 0.0032 and 0.0018 for the respective airfoils. Using Figure 4.38 and the transition points of $0.2C$ and $0.65C$, values of C_f of 0.0026 and 0.0014, respectively, are obtained corresponding to C_d values of 0.0052 and 0.0028. The difference between these values and the experimental results may be attributable to errors in the estimated transition locations. Most likely, however, the difference is attributable to form drag. For both airfoils, the differences are close to the estimates of form drag that one obtains from examining the increase in total C_d with thickness ratio.

The favorable pressure distribution of the series-66 airfoils undoubtedly delays transition, thereby reducing the skin friction drag. For a particular airfoil, however,

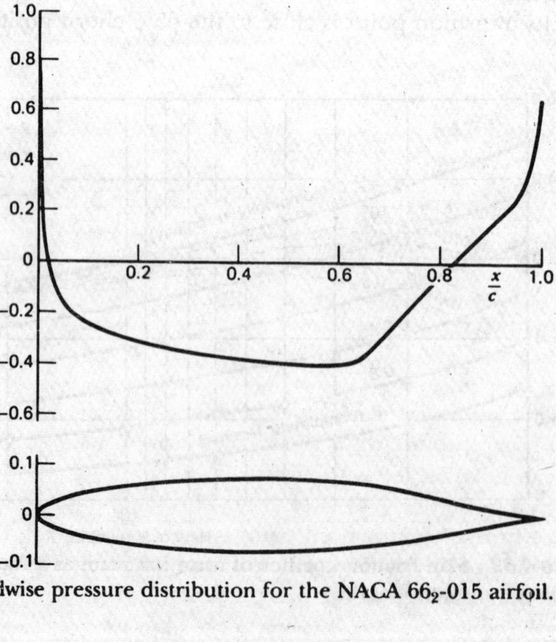

Figure 4.40 Chordwise pressure distribution for the NACA 66_2-015 airfoil.

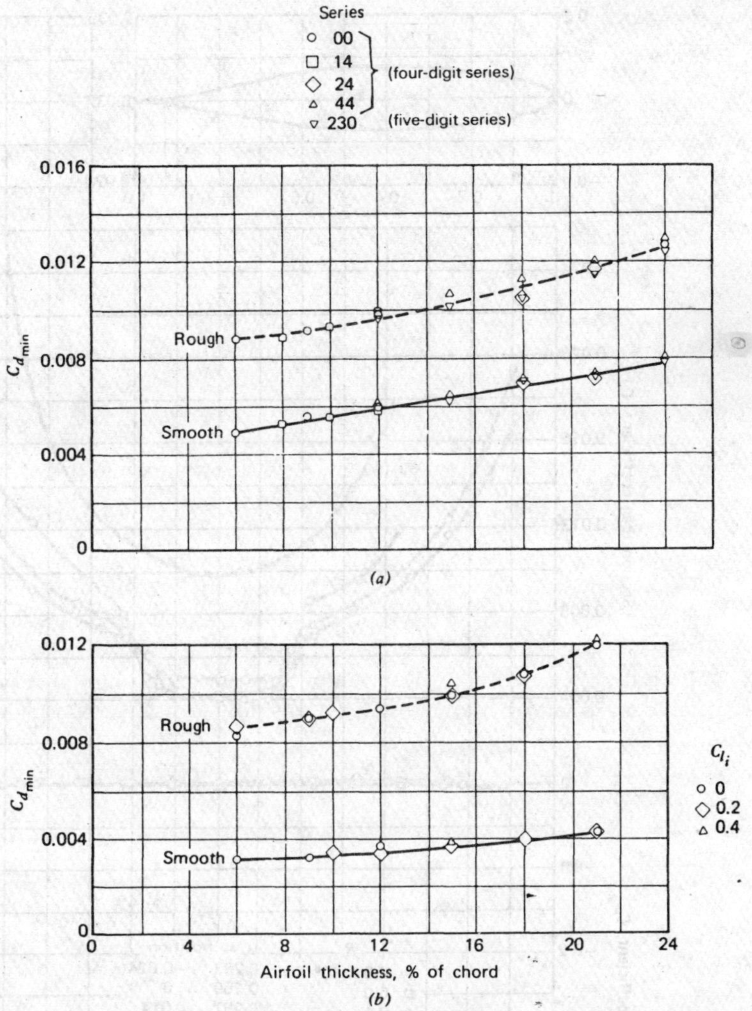

Figure 4.41 Variation of section $C_{d_{min}}$ with thickness ratio for conventional and laminar flow NACA airfoils. (*a*) NACA four- and five-digit series. (*b*) NACA 66-series.

extensive laminar flow can only be maintained over a limited range of C_l values and for Reynolds numbers that are not too large. The C_d versus C_l curve, known as the *drag polar*, for a laminar flow airfoil has the rather unusual shape typified by Figure 4.42. This drag bucket results from the fact that for C_l values between approximately + or -0.2, the chordwise pressure distribution is sufficiently favorable to maintain laminar flow over most of the airfoil. Without this "bucket," the drag curve extrapolates to a C_d value at a zero C_l close to that for a more conventional airfoil having this same thickness.

With careful attention to surface waviness and roughness, appreciable laminar flow can be achieved with airfoils up to Reynolds numbers in excess of 20 million, as shown by Figure 4.43 (Ref. 3.1). This same figure emphasizes the importance of surface finish. Unimproved paint is seen to be rough enough to cause premature transition at a Reynolds number of approximately 20×10^6. The result is a doubling in the drag coefficient for this particular airfoil.

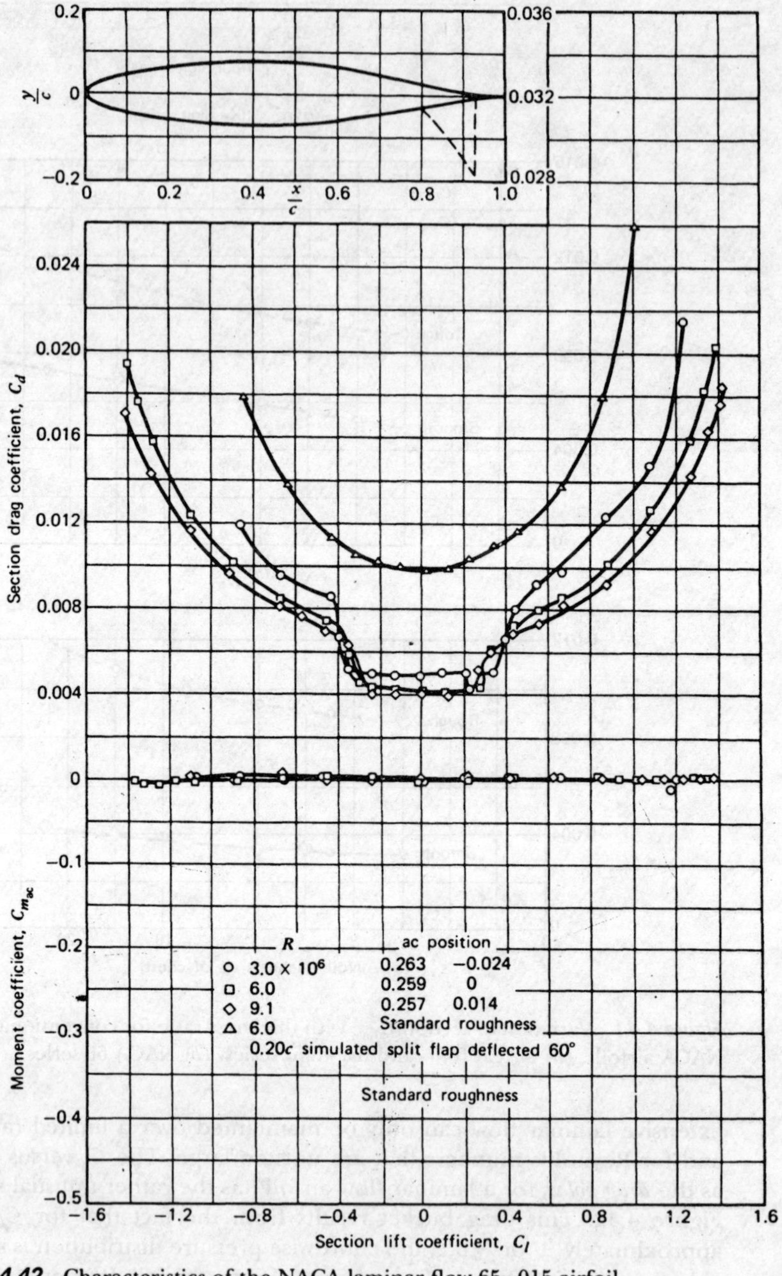

Figure 4.42 Characteristics of the NACA laminar flow 65_2-015 airfoil.

One has to be somewhat careful in interpreting this figure. At first glance, it might appear that transition is being significantly delayed to a Reynolds number of 60 × 10^6, since the drag coefficient is nearly constant up to this Reynolds number. A closer look shows the C_d to decrease up to an R of approximately 32 × 10^6. It then increases up until an R of approximately 54 × 10^6. Above this value of R, it appears C_d is tending to decrease.

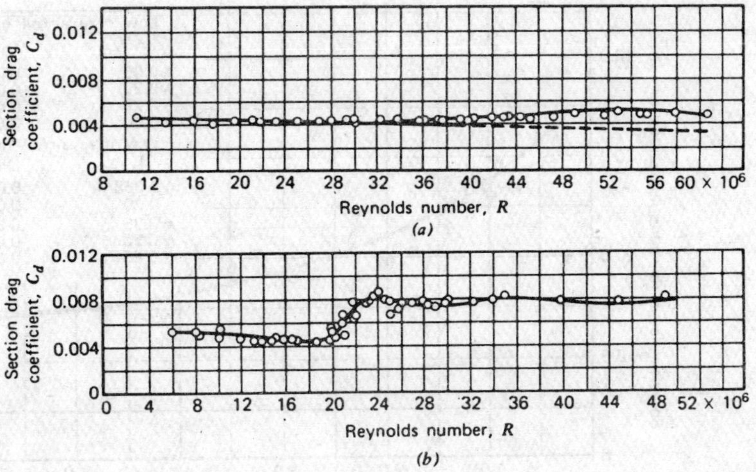

Figure 4.43 Variation of drag coefficient with Reynolds number for a 60-in chord model of the NACA $65_{(421)}$-420 airfoil for two surface conditions. (*a*) Smooth condition. (*b*) Lacquer camouflage unimproved after painting.

Obviously, from Figure 4.38, a constant C_d as R increases requires that the transition point move forward. This is assuming that the form drag is not dependent on R. This is a valid assumption; if anything, the form C_d tends to decrease with R.

It is difficult to divide the total drag into form and skin friction drag because of the dependence of the skin friction drag on the transition location. However, based on the potential flow pressure distribution, it is reasonable to assume that transition occurs at around the 50% chord point at the lower Reynolds numbers. With this assumption, the same form drag coefficient is obtained at R values of 12×10^6 and 30×10^6, that is, a form C_d, of 0.0013. for the same transition location and form C_d, Figure 4.38 leads to a predicted C_d of 0.0036. This is close to what one might expect if the data for Figure 4.43*a* are extrapolated beyond an R of 32 million.

Using the form C_d of 0.0013 and Figure 4.38, the peak C_d of 0.0050 at an R of 54×10^6 leads to a transition location at this higher Reynolds number of 18% of the chord. Thus, it is concluded that the shape of the $65_{(421)}$-420 airfoil is able to stabilize the laminar boundary layer up to the midchord point for Reynolds numbers as high as 30 million. For higher Reynolds numbers, the transition point moves progressively forward.

The size of roughness that can be tolerated without causing transition can be estimated from Figure 4.44 (Ref. 3.1). It is somewhat surprising to find that the results do not depend significantly on the chordwise position of the roughness. In fact, it appears that the downstream positions are less tolerant to roughness height than positions near the leading edge.

Since one can never be sure of the shape of a particle, based on Figure 4.44, a value of R_{kt} of 1400 is recommended as being reasonable. In the case of Figure 4.43, this criterion leads to a roughness as small as 0.004 in. in height as the cause of the drag rise at an R value of 20 million.

Figure 4.45 is a convenient graph for quickly determining Reynolds numbers at a given speed and altitude. For example, a typical light airplane operating at 10,000 ft at a speed of 150 mph has a unit Reynolds number of 1.1×10^6 or, for a chord of 5.5 ft, a Reynolds number of 6.05×10^6. A jet transport cruising at 35,000 ft at

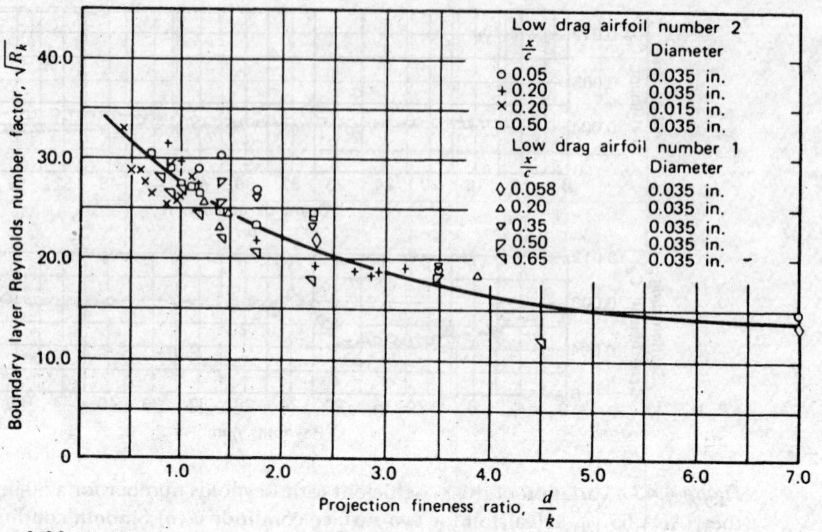

Figure 4.44 Variation of boundary layer Reynolds number with projection fineness ratio for two low drag airfoils. (R_k = transition Reynolds number based on height of protuberance, which will cause transition and local velocity outside of boundary.) (Ref. 3.1, reprinted by permission of Dover Publications, Inc.)

500 mph has a unit R of 1.8×10^6. This results in an R of 27×10^6 for a chord length of 15 ft.

For the light plane, an R_{kt} of 1400 gives an allowable roughness height of 0.015 in. A height of 0.009 in. or less should not cause transition on the jet transport's wing. These may not be difficult criteria to meet for a wind tunnel model or an isolated panel. On an operational, full-scale aircraft with rivets, access panels, deicers, gas caps, wheel-well covers, and the like, the achievement of this degree of smoothness is a real challenge. Even if such smoothness is attained, a few bugs smashed on the leading edge can easily destroy the aerodynamic cleanliness.

An excellent summary of the more recent work undertaken by NASA and others to achieve laminar flow is found in Reference 4.21. Part 3 of this reference contains a number of papers relating to maintaining natural laminar flow on airfoils and fuselage shapes.

One of the papers in this collection, (Ref. 4.22), reported on flight test measurements obtained with a glove wrapped around the wing of a Boeing 757 transport immediately outboard of the right engine. This glove, which adds material to the existing wing to modify its contour, was designed to promote extensive laminar flow on both the upper and lower surfaces at high altitudes and Mach numbers. In order to avoid surface contamination of bugs during the climbout, paper was wrapped around the glove and then torn away at altitude. This same technique was used for tests of the North American P-51 Mustang fighter of World War II fame, which utilized a laminar-flow airfoil. Tests on the 757 were conducted up to Mach numbers of approximately 0.82 and altitudes of 41,000 ft. The design condition for the glove was for an altitude of 40,500 ft, a C_L of 0.53, and a Mach number of 0.80.

The extent of laminar flow measured on the upper and lower surfaces of the glove at conditions to either side of the design C_L is shown in Figure 4.46 as taken from the reference. The maximum extent of laminar flow on the upper surface was approximately 29% of chord at an altitude of 40,761 ft and a Mach number of

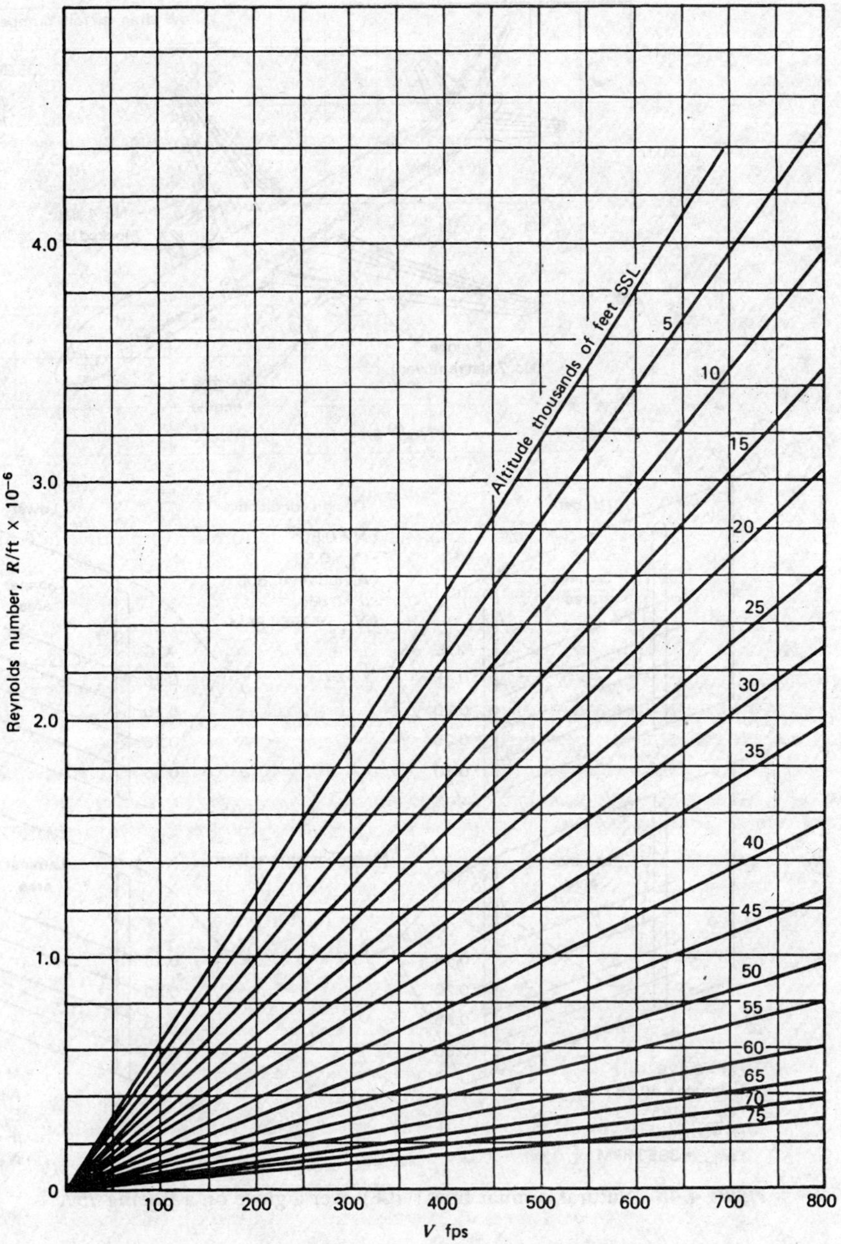

Figure 4.45 Reynolds number as a function of velocity and altitude.

0.825. With the chord of the glove equaling approximately 6 ft (with a span of 10 ft), these numbers translate to a Reynolds number based on the chord of $R = 9.0 \times 10^6$ and a transition Reynolds number of $R = 2.6 \times 10^6$.

Reference 4.22 concluded that there was little effect of engine noise on NLF on the upper surface and only a 2 or 3% forward movement of the transition point on the lower surface when going from low power to high power. Sweep has a strong influence on the stability of the boundary layer which was shown by slipping the

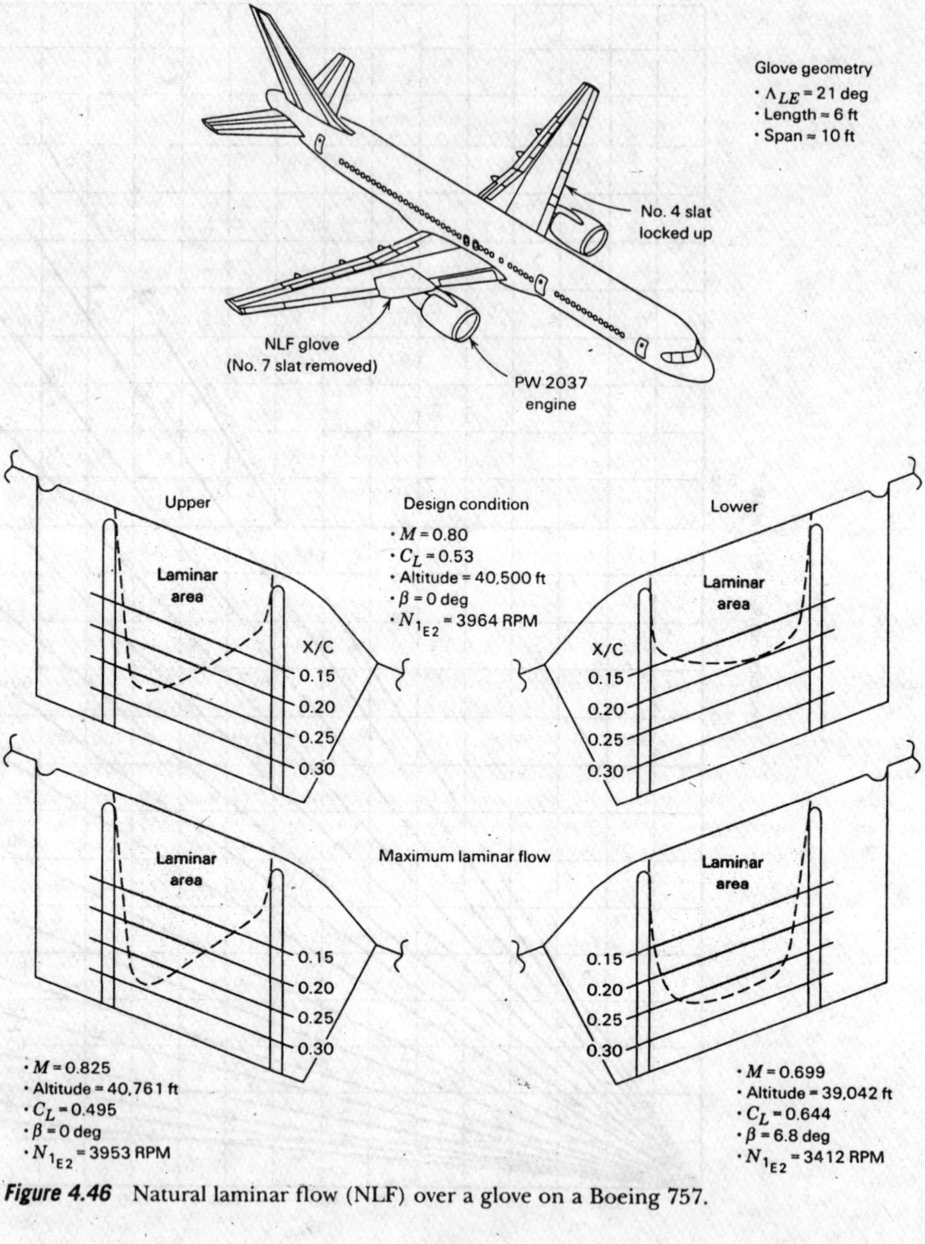

Figure 4.46 Natural laminar flow (NLF) over a glove on a Boeing 757.

test airplane so as to change the effective cross-flow on the wing. Generally, the effect of sweep is deleterious to the stability of the laminar boundary layer and was shown to be the cause of transition at most flight conditions. It is interesting to note that in one test the protective paper was omitted with the result that early transition was found because of smashed bugs found on the leading edge. This problem of climbing through the "bug layer" is discussed in the next section on LFC.

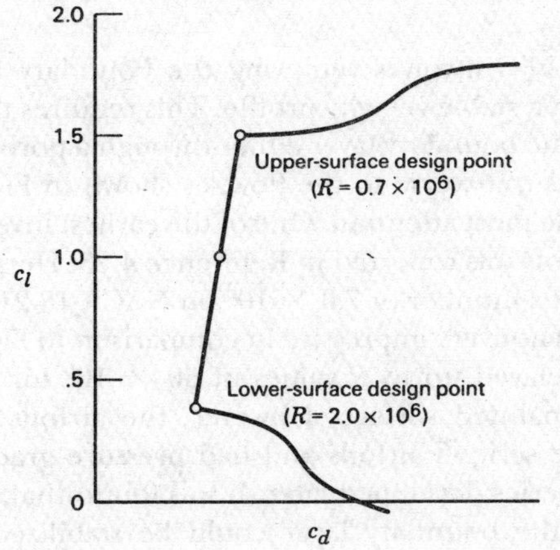

Figure 4.47 Criteria for a multipoint airfoil design.

Reference 4.23 by Maughmer and Somers is significant because it demonstrates clearly what can be accomplished with modern numerical airfoil analysis. Here, a multipoint-design of an airfoil was undertaken where performance was specified for the four points shown in Figure 4.47 taken from the reference. At the lower lift coefficients, the resulting airfoil was shown to achieve a low drag with NLF. As the lift coefficient increases, the airfoil was designed so that the transition point moves forward reaching the leading edge at $C_{L_{max}}$. This is important to the operation of an airplane since, with this airfoil, its stalling characteristics will not be affected by surface contamination.

To demonstrate that airfoil design can make a difference on an airplane's performance, Maughmer and Somers include in their paper a table, reproduced here as Table 4.5, with the gains to be realized with the NLF airfoil as compared to the NACA 23015 airfoil.

Some additional observations regarding NLF can be found in Reference 4.24. Holmes and Obara examined the test results on many general aviation airplanes and came to several interesting conclusions. First, somewhat contradictory to the Jetstar results to be discussed in the next section, it was found using a T-34C airplane with an NLF glove that the transition point was not necessarily affected by flying through clouds. If no mist was observed on the windshield, the transition point was not affected. However, if mist was observed, the transition point moved forward to the leading edge. Once clear of the clouds, laminar flow was quickly reestablished.

Reference 4.24 also notes that significant regions of laminar flow were presented in the propeller slipstream and that no premature transition was attributable to surface waviness. Based on the rather extensive research program reported by Holmes and Obara, it appears that transition Reynolds numbers up to 11×10^6 can be attained on modern airplane surfaces using NLF design methods.

Table 4.5 Performance Gains Achieved with an NLF Airfoil

	NACA 23015	NLF(1)-1015	% Gain
Endurance	72 hr	93 hr	30
Range	18,000 mi.	21,000 mi.	14

Laminar Flow Control

An active method of providing LFC involves removing the boundary layer as it develops so as to keep it thin with a stable velocity profile. This requires that power be expended to apply suction to the boundary layer either through a porous surface or across closely spaced thin slots transverse to the flow, as shown in Figure 4.48. The latter method has received the most attention. One of the earliest investigations of LFC using discrete spanwise slots was reported in Reference 4.25. Here, laminar flow was achieved up to a Reynolds number of 7.0×10^6 on NACA 18-212, 27-215, and 0007-34 airfoils. This result is not very impressive in comparison to Figure 4.43, where transition is apparently delayed up to R values of 30×10^6 for a smooth surface and 20×10^6 for the painted surface. However, the airfoils tested by Reference 4.18 were prior to the series-6 airfoils and had pressure gradients less favorable than the laminar flow series developed later. It was found that, with only a small expenditure of power, the boundary layer could be stabilized over an extensive region having an adverse pressure gradient. Somewhat discouraging was the fact that the use of suction did not reduce the sensitivity of transition to roughness.

Flight testing performed in the mid-1960s provided more encouraging results, as reported in Reference 4.26. Two WB-66 airplanes were modified and redesignated X-21A. These airplanes had 30° swept wings with an aspect ratio of 7. The boundary layer was removed by approximately 120 slots on each surface. The slots varied in width from about 0.0035 to 0.01 in.

Difficulties were encountered with instabilities in the skewed boundary layer along the swept leading edge produced by the spanwise flow. However, the use of fences and chordwise suction slots spaced along the leading edge alleviated this problem. The final result was the attainment of full-chord laminar flow at a Reynolds number of 45.7×10^6.

The adoption of a powered LFC system represents a challenging exercise in systems analysis and design. The saving in drag must be measured against the weight and initial cost of the ducting, pumps, and double skin required to remove the boundary layer. According to Reference 4.26, performance analyses showed that the required engine size for a jet transport in the 300,000-lb class is smaller than that for the turbulent counterpart. This smaller engine results in a weight saving that offsets the weight penalty of the pumping equipment. With both the inner and outer skins contributing to the wing's structural integrity, the weight of all of the pumping equipment is estimated at between 1.3 and 1.4 psf. Considering the

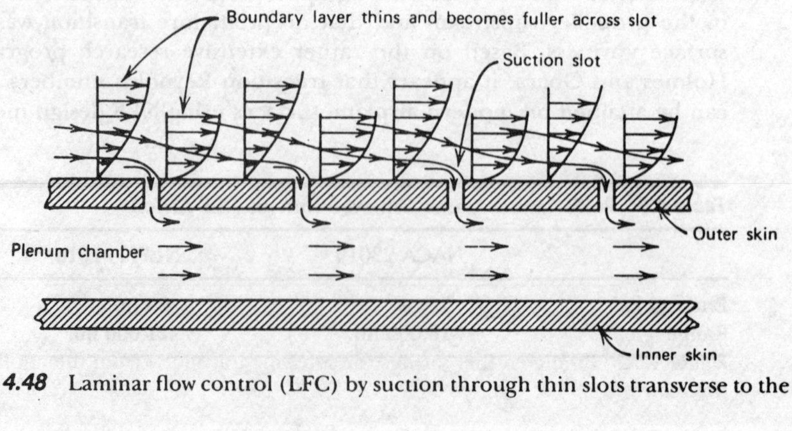

Figure 4.48 Laminar flow control (LFC) by suction through thin slots transverse to the flow.

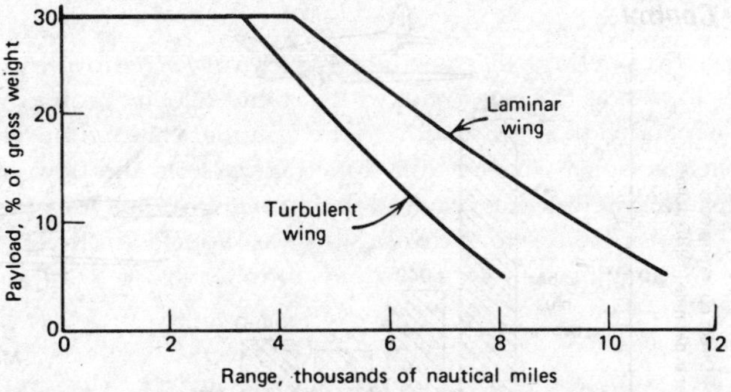

Figure 4.49 Performance gains from laminar flow control. (Ref. 4.26, reprinted by permission of the American Institute of Aeronautics and Astronautics)

weight, drag, and specific fuel consumption, an optimized design incorporating LFC shows an increase of one-third in the range for a fixed payload or in the payload for a fixed range at a design range of 5000 nmi. These predicted performance gains are shown in Figure 4.49 (taken from Ref. 4.26).

Figure 4.49 was produced in the middle 60's. However, the benefits of laminar flow shown on that graph are believed to be true today. Figure 4.50 taken from Reference 4.27 shows fuel savings approaching 30% for a jet transport with LFC and between 5 and 15% for NLF or a combination of LFC and NLF.

If these substantial benefits are there to be realized, why haven't they been? First, and possibly foremost, no manufacturer has been willing to gamble the large investment it will take to develop a production transport airplane with LFC because it is still in the research stage. The X-21A program was disappointing and made people wary of the practical problems of stabilizing a laminar boundary on a swept wing having a surface made to production standards and operating in an adverse environment.

Another argument against the adoption of LFC is the fact that airplanes have seen dramatic improvements in their direct operating costs since the era of the

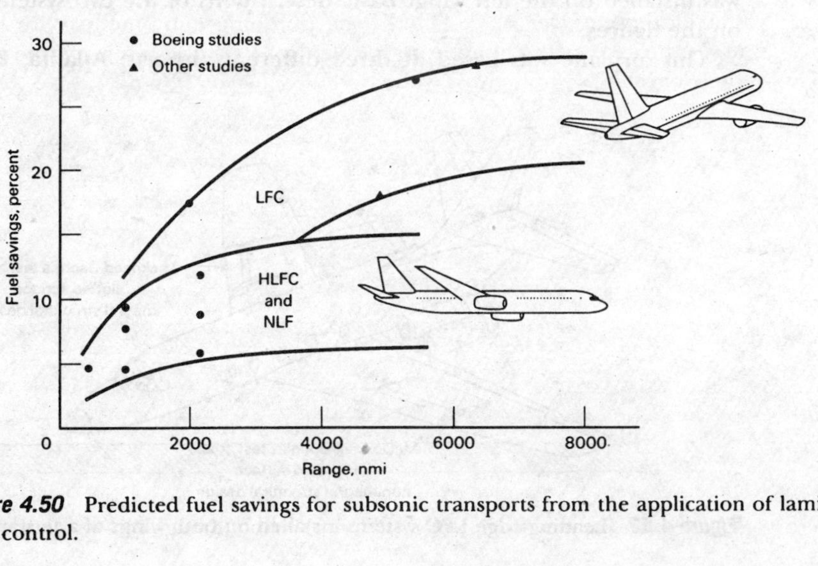

Figure 4.50 Predicted fuel savings for subsonic transports from the application of laminar flow control.

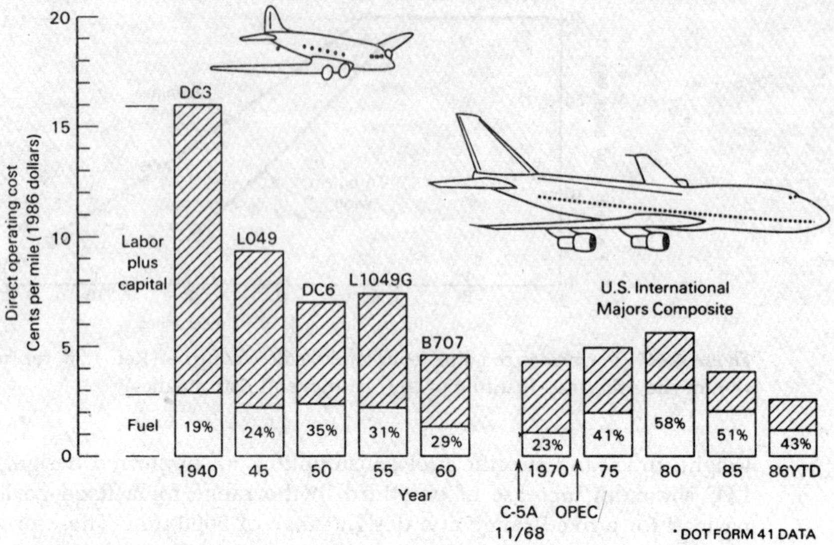

Figure 4.51 History of operating costs for U.S. air transports.

DC-3. This is shown in Figure 4.51 taken from Reference 4.27. The added costs and uncertainties associated with LFC may not be worth it. However, in the past decade, fuel costs have risen dramatically, particularly as a percentage of the direct operating cost, so the time may be approaching for the air transport community to take another look at LFC.

One of the most ambitious programs intended to demonstrate the feasibility of LFC was the *Jetstar LFC Leading-Edge Flight Test Program.* The Lockheed Jetstar is a 4-engine, jet, executive transport and for this program was modified as pictured in Figures 4.52 and 4.53 (*a*) and (*b*). The leading edges of the wing back to the main spar were modified with two different LFC systems. The middle portion of the leading edge of the right wing was modified with an LFC system designed by McDonnell-Douglas Co. whereas a system designed by the Lockheed-Georgia Co. was installed on the left wing. Basic descriptions of the two systems are included on the figures.

This airplane was based at three different airports: Atlanta, Pittsburgh, and

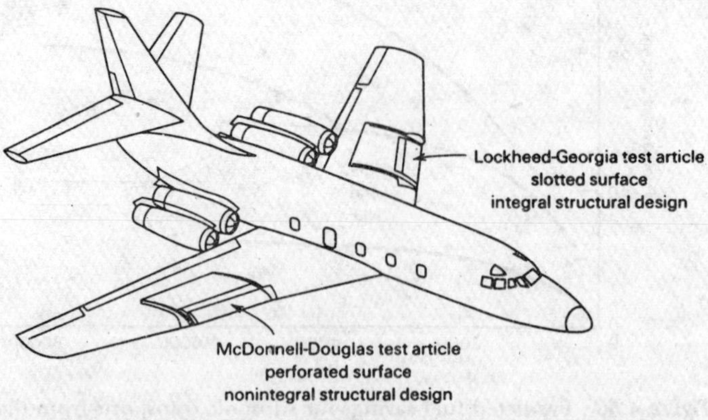

Figure 4.52 Leading-edge LFC systems installed on both wings of a Jetstar.

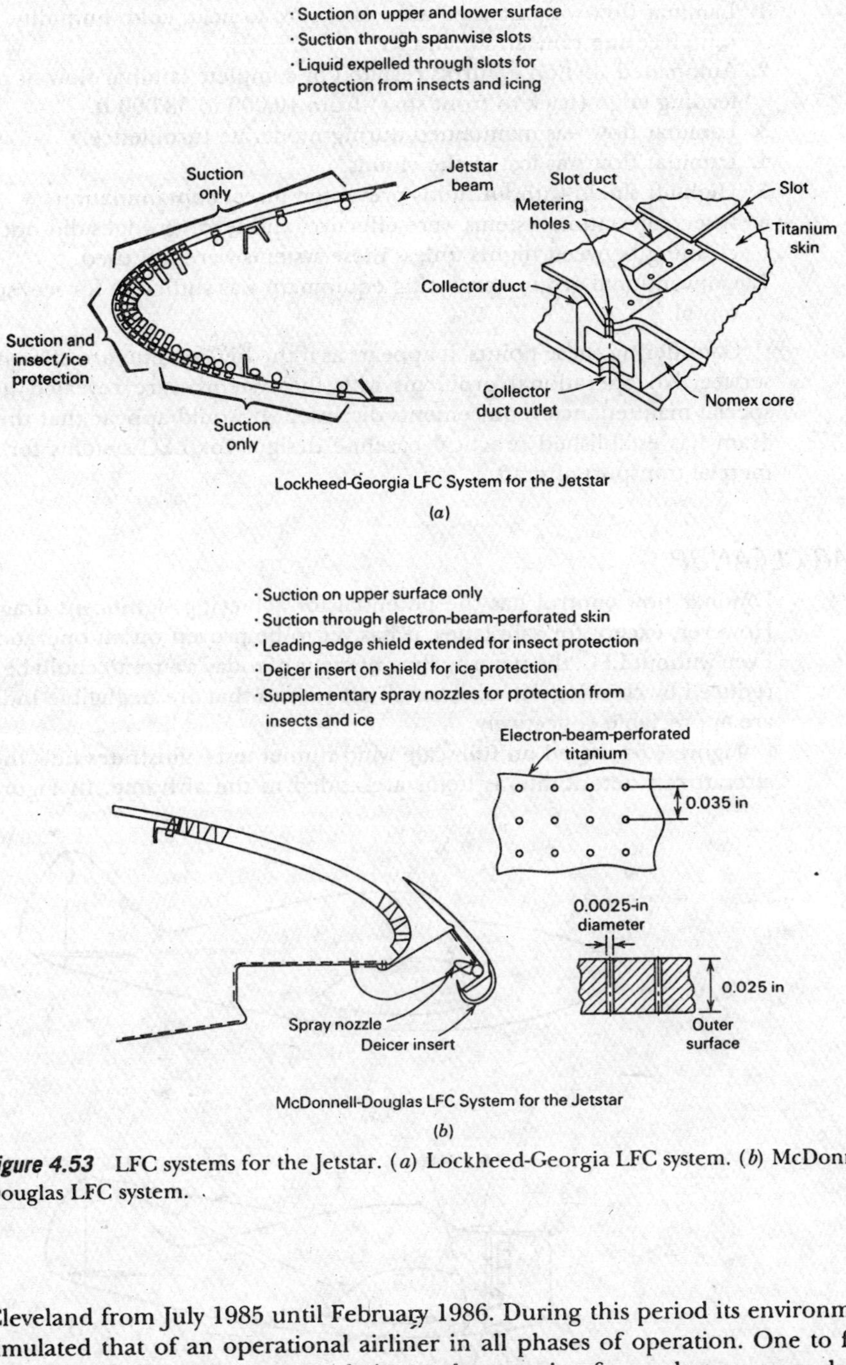

Figure 4.53 LFC systems for the Jetstar. (*a*) Lockheed-Georgia LFC system. (*b*) McDonnell-Douglas LFC system.

Cleveland from July 1985 until February 1986. During this period its environment simulated that of an operational airliner in all phases of operation. One to four scheduled flights a day were made in weather ranging from a hot, summer day to severe winter weather. Essentially five systems were evaluated during this program: suction, high-lift/shield, wetting, purge, and anti-icing. Purge refers to the pumping of a cleaning fluid out of the suction holes to keep them dirt-free.

The conclusions reached from this study based, in part, on over 2000 data points are as follows:

1. Laminar flow was obtained after exposure to heat, cold, humidity, insects, rain, freezing rain, snow, and ice.
2. Automated suction controls resulted in complete laminar flow of perforated leading edge (back to front spar) from 10,000 to 38,000 ft.
3. Laminar flow was maintained during moderate turbulence.
4. Laminar flow was lost in the clouds.
5. High-lift shield without fluids prevented insect contamination.
6. Insect alleviation systems were effective, and leading edges did not require cleaning between flights unless these systems were not used.
7. Conventional ground anti-icing equipment was sufficient for ice/snow removal.

Considering these points, it appears as if the LFC systems are effective for airline service. No operational problems with the systems were revealed nor were any special maintenance requirements disclosed. It would appear that the Jetstar program has established practical baseline designs for LFC systems for future commercial transport aircraft.

DRAG CLEANUP

Laminar flow control has the potential for achieving significant drag reductions. However, except for sailplanes, it has yet to be proven on an operational aircraft. Even without LFC, the parasite drag of many of today's aircraft could be significantly reduced by cleaning up many small drag items that are negligible individually but are appreciable collectively.

Figure 4.54, based on full-scale wind tunnel tests, illustrates how the drag of an aircraft can deteriorate as items are added to the airframe. In Figure 4.54*a*, the

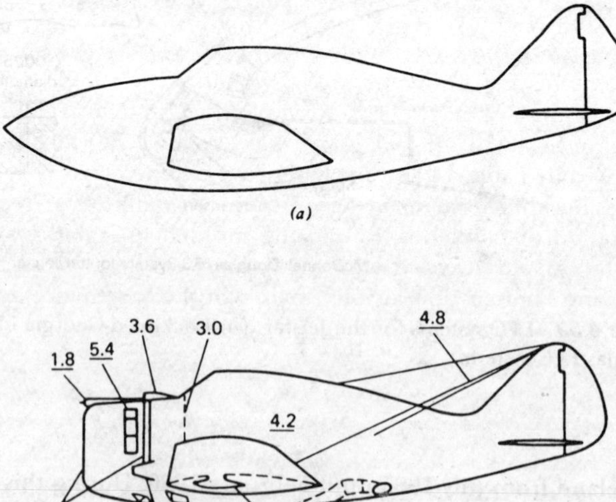

Figure 4.54 Drag penalties for an airplane. (*a*) Airplane in faired and sealed condition. (*b*) Airplane in service condition (numbers indicate drag increments in percent of total drag of clean airplane).

Table 4.6 Drag Items as Shown in Figure 4.54 (Reference 4.14)

Power plant installation	
Open cowling inlet and exit	18.6%
Unfaired carburetor airscoop	3.6%
Accessory cooling airflow	3.0%
Exhaust stacks and holes	3.6%
Intercooler	6.6%
Oil cooler	10.2%
Total	45.6%
Other items for service condition	
Remove seals from cowl flaps	5.4%
Opening case and link ejector	1.8%
Opening seals around landing gear doors	1.2%
Sanded walkway	4.2%
Radio aerials	4.8%
Guns and blast tubes	1.8%
Total	19.2%

airplane is shown in the faired and sealed condition. Then, as the items tabulated in Table 4.6 were added, drag increments were measured. These are expressed as a percentage of the original, clean airplane drag.

Table 4.6 shows that the drag of the original, clean airplane is increased by nearly 65% by the total effect of these drag items. Some of this additional drag is, of course, necessary, but more than half of it is not. Additional tests and analysis of this particular airplane showed that the drag of the power plant items could be reduced to 26.6% of the initial drag.

The moral of the foregoing and other material contained in this chapter is that, with regard to drag, attention should be paid to detail. Surfaces should be smooth and protuberances streamlined or avoided if possible. Tight seals should be provided around wheel wells, door openings, and other cutouts. It is exactly this attention to detail (or lack of it) that explains the wide disparity in the C_F values tabulated in Table 4.2 for airplanes of the same class.

Possibly the ultimate in aerodynamic cleanliness is represented by the latest generation of sailplanes. Employing molded fiberglass or other types of plastics, ultrasmooth surfaces are achieved. Using very high aspect ratios, ranging from 10 to 36, and laminar flow airfoils, mainly of the Wortmann design (Ref. 4.31), lift-to-drag ratios as high as 40 have been accomplished.

TOTAL AIRPLANE DRAG

To this point, this chapter has considered the separate sources of drag that contribute to the total drag of an airplane. Let us now consider the behavior of the total drag of an airplane as a function of airspeed and altitude. To do this it will be assumed that the drag is composed only of the parasite and induced drag. Thus the total drag is given by

$$D = q(f + SC_{D_i}) \tag{4.56}$$

Using Equation 4.33 for the induced drag coefficient and substituting the definition of lift coefficient and aspect ratio leads to the following for the total drag:

$$D = \frac{\rho f}{2} V^2 + \frac{2}{\rho \pi e}\left(\frac{W}{b}\right)^2 \frac{1}{V^2} \tag{4.57}$$

Thus, the drag is composed of two parts: the parasite drag, which varies directly with the square of the airspeed and the induced drag, which varies inversely with V^2. Differentiating Equation 4.57 with respect to V and equating the result to zero results in the velocity for minimum drag.

$$V_{\min} = \left[\frac{4}{\pi e f}\left(\frac{W}{\rho b}\right)^2\right]^{1/4} \tag{4.58}$$

The minimum drag is then obtained by substituting this value for V into Equation 4.57.

$$D_{\min} = 2\left(\frac{W}{b}\right)\sqrt{\frac{f}{\pi e}} \tag{4.59}$$

It is interesting to note that the minimum drag is independent of density and depends on the span loading, W/b, instead of the wing loading, W/S.

For a "ballpark" number, one can usually estimate a reasonable value for the lift-to-drag of a given airplane from its appearance. Assuming the lift equal to the weight, L/D can be obtained from the above as,

$$\frac{L}{D} = \sqrt{\frac{\pi e A}{4 C_f}\left(\frac{S}{S_w}\right)} \tag{4.60}$$

In the above, S/S_w can be no greater than 2.0 (a flying wing) and Oswald's efficiency factor, e, will generally be less than 1. Thus, L/D will be equal to or less than

$$\frac{L}{D} \leq 0.63\sqrt{\frac{A}{C_f}} \tag{4.61}$$

For a typical light airplane, from Table 4.2, C_f is approximately 0.01 and the aspect ratio is between 5 and 7. Thus, for this class of airplane, one would not expect L/D to be any higher than approximately 15. Considering that the ratio of wetted area to planform area is probably around 4 or 5, an L/D of 10 is more reasonable in this case. On the other end of the spectrum, sailplanes with high aspect ratios and extensive laminar flow can achieve L/D values of 50 or more. The maximum L/D for a subsonic jet transport will be approximately 25.

D, $D_{\min}$, and $V_{\min}$ can be combined and expressed in a general manner as

$$\left(\frac{D}{D_{\min}}\right) = \frac{1}{2}\left[\left(\frac{V}{V_{\min}}\right)^2 + \frac{1}{\left(\frac{v}{v_{\min}}\right)^2}\right] \tag{4.62}$$

This first term on the right-hand side of Equation 4.62 represents the parasite drag and the second term is proportional to the induced drag. Both of these terms and the total drag are represented in Figure 4.55.

Computer Exercise 4.2 "POLAR"

Write a program to calculate the total (induced + parasite) drag of an airplane given the operating conditions and airplane parameters including weight, wing area, equivalent flat-plate area, and Oswald's efficiency factor. Design the program to produce a data file of true airspeed and drag at a given altitude.

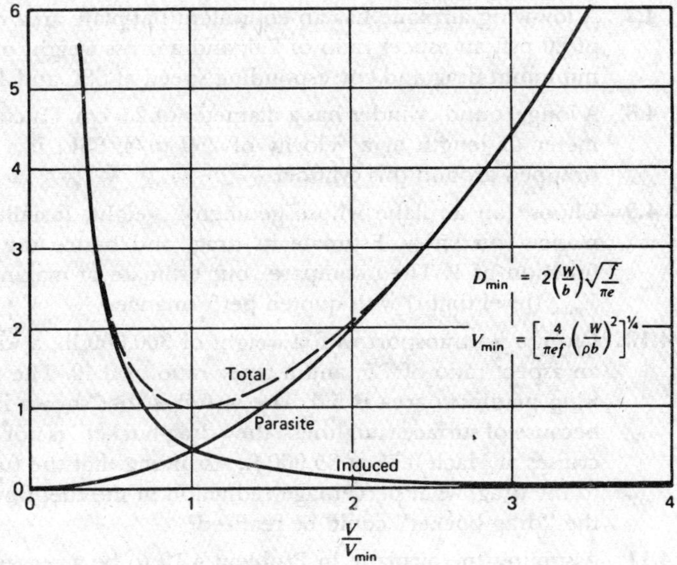

Figure 4.55 Generalized Drag Curves.

PROBLEMS

4.1 A flat plate aligned with the flow has a length of 4 m and a width of 10 m. Calculate its skin friction drag at an airspeed of 40 m/s for SSL conditions. Assume a transition Reynolds number of 1×10^6.

4.2 For the plate in Problem 4.1, calculate the displacement thickness 1 m back from the leading edge.

4.3 Estimate the roughness height that would cause premature transition on the plate in Problem 4.1.

4.4 A "turbulence detector" uses a cylinder operating just below its critical Reynolds number. Stimulated by turbulence, its drag drops suddenly. What size cylinder should be used to detect turbulence in an airflow having a velocity of 120 mph, SSL conditions?

4.5 The preliminary design of a light, twin-engine, propeller-driven airplane is being undertaken. It will have a rectangular wing with an aspect ratio of 7, a wing loading of 960 Pa, and a gross weight of 17,800 N. Assuming a relatively clean airplane, estimate its drag at a speed of 90 m/s.

4.6 An airplane has four flap hinge brackets. Each bracket projects vertically below the wing a distance of 5 cm and is 0.6 cm thick, as shown.

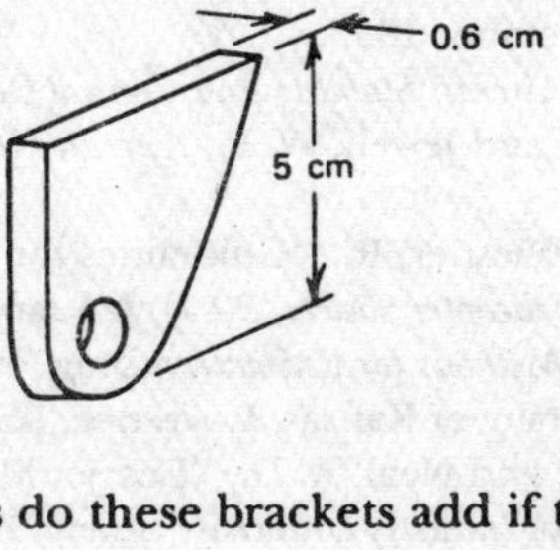

How many drag counts do these brackets add if the wing area is 18.6 m^2?

4.7 A low-wing airplane has an equivalent flat-plate area of 8 ft^2, a wing loading of 20 psf, an aspect ratio of 7.0, and a gross weight of 5000 lb. Calculate its minimum drag and corresponding speed at SSL and 10,000 ft altitude.

4.8 A long, round cylinder has a diameter of 2.5 cm. Calculate the drag saved per meter of length at a velocity of 250 m/s, SSL, if a streamlined fairing is wrapped around this cylinder.

4.9 Choose an airplane whose geometry, weight, installed power, and performance you know. Estimate its drag, and hence the power required, as a function of *V*. Then compare your estimate of maximum rate of climb and V_{max} (level flight) with quoted performance.

4.10 Given a jet transport with a weight of 300,000 lb, a wing loading of 100 psf, an aspect ratio of 7.5, and a taper ratio of 0.40. The ratio of wetted area to wing planform area is 5.0. The airfoil is that shown in Figure 3.9; however, because of surface roughness, the "drag bucket" is not achieved. The airplane cruises at Mach 0.75 at 35,000 ft. Assuming that the fuel flow is proportional to the drag, what percentage reduction in the fuel flow would be achieved if the "drag bucket" could be realized?

4.11 Assuming the airplane in Problem 4.10 to be a second-generation jet transport, how much fuel percentage-wise can be saved by the use of winglets?

4.12 The drag coefficient of an airplane is a function only of its lift coefficient (neglecting Mach number and Reynolds number effects).
(a) If the total drag of the airplane in Problem 4.10 equals 14,000 lb, at what operating condition at 15,000 ft can you calculate the drag and what the drag will equal at this condition?
(b) If the minimum drag of the airplane equals 12,000 lb at 40,000 ft at a Mach number of 0.45, what will the drag be at 15,000 ft at a Mach number of 0.7?

4.13 This is an open-ended problem. A canard configuration is touted by many because its trim drag appears to be less than that of a comparable, conventional airplane. Investigate this claim.

REFERENCES

4.1 Kuethe, A. M., and Chow, C., *Foundations of Aerodynamics, 3rd Ed.*, John Wiley & Sons, Inc., New York, 1976.

4.2 Oswald, W. Bailey, *General Formulas and Charts for the Calculation of Airplane Performance*, NACA Report 408, 1933.

4.3 Raspet, August, "Application of Sailplane Performance Analysis to Airplanes," *Aeronautical Engineering Review, 13(8)*, August 1954.

4.4 Hoerner, S. R., *Fluid-Dynamic Drag*, published by the author, Midland Park, NJ, 1965.

4.5 Delany, N. K., and Sorensen, N. E., *Low-speed Drag of Cylinders of Various Shapes*, NACA TN 3038, November 1953.

4.6 Teper, G. L., *Aircraft Stability and Control Data*, NASA CR-96008, April 1969.

4.7 Heffley, R. K., and Jewell, W. F., *Aircraft Handling Qualities Data*, NASA CR-2144, December 1972.

4.8 Keys, C., and Wiesner, R., "Guidelines for Reducing Helicopter Parasite Drag," *J. of the American Helicopter Society, 20(1)*, January 1975.

4.9 Roskam, Jan, *Methods for Estimating Drag Polars of Subsonic Airplanes*, published by the author, University of Kansas, Lawrence, KS, 1971.

4.10 Ross, Richard, and Neal, R. D., "Learjet Model 25 Drag Analysis," a paper in *Proceedings of the NASA-Industry-University General Aviation Drag Reduction Workshop*, Jan Roskam, editor, Lawrence, KS, July 14–16, 1975.

4.11 Anderson, A. A., "General Overview of Drag," a paper in *Proceedings of the NASA-Industry-University General Aviation Drag Reduction Workshop,* Jan Roskam, editor, Lawrence, KS, July 14–16, 1975.

4.12 Anonymous, *Installation Design for Engine Cooling,* Avco Lycoming Division, Avco Corporation, an in-house manual available from Lycoming on request.

4.13 Frass, A. P., *Aircraft Power Plants,* McGraw-Hill, New York, 1943.

4.14 McKinney, M. O., "Summary of Drag Clean-Up Tests in NASA Langley Full-Scale Tunnel," a paper in *Proceedings of the NASA-Industry-University General Aviation Drag Reduction Workshop,* Jan Roskam, editor, Lawrence, KS, July 14–16, 1975.

4.15 Whitcomb, R. T., *A Design Approach and Selected Wind-Tunnel Results at High Subsonic Speeds for Wing-tip Mounted Winglets,* NASA TN D-8260, July 1976.

4.16 Flechner, S. G., Jacobs, P. F., and Whitcomb, R. T., *A High Subsonic Speed Wind-Tunnel Investigation of Winglets on a Representative Second-Generation Jet Transport Wing,* NASA TN D-8264, November 1976.

4.17 Heyson, H. H., Riebe, G. D., and Fulton, C. L., *Theoretical Parametric Study of the Relative Advantages of Winglets and Wing-Tip Extensions,* NASA TP 1020, September 1977; *Industry-University General Aviation Drag Reduction Workshop,* Jan Roskam, editor, Lawrence, KS, July 14–16, 1975.

4.18 van Dam, C. P., "Induced Drag Characteristics of Crescent-Moon-Shaped Wings," *AIAA J. of Aircraft, 24(2),* February 1987.

4.19 Smith, S. C., Kroo, I. M., "A Closer Look at the Induced Drag of Crescent-Shaped Wings," AIAA Paper 90-3063, AIAA 8th Applied Aerodynamics Conference, August 20–22, 1990.

4.20 van Dam, C. P., Vijgen, P. M., Holmes, H. W., "Wind-tunnel Investigation on the Effect of the Crescent Planform Shape on Drag," AIAA Paper 90-0300, AIAA Annual Conference, Reno, Nev., January 8–11, 1990.

4.21 "Research in Natural Laminar Flow and Laminar-Flow Control," NASA Conference Publication 2487, Parts 1, 2, and 3, *Proceedings of symposium at NASA Langley Research Center,* March 16–19, 1987.

4.22 Runyan, L. J., Bielak, G. W., Chen, A. W., Rozendaal, R. A., "757 NLF Glove Flight Test Results," paper contained in Ref. 4.21.

4.23 Maughmer, M. M., and Somers, D. N., "The Design of an Airfoil for a High-Altitude, Long-Endurance Remotely Piloted Vehicle," Paper contained in Ref. 4.21.

4.24 Holmes, B. J., Obara, C. J., "Observations and Implications of Natural Laminar Flow on Practical Airplanes," *AIAA J. of Aircraft, 20(12),* December 1983.

4.25 Loftin, J. K., and Burrows, D. L., *Investigations Relating to the Extension of Laminar Flow by Means of Boundary-Layer Suction Through Slots,* NACA TN 1961, October 1949.

4.26 Kosin, R. E., "Laminar Flow Control by Suction as Applied to the X-21A Airplane," *AIAA J. of Aircraft, 2(5),* September–October 1965.

4.27 Kerschner, M. E., "Laminar Flow: Challenge and Potential," paper contained in Reference 4.21.

4.28 Fisher, D. F., and Fischer, M. C., "Development Flight Tests of Jetstar LFC Leading-Edge Flight Test Experiment," paper contained in Reference 4.21.

4.29 Powell, A. G., "The Right Wing of the L. E. F. T. Airplane," paper contained in Reference 4.21.

4.30 Maddalon, D. V., Fisher, D. F., Jennett, L. A., Fischer, M. C., "Simulated Airline Service Experience with Laminar-Flow Control," paper contained in Reference 4.21.

4.31 McMasters, John H., and Palmer, G. M., "Possible Applications of Soaring Technology to Drag Reduction in Powered General Aviation Aircraft," a paper in *Proceedings of the NASA-Industry-University General Aviation Drag Reduction Workshop,* Jan Roskam, editor, Lawrence, KS, July 14–16, 1975.

5 LIFT AND DRAG AT HIGH MACH NUMBERS

The preceding material on lift and drag was limited primarily to incompressible flows; that is, to Mach numbers less than approximately 0.4. Compressibility becomes more and more important as the Mach number increases. In the vicinity of a Mach number of unity, airfoils and wings undergo a radical change in their behavior. It is not too surprising, therefore, to find that the equations covering the flow of air undergo a similar change at around $M_\infty = 1.0$.

Many textbooks are devoted entirely to the subject of compressible flows. References 5.1 land 5.2 are two such examples. Here, the several equations and techniques for the study of gas dynamics are developed in considerable detail. An excellent, lucid, qualitative explanation of compressibility effects on wings and airfoils is found in Reference 5.3.

QUALITATIVE BEHAVIOR OF AIRFOILS AS A FUNCTION OF MACH NUMBER

We will consider three regimes of flow around an airfoil. In the first, the flow is everywhere subsonic with a relative high Mach number. The second regime is referred to as transonic flow. Here, the free-stream Mach number is less than unity, but sufficiently high so that the flow locally, as it accelerates over the airfoil, exceeds the local speed of sound; that is, locally the flow becomes supersonic. The lowest free-stream Mach number at which the local flow at some point on the airfoil becomes supersonic is known as the *critical Mach number.* The third regime is the supersonic flow regime, in which the free-stream Mach number exceeds unity. Even here, a small region of subsonic flow may exist near the leading edge of the airfoil immediately behind a shock wave depending on the bluntness of the leading edge. The sharper the leading edge, the smaller is the extent of the subsonic flow region.

Subsonic Flow at High Mach Numbers

Since weak pressure disturbances propagate at the speed of sound, the time that a fluid particle ahead of a moving body is influenced by the pressure field around the body is proportional to the difference between the acoustic velocity, a, and the speed of the body, V. As V increases to a (i.e., as M_∞ approaches unity), the fluid is displaced less and less ahead of the body. Thus, the streamline pattern around an airfoil and hence, its pressure distribution can be expected to change with M_∞, even though the flow is subsonic everywhere.

As long as the flow remains entirely subsonic, the effect of M_∞ on airfoil characteristics can be estimated by the use of a factor, β, where β is defined as

$$\beta = \sqrt{1 - M_\infty^2} \tag{5.1}$$

β is known as the *Prandtl-Glauert compressibility correction factor.*

In a later, more complete treatment of β, it will be noted that the local pressure coefficient at a given point on an airfoil in subsonic compressible flow C_{p_c}, is related to the pressure coefficient in incompressible flow, C_{p_i}, by

$$C_{p_c} = \frac{C_{p_i}}{\beta} \tag{5.2}$$

It can easily be shown that

$$C_l = \int_0^1 (C_{p_l} - C_{p_u})\,dx \tag{5.3}$$

and

$$C_{m_{l_e}} = -\int_0^1 x(C_{p_l} - C_{p_u})\,dx \tag{5.4}$$

where x is the dimensionless distance along the airfoil chord and the subscripts l and u refer to lower and upper surfaces, respectively. Thus, it follows from Equations 5.2, 5.3, and 5.4 that the lift and moment coefficients for compressible flow are related to those for incompressible flow in a manner similar to Equation 5.2.

$$C_{l_c} = \frac{C_{l_i}}{\beta} \tag{5.5}$$

$$C_{M_c} = \frac{C_{M_i}}{\beta} \tag{5.6}$$

Notice from the use of Equations 3.11 and 3.12 that neither the center of pressure, X_{cp}, nor the location of the aerodynamic center, X_{ac}, varies with Mach number in the purely subsonic regime.

Obviously, the lift curve slope, C_{l_α}, also obeys Equation 5.5.

$$Cl_{\alpha_c} = \frac{C_{l_{\alpha_i}}}{\beta} \tag{5.7}$$

This relationship is presented graphically in Figure 5.1 together with the corresponding supersonic relationship, which will be discussed later. However, it must

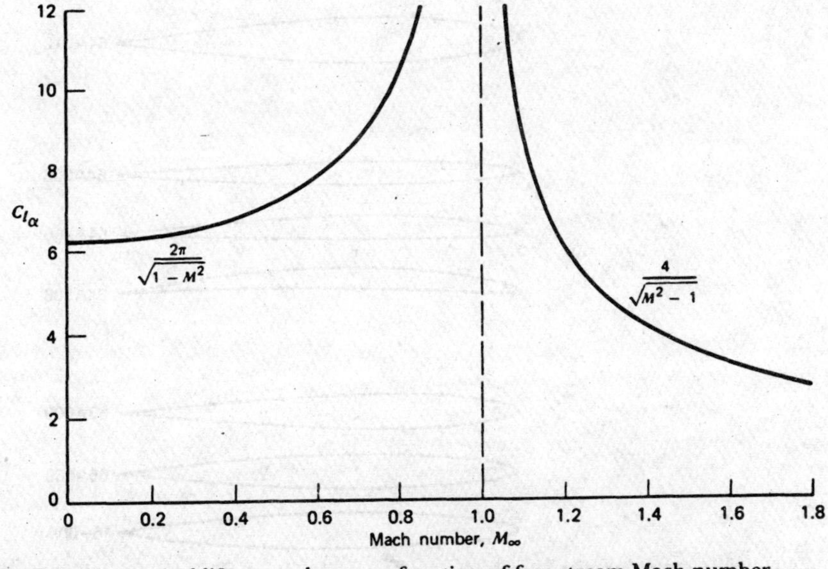

Figure 5.1 Theoretical lift curve slope as a function of free-stream Mach number.

be used with caution. First, the theoretical basis on which it rests is valid only for Mach numbers less than critical. Second, by comparison with experiment, the ratio C_{l_r}/C_{l_i} is overestimated by Equation 5.5 in some cases and underestimated for others, depending on the airfoil geometry.

Reference 5.4 presents data on nine different airfoils at Mach numbers up to 1.0. These airfoils vary in thickness, design lift coefficient, and thickness distribution; they are illustrated in Figure 5.2. Pressure distribution measurements were made to determine lift and pitching moment, and wake surveys were taken for determination of drag. Unfortunately, it is difficult to generalize on the data, and they are too voluminous to present here. A sample of the data is presented in Figure 5.3 for the 64A009 airfoil (taken from Ref. 5.4). The normal force coefficient, C_n, is defined as the force normal to the chord line (obtained by integrating the normal pressure around the airfoil contour) divided by the product of the free-stream dynamic pressure and the airfoil chord. C_d is the usual drag coefficient and is composed of the skin friction drag and the component of C_n in the drag direction. The lift coefficient is slightly less than C_n and can be obtained from

$$C_l = C_n \cos \alpha \tag{5.8}$$

Estimated critical Mach numbers are indicated by arrows in Figure 5.3 and were obtained from calculated graphs found in Reference 3.13. An example of such graphs is presented in Figures 5.4a and b. The results of Figure 5.4*a* apply approximately to the airfoils of Figure 5.3 and were used to obtain the M_{cr} values shown there. The 64Axxx airfoils are similar to the 64-xxx airfoils except that the rear portion of the 64Axxx airfoils are less curved than the corresponding surfaces of the 64-xxx airfoils.

Observe that the thinner symmetrical airfoils, as one might guess, have the higher critical Mach numbers at a C_l of zero. However, the rate at which M_{cr}

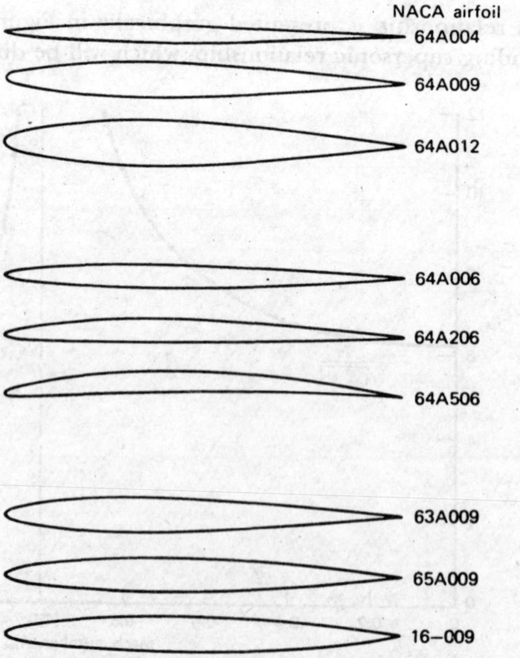

Figure 5.2 Airfoil profiles.

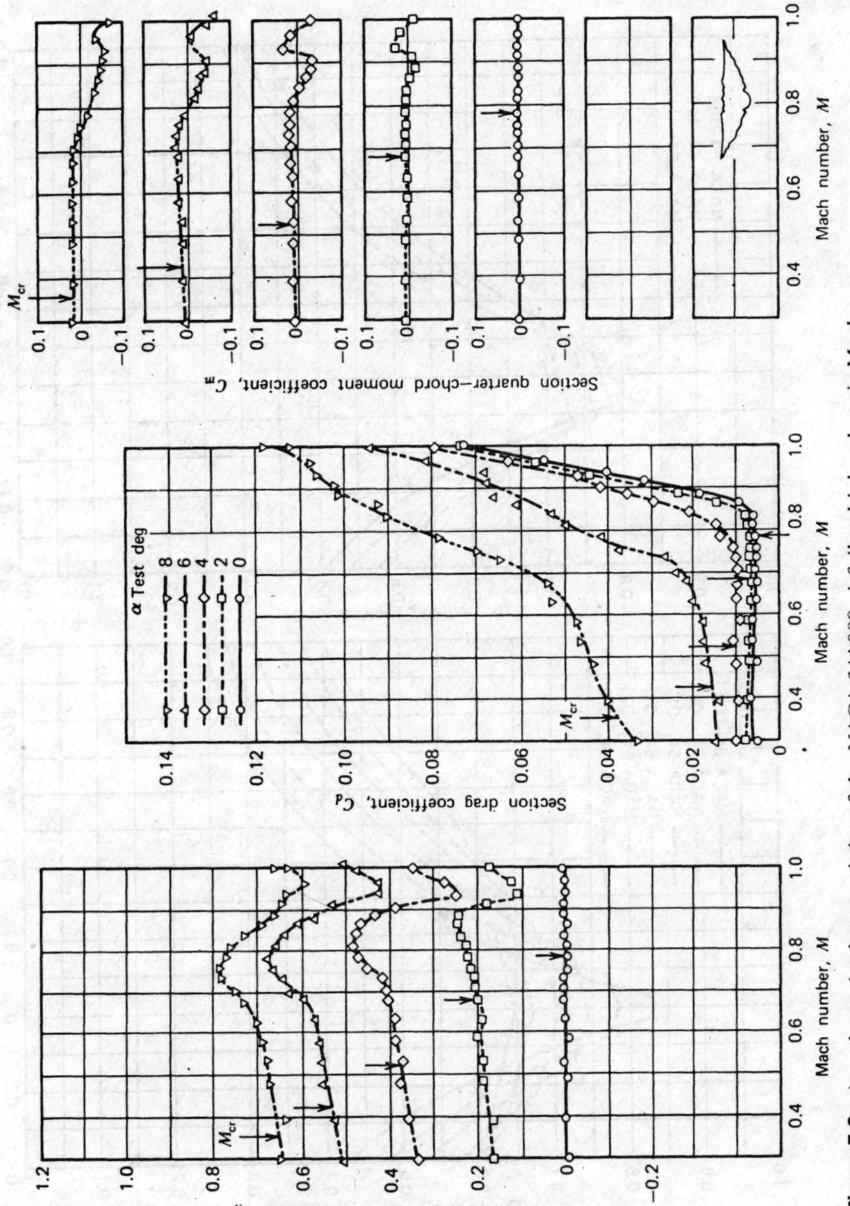

Figure 5.3 Aerodynamic characteristics of the NACA 64A009 airfoil at high subsonic Mach numbers.

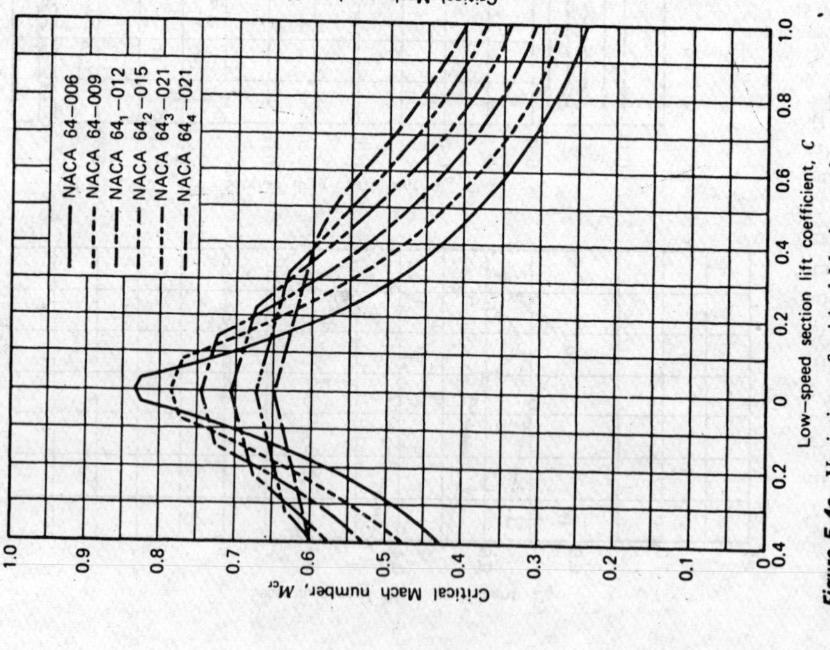

Figure 5.4a Variation of critical Mach number with low-speed section lift coefficient for several NACA 64-series symmetrical airfoil sections of various thicknesses.

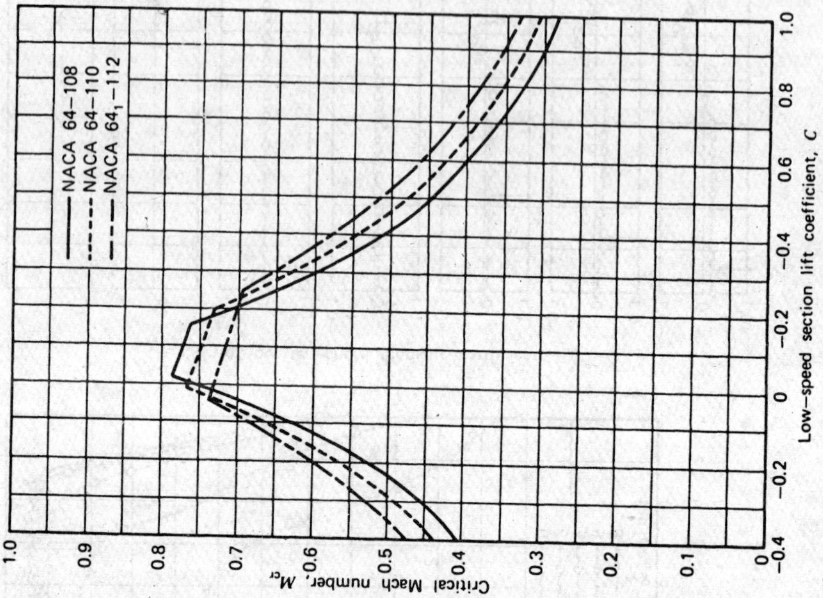

Figure 5.4b Variation of Critical Mach number with low-speed section lift coefficient for several NACA 64-series airfoil sections of various thicknesses, cambered for a design lift coefficient of 0.1.

decreases with C_l is greater for the thinner airfoils. Thus, the thicker airfoils become relatively more favorable as C_l increases. As shown in Figure 5.4*b*, camber results in shifting the peak M_{cr} to the right. As a function of thickness, the curves for the cambered airfoils are similar in appearance to those for the symmetrical airfoils.

It can be seen from Figure 5.3 that Equation 5.5 holds in a qualitative sense. At a given angle of attack, the lift coefficient increases with Mach number; however, the increase is not as great as Equation 5.5 predicts. For example, at an angle of attack of 6° and a Mach number of 0.3, C_l is equal to 0.51. Therefore, at this same angle of attack, one would predict a C_l of 0.61 at a Mach number of 0.6. Experimentally, however, C_l equals only 0.57 at the higher M value.

Figure 5.5, also based on the data of Reference 5.4, presents the variation with Mach number of the slope of the normal force coefficient curve for 4, 6, 9, and 12% thick airfoils. The theoretical variation of $C_{n\alpha}$ with M, matched to the experiment at an M of 0.3, is also included. Again, the Glauert correction is seen to be too high by comparison to the experimental results. Contrary to these observations, Reference 5.2 states that Equation 5.2 underestimates the effect of Mach number and presents a comparison between theory and experiment for a 4412 airfoil to substantiate the statement.

Reference 5.5 presents a graph similar to Figure 5.5 for symmetrical airfoils varying in thickness from 6 to 18%. The results are somewhat similar except that, at the lower Mach numbers, below approximately 0.8, the trend of $C_{l\alpha}$ with thickness is reversed. Both graphs show $C_{l\alpha}$ continuing to increase with a Mach number above the critical Mach number. Unlike Figure 5.5, the results presented in Reference 5.5 show a closer agreement with the Prandtl-Glauert factor for the lower thickness ratios.

Reference 5.6 is a voluminous collection of data pertaining to aircraft and missiles. Subsonic and supersonic data are given for airfoil sections, wings, bodies, and wing-body combinations. Any practicing aeronautical engineer should be aware of its existence and have access to the wealth of material contained therein. In Section 4 of this reference, the Prandtl-Glauert factor is used up to the critical Mach number. Isolated examples given in this reference using β show reasonably good

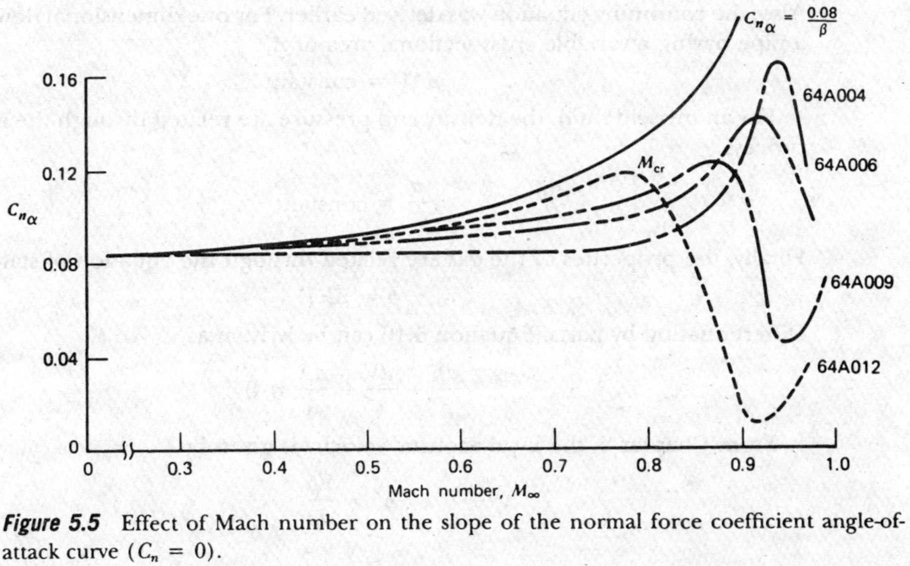

Figure 5.5 Effect of Mach number on the slope of the normal force coefficient angle-of-attack curve ($C_n = 0$).

agreement with test results. Thus, in the absence of reliable data, it is recommended that the Prandtl-Glauert compressibility correction be used, but with caution, keeping in mind discrepancies such as those shown in Figure 5.5.

From Figures 5.3 and 5.5, it is interesting to note that nothing drastic happens to the lift or drag when the critical Mach number is attained. Indeed, the lift appears to increase at a faster rate with Mach number for M values higher than M_{cr}. Only when M_{cr} is exceeded by as much as 0.2 to 0.4 does the normal force coefficient drop suddenly with increasing M_∞. The same general behavior is observed for C_d, except that the increments in M_∞ above M_{cr} where the C_d curves suddenly bend upward are somewhat less than those for the breaks in the C_n curves.

The value of M_∞ above which C_d increases rapidly with Mach number is known as the *drag-divergence Mach number.* A reliable determination of this number is of obvious importance in estimating the performance of an airplane such as a jet transport, designed to operate at high subsonic Mach numbers.

FUNDAMENTALS OF GAS DYNAMICS

Before proceeding further into the question of airfoil characteristics at Mach numbers higher than M_{cr}, it is necessary to develop some basic relationships relating to compressible subsonic and supersonic flows.

One-Dimensional Isentropic Flow

We will begin by considering briefly a reversible, adiabatic flow where the state of the flow is a function only of the position along the flow direction as, for example, a uniform flow through a duct. This simple case illustrates some of the pronounced differences between subsonic and supersonic flows. Applying the momentum theorem to a differential fluid element, Euler's equation of motion along a streamline was derived in Chapter 2 and is again stated here.

$$V\,dV + \frac{dp}{\rho} = 0 \tag{5.9}$$

Also, the continuity equation was derived earlier. For one-dimensional flow through a pipe having a variable cross-sectional area of A,

$$\rho A V = \text{constant} \tag{5.10}$$

For an inviscid fluid, the density and pressure are related through the isentropic process

$$\frac{p}{\rho^\gamma} = \text{constant} \tag{5.11}$$

Finally, the properties of the gas are related through the equation of state.

$$p = \rho R T \tag{5.12}$$

Differentiating by parts, Equation 5.10 can be written as

$$\frac{d\rho}{\rho} + \frac{dV}{V} + \frac{dA}{A} = 0 \tag{5.13}$$

From Chapter 2, the local acoustic velocity is given by

$$a^2 = \frac{dp}{d\rho} \tag{5.14}$$

Defining the local Mach number, M, as V/a and substituting Equations 5.13 and 5.14 into Equation 5.9 leads to a relationship between u and A.

$$\frac{dV}{ds} = \frac{-1}{1 - M^2} \frac{V}{A} \frac{dA}{ds} \tag{5.15}$$

Since V and A are both positive, we arrive at the surprising result (at least to those who have never seen it) that, for supersonic flow through a duct, an increase in cross-sectional area in the direction of flow will cause the flow to accelerate. Also, Equation 5.15 shows that a Mach number of unity can only occur if $dA/ds = 0$ since, for $M = 1$, dV/ds will be finite only if the cross-sectional area does not change with distance along the duct. This does not mean that M must equal unity when dA/ds equals zero, but instead that dA/ds equal to zero is a necessary condition for $M = 1$.

Consider flow from a reservoir through a converging-diverging nozzle, as pictured in Figure 5.6. Such a nozzle is referred to as a *Laval nozzle*. If the reservoir pressure, p_0, is sufficiently high relative to the exit pressure, p_E, the flow will accelerate to Mach 1 at the throat. Beyond the throat, with dA/ds positive, the flow will continue to accelerate, thereby producing a supersonic flow. Downstream of the throat, pressure and density decrease as the velocity increases with the increasing area.

The compressible Bernoulli equation governing one-dimensional isentropic flow was derived earlier. In terms of the local acoustic velocity,

$$\frac{V^2}{2} + \frac{a^2}{\gamma - 1} = \text{constant}$$

For this case of flow from a reservoir,

$$\frac{V^2}{2} + \frac{a^2}{\gamma - 1} = \frac{{a_0}^2}{\gamma - 1} \tag{5.16}$$

Dividing this equation through by a^2 and using the isentropic relationships among p, ρ, and T leads to these three quantities as a function of the local Mach number.

$$\frac{T}{T_0} = \left(1 + \frac{\gamma - 1}{2} M^2\right)^{-1} \tag{5.17a}$$

$$\frac{p}{p_0} = \left(1 + \frac{\gamma - 1}{2} M^2\right)^{-\gamma/(\gamma-1)} \tag{5.17b}$$

$$\frac{\rho}{\rho_0} = \left(1 + \frac{\gamma - 1}{2} M^2\right)^{-1/(\gamma-1)} \tag{5.17c}$$

These relationships, presented graphically in Figure 5.7, are valid for Mach numbers greater than unity if the flow is shockless. The subject of shock waves will be treated later.

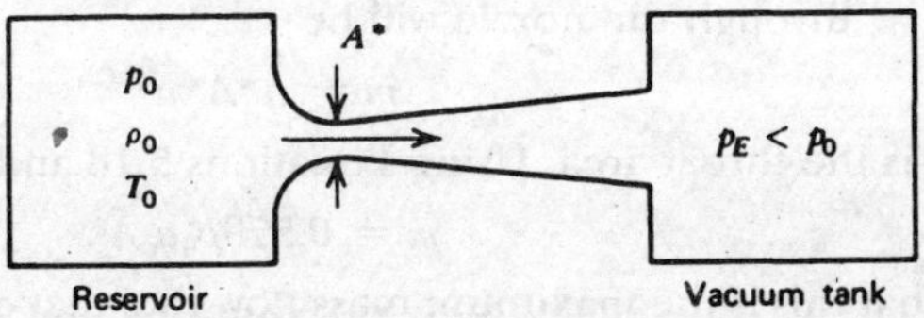

Figure 5.6 Flow from a reservoir through a converging–diverging nozzle.

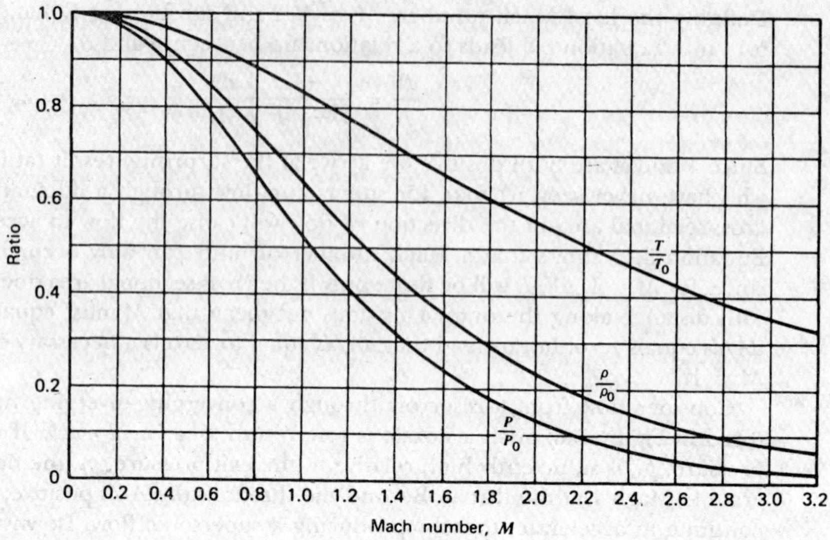

Figure 5.7. Isentropic flow. Pressure, density, and temperature as a function of Mach number.

At the throat the local velocity and the local acoustic velocity are equal. Designating this velocity by a^*, Equation 5.16 can be written as

$$\frac{a^{*2}}{2} + \frac{a^{*2}}{\gamma - 1} = \frac{a_0^2}{\gamma - 1}$$

or

$$\frac{a^*}{a_0} = \left[\frac{2}{(\gamma + 1)}\right]^{1/2} = 0.913 \tag{5.18}$$

Using Equations 5.11 and 5.14, it follows that

$$\frac{p^*}{p_0} = \left[\frac{2}{(\gamma + 1)}\right]^{\gamma/(\gamma-1)} = 0.528 \tag{5.19}$$

$$\frac{\rho^*}{\rho_0} = \left[\frac{2}{(\gamma + 1)}\right]^{1/(\gamma-1)} = 0.634 \tag{5.20}$$

In these equations, the superscript * refers to the throat.

Equation 5.19 shows that the airflow from a reservoir will reach Mach 1 if the reservoir pressure exceeds the exit pressure by a factor of at least 1.894. The mass flow rate, m, through the nozzle will be

$$m = \rho^* A^* a^* \tag{5.21}$$

where A^* is the throat area. Using Equations 5.18 and 5.20, this becomes

$$m = 0.579 \rho_0 a_0 A^* \tag{5.22}$$

Observe that this is the maximum mass flow rate that can be obtained from a given reservoir independent of the exit pressure. For example, consider two tanks connected together through a nozzle having a throat area of 1 m^2, as pictured in Figure 5.6. Assume that the air in both tanks is at standard sea level conditions. We will now begin to lower the pressure in one tank, causing air to flow from the other tank into the one with the vacuum. As the pressure in the vacuum tank is gradually reduced, the mass flow through the pipe will increase continuously, assuming the volume in the other tank, or reservoir, is sufficiently large so that its pressure and density do not change significantly. The pressure drop along the pipe resulting from skin friction will exactly equal the pressure difference between the two tanks. However, when the pressure in the vacuum tank is reduced to 53.5 kN/m^2 (from Eq. 5.19), a value of $M = 1$ occurs at the throat. The nozzle, or flow, is then said to be "choked," since a further reduction in the pressure downstream of the throat will not result in any further increase in the mass flow. From Equation 5.22, this critical mass flow will equal 242.3 kg/s.

If we assume that the flow beyond the throat is still isentropic, Equations 5.17, 5.18, 5.20, and 5.21 can be combined to give

$$\left(\frac{A}{A^*}\right)^2 = \frac{\gamma - 1}{2} \frac{(2/\gamma + 1)^{(\gamma+1)/(\gamma-1)}}{[1 - (p/p_0)^{(\gamma-1)/\gamma}](p/p_0)^{2/\gamma}} \tag{5.23}$$

This is known as *St. Venant's equation.*

Substituting for the local pressure ratio in terms of Mach number, this can also be written as

$$\left(\frac{A}{A^*}\right)^2 = \frac{1}{M^2}\left[\frac{2}{\gamma + 1}\left(1 + \frac{\gamma - 1}{2} M^2\right)\right]^{(\gamma+1)/(\gamma-1)} \tag{5.24}$$

Since p, ρ, and T are related through the adiabatic process and the equation of state, it follows from the foregoing that p, ρ, and T are all uniquely related to their corresponding reservoir values by the ratio of A to A^*. This obviously raises some problems since, in the example of Figure 5.6, the pressure at the exit into the vacuum tank does not necessarily have to match the pressure from Equation 5.23 corresponding to the area of the duct at its connection to the vacuum tank. Some nonisentropic mechanism must exist that will allow the pressure to adjust to exit conditions. This leads us to the concept of a shock wave.

Normal Shock Waves

Figure 5.8 pictures a shock wave normal to a one-dimensional flow through a duct having a unit cross-sectional area. The *shock wave* is a surface across which the flow properties p, ρ, and T, and V change discontinuously. We will now examine the equations governing the flow to see if such a standing wave is possible, and to determine the relationships between the upstream and downstream fluid proper-

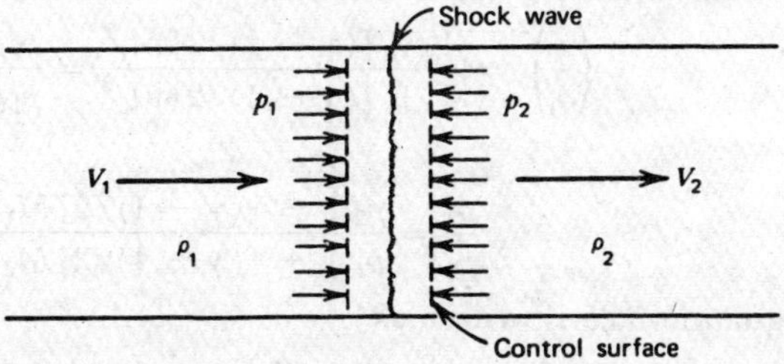

Figure 5.8 Flow through a normal shock wave.

ties. To begin, the equations governing the conservation of mass and momentum must hold.

continuity

$$\rho_1 V_1 = \rho_2 V_2 = m \tag{5.25}$$

momentum

$$p_1 - p_2 = m(V_2 - V_1) \tag{5.26}$$

In addition, the equation of state, Equation 2.1, must also hold. It is repeated here for convenience.

equation of state

$$p = \rho RT \tag{5.27}$$

A fourth relationship, which has not been used as yet, is the *energy equation.*

energy

$$C_p T_1 + \frac{V_1^2}{2} = C_p T_2 + \frac{V_2^2}{2} = C_p T_0 \tag{5.28}$$

This equation, derived in Reference 5.5, applies to adiabatic flows where no heat is added to the flow. C_p is the specific heat at constant pressure. The product C_pT is the enthalpy of the flow per unit mass. Thus, Equation 5.28 states that the sum of the enthalpy and kinetic energy per unit mass of an adiabatic flow remains constant. C_p, R, γ, and the specific heat at constant volume, C_v, are all interrelated.

$$R = C_p - C_v$$

$$\gamma = \frac{C_p}{C_v} \tag{5.29}$$

$$C_p = \frac{R\gamma}{(\gamma - 1)}$$

If Equation 5.29 is substituted into the energy equation (Equation 5.28), it is interesting to note that one obtains the compressible Bernoulli equation (Equation 5.16). Thus, Equation 5.16 and the energy equation are equivalent for isentropic flow. However, across the shock wave the flow is not a reversible, adiabatic process, so the changes in state are not related by Equation 5.11.

To see how p, ρ, T, and M change across a normal shock wave, we begin by substituting Equation 5.25 into Equation 5.26 so that

$$p_1 + \rho_1 V_1^2 = p^2 + \rho V_2^2$$

Since $a^2 = \gamma p/\rho$, it follows that

$$p_l + \gamma p_1 M_1^2 = p_2 + \gamma p_2 M_2^2$$

or

$$\frac{p_2}{p_2} = \frac{1 + \gamma M_1^2}{1 + \gamma M_2^2} \tag{5.30}$$

Manipulating Equation 5.28 in a similar manner, it follows that

$$\left(\frac{a_2}{a_1}\right)^2 = \frac{1 + [(\gamma - 1)/2] M_1^2}{1 + [(\gamma - 1)/2] M_2^2} = \frac{p_2 \rho_1}{p_1 \rho_2} = \frac{T_2}{T_1} \tag{5.31}$$

Thus,

$$\frac{p_2}{p_1} = \frac{\rho_2}{\rho_1} \frac{1 + [(\gamma - 1)/2] M_1^2}{1 + [(\gamma - 1)/2] M_2^2} \tag{5.32}$$

Next, Equation 5.25 is written as

$$\frac{\rho_1}{\rho_2} \frac{M_1}{M_2} \frac{a_1}{a_2} = 1.0$$

Substituting Equations 5.30, 5.31, and 5.32 into the preceding equations leads to an implicit relationship for M_2 as a function of M_1.

$$\frac{M_1\sqrt{1+[(\gamma-1)/2]M_1^2}}{1+\gamma M_1^2}=\frac{M_2\sqrt{1+[(\gamma-1)/2]M_2^2}}{1+\gamma M_2^2}=f(M) \quad (5.33)$$

Obviously, one solution of the above is $M_2 = M_1$, in which case $p_2 = p_1$, $\rho_2 = \rho_1$, and $T_2 = T$, so that there is no discontinuity in the flow and the solution is trivial. The other solution is apparent from Figure 5.9, where $f(M)$ is presented as a function of M. It is seen that the same value for $f(M)$ is obtained from two different values of M, one greater and one less than unity. For example, if M_1 is equal to 2.0, Equation 5.33 would be satisfied by an M_2 value of approximately 0.57. One might, of course, say that a value of M_1 equal to 0.57 with M_2 equal to 2.0 would also satisfy Equation 5.33 which, indeed, would be the case. However, it can be argued on the basis of the second law of thermodynamics that the flow ahead of the shock wave must be supersonic (see Ref. 5.6, p. 234). An entropy loss, in violation of the second law, will occur if M_1 is less than unity. Therefore, in Figure 5.9, M_1 is greater than unity and M_2 is the value of M less than unity for which

$$f(M_2) = f(M_1)$$

Notice from Equation 5.33 that $f(M)$ approaches a value of $(\gamma - 1)^{1/2}/\gamma$, or approximately 0.319, as $M_1 \rightarrow \infty$. Thus, behind a normal shock wave, the Mach number has a lower limit of approximately 0.38.

Since there is an entropy gain across the normal shock wave, a loss occurs in the total, or reservoir, pressure as the flow passes through the wave. With some algebraic manipulation of the energy equation and application of the isentropic relationships before and after the wave (but not across it), the following equation can be obtained:

$$\frac{p_{0_2}}{p_{0_1}}=\frac{p_2}{p_1}\left[\frac{1+[(\gamma-1)/2]M_2^2}{1+[(\gamma-1)/2]M_1^2}\right]^{\gamma(\gamma-1)} \quad (5.34)$$

p_{0_2} and p_{0_1} are the reservoir pressures behind and ahead of the wave, respectively.

Equations 5.30 to 5.34 are unwieldy to use because of their implicit nature. However, after some algebraic manipulation, they can all be reduced to explicit

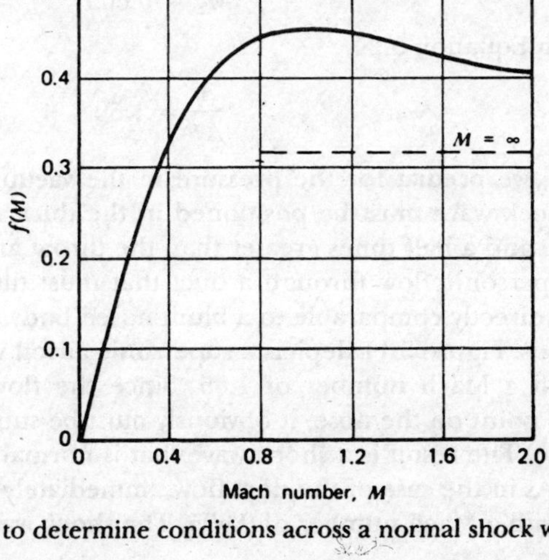

Figure 5.9 Function to determine conditions across a normal shock wave.

functions of M_1. These can be found in Reference 5.7 and are repeated here. Following the lead of Reference 5.7, the second form of each equation is for $\gamma = 7/5$.

$$\frac{p_2}{p_1} = \frac{2\gamma M_1^2 - (\gamma - 1)}{\gamma + 1} = \frac{7M_1^{\,2} - 1}{6} \tag{5.35}$$

$$\frac{\rho_2}{\rho_1} = \frac{(\gamma + 1)\, M_1^{\,2}}{(\gamma - 1)\, M_1^{\,2} + 2} = \frac{6\, M_1^{\,2}}{M_1^{\,2} + 5} \tag{5.36}$$

$$M_2^{\,2} = \frac{(\gamma - 1)\, M_1^{\,2} + 2}{2\gamma M_1^{\,2} - (\gamma - 1)} = \frac{M_1^{\,2} + 5}{7M_1^{\,2} - 1} \tag{5.37}$$

$$\frac{p_{0_2}}{p_{0_1}} = \left[\frac{(\gamma + 1)\, M_1^{\,2}}{(\gamma - 1)\, M_1^{\,2} + 2}\right]^{\gamma/(\gamma - 1)} \cdot \left[\frac{\gamma + 1}{2\gamma M_1^{\,2} - (\gamma - 1)}\right]^{1/(\gamma - 1)} \tag{5.38}$$

$$= \left(\frac{6M_1^{\,2}}{M_1^{\,2} + 5}\right)^{7/2} \left(\frac{6}{7M_1^{\,2} - 1}\right)^{5/2}$$

Figure 5.10 presents these relationships graphically and can be used for approximate calculations.

Let us now return to the problem of flow in the duct illustrated by Figure 5.8. Suppose the pressure in the vacuum tank is lowered to a value of 80 kN/m^2. Furthermore, let us assume that the area of the duct entering the second tank is large so that, as the air enters the tank, its velocity, and hence, dynamic pressure, is low. Thus, 80 kN/m^2 would represent approximately the reservoir pressure downstream of a normal shock in the duct. The upstream reservoir pressure is equal to the standard sea level value of 101.3 kN/m^2. Thus,

$$\frac{p_{0_2}}{p_{0_1}} = 0.790$$

From Figure 5.10, the Mach number, M_1, just upstream of the normal shock wave, equals 1.85. In addition,

$$\frac{p_2}{p_1} = 3.85$$

$$\frac{\rho_2}{\rho_1} = 3.43$$

$$M_2 = 0.605$$

Thus, from Equation 5.24,

$$\frac{A}{A^*} = 1.50$$

Therefore, we predict for the pressure in the vacuum tank of 80 kN/m^2 that a normal shock wave must be positioned in the duct at a location where the duct area is one and a half times greater than the throat area.

This supersonic flow through a duct that must ultimately come to rest in the vacuum is directly comparable to a blunt-nosed body, or airfoil, traveling at supersonic speeds. Figure 5.11 depicts a supersonic airfoil with a rounded leading edge traveling at a Mach number of 1.85. Since the flow must come to rest at the stagnation point on the nose, it obviously must be subsonic for some extent ahead of the nose. The result is a shock wave that is normal to the flow in the vicinity of the nose. As in the case of the duct flow, immediately behind this wave, the flow is subsonic with a Mach number of 0.605. The shock wave, positioned away from the

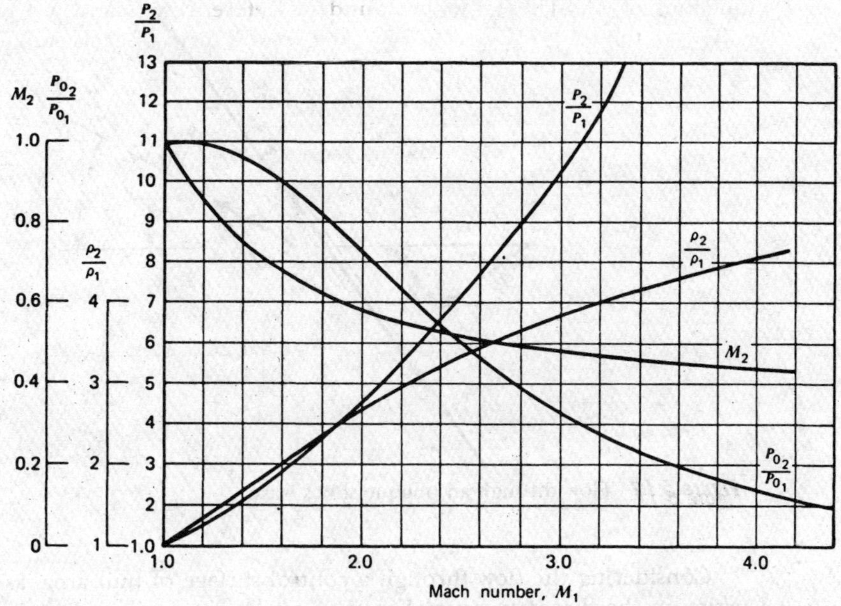

Figure 5.10 Pressure, density, and Mach number changes across a normal shock wave.

nose some small distance, is referred to as a "detached" shock wave, since it is detached from the surface.

Oblique Shock Waves

Generally, a shock wave is not normal to the flow. In Figure 5.11, for example, the wave becomes oblique to the flow as one moves away from the nose. Let us therefore examine this more general case in the same manner as we did the normal shock wave.

Figure 5.12 pictures a flow passing through an oblique shock wave that is at an angle of θ relative to the incoming velocity vector. θ is equal to 90° for a normal shock. We will assume that the flow is deflected through the angle δ as it passes through the wave, with δ being in the direction shown in the figure.

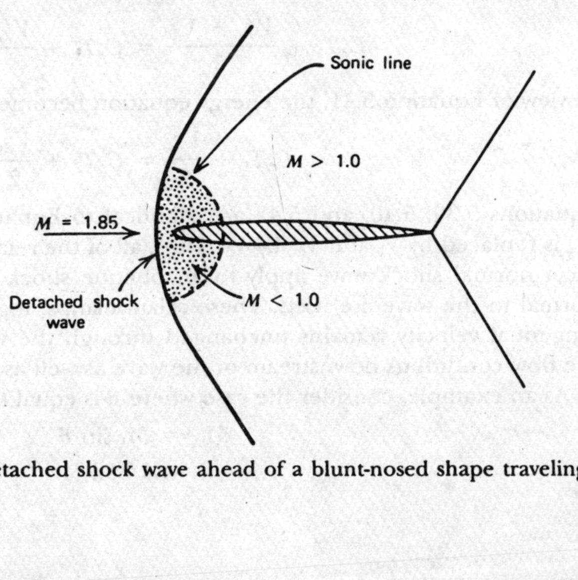

Figure 5.11 A detached shock wave ahead of a blunt-nosed shape traveling at supersonic speed.

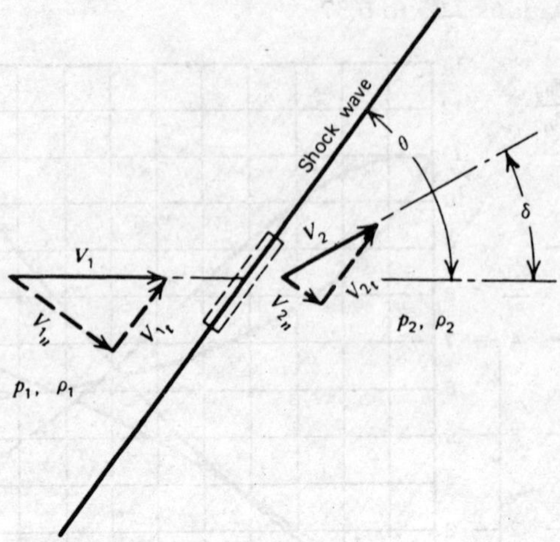

Figure 5.12 Flow through an oblique shock wave.

Considering the flow through a control surface of unit area, as shown, we can write, in the direction normal to the wave, two equations directly comparable to Equations 5.25 and 5.26.

continuity

$$\rho_1 V_{1_n} = \rho V_{2_n} = m \tag{5.39}$$

momentum

$$p_1 - p_2 = m(V_{2_n} - V_{1_n}) \tag{5.40}$$

In the direction tangential to the wave, the momentum theorem gives

momentum

$$0 = M(V_{2_t} - V_{1_t})$$

or

$$V_{1_T} = V_{2_t} = V_t \tag{5.41}$$

Again, the energy equation holds.

energy

$$C_p T_1 + \frac{V_{1_n}^2 + V_{1_t}^2}{2} = C_p T_2 + \frac{V_{2_n}^2 + V_{2_t}^2}{2}$$

In view of Equation 5.41, the energy equation becomes

$$C_p T_1 + \frac{V_{1_n}^2}{2} = C_p T_2 + \frac{V_{2_n}^2}{2} \tag{5.42}$$

Equations 5.39, 5.40, and 5.42 are identical to Equations 5.25, 5.26, and 5.28 if V_{1_n} is replaced by V_1 and V_{2_n} by V_2. Thus, all of the relationships previously derived for a normal shock wave apply to an oblique shock wave if the Mach numbers normal to the wave are used. These relationships, together with the fact that the tangential velocity remains unchanged through the wave, allow us to determine the flow conditions downstream of the wave as well as the angles θ and δ.

As an example, consider the case where θ is equal to 50° and M_1 equals 2.0.

$$M_{1_n} = M_1 \sin\theta$$
$$= 1.532$$

From Equations 5.35 to 5.37,

$$\frac{p_2}{p_1} = 2.57$$

$$\frac{\rho_2}{\rho_1} = 1.92$$

$$M_{2_n} = 0.69$$

Now we must be careful, because the tangential velocity is constant across the wave, *not* the tangential Mach number. To obtain M_{2_t}, we write

$$\begin{aligned} M_{2_t} &= \frac{V_t}{a_2} \\ &= M_{t_1}\frac{a_1}{a_2} \\ &= M_1 \cos\theta \sqrt{\frac{\rho_2/\rho}{p_2/p_1}} \end{aligned}$$

or, in this case,

$$M_{2_t} = 1.11$$

The turning angle, δ, can now be determined from the geometry of Figure 5.12.

$$\delta = \theta - \frac{\pi}{2} + \tan^{-1}\frac{M_{2_t}}{M_{2_n}} \tag{5.43}$$

or, in this example,

$$\delta = 18.1°$$

Notice for this example that

$$\begin{aligned} M_2 &= \sqrt{M_{2n}^2 + M_{2_t}^2} \\ &= 1.31 \end{aligned}$$

Thus, the flow is still supersonic after it has passed through the wave, unlike the flow through a normal shock wave. (As an exercise, repeat the foregoing example, but with a θ of 76.5°.)

Surprisingly, the same turning angle is obtained for the same upstream Mach number but different wave angle, θ. For this steeper wave, which is more like a normal shock, the flow becomes subsonic behind the wave, M_2 being equal to 0.69.

The deflection angle, δ, as a function of θ, for a constant Mach number will appear as shown qualitatively in Figure 5.13. For a given M_1, a maximum deflection angle exists with a corresponding shock wave angle. For deflections less than the maximum, two different θ values can accomplish the same deflection. The oblique shock waves corresponding to the higher θ values are referred to as *strong waves,* while the shock waves having the lower θ values are known as *weak waves.* There appears to be no analytical reason for rejecting either possible family of waves but, experimentally, one finds only the *weak* oblique shock waves. Thus, the flow tends to remain supersonic through the wave unless it has no other choice. If, for a given M_1, the boundary of the airfoil requires a turning greater than δ_{max}, the wave will become detached, as illustrated in Figure 5.11. The flow then becomes subsonic just behind the normal part of the wave and navigates around the blunt nose under

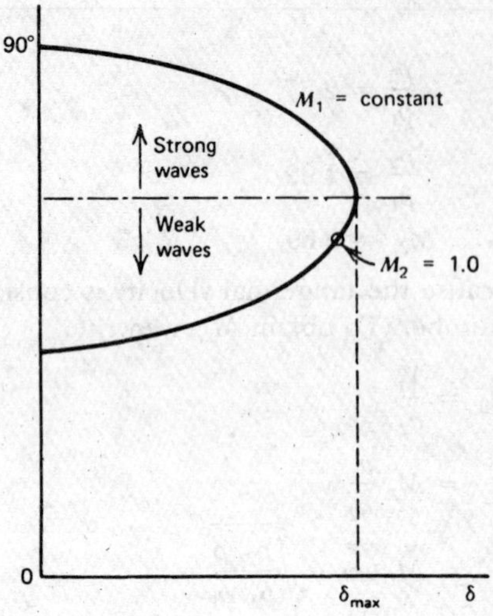

Figure 5.13 Relationship between the shock wave angle, θ, and the deflection angle, δ.

the influence of pressure gradients propagated ahead of the airfoil in the subsonic flow region. It then accelerates downstream, again attaining supersonic speeds.

An explicit relationship for δ as a function of M_1 and θ can be obtained by applying the equations for the normal shock wave to the Mach number normal components of the oblique shock wave, $M_1 \sin \theta$ and $M_2 \sin (\theta - \delta)$. After a considerable amount of algebraic reduction, one obtains the result

$$\tan \delta = \frac{M_1^2 \sin 2\theta - 2 \cot \theta}{2 + M_1^2(\gamma + \cos 2\theta)}$$
$$= 5 \frac{M_1^2 \sin 2\theta - 2 \cot \theta}{10 + M_1^2(7 + 5 \cos 2\theta)}$$
$$(\text{for } \gamma = 7/5) \qquad (5.44)$$

This equation is presented graphically in Figure 5.14 (taken from Ref. 5.7) for a range of Mach numbers and shock wave angles from 0 to 90°. θ values lying below the broken line correspond to weak oblique shock waves. This dividing line is close to but slightly below the solid line through the maximum deflection angles.

Expansion Waves

Let us consider the two supersonic flows pictured in Figure 5.15*a* and 5.15*b*. When the flow is turned by a surface concave to the flow, as in Figure 5.15*a*, we have seen that an oblique shock originating from the bend in the surface will compress the flow and turn it through the angle, δ. The question then posed is, how is the flow turned around a bend convex to the flow, as shown in Figure 5.15*b*. As suggested by the figure, this is accomplished through a continuous ensemble of weak expansion waves, known as an *expansion fan*.

To examine the flow relationships in this case, we take an approach similar to that for oblique shock waves. Consider supersonic flow through a single, weak wave, known as a *Mach wave*, as illustrated in Figure 5.16. The wave represents a limiting case of zero entropy gain across the wave. Hence, the turning and velocity changes are shown as differentials instead of as finite changes. Since the wave is a weak

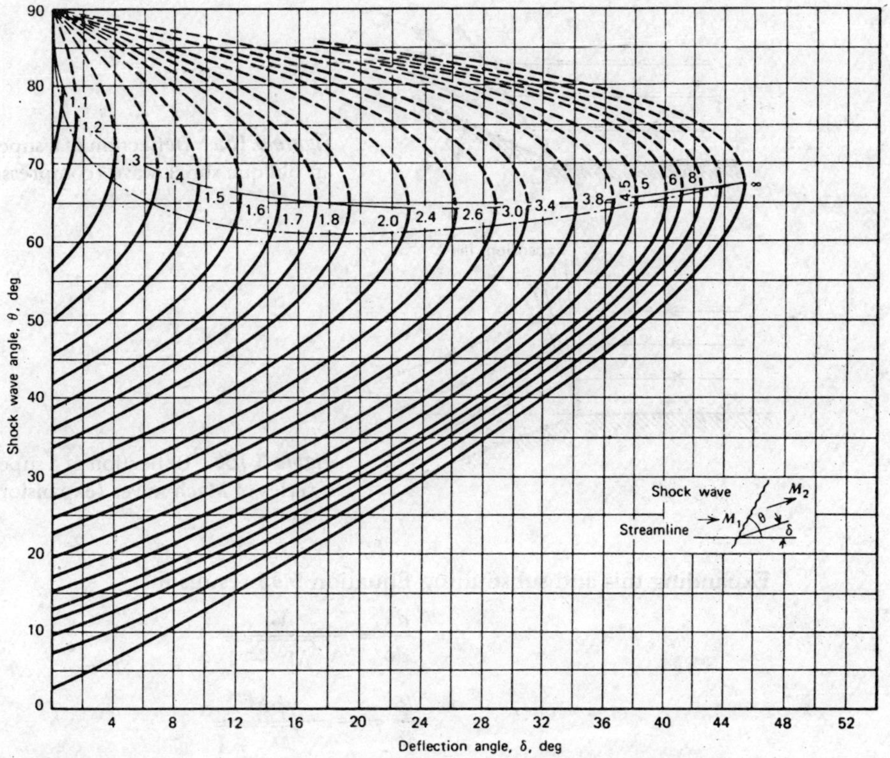

Figure 5.14 Variation of shock wave angle with flow deflection angle of various upstream Mach numbers.

wave, it propagates normal to itself at the acoustic velocity, a, which added vectorially to the free-stream velocity, V, defines the angle of the wave μ.

$$\mu = \sin^{-1}\frac{a}{V}$$

$$= \sin^{-1}\frac{1}{M} \tag{5.45}$$

Applying momentum principles across the wave, as done previously for the oblique shock wave, results in

$$dV_t = 0$$

and

$$-dp = \rho a[(V + dV)\sin(\mu + d\delta) - a]$$

This reduces to

$$-dp = \rho a^2\left[\frac{dV}{V} + \sqrt{M^2 - 1}\, d\delta\right]$$

Since the tangential velocity component is unchanged across the wave, it follows that

$$\frac{a}{\tan\mu} = (V + dV)\cos(\mu + d\delta)$$

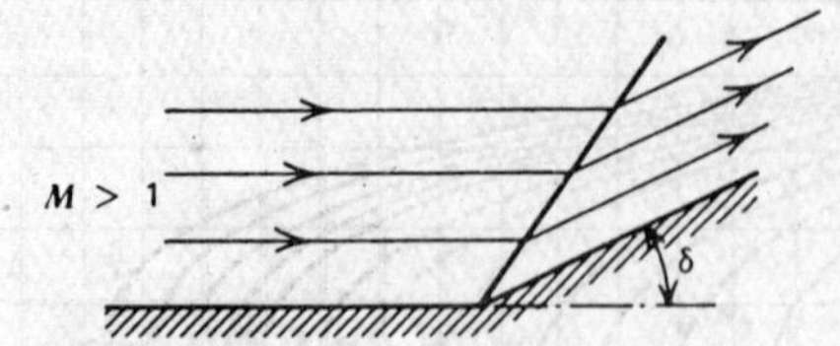

Figure 5.15a Deflection of a supersonic flow by an oblique shock wave (compression).

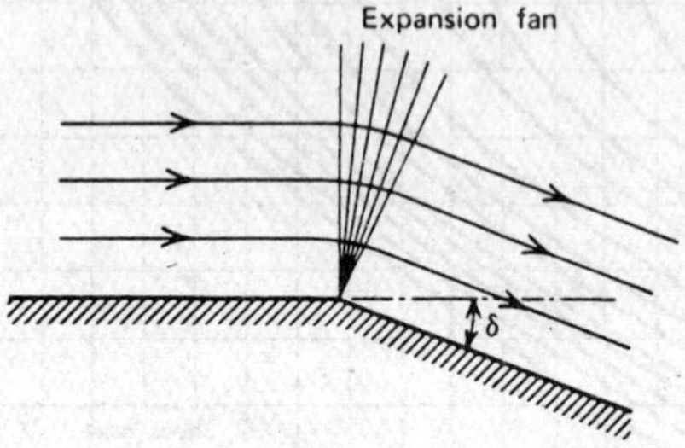

Figure 5.15b Deflection of a supersonic flow by a series of Mach waves (expansion).

Expanding this and substituting Equation 5.45 results in

$$\frac{dV}{d\delta} = \frac{V}{\sqrt{M^2 - 1}} \tag{5.46}$$

$$\frac{dp}{d\delta} = -\frac{\gamma p M^2}{\sqrt{M^2 - 1}} \tag{5.47}$$

Thus, this weak wave, deflecting the flow in the direction shown in Figure 5.16, results in an expansion of the flow, since $dp/d\,\delta$ is negative. It is also possible for small deflections in the opposite direction to produce a compression with a Mach wave. This represents a limiting case of an oblique shock wave.

The expansion fan shown in Figure 5.15*b* represents a continuous distribution of Mach waves. Each wave deflects the flow a small amount, so that the integrated effect produces the total deflection, δ. The changes in the flow can be related to the total deflection by integrating Equation 5.46. The energy equation is used to relate the local sonic velocity to V. It is convenient in so doing to let $\delta = 0$ at $M = 1.0$. This corresponds to $V = a^*$ for a given set of reservoir conditions. Therefore,

$$\delta = \int_{a^*}^{V} \frac{\sqrt{M^2 - 1}}{V}\, dV$$

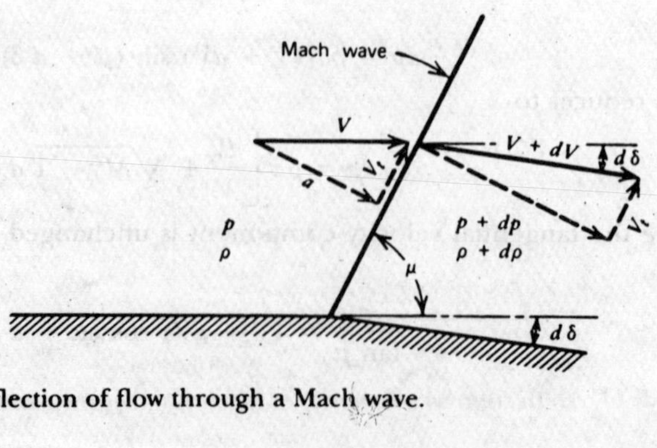

Figure 5.16 Deflection of flow through a Mach wave.

The details of performing this integration will not be presented here. They can be found in several texts and in Reference 5.7. The final expression for δ becomes

$$\delta = \sqrt{\frac{\gamma + 1}{\gamma - 1}} \tan^{-1} \sqrt{\frac{\gamma - 1}{\gamma + 1}(M^2 - 1)} - \cos^{-1}\frac{1}{M} \tag{5.48}$$

This relationship is presented graphically in Figure 5.17 and is referred to as *Prandtl-Meyer flow*. To use this graph one relates a given flow state back to the $M = 1$ condition. For example, suppose the local Mach is equal to 3.0. This means that, relative to $M = 1$, the flow has already been deflected through an angle of approximately 50°. Suppose the flow is turned an additional 50°. Relative to $M = 1$, this gives a total deflection of 100°. Thus, one enters Figure 5.17 with this value of δ to determine a final Mach number slightly in excess of 9.0. Since the Prandtl-Meyer flow is isentropic, the flow state is determined completely by the reservoir conditions and the local Mach number (Eq. 5.17).

Computer Exercise 5.1 "SUPER"

Write four subroutines to do the following calculations:

1. Isentropic flow relationships given by Equations 5.17–5.20 with M as the input.
2. Normal shock relationships given by Equations 5.35–5.38 with M as input.
3. Oblique shock relationships for the weak wave family with M and deflection angle, δ, as input. An iteration can be used by initially assuming $\theta = \delta$ and then incrementing θ until the calculated δ is equal to the desired δ within a specified tolerance.
4. Prandtl Meyer relationship, Equation 5.48, with M as the input.

These subroutines will be used later in a program to predict the lift, wave drag, and moment coefficients of a supersonic airfoil. At this point, write a main program, which will call and execute the desired subroutine.

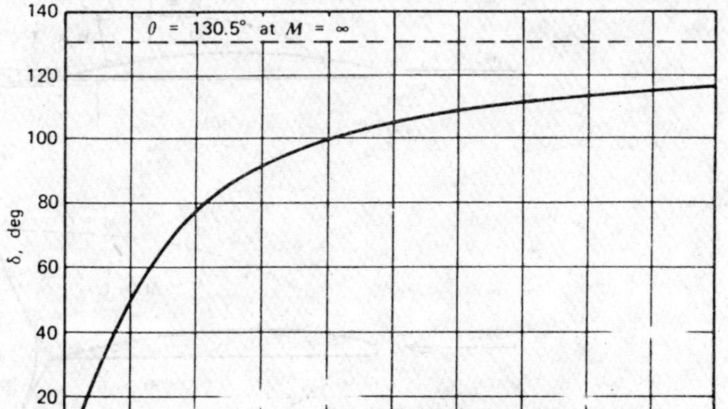

Figure 5.17 Deflection as a function of Mach number for expansion flows.

TRANSONIC AIRFOILS

An airfoil operating at high, but subsonic, Mach numbers is pictured in Figure 5.18. If the free-stream Mach number is sufficiently high, the local flow as it progresses back along the upper surface will reach a point where the local Mach is equal to, or greater than, unity. As the flow continues along the concave surface, a region of supersonic flow develops. However, as the flow approaches the trailing edge, it must eventually become subsonic again. As we saw in the preceding section, this can occur only through a shock wave.

If the compression to subsonic flow occurs before the trailing edge, as shown in Figure 5.18*a*, there is no deflection of the flow as it passes through the wave, so that the shock wave is normal to the flow. As the free-stream Mach number is increased, a similar transonic region will develop on the lower surface, as shown in Figure 5.18*b*. Immediately behind both normal shocks, the boundary layer will separate because of the strong positive pressure gradients. This results in a loss in lift and a sharp increase in the drag. Increasing M_∞ still further will result in the shock waves on both the upper and lower surfaces moving to the trailing edge. Here, they become oblique shocks in order to turn the flow from both surfaces in the free-stream direction. Because of the developing boundary layer, as the shocks move toward the trailing edge, they assume a bifurcated or λ form, as shown. Here, within the boundary layer, compression begins initially through an oblique shock and continues through to a normal shock.

Increasing the Mach number also causes the sonic line, defining the forward extent of the supersonic flow, to move forward. This line, shown dashed in Figure 5.18, is a constant pressure surface along which $M = 1$.

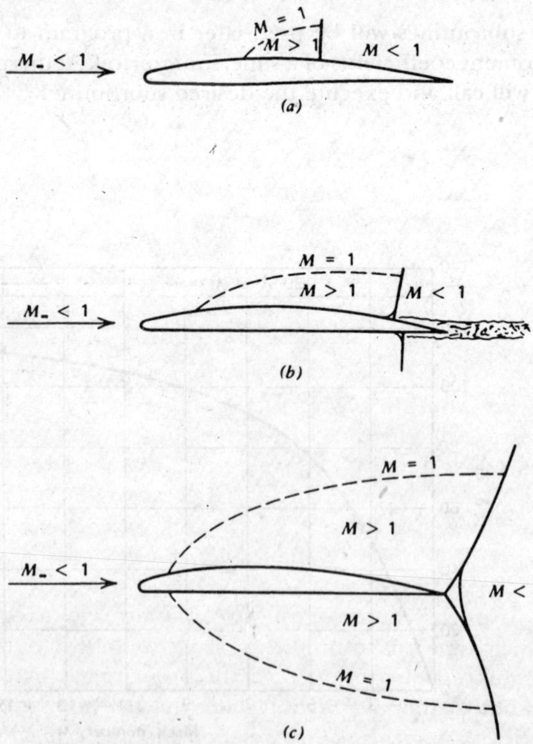

Figure 5.18 Airfoils in transonic flow.

A detailed treatment of analytical methods for predicting airfoil characteristics is beyond the scope of this text. However, an interesting aspect of transonic airfoil behavior, discussed in Reference 5.8, is the limiting Mach number concept, which also leads to a limit on pressure coefficients. This particular reference presents semiempirical methods for estimating two-dimensional and three-dimensional values of C_L and C_D through the transonic regime.

Combining Equations 5.35 and 5.17*b*, the pressure, p_2, immediately downstream of the normal shock can be written in terms of upstream reservoir pressure, p_0, and the local Mach number just ahead of the normal shock.

$$\frac{p_2}{p_{0_1}} = \left(1 + \frac{\gamma - 1}{2} M^2\right)^{-\gamma/(\gamma-1)} \left(\frac{2\gamma M^2 - \gamma + 1}{\gamma + 1}\right) \tag{5.49}$$

This ratio reaches a maximum value at a Mach number denoted as the limiting Mach number, M_{limit}, and given by

$$M_{\text{limit}} = \sqrt{\frac{\gamma + 3}{2}} = 1.483 \ (\gamma = 1.4) \tag{5.50}$$

Laitone argues in Reference 5.9 that the normal shock will be positioned on the surface of a transonic airfoil at the location where the local Mach number equals the limiting Mach number and thus, assuring the maximum positive pressure downstream of the shock wave.

This limit on the local Mach number leads to a minimum pressure coefficient that can be attained on an airfoil surface ahead of the shock wave. C_p is defined as

$$C_p = \frac{p - p_\infty}{(1/2)\rho_\infty V_\infty^2} \tag{5.51}$$

which can be written as

$$C_p = \frac{2p_\infty}{\rho_\infty V_\infty^2}\left(\frac{p}{p_0}\frac{p_0}{p_\infty} - 1\right) \tag{5.52}$$

The ratio of the local pressure to the reservoir pressure, p_0, is a function of the local M, according to Equation 5.17*b*, and decreases monotonically with M. When the local M reaches M_{limit}, this ratio attains a minimum value of 0.279. Using Equation 5.17*b* also to relate the free-stream static pressure, p_∞, to p_0, a limiting value for C_p is obtained as a function of the free-stream Mach number.

$$C_{p_{\text{limit}}} = \frac{2}{\gamma M_\infty^2}\left[0.279\left(1 + \frac{\gamma - 1}{2} M^2\right)^{\gamma/(\gamma-1)} - 1\right] \tag{5.53}$$

This relationship is presented graphically in Figure 5.19. The limiting value of C_p is seen to decrease rapidly in magnitude as M_∞ increases. $C_{p_{cr}}$ is also presented on this same figure and is a value of C_p necessary to achieve local sonic flow. The value of M_∞ corresponding to $C_{p_{cr}}$ is equal to M_{cr}, the critical Mach number. $C_{p_{cr}}$ is obtained from Equation 5.52 by setting the local M equal to unity to obtain p/p_0. The result is identical to Equation 5.53 except for replacing the constant 0.279 by 0.528.

Before discussing the significance of these relationships, let us return to Equation 5.2, which allows us to predict C_p at subsonic Mach numbers based on predictions for incompressible flow. If C_{p_c} is the pressure coefficient at a given Mach number, the Prandtl-Glauert correction states that, for the same geometry, C_p at $M = 0$ will equal βC_{p_c}. Using this scaling relationship, the critical value for the incompressible

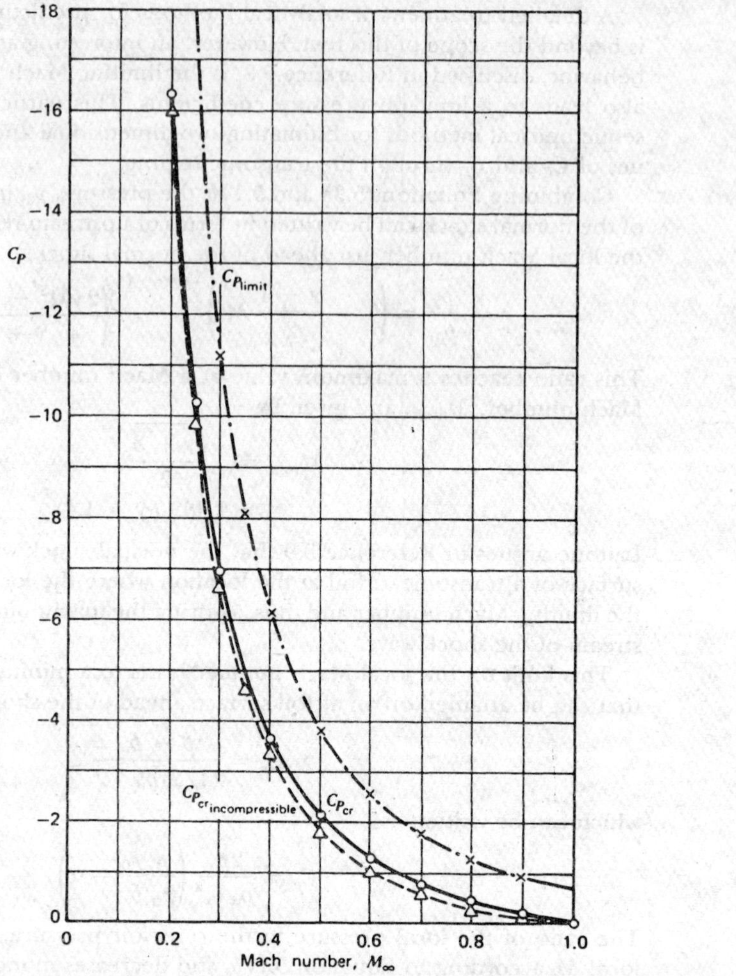

Figure 5.19 Limiting and critical pressure coefficients as a function of free-stream Mach number.

C_p can be calculated from the compressible $C_{p_{cr}}$. This result is also presented in Figure 5.19.

As an example of the use of Figure 5.19, consider the Liebeck airfoil in Figure 3.46. The minimum C_p at $M_\infty = 0$ for this airfoil is approximately -2.8. Hence, its critical Mach number is estimated from the lower curve of Figure 5.19 to be approximately 0.43. Its limiting Mach number, based on $C_p = -2.8$, would be 0.57. However, at this Mach number, the Prandtl-Glauert factor, β, equals 0.82, so that the minimum C_p at this Mach number is estimated to be -3.4. A second iteration on M_{limit} then gives a value of 0.52. Continuing this iterative procedure, a value of $M_{\text{limit}} = 0.53$ is finally obtained.

Next, consider the chordwise pressure distributions presented in Figure 5.20. Here, C_p as a function of chordwise position is presented for the NACA 64A010 airfoil for free-stream Mach numbers of 0.31, 0.71, and 0.85, all at a constant angle of attack of 6.2° (Ref. 5.10). Only the pressure distributions over the upper surface are shown. The critical Mach number for this airfoil, corresponding to the mini-

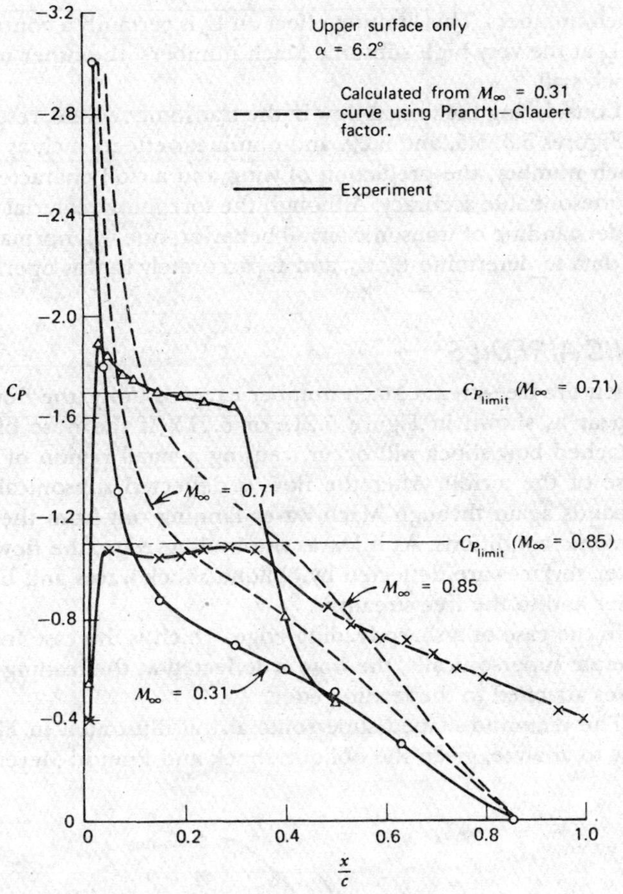

Figure 5.20 Pressure distributions for NACA 64A010 airfoil.

mum C_p of approximately -3.0, is approximately $M_{cr} = 0.43$. Thus, at $M_\infty = 0.31$, this airfoil is operating in the subsonic regime. At $M_\infty = 0.71$, the flow is transonic and theoretically limited to a C_p of -1.7. Near the nose this value is exceeded slightly. However, the experimental values of C_p are indeed nearly constant and equal to $C_{p_{limit}}$ over the leading 30% of the chord. At between the 30 and 40% chord locations, a normal shock compresses the flow, and the pressure rises over the aft 60% to equal approximately the subsonic distribution of C_p over this region. Using the preceding relationships for isentropic flow, the pressure rises across a normal shock, and with the limiting Mach number of 1.483, one would expect an increase in C_p of 1.56. The experiment shows a value of around 1.2. This smaller value may be the result of flow separation downstream of the shock.

The results at $M_\infty = 0.85$ are somewhat similar. Over the forward 35% of the chord, C_p is nearly constant and approximately equal to $C_{p_{limit}}$. Behind the normal shock, at around the 40% chord location, the increase in C_p is only approximately 0.25 as compared to an expected increment of approximately 1.24. Here, the separation after the shock is probably more pronounced, as evidenced by the negative C_p values all the way to the trailing edge.

Notice the appreciable reduction in the area under the C_p curve for $M_\infty = 0.85$ as compared to $M_\infty = 0.71$ because of the difference in the $C_{p_{limit}}$ values of the two

Mach numbers. This limiting effect on C_p is certainly a contribution to the decrease in C_l at the very high subsonic Mach numbers, the other major contributor being shock stall.

Considering the mixed flow in the transonic regime, results such as those shown in Figures 5.3, 5.5, and 5.20, and nonlinear effects such as shock stall and limiting Mach number, the prediction of wing and airfoil characteristics is a difficult task of questionable accuracy. Although the foregoing material may help to provide an understanding of transonic airfoil behavior, one will normally resort to experimental data to determine C_l, C_d, and C_m accurately in this operating regime.

SUPERSONIC AIRFOILS

When the free-stream Mach number exceeds unity, the flow around an airfoil will appear as shown in Figure 5.21*a* or 5.21*b*. If the nose of the airfoil is blunt, a detached bow shock will occur, causing a small region of subsonic flow over the nose of the airfoil. After the flow is deflected subsonically around the nose, it expands again through Mach waves fanning out from the convex surfaces to supersonic conditions. As it leaves the trailing edge, the flows along the upper and lower surfaces are deflected by oblique shock waves and become parallel to each other and to the free stream.

In the case of a sharp leading edge, which is the case for an airfoil designed to operate supersonically, the flow is deflected at the leading edge by oblique shock waves attached to the leading edge.

The diamond-shaped supersonic airfoil illustrated in Figure 5.21*b* is relatively easy to analyze, given the oblique shock and Prandtl-Meyer flow relationships. To

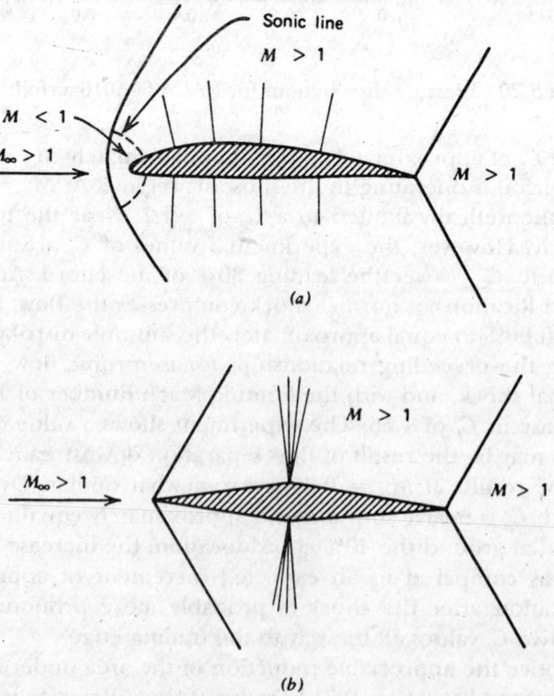

Figure 5.21 Supersonic airfoils. (*a*) Blunt-nosed airfoil. (*b*) Sharp-nosed airfoil.

begin, since pressure distributions cannot be propagated ahead, the flow will be uniform until it is deflected by the oblique shock waves above and below the leading edge. The streamlines, after passing through the oblique shocks, will remain parallel and straight until they are turned through the expansion fan, after which they are again straight and parallel until they are deflected to approximately the freestream direction by the oblique shock waves from the trailing edge. This flow is illustrated in detail in Figure 5.22.

The flow from the trailing edge does not necessarily have to satisfy the Kutta conditions, as in the subsonic case. Instead, the final deflection and hence, the strength of the trailing olique shock waves, is fixed by stipulating that the pressure and flow directions be the same for the flows from the upper and lower surfaces as they meet behind the trailing edge.

As an example, consider a supersonic airfoil having the shape of a symmetrical wedge as shown in Figure 5.22. Consider the case where the airfoil has a maximum thickness ratio of 10% and the chord line is at an angle of attack of 2° in a Mach 2 flow.

Let us begin by considering conditions of the upper surface across the oblique shock wave at the leading edge. The angle shown in Figure 5.22 for this case is equal to 11.42°, or half of this angle is 5.71°. Thus, for a 2-degree angle of attack, the oblique shock at the leading edge must turn the flow through an angle of 3.71°. From Figure 5.14, this deflection angle and a Mach number of 2 results in a shock wave angle, θ, of approximately 33° for the weak wave family. Using the programs developed in Computer Exercise 5.1, a more exact value of 33.14° is obtained. Additionally, using the Mach number normal to the wave and the relationships across a normal shock wave, the following numbers can be determined:

$$M_2 = 1.867$$

$$\frac{P_2}{P_1} = 1.228$$

$$\frac{p_{0_2}}{p_{0_1}} = 0.9991$$

In going from region 2 to region 3 on the upper surface, the flow must be turned through an additional 11.42°. This turning is done by an expansion fan of Mach waves in an isentropic process. For the M_2 value of 1.861, from Figure 5.17

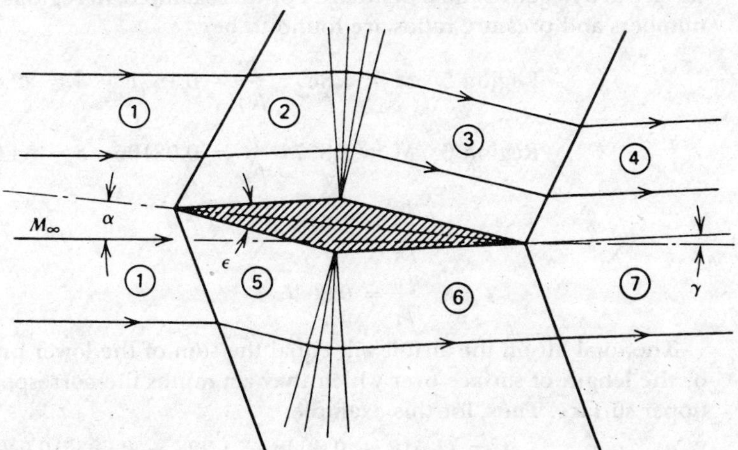

Figure 5.22 A symmetric wedge airfoil in supersonic flow.

(Eq. 5.48), the flow in region 2 is already "turned" 22.64°. Thus, in region 3, the flow must be deflected to a total angle of 34.06°, in expanding isentropically from Mach 1. Using an iterative computer program for Equation 5.48, an M of 2.291 is determined for region 3.

For flow expanded isentropically from a reservoir to the Mach numbers given above, the following values can be calculated.

$$\text{Region 2} \quad M = 1.867 \quad \frac{p}{p_{0_2}} = 0.1571 \quad \delta = 22.64°$$

$$\text{Region 3} \quad M = 1.867 \quad \frac{p}{p_{0_2}} = 0.08106 \quad \delta = 34.06°$$

where p_{0_2} is the reservoir pressure behind the oblique shock wave from the leading edge of the upper surface.

In order to integrate the above pressures over the surface of the airfoil, it is convenient to relate them to the free-stream static pressure, p_1, ahead of the airfoil. This is done by writing

$$\frac{p}{p_1} = \left(\frac{p}{p_{0_2}}\right)\left(\frac{p_{0_2}}{p_{0_1}}\right)\left(\frac{p_{0_1}}{p_1}\right)$$

$\frac{p_1}{p_{0_1}}$ is obtained by expanding isentropically from a reservoir to Mach 2.

$$\frac{p_1}{p_{0_1}} = 0.1278$$

Thus, in region 2,

$$\frac{p_2}{p_1} = \frac{(0.1571)(0.9991)}{(0.1278)} = 1.228$$

This agrees with the pressure ratio obtained from conditions across the oblique shock wave.

In region 3,

$$\frac{p_3}{p_1} = \frac{(0.08106)(0.9991)}{(0.1278)} = 0.6337$$

The lower surface is treated in a similar way and can be visualized as an upper surface at a negative angle of attack. For this example, in regions 5 and 6, the Mach numbers and pressure ratios are found to be

$$\text{Region 5} \quad M = 1.867 \quad \frac{p}{p_{0_2}} = 0.1571 \quad \delta = 22.64°$$

$$\text{Region 6} \quad M = 1.867 \quad \frac{p}{p_{0_2}} = 0.08106 \quad \delta = 34.06°$$

$$\frac{p_5}{p_1} = 1.518$$

$$\frac{p_6}{p_1} = 0.8086$$

The total lift on the airfoil will equal the sum of the lower pressures multiplied by the length of surface over which they act minus the corresponding sum on the upper surface. Thus, for this example,

$$L = (1.518 + 0.8086 - 1.228 - 0.6337)0.5p_1$$

The lift coefficient therefore becomes

$$C_l = \frac{L}{\frac{1}{2}\rho_1 V_1^2 c} = \frac{0.2325}{\frac{1}{2}\rho_1 V_1^2}$$

But $a_1^2 = \frac{\gamma p_1}{\rho_1}$. Therefore,

$$C_l = \frac{(2)(0.2325)}{\gamma M_1^2}$$

or, for $\gamma = 1.4$ and $M_1 = 2.0$, C_l becomes

$$C_l = 0.0830$$

Unlike the two-dimensional, inviscid, subsonic flow, a drag known as *wave drag* exists for the supersonic case. This drag can be obtained by resolving the integral of the normal pressure forces over the body in the drag direction.

For the symmetrical wedge pictured in Figure 5.22, the wave drag is therefore

$$D = [(p_2 - p_6)\sin 3.71° + (p_5 - p_3)\sin 7.71°]\frac{c}{2}$$

or in dimensionless form,

$$C_d = \frac{2\left(\frac{1}{2}\right)}{\gamma M_1^2}[(1.228 - 0.8086)(0.06470) + (1.518 - 0.6337)(0.1342)]$$
$$= 0.0260$$

The moment about the leading edge will, of course, equal the sum of the moments contributed by each lift component. Thus,

$$M = \frac{1}{2}\left[(p_2 - p_5)\left(\frac{1}{4}\right) + (p_3 - p_6)\left(\frac{3}{4}\right)\right]$$

or in coefficient form,

$$C_{mLE} = \frac{\left(\frac{p_2}{p_1} - \frac{p_5}{p_1}\right)\left(\frac{1}{4}\right) + \left(\frac{p_3}{p_1} - \frac{p_6}{p_1}\right)\left(\frac{3}{4}\right)}{\gamma M_1^2}$$

For this example,

$$C_{mLE} = -0.03637$$

Observe that the characteristics of the airfoil have been determined at this point without any reference to the flow downstream of the trailing edge. Unlike the subsonic airfoil, there is no need to impose a Kutta condition in order to determine the airfoil life since conditions at the trailing edge cannot affect the flow upstream. In order to know the conditions downstream of the airfoil, a value for the angle δ in Figure 5.22 is assumed. This then determines the deflection, which will be required by the oblique shock waves on the upper and lower surfaces. Knowing the Mach numbers and static pressures immediately upstream of each shock wave then leads to a prediction of the static pressures downstream of each shock. The angle, γ, is then changed until the static pressures are equal. For this example, a value of γ of zero leads to static pressures downstream of both oblique shocks, which are nearly equal to the static pressure ahead of the airfoil with downstream Mach numbers of approximately 1.99.

Computer Exercise 5.2 "WEDGE"

Using the subroutines written for Computer Exercise 5.1, write a program to predict the characteristics of a supersonic airfoil following the example just presented. Define the upper and lower surfaces of the airfoil by connected straight line segments where the coordinates of the nodes are read from a file and are relative to the chord line. Check your program against the example of the 10% symmetrical wedge at a 2° angle of attack. Then run it for a 10% thick, symmetrical, convex airfoil at 1, 2, 3, 4, and 5° angle of attack where both surfaces are circular arcs, each approximated by 50 straight line segments.

LINEARIZED COMPRESSIBLE POTENTIAL FLOW

The foregoing treatment based on Prandtl-Meyer and oblique shock relationships is somewhat tedious to apply. Also, the general behavior of supersonic airfoils is not disclosed by this approach. Therefore, we will now consider a linearized solution that holds for slender profiles and for Mach numbers that are not too close to unity or not too high.

Assuming that the free-stream velocity is only perturbed by the presence of a slender body at a small angle of attack, the x and y components of the local velocity can be written as

$$\begin{aligned} V_x &= V_\infty + u \\ V_y &= v \end{aligned} \tag{5.54}$$

where

$$\frac{u}{V_\infty}, \frac{v}{V_\infty} \ll 1$$

We now define a *perturbation velocity potential*, ϕ, such that

$$\begin{aligned} u &= \frac{\partial \phi}{\partial x} \\ v &= \frac{\partial \phi}{\partial y}. \end{aligned} \tag{5.55}$$

If Equations 5.54 and 5.55 are substituted into the equations of fluid motion together with isentropic relationships, the following linearized equation is obtained for the perturbation velocity potential.

$$(1 - M_\infty^{\,2})\frac{\partial^2 \phi}{\partial x^2} + \frac{\partial^2 \phi}{\partial y^2} = 0 \tag{5.56}$$

To arrive at this equation, Reference 5.3 shows that the following must hold:

$$M_\infty^{\,2}(u/V_\infty) \ll 1 \tag{5.57a}$$

$$\frac{u}{V_\infty} \ll (1 - M_\infty^{\,2}) \tag{5.57b}$$

Thus, from Equation 5.57a, M_∞ cannot be too large (i.e., the application of Eq. 5.56 to hypersonic flow is questionable). On the other hand, from Equation 5.57b, M_∞ is restrained from becoming close to unity, so that the application of Equation 5.56 to transonic flows is ruled out.

Subsonic Flow

For M_∞ values less than unity, Equation 5.56 is of the elliptic form. In this case, a disturbance at any point in the flow affects the flow at all other points. A solution to Equation 5.56 for the subsonic case can be obtained in terms of the solution for $M_\infty = 0$. This latter solution has been discussed in previous chapters.

Let $\phi_i(x, y)$ be a solution to Equation 5.56 for $M_\infty = 0$. Now consider a function, ϕ_c, given by

$$\phi_c = \frac{\phi_i(x, \beta y)}{\beta} \tag{5.58}$$

where as before β is defined as

$$\beta^2 = 1 - M_\infty^2$$

Thus,

$$\frac{\partial^2 \phi_c}{\partial x^2} = \frac{1}{\beta}\frac{\partial^2 \phi_i}{\partial x^2}$$

$$\frac{\partial^2 \phi_c}{\partial y^2} = \beta\frac{\partial^2 \phi_i}{\partial y^2}$$

Substituting this into the left side of Equation 5.56 gives

$$\beta\left(\frac{\partial^2 \phi_i}{\partial x^2} + \frac{\partial^2 \phi_i}{\partial y^2}\right)$$

Since the terms within the parentheses are equal to zero, it follows that Equation 5.58 is a solution of Equation 5.56.

Now consider a body contour $Y(x)$. At any point along the contour, the following boundary condition must hold.

$$\frac{dY}{dx} = \frac{v}{V_\infty + u} \tag{5.59}$$

$$\simeq \frac{v}{V_\infty}$$

Equation 5.59 holds to the first order in the perturbation velocities.

Relating v to the incompressible perturbation velocity potential leads to

$$\frac{dY}{dx} = \frac{1}{V_\infty}\left[\frac{\partial \phi_i}{\partial y}(x, y)\right] \tag{5.60}$$

In the compressible case,

$$\frac{dY}{dx} = \frac{1}{V_\infty}\left[\frac{\partial \phi_i}{\partial y}(x, \beta y)\right] \tag{5.61}$$

$\partial\phi_i/\partial y$ can be expanded in a Maclaurin series to give

$$\frac{\partial \phi_i}{\partial y}(x, y) = \frac{\partial \phi_i}{\partial y}(x, 0) + y\frac{\partial^2 \phi_i}{\partial y^2}(x, 0) + \frac{y^2}{2}\frac{\partial^3 \phi_i}{\partial y^3}(x, 0) + \ldots$$

Since y and the derivatives of ϕ_i are assumed to be small, to a first order,

$$\frac{\partial \phi_i(x, y)}{\partial y} = \frac{\partial \phi_i(x, 0)}{\partial y} \tag{5.62}$$

Thus, by comparing Equations 5.60 and 5.61, it follows that the body contour for which ϕ_c holds is the same (to a first order) as that for ϕ_i.

We are now in a position to determine the pressure distribution for a given slender body shape as a function of Mach number. Along a streamline the resultant velocity, U, in terms of the perturbation velocities, can be written as

$$U = [(V_\infty + u)^2 + v^2]^{1/2}$$
$$= V_\infty + u \text{ (to a first order)} \tag{5.63}$$

Euler's equation along a streamline was derived earlier in differential form. Expressed in finite difference form, it can be written as

$$U\,\Delta U + \frac{\Delta p}{\rho} = 0$$

Using Equation 5.63, this becomes

$$(V_\infty + u)\,u + \frac{\Delta p}{\rho} = 0$$

Finally,

$$C_p = \frac{p - p_\infty}{(1/2)\rho V_\infty^{\,2}}$$
$$= \frac{\Delta p}{(1/2)\rho V_\infty^{\,2}}$$
$$= -2\frac{u}{V_\infty} \tag{5.64}$$

Since $u = \partial\phi/\partial x$, it follows that from Equations 5.64 and 5.58, the pressure distribution over a slender body at a finite subsonic Mach number is related to the pressure distribution over the *same* body at $M = 0$ by

$$C_{p_c} = \frac{C_{p_i}}{\beta} \tag{5.65}$$

This was assumed earlier in this chapter as Equation 5.2.

Thus, to predict the lift and moment on a two-dimensional shape such as an airfoil, one simply calculates these quantities in coefficient form for the incompressible case and then multiplies the results by the factor $1/\beta$.

The three-dimensional case is somewhat more complicated but not much. Here,

$$\phi_c = \frac{1}{\beta^2}\,\phi_i(x, \beta y, \beta z) \tag{5.66}$$

Hence, to find the compressible flow past a three-dimensional body with coordinates of x, y, and z, one solves for the incompressible flow around a body having the coordinates x, βy, and βz. The pressure coefficients are then related by

$$C_{p_c} = \frac{C_{p_i}}{\beta^2} \tag{5.67}$$

Supersonic Flow (Ackeret Theory)

If M is greater than unity, Equation 5.56 changes to a hyperbolic partial differential equation, specifically, to the following wave equation:

$$(M_\infty^{\,2} - 1)\frac{\partial^2\phi}{\partial x^2} = \frac{\partial^2\phi}{\partial y^2} \tag{5.68}$$

Letting

$$B = \sqrt{M_\infty^{\,2} - 1}$$

a general solution of Equation 5.68 can be written

$$\phi = f(x - By) + g(x + By) \tag{5.69}$$

where f and g represent arbitrary functions of their arguments. As an exercise, verify that Equation 5.69 satisfies Equation 5.68. ϕ is seen to be constant along families of straight lines defined by

$$x - By = \text{constant} \tag{5.70a}$$

$$x + By = \text{constant} \tag{5.70b}$$

The slope of the lines represented by Equation 5.70a is

$$\frac{dy}{dx} = \frac{1}{B} = (M^2 - 1)^{-1/2}$$

But this is the tangent of the Mach wave angle as defined by Equation 5.45. Thus, ϕ is constant along a Mach wave. In the case of Equation 5.70b,

$$\frac{dy}{dx} = -\frac{1}{B}$$

On the upper surface of a body, this would correspond to a disturbance being propagated forward in the flow, which is physically impossible in a supersonic flow. Thus, Equation 5.70b is ruled out for the upper surface. However, on the lower surface of a body, $g(x + By)$ is a physically valid flow and represents a disturbance being propagated rearward along a Mach wave. Similarly, $f(x - By)$ is not allowed as a solution on the lower surface of a body. The net result is pictured in Figure 5.23, where it is seen that $f(x - By)$ and $g(x + By)$ are solutions to ϕ on the upper and lower surfaces, respectively. Since ϕ is constant along Mach waves emanating from disturbances from the upper and lower surfaces, it follows that the properties of the flow (velocity and state) are also constant along these waves.

Along the surface of the body, the flow must be tangent to the body. Thus, if $Y(x)$ represents the body surface, it follows, to a first order in the perturbation velocities, that

$$\frac{1}{V_\infty}\left(\frac{\partial \phi}{\partial y}\right) = \left(\frac{dY}{dx}\right) \tag{5.71}$$

Consider the upper surface and let $x - By = z$. Then

$$\left(\frac{\partial \phi}{\partial y}\right)_u = \frac{df}{dz}\frac{\partial z}{\partial y} = -Bf' \tag{5.72}$$

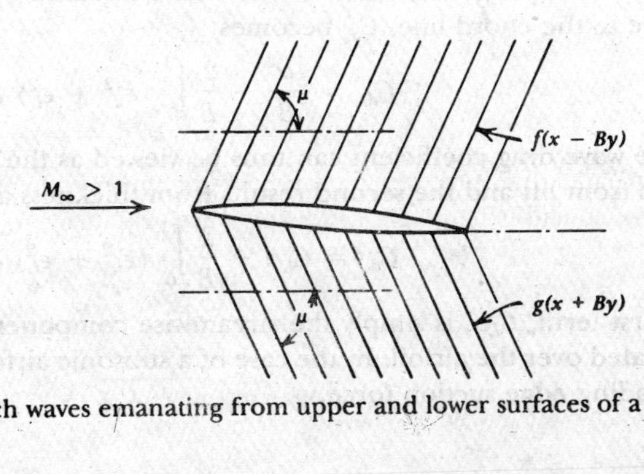

Figure 5.23 Mach waves emanating from upper and lower surfaces of a body.

where f' denotes df/dz. In addition, from Equation 5.64,

$$C_{p_u} = -\frac{2}{V_\infty}\frac{\partial \phi}{\partial x} = \frac{2}{V_\infty} f' \tag{5.73}$$

Thus, combining Equations 5.71 to 5.73,

$$C_{p_u} = \frac{2}{B}\left(\frac{dY}{dx}\right)_u \tag{5.74}$$

Similarly, on the lower surface,

$$C_{p_l} = -\frac{2}{B}\left(\frac{dY}{dx}\right)_l \tag{5.75}$$

Thus, according to Ackeret's linearized theory for supersonic flow around a slender body, the pressure locally on the body is determined by the slope of the surface at the particular location in question.

This simple result leads quickly to some interesting conclusions regarding the characteristics of thin, supersonic airfoils at low angles of attack. Since, for a unit chord,

$$C_l = \int_0^1 (C_{p_l} - C_{p_u})\, dx$$

you can quickly verify that

$$C_l = \frac{4\alpha}{B} \tag{5.76}$$

Thus, within the limitations of the linearized theory, the section lift coefficient of a supersonic airfoil depends only on its angle of attack. Camber is predicted to have no effect on C_l.

The wave drag coefficient is obtained by integrating the component of C_p in the drag direction around the airfoil.

$$C_{d_w} = \oint C_p \frac{dy}{dx}\, dx = \frac{2}{B}\left[\int_0^1 \left(\frac{dy}{dx}\right)_u^2 dx + \int_0^1 \left(\frac{dy}{dx}\right)_l^2 dx\right] \tag{5.77}$$

If dy/dx is expressed in the form

$$\frac{dy}{dx} = -\alpha + \epsilon$$

where α is the angle of attack of the chord line and ϵ is the slope of the surface relative to the chord line, C_{d_w} becomes

$$C_{d_w} = \frac{4\alpha^2}{B} + \frac{2}{B}\int_0^1 (\epsilon_u^2 + \epsilon_l^2)\, dx \tag{5.78}$$

The wave drag coefficient can thus be viewed as the sum of two terms: the first results from lift and the second results from thickness and camber.

$$C_{d_w} = C_L\alpha + \frac{2}{B}\int_0^1 (\epsilon_u^2 + \epsilon_l^2)\, dx$$

The first term, $C_L\alpha$, is simply the streamwise component of the normal pressures integrated over the airfoil. In the case of a subsonic airfoil, this term is canceled by the leading edge suction force.

The pitching moment coefficient about the leading edge of a thin, supersonic airfoil can be written

$$C_{m_{LE}} = \int_0^1 C_{p_l} x\,dx - \int_0^1 C_{p_u} x\,dx$$
$$= \frac{2}{B}\int_0^1 \left[\left(\frac{dy}{dx}\right)_u + \left(\frac{dy}{dx}\right)_l\right] x\,dx$$
$$= -\frac{2}{B}\alpha + \frac{2}{B}\int_0^1 (\epsilon_u + \epsilon_l)\, x\,dx \qquad (5.79)$$

$$C_m = -\frac{2}{\sqrt{M^2 - 1}}\alpha \qquad (5.80)$$

The center of pressure for a symmetrical airfoil in supersonic flow is thus predicted to be at the midchord point.

Figure 5.24 (taken from Ref. 5.5) provides a comparison between the linear theory and experiment for a 10% thick biconvex airfoil. This figure shows fairly good agreement of the theory with experiment with the differences being of the same order as those in Table 5.1.

Table 5.1 compares the results of the linearized theory with the more exact predictions made earlier for the symmetrical wedge airfoil pictured in Figure 5.22. In this particular case, the linearized theory is seen to be somewhat optimistic with regard to lift and drag and predicts the center of pressure to be farther aft than the position obtained from the more exact calculations. Nevertheless, the Ackeret theory is valuable for predicting trends. For example, *for symmetrical airfoils,* the expressons of C_{d_w} and C_m reduce to

$$C_{d_w} = \frac{4}{\sqrt{M^2 - 1}}\left[\alpha^2 + \left(\frac{t}{c}\right)^2\right] \qquad (5.81)$$

THREE-DIMENSIONAL WINGS

Wings designed to operate at high speeds are generally thin and employ sweepback in order to increase the critical Mach number. In some instances the sweep is variable to accommodate operation at both low and high speeds. Many airplanes for which the primary mission involves supersonic flight employ delta planforms. Figure 5.25 illustrates various types of planforms utilized on high-speed airplanes. The swept wing is common to all subsonic jet transports and to subsonic military airplanes. It is also used on supersonic airplanes but with much lower aspect ratios than are found on subsonic transports. The delta wing and the swing-wing, or variable-sweep, are employed primarily on supersonic airplanes. The best known airplane that uses the ogee planform is the Concorde, the only operational supersonic transport. For subsonic applications, these configurations incur both aerodynamic, weight, and cost penalties.

Table 5.2 lists a number of airplanes selected from the 1991–92 issue of Reference 5.11 for which the sweepback and operating Mach number could be obtained. There are generally four types of airplanes included in the table: high subsonic airplanes with swept wings, supersonic airplanes with swept wings, supersonic airplanes with variable sweep, and supersonic airplanes with delta wings. This data is graphed in Figure 5.26, which presents the Mach number normal to the leading edge as a function of the free-stream Mach number. Admittedly, the sample is small, but it is interesting to observe that all of the high-subsonic airplanes designed

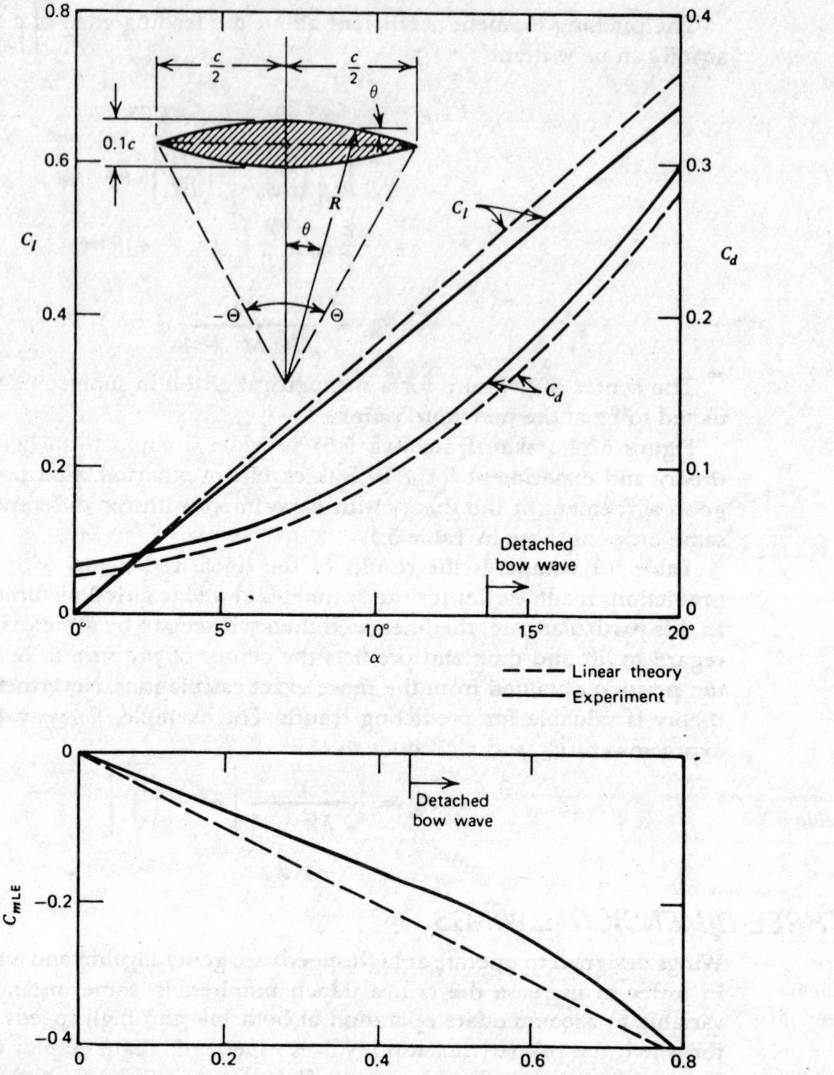

Figure 5.24 Comparison between measured aerodynamic characteristics (Ferri 1939) and those predicted by linear theory at $M = 2.13$. Airfoil is biconvex.

Table 5.1 Predicted Characteristics of a Symmetrical, 10% thick, Double-Wedge Airfoil at a 2° Angle of Attack and a Mach Number of 2.0

	Using Oblique Shock Wave and Prandtl-Meyer Relationships	Linearized Ackeret Theory
C_l	0.0830	0.0806
C_{d_w}	0.0260	0.0259
C_m	−0.0364	−0.0403
Center of pressure	0.4386	0.5000

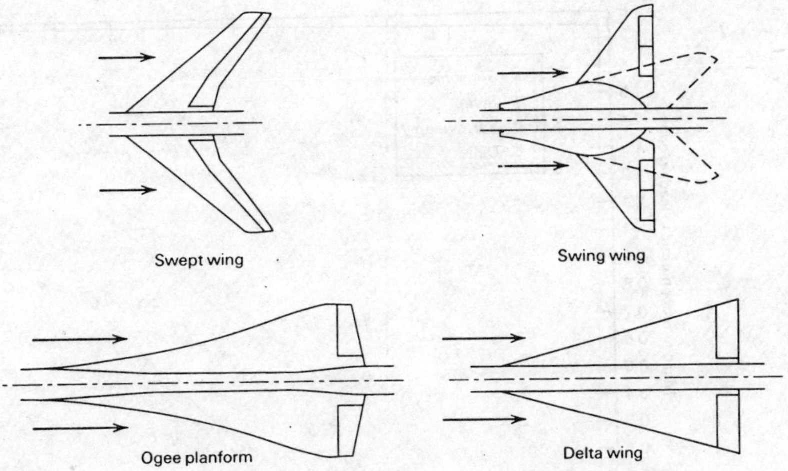

Figure 5.25 Types of swept planform shapes to operate at high Mach numbers.

to operate at around Mach 0.8–0.9 employ sweepback, which reduces the normal Mach number to approximately 0.7. It is also interesting to note that the design normal Mach number increases progressively in going from swing-wings to delta wings to fixed-swept wings.

The sweep angles for the fixed-swept airplanes are probably a compromise between supersonic and subsonic flight. For landing, swing-wings are brought forward to a point where there is practically no sweep. In high-speed flight, the wings can be swept back to an angle that provides subsonic flow normal to the leading edge.

Table 5.2 Characteristics of Swept-Wing Airplanes

Airplane	Sweep	Mach No.	Mach No. Normal to Leading Edge	Wing Type
British Aerospace Hawk	26.00	0.88	0.79	Swept
Boeing E-6a	36.00	0.79	0.64	Swept
Boeing 747	41.00	0.90	0.68	Swept
Boeing 757	30.00	0.80	0.69	Swept
Boeing 777	35.00	0.83	0.68	Swept
Cessna 750 Citation X	38.00	0.90	0.71	Swept
Airbus A300	31.00	0.82	0.70	Swept
McDonnell-Douglas F-15	45.00	2.50	1.77	Swept
McDonnell-Douglas F-18	26.00	1.80 +	1.62	Swept
Dassault Mirage F1	50.00	2.20	1.41	Swept
GD F-16	40.00	2.00	1.53	Swept
MiG-23	72.00	2.35	0.73	Variable Sweep
Tupolev TU-160	66.00	1.88	0.76	Variable Sweep
Panavia Tornado	68.00	2.20	0.82	Variable Sweep
Grumman F-14A	68.00	2.34	0.88	Variable Sweep
Dassault Mirage 2000N	60.00	2.20	1.10	Delta
Dassault Rafale C	47.00	2.00 E	1.36	Delta-canard
Eurofighter EFA	52.00	1.80 +	1.11	Delta-canard

E = estimated
\+ means M in excess of number

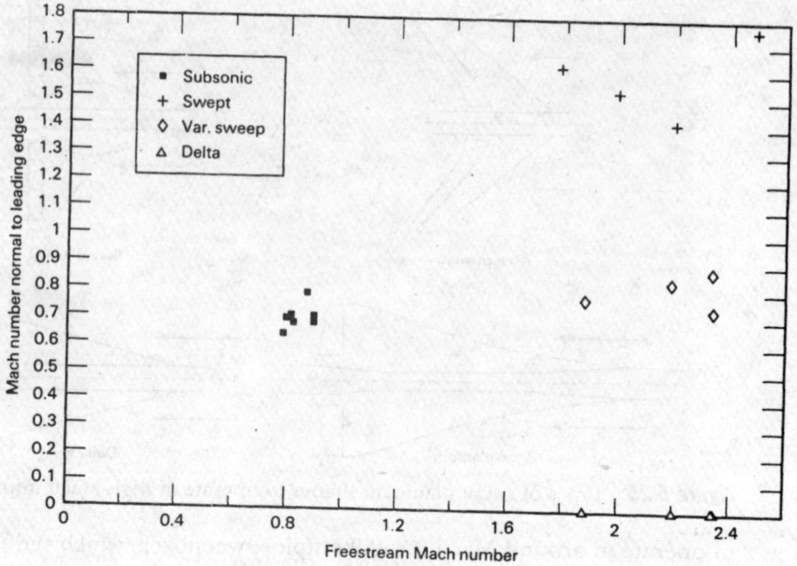

Figure 5.26 Mach number normal to leading edge as related to free-stream Mach number.

Figure 5.27a Lockheed F-117A stealth fighter. (Courtesy Lockheed Advanced Development Co.)

Figure 5.27b YF-22A and F-22 advanced tactical fighter. (Courtesy Lockheed Advanced Development Co.)

Figure 5.27c

Figures 5.27 (*a*) and (*b*) present two modern aircraft, which were designed not simply to satisfy aerodynamic performance requirements but stealth requirements as well. The specifics of each aircraft's performance are classified, but generally, it can be said that the Lockheed F-117A is designed to operate at high subsonic speeds, whereas the F-22 is supersonic. The sweepback of the F-117A is approximately 67° whereas the F-22, with a modified diamond planform, has a sweepback of 42°. This was decreased from the YF-22A prototype value of 48° and, according to Reference 5.12, was done to improve aerodynamic performance.

The F-22 is not yet operational (at the time of this writing), although the prototype has undergone extensive flight testing, which apparently has met or exceeded predictions. The F-117A, which played a major role in the Persian Gulf War, was developed and flew in only 31 months under the strictest security, flying for almost 8 y before its existence became public. The F-117A is an unusual design with its exterior being composed of flat panels, not exactly what the aerodynamicist would desire. Their purpose is, of course, to reflect radar signals away from, and not return them to, the radar that is transmitting the signals.

The aerodynamic design of the F-22, while still a stealth aircraft, is certainly more pleasing to the aerodynamicist. According to Reference 5.12, some of the key aerodynamic elements in its design include:

1. Blended wing-body with internal weapons bays and sufficient fuel volume for meeting long endurance missions.
2. Modified diamond wing (848 ft^2) with a wing span (44.6 ft) compatible with existing aircraft shelters. (*Author's note:* This is another example where reasons other than aeronautical can contribute to an airplane's shape.)
3. Constant chord, full leading edge flaps
4. Two-dimensional, convergent–divergent exhaust nozzles with independent throat and exit area actuation and pitch axis thrust vectoring.
5. Free-stream fixed geometry supersonic inlets with swept cowl lips, boundary layer bleed and overboard bypass systems, and a relatively long subsonic diffuser having 100% line-of-sight RF blockage.
6. All exterior edge angles aligned with either the wing leading or trailing edge angles.

Characteristics of Sweptback Wings

Qualitatively, the effect of wing sweep can be seen by referring to Figure 5.28. In Figure 5.28*a*, a wing section is shown extending from one wall of a wind tunnel to the other. The test section velocity is denoted by V_n. Imagine that the wing in Figure 5.28*a* is only a section of an infinitely long wing that is being drawn through contoured slots in the tunnel walls at a velocity of v. Obviously, the pressure distribution around the section does not depend on v. V_n and v combine vectorially, as shown in Figure 5.28*b*, to give a velocity of V_∞ relative to the wing. As shown in Figure 5.28*c*, this is equivalent to a swept wing of an infinite aspect ratio with a sweep angle of Λ and a free-stream velocity of V_∞. Thus, the chordwise C_p distribution of such a wing depends only on the component of V_∞ normal to the span, which is given by

$$V_n = V_\infty \cos \Omega \tag{5.82}$$

Based on this velocity, a pressure coefficient can be defined as

$$\begin{aligned} C_{p_n} &= \frac{p - p_\infty}{(1/2)\rho_\infty V_\infty^2 \cos^2 \Omega} \\ &= \frac{(p/p_\infty) - 1}{(1/2)\gamma M_{\infty_n}^2} \end{aligned} \tag{5.83}$$

The Prandtl-Glauert transformation can then be applied to C_{p_n} using M_n to account for compressibility. In practice, a swept wing has a finite length. Near the apex of the wing and at the tips, a three-dimensional flow effect will be encountered. Indeed, one cannot test a two-dimensional swept wing in a wind tunnel (except in the manner shown in Figure 5.28*a*). For example, a wing placed wall to wall and

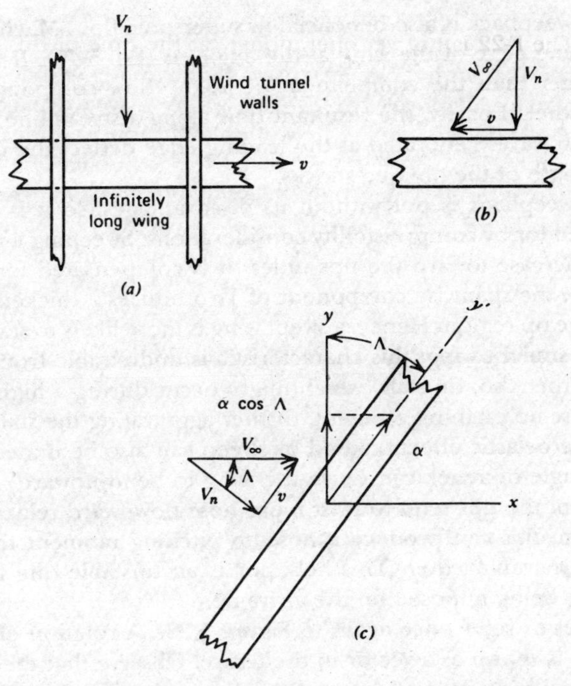

Figure 5.28 Effect of sweepback.

yawed in a wind tunnel models a saw-toothed planform instead of an infinitely long swept wing. As illustrated in Figure 5.29, this results from the fact that the flow must be parallel to the wind tunnel walls at the walls. This can only be satisfied by assuming an image system of wings having alternating sweep, as shown.

The effect of sweepback on critical Mach number can be estimated using Equation 5.83. For example, suppose a straight wing has a certain chordwise and spanwise C_p distribution that produces a given lift. At some point on the wing, suppose that a minimum C_p is equal to -0.5. According to Figure 5.19, its M_{cr} value would equal 0.71. Now suppose the same wing were swept back 45° and its twist, camber, and angle of attack were adjusted to give the same chordwise C_p distribution (based on V_∞) at each spanwise station as for the unswept wing. The total lift for the two wings would then be the same. For the swept wing the minimum C_p based on V_n becomes $C_{p_n} = -1.0$. Thus, according to Figure 5.19, $m_{n_{cr}} = 0.605$. This corresponds to a free-stream critical Mach number, $M_{\infty_{cr}}$, of 0.86. Therefore, sweeping the wing back 45° for the same wing area and lift has increased the critical Mach number from 0.71 to 0.86.

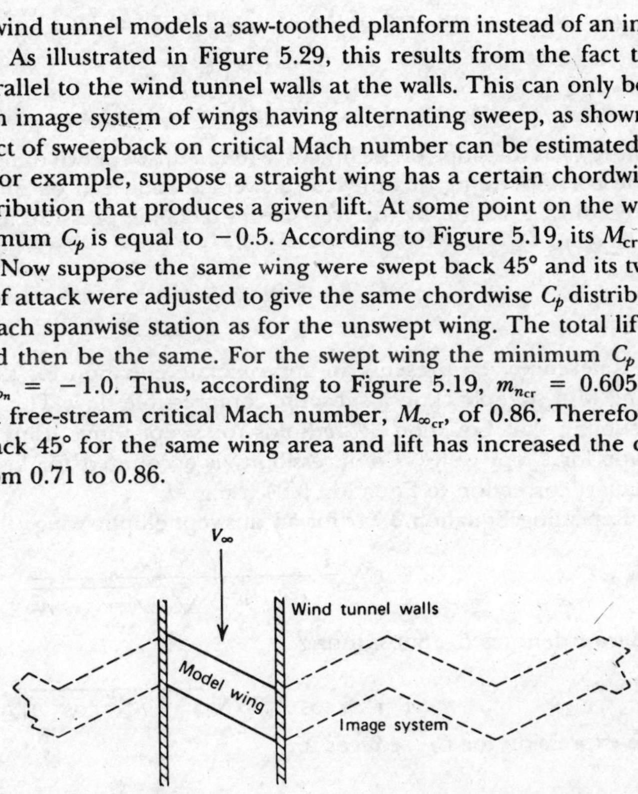

Figure 5.29 Swept wing in a wind tunnel.

Sweepback is also beneficial in supersonic flow. Mach waves propagate from the leading edge of the wing at the angle, μ (Eq. 5.45). If the sweepback angle, Λ, is greater than the complement of μ, the flow component normal to the span is subsonic. Locally, the resultant flow along a streamline is still supersonic, but the Mach waves generated at the leading edge deflect the flow, thereby lessening the strength of the oblique shocks.

Sweepback is not without its disadvantages, so it is normally used only when called for by compressibility considerations. Sweeping a wing will cause the loading to increase toward the tips unless it is compensated for by washout. At the same time, the spanwise component of V_∞ produces a thickening of the boundary layer in the tip region. Hence, a swept wing is more likely to stall outboard by comparison to a straight wing; this characteristic is undesirable from the standpoint of lateral control. Also, tip stall (which might occur during a high-speed pull-up) can cause a nose-up pitching moment, further aggravating the stall.

Aeroelastic effects caused by sweep can also be undesirable. A sudden increase in angle of attack can cause the wing to bend upward. As it does, because of the sweep, the tips tend to twist more nose downward relative to the rest of the wing. Again, this can produce a nose-up pitching moment that increases the angle of attack even further. This behavior is an unstable one that can lead to excessive loads being imposed on the airframe.

Let us refer once again to Figure 5.28*c*. A rotation of α about a line along the wing is shown as a vector in the figure. Observe that the component of this vector normal to V_∞ is equal to $\alpha \cos \Lambda$. Thus, an angle of attack of α relative to V_n results in a smaller angle of attack of $\alpha \cos \Omega$ relative to V.

The lift on a unit area of the wing will be given by

$$\begin{aligned} L &= \tfrac{1}{2}\rho V_n^{\,2} C_{l\dot{\alpha}}\alpha \\ &= \tfrac{1}{2}\rho (V_\infty \cos \Lambda)^2 C_{l_\alpha}\alpha \end{aligned}$$

where C_{l_α} is the slope of the lift curve for an unswept two-dimensional airfoil section. The corresponding quantity for a swept section can be obtained by dividing the preceding equation by the free-stream dynamic pressure and the angle of attack relative to V_∞.

$$\begin{aligned} C_{l_\alpha}\,(\Lambda \neq 0) &= L/(\tfrac{1}{2}\rho\, V_\infty^{\,2}\alpha \cos \Lambda) \\ &= C_{l_\alpha} \cos \Lambda \end{aligned} \tag{5.84}$$

Reference 3.35 presents an approximate equation for the lift curve slope of a wing with sweepback in a subsonic compressible flow. The equation is derived by assuming that Equation 3.74*b* holds for swept wings using the section lift curve slope for swept wings. Compressibility is accounted for by applying the Prandtl-Glauert correction to Equation 5.84 using M_{∞_n}.

Repeating Equation 3.74*b* for an unswept elliptic wing,

$$C_{L_\alpha} = \frac{aA}{(\alpha/\pi) + \sqrt{(a/\pi)^2 + A^2}}$$

where a denotes C_{l_α}. Substituting

$$a = a_0 \cos \Lambda_{1/2}/\sqrt{1 - M_\infty^{\,2} \cos^2 \Lambda_{1/2}}$$

the expression for C_{L_α} reduces to

$$C_{L_\alpha} = \frac{a_0 A}{(a_0/\pi) + \sqrt{A/\cos \Lambda_{1/2})^2 + (a_0/\pi)^2 - (AM_\infty)^2}} \tag{5.85}$$

where $a_0 = C_{l_\alpha}$ for $\Lambda_{1/2} = 0$. Note that a subscript $\frac{1}{2}$ has been added to Λ to indicated that $\Lambda_{1/2}$ should be measured relative to a line through the midchord points. It is argued in Reference 3.35 that the use of $\Lambda_{1/2}$ makes Equation 5.85 independent of taper ratio, λ. This conclusion appears to be supported by Figure 5.30 (taken from Ref. 3.35). Here, Equation 5.85 (for $M = 0$) is seen to compare favorably with several lifting surface calculations for elliptical and tapered wings.

A tabulation of subsonic lift and moment characteristics for tapered sweptback wings is presented in Table 5.3. Taken from Reference 5.13, these results are based on a numerical lifting surface theory. They should be more accurate than Equation 5.85, although not as convenient to use. As an example in the use of this table, let

$$\beta A = 5.0$$
$$\beta = 0.6$$
$$A \tan \Lambda_{1/2} = 4$$

It then follows that

$$A = 8.333$$
$$\Lambda_{1/2} = 25.6°$$
$$\beta C_{L_\alpha} = 3.43$$
$$C_{L_\alpha} = 5.717$$

For this example, Equation 5.85 gives a C_{L_α} value of 6.017, which is about 5% higher than the lifting surface theory result. Of course, both results can be expected to be somewhat high by comparison to experiment, since the theoretical value for a_0 of $2\pi C_l/\text{rad}$ is a few percent higher than that found experimentally for most airfoils.

Figure 5.31 (from Ref. 5.14) presents some experimental results on C_{L_α} at low Mach numbers. Generally, the trends shown on this figure confirm the predictions of Figure 5.30. The two figures are not directly comparable, since the sweep angles of Figure 5.31 are relative to the leading edge instead of to mischord. To go from

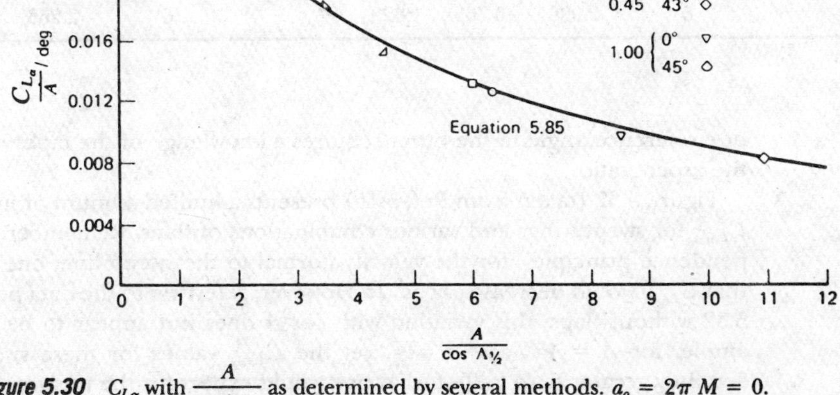

Figure 5.30 C_{L_α} with $\dfrac{A}{\cos\Lambda_{1/2}}$ as determined by several methods. $a_o = 2\pi$ $M = 0$.

Table 5.3 Subsonic Theoretical Lift Slopes, Pitching Moments, and Aerodynamic Centers for Wings of Varying Sweep and Taper (M about leading edge at midspan, $\bar{c}$ = geometric mean chord)

λ	βA	$A \tan \Lambda_{1/2}$	$-\beta \dfrac{\partial C_M}{\partial \alpha}$	$\dfrac{1}{A}\dfrac{\partial C_L}{\partial \alpha}$	$\dfrac{x_{ac}}{\bar{c}}$	λ	βA	$A \tan \Lambda_{1/2}$	$-\beta \dfrac{\partial C_M}{\partial \alpha}$	$\dfrac{1}{A}\dfrac{\partial C_L}{\partial \alpha}$	$\dfrac{x_{ac}}{\bar{c}}$
1.0	8.0	0	1.110	0.574	0.242	0.25	8.0	0	2.465	0.594	0.418
		2	3.156	0.563	0.701			2	4.433	0.582	0.951
		4	5.025	0.532	1.181			4	6.155	0.551	1.397
		6	6.579	0.490	1.677			6	7.510	0.508	1.848
	5.0	0	0.934	0.791	0.236		5.0	0	2.100	0.816	0.515
		2	2.628	0.761	0.691			2	3.788	0.786	0.963
		4	4.030	0.686	1.175			4	5.062	0.711	1.423
		6	5.032	0.599	1.680			6	5.859	0.622	1.882
	3.0	0	0.707	1.049	0.225		3.0	0	1.623	1.071	0.505
		2	1.998	0.982	0.678			2	2.064	1.009	0.980
		4	2.916	0.830	1.171			4	3.750	0.861	1.452
		6	3.446	0.682	1.684			6	4.078	0.712	1.910
	1.5	0	0.393	1.348	0.195		1.5	0	0.978	1.351	0.483
		2	1.230	1.232	0.665			2	1.891	1.254	1.006
		4	1.676	0.952	1.174			4	2.203	0.990	1.484
		6	1.869	0.738	1.689			6	2.241	0.774	1.931
0.5	8.0	0	1.888	0.592	0.399	0	8.0	0	3.092	0.568	0.681
		2	3.917	0.580	0.844			2	4.818	0.557	1.081
		4	5.728	0.548	1.306			4	6.296	0.529	1.488
		6	7.188	0.505	1.779			6	7.433	0.491	1.894
	5.0	0	1.600	0.814	0.393		5.0	0	2.636	0.773	0.682
		2	3.312	0.784	0.845			2	4.143	0.746	1.111
		4	4.663	0.708	1.317			4	5.218	0.680	1.536
		6	5.563	0.619	1.799			6	5.841	0.600	1.948
	3.0	0	1.224	1.072	0.381		3.0	0	2.066	1.009	0.683
		2	2.561	1.007	0.848			2	3.286	0.953	1.149
		4	3.419	0.856	1.331			4	3.906	0.821	1.585
		6	3.845	0.706	1.816			6	4.092	0.685	1.991
	1.5	0	0.714	1.360	0.350		1.5	0	1.310	1.281	0.682
		2	1.609	1.254	0.855			2	2.150	1.190	1.205
		4	1.989	0.982	1.350			4	2.325	0.947	1.636
		6	2.099	0.765	1.829			6	2.263	0.746	2.022

one reference angle to the other requires a knowledge of the taper ratio as well as the aspect ratio.

Figure 5.32 (taken from Ref. 5.15) presents a limited amount of information on $C_{L_{max}}$ for swept wings and various combinations of flaps. Remembering the "independence principle" for the velocity normal to the sweep line, one might expect that $C_{L_{max}}$ would decrease as $\cos^2 \Omega$. However, referring to the data points in Figure 5.32 without flaps, this variation with $\cos^2 \Lambda$ does not appear to be valid. For example, for $\Lambda = 45°$, $\cos^2 \Lambda = \frac{1}{2}$, yet the $C_{L_{max}}$ values for these swept wings are certainly greater than half of what one would expect for the unswept wings. Three-

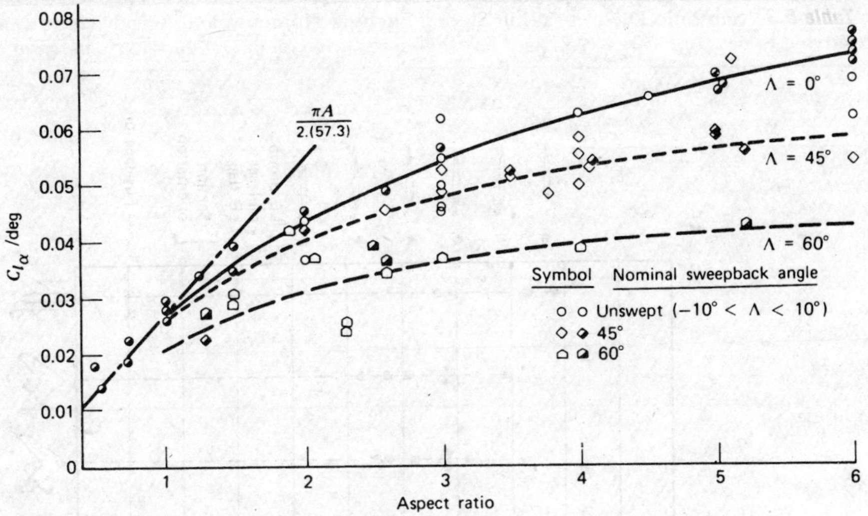

Figure 5.31 Lift curve slopes for swept, tapered wings.

dimensional effects such as spanwise flow and leading edge vortices undoubtedly play an important role in determining the stalling characteristics of a swept wing. Figure 5.33 indicates that, if anything, there is a tendency for $C_{L_{max}}$ to increase with sweepback (or with sweep forward). Admittedly, this figure includes other factors affecting $C_{L_{max}}$, but the general impression that it portrays is probably valid; that is, sweep has little effect on $C_{L_{max}}$.

The effect of sweepback on drag can be estimated by reference to Figures 5.34 and 5.35. Figure 5.34 presents the relative effect of sweepback on induced drag for three planforms having taper ratios of 0.3, 0.6, and 1.0, all with an aspect ratio of 6.0. This graph was prepared using a vortex lattice model with 80 spanwise, and 3 chordwise, panels. The efficiency factors shown here exceed unity by a small amount because of the numerical methods used and because of the fact that Equation 4.19 for the minimum C_{D_i} is based on lifting line theory. These graphs apply only to $M = 0$, so they must be used in conjunction with the three-dimensional, Prandtl–Glauert transformation. The trends shown by Figure 5.34 are reasonable in view of the fact that sweeping a wing shifts the loading outboard. Thus, the loading for a taper ratio of 0.3 at a sweepback angle of, say, 25° is similar to that for a taper ratio of 0.6 with no sweepback. With this in mind, wings having a taper ratio equal to or greater than optimum for their aspect ratio will suffer increased induced drag as sweepback is increased. For taper ratios lower than optimum, sweepback can reduce the induced drag to a point.

Figure 5.35 indicates that there is little, if any, effect of sweepback on the minimum drag coefficient. Thus, one can estimate $C_{D_{min}}$ on the basis of two-dimensional airfoil measurements. Usually these drag measurements are taken at low Mach numbers so that it is necessary to correct the skin friction part of $C_{D_{min}}$ for compressibility effects. Such a correction is given by the graph of Figure 5.36 (Ref. 5.1) for Mach numbers as high as 10.0.

An empirical equation that closely fits the graph of Figure 5.36 is

$$\frac{C_f}{C_{f_i}} = (1 + 0.104\,M^2)^{-0.773} \tag{5.86}$$

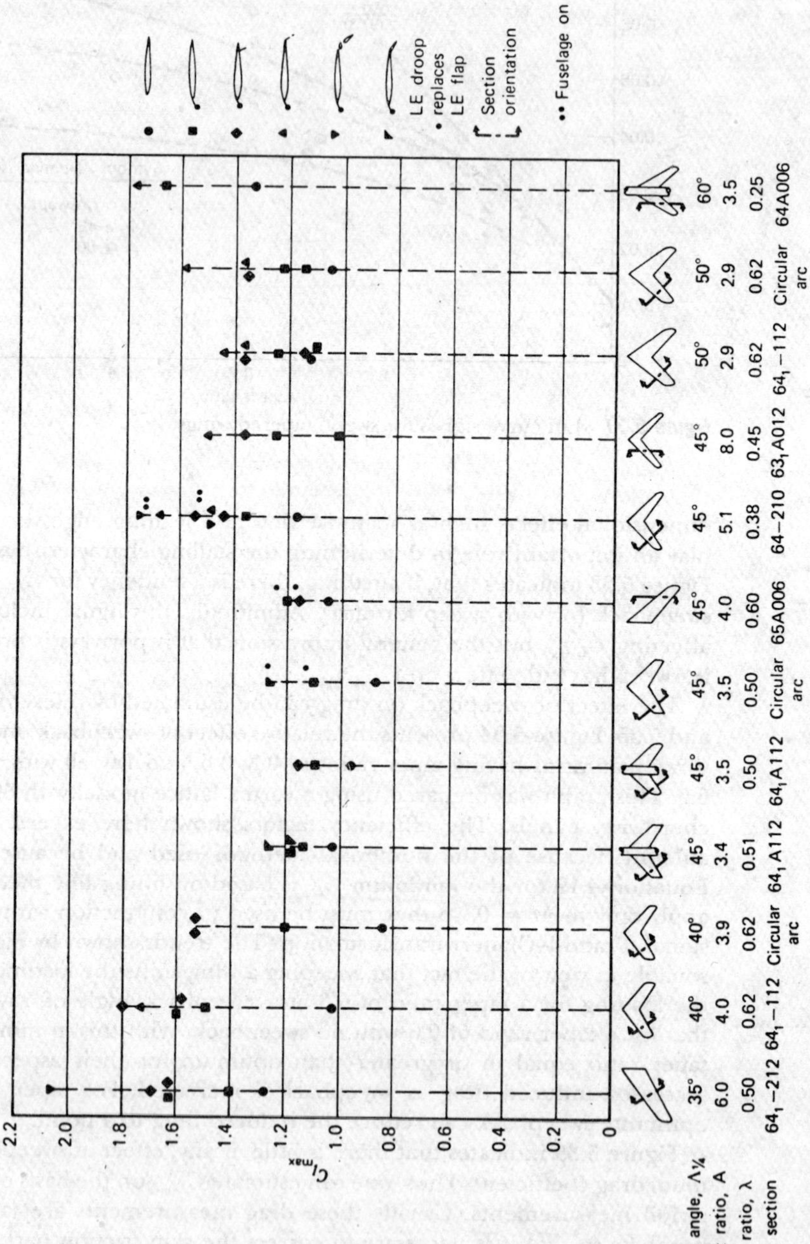

Figure 5.32 Maximum lift coefficients obtained with various types of trailing edge flaps.

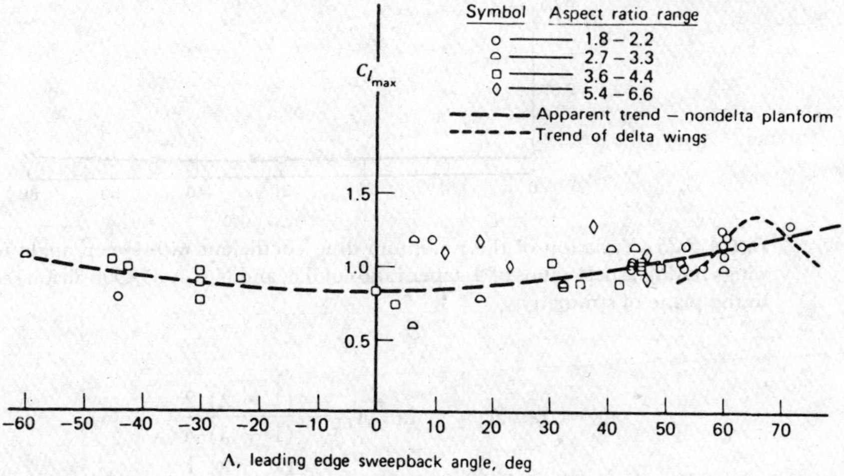

Figure 5.33 Variation of $C_{L_{max}}$ with sweepback for wings having planforms other than delta.

As an example in the use of the foregoing relationships for swept wings, consider a swept wing having the following geometry and operating at a free-stream Mach number of 0.7 at 30,000 ft.

$$A = 7.0$$
$$\lambda = 0.5$$
$$\Lambda_{LE} = 30° \text{ (leading edge)}$$

64A009 airfoil (defined normal to michord line)
(see Figure 5.3 for two-dimensional characteristics)
wing loading = 100 psf

To use Figure 5.30 or Equation 5.85 we need $\Lambda_{1/2}$. For a linearly tapered swept wing, Λ_{LE} and $\Lambda_{1/2}$ can easily be shown to be related by

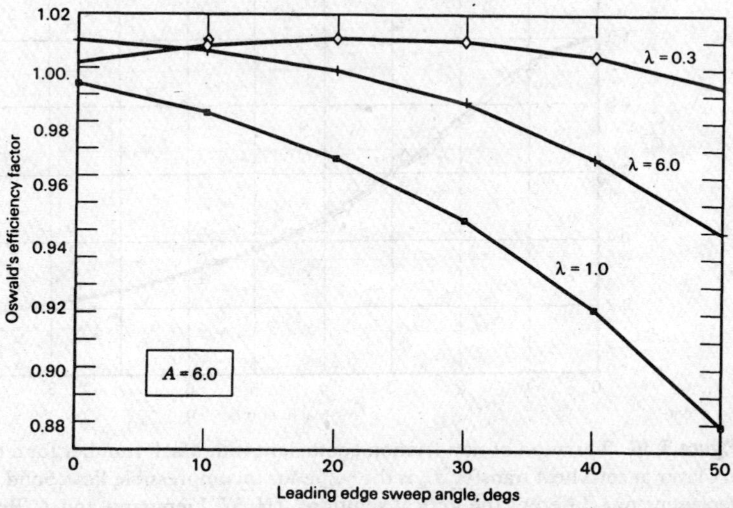

Figure 5.34 Effect of sweep on Oswald's efficiency factor for linearly tapered wings.

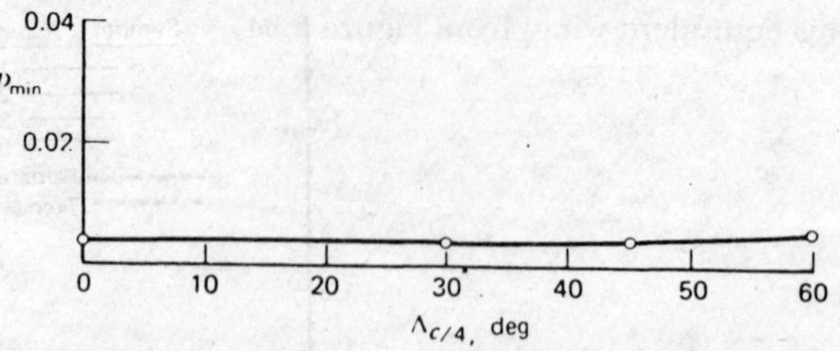

Figure 5.35 Variation of the minimum drag coefficient with sweep angle for a family of wings having aspect ratios of 4, taper ratios of 0.6, and NACA 65A006 airfoil sections parallel to the plane of symmetry.

$$\tan \Lambda_{1/2} = \tan \Lambda_{LE} - \frac{(1-\lambda)}{(1+\lambda)}\frac{2}{A} \quad \text{(a)} \tag{5.87}$$

$$\tan \Lambda_{1/4} = \tan \Lambda_{LE} - \frac{(1-\lambda)}{(1+\lambda)}\frac{1}{A} \quad \text{(b)}$$

Thus, $\Lambda_{1/2} = 25.7°$.

The incompressible slope of the lift curve, a_0, for this particular airfoil equals approximately 6.2 C_l/rad. Thus, for $M_\infty = 0.7$, using Equation 5.85, $C_{L_\alpha} = 5.21$ C_L/rad.

According to Equation 5.66, the equivalent wing in incompressible flow will have its y dimensions decreased by the factor β. In this case, $\beta = 0.714$. Hence, the equivalent wing's geometry becomes

$$A_i = 5.0$$
$$\Lambda_{i1/4} = 36.6°$$

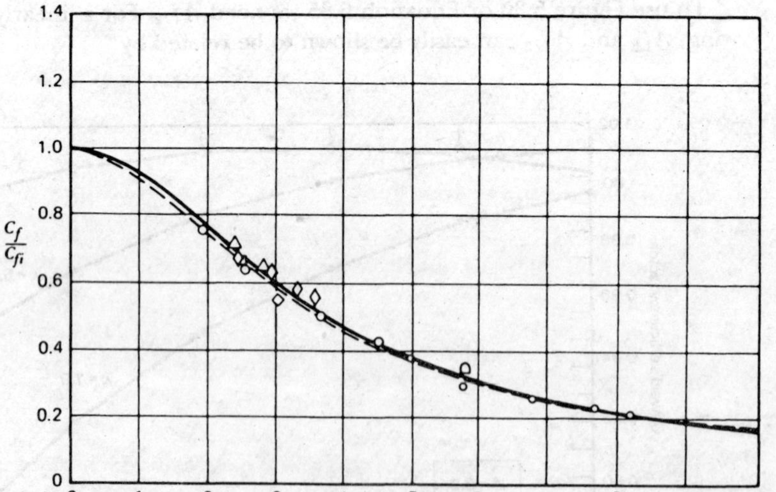

Figure 5.36 Variation of skin friction coefficient with Mach number for a turbulent boundary layer at zero heat transfer. C_{fi} is the value for incompressible flow. Solid and dashed lines represent two different theoretical solutions. (H. W. Liepmann and A. Roshko, *Elements of Gas Dynamics,* John Wiley & Sons, Inc., 1957.

For this equivalent wing, from Figure 5.34.

$$\frac{C_{Di\,\text{elliptical}}}{C_{D_i}} = 0.98$$

For the given wing loading and operating conditions,

$$C_L = \frac{W/S}{q}$$
$$= 0.462$$

Thus, for the given wing

$$C_{D_i} = \frac{1}{0.98}\frac{C_L^2}{\pi A}$$
$$= \frac{1}{0.98}\frac{(0.462)^2}{\pi(7)}$$
$$= 0.00990$$

We do not need to correct the profile drag for compressibility effects in this example, since Figure 5.3 provides us with the section C_d as a function of M and C_l. Assuming that the section C_l is equal approximately to the wing C_L of 0.462, C_d is read from Figure 5.3 to be 0.015 for an M_∞ of 0.7. According to Figure 5.35, this does not need to be corrected for sweep. The total wing C_D at the operating C_L and Mach number is thus estimated to equal 0.0249.

Before leaving this example, it is of interest to generalize on the calculation of C_{D_i}.

$$A_i = \beta A_c$$
$$C_{L_i} = \beta^2 C_{L_c}$$

Thus,

$$C_{D_{i_i}} = \frac{C_{L_i}^2}{\pi A_i}$$
$$= \beta^3 \frac{C_{L_c}^2}{\pi A_c}$$
$$= \beta^3 C_{D_{i_c}} \tag{5.88}$$

This result can also be obtained by applying the three-dimensional Prandtl-Glauert transformation to the relationship

$$C_{D_i} = C_L \alpha_i$$

α_i, the induced angle of attack, is proportional to the vertical velocity associated with $\phi(x, \beta y, \beta z)$. Hence, α_i varies as $1/\beta$.

SUPERSONIC WINGS

This section deals with wings operating at free-stream Mach numbers greater than unity. Since the governing flow equations change their form in going from subsonic to supersonic flow, the behavior of a wing changes its characteristics also. The induced drag, lift curve slope, and center of pressure are all affected significantly when M_∞ exceeds unity.

There are many variations possible in supersonic wing configurations, so that it is almost impossible (at least with the current state of the art) to present a universal approach that will satisfy the intended level of this text. Instead, we will examine two specific cases. The first case is that of swept wings (including delta wings) with subsonic leading edges and supersonic trailing edges; the second case is that of swept or unswept wings with supersonic leading and trailing edges. The term "subsonic leading edge" may be somewhat misleading. It refers to the case where the leading edge lies within the Mach cone generated at the apex. The Mach number normal to the edge is subsonic, so that even though such a wing is supersonic, because of the independence of the normal flow, it exhibits some of the characteristics of a subsonic wing (such as a leading edge suction force).

No attempt will be made to develop the theory of supersonic wings. Instead, a brief description will be offered of each theoretical approach together with the limitations. Results will then be presented that permit estimations of wing behavior in supersonic flow. We begin with the case of subsonic leading edges.

Subsonic Leading Edges

Reference 5.17 is a linearized treatment of delta wings with subsonic leading edges. The method is valid for large values of the apex angle up to and coincident with the Mach angle. Reference 5.23 extends the method of Reference 5.17 to include swept wings with pointed tips, but with trailing edges that are also swept. The trailing edges in Reference 5.17 are restricted only by the requirement that they be supersonic. Hence, their sweep angle must be less than the "sweep" angle of the Mach wave. Thus, swept forward trailing edges are allowed. It is interesting to note that any triangular wing having subsonic leading edges and supersonic trailing edges will have the same pressure distribution along a ray from the apex as any other triangular wing having the same leading edge sweep angle and operating at the same Mach number. Thus, if one finds the pressure distribution over a "basic" delta wing, the effect of sweeping its trailing edges can be readily determined. The conclusion follows from the fact that disturbances cannot propagate ahead of the Mach cone in linearized supersonic flow. Figure 5.37 illustrates the foregoing principle.

For delta wings only, the following expression was obtained by Brown (Ref. 5.17) for the lift curve slope:

$$C_{L_\alpha} = \frac{2\pi^2 \tan \epsilon}{\pi + \lambda} \tag{5.89}$$

λ is a function of the ratio of apex angle tangent to that of the Mach angle; it is presented in Figure 5.38. Note that when the leading edge of the wing and the Mach line are coincident, C_{L_α} reduces to Equation 5.76; that is, the slope of the lift curves are the same for a supersonic two-dimensional airfoil and a delta wing, the leading edge of which is coincident with the Mach line. This also holds if the leading edge is supersonic. Thus, for values of $\tan \epsilon/\tan \mu$ greater than unity, the value of the λ function in Figure 5.38 is constant and equal to 1.793. Note also that Equation 5.89 reduces to the results from slender wing theory as the apex angle approaches zero.

For subsonic leading edges, Brown shows that a leading edge suction force will exist. As a result, an induced drag is obtained that is less than the streamwise component of the wing's normal force.

$$C_{D_i} = \frac{{C_L}^2}{\pi A}\left[2\left(1 + \frac{\lambda}{\pi}\right) - \sqrt{1 - B^2 \tan^2 \epsilon}\right] \tag{5.90}$$

It is emphasized that Equations 5.89 and 5.90 hold only for delta wings with subsonic leading edges. λ in these particular equations refers to the function shown

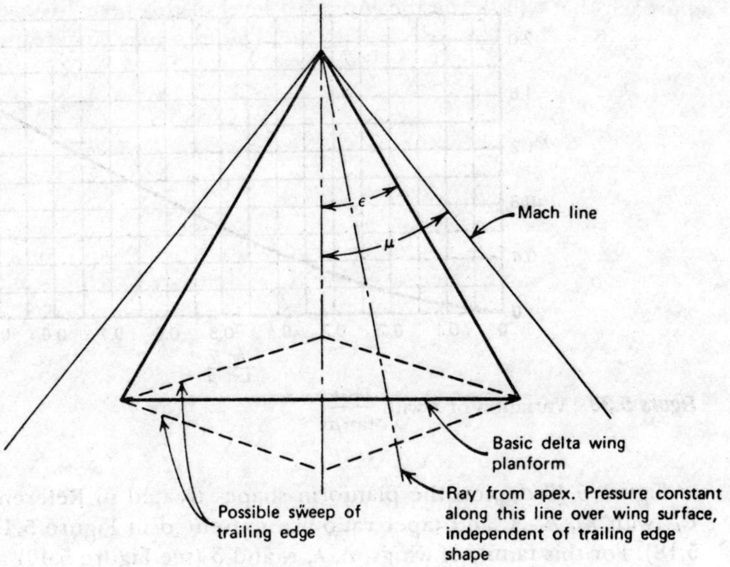

Figure 5.37 Supersonic delta wing.

in Figure 5.38 and does *not* stand for the taper ratio. The taper ratio of a delta wing is zero. Since $\tan\mu = 1/B$ and λ is a function of $\tan\epsilon/\tan\mu$, we can write

$$\frac{C_{D_i}}{(C_L^2/\pi A)} = 2\left(1 + \frac{\lambda}{\pi}\right) - \sqrt{1 - \left(\frac{\tan\epsilon}{\tan\mu}\right)^2} \qquad (5.91)$$
$$= f\left(\frac{\tan\epsilon}{\tan\mu}\right)$$

Referencing $C_{L\alpha}$ to the two-dimensional value, we can also write

$$\frac{C_{L\alpha}}{C_{L\alpha\text{2-D}}} = \frac{B}{4}C_{L\alpha}$$
$$= \frac{\pi^2}{2(\pi + \lambda)}\left(\frac{\tan\epsilon}{\tan\mu}\right) \qquad (5.92)$$
$$= g\left(\frac{\tan\epsilon}{\tan\mu}\right)$$

The functions, f and g, are presented in Figure 5.39. With regard to the drag, note that the leading edge suction force vanishes when the leading edge and Mach line are coincident. Thus, for μ values equal to or less than ϵ, the drag is simply equal to the streamwise component of the wing normal force. In coefficient form,

$$C_{D_i} = C_L \tan\alpha$$

But

$$C_L = \frac{4}{B} = 4\tan\mu \qquad \text{and} \qquad A = 4\tan\epsilon$$

Thus, for $\mu \le \epsilon$,

$$\frac{C_D}{(C_L^2/\pi A)} = \pi$$

This is the value shown in Figure 5.39 for $\tan\epsilon/\tan\mu = 1$.

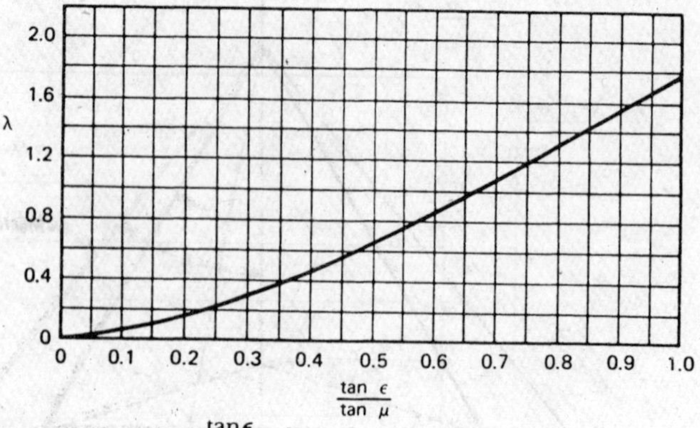

Figure 5.38 Variation of λ with $\frac{\tan\epsilon}{\tan\mu}$.

Figure 5.40 depicts the planform shapes treated in Reference 5.18. Trends of $C_{L\alpha}$ with M, A, Λ, and taper ratio are presented in Figure 5.41 (taken from Ref. 5.18). For this family of wings, A, λ, ϵ, and δ (see Figure 5.40) are related by

$$\tan\delta = \frac{(1+\lambda)A\tan\epsilon}{(1+\lambda)A - 4(1-\lambda)\tan\epsilon} \tag{5.93}$$

Before leaving the subject of wings with subsonic leading edges, it should be emphasized once again that such wings can develop leading edge suction forces. Thus, to prevent leading edge separation in order to maintain this suction force, it is beneficial (from a drag standpoint) to round the leading edges of these wings, even though they are operating at a supersonic free-stream Mach number. Experimental proof of this (Ref. 5.19) is offered by Figure 5.42*a*. Here, the lift-to-drag ratio as a function of C_L is presented for a delta wing with its leading edges lying within the Mach cone. Three different airfoil sections were tested, including one with a rounded leading edge. This latter section is seen to have a maximum L/D value approximately 8% higher than the other two.

Other data from the same reference are presented in Figure 5.42*b* to 5.42*f*, which shows the effects of aspect ratio and sweep on lift, moment, and drag for a family of tapered wings. These wings all have a taper ratio of 0.5 and employ the cambered wedge section having a maximum thickness ratio of 5%, pictured in Figure 5.42*b*.

The moment coefficients for this set of graphs are about the centroid of the planform area, with the mean aerodynamic chord as the reference length. If the aerodynamic center of a wing is a distance of x ahead of the centroid and M is the moment about the centroid, then

$$M_{ac} = M - xL$$

In coefficient form,

$$C_{M_{ac}} = C_M - C_L\frac{x}{c}$$

Differentiating with respect to C_L and recalling that, by definition, $dC_{M_{ac}}/dC_L = 0$, gives

$$\frac{x}{c} = \frac{dC_M}{dC_L}$$

Thus, Figure 5.42*d* and 5.42*e* represent, in effect, the distance of the aerodynamic center ahead of the centroid.

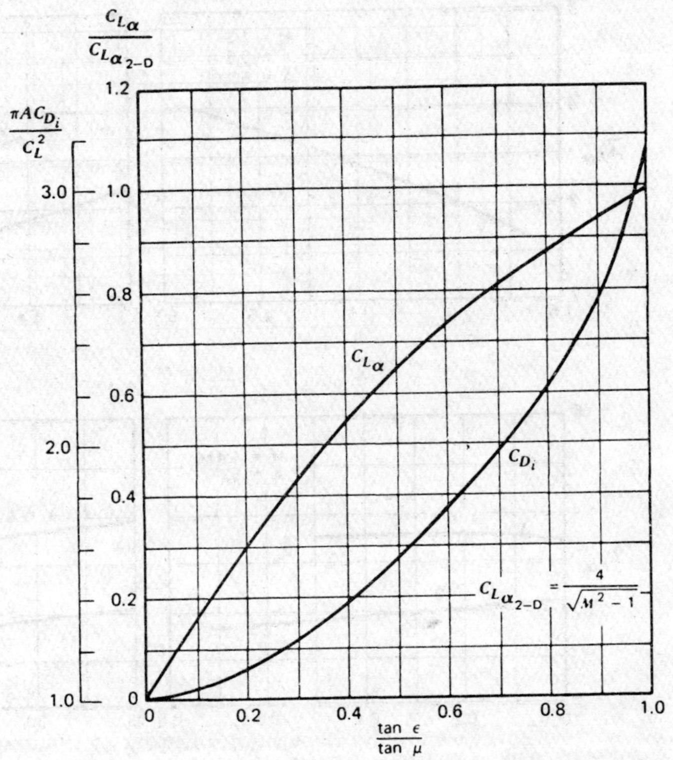

Figure 5.39 Lift curve slope and induced drag coefficient for delta wings with subsonic leading edges.

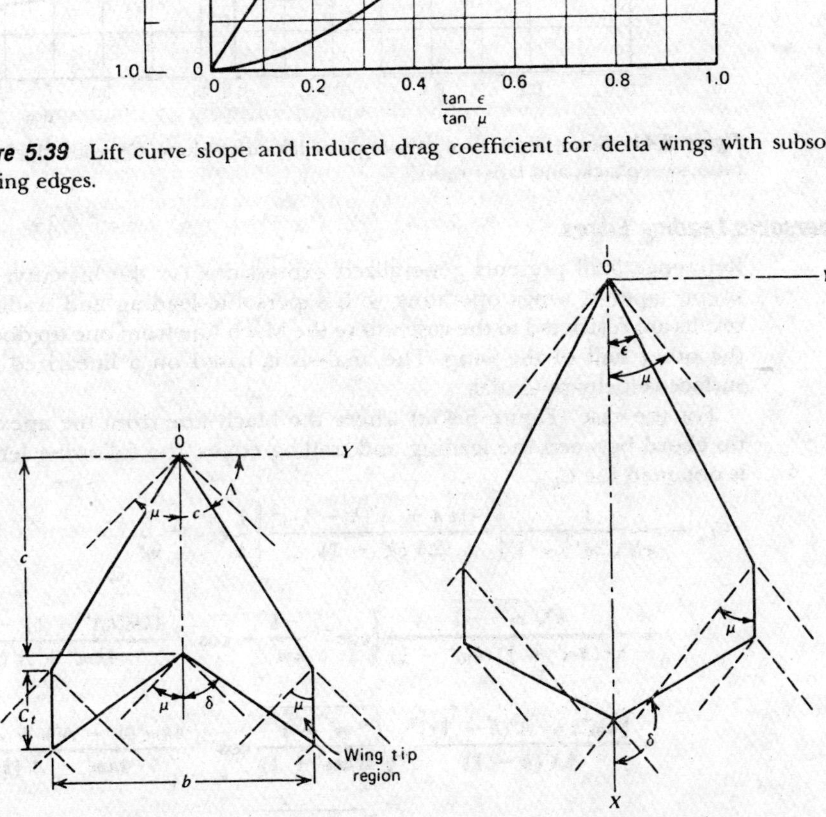

Figure 5.40 Supersonic wings and subsonic leading edges. (*a*) Sweptback trailing edge. (*b*) Sweptforward trailing edge.

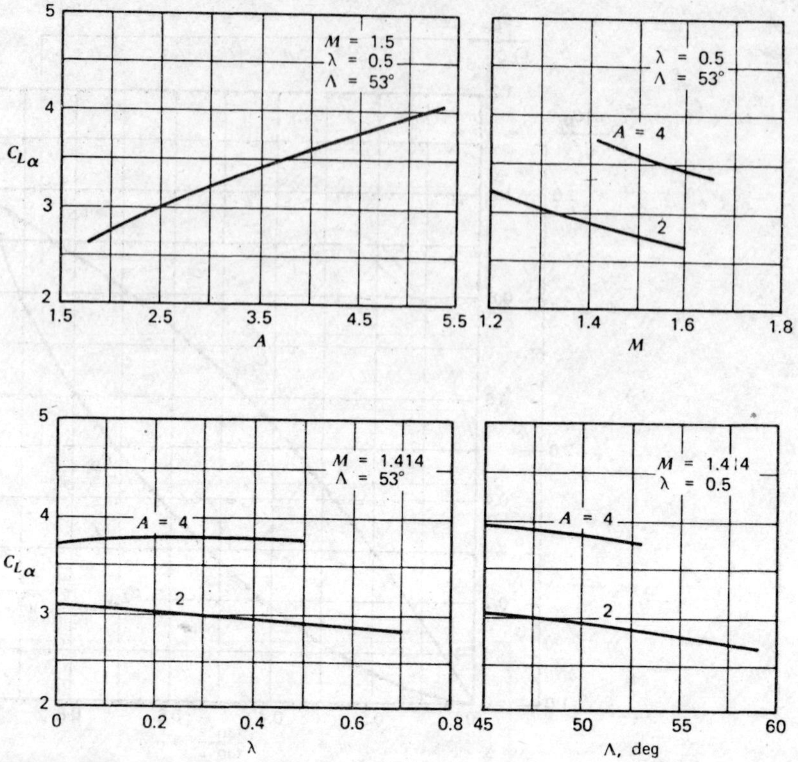

Figure 5.41 Some illustrative variations of lift curve slope $C_{L\alpha}$ with Mach number, aspect ratio, sweepback, and taper ratio.

Supersonic Leading Edges

Reference 5.20 presents generalized expressions for the lift curve slope of thin, swept, tapered wings operating with supersonic leading and trailing edges. The results are restricted to the case where the Mach line from one tip does not intersect the other half of the wing. The analysis is based on a linearized theory for the surface velocity potential.

For the case (Figure 5.43*a*) where the Mach line from the apex intersects the tip chord between the leading and trailing edges, the following lengthy equation is obtained for $C_{L\alpha}$.

$$C_{l\alpha} = \frac{1}{\pi B\sqrt{m'^2 - 1}} \left\{ \frac{[4m'k + A'(k-1)]^2}{2A'(k^2-1)} \left[\frac{1}{k}\cos^{-1}\frac{1}{m'} \right.\right.$$

$$+ \left.\frac{k\sqrt{m'^2-1}}{\sqrt{(km'+1)(km'-1)}}\left(\cos^{-1}\frac{-1}{km'} - \cos^{-1}\frac{4km'(A'-1) - A'(k+3)}{4km' + A'(k-1)}\right)\right]$$

$$- \frac{[4m'k - A'(k-1)]^2}{4A'(k-1)}\sqrt{\frac{m'+1}{k(km'+1)}}\cos^{-1}\frac{4km'(1-A') + A'(3k+1)}{4km' - A'(k-1)}$$

$$\left. + \frac{[4m'k + A'(1+3k)]^2}{4A'(k+1)}\sqrt{\frac{m'-1}{k(km'+1)}}\cos^{-1}\frac{4km'(A'-1) + A'(k-1)}{4km' + A'(3k+1)} \right\} \tag{5.94}$$

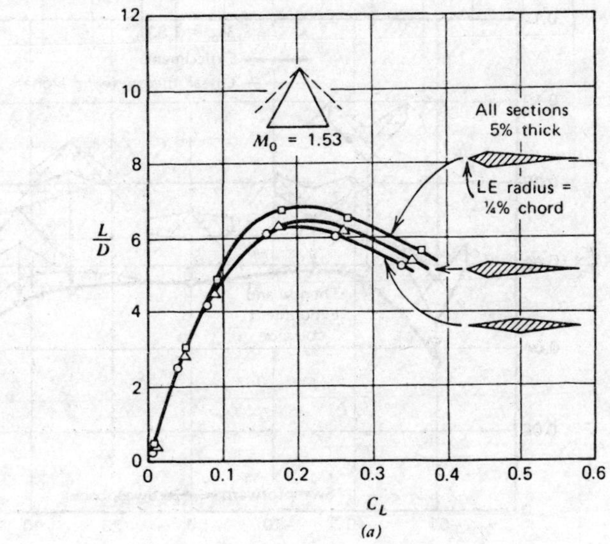

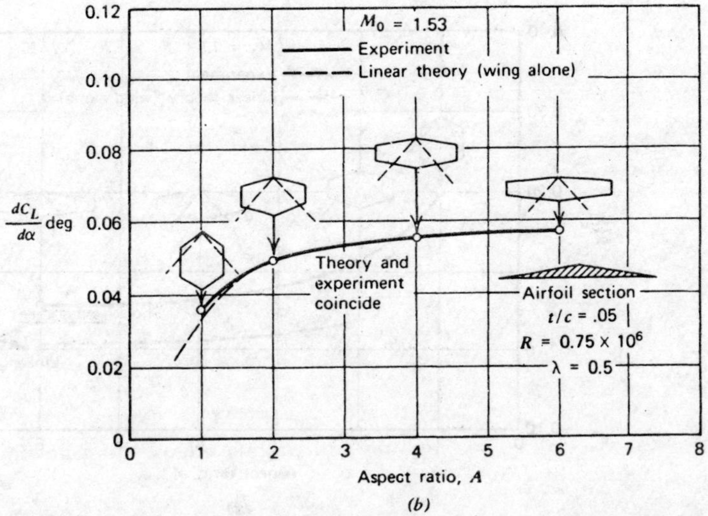

Figure 5.42 Supersonic wing characteristics. (*a*) Effect of wing section on lift–drag ratio of triangular wings. (*b*) Effect of aspect ratio on lift curve slope. (*c*) Effect of sweep on lift curve slope. (*d*) Effect of aspect ratio on moment curve slope. (*e*) Effect of sweep on moment curve slope. (*f*) Effect of sweep on minimum drag.

If the Mach line from the apex intersects the trailing edge inboard of the tip, as shown in Figure 5.43*b*, C_{L_α} is given by

$$C_{L_\alpha} = \frac{1}{B\pi\sqrt{m'^2 - 1}}\left\{\frac{[4m'k + A'(k-1)]^2}{2A'(k^2-1)}\left[\frac{1}{k}\cos^{-1}\frac{1}{m'} + \frac{k\sqrt{m'^2-1}}{\sqrt{(km'-1)(km'+1)}}\cos^{-1}\frac{-1}{km'}\right] - \frac{\pi[4km' - A'(k-1)]^2}{4A'(k-1)}\sqrt{\frac{m'+1}{k(km'+1)}}\right\} \tag{5.95}$$

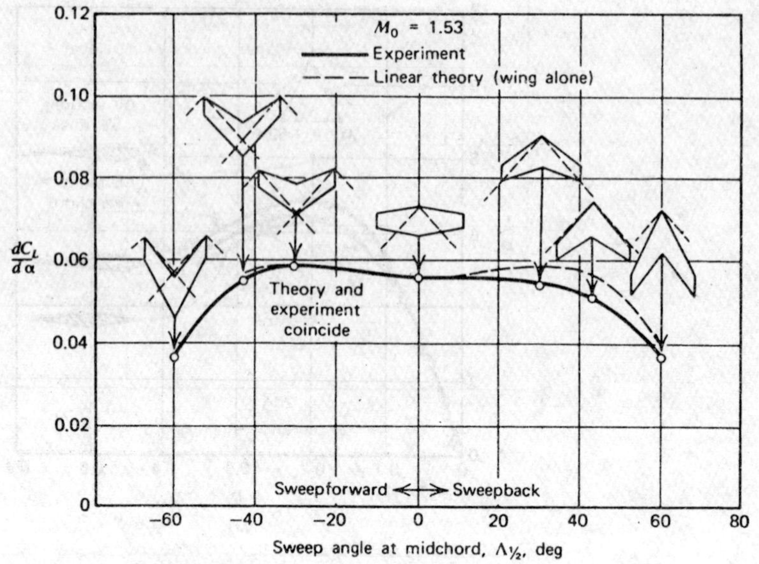

(c)

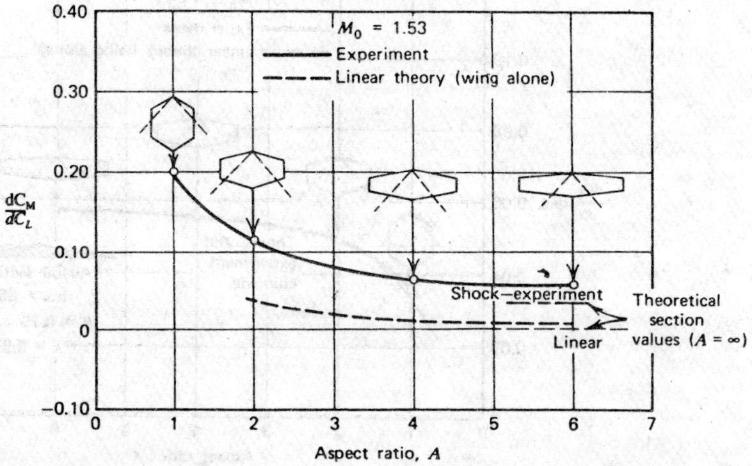

(d)

Figure 5.42 (*continued*)

In applying Equations 5.94 and 5.95, care must be taken to retain the signs of quantities under the radical. For example, if x and y are two arbitrary positive quantities,

$$\sqrt{(-x)(-y)} = \sqrt{(-1)^2 xy} = -\sqrt{xy}$$

In order to use Equations 5.94 and 5.95, the following quantities are defined:

$$k = \frac{\cot \Lambda_{TE}}{\cot \Lambda}$$

$$= \frac{A(1+\lambda)}{A(1+\lambda) - 4m(1-\lambda)}$$

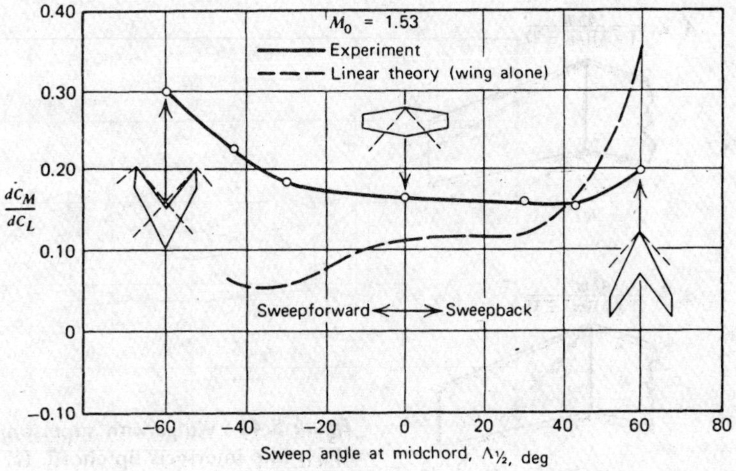

(e)

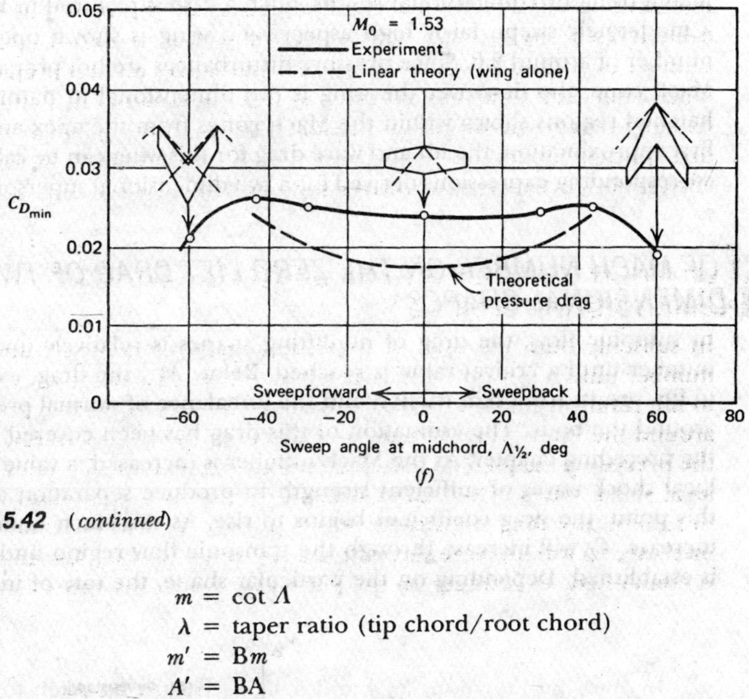

(f)

Figure 5.42 (*continued*)

$$m = \cot \Lambda$$
$$\lambda = \text{taper ratio (tip chord/root chord)}$$
$$m' = Bm$$
$$A' = BA$$

Since there is no leading edge suction for a wing with a supersonic leading edge, its leading edges should be sharp to reduce the drag. The drag caused by lift in this case is given simply by

$$C_{D_i} = C_L \tan \alpha \qquad (5.96)$$

To this must be added the wave drag and skin friction drag to obtain the total C_D.

A delta wing having the leading edge ahead of the Mach cone from the apex is a relatively simple case to treat. For this configuration, the expression for C_{L_α} is identical to that obtained for a two-dimensional airfoil (Equation 5.76).

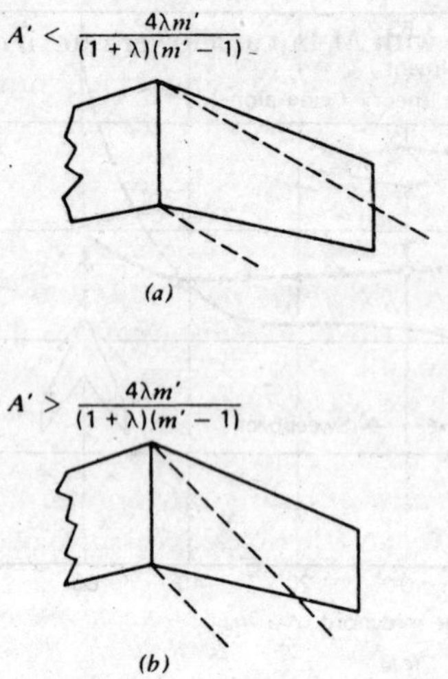

Figure 5.43 Wings with supersonic leading edges. (*a*) Mach line intersects tip chord. (*b*) Mach line intersects trailing edge.

For certain extreme cases, the lift of a supersonic wing can be quickly approximated using two-dimensional results. Such a case is pictured in Figure 5.44. Here, a moderately swept, fairly high aspect ratio wing is shown operating at a Mach number of around 2.0. Since pressure disturbances are not propagated outside the Mach cone, the flow over the wing is two dimensional in nature, except for the hatched regions shown within the Mach cones from the apex and tips. Thus, as a first approximation, the lift and wave drag for this wing can be calculated using the corresponding expressions derived for a two-dimensional supersonic airfoil section.

EFFECT OF MACH NUMBER ON THE ZERO LIFT DRAG OF TWO- AND THREE-DIMENSIONAL SHAPES

In subsonic flow, the drag of nonlifting shapes is relatively unaffected by Mach number until a critical value is reached. Below M_{cr}, the drag, excluding that due to lift, results from skin friction and the unbalance of normal pressures integrated around the body. The estimation of this drag has been covered in some detail in the preceding chapter. As the Mach number is increased, a value is reached where local shock waves of sufficient strength to produce separation are generated. At this point, the drag coefficient begins to rise. As the Mach number continues to increase, C_D will increase through the transonic flow region until supersonic flow is established. Depending on the particular shape, the rate of increase of C_D with

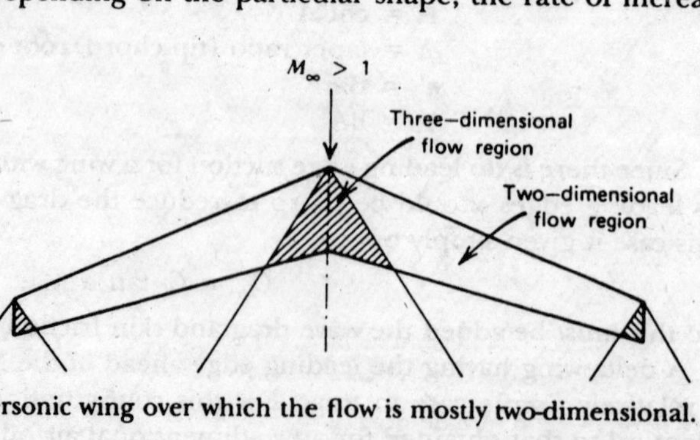

Figure 5.44 Supersonic wing over which the flow is mostly two-dimensional.

M diminishes. C_D may continue to increase with M, but at a lower rate, it can remain fairly constant, or it can actually decrease with increasing Mach number. The behavior of C_D in the supersonic flow regime depends on the composition of the drag. Excluding the drag caused by lift, the remainder of the drag is composed of skin friction drag, wave drag, and base drag.

Base drag is a term not yet used. It refers to the drag produced by the pressure acting on the blunt rear end (base) of a body, such as that pictured in Figure 5.45. The base drag was not stressed in Chapter 4 since, in subsonic flow, the shape of the base affects the flow over the rest of the body ahead of it, so the base was viewed as simply an integral part of the overall pressure drag. In supersonic flow, however, the base does not affect the flow ahead of it, so it is convenient to treat it separately.

An upper limit on the base drag can be easily obtained by noting that the base pressure can never be less than zero. Hence, the base pressure coefficient is bounded by

$$C_{p_B} = \frac{p_B - p_0}{(1/2)\rho_0 V^2}$$

$$> \frac{-p_0}{(1.2)\rho V^2}$$

$$> -\frac{2}{\gamma M^2}$$

Therefore, the base drag coefficient, *based on the base area,* must satisfy the inequality

$$C_{D_B} < \frac{2}{\gamma M^2} \tag{5.97}$$

There appears to be no accepted method available for calculating C_{D_B}. Generally, the base drag is affected by the thickness of the boundary layer just ahead of the base. Hence, C_{D_B} depends on the body shape, surface condition, and Reynolds number, as well as on the Mach number. In the supersonic regime, experimental data presented as a ratio to the upper limit (Eq. 5.97) show the trends pictured in Figure 5.46.

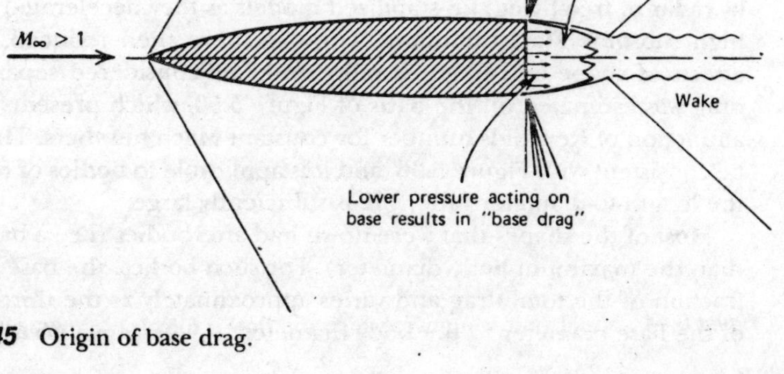

Figure 5.45 Origin of base drag.

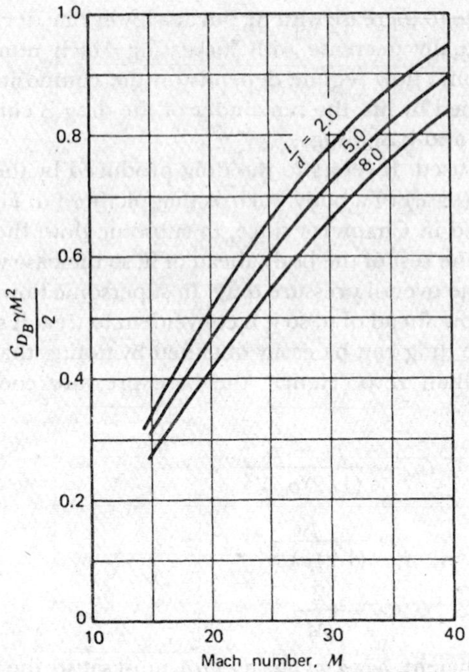

Figure 5.46 Ratio of base C_D to upper limit as a function of M and body-length–diameter ratio.

The total drag coefficients of some basic shapes are presented in Figures 5.47 to 5.49. These curves are based on data from a number of sources. In Figure 5.47, C_D as a function of M is presented for circular disks and 2-dimensional flat plates, while Figure 5.48 presents the corresponding graphs for spheres and cylinders. In both cases, for the two-dimensional bluff shapes, C_d peaks at a Mach number of unity and then decreases, with increasing M reaching a value at approximately $M = 2$, which is equal to or slightly less than the low-speed value. Above $M = 2$, C_D remains nearly constant as M increases. The three-dimensional values, however, begin to rise at an M of approximately 0.7 and continue to rise until an M of approximately 2.0 is reached.

Figure 5.49 presents the drag of various conical heads having different apex angles as a function of Mach number. This is only the drag resulting from the pressure on the forward surface of the cone.

Drag data on a number of bodies of revolution are presented in Reference 5.21. Most of these data were derived by differentiating the velocity time history obtained by radar of free-flying, fin-stabilized models as they decelerated from supersonic to high subsonic Mach numbers. The data were then reduced, assuming that the effects of shape and fineness ratio could be considered separately. Skin friction drag was estimated on the basis of Figure 5.50, which presents C_f for flat plates as a function of Reynolds number for constant Mach numbers. This figure appears to be consistent with Figure 5.36, and it is applicable to bodies of revolution, provided the length-to-diameter ratio, l/d, is sufficiently large.

Most of the shapes that were flown had afterbodies (i.e., a base diameter smaller than the maximum body diameter). For such bodies, the base drag is only a small fraction of the total drag and varies approximately as the third power of the ratio of the base diameter to the body diameter. *(Text continued on p. 274.)*

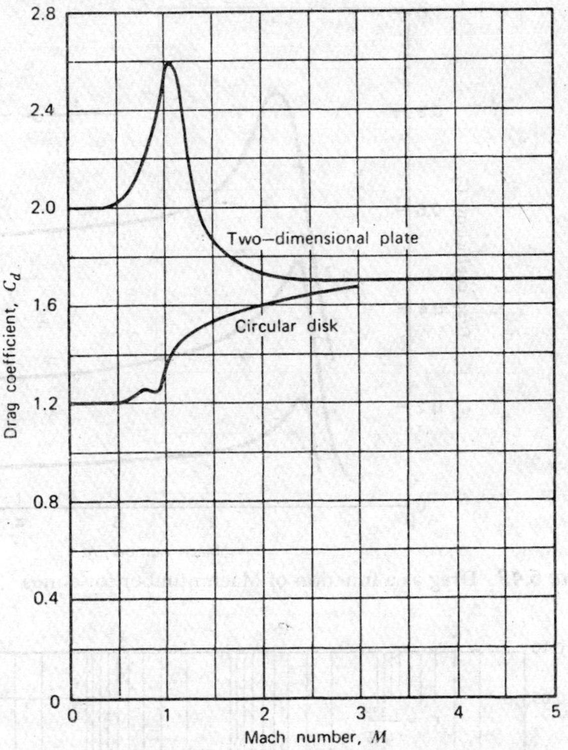

Figure 5.47 Drag of flat plates and discs normal to the flow as a function of Mach number.

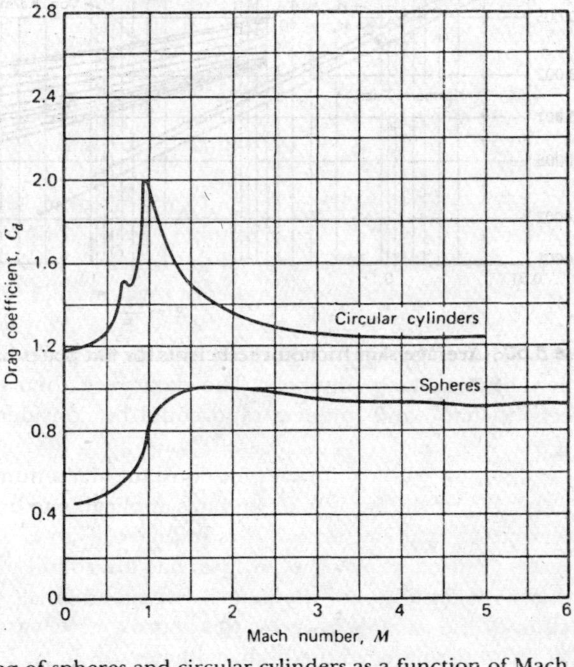

Figure 5.48 Drag of spheres and circular cylinders as a function of Mach number.

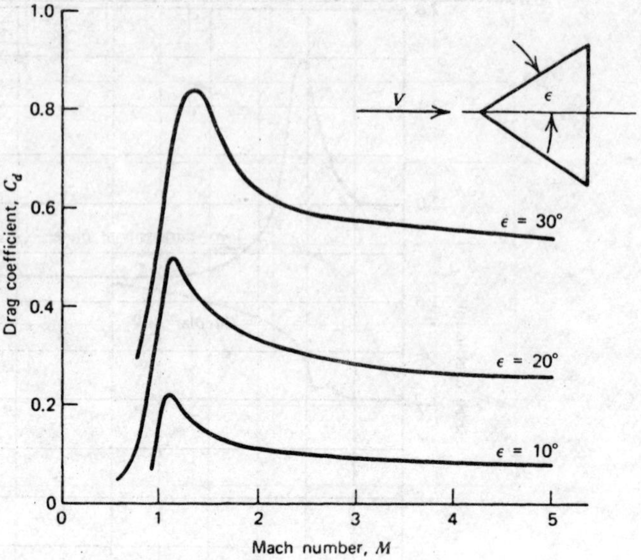

Figure 5.49 Drag as a function of Mach number for cones.

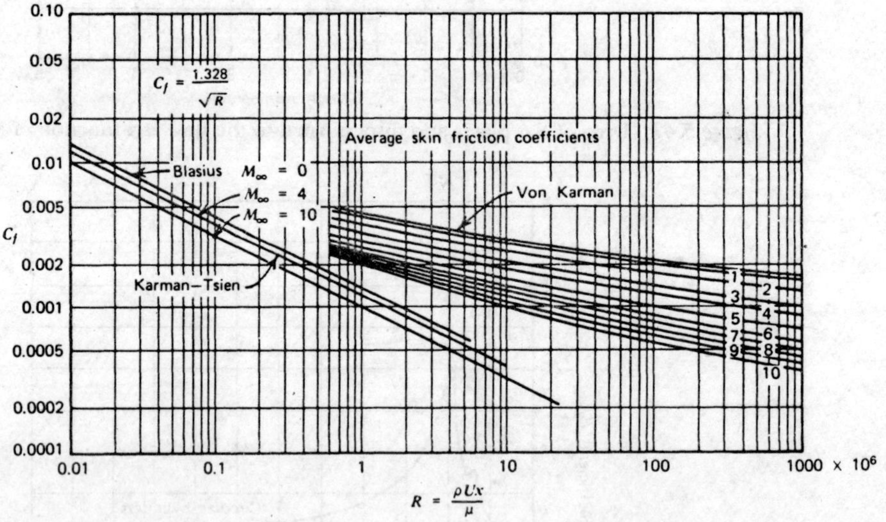

Figure 5.50 Average skin friction coefficients for flat plates based on wetted area.

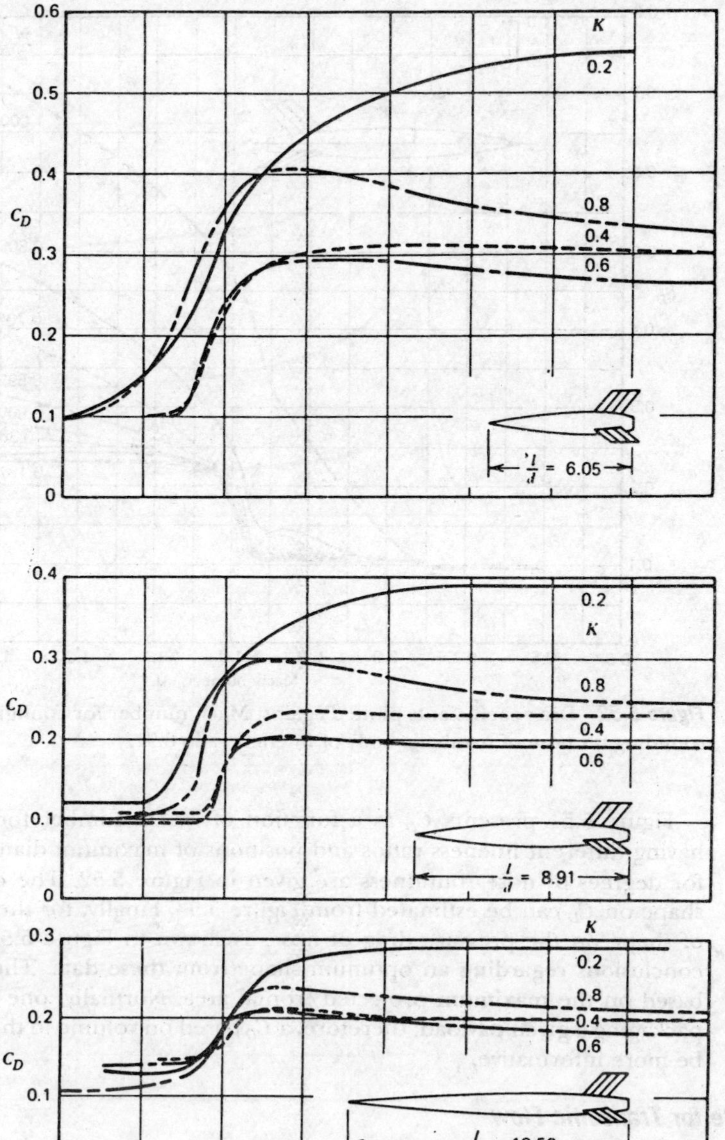

Figure 5.51 **Drag coefficients of parabolic bodies showing effects of fineness ratio and position of maximum diameter. (k = position of maximum diameter from nose as fraction of body length).**

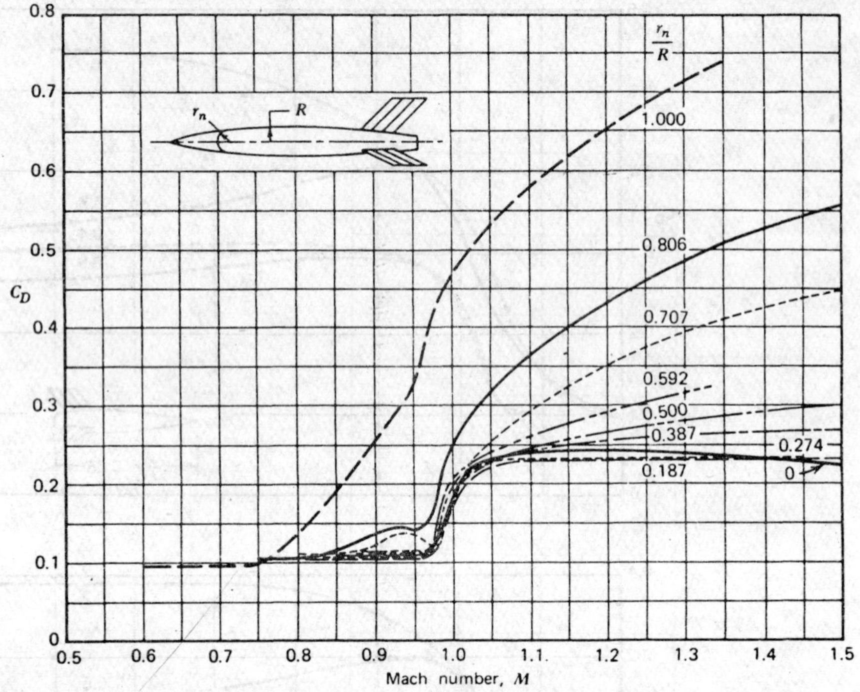

Figure 5.52 Drag coefficients plotted against Mach number for configurations obtained by rounding off nose of parabolic body of fineness ratio 8.91.

Figure 5.51 presents C_D as a function of Mach number for parabolic bodies having different fineness ratios and positions of maximum diameter. Similar data for degrees of nose roundness are given in Figure 5.52. The effect of afterbody shape on C_D can be estimated from Figure 5.53. Finally, for these data, the effect of shape on the pressure drag of noses is shown in Figure 5.54. Draw your own conclusions regarding an optimum shape from these data. The C_D values are all based on the maximum projected frontal area. Normally, one is concerned with packaging a given payload; therefore, a C_D based on volume to the 2/3 power might be more informative.

Area Rule for Transonic Flow

Reference 5.22 represents a significant contribution to the aerodynamics of high-speed aircraft. Whitcomb experimentally investigated the zero lift drag of wing-body combinations through the transonic flow regime. Based on analyses by Hayes (Ref. 5.23), Busemann (Ref. 5.24), and others, Whitcomb formulated some general guidelines for the design of wing-body combinations to have minimum wave drag. These are reflected in most of today's aircraft designed to operate near or in excess of Mach 1. These guidelines are included in the general designation "area rule."

The essence of the area rule is contained in the data of Figures 5.55 to 5.58 (taken from Whitcomb's original NACA report). These figures present C_D at zero lift as a function of M for a cylindrical body alone and in combination with a triangular wing and a tapered swept wing. C_{D_0} is the total C_D measured for zero lift, while ΔC_{D_0} is obtained by simply subtracting C_{D_0} at M = 0.85 from the other C_{D_0}

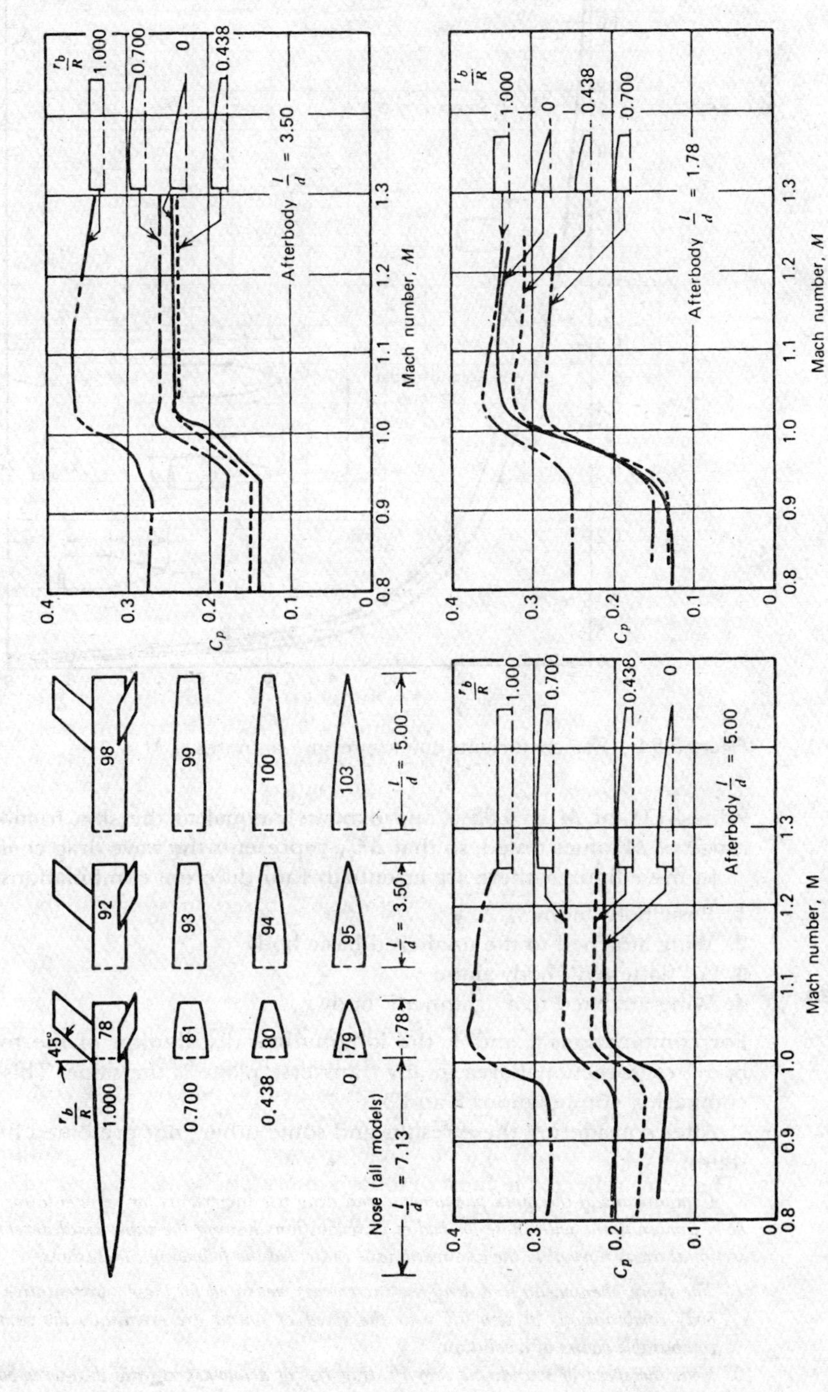

Figure 5.53 Drag coefficients for 12 bodies having identical noses and different afterbodies.

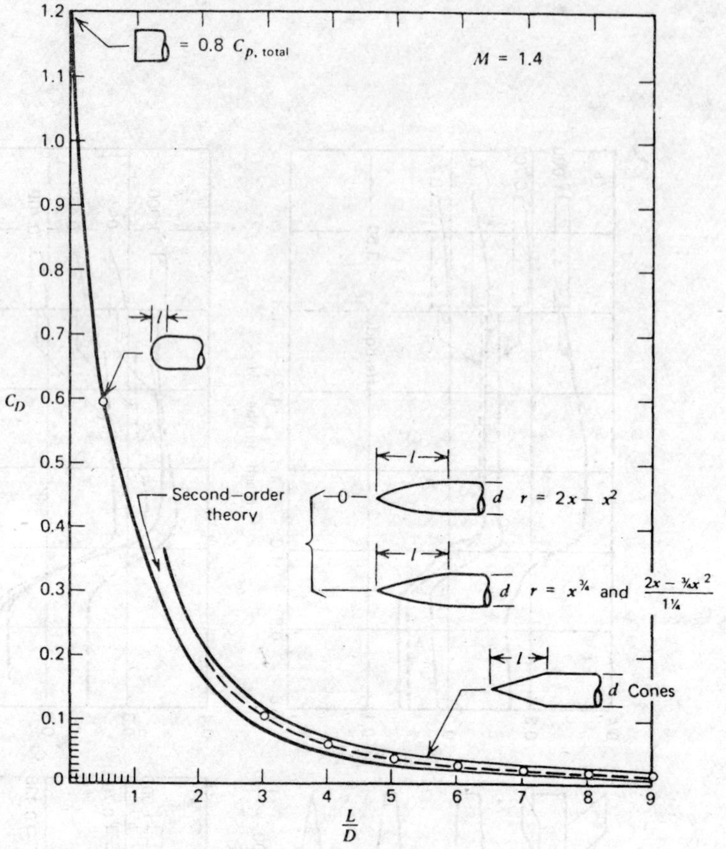

Figure 5.54 Drag coefficients due to pressure on noses at $M = 1.4$.

values. C_{D_0} at $M = 0.85$ is approximately equal to the skin friction C_D over the range of M values tested, so that ΔC_{D_0} represents the wave drag coefficient.

In these figures, there are essentially four different combinations.

1. Basic body alone
2. Wing attached to the unaltered basic body
3. A "flattened" body alone
4. Wing attached to a "thinned" body

For combinations 1 and 4, the longitudinal distribution of the total (wing plus body) cross-sectional area in any transverse plane is the same. This is also true in comparing combinations 2 and 3.

After considering these results and some others not presented here, Whitcomb states:

Comparisons of the shock phenomena and drag-rise increments for representative wing and central-body combinations with those for bodies of revolutions having the same axial developments and cross-sectional areas normal to the airstream have indicated the following conclusions:

1. *The shock phenomena and drag-rise increments measured for these representative wing and central-body combinations at zero lift near the speed of sound are essentially the same as those for the comparable bodies of revolution.*
2. *Near the speed of sound, the zero-lift drag rise of a low-aspect-ratio thin-wing-body combination is primarily dependent on the axial development of the cross-sectional areas normal to the airstream.*

(Text continued on p. 281.)

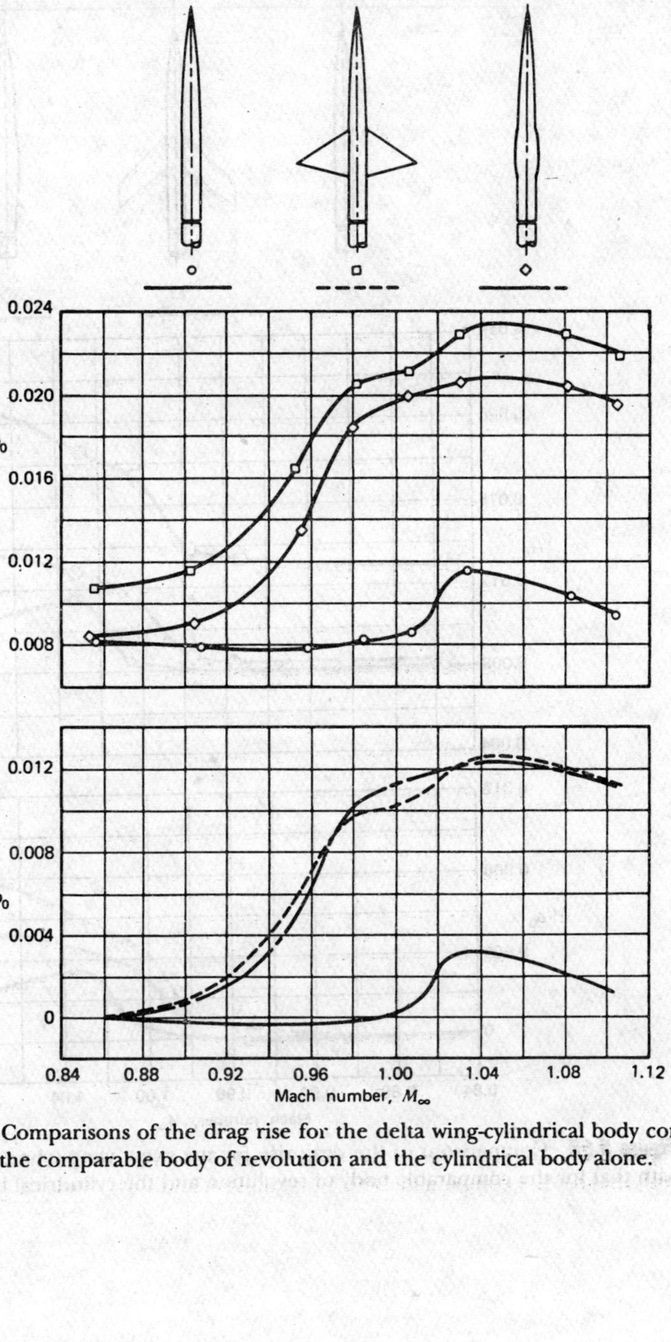

Figure 5.55 Comparisons of the drag rise for the delta wing-cylindrical body combination with that for the comparable body of revolution and the cylindrical body alone.

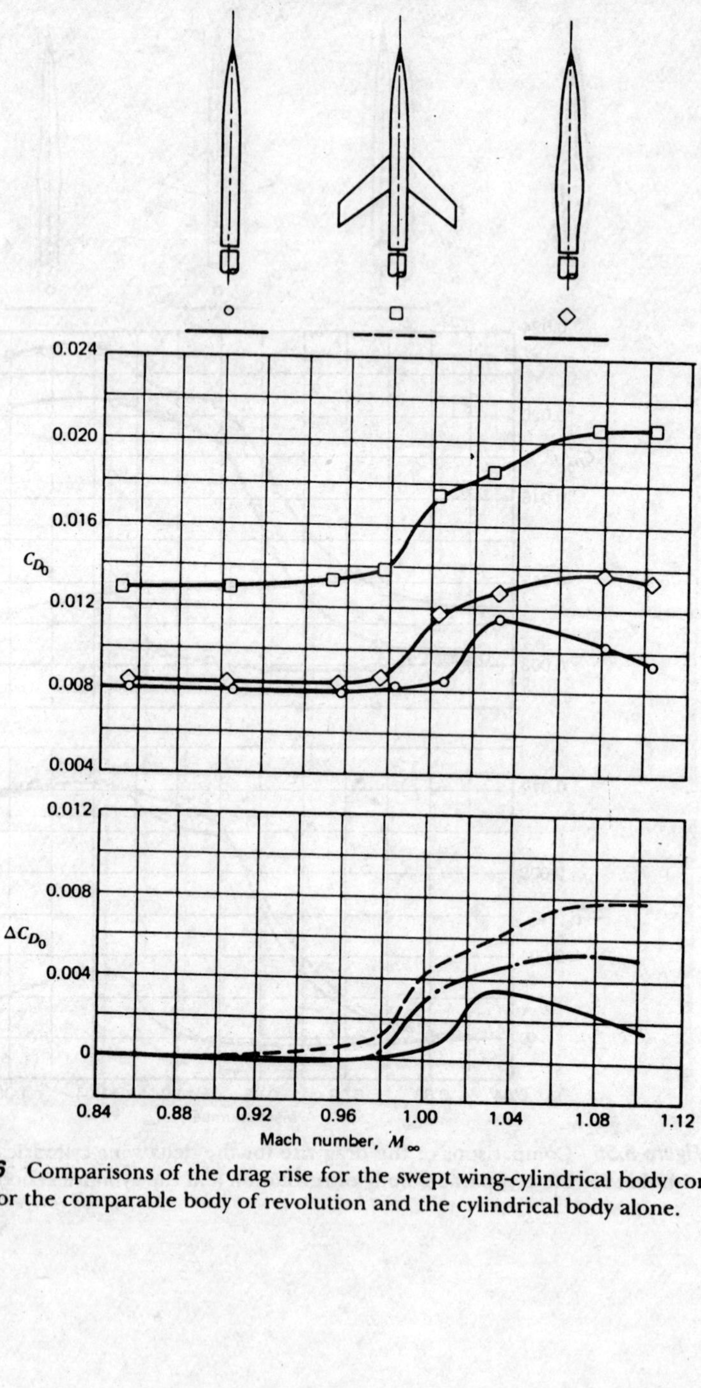

Figure 5.56 Comparisons of the drag rise for the swept wing-cylindrical body combination with that for the comparable body of revolution and the cylindrical body alone.

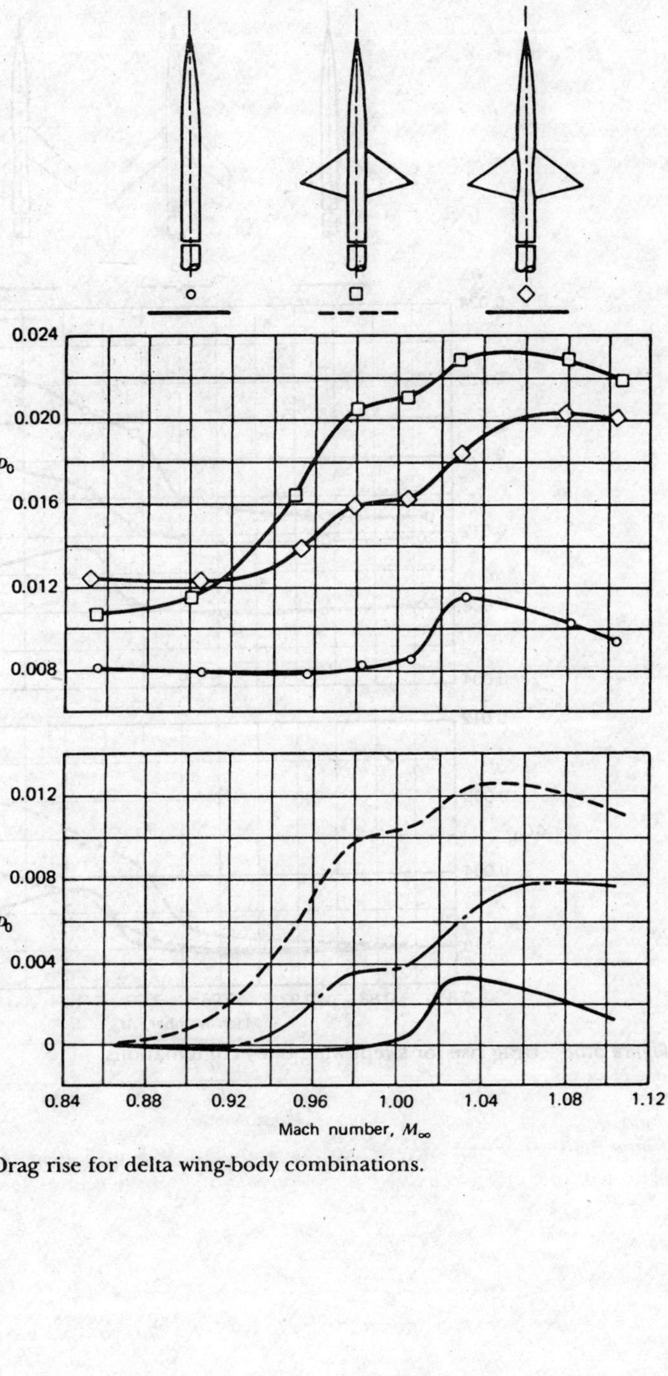

Figure 5.57 Drag rise for delta wing-body combinations.

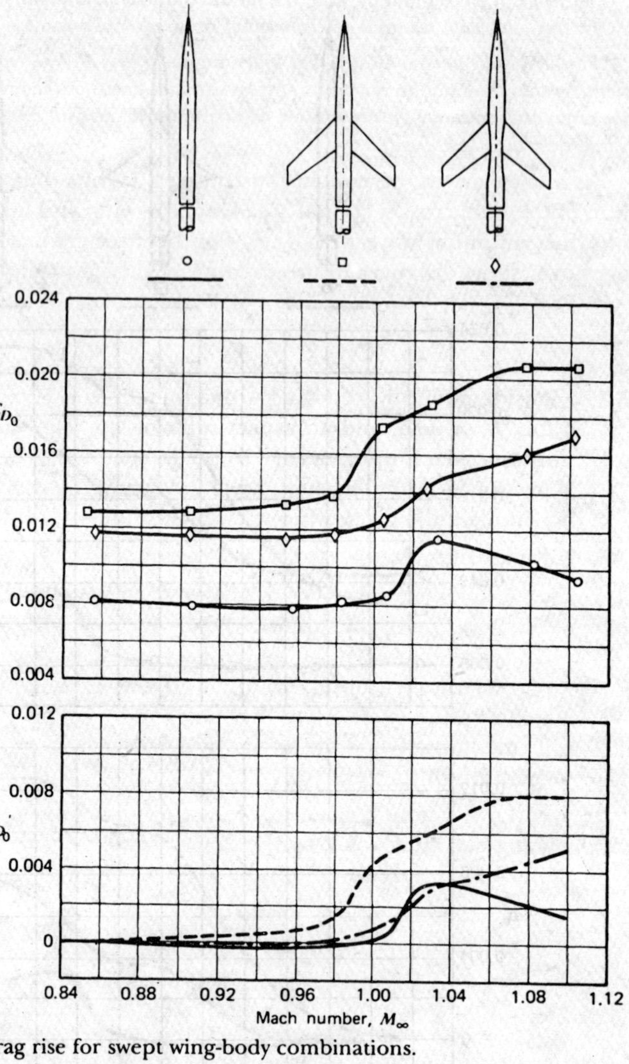

Figure 5.58 Drag rise for swept wing-body combinations.

Therefore, it follows that the drag rise for any such configuration is approximately the same as that for any other with the same development of cross-sectional areas.

Further results have indicated that indenting the bodies of three representative wing-body combinations, so that the axial developments of cross-sectional areas for the combinations were the same as for the original body alone, greatly reduced or eliminated the zero-lift drag-rise increments associated with wings near the speed of sound.

As a corollary to the conclusions stated by Whitcomb, the longitudinal distribution of the total cross-sectional area from the wing and fuselage should be a smooth one for minimum wave drag (i.e., significant increases or decreases of the total area over short distances in the streamwise direction should be avoided).

PROBLEMS

5.1 Consider an untwisted wing with a midchord sweep angle, $\Lambda_{1/2}$, of 40°, a taper ratio, λ, of 0.35, and an aspect ratio of 4.5. The wing characteristics at high subsonic Mach numbers are to be determined by testing equivalent wings in a low-speed tunnel having $M \simeq 0.2$. What is the shape of the wing to be tested to simulate the given wing at a Mach number of 0.7?

5.2 Given a thin circular arc airfoil having a 5% camber ratio and operating at zero angle of attack, what is its critical Mach number?

5.3 A 56.63 m^3 reservoir supplies the air for a blowdown supersonic wind tunnel that operates at $M = 2.0$ in a test section that measures 15.24 × 15.24 cm. If the reservoir is compressed initially to an absolute pressure of 2×10^6 Pa with a temperature of 20°C and the valve suddenly opened to the tunnel, how long will the tunnel operate at its design Mach number? Assume an adiabatic expansion in the reservoir. The test section exits to SSL conditions.

5.4 Redo the example given in the text for the airfoil pictured in Figure 5.22 at an angle of attack of 4° at $M = 2.0$. Use both Ackeret's theory and the oblique shock expansion wave method.

5.5 Using Ackeret's theory, predict C_l, C_d, and C_M for a convex airfoil that is flat on the bottom with a circular arc on the top. $t/c = 0.10$, $M = 1.5$, and $\alpha = 2°$.

5.6 Prove that Equations 5.66 and 5.67 satisfy Equation 5.64 and the following: $\beta^2 \phi_{xx} + \phi_{yy} + \phi_{zz} = 0$.

5.7 Check a point on Figure 5.24 using both Ackeret's theory and the oblique shock expansion wave method. As a class exercise, each student (or team) should take a different point in order to check the entire figure collectively.

5.8 Estimate the slope of the wing lift curve, position of the aerodynamic center, and $dC_{D_i}/dC_L{}^2$ for the Vought A7-D at $M = 0.88$.

5.9 Given a thin, flat delta wing with $A = 2.0$, calculate C_L and C_D for $\alpha = 20°$ for $M = 0.9$. Include an estimate of skin friction drag. Assume SSL and b = 30 ft. wing span of 30 ft.

5.10 Repeat Problem 5.9 for $\alpha = 5°$, $M = 2.0$.

5.11 Consider a thin, swept, tapered wing with $A = 4.0$, $\lambda = 0.5$, and $\Lambda_{LE} = 25°$. The wing loading equals 110 psf and the wing section is that described for Problem 5.5, with a slightly rounded leading edge. The mean geometric chord equals 10 ft. For the wing alone, estimate $C_{L_{max}}$ without flaps at SSL conditions. Then, beginning at the stalling speed, estimate C_D and, hence, L/D as a function of true airspeed up to a Mach number 20% higher than that which will make the Mach cone coincident with the leading edge. Use SSL conditions for M values below 0.5 and a standard altitude of 25,000 for higher Mach numbers.

5.12 Given a thin delta wing with $\Lambda_{LE} = 45°$ operating at $M = 2.0$, calculate $C_{L\alpha}$ and dC_D/dC_L^2 at $\alpha = 0$. Repeat for a tapered wing with the same A, leading edge sweep, and $\lambda = 0.5$.

5.13 Estimate the total drag of the Concorde (Figure 5.25) at its design cruising speed and altitude. Assume that the fuselage is area-ruled so that the zero lift wave drag can be neglected. Use a wing planform area of 358 m^2 and a weight of 1434 kN. The wing span equals 25.56 m. Approximate the wing as a delta wing having the same A as the actual wing.

5.14 This is an open-ended problem. A supersonic airfoil is designed to operate at $M = 2.5$ at a lift coefficient of 0.1. Design an airfoil for this application to minimize wave drag. The thickness ratio must be at least 10% for structural reasons.

REFERENCES

5.1 Liepmann, H. W., and Roshko, A., *Elements of Gas Dynamics,* John Wiley, New York, 1957.

5.2 Shapiro, A. H., *The Dynamics and Thermodynamics of Compressible Fluid Flow,* Vols. I and II, Ronald Press, New York, 1953.

5.3 Clancy, L. J., *Aerodynamics,* A Halsted Press Book, John Wiley, New York, 1975.

5.4 Daley, B. N., and Dick, R. S., *Effect of Thickness, Camber, and Thickness Distribution on Airfoil Characteristics at Mach Numbers Up to 1.0,* NACA TN 3607, March 1956.

5.5 Kuethe, A. M., and Chow, C., *Foundations of Aerodynamics, 3rd edition* (first two editions by A. M. Kuethe and J. D. Schetzer), John Wiley, New York, 1976.

5.6 *USAF Stability and Control Datcom,* Flight Control Division, Air Force Flight Dynamics Laboratory, Wright-Patterson Air Force Base, OH, October 1960 (revised January 1975).

5.7 Ames Research Staff, *Equations, Tables and Charts for Compressible Flow,* NACA Report 1135, 1953.

5.8 Axelson, J. A., "Estimation of Transonic Aerodynamics to High Angles of Attack," *AIAA J. of Aircraft, 14(6),* pp. 553–559, June 1977.

5.9 Laitone, E. V., *Limiting Velocity by Momentum Relations for Hydrofoils Near the Surface and Airfoils in Near Sonic Flow,* Proceedings of the Second U.S. National Congress of Applied Mechanics, pp. 751–753, June 14–18, 1954.

5.10 Stivers, L. S., Jr., *Effects of Subsonic Mach Number on the Forces and Pressure Distributions on Four NACA 64A-series Airfoil Sections at Angles of Attack as High as 28,* NACA TN 3162, March 1954.

5.11 *Jane's All the World's Aircraft,* published annually by Paulton House, London. Taylor, John W. R., Ed.

5.12 Mullin, S. N., *The Evolution of the F-22 Advanced Tactical Fighter,* 1992 Wright Brothers Lecture, AIAA Aircraft Design Systems Meeting, Hilton Head, SC, August 1992.

5.13 Garner, H. C., and Inch, S. M., *Subsonic Theoretical Lift-Curve Slope Aerodynamic Centre and Spanwise Loading for Arbitrary Aspect Ratio, Taper Ratio and Sweepback,* Aeronautical Research Council C. P. No. 1137, 1971.

5.14 Razak, Kenneth, and Snyder, M. H., *A Review of the Planform Effects of the Low-Speed Aerodynamic Characteristics of Triangular and Modified Triangular Wings,* NASA CR-421, April 1966.

5.15 Furlong, G. C., and McHugh, J. G., *A Summary and Analysis of the Low-Speed Longitudinal Characteristics of Swept Wings at High Reynolds Number,* NACA Report 1339, 1957.

5.16 Hicks, R. M., and Hopkins, E. J., *Effects of Spanwise Variation of Leading-Edge Sweep on the Lift, Drag, and Pitching Moment of a Wing-Body Combination at Mach Numbers from 0.7 to 2.94*, NASA TB D-2236, April 1964.

5.17 Brown C. E., *Theoretical Lift and Drag of Thin Triangular Wings at Supersonic Speeds*, NACA Report 839, 1946.

5.18 Malvestuto, F. S., Margolis, Kenneth, and Ribner, H. S., *Theoretical Lift and Damping in Roll at Supersonic Speeds of Thin Sweptback Tapered Wings with Streamwise Tips, Subsonic Leading Edges, and Supersonic Trailing Edges*, NACA Report 970, 1950.

5.19 Vicenti, W. G., *Comparison Between Theory and Experiment for Wings at Supersonic Speeds*, NACA Report 1033, 1951.

5.20 Harmon, S. M., and Jeffreys, Isabella, *Theoretical Lift and Damping in Roll of Thin Wings with Arbitrary Sweep and Taper at Supersonic Speeds-Supersonic Leading and Trailing Edges*, NACA TN 2114, May 1950.

5.21 Stoney, W. E., *Collection of Zero-Lift Drag Data on Bodies of Revolution from Free-Flight Investigations*, NACA TN 4201, January 1958.

5.22 Whitcomb, R. T., *A Study of the Zero-Lift Drag-Rise Characteristics of Wing-Body Combinations Near the Speed of Sound*, NACA Report 1273, 1956.

5.23 Hayes, W. D., *Linearized Supersonic Flow*, Report No. AL-222, North American Aviation, Inc. (now Rockwell International), June 28, 1947.

5.24 Busemann, Adolf, *Application of Transonic Similarity*, NACA TN 2687, 1952.

6 THE PRODUCTION OF THRUST

A BRIEF HISTORY OF THE PISTON ENGINE

The evolution of today's airplane has depended significantly on the development of power plant technology. As pointed out by Torell (Ref. 6.1),

> *Each successive generation of transport aircraft has been paced by the development of engines with greater power and improved efficiency, and more power for each pound of weight.*

From 1903 until the early 1940s, the piston engine-propeller combination was the only type of power plant used to any extent for airplane propulsion. The earliest piston engine designs were water-cooled, in-line designs requiring bulky, high-drag radiators. The Wright Brothers' first engine was of this type. This engine had four upright cylinders, weighed 890 N (200 lb), and developed approximately 8.95 kW (12 hp). The liquid-cooled airplane piston engine remained in extensive use through World War II and evolved into powerful, efficient configurations typified by the Rolls-Royce Merlin. This ethylene glycol-cooled V-12 design, weighing 6360 N (1430 lb), delivered 843 kW (1130 hp) at 3000 rpm.

Air-cooled airplane piston engines trace their lineage from the French Gnome rotary-radial design built in 1908. For this configuration, the propeller was attached to the cylinders which, displaced radially, rotated around the engine centerline. The first engines of this type delivered approximately 52 to 60 kW (70 to 80 hp) for a weight of approximately 1023 N (230 lb). The power-to-weight ratio of the rotary-radial engine was an improvement over the contemporary water-cooled designs; however, the fuel and oil consumption was high for the rotary-radial designs. Also, gyroscopic effects resulting from their high angular momentum were severe under maneuvering conditions.

The rotary-radial engine was superseded by the static, or stationary, radial configuration. In the 1920s, a competitive series of static-radial piston engines were produced by the Wright Co. and the Pratt & Whitney Co. The first of these was the Wright Whirlwind. It was a Wright J-5C Whirlwind, nine-cylinder radial engine weighing approximately 2224 N (500 lb) and developing 164 kW (220 hp); that powered the Ryan monoplane "Spirit of St. Louis" in which Colonel Charles Lindbergh made his epic crossing of the Atlantic in May 1927. It was also the Wright Whirlwind, three of them each rated at 224 kW (300 hp), that took Admiral Richard Byrd to Antarctica in a Ford Tri-motor in 1929.

Pratt & Whitney's first radial engine, developed shortly after the Whirlwind, was designated the Wasp. This engine, weighing 2891 N (650 lb), developed 298 kW (400 hp) at 1900 rpm. The Wasp was followed by the more powerful Hornet and then by the Twin-Wasp, a radial engine with two rows of radially displaced cylinders. The Wright Cyclone was a similar design that first powered one of the most famous air transports of all time, the Douglas DC-3. Succeeding versions of the DC-3 were also powered with the Twin-Wasp designated the R-1830. The 1830 refers to the engine displacement in cubic inches. More than 10,000 of the Wasp-powered DC-3's were built, and many are still flying today.

Horizontally opposed, air-cooled piston engines for light aircraft appeared around 1932. These designs are still produced in large quantities today by Continental Motors and Avco Lycoming. Production of the larger radial engines ceased in the late 1950s with the development of the turboprop and turbojet engines.

PISTON ENGINE CHARACTERISTICS

An aircraft piston engine is similar to an automobile engine with a few differences. First, an aircraft engine is designed with weight as a primary consideration. Thus, the weight-to-power ratio is generally lower for an aircraft engine when compared with an automobile engine of comparable size. Weight-to-power ratios for various aircraft piston engine configurations are presented in Figure 6.1 as a function of engine power. The ratio is seen to improve as the engine gets larger. Turboprop engines are also included in this figure and are seen to have weight-to-power ratios that are less than half of those for the piston engines. The turboprop engine will be discussed in more detail later. Notice for the piston engines that the weight-to-power ratio trend is about the same for all of the engine configurations. The horizontally opposed, air-cooled configuration appears to be somewhat better than the rest, but this may be because most of these engines were designed and built at a later date than the radial or in-line engines.

Today's airplane piston engine is a very reliable piece of machinery. With recommended major overhaul periods of up to 2000 hr, one can get 400,000 to 800,000 km on an engine (depending on the cruising speed) before major components are replaced. An airplane engine has two spark plugs on each cylinder that are fired independently from engine-driven magnetos. Before taking off, a pilot checks to see that the engine will run smoothly on either "mag" alone. Many airplane engine systems also incorporate an additional fuel pump that is electrically driven, independent of the engine.

Since it must operate over a range of density altitudes, an airplane engine has a

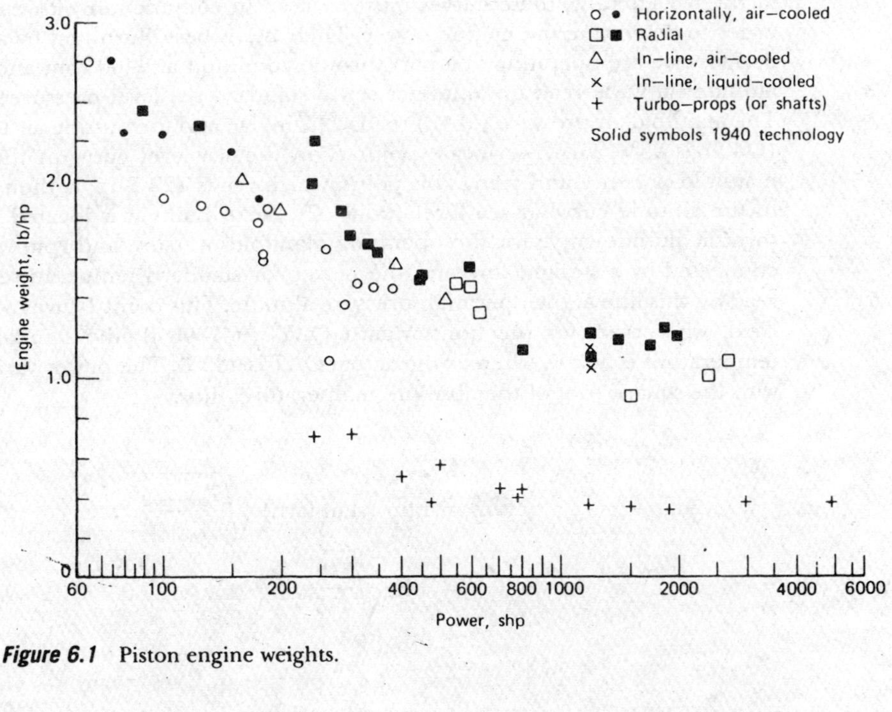

Figure 6.1 Piston engine weights.

manual mixture control. At low altitudes, the mixture is set relatively rich and is leaned at the higher altitudes, where the air is less dense. During a continuous climb, where a large amount of power is required, the pilot will also set the mixture on the rich side in order to provide better cooling. The richer the mixture, the cooler the exhaust gases will be. Running too rich, however, can result in a loss of power and premature spark plug fouling. Therefore, many airplanes are equipped with an exhaust gas temperature (EGT) gage that allows the pilot to set the mixture control more accurately.

Carbureted airplane engines, as opposed to fuel-injected engines, have a carburetor heat control. In the "on" position, this control provides heated air to the carburetor in order to avoid the buildup of ice in the venturi. Since the air expands in the carburetor throat, the temperature in this region can be below freezing even when the outside air temperature is above freezing. If a pilot is flying through rain or heavy clouds at temperatures close to freezing, he or she can experience carburetor icing with an attendant loss (possibly complete!) of power unless he or she pulls on the carburetor heat.

An airplane piston engine is operated with primary reference to two gages: the tachometer (which indicates the engine rpm), and the manifold pressure gage (which measures the absolute pressure within the intake manifold). These two quantities, at a given density altitude, determine the engine power. To develop this point further, consider the characteristics of the Lycoming 0-360-A engine, which is installed in the Piper Cherokee 180 used previously as an example. A photograph of this engine is presented in Figure 6.2.

This horizontally opposed, four-cylinder, air-cooled engine is rated at 180 bhp at 2700 rpm. On a relative basis, its performance is typical of piston engines and will be discussed now in some detail. Operating curves for this engine are presented in Figure 6.3. Its performance at sea level is on the left and the altitude performance is on the right. These curves are given in English units, as prepared by the manufacturer. Generally, these curves must be used in conjunction with each other in order to determine the engine power. Their use is best illustrated by an example. Suppose we are operating at a part throttle condition at 2400 rpm and a pressure altitude (altitude read on altimeter set to standard sea level pressure) of 1800 ft. The manifold air pressure (MAP) reads 23.2 in. Hg and the outside air temperature (OAT) is 25°F. First, we locate point B on the sea level curve at the operating manifold pressure and rpm. This point, which reads 129 Bhp, is then transferred to the altitude curve at sea level (point C). Next, point A is located on the full-throttle altitude curve for the operating manifold pressure and rpm. A and C are connected by a straight line and the power for standard temperature conditions read on this line at the operating pressure altitude. This point D gives a bhp of 133. Next, we correct for the nonstandard OAT. At 1800 ft, the standard absolute temperature is 513°R, whereas the actual OAT is 485°R. The power varies inversely with the square root of the absolute temperature; thus,

$$\begin{aligned} \text{bhp} &= \text{bhp (standard)} \sqrt{\frac{T_{\text{standard}}}{T_{\text{actual}}}} \\ &= 133 \sqrt{\frac{513}{485}} \\ &= 137 \text{ hp} \end{aligned} \tag{6.1}$$

SPECIFICATIONS AND DESCRIPTION

TYPE—Four-cylinder, direct drive, horizontally opposed, wet sump, air-cooled engine

	0-360-A
FAA Type certificate	286
Takeoff rating, O-360-C2D only	
hp, rpm and manifold pressure	180 @ 2900 and 28 in. hg
Rated hp and rpm	180 @ 2700
Cruising rpm, 75% rated	135 @ 2450
65% rated	117 @ 2350
Bore, in.	5.125
Stroke, in.	4.375
Displacement, in.^{3}	361.0
Compression ratio	8.5:1
Cylinder head temperature, max	500°
Cylinder base temperature, max	325°
Fuel octane, aviation grade, min	91/96
Valve rocker clearance (hydraulic tappets collapsed)	0.028 to 0.080
Oil sump capacity, qt	8
Oil pressure, idling psi	25
normal psi	65 to 90
start and warmup psi	100
Spark occurs, deg BTC	25
Spark plug gap, fine wire	0.015 to 0.018
massive wire	0.018 to 0.022
Firing order	1-3-2-4
Standard engines (dry weight)	
(Includes 12-V–20-amp generator and 12-volt starter)	285 lb

Figure 6.2 Lycoming Aircraft Engine. Three-quarter right front view. (Courtesy, Lycoming Division, Avco Corp.)

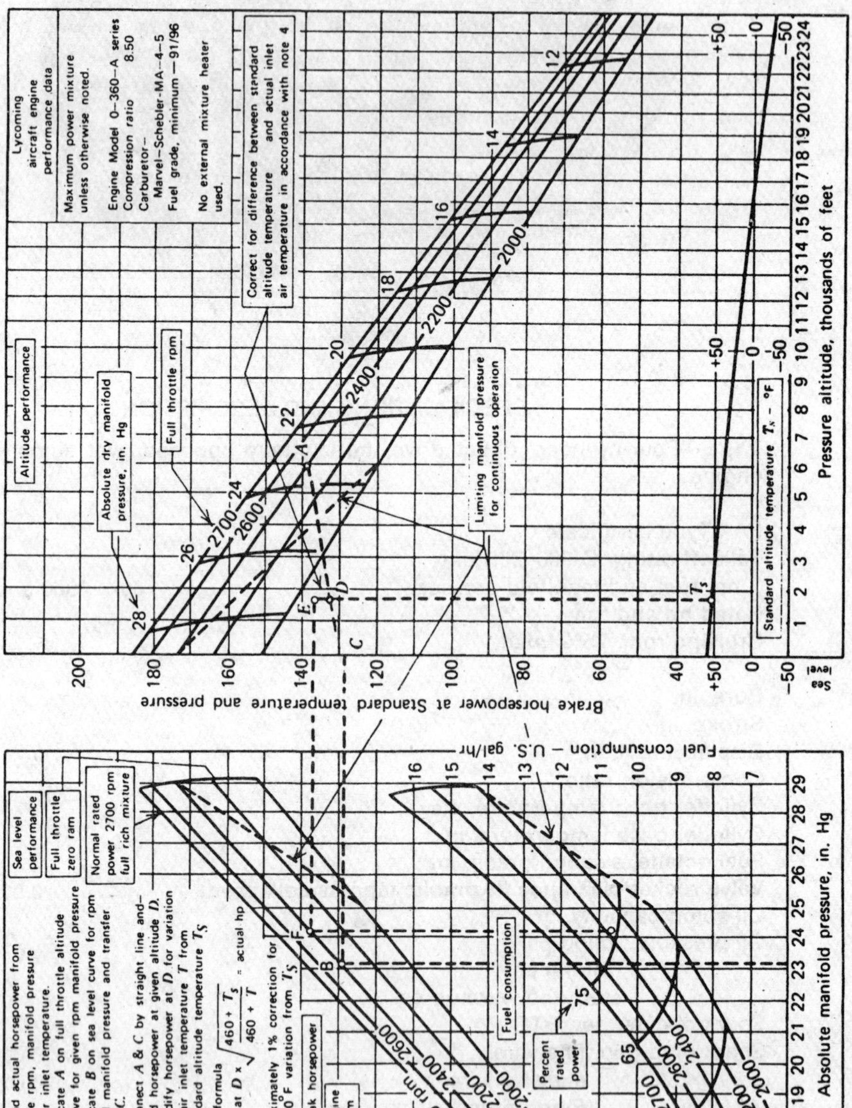

Figure 6.3 Lycoming Aircraft engine performance data. (Courtesy, Lycoming Division, Avco Corp.)

The engine is developing 137 bhp under these conditions. To determine the fuel consumption, we enter the sea level curve with this power and the engine rpm (point F). We then read the fuel consumption at the same rpm and manifold pressure corresponding to point F. In this case, a consumption of 10.8 gph is determined.

These curves are typical of operating curves for nonsupercharged, reciprocating engines. It is interesting to note that all wide-open throttle (WOT) power curves at a constant rpm decrease linearly with density ratio approaching zero at a σ of approximately 0.1 corresponding to a standard altitude of approximately 18,000 m (59,000 ft).

Fuel consumption for a piston engine is frequently given as a brake specific fuel consumption (BSFC). This quantity is measured in pounds per brake horsepower hour. For the example just covered, since gasoline weighs 6 lb/gal,

$$\begin{aligned}\text{BSFC} &= \frac{\text{gph (lb/gal)}}{\text{bhp}} \\ &= \frac{(10.8)(6)}{137} \\ &= 0.47 \text{ lb/bhp-hr}\end{aligned}$$

Referring to Figure 6.4, the value 0.47 is seen to be reasonable. This figure presents BSFC as a function of engine size for both piston and turboshaft engines. Notice that BSFC tends to improve as engine power increases. Also observe that there is little difference in BSFC for the different piston engine types. Some gains (~6%) appear to be realized by the use of fuel injection. Turboshaft engines, to be discussed later, have specific fuel consumptions that are approximately 25% higher than those for a piston engine of the same power.

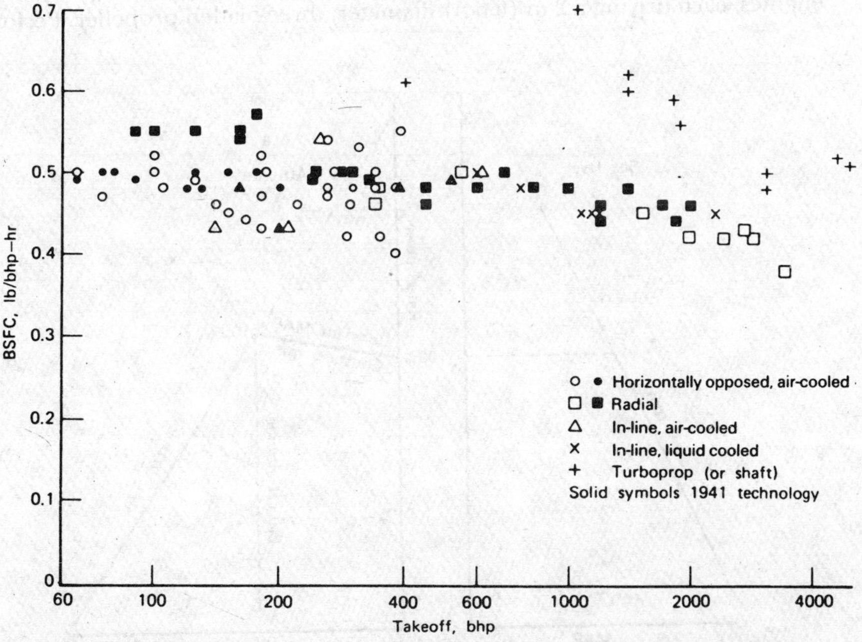

Figure 6.4 Brake specific fuel consumption for piston and turboshaft engines.

In the SI system of units, BSFC is expressed in Newtons per kilowatt-hour. In this system, the typical value of 0.5 lb/bhp-hr becomes 2.98 N/kWhr.

Supercharged Engines

The altitude performance of a piston engine can be improved by supercharging. This involves compressing the air entering the intake manifold by means of a compressor. In earlier supercharged engines, this compressor was driven by a gear train from the engine crankshaft; hence, the term *gear-driven supercharger.* A typical performance curve for a gear-driven supercharged engine is presented in Figure 6.5. Such an engine is generally limited by the manifold absolute pressure (MAP), so the pilot cannot operate at full throttle at sea level. As the pilot climbs to altitude, the throttle is opened progressively, holding a constant MAP. The engine power will increase slightly until an altitude, at which the throttle is wide open, is reached. Above this altitude, known as the *critical altitude,* the power decreases linearly with density ratio in the same relative way as a nonsupercharged engine.

Today's supercharged engines employ a turbine-driven compressor powered by the engine's exhaust. This configuration is referred to as an *exhaust turbosupercharger.* The advantage of this type of supercharger as compared to the gear-driven type is twofold. First, the compressor does not extract power from the engine, but uses energy that would normally be wasted. Second, the turbosupercharger is able to maintain sea level-rated power up to much higher altitudes than the gear-driven supercharger.

Turbosupercharged engines are equipped with a regulating system that maintains an approximately constant manifold pressure independent of altitude. A density and pressure controller regulates the position of a waste gate, or bypass valve, which regulates the amount of exhaust gases through the turbine.

A pressurized aircraft used by the general aviation industry is the Piper Navajo, pictured in Figure 6.6. This aircraft uses two Lycoming TIO-540 turbocharged engines, each driving a 2-m (6.6-ft) diameter, three-bladed propeller. Performance

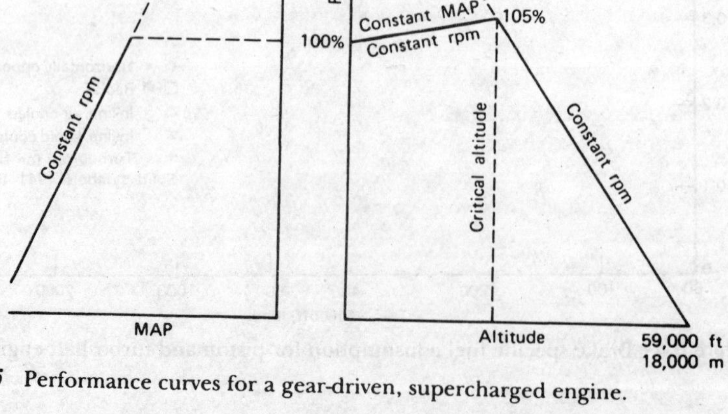

Figure 6.5 Performance curves for a gear-driven, supercharged engine.

Figure 6.6 Pressurized Navajo. (Courtesy, Piper Aircraft)

and fuel flow curves for this engine are presented in Figures 6.7 and 6.8. As shown, the engine is able to maintain its rated power up to an altitude of approximately 7300 m (24,000 ft). The use of these curves is straightforward, so they do not need any detailed explanation. Correction for nonstandard temperature is the same as that for a nonsupercharged engine.

PROPELLER ANALYSIS

The efficiency of a piston engine-propeller combination depends on a proper match of the propeller to the engine as well as a match of the two to the airframe. Understanding propeller behavior is important, even though we are currently in the "jet age." First, because of their cost, it is doubtful that gas turbine engines will be used in the smaller, general aviation airplanes in the foreseeable future. Second, there is currently a renewed interest in the turboprop engine because of its lower fuel consumption compared to turbojet or turbofan engines.

Momentum Theory

The classical momentum theory provides a basic understanding of several aspects of propeller performance. Referring to Figure 6.9, the propeller is approximated by an infinitely thin "actuator" disc across which the static pressure increases discontinuously. The assumptions inherent in this model are the following:

1. The velocity is constant over the disc.
2. The pressure is uniform over the disc.
3. Rotation imparted to the flow as it passes through the propeller is neglected.
4. The flow passing through the propeller can be separated from the rest of the flow by a well-defined streamtube.
5. The flow is incompressible.

The transverse plane 1 is far ahead of the propeller, while plane 4 is far downstream. Planes 2 and 3 are just upstream and downstream of the propeller, respectively. In planes 1 and 4, all streamlines are parallel so that the static pressure is constant and equal to the free-stream static pressure p_0.

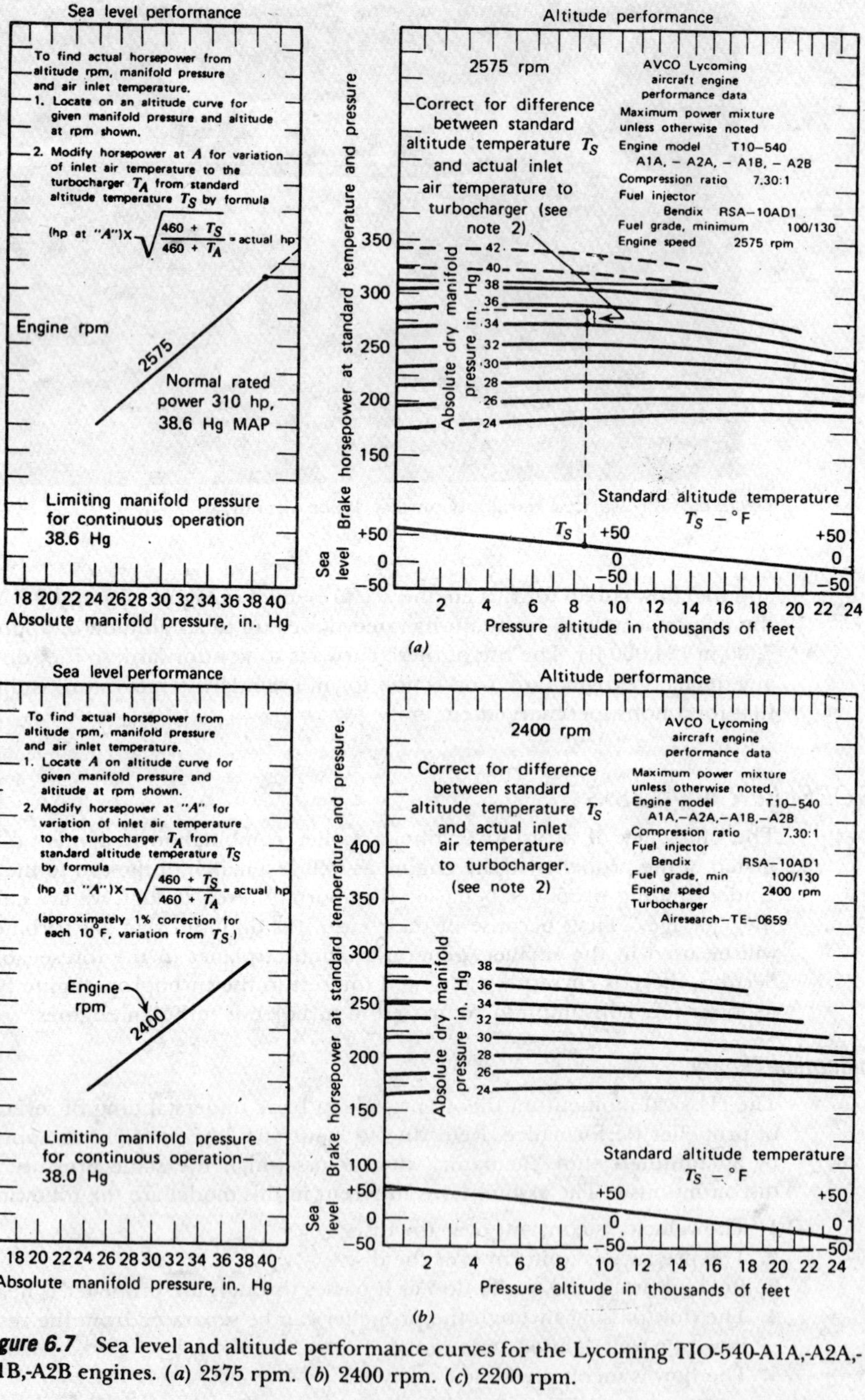

Figure 6.7 Sea level and altitude performance curves for the Lycoming TIO-540-A1A,-A2A,-A1B,-A2B engines. (*a*) 2575 rpm. (*b*) 2400 rpm. (*c*) 2200 rpm.

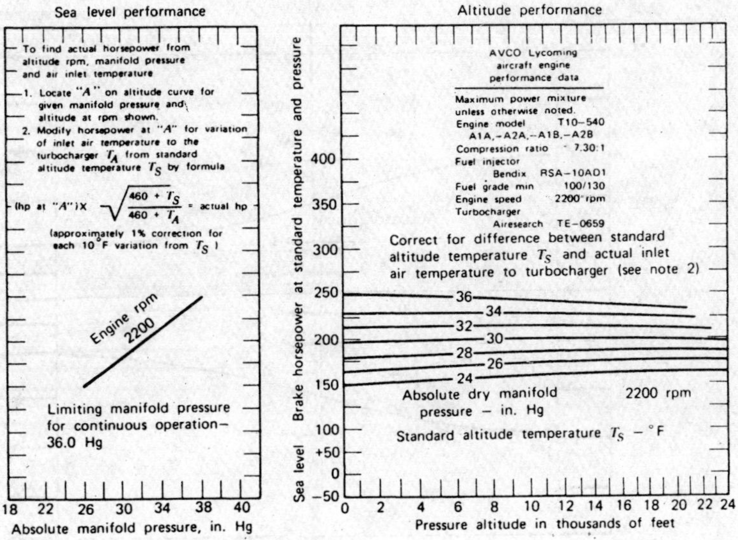

(c)

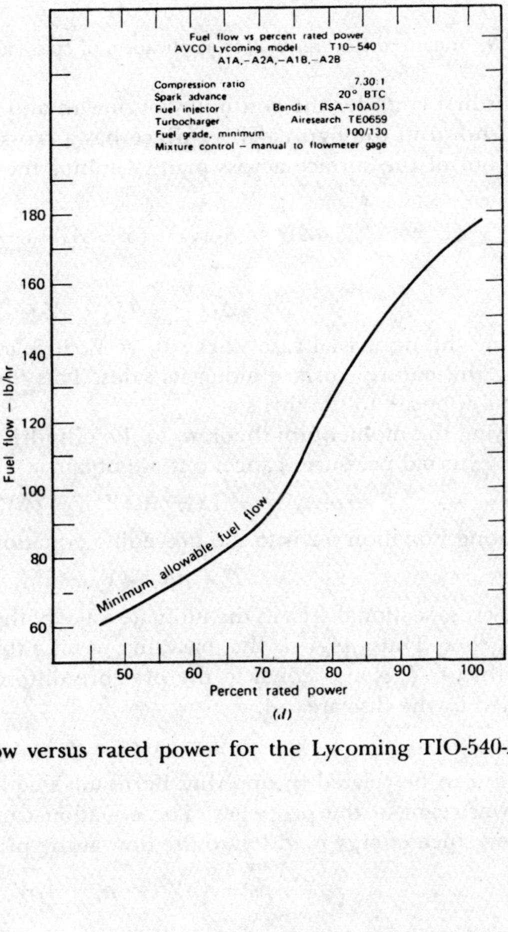

(d)

Figure 6.8 Fuel flow versus rated power for the Lycoming TIO-540-A1A,-A2A,-A1B,-A2B engines.

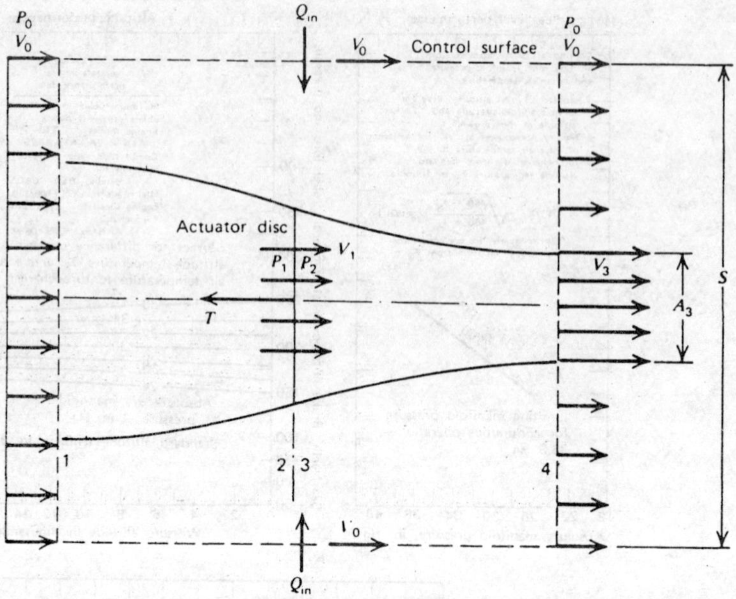

Figure 6.9 Idealized flow model for application of classical momentum theory.

Let us first consider the continuity of flow in and out of the cylindrical control surface shown in the figure. This surface has a cross-sectional area of S. The flux passing out of the surface across plane 4 minus the flux entering across plane 1 will be

$$\Delta Q = A_3 V_3 + (S - A_3) V_0 - S V_0$$

or

$$\Delta Q = A_3 (V_3 - V_0) \tag{6.2}$$

Assume the nontrivial case where $V_3 \neq V_0$, it follows that the flux ΔQ must be entering the control surface along its sides. This flux has a velocity of V_0 in the direction opposite to the thrust.

Applying the momentum theorem to the cylindrical control surface and noting that the external pressures cancel out, we obtain

$$T = \rho[A_3 V_3^2 + (S - A_3) V^2] - \rho S V_0^2 - \rho\, \Delta Q V_0$$

Substituting Equation 6.2 into the preceding equation gives

$$T = \rho A_3 V_3 (V_3 - V_0) \tag{6.3}$$

A_3 is the cross-sectional area in the ultimate wake of the streamtube passing through the propeller. Thus, $\rho A_3 V_3$ is the mass flux passing through the propeller.

The thrust, T, is also equal to the pressure difference across the actuator disc multiplied by the disc area, A.

$$T = A(p_2 - p_1) \tag{6.4}$$

p_1 and p_2 can be related by applying Bernoulli's equation ahead of the propeller and downstream of the propeller. The equation cannot be applied through the propeller, since energy is added to the flow at the propeller.

$$p_0 + \tfrac{1}{2}\rho V^2 = p_1 + \tfrac{1}{2}\rho V_1^2 \tag{6.5}$$

$$p_0 + \tfrac{1}{2}\rho V_3^2 = p_2 + \tfrac{1}{2}\rho V_2^2 \tag{6.6}$$

Subtracting Equation 6.5 from Equation 6.6 and noting that the velocity is continuous through the propeller gives

$$p_2 - p_1 = \tfrac{1}{2}\rho(V_3^2 - V_0^2) \tag{6.7}$$

Using the fact from continuity that $A_3 V_3 = A V_1$ and combining Equations 6.3, 6.4, and 6.7 results in the well-known relation

$$V_1 = \frac{(V_3 + V_0)}{2} \tag{6.8}$$

In words, the velocity through the propeller equals the average of the velocity far ahead of and far behind the propeller.

Let us now write

$$V_3 = V_0 + 2w \tag{6.9}$$

where w is the propeller-induced velocity. It follows that

$$V_1 = V_0 + w$$

so that

$$T = 2\rho A(V_0 + w)w \tag{6.10}$$

This is easily remembered, since $\rho A(V_0 + w)$ is the mass flux through the propeller and $2w$ is the total increase in the velocity of the flow.

Applying the energy theorem to this system gives for the power, P, added to the flow,

$$P = \tfrac{1}{2}\rho A(V + w)[(V_0 + 2w)^2 - V^2]$$

or

$$P = 2\rho A w(V_0 + w)^2 \tag{6.11}$$

Using Equation 6.10, this becomes

$$P = T(V_0 + w) \tag{6.12}$$

This important result states that the power required by the propeller equals the product of its thrust and the velocity *through* the propeller. This can be divided into two parts. The first part is defined as the *useful power.*

$$P_{use} = TV_0 \tag{6.13}$$

The second part is known as the *induced power.*

$$P_i = Tw \tag{6.14}$$

Equation 6.10 can be solved for the induced velocity to give

$$w = \frac{1}{2}\left[-V_0 + \sqrt{V_0^2 + \left(\frac{2T}{\rho A}\right)}\right] \tag{6.15}$$

For the static case where V equals zero,

$$w_0 = \sqrt{\frac{T}{2\rho A}} \tag{6.16}$$

and

$$P_{i0} = \frac{T^{3/2}}{\sqrt{2\rho A}} \tag{6.17}$$

As an example, consider a propeller 2 m (6.6 ft) in diameter driven by a 150 kW engine (201 hp). *The maximum* static thrust that one might expect from this propeller can be calculated by solving Equation 6.17 for T.

$$T = P_{io}^{2/3}(2\rho A)^{1/3} \tag{6.18}$$

Since 1 W = 1.00 m-N/s,

$$P_{io} = 150{,}000 \text{ m-N/s}$$
$$(110{,}550 \text{ ft-lb/sec})$$

For standard sea level conditions, $\rho = 1.226\ \text{kg/m}^2$. Thus,

$$T = (150{,}000)^{2/3}(2 \times 1.226 \times 3.14)^{1/3}$$
$$= 5576 \text{ N } (1254 \text{ lb})$$

This value of T represents an upper limit that is not attainable in practice, since the momentum theory neglects profile drag of the propeller blades. Also, additional induced losses occur near the tips of the blades. Since the pressure difference across the blades must vanish at the tips, a trailing vortex system, helical in shape, is generated by the propeller in a manner similar to a finite wing.

In forward flight, an ideal efficiency, η_i, can be defined as the ratio of the useful power to the total power given by Equation 6.12.

$$\eta_i = \frac{TV}{T(V + w)}$$

or

$$\eta_i = \frac{1}{1 + (w/V)} \tag{6.19}$$

Using Equation 6.15, w/V can be written as

$$\frac{w}{V_0} = \frac{1}{2}(-1 + \sqrt{1 + T_c}) \tag{6.20}$$

where T_c is a thrust coefficient defined by

$$T_c = \frac{T}{qA} \tag{6.21}$$

Thus, η_i becomes

$$\eta_i = \frac{2}{1 + \sqrt{1 + T_c}} \tag{6.22}$$

The thrust of a propeller divided by its disc area is referred to as the *disc loading*. As this loading approaches zero, the ideal efficiency is seen to approach unity.

As an example, let us again consider the Cherokee 180 having a propeller diameter of 1.88 m (6.17 ft). At a cruising speed of 60.4 m/s (135 mph) at standard sea level, its drag will equal approximately 1390 N (312 lb). Thus, $T_c = 0.224$ giving an η_i of 0.95. As we will see later, the actual propeller efficiency is more like 0.83. As with the static thrust, η_i given by the momentum theory is optimistic and represents an upper limit that is really not attainable.

Although the momentum theory is not too accurate with regard to predicting power, it is useful for estimating the induced velocity. An interesting and easily remembered relationship is the following: The dynamic pressure in the ultimate wake of a propeller is equal to the sum of the free-stream dynamic pressure and the disc loading. Proof of this statement is left to you.

Computer Exercise 6.1 "MOMENTUM"

Write a computer program to solve the relationships for the momentum theory of propellers. Design the program to allow for either the thrust or power to be specified. The input should be

(a) Propeller diameter
(b) Free-stream velocity
(c) (1) Thrust required *or* (2) Power available
(d) Altitude

The output should be the induced velocity, the ideal efficiency, and the ideal (1) Power required or (2) Thrust available.

If the power, instead of thrust, is read into the program, then an iteration will have to be done on the thrust until the correct power is obtained For this, take the initial thrust and induced velocity equal to their static values corresponding to the given power. Then recognize that $\Delta V/\Delta P \simeq V/2$.

Blade Element Theories

In order to design a propeller or to predict the performance of an existing propeller more accurately, it is necessary to examine the aerodynamics of the blade in detail. Figure 6.10*a* presents the front view of a three-bladed propeller that is rotating with an angular velocity of ω rad/s and advancing through the air with velocity of V. Two cylindrical surfaces concentric with the axis of rotation and a differential distance of dr apart cut the propeller blade at a radius of r from the axis. The blade element thus defined is illustrated in Figure 6.10*b*. Here, we are looking in along the blade. The section is moving to the right (due to rotation) and toward the top of the page as the propeller advances into the air. The velocities influencing the element are shown relative to the element.

For the following analysis, the pitch angle, β, of the section is defined relative to the zero lift line of the airfoil section. In this regard, however, one must be careful, since propeller pitch angles are frequently tabulated with respect to the chord line or to a flat lower surface.

The pitch of a propeller has reference to the corresponding quantity for the ordinary screw. In fact, the early literature refers to propellers as "airscrews." If the propeller "screws" itself through the air without slipping, the distance it would move forward in one revolution is the pitch, p. From Figure 6.10*b*;

$$p = 2\pi r \tan \beta \tag{6.23}$$

Propellers are sometimes categorized by their pitch-diameter ratios. Thus,

$$\frac{p}{D} = \pi x \tan \beta$$

where $x = r/R$, the relative radius of the blade section.

A constant pitch propeller is one whose pitch does not vary with radius. For such a propeller,

$$\beta = \tan^{-1} \frac{p/D}{\pi x} \tag{6.24}$$

At the tip, x equals unity so that

$$\beta(x = 1) = \tan^{-1} \frac{p/D}{\pi}$$

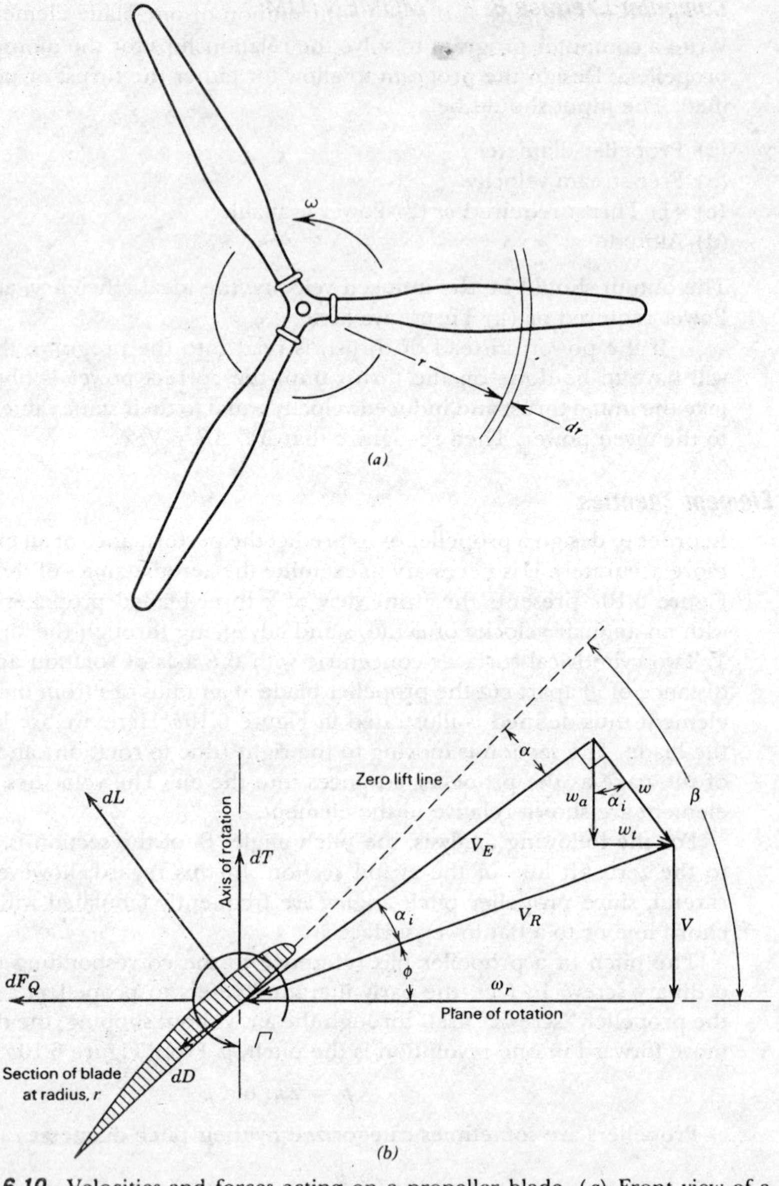

Figure 6.10 Velocities and forces acting on a propeller blade. (*a*) Front view of a three-bladed propeller. (*b*) Blade Element as seen looking in along the blade toward the hub.

The terms *constant pitch, fixed pitch,* and *variable pitch* are somewhat confusing. "Constant pitch" refers to the propeller geometry as just defined. "Fixed" or "variable" pitch refers to whether or not the whole blade can rotate about an axis along the blade (feathering axis) in order to vary the pitch angles of the blade sections all along the blade. Some propellers are equipped with governors to maintain a constant rpm as the engine throttle is varied. This is done by increasing the blade pitch angles as the propeller rpm tends to increase due to increased power or, vice versa, by decreasing the pitch for reduced power. Such a propeller is called a "constant speed" propeller.

Referring to Figure 6.10*b*, the contribution of one-blade element to the thrust, T, and torque, Q, will be

$$dT = dL\cos(\phi + \alpha_i) - dD\sin(\phi + \alpha_i) \tag{6.25a}$$

$$dQ = r[dL\sin(\phi + \alpha_i) + dD\cos(\phi + \alpha_i)] \tag{6.25b}$$

dL and dD are the differential lift and drag forces, respectively. Similar to finite wing theory, α_i is an induced angle of attack resulting from the induced velocity, w. dL and dD can be calculated by

$$dL = \tfrac{1}{2}\rho V_E^2 cC_l dr \tag{6.26a}$$

$$dD = \tfrac{1}{2}\rho V_E^2 cC_d dr \tag{6.26b}$$

The chord, c, is usually a function of the local radius, r. The section C_d is primarily a function of the section C_l. It can also depend on the local Reynolds and Mach numbers. C_l can be found from

$$C_l = \alpha(\beta - \phi - \alpha_i) \tag{6.27}$$

We are now at somewhat of a dilemma. We need α_i, which is a function of w, in order to get the blade loading. But w depends in turn on the blade loading.

Momentum-Blade Element Theory

The *momentum-blade element theory* is one means around this difficulty. If we assume α_i and the drag-to-lift ratio to be small, then $V_E \simeq V_R$ and Equation 6.25 can be written approximately for B blades as

$$dT = \frac{B\rho}{2} V_R^2 ca(\beta - \phi - \alpha_i)\cos\phi\, dr$$

Applying momentum principles to the differential annulus and letting $w \simeq V_R\alpha_i$, we can also write for dT:

$$dT = \rho(2\pi r\, dr)\,(V + V_R\alpha_i\cos\phi)2V_R\alpha_i\cos\phi$$

Equating these two expressions for dT/dr gives the following quadratic for α_i.

$$\alpha_i^2 + \alpha_i\left(\frac{\lambda}{x} + \frac{\sigma a V_R}{8x^2 V_T}\right) - \frac{\sigma a V_R}{8x^2 V_T}(\beta - \phi) = 0$$

where

$$\lambda = \frac{V}{\omega R} \qquad V_R = V_T\sqrt{x^2 + \lambda^2}$$

$$\sigma = \frac{Bc}{\pi R} \qquad \phi = \tan^{-1}\frac{\lambda}{x}$$

$$V_T = \omega R \qquad x = \frac{r}{R}$$

The induced angle of attack then becomes

$$\alpha_i = \frac{1}{2}\left\{-\left(\frac{\lambda}{x} + \frac{\sigma a V_R}{8x^2 V_T}\right) + \left[\left(\frac{\lambda}{x} + \frac{\sigma a V_R}{8x^2 V_T}\right)^2 + \frac{\sigma a V_R}{2x^2 V_T}(\beta - \phi)\right]^{1/2}\right\} \tag{6.28}$$

Given the geometry, forward speed, and rotational speed of a propeller, Equation 6.28 can be solved for α_i. Equations 6.25*a* and 6.25*b* can then be numerically integrated using Equations 6.26 and 6.27 to give the thrust and torque.

The thrust and power of a propeller are normally expressed in coefficient form. These thrust and power coefficients are defined in various ways, depending on what particular reference areas and velocities are used. Test results on propellers almost always define the thrust coefficient, C_T, and power coefficient, C_P, as follows:

$$C_T = \frac{T}{\rho n^2 D^4} \tag{6.29a}$$

$$C_P = \frac{P}{\rho n^3 D^5} \tag{6.29b}$$

where n is the rotational speed in revolutions per second and D is the propeller diameter. The thrust, power, ρ, and D must be in consistent units. For this convention, one might say that nD is the reference velocity and D^2 is the reference area.

One would expect these dimensionless coefficients to be a function only of the flow geometry (excluding scale effects such as Mach number and Reynolds number). From Figure 6.10*b*, the angle of the resultant flow, ϕ, is seen to be determined by the ratio of V to ωr.

$$\phi = \tan^{-1} \frac{V}{\omega r}$$

This can be written as

$$\phi = \tan^{-1} \frac{J}{\pi x} \tag{6.30}$$

The quantity, J, is called the *advance ratio* and is defined by

$$J = \frac{V}{nD} \tag{6.31}$$

Thus, C_T and C_p are functions of J.

In a dimensionless form, Equations 6.25 and 6.26 can be combined and expressed as

$$C_T = \frac{\pi}{8} \int_{x_h}^{1} (J^2 + \pi^2 x^2) \sigma [C_l \cos(\phi + \alpha_i) - C_d \sin(\phi + \alpha_i)] \, dx \tag{6.32a}$$

and, since $P = \omega Q$,

$$C_P = \frac{\pi}{8} \int_{x_h}^{1} \pi x (J^2 + \pi^2 x^2) \sigma [C_l \sin(\phi + \alpha_i) + C_d \cos(\phi + \alpha_i)] \, dx \tag{6.32b}$$

x_h is the hub station where the blade begins. x_h is rather arbitrary, but C_T and C_P are not too sensitive to its value.

To reiterate, one would be given D, V, ρ, and n. Also, c and β would be given as a function of x. At a given station, x, α_i is calculated from Equation 6.28. This is followed in order by C_l and C_d and finally by dC_T/dx and dC_p/dx. These are then integrated from x_h to 1 to give C_T and C_p.

Given J and having calculated C_T and C_p, one can now calculate the propeller efficiency. The useful power is defined as TV and P is, of course, the input power. Thus,

$$\eta = \frac{TV}{P} \tag{6.33}$$

In terms of C_T, C_p, and J, this becomes

$$\eta = \frac{C_T J}{C_p} \tag{6.34}$$

The combined blade element-momentum theory is presented here mainly to make the reader aware of the theory, which can be found in the literature. However, with the advent of computers, there is little reason to use this theory, which is approximate in the treatment of induced effects. No account is taken of flow rotation, and the blade loading is not predicted to vanish at the tips of the blades. Vortex theory, which is presented next, is considerably more precise in the treatment of induced effects and is found to give close agreement with experimental results when profile drag is included.

Vortex Theory

Other blade element theories differ from the combined blade element-momentum theory in the way in which the induced velocities are determined. Numerically based methods use vortex lattice models analogous to the lifting surface model for the finite wing. The geometry of the vortex wake is either prescribed or uses vortices that are free to align themselves with the resultant flow that they produce. This latter, free-wake model is more precise than one using a prescribed wake but requires an inordinate amount of computer time to accomplish.

The vortex theory begins with a consideration of the optimum propeller, which leads to the so-called Betz condition. This condition can be derived by considering conditions in the ultimate wake, or slipstream, of the propeller. Figure 6.10*b* presents the blade geometry and velocities at the propeller. Figure 6.11 presents similar quantities but in the ultimate wake far downstream of the propeller. The notation used on this figure is similar to Betz's. Referring to Figure 6.11, Ωr is the velocity due to rotation and V is the advance velocity of the propeller. w is the velocity induced by the trailing vortex system whereas ωr and v are the tangential and axial components, respectively, of w. These induced velocities are approximately twice their respective values given in Figure 6.10*b*. w_o is a fictitious velocity referred to as the impact velocity. If one moves along the helical surface parallel to the vortex sheet, because of the symmetry, the velocity potential, ϕ, will remain unchanged. Thus, grad ϕ in this direction is zero, which means that the velocity induced by the vortex system must be normal to the sheet since there can be no component tangential to the sheet.

In the same manner as a wing, the aerodynamic lift on a propeller blade can be related to a bound circulation, Γ, around the blade. Now imagine that the induced velocity in the ultimate wake is the reaction to a loaded lifting line in the ultimate wake under the influence of the velocities shown in Figure 6.11. If $\Delta\Gamma$ is an increment in Γ with corresponding increments in the thrust and torque, ΔT and ΔQ, then a local efficiency, which is a function of r, can be defined as

$$k = \frac{V\Delta T}{\Omega\Delta Q} \tag{6.35}$$

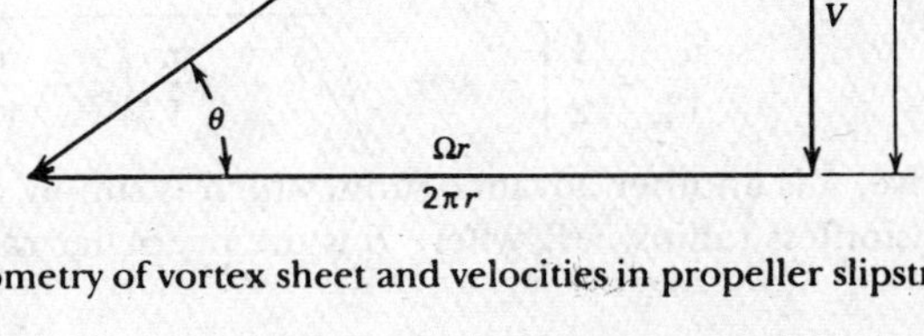

Figure 6.11 Geometry of vortex sheet and velocities in propeller slipstream far downstream.

Now, for the optimum propeller, k must be equal to a constant, independent of r. Otherwise, one could increase Γ where k was highest and decrease Γ where k was lowest, thereby increasing the overall efficiency.

From the Kutta–Joukowski relationship applied to Figure 6.11,

$$\Delta T = \rho\,(\Omega - \omega)\; r\Delta\Gamma$$
$$\Delta Q = \rho\,(V + v)\; r\Delta\Gamma$$

which leads to

$$k = \frac{V(\Omega - \omega)}{\Omega\,(V + v)}$$

But from the geometry of Figure 6.11,

$$p = 2\pi r \tan\theta = 2\pi r \frac{(V + v)}{(\Omega - \omega)\; r}$$

Thus, substituting in the above for k, leads to the Betz condition that p, the pitch of the trailing vortex sheet in the ultimate wake of a propeller, is a constant for the optimum propeller. A corollary to this is the fact that the impact velocity, w_o, must be a constant.

Goldstein's classical vortex theory is akin to a prescribed wake model in that it satisfies the Betz condition. The vortex sheet trailing from a propeller blade is taken to lie along a helical surface of constant pitch. For the optimum propeller, the velocity normal to the sheet must be zero. This is the same condition that is satisfied by a solid helical surface translating through the fluid. Thus, Goldstein solved this equivalent potential flow problem in order to predict the spanwise loading over the blade. His theory is described in detail in Reference 3.12 and is briefly outlined here.

Referring to the plane of the propeller (Figure 6.10*b*), it is assumed that the normality condition between the resultant velocity, V_E, and the induced velocity, w, also holds here. Actually, the wake is contracting immediately downstream of the propeller as the induced velocity increases so that normality at the plane of the propeller is not as easily justified as it is in the ultimate wake. Let it suffice to say that studies have been performed that support normality at the plane of the propeller.

In Figure 6.10*b*, the resultant induced velocity, w, is composed of a tangential component, w_t, and an axial component, w_a. w_t and w_a are related by

$$\frac{V + w_a}{\omega r - w_t} = \frac{w_t}{w_a}$$

This can be solved for w_a as a function of w_t.

$$w_a = \frac{1}{2}\left[-V + \sqrt{V^2 + 4w_t\,(\omega r - w_t)}\right]$$

It is convenient to express all velocities in terms of the tip speed caused by rotation, $V_T = \omega R$. V is not used as the reference velocity since, in the static case, V will be zero. In terms of V_T, Equation 6.37 becomes

$$\frac{w_a}{V_T} = \frac{1}{2}\left[-\lambda + \sqrt{\lambda^2 + 4\frac{w_t}{V_T}\left(x - \frac{w_t}{V_T}\right)}\right] \qquad (6.35)$$

In the above, λ is another advance ratio, which is simply defined as V/V_T, and x is the dimensionless radius, r/R, where R is the propeller radius.

Goldstein's vortex theory relates w_t to the bound circulation, Γ, around any blade station by

$$B\Gamma = 4\pi r\kappa w_t \tag{6.36}$$

κ is known as *Goldstein's kappa factor.* This factor is not expressible in a closed form, but it is available in graphical form in the literature. An approximation of κ is *Prandtl's tip loss factor, F,* which becomes more exact as J becomes smaller or as the number of blades increases. Thus,

$$B\Gamma \simeq 4\pi rFw_t \tag{6.37}$$

where

$$F = \frac{2}{\pi}\cos^{-1}\exp\left[-\frac{B(1-x)}{2\sin\phi_T}\right] \tag{6.38}$$

ϕ_T is the helix angle of the propeller's helical trailing vortex system at the tip. For a lightly loaded propeller,

$$\phi_T = \tan^{-1}\lambda \tag{6.39}$$

However, the lift must vanish at the tip of a propeller blade, which generally means that the local angle of attack at the tip must be zero. Thus, another expression for ϕ_T can be obtained by reference to Figure 6.9.

$$\phi_T = \beta_T \tag{6.40}$$

For most cases, Equation 6.40 is preferred to Equation 6.39.

From the Kutta-Joukowski theorem,

$$L = \rho V\Gamma$$

Thus,

$$\Gamma = \tfrac{1}{2}cC_lV_e \tag{6.41}$$

Substituting Equation 6.41 into Equation 6.37, the result can be expressed as

$$\sigma C_l\frac{V_e}{V_T} = 8xF\frac{w_t}{V_T} \tag{6.42}$$

C_l can be calculated from

$$C_l = a\,(\beta - \phi - \alpha_i) \tag{6.43}$$

where

$$\alpha_i = \tan^{-1}\left(\frac{w_t}{w_a}\right) - \phi$$

$\dfrac{V_e}{V_t}$ is obtained from

$$\frac{V_e}{V_T} = \sqrt{\left(\lambda + \frac{w_a}{V_T}\right)^2 + \left(x - \frac{w_t}{V_T}\right)^2} \tag{6.44}$$

The induced angle of attack can be found from

$$\alpha_i = \tan^{-1}\left(\frac{V + w_a}{\omega r - w_t}\right) - \phi \tag{6.45}$$

Equations 6.35–6.45 can be solved iteratively for $w_t/V_T \cdot C_l$, and the other quantities can then be found to evaluate C_T and C_p from relationships similar to Equations 6.32*a* and 6.32*b*.

$$C_T = \frac{\pi^3}{8}\int_{x_h}^{1.0}\left(\frac{V_e}{V_T}\right)^2 \sigma\,[C_l\cos(\phi + \alpha_i) - C_d\sin(\phi + \alpha_i)]\,dx \qquad (6.46a)$$

$$C_p = \frac{\pi^3}{8}\int_{x_h}^{1.0} \pi x\left(\frac{V_e}{V_T}\right)^2 \sigma\,[C_l\sin(\phi + \alpha_i) + C_d\cos(\phi + \alpha_i)]\,dx \qquad (6.46b)$$

There are several refinements to the classical vortex theory, which improve its accuracy. The first is the inclusion of profile drag, which has already been used in formulating the expressions for C_T and C_p in the combined blade element-momentum theory.

The second correction, which is small for a typical airplane propeller, results from the finite thickness of a blade. From continuity, the axial component of the flow, $V + w_a$, must increase in magnitude and then decrease as it passes through the propeller. This results in a flow path that effectively decreases the angle of attack of the blade section. Approximating the blade section by an ellipse, Reference 3.3 obtains a closed-form expression for this correction to α as a function of the section thickness ratio given by

$$\Delta\alpha = \frac{4\lambda\sigma}{15(\lambda^2 + x^2)}\frac{t_{max}}{c} \qquad (6.47)$$

The third correction is to the camber and results from the tangential component of induced velocity, which increases from zero to its final value through the propeller. This results in a curvature of the flow, which effectively reduces the camber of each blade section. Based on the derivation presented in Reference 3.3, a reduction to the section lift coefficient at zero angle of attach can be obtained as

$$\Delta C_l = \frac{a\Delta\theta}{4} \qquad (6.48)$$

where the angle $\Delta\theta$ is found from

$$\Delta\theta = \tan^{-1}\left(\frac{V + w_a}{\omega r - 2w_t}\right) - \tan^{-1}\left(\frac{V + w_a}{\omega r}\right) \qquad (6.49)$$

Given the propeller blade geometry and operating condition, J, at a particular x value, the iteration for w_t proceeds as outlined below:

1. Assume $\frac{w_t}{V_T} = 0$. Then calculate in order.
2. $\frac{w_a}{V_T}$ from Equation 6.35.
3. Corrections to α and C_{l_0}.
4. α_i from Equation 6.45.
5. α from $\beta - \alpha_i - \phi - \Delta\alpha$.
6. C_l from $a\alpha + C_{l_0} - \Delta C_{l_o}$.
7. $\frac{V_E}{V_T}$ from Equation 6.44.
8. Γ from Equation 6.41.
9. w_t from Equation 6.37.

The result from Step 9 is then returned to Step 2 until the absolute difference between the calculated value of α_i and that from the previous calculation reduces to a desired value (like 0.00001 rad).

As an alternative to the above iteration, a solution for α_i can be obtained directly if the angle is assumed to be small. In this case, Reference 3.3 shows that α_i can be solved directly by assuming that

$$w_t = V_R \alpha_i \sin(\phi + \alpha_i) \tag{6.50}$$

$$w_a = V_R \alpha_i \cos(\phi + \alpha_i) \tag{6.51}$$

The induced angle of attack can then be determined approximately from

$$\alpha_i = \frac{1}{2}\left(-X + \sqrt{X^2 + 4Y}\right) \tag{6.52}$$

where

$$X = \tan\phi + \frac{a\sigma}{8x\mathrm{F}\cos\phi} \tag{6.53}$$

$$Y = \frac{a\sigma(\beta - \phi)}{8xF\cos\phi} \tag{6.54}$$

As an example in the use of the vortex, consider the three-bladed propeller having the geometry shown in Figure 6.12. Wind tunnel testing of this particular propeller, designated 5868-R6, Clark-Y section, three blades, is reported in Reference 6.3. These results are presented in Figures 6.13, 6.14, and 6.15.

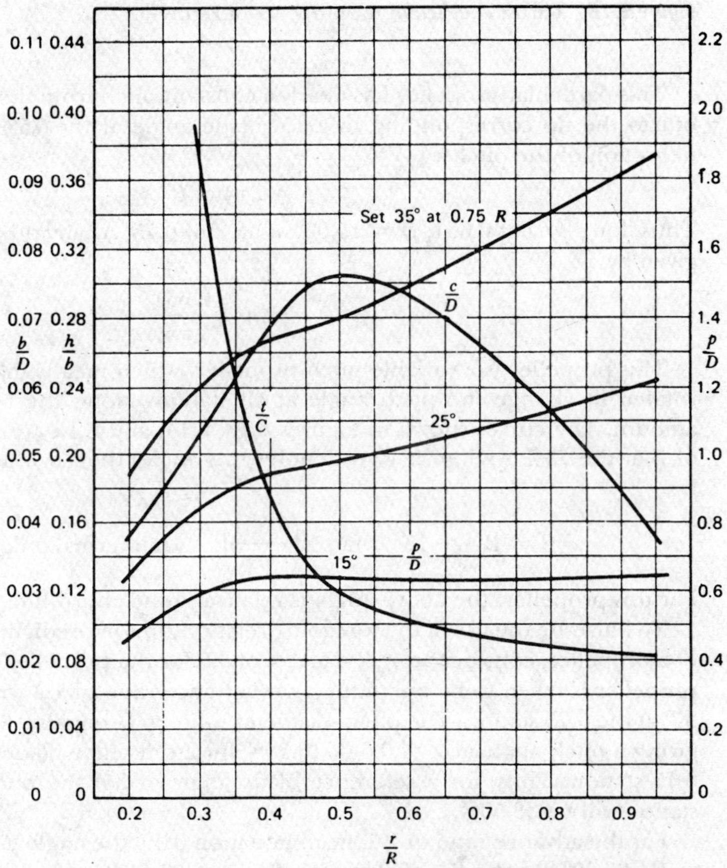

Figure 6.12 Geometry of example propeller 5868-9, Clark-Y section, three blades.

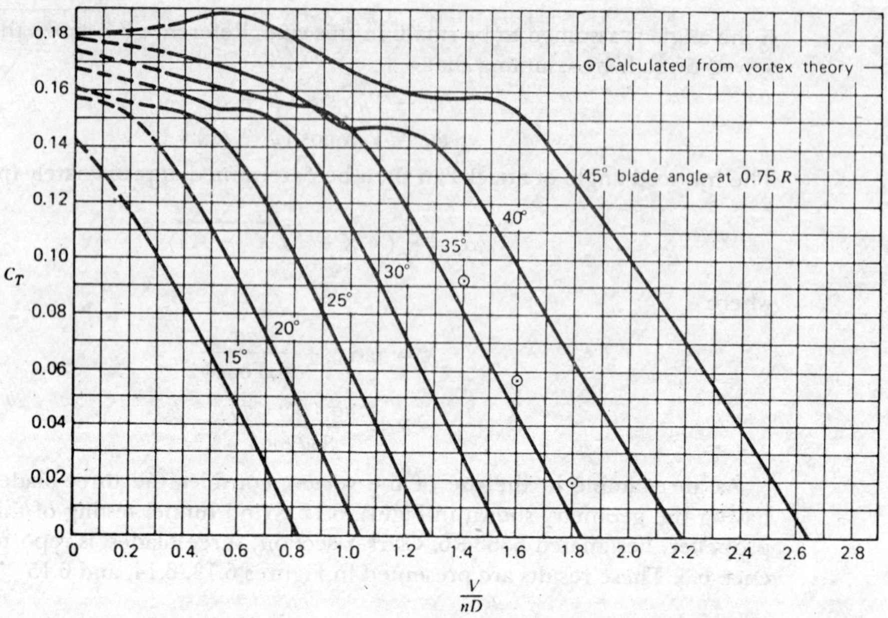

Figure 6.13 Thrust coefficient for propeller of Figure 6.12.

This particular propeller has nearly a constant pitch from the 35% radius station out to the tip corresponding to a 15° blade angle at the 75% station. From the definition of the pitch

$$p = 2\pi r \tan\beta \tag{6.55}$$

Thus, for a p/D of 0.631, $\beta = 15.00°$ at an x of 0.75. At any other station, β will be given by

$$\beta = \tan^{-1}\frac{0.631}{\pi x} \tag{6.56}$$

This propeller is a variable pitch propeller, which means that the blade can be rotated to change the pitch angle at all stations along the blade by a constant amount. The curves shown in Figures 6.13, 6.14, and 6.15 are for different values of β at the 75% station. If $\beta_{0.75}$ denotes this angle, then β will generally be given by

$$\beta = \beta_{0.75} - 15 + \tan^{-1}\left(\frac{0.631}{\pi x}\right) \quad \text{(all terms in degrees)} \tag{6.57}$$

For this propeller, the above angle is relative to the chord line.

To illustrate the use of the foregoing relationships in predicting the performance of a propeller, consider an r/R value, x, of 0.6 for the propeller of Figure 6.12. The propeller is taken to be operating at an advance ratio of 1.4 with a blade angle of 35° at the 0.75 station. The pitch–diameter ratio, p/D, equals 0.631 along the blade, giving a pitch angle at $x = 0.6$ of 18.508° for a reference blade angle of 15° at the 0.75 station. Thus, for a reference blade angle of 35° the pitch angle at the 0.6 station will be 38.508°.

For the advance ratio of 1.4, from Equation 6.30, the angle ϕ at $x = 0.6$ is found to be 36.60°. At the tip of the propeller, $x = 1.0$, the blade pitch angle will be 31.36°. This blade pitch and the previous one at $x = 0.6$ are relative to the chord

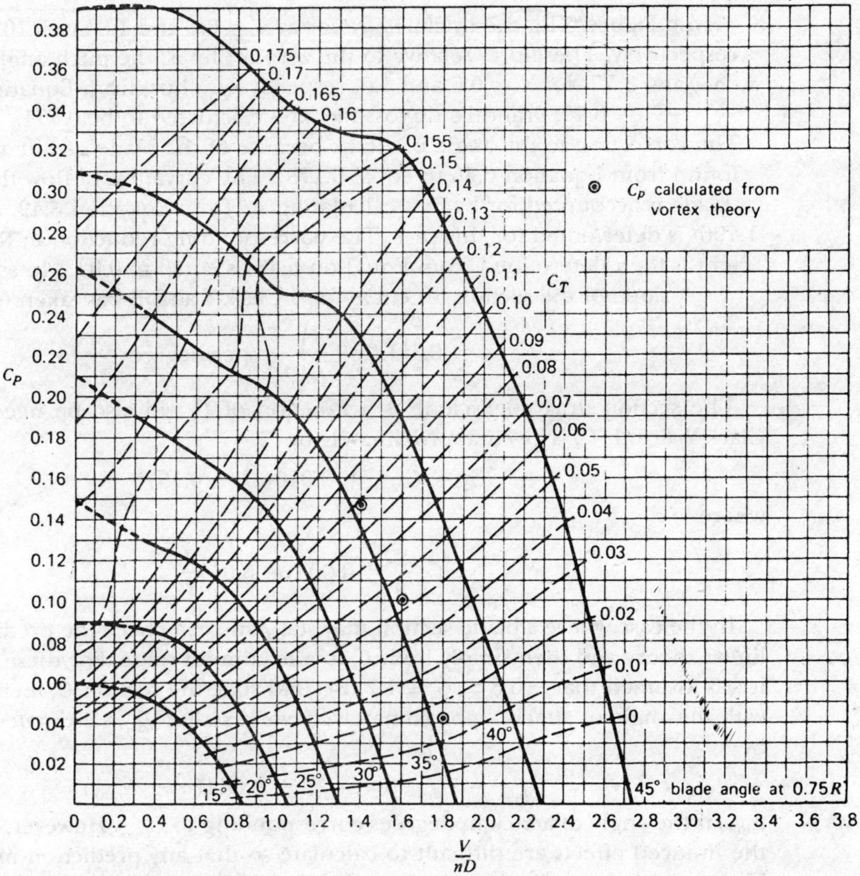

Figure 6.14 Power coefficient for propeller of Figure 6.12.

line and must be adjusted to refer to the zero lift line. The angle between the zero lift line and the chord line for this airfoil is equal approximately to

$$\alpha_{0l} = 46\frac{t}{C} \tag{6.58}$$

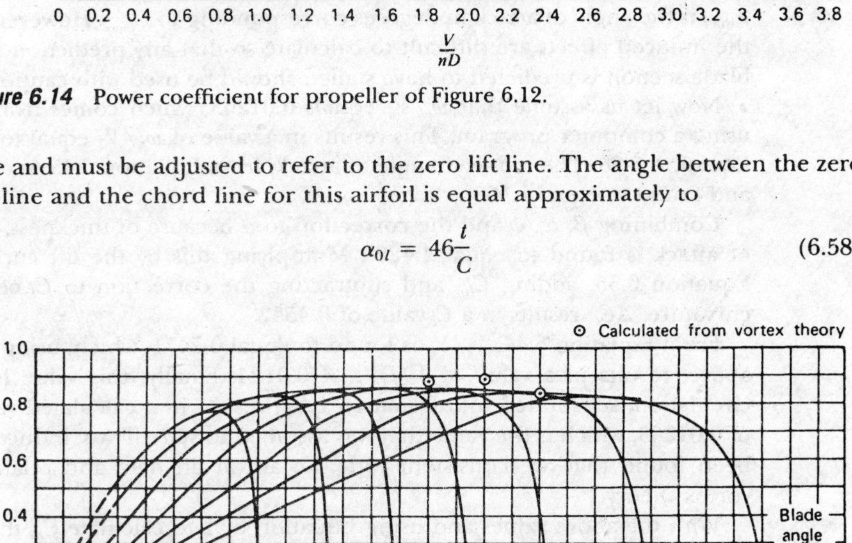

Figure 6.15 Efficiency for propeller of Figure 6.12.

From Figure 6.12, the thickness ratios at $x = 0.6$ and 1.0 are 0.103 and 0.082, respectively. Therefore, relative to the zero lift lines, the pitch angles become 4.738° and 3.77° for $x = 0.6$ and 1.0, respectively. Thus, from Equations 6.38 and 6.40, Prandtl's tip loss factor is calculated to be 0.771.

The correction to the angle of attack because of the blade's finite thickness is found from Equation 6.46 to equal 0.175°. The curvature of flow through the blade is accounted for by first calculating $\Delta\theta$ from Equation 6.49. A value of 1.196° is determined for this case. The corresponding reduction in the lift coefficient is then determined from Equation 6.48 as equal to 0.033. In so doing, the slope of the section lift curve for a Clark-Y airfoil was taken to be

$$a = 0.1\left(1.0 + \frac{t}{C}\right) \quad \text{per degree} \tag{6.59}$$

The section drag coefficient as a function of C_l will also be needed. For the Clark-Y airfoil, C_d is given approximately by

$$C_d = C_{d_{\min}} + 0.010(C_l - 0.15)^2 \tag{6.60}$$

where

$$C_{d_{\min}} = 0.004 + 0.017\frac{t}{C} \tag{6.61}$$

In the case where a blade section may stall, one can check the predicted C_l from linear theory and then simply limit C_l to a maximum value. For these calculations, it was assumed that $-0.8 \leq C_l \leq 1.2$. Beyond stall, the section C_d increases rapidly with the angle of attack. Approximately, beyond stall, $C_{d_{\min}}$ is estimated to be given by

$$C_{d_{\min}} = C_{d_{\min}}\,(@C_{l_{\max}}) + 0.044(\alpha - \alpha_{\max}) \tag{6.62}$$

$\alpha_{\max}$ is the angle of attack in degrees corresponding to $C_{l_{\max}}$. However, beyond stall, the induced effects are difficult to calculate so that any prediction made where a blade section is predicted to have stalled should be used with caution.

Now let us assume that w_t/V_T equals 0.01263, which comes from an iteration using a computer program. This results in a value of w_a/V_T equal to 0.01606 from Equation 6.38. The induced angle of attack, from Equation 6.45, is calculated next and found to equal 1.566°.

Combining β, α_i, ϕ and the correction to α because of thickness, $\Delta\alpha$, the angle of attack is found to equal 4.426°. Multiplying this by the lift curve slope from Equation 6.55, adding C_{l_0}, and subtracting the correction to C_l because of flow curvature, ΔC_l, results in a C_l value of 0.4552.

From Equation 6.44, V_e/V_T is found to equal 0.7471, which, using the value of F above, results in a value of Γ/DV_T of 0.0124. Finally, this value for the bound circulation, substituted into Equation 6.37, results in a calculated value for w_t/V_T of 0.01263, which is the value that was assumed to start. Thus, induced effects have been found that are consistent with the airfoil lift data and relationships from vortex theory.

With the above values and using Equation 6.57 to calculate C_d, the gradients of C_T and C_P can be found from Equation 6.46*a* and 6.46*b*.

$$\frac{dC_T}{dx} = 0.1064$$

$$\frac{dC_P}{dx} = 0.1635$$

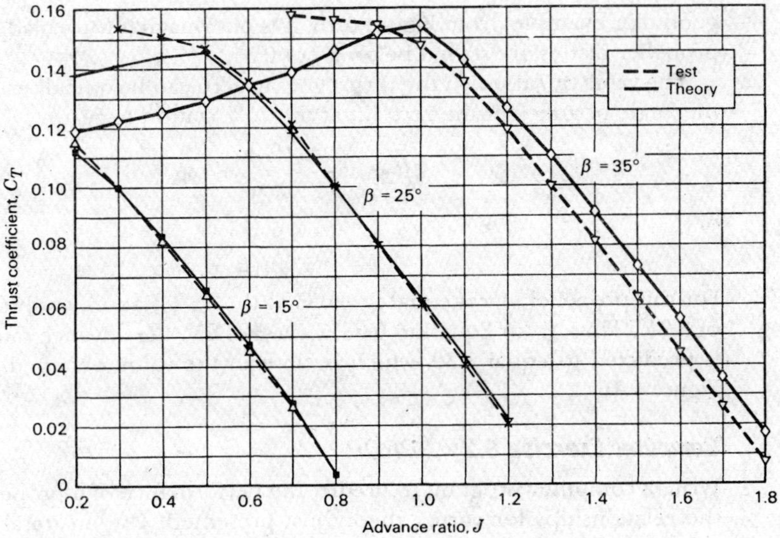

Figure 6.16 Comparison of prediction of thrust from vortex theory with experiment for propeller of Figure 6.12.

In this manner, the graphs of Figures 6.16 and 6.17 were produced. From these figures it can be seen that the predictions of C_T and C_P agree closely with the experimental results, particularly at the lower blade angles, for the range of J values for which the blade is not stalled. As J is reduced, the angles of attack of the sections increase until a value of J is reached below which some portion of the blade is stalled. Below this value, the agreement of the theory with the test results is not too

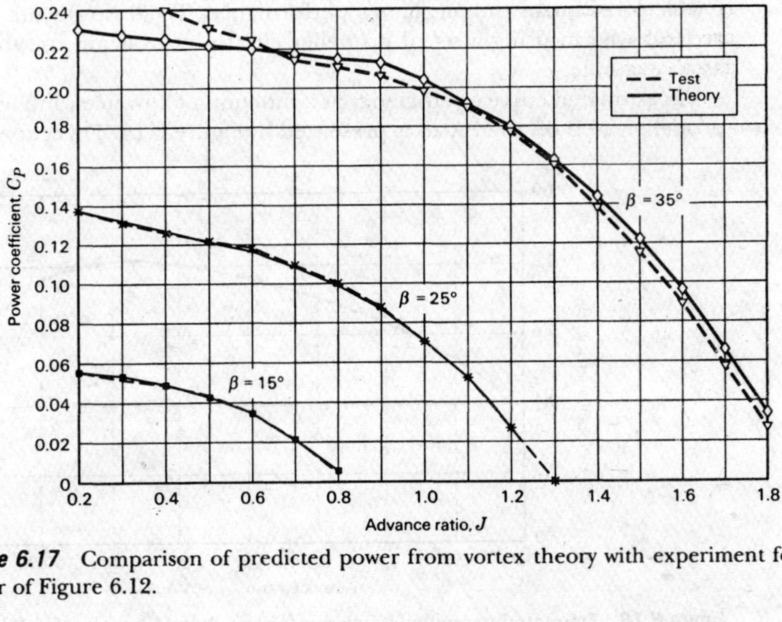

Figure 6.17 Comparison of predicted power from vortex theory with experiment for propeller of Figure 6.12.

good. For example, from Figure 6.16, it is obvious that for a blade angle of 25° the propeller blades are stalled below a J of 0.5.

The value of J at which the thrust goes to zero can be quickly estimated by setting the angle of attack of the zero lift at the 75% station to zero.

$$\beta - \tan^{-1}\left(\frac{J}{\pi x}\right) + \alpha_{0l} = 0$$

or

$$J = \pi x \tan(\beta + \alpha_{0l}) \tag{6.63}$$

For this propeller, t/c is equal to 0.085 at the 75% station resulting in an α_{0l} value of 3.91°. Thus, from Equation 6.63, for a β of 25°, the advance ratio for zero thrust is predicted to equal 1.30, which is close to the value obtained by extrapolating Figure 6.16.

Computer Exercise 6.2 "PROP"

Write a computer program to predict the performance of a propeller according to the relationships for vortex theory just presented. Use an iterative technique to solve for w_t and include corrections for flow curvature and finite thickness. As input, the program should be capable of reading a table of data for the propeller and calculating β from a given p/D value. You may also wish to have the program accept functional relationships (like linear) to define the propeller geometry. The program should be able to loop on J and β.

Run the program for the 5868-9 propeller and compare your results with Figures 6.16 and 6.17.

Practical Use of Propeller Charts

The practicing aerodynamicist will normally have available both engine and propeller operating curves as supplied by the respective manufacturers. Using these curves together with a knowledge of the airplane's aerodynamic characteristics, one is able to estimate the airplane's performance. To illustrate the procedures that are followed in using a set of propeller charts, let us again use the Cherokee 180 as an example.

An estimated curve of efficiency as a function of advance ratio for the fixed pitch propeller used on the PA-28 is presented in Figure 6.18. This curve is applicable to

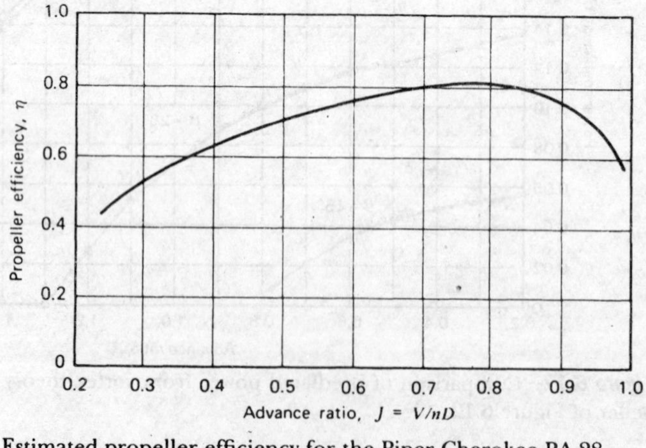

Figure 6.18 Estimated propeller efficiency for the Piper Cherokee PA-28.

the aircraft pictured in Figure 3.59 with the engine operating curves of Figure 6.3. This particular propeller has a diameter of 1.88 m (6.17 ft).

As an example in the use of the engine performance charts together with the graph of propeller efficiency, assume that in steady, level flight, the pilot of a PA-28 reads a manifold pressure of 24 in., an rpm of 2400, a pressure altitude of 3000 ft, an OAT of 65°F, and an indicated airspeed of 127 mph. From this information, together with Figure 6.18, one can estimate the drag of the airplane at this indicated airspeed and density altitude.

From Figure 2.3 at 914 m (3000 ft),

$$\frac{p}{p_0} = 0.90$$

Thus, p = 91,163 Pa (1904 psf). Furthermore, air obeys closely the equation of state for a perfect gas.

$$\frac{P}{\rho T} = \text{constant} = R \tag{6.64}$$

where T is the absolute temperature.

Using standard sea level values for p_0, ρ_0, and T_0, the preceding constant is seen to be

$$R = 287.1\ (\text{m/s})^2/°\text{R}\ [1717\ (\text{fps})^2/°\text{K}]$$

Thus, for this example, T = 292°K (525 °R), so that

$$\rho = 1.087\ \text{kg/m}^3\ (0.00211\ \text{slugs/ft}^3)$$

This corresponds to a σ of 0.888. Thus, the density altitude is found to be 1220 m (4005 ft) and the true airspeed is calculated to be 60.4 m/s (135 mph or 198 fps).

The propeller advance ratio, defined by Equation 6.31, is

$$J = 0.802$$

For this value of J, a propeller efficiency of 0.81 is read from Figure 6.17.

One can verify that the engine power for these operating conditions, from Figure 6.3, is equal to 138 bhp. Therefore, from Equation 6.33, knowing η, P, and V, the propeller thrust can be calculated as

$$T = \frac{\eta P}{V}$$
$$= 310\ \text{lb}\ (1380\ \text{N})$$

In steady, level flight, the propeller thrust and airplane drag must be equal. Thus, 310 lb is the drag of the airplane at this particular density altitude and airspeed.

For analyzing a variable pitch propeller a set of curves for different blade pitch angles is required. These are given in Figures 6.19 and 6.20 for the propeller installed on the Piper PA-28R, the Cherokee Arrow. Here we are given both η and C_p as a function of J. To illustrate the use of such graphs, let us assume that they apply to the preceding example for the PA-28. Here, ρ = 0.00211 slugs/ft^3, D = 6.17 ft, V = 198 fps, n = 40 rps, J = 0.802, and hp = 139. Thus,

$$c_p = \frac{P}{\rho n^3 D^5}$$
$$= 0.0633$$

From Figure 6.20 for the preceding C_P and a J of 0.802, the blade pitch angle must be equal to 24°. Entering Figure 6.19 with this β and J results in an efficiency, η, of 0.83.

A well-designed propeller, or one carefully selected to match the engine and

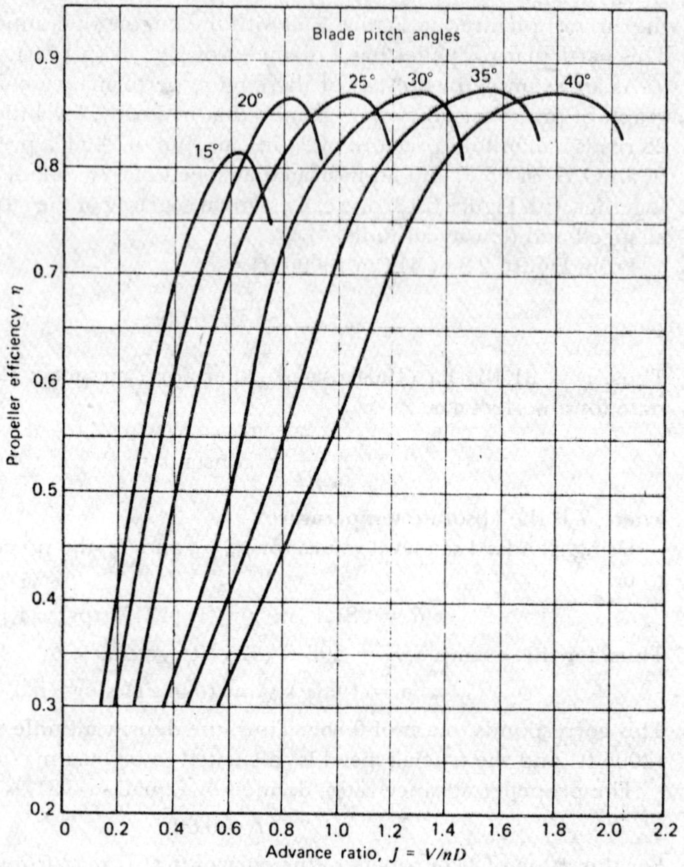

Figure 6.19 Estimated propeller efficiency for the Piper Cherokee Arrow PA-28R.

airplane on which it is to operate, can be expected to have a cruise efficiency of approximately 85%. At low speeds, however (e.g., during the takeoff roll), the efficiency is difficult to estimate. At zero forward speed, the efficiency of a propeller is zero by definition, even though its thrust is not zero. In fact, for the same shaft power, a variable pitch propeller will produce the most thrust at zero advance velocity (i.e., its static thrust is greater than the thrust produced in forward flight).

Figures 6.21 and 6.22 may be used to estimate the thrust attainable from a variable pitch propeller at low forward speeds. The static thrust is first obtained from Figure 6.22 and then reduced by the factor from Figure 6.21 to give the thrust in forward flight. These curves apply only to a constant speed propeller, which will allow the engine to develop its rated power regardless of forward speed. As an example of the use of these figures, consider a propeller having a diameter of 6.2 ft, turning at 2700 rpm, and absorbing 200 hp. The power loading for this propeller is

$$\frac{\text{hp}}{A} = 6.62 \text{ hp/ft}^2$$

Hence, from Figure 6.21, the static thrust to power loading should be

$$\frac{T_0}{\text{hp}} = 4.9$$

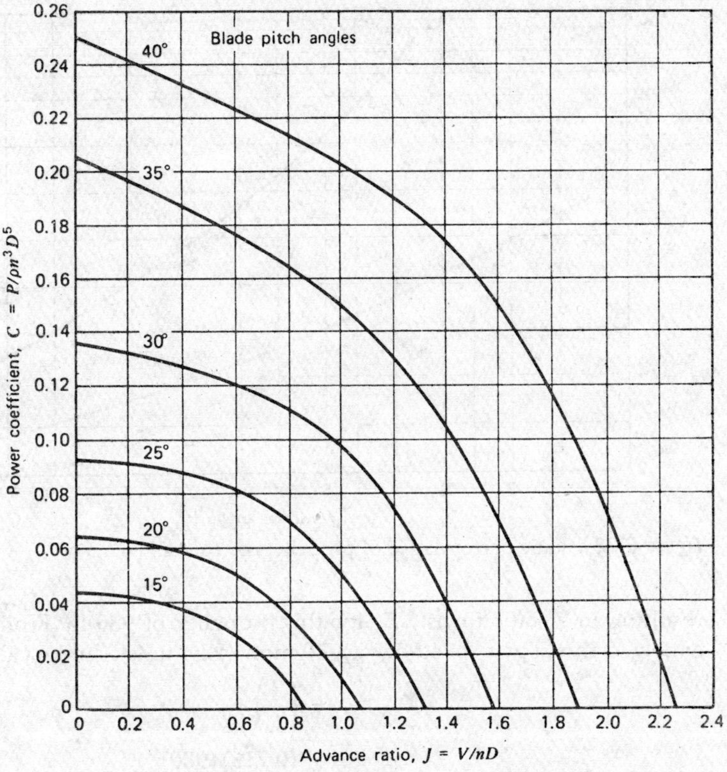

Figure 6.20 Estimated propeller power coefficients for the Piper Cherokee Arrow PA-28R.

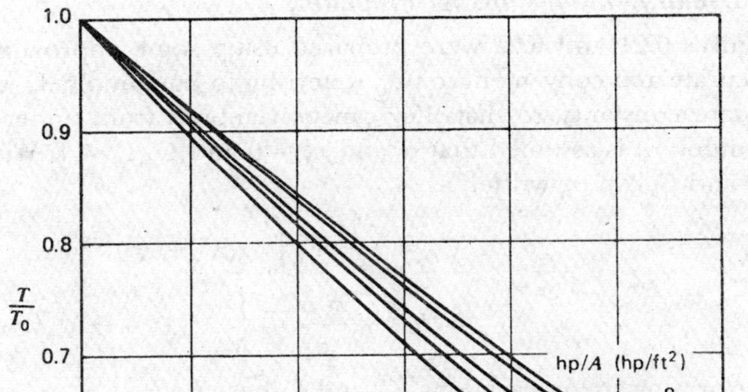

Figure 6.21 Decrease of thrust with velocity for different power loadings.

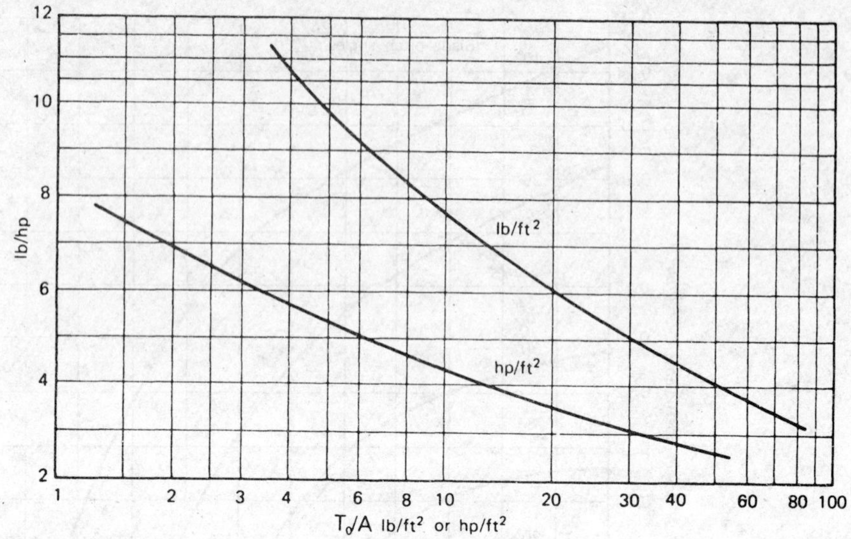

Figure 6.22 Static performance of propellers or rotors.

resulting in a static thrust, T_0, for this propeller of 980 lb. From Figure 6.21, the expected thrust at a speed of, say, 50 mph (22.4 m/s) can be calculated as

$$\begin{aligned} T &= \left(\frac{T}{T_0}\right) T_0 \\ &= (0.715)(980) \\ &= 700 \text{ lb} \end{aligned}$$

Approximate Useful Relationships for Propellers

Figures 6.21 and 6.22 were prepared using some approximations that are fairly accurate and convenient to use. Referring to Equation 6.32, assume that σ, C_l, and C_d are constants, so that they can be removed from under the integral sign. In addition, it is assumed that α_i and $x_h \simeq 0$ and $C_d/C_l \ll 1$. With these assumptions, C_T and C_p can be written as

$$C_T = \frac{\pi^3 \sigma}{8} \overline{C_l} \int_0^1 x(\lambda^2 + x^2)^{1/2}\, dx \tag{6.65a}$$

$$C_P = JC_T + \frac{\pi^4 \sigma}{8} \overline{C_d} \int_0^1 x^2(\lambda^2 + x^2)^{1/2}\, dx \tag{6.65b}$$

Performing the integrations, C_T and C_p become

$$C_T = \frac{\pi^3 \sigma}{24} \overline{C_l} f(\lambda) \tag{6.66a}$$

$$C_P = JC_T + \frac{\pi^4 \sigma}{32} \overline{C_d}\, g(\lambda) \tag{6.66b}$$

where

$$f(\lambda) = (1 + \lambda^2)^{3/2} - \lambda^3$$

$$g(\lambda) = \frac{1}{2}\left[(1 + \lambda^2)^{1/2}(2 + \lambda^2) - \lambda^4 \log \frac{1 + \sqrt{1 + \lambda^2}}{\lambda}\right]$$

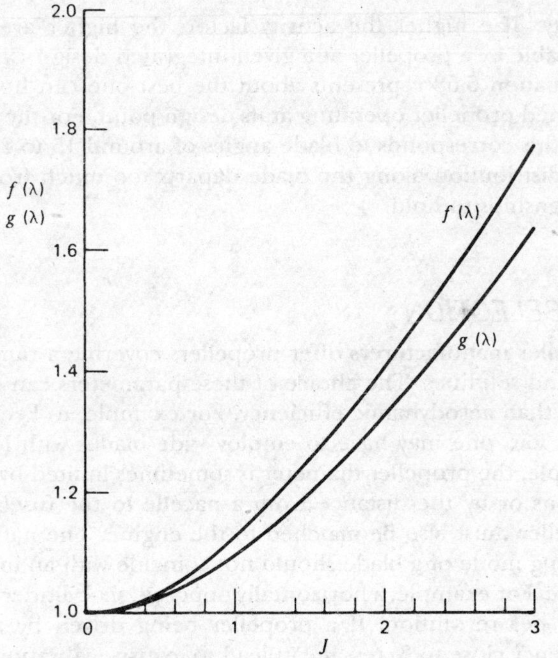

Figure 6.23 Functions for approximating C_T and C_P.

$f(\lambda)$ and $g(\lambda)$ are given as a function of J in Figure 6.23. $\overline{C_l}$ and $\overline{C_d}$ indicate average values of these quantities as defined by Equation 6.65.

The term JC_T in the expression for C_p simply represents the useful power. The remaining term in C_p is the profile power, or the power required to overcome the profile drag of the blades. The induced power is missing, since α_i was assumed to be zero. Experience shows that the induced power is typically 12% higher than the ideal value given by Equations 6.14 and 6.15. Thus, in coefficient form,

$$C_{p_i} \cong 1.12 C_T \frac{w}{nD} = 1.12 \frac{C_T}{2}\left[-J + \left(J^2 + \frac{8C_T}{\pi}\right)^{1/2}\right] \tag{6.67}$$

C_p then becomes approximately

$$C_p = C_T J + C_{p_i} + \frac{\pi^4 \sigma}{32} \overline{C_d}\, g(\lambda) \tag{6.68}$$

The average value of σ is referred to as *propeller solidity* and is equal to the ratio of blade area to disc area.

Propeller designs are sometimes identified by an "integrated design lift coefficient" and an activity factor. These are defined by

$$C_{L_d} = 3 \int_0^1 C_{l_d} x^2\, dx \tag{6.69}$$

$$\text{AF} = \frac{100{,}000}{16} \int_0^1 \left(\frac{c}{D}\right) x^3\, dx \tag{6.70}$$

The integrated design lift coefficient represents the average of the section design lift coefficient weighted by x^2. The activity factor is simply another measure of the solidity. The higher the activity factor, the higher are the values of C_T and C_p attainable by a propeller at a given integrated design C_L.

Equation 6.68 represents about the best one can hope to achieve with a well-designed propeller operating at its design point. For the propeller shown in Figure 6.12, this corresponds to blade angles of around 15 to 25°. Beyond this range, the twist distribution along the blade departs too much from the optimum for these relationships to hold.

PROPELLER SELECTION

Propeller manufacturers offer propellers covering a range of diameters, pitch values, and solidities. The choice of these parameters can depend on considerations other than aerodynamic efficiency. For example, to keep the noise level of a propeller low, one may have to employ wide blades with low tip speeds. As another example, the propeller diameter is sometimes limited by ground clearance considerations or by the distance from a nacelle to the fuselage. The dynamics of the propeller must also be matched to the engine. The natural frequency of the first bending mode of a blade should not coincide with an impulse frequency from the engine. For example, a horizontally opposed, six-cylinder engine has three torsional peaks per revolution. If a propeller being driven by this engine has a natural frequency close to 3/rev, it can lead to exessive vibration and fatigue stresses.

Aerodynamically, one strives to select a propeller that provides a high efficiency for cruise and a high static thrust for takeoff. These two requirements are easier to satisfy with a variable pitch propeller. A fixed pitch propeller is usually a compromise between these two operating regimes.

Given the results of a series of propeller tests, such as Figures 6.13 and 6.14, one can utilize these data to select the best propeller diameter and blade angle to match a given airplane-engine combination. One approach that is sometimes used is based on a coefficient C_s, the speed power coefficient, defined by

$$C_s = \left(\frac{\rho V^5}{Pn^2}\right)^{1/5} \tag{6.71}$$

Knowing C_p as a function of J, C_s can be calculated from

$$C_s = \frac{J}{C_p^{1/5}} \tag{6.72}$$

The advantage of C_s is that it does not contain the diameter in its definition.

Figure 6.24 presents J as a function of C_s for the same propeller for which Figure 6.13 and 6.14 hold. A maximum efficiency line is also shown in Figure 6.24. The use of this graph is best illustrated with an example. The problem will be to select the optimum diameter for this propeller if it is to be installed on a Cherokee 180. Consider the selection of a propeller to absorb 75% of the maximum power of 180 bhp at 2500 rpm at standard sea level conditions. Using a value for f of 0.5 m^2 (5.38 ft^2) and an e of 0.6, C_D can be calculated as a function of V. C_T and C_D are then related by $(T = D)$.

$$C_T = C_D \frac{SJ^2}{2D^2} \tag{6.73}$$

Assume a value for V of 130 mph leads to a C_s of 1.360. From the maximum efficiency line in Figure 6.24, a J of 0.76 and a β of 20° are obtained. These values in turn lead to a C_T value of 0.0573, so obviously 130 mph will not be the trim

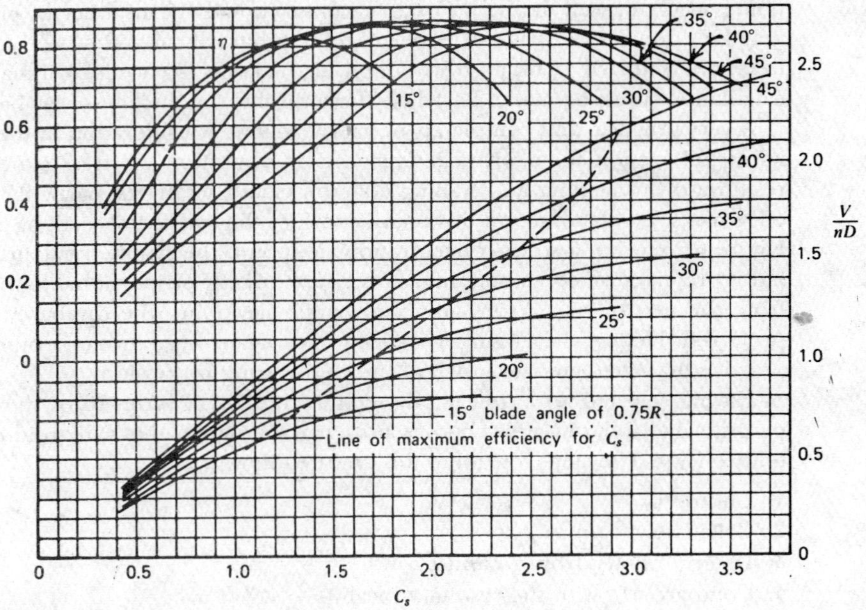

Figure 6.24 Design chart for propeller of Figure 6.12.

speed for the optimum propeller at this power and rpm. By iteration, one obtains a trim speed of 132 mph and the following:

$$J = 0.76$$
$$\beta = 20°$$
$$C_T = 0.0592$$
$$\eta = 0.84$$
$$D = 6.1 \text{ ft}$$

DESIGN OF A NEW PROPELLER

This section deals mainly with the aerodynamic considerations of designing a new propeller. The optimum blade loading is prescribed by the Betz condition which requires the trailing vortex system to lie along a helical surface in the ultimate wake. This condition will be met if

$$\begin{aligned} \omega r \tan(\phi + \alpha_i) &= \text{constant} \\ &= V + w_0 \end{aligned} \tag{6.74}$$

w_0 is a fictitious velocity called the impact velocity. Given the design advance ratio, one can arbitrarily choose a value of $w_0/\omega R$. From the geometry of Figure 6.10*b*, it follows that

$$\frac{w_t}{V_T} = \frac{w_0}{V_T} \sin(\phi + \alpha_i) \cos(\phi + \alpha_i) \tag{6.75}$$

where

$$\phi + \alpha_i = \tan^{-1}\left(\frac{V + w_0}{xV_T}\right) \tag{6.76}$$

Substituting Equation 6.75 into Equation 6.42 leads to the product σC_l as a function of x. One must then decide how to choose between σ and C_l. The procedure for doing so is not well defined. First, one must choose the number of blades. This may be done on the basis of experience or arbitrarily as a first step in a design iteration. Similarly, a radial distribution of thickness is chosen. Ultimately, stress calculations must be made. Based on these results, the thickness may be changed.

A very practical and completely nonaerodynamic consideration in the choice of an airfoil section for a propeller blade is the question of stress concentrations resulting from leading and trailing edge nicks and scratches, particularly leading edge nicks. To elaborate on this point, consider the two airfoil sections pictured in Figure 6.25. From a stress-concentration viewpoint, the symmetrical airfoil on the left is preferred since, in bending, stresses are directly proportional to the distance from the neutral axis. From an aerodynamic viewpoint, the cambered section is preferred. Hence, the engineer is faced once again with another compromise, a practice that characterizes much of the engineering profession.

Having selected an airfoil family such as the NACA series-16 or the newer supercritical airfoil, one now chooses at each x a design C_l that will avoid compressibility effects. The steps for doing this are the following:

1. Choose C_l.
2. Calculate c from σC_l.
3. Determine M_{cr} from t/c and C_l.
4. Compare M_{cr} with the resultant local M.
5. If M_{cr} is less than M, decrease C_l and repeat.

If Mach number is not a consideration in the design, then one can choose C_{l_i} to give the lowest C_d to C_l ratio for the chosen airfoil family.

Having determined the radial distribution of c (and hence σ) and C_l, the corresponding C_d values are calculated. These, together with $(\phi + \alpha_i)$, are substituted into Equations 6.32*a* and 6.32*b* to determine thrust and power. The entire design process is performed with different w_0 values until the desired value of C_T or C_p is achieved. Generally, increasing w_0 will increase either of these coefficients.

Most propellers are designed to operate immediately in front of a fuselage or nacelle. The inflow velocity in this case is no longer a constant but is, indeed, a function of x, the dimensionless radial station. This three-dimensional flow field can be determined by the potential flow methods presented in Chapter 2. With V a function of x, the resultant flow angle, ϕ, becomes

$$\phi = \tan^{-1}\frac{V(x)}{wr} \tag{6.77}$$

or

$$\phi = \tan^{-1}\left(\frac{V(x)/V_0}{x}\frac{J}{\pi}\right) \tag{6.78}$$

For this case of a nonuniform, *potential* inflow, the Betz condition is *not*

$$V(x) + w_0 = \text{constant}$$

Instead, one should impose only

$$w_0 = \text{constant} \tag{6.79}$$

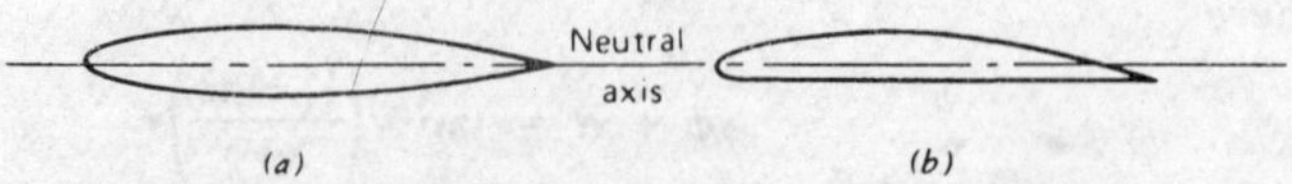

Figure 6.25 Susceptibility of airfoil shapes to leading and trailing edge stress concentrations. (*a*) Edges close to neutral axis. (*b*) Edges removed from neutral axis.

Equation 6.79 follows from superimposing the potential flow from the propeller on that produced by the body. In the ultimate wake, $V(x)$ will approach V_0 and w_0 will approach $2w_0$, so that the Betz condition is again satisfied.

A BRIEF HISTORY OF THE TURBOJET

One might argue that the turbojet engine had its beginnings with the turbosupercharger, since the latter has an exhaust-driven turbine that drives a compressor to supply air to the engine. These are the essential ingredients of a turbojet engine. Dr. Sanford A. Moss is generally credited with developing the turbosupercharger, at least in this country. In 1918, Moss successfully tested his turbosupercharger atop Pikes Peak. Two years later, a La Pere biplane equipped with a turbosupercharger set a world altitude record of over 10,000 m.

In 1930, Frank Whittle (later to become Sir Frank Whittle) received a patent for a turbojet engine. Unfortunately, he was unable to gain support for the development of his design. It was not until 1935, when a young German aeronautical engineering student, Hans von Ohain, received a German patent on a jet engine, that development work began in earnest on the turbojet engine. On August 27, 1939 (some references say June 1939), the first turbojet engine was flown in a Heinkel He178. This engine designed by von Ohain delivered 4900 N (1100 lb) of thrust. It was not until May 1941 that Whittle's engine was flown in England.

German jet engine development progressed rapidly. By 1944, both BMW and Junkers turbojet engines were introduced into the Luftwaffe. One can imagine the astonishment of the allied aircrews upon first seeing propellerless airplanes zip by them at incredible speeds of over 500 mph.

On October 1, 1942, the first American jet-propelled airplane, the Bell Airacomet, was flown. This twin-engine airplane was powered by an American copy of Whittle's engine built by the General Electric Co. Designated the "I-A," the engine weighed approximately 4450 N (1000 lb) with a thrust-to-weight ratio of 1.25. The first production American jet aircraft, the Lockheed F-80, first flew in January 1944. In production form, it was powered by the J33 engine, which delivered a thrust of approximately 17,800 N (4000 lb) at a weight of 8900 N (2000 lb).

General Electric's J47 was the first turbojet power plant certified in the United States for commercial aviation in 1949. The world's first commercial jet transport to fly, however, on July 27, 1949, was the British-built de Havilland Comet powered by four de Havilland Ghost 50 Mk1 turbojets. This engine, incorporating a centrifugal compressor, developed 19,800 N (4450 lb) of thrust. The Comet must be recognized as one of the most famous airplanes in history, because it truly ushered in the age of jet transportation. Unfortunately, its career was short-lived after three of the nine that had been built broke up in the air. An exhaustive investigation showed the cause to be fuselage structural fatigue because of repeated pressurizations. Despite its tragic demise, the Comet proved the feasibility of commercial jet transportation and paved the way for Boeing's successful 707. This airplane first took to the air on July 15, 1954, powered by four Pratt & Whitney JT3 (military designation J57) turbojet engines. Each engine developed a static thrust of approximately 57,800 N (13,000 lb) with a dry weight of 18,200 N (4100 lb).

A historical note of interest is the following quotation taken from a report by the Gas Turbine Committee of the U.S. National Academy of Sciences in 1940.

> *. . . Even considering the improvements possible . . . the gas turbine could hardly be considered a feasible application to airplanes mainly because of the difficulty with the stringent weight requirements . . . .*

This conclusion, made by a panel of eminent persons, including Dr. Theodore von Karman, is a sobering reminder to any engineer not to be too absolute.

DESCRIPTION OF THE GAS TURBINE ENGINE

Basically, the gas turbine engine consists of a compressor, a combustion chamber, and a turbine. The combination of these basic components is referred to as the *gas generator* or *core engine*. Other components are then added to make the complete engine. It is beyond the scope of this text to delve into the details of gas turbine engine design. However, the various types of gas turbine engines will be described, and their operating characteristics will be discussed in some detail.

Beginning with the core engine, the turbojet engine pictured in Figure 6.26 is obtained by adding an engine air inlet and a jet nozzle. As the air enters the inlet, it is diffused and compressed slightly. It then passes through a number of blade rows that are alternately rotating and stationary. The collection of rotating blades is referred to as the *rotor*; the assembly of stationary blades is called the *stator*. This particular compressor configuration is known as an *axial-flow compressor* and is the type used on all of today's larger gas turbine engines. Early gas turbine engines, such as Whittle's engine, employed a centrifugal compressor, as shown in Figure 6.27. Here, the air enters a rotating blade row near the center and is turned radially outward. As the air flows out through the rotating blade passage, it acquires a tangential velocity component and is compressed. A scroll or radial diffuser collects the compressed air and delivers it to the combustion chamber. Centrifugal compressors were used on the early turbojet engines simply because their design was better understood at the time. As jet engine development progressed, centrifugal compressors were abandoned in favor of the more efficient axial-flow compressors. The axial-flow compressor also presents a smaller frontal area than its centrifugal counterpart and is capable of achieving a higher pressure ration.

Smaller sizes of gas turbine engines still favor the centrifugal compressor. Figure 6.28 is a cutaway drawing of the Garrett TPE 331/T76 turboprop engine. The compressor section of this engine consists of two stages of radial impellers made of forged titanium.

After the compressed air leaves the compressor section, it enters the combustor, or burner, section. Atomized fuel is sprayed through fuel nozzles and the resulting air-fuel mixture is burned. Typically, the ratio of air to fuel by weight is about 60:1. However, only approximately 25% of the air is used to support combustion. The

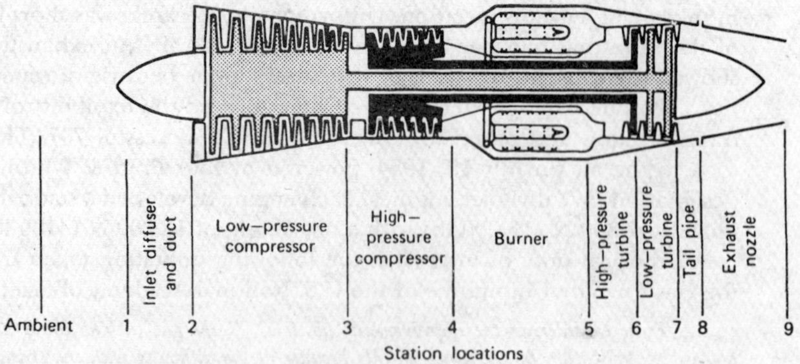

Figure 6.26 Typical dual rotor (two-spool) turbojet.

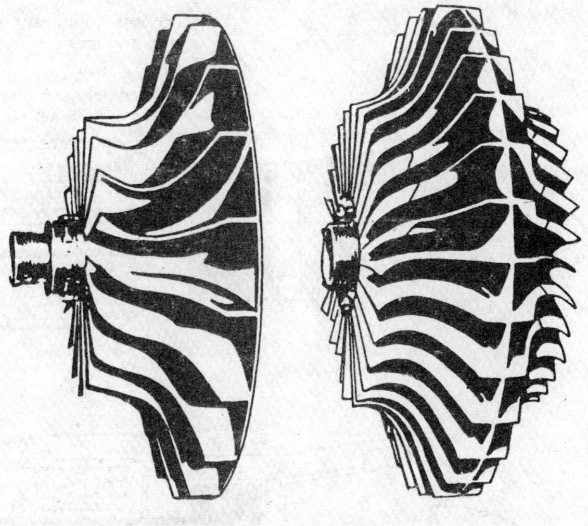

Figure 6.27 Typical centrifugal flow compressor impellers. (*a*) Single-entry impeller. (*b*) Double-entry impeller. (Courtesy, General Electric Co.)

remainder bypasses the fuel nozzles and mixes downstream of the burner to cool the hot gases before they enter the turbine.

The mixed air, still very hot (about 1100°C), expands through the turbine stages which are composed of rotating and stationary blade rows. The turbines extract energy from the moving gases, thereby furnishing the power required to drive the compressor. Nearly 75% of the combustion energy is required to drive the compressor. The remaining 25% represents the kinetic energy of the exhaust, which provides the thrust. For example, in the General Electric CF6-6 turbofan engine [180,000 N (40,000 lb) thrust class], the turbine develops approximately 65,600 kW (88,000 shp) to drive the high- and low-pressure compressors.

Figure 6.28 Cutaway of Garrett TPE 331/76. (Courtesy, The Garrett Corp.)

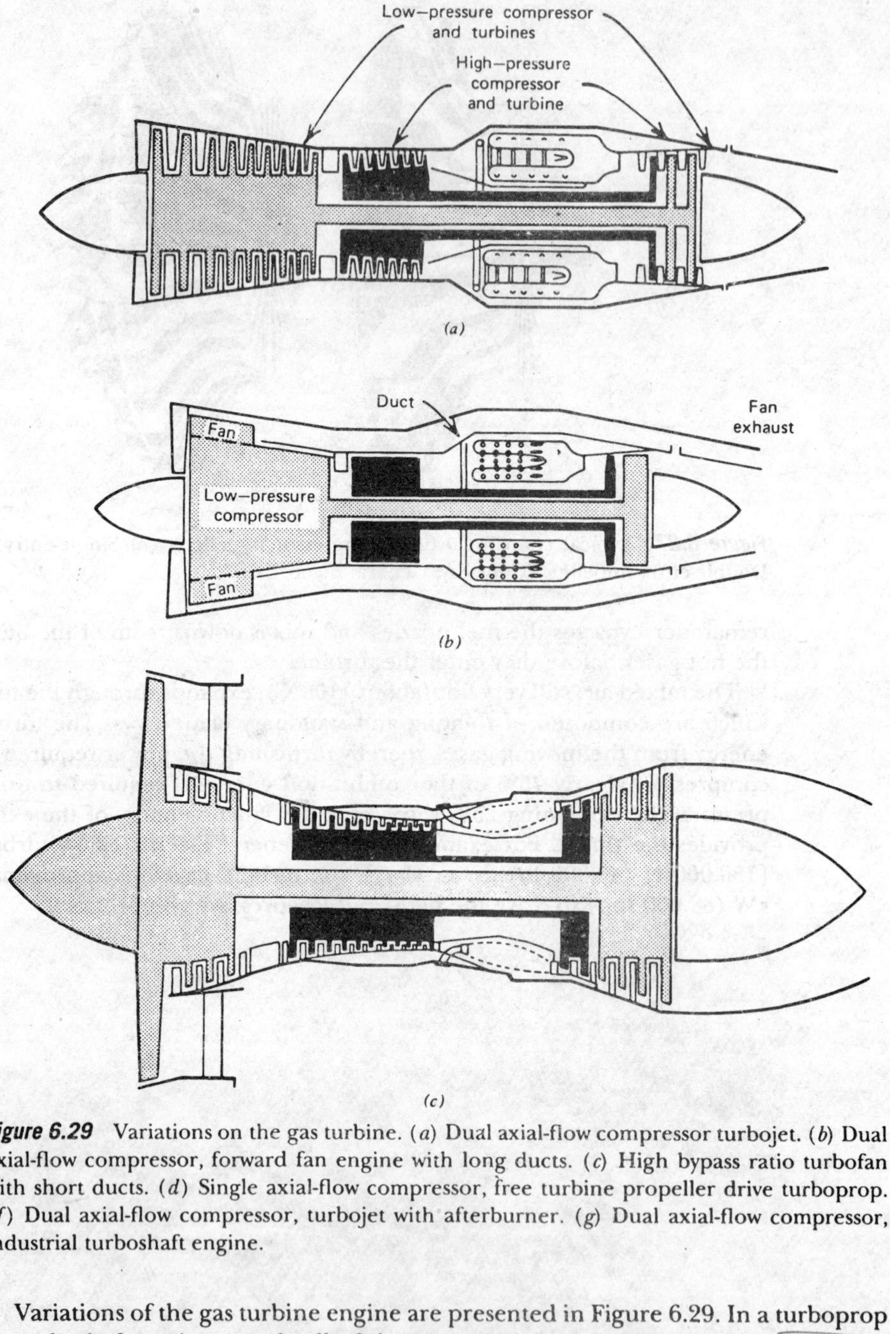

Figure 6.29 Variations on the gas turbine. (*a*) Dual axial-flow compressor turbojet. (*b*) Dual axial-flow compressor, forward fan engine with long ducts. (*c*) High bypass ratio turbofan with short ducts. (*d*) Single axial-flow compressor, free turbine propeller drive turboprop. (*f*) Dual axial-flow compressor, turbojet with afterburner. (*g*) Dual axial-flow compressor, industrial turboshaft engine.

Variations of the gas turbine engine are presented in Figure 6.29. In a turboprop or turboshaft engine, nearly all of the energy of the hot gases is extracted by the turbines, leaving only a small residual thrust. The extracted energy in excess of that required to drive the compressor is then used to provide shaft power to turn the propeller or a power-output shaft in general. Turboshaft engines power most of today's helicopters and are used extensively by the electric utilities to satisfy peak power load demands.

A "spool" refers to one or more compressor and turbine stages connected to the same shaft and thus rotating at the same speed. Gas turbine engines generally

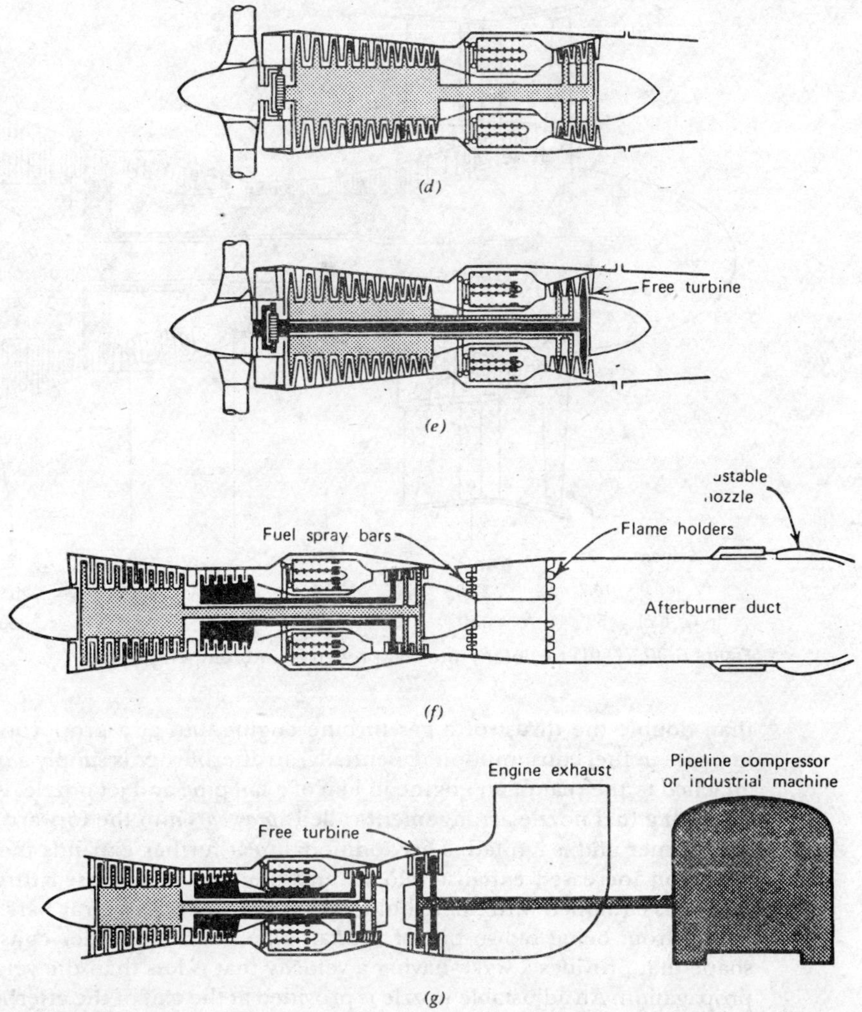

(d) (e) (f) (g)

Figure 6.29 *(continued)*

use one or two spools and are referred to as single or dual compressor engines. A turboshaft engine may incorporate a free turbine that is independent of any compressor stage and is used solely to drive the shaft. Since the rotational speed of a turbine wheel is of the order of 10,000 rpm, a reduction gear is required between the turbine shaft and the power output shaft.

A turboprop produces a small amount of jet thrust in addition to the shaft power that it develops; these engines are rated statically in terms of an equivalent shaft horsepower (eshp). This rating is obtained by assuming that 1 shp produces 2.5 lb of thrust. For example, the dash 11 model of the engine shown in Figure 6.28 has ratings of 1000 shp and 1045 eshp. From the definition of eshp, this engine therefore produces a static thrust from the turbine exhaust of approximately 113 lb.

A turbojet engine equipped with an afterburner is pictured in Figure 6.29*f*. Since only 25% or so of the air is used to support combustion in the burner section, there is sufficient oxygen in the turbine exhaust to support additional burning in the afterburner. Both turbofans and turbojets can be equipped with afterburners to provide additional thrust for a limited period of time. Afterburning can more

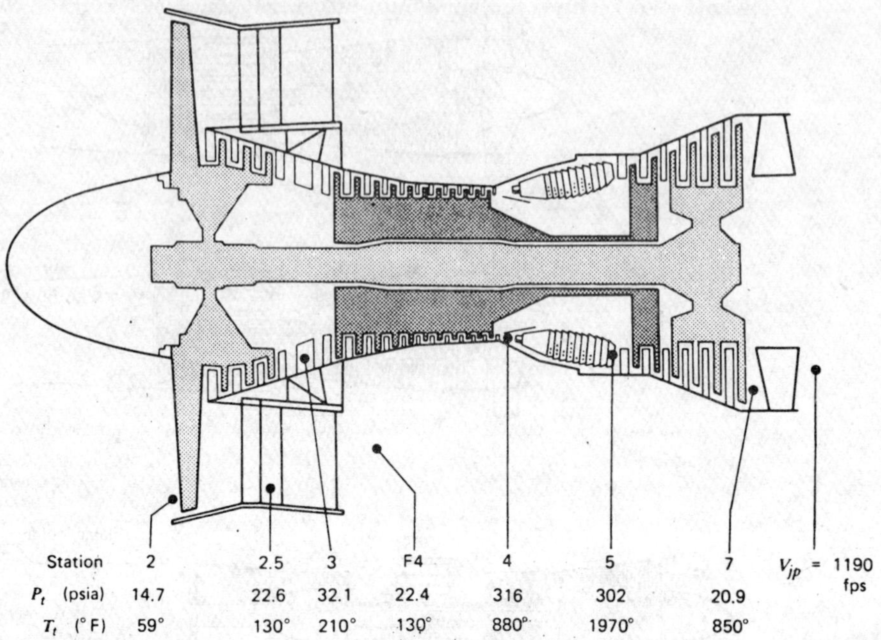

Figure 6.30 JT9D turbofan, internal pressures and temperatures.

than double the thrust of a gas turbine engine, but at a proportionately greater increase in fuel consumption. Essentially, an afterburner is simply a huge stovepipe attached to the rear of an engine in lieu of a tail pipe and jet nozzle. Fuel is injected through a fuel nozzle arrangement called *spray bars* into the forward section of the afterburner and is ignited. This additional heat further expands the exhaust, providing an increased exhaust velocity and thereby an increased thrust. The afterburner is equipped with flame holders downstream of the spray bars to prevent the flames from being blown out of the tail pipe. A flame holder consists of a blunt shape that provides a wake having a velocity that is less than the velocity for flame propagation. An adjustable nozzle is provided at the exit of the afterburner in order to match the exit area to the engine operating condition.

Two different types of turbofan engines are shown in Figure 6.29*b* and 6.29*c*; the forward fan with a short duct and the forward fan with a long duct. These engines are referred to as *bypass engines,* since part of the air entering the engine bypasses the gas generator to go through the fan. The ratio by weight of the air that passes through the fan (secondary flow) to the air that passes through the gas generator (primary flow) is called the *bypass ratio.* Early turbofan engines had bypass ratios of around 1:1; the latest engines have ratios of about 5:1. One such engine, Pratt & Whitney's JT9D turbofan, is shown in Figure 6.30. Included on the figure are temperatures and absolute pressures throughout the engine for static operation at standard sea level conditions.

ENGINE RATINGS

An engine rating specifies the thrust that an engine can (or is allowed) to develop in a particular operating mode. For commercial certification, these ratings are defined as follows:

Takeoff (Wet) This is the maximum thrust available for takeoff for engines that use water injection. The rating is selected by actuating the water injection system and setting the aircraft throttle to obtain the computed "wet" takeoff thrust. The rating is restricted to takeoff, is time limited to 5 min, and has altitude and ambient air or water temperature limitations.

Takeoff (Dry) This is the maximum thrust available without the use of water injection. The rating is selected by setting the aircraft throttle to obtain the computed takeoff (dry) thrust for the prevailing conditions of ambient temperature and barometric pressure. The rating is time limited to 5 min and is to be used only for takeoff and, as required, for reverse thrust operations during landing.

Maximum Continuous This rating is the maximum thrust that may be used continuously, and is intended only for emergency use at the discretion of the pilot.

Maximum Climb Maximum climb thrust is the maximum thrust approved for normal climb. On some engines, maximum continuous and maximum climb thrusts are the same. For commercial engines, the term formerly used, normal rated thrust, has been replaced by the more appropriate term, maximum climb thrust.

Maximum Cruise This is the maximum thrust approved for cruising.

Flat Rating

Engines that must be operated at "part throttle" at standard ambient conditions to avoid exceeding a rated thrust are referred to as "flat-rated" engines. This refers to the shape of the thrust versus the ambient temperature curve. For example, the General Electric Company's CF6-6 high bypass turbofan engine is flat rated up to an ambient temperature of 31°C at sea level, or 16°C higher than a standard day. Thus, its thrust as a function of ambient temperature varies, as shown in Figure 6.31. At full throttle, the thrust is seen to decrease with increasing temperature. Therefore, by flat rating an engine out to a temperature higher than standard, one is able to maintain rated thrust on a hot day.

SOME CONSIDERATIONS RELATING TO GAS TURBINE PERFORMANCE

To understand, at least qualitatively, why a particular configuration of gas turbine engine performs as it does, let us consider a few basic principles. The ideal ther-

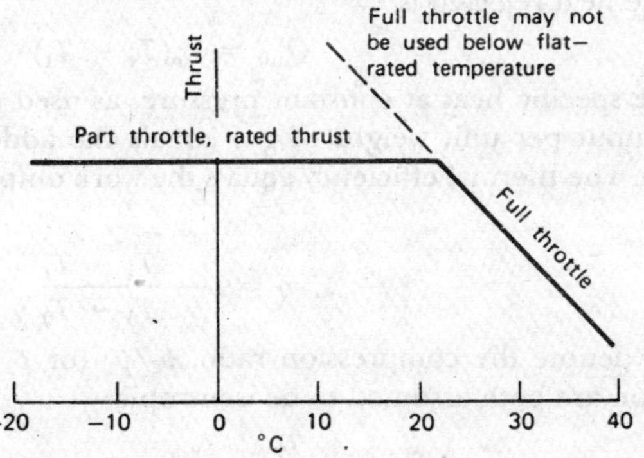

Figure 6.31 Thrust curve for a flat-rated engine.

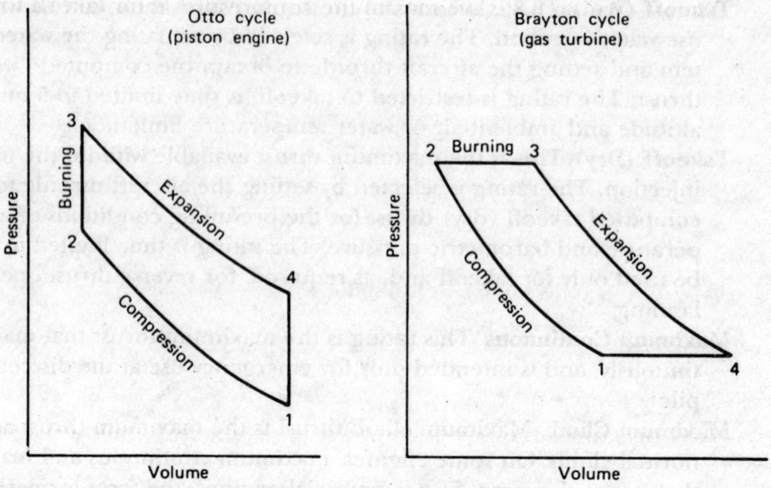

Figure 6.32 A comparison between the Otto and Brayton cycles.

modynamic cycle for the gas turbine engine is shown in Figure 6.32, where it is compared to the cycle for the piston engine. The Otto cycle, which approximates the piston engine thermodynamics, consists of an isentropic compression of the gas followed by a rapid combustion at nearly constant volume. The gas then expands isentropically, forcing the piston ahead of it. Unlike the piston engine, the gas turbine engine involves a continuous flow of the working gas. The Brayton or constant pressure cycle, which approximates the actual gas turbine cycle, begins with an isentropic compression of the air from ambient conditions. Part of this compression occurs prior to the compressor stages as the air enters the engine inlet. Following the compression, burning occurs at constant pressure, resulting in increased volume and total temperature. The air then expands isentropically through the turbines and jet nozzle to the ambient static pressure. In a turboprop or turboshaft engine, nearly all of the expansion occurs within the turbines in order to drive the compressor and produce shaft power. In a turbojet engine, an appreciable amount of expansion occurs after the turbines in order to produce the high-momentum jet.

The heat that is added to the flow per unit weight of gas is given by

$$Q_{\text{in}} = C_p(T_3 - T_2) \tag{6.80}$$

while the heat rejected is

$$Q_{\text{out}} = C_p(T_4 - T_1) \tag{6.81}$$

C_p is the specific heat at constant pressure, as used previously in Chapter 5. The work output per unit weight of gas equals the added heat minus that which is rejected. The thermal efficiency equals the work output divided by the added heat. Thus,

$$\eta = 1 - \frac{T_4 - T_1}{T_3 - T_2} \tag{6.82}$$

Let r denote the compression ratio, p_2/p_1 (or p_3/p_4). Since compression and expansion are both assumed to be isentropic,

$$\frac{T_2}{T_1} = \frac{T_3}{T_4} = r^{(\gamma-1)/\gamma} \tag{6.83}$$

Thus, Equation 6.82, in terms of the compression ratio, can be written as

$$\eta = 1 - \frac{1}{r^{(\gamma-1)/\gamma}} \tag{6.84}$$

As stated previously, the compression ratio r is achieved partly in the inlet (ram pressure), and the remainder is achieved through the compressor. The pressure increase across the compressor, at a constant rpm, as a first approximation, is proportional to the mass density, ρ_c, just ahead of the compressor.

$$\Delta p \propto \rho_c$$

If r_0 denotes the value of r for static sea level operation,

$$p = p_0(r_0 - 1)\frac{\rho_c}{\rho_0} \tag{6.85}$$

ρ_0 is, of course, the standard sea level value of mass density.

For isentropic compression in the inlet up to the compressor, the ambient mass density and ρ_c are related by

$$\frac{\rho_c}{\rho_\infty} = \left[\frac{1 + (\gamma - 1/2)M_\infty{}^2}{1 + (\gamma - 1/2)M^2}\right]^{1/(\gamma-1)} \tag{6.86}$$

where M_∞ is the free-stream Mach number and M is the local Mach number just ahead of the compressor. The pressure ratio, r, thus becomes

$$r = \frac{p_c + \Delta p}{p_\infty}$$

or

$$r = \frac{p_c}{p_\infty} + \frac{(r_0 - 1)}{\theta}\frac{\rho_c}{\rho_\infty} \tag{6.87}$$

p_c/p_∞ is given by

$$\frac{\rho_c}{\rho_\infty} = \left[\frac{1 + (\gamma - 1/2)M_\infty{}^2}{1 + (\gamma - 1/2)M^2}\right]^{\gamma/\gamma-1} \tag{6.88}$$

Thus, r finally becomes

$$r = [f(M, M_\infty)]^{\gamma/(\gamma-1)} + \frac{r_0 - 1}{\theta}[f(M, M_\infty)]^{1/(\gamma-1)} \tag{6.89}$$

where

$$f(M, M_\infty) = \frac{1 + (\gamma - 1/2)M_\infty{}^2}{1 + (\gamma - 1/2)M^2}$$

Equation 6.89 is substituted into Equation 6.84 and an expression for the thermal efficiency results that is a function of δ, r_0, M_∞, and M.

The effect of pressure ratio, altitude, and free-stream Mach number on the ideal thermal efficiency is shown in Figure 6.33. This figure assumes the ratio of M to M_∞ just before the compressor to equal approximately zero. This is a fairly reasonable assumption, since values of this ratio up to at least 0.4 affect η by less than 1%. Figure 6.33 shows the effect of varying one parameter at a time while keeping the other two parameters at their normal values. Increasing M_∞ from zero to 0.8 is seen to result in a 7% improvement in η. The efficiency also improves with altitude, increasing by approximately 6% in going from sea level to 40,000 ft (12,200 m). Doubling the pressure ratio, r_0, from 10 to 20 results in a 16% improvement in η.

With regard to the production of thrust, η does not tell the whole story. η is

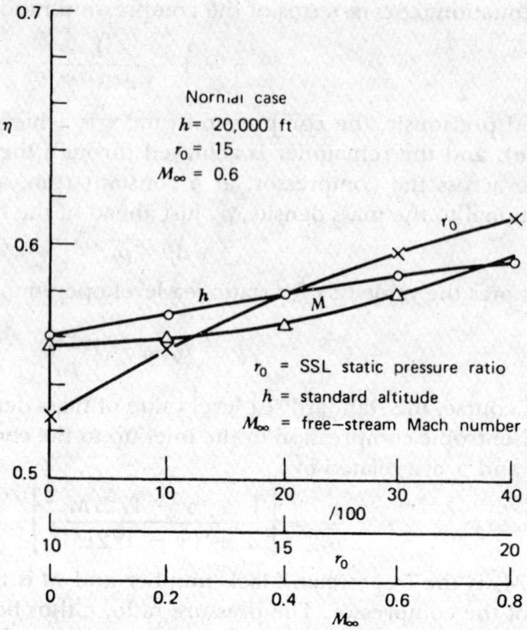

Figure 6.33 Effect of pressure ratio, altitude, and Mach number on the ideal thermal efficiency of a gas turbine engine.

simply a measure of how efficiently the air passing through the engine is being used. The heat added to the flow per unit weight is given by

$$Q_{in} = \eta C_p(T_3 - T_2)$$

Thus, for the same efficiency, if T_3 is increased or the mass flow increased, the thrust will be increased.

QUALITATIVE COMPARISON OF THE PERFORMANCE OF TURBOJET, TURBOFAN, AND TURBOPROP ENGINES

Figure 6.34*a*, 6.34*b*, and 6.34*c* presents a qualitative comparison of the turbojet, turbofan, and turboprop engines, each having the same core engine.

The specific fuel consumption for a turbojet or turbofan engine is expressed as a *thrust specific fuel consumption* (TSFC). In the English system of units, one states TSFC as pounds of fuel per hour per pound of thrust, so that TSFC actually has the dimensions of 1/time. Thus, its numerical value is the same in the SI system as in the English system. In the SI system, TSFC is given as N/hr/N. The characteristics of the three engines are seen to be quite different with the turbofan, not surprisingly, lying between the turboprop and turbojet. The relative differences in these curves are explained mainly by the momentum and energy considerations undertaken previously for the propeller. "Disc loadings" for turbojet engines are of the order of 81,400 Pa (1700 psf), while turbofans operate at approximately half of this loading and propellers at only approximately 4% of the disc loading for a turbojet. If we assume that the core engine is delivering the same power to each engine configuration then, from Equation 6.17, for static thrust one obtains

$$F_0 \propto \left(\frac{F_0}{A}\right)^{-1/2}$$

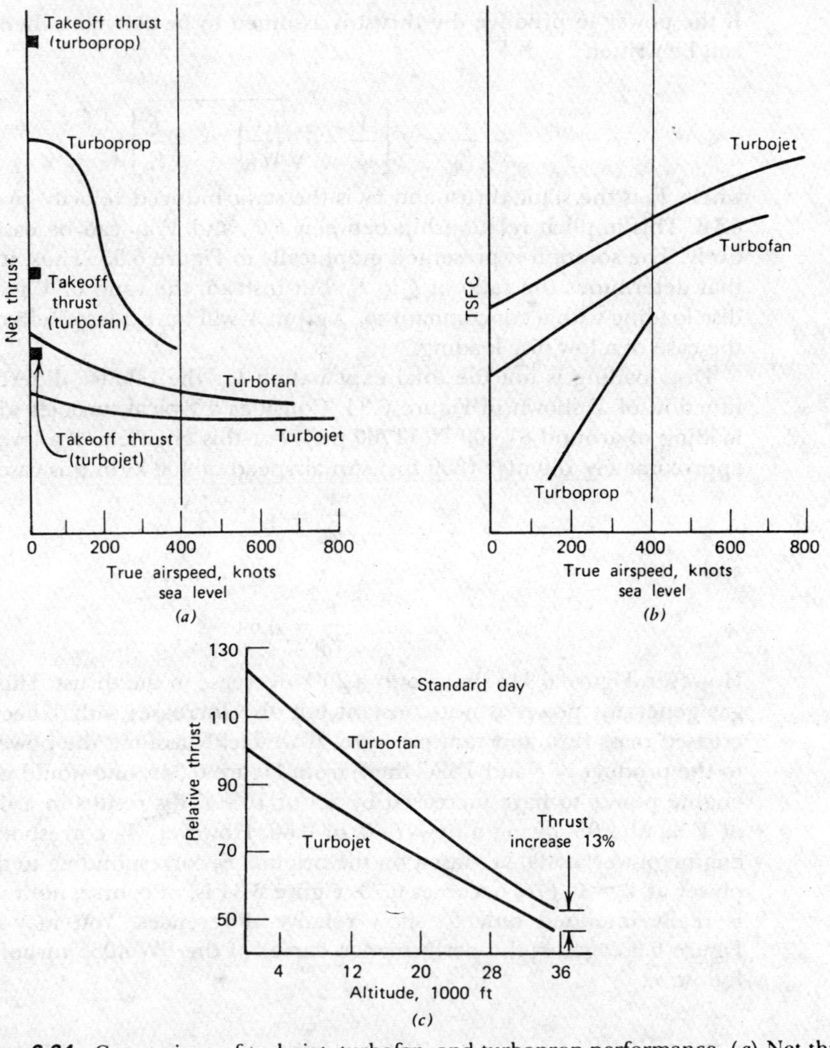

Figure 6.34 Comparison of turbojet, turbofan, and turboprop performance. (*a*) Net thrust at sea level. (*b*) Thrust specific fuel consumption. (*c*) Relative maximum continuous thrust during climb.

Note that thrust for a turbojet engine is denoted by F instead of T, since T is understood to refer to temperature when working with a gas turbine. Thus, with its appreciably lower disc loading, one would expect the static thrust of a turboprop to be significantly higher than the corresponding turbojet, possibly even more so than that shown in Figure 6.34 (taken from Ref. 6.6).

The rapid decrease in thrust with airspeed for the turboprop and the more gradual changes for the turbofan and turbojet engines are also explained in part by the relative disc loadings. Combining Equations 6.13, 6.14, and 6.15 gives

$$P = \frac{F}{2}\left[V + \sqrt{V^2 + \left(\frac{2F}{\rho A}\right)} \right] \tag{6.90}$$

If the power to produce the thrust is assumed to be constant, then Equation 6.90 can be written

$$\frac{F}{F_0} = 2\left[\frac{V}{w_0} + \sqrt{\left(\frac{V}{w_0}\right)^2 + 4\frac{F}{F_0}}\right]^{-1} \tag{6.91}$$

where F_0 is the static thrust and w_0 is the static-induced velocity given by Equation 6.16. This implicit relationship between F/F_0 and V/w_0 can be easily solved iteratively. The solution is presented graphically in Figure 6.35. Thus, it is not V per se that determines the ratio of F to F_0, but instead, the ratio of V to w_0. For a high disc loading with a concomitant w_0, a given V will have a lesser effect on F than for the case of a low disc loading.

Disc loading is not the total explanation for the relative differences in T as a function of V shown in Figure 6.34. Consider a typical turbojet with a static disc loading of around 81,400 Pa (1700 psf). For this engine at sea level, w_0 will equal approximately 180 m/s (600 fps). An airspeed of 400 kt in this case gives

$$\frac{V}{w_0} = 1.13$$

and

$$\frac{F}{F_0} = 0.64$$

However, Figure 6.34*a* shows only a 20% decrease in the thrust. This is because the gas generator power is not constant but also increases with V because of the increased mass flow and ram pressure. If we tacitly assume the power proportional to the product of F and TSFC then, from Figure 6.34*b*, one would predict the core engine power to have increased by about 10%. This results in a decreased value of V/w_0 of 1.08, giving a new T/T_0 of 0.66. However, T_0 corresponds to the core engine power at 400 kt. Based on the original F_0 corresponding to the core engine power at $V = 0$, F/F_0 becomes 0.73. Figure 6.34 is, of course, not too accurate and is really intended only to show relative differences. You may wish to check Figure 6.35 against the performance curves of the PW 4056 turbofan engine that follow.

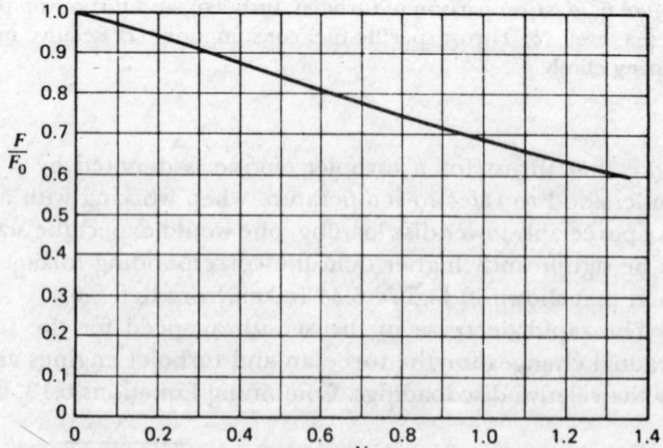

Figure 6.35 Predicted variation of thrust with forward velocity for a constant power.

SPECIFIC ENGINE CHARACTERISTICS AND PERFORMANCE

For illustrative purposes, this section will consider the characteristics and performance of specific engines.

Turbojet

Most modern engine developments incorporate the turbofan configuration. However, for completeness, the characteristics of a pure turbojet engine will be presented; namely, the Pratt & Whitney JT4A-3, which has been used extensively in the past. This engine, installed on the Boeing 707-320 and McDonnell-Douglas DC-8-20, is a two-spool engine. The low-pressure compressor section has eight stages, with seven stages in the high-pressure section. The turbine has two low-pressure stages and one high-pressure stage. Other characteristics of this engine are presented in Table 6.1. This particular model of the JT4 engine has a takeoff thrust-to-dry weight ratio of 3.15. Later versions of this engine, such as the JT4A-12, develop T/W ratios of 3.58.

The net thrust and fuel consumption curves for this engine are reproduced from the manufacturer's installation handbook in Figure 6.36*a*, 6.36*b*, 6.36*c*, and 6.36*d*. At this point, a definition of net thrust is needed. To do this we first define gross thrust, F_g, as the product of the mass flow rate in the jet exhaust and the velocity attained by the jet after expanding to ambient static pressure.

$$F_g = m_j V_j$$

The net thrust, F_n, is then defined by

$$F_n = F_g - m_i V_{am}$$

where m_i is the inlet mass flow and V_{am} is the velocity of the ambient air. For static operation, F_g and F_n are equal. Net thrust for the takeoff rating of this engine is presented in Figure 6.37*a*, 6.37*b*, and 6.37*c* for speeds of 0, 100, and 200 kt and altitudes from sea level to 14,000 ft.

The curves of Figure 6.36 are for standard atmospheric conditions. One rarely finds a standard day, so it is usually necessary to correct engine performance for deviations from the standard. Without delving into the details of compressor design, one can argue that, for the same flow geometry (ratio of rotor speed to axial velocity and M_∞), the pressure increase across the compressor can be written as

$$\Delta p \propto \rho_\infty N^2$$

where N is the rotor angular velocity.

Table 6.1 Characteristics of the JT4A-3 Engine

Type—turbojet	
SSL static thrust	
Dry takeoff	70,300 N (15,800 lb)
Maximum continuous	55,600 N (12,500 lb)
TSFC	
Dry takeoff	0.780/hr
Maximum continuous	0.740/hr
Gas generator at dry takeoff	
Total airflow	1,108 N/s (249 lb/sec)
Overall pressure ratio	11.8
Engine dry weight	22,239 N (5020 lb)
Engine diameter	1.09 M (43 in.)
Engine length	3.66 M (144.1 in.)

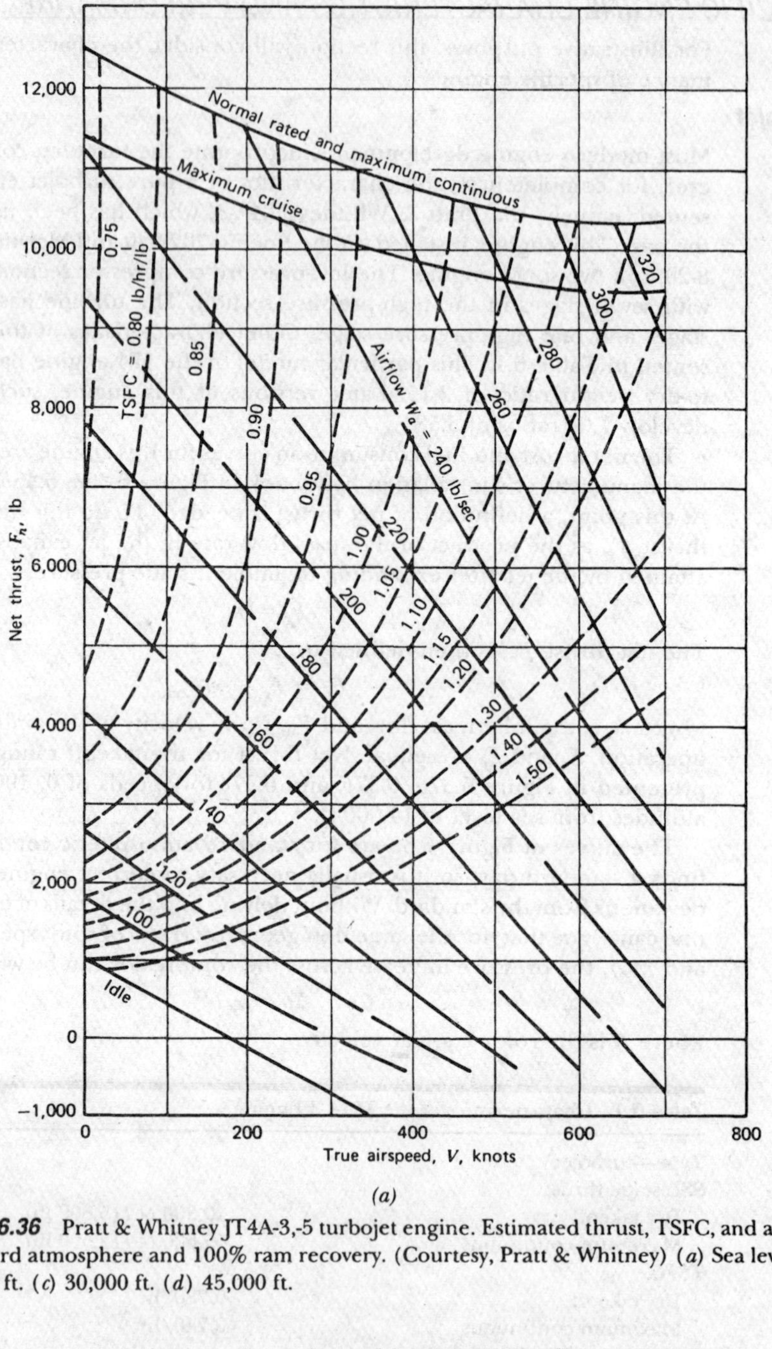

(*a*)

Figure 6.36 Pratt & Whitney JT4A-3,-5 turbojet engine. Estimated thrust, TSFC, and airflow. Standard atmosphere and 100% ram recovery. (Courtesy, Pratt & Whitney) (*a*) Sea level. (*b*) 15,000 ft. (*c*) 30,000 ft. (*d*) 45,000 ft.

If Δp is expressed as a ratio to the ambient pressure, then

$$\frac{\Delta p}{p_\infty} \propto \frac{N^2}{p_\infty/\rho_\infty}$$

or

$$\frac{\Delta p}{p_\infty} \propto \frac{N^2}{T_\infty}$$

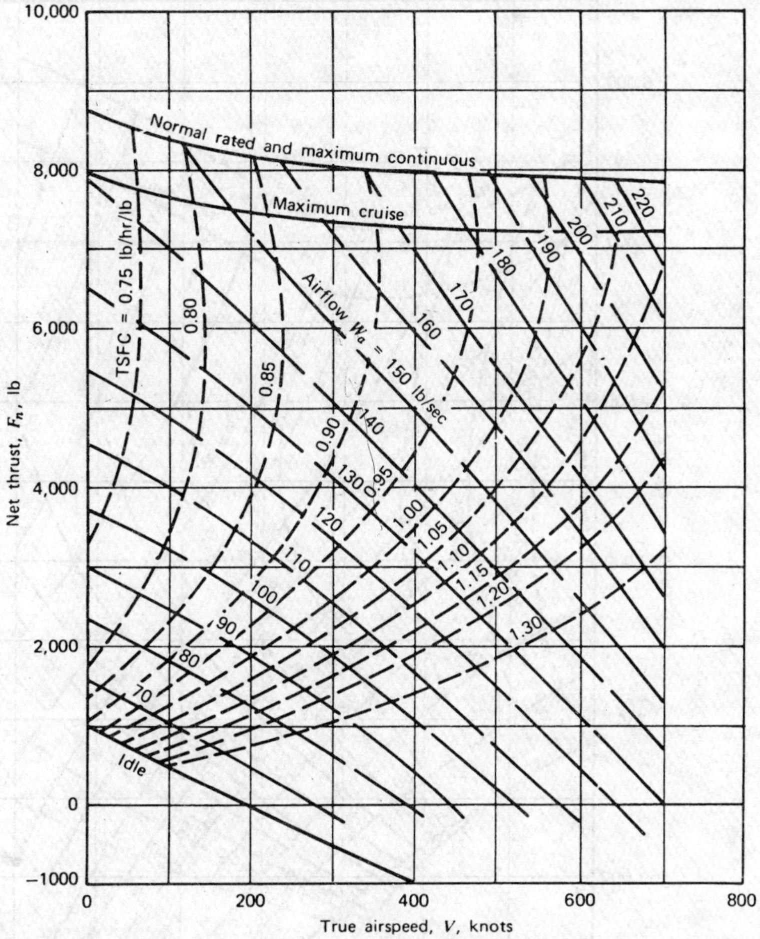

(b) JT4A, 15,000 ft.

Figure 6.36 *(continued)*

Thus, if N denotes the rpm of a compressor operating at an ambient temperature of T_∞, the rpm required to deliver the same pressure ratio at standard sea level conditions is known as the corrected rpm, N_c, given by

$$N_c = \frac{N}{\sqrt{\theta}} \tag{6.92}$$

θ being the ratio of the absolute temperature to the standard absolute temperature at sea level.

Similarly, one can say that the thrust, F, must be proportional to Δp, or p_∞, for a constant pressure ratio. Thus, the corrected thrust, F_c, corresponding to the corrected rpm, is defined by

$$F_c = \frac{F}{\delta} \tag{6.93}$$

where δ is the ratio of the ambient pressure to standard sea level pressure.

Similarly, corrected values for fuel flow, airflow, and exhaust gas temperature (EGT) are defined by

$$W_{fc} = \frac{W_f}{\delta\sqrt{\theta}} \tag{6.94}$$

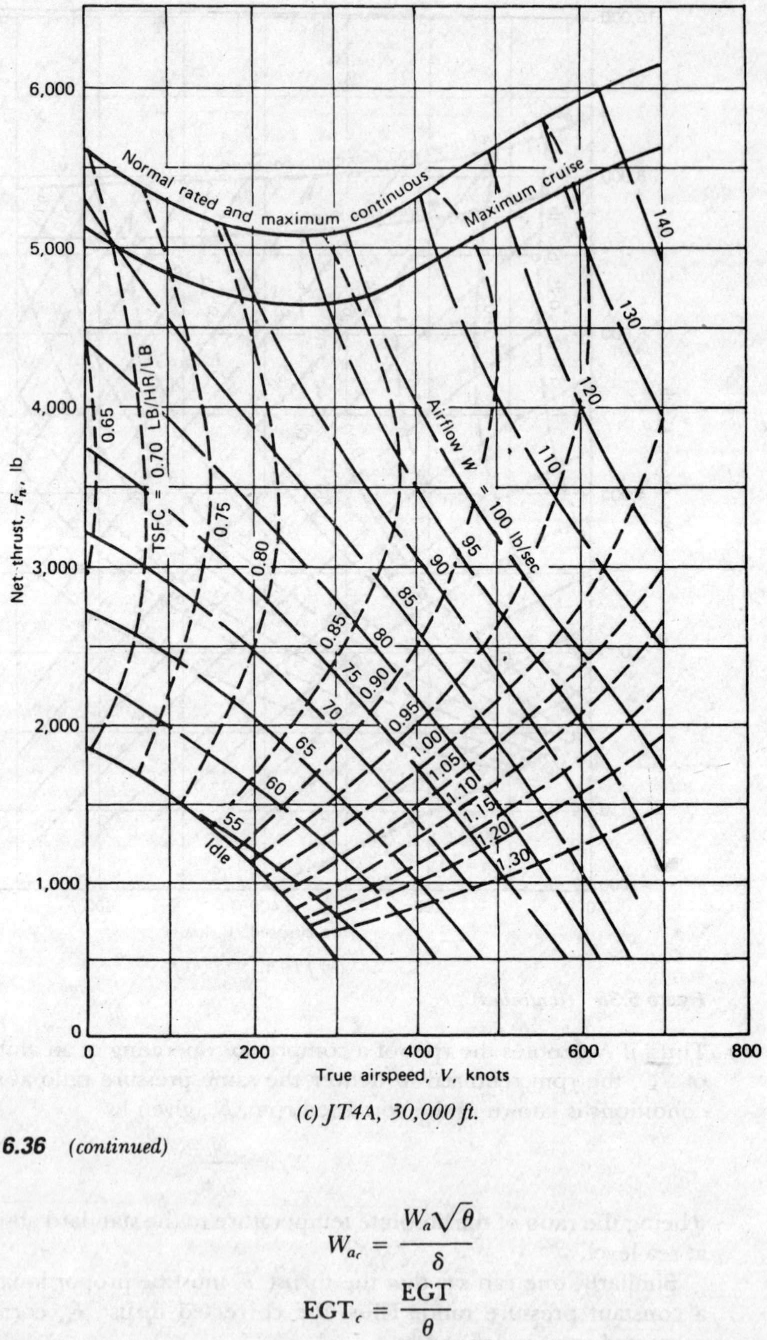

(c) JT4A, 30,000 ft.

Figure 6.36 *(continued)*

$$W_{a_c} = \frac{W_a\sqrt{\theta}}{\delta} \tag{6.95}$$

$$\text{EGT}_c = \frac{\text{EGT}}{\theta} \tag{6.96}$$

A more elegant derivation of these corrected parameters, based on Buckingham's π theorem of dimensional analysis, can be found in Reference 6.9.

Excluding scale effects, the important point is made that the corrected thrust of a gas turbine engine is a unique function of the corrected values of N, W_a, and W_f. These, in turn, assure a constant value of the pressure ratio.

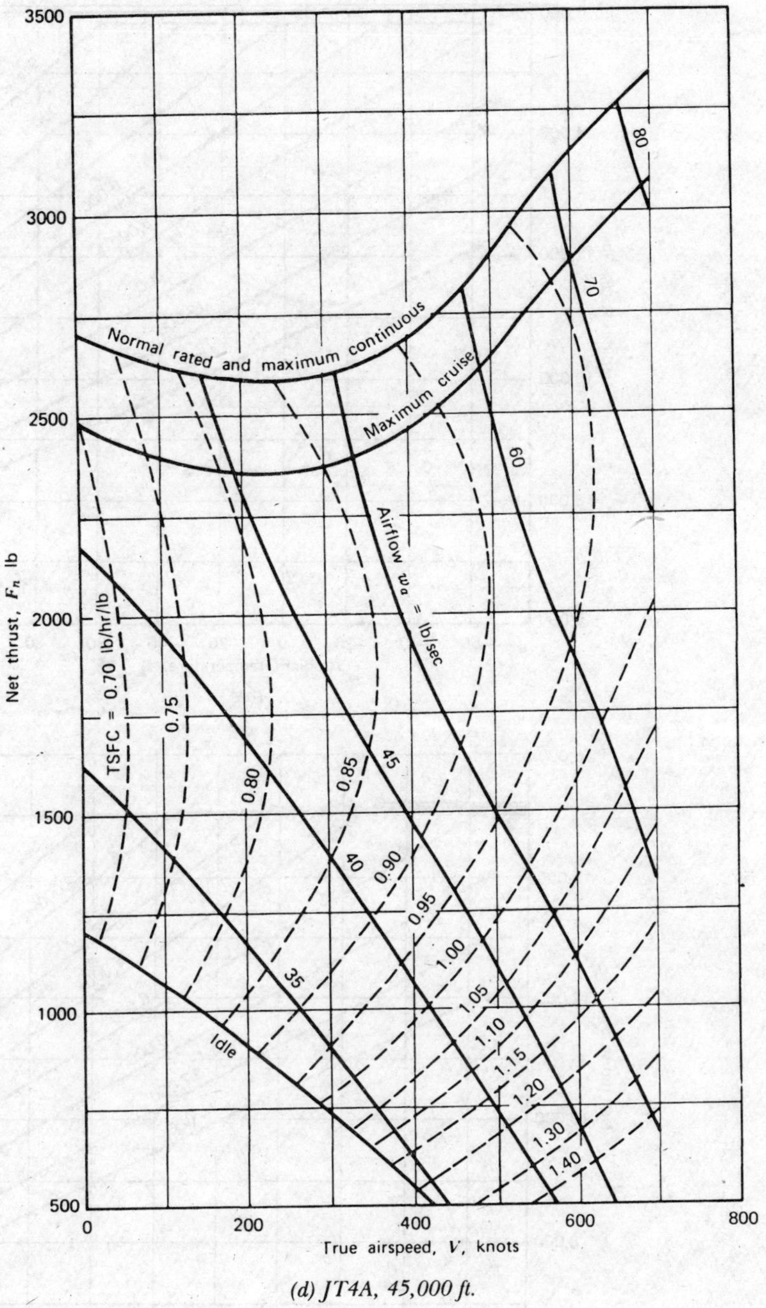

(d) JT4A, 45,000 ft.

Figure 6.36 *(continued)*

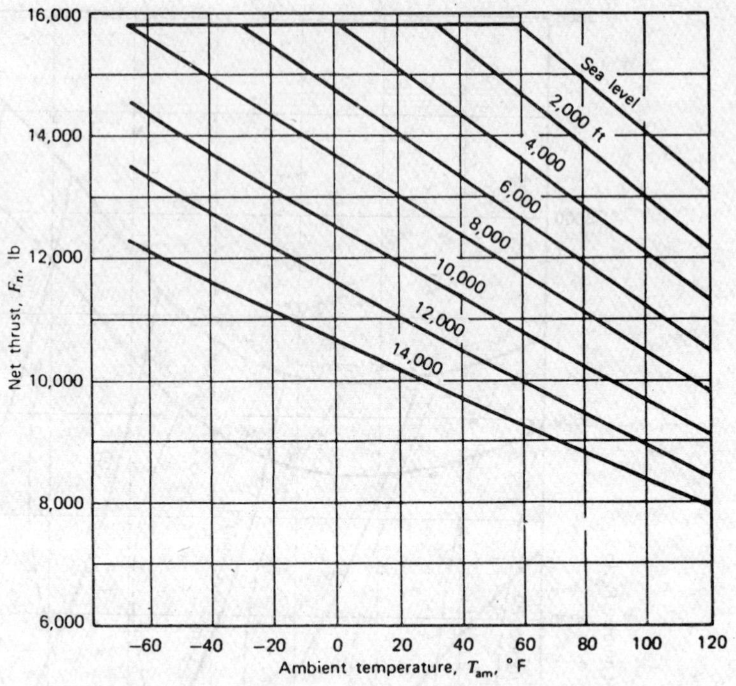

(*a*)

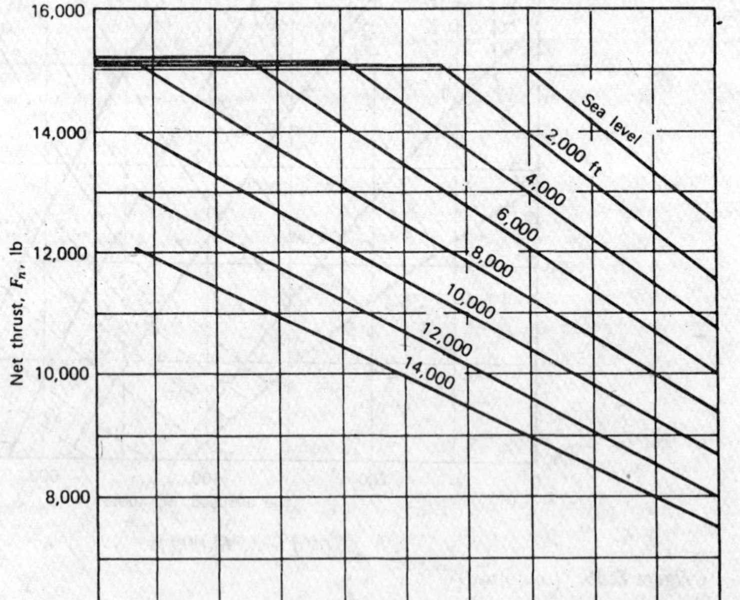

(*b*)

Figure 6.37 Pratt & Whitney JT4A-3,-5 turbojet engines. Estimated thrust on runway during takeoff. 100% ram recovery. (Courtesy, Pratt & Whitney) (*a*) Zero knots. (*b*) One hundred knots. (*c*) Two hundred knots.

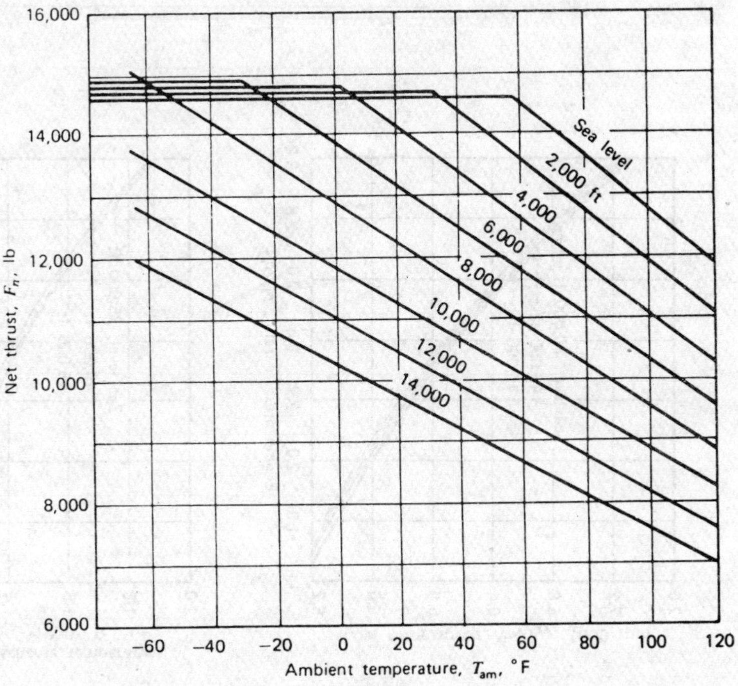

(c)

Figure 6.37 *(continued)*

In practice, the pressure ratio used to monitor the corrected thrust is referred to as the *engine pressure ratio* (EPR) defined by

$$\text{EPR} = \frac{p_{t7}}{p_{t2}} \tag{6.97}$$

The subscript t refers to the total stagnation pressure, the 7 and 2 refer to the engine stations shown in Figure 6.26. Thus, EPR is the ratio of the total pressure at the turbine nozzle to the total pressure at the compressor inlet.

θ and δ, used to correct the operating parameters, are also based on the total temperature and pressure, respectively, at the compressor inlet.

$$\theta_{T2} = \frac{T_{t2}}{T_0} \tag{6.98}$$

$$\delta_{T2} = \frac{p_{t2}}{p_0} \tag{6.99}$$

where T_0 and p_0 are the standard sea level values of temperature and pressure. Assuming 100% ram pressure recovery, (i.e., that $M = 0$ at station 2), θ_{T2} and δ_{T2} can be calculated from

$$\theta_{T2} = \theta\,[1 + (\gamma - 1)\, M_\infty^2/2] \tag{6.100}$$

$$\delta_{T2} = \delta\,[1 + (\gamma - 1)\, M_\infty^2/2]^{\gamma/\gamma - 1} \tag{6.101}$$

The operating curves for the JT4A-3 turbojet are presented in Figures 6.38 and 6.39. Turbine discharge temperature, compressor speeds, and fuel flow are presented in Figure 6.38 as a function of EPR. Figure 6.39 shows the net thrust as a function of Mach number for constant values of EPR. All of the curves presented

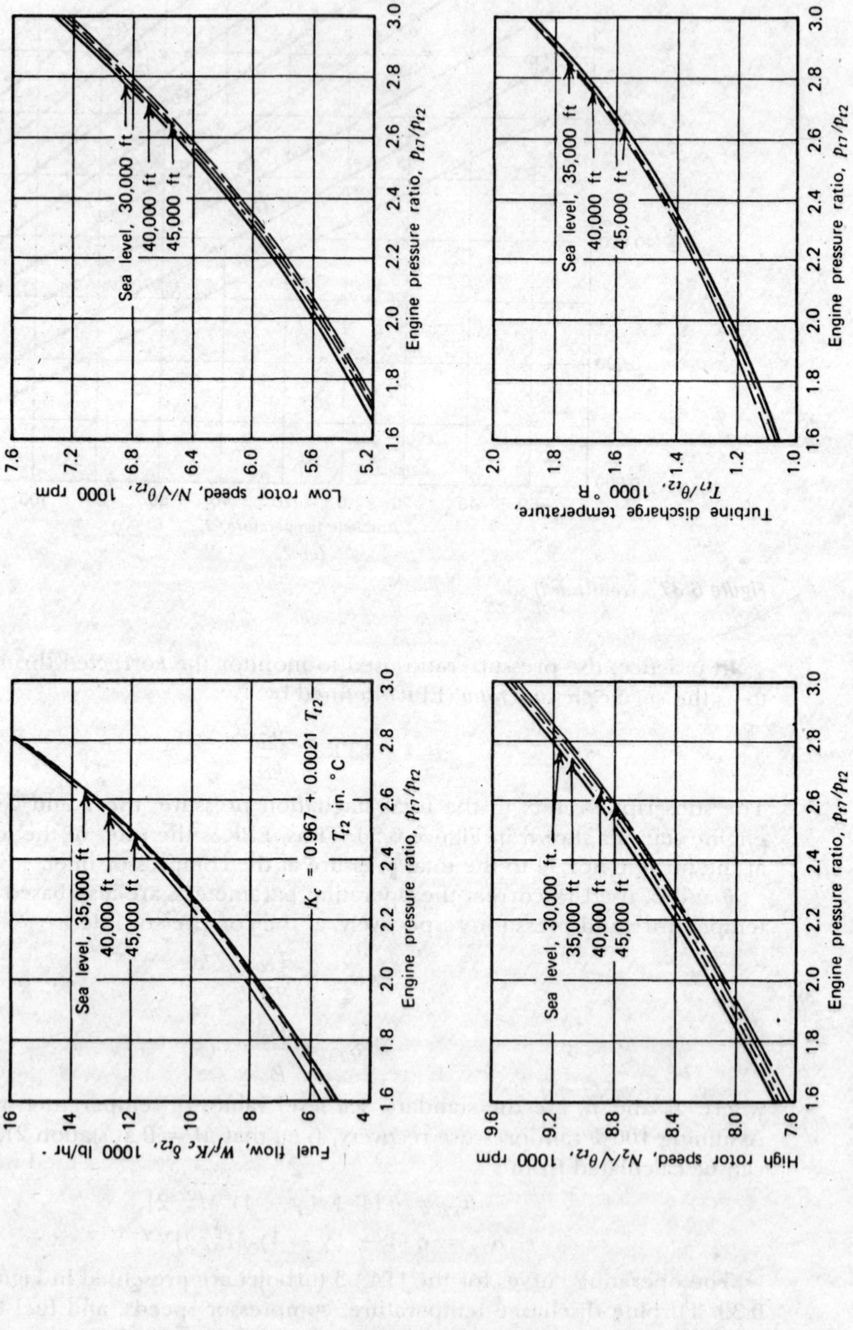

Figure 6.38 Operating curves for the JT4A-3 turbojet. (Courtesy, Pratt & Whitney).

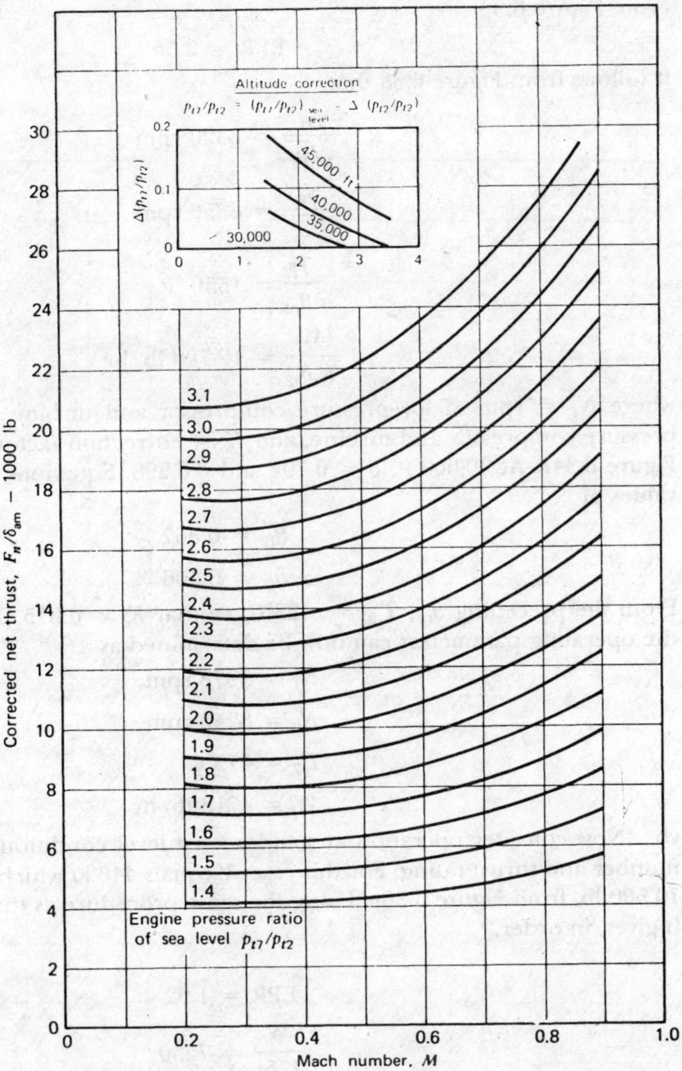

Figure 6.39 Net thrust for the JT4A-3 turbojet. 100% ram recovery. Standard exhaust nozzle. No airbleed. No power extraction.

thus far for the JT4A-3 engine assume 100% ram recovery (no inlet duct loss) and a standard nozzle installation prescribed by the manufacturer. They also assume zero power extraction or compressor air bleed. In an actual airplane installation, corrections must be made for these factors. The details of these corrections are too lengthy to be presented here.

As an example of the use of the performance curves presented thus far for the JT4A-3 engine, consider its operation at an airspeed of 400 kt at an altitude of 30,000 ft. For the maximum continuous thrust rating, a net thrust of 5300 lb is read from Figure 6.36*f*. Thus, for this altitude and airspeed,

$$\frac{F_n}{\delta_{am}} = 17{,}785$$

$$M_\infty = 0.679$$

From Figure 6.39,

$$\text{EPR} = 2.56$$

It follows from Figure 6.38 that

$$\frac{N_1}{\sqrt{\theta_{t2}}} = 6420 \text{ rpm}$$

$$\frac{N_2}{\sqrt{\theta_{t2}}} = 8850 \text{ rpm}$$

$$\frac{T_{t7}}{\theta_{t2}} = 1550\ °\text{R}$$

$$\frac{W_f}{K_c \delta_{t2}} = 12{,}700 \text{ lb/hr}$$

where N_1 = rpm of low-pressure compressor and turbine, N_2 = rpm of high-pressure compressor and turbine, and K_c = correction factor yet to be read from Figure 6.34*a*. At 30,000 ft, $\theta = 0.794$ and δ 0.298. Equations 6.100 and 6.101 give values of

$$\theta_{t2} = 0.867$$

$$\delta_{t2} = 0.406$$

From the preceding θ_{t2}, $T_{t2} = -23°\text{C}$, so that $K_c = 0.915$. The actual values for the operating parameters can now be determined as

$$N_1 = 5978 \text{ rpm}$$

$$N_2 = 8240 \text{ rpm}$$

$$T_{t7} = 473°\text{C}$$

$$W_f = 4718 \text{ lb/hr}$$

vb Now consider operation at standard sea level conditions at this same Mach number and thrust rating. For this case, V equals 448 kt which gives a net thrust of 10,500 lb. from Figure 6.36*a*. Using the same procedure as that followed at 30,000 ft gives, in order,

$$\text{EPR} = 1.92$$

$$\frac{W_f}{K_c \delta_{t2}} = 7250$$

$$\frac{N_1}{\sqrt{\theta_{t2}}} = 5490$$

$$\frac{N_2}{\sqrt{\theta_{t2}}} = 8070$$

$$\frac{T_{7t}}{\theta_{t2}} = 1190$$

$$\delta_{t2} = 1.362$$

$$\theta_{t2} = 1.092$$

Hence,

$$N_1 = 5737 \text{ rpm}$$

$$N_2 = 8433 \text{ rpm}$$

$$T_{t7} = 449°C$$

$$W_f = 10{,}418 \text{ lb/hr}$$

Note that the engine rotational speeds and exhaust gas temperature are approximately the same in both cases. Indeed, if other speeds and altitudes at the maximum thrust rating are examined, N_1, N_2, and T_{t7} values approximately equal to those just calculated are found. Thus, the thrust available from a turbojet engine at a given speed and altitude depends on the maximum stress and temperature levels that can be tolerated by the engine materials. As a result, the net thrust of a turbojet will not decrease with altitude in proportion to the density ratio, as with piston engines. As a rough approximation, one can assume F_n to be proportional to σ but, in practice, F_n will not decrease with altitude as rapidly as this approximation predicts. For the example just presented, one might predict a net thrust at 30,000 ft at 400 kt based on σ and the standard sea level value of F_n of 3941 lb. This value is 25.6% lower than the rated value previously noted. To illustrate further the accuracy of the approximation, Figure 6.40 presents the rated maximum continuous thrust at 200,400 and 600 kt as a function of altitude and compares this thrust with that obtained by multiplying the sea level values by σ. The approximation to F_n is seen to improve for the lower airspeeds and certainly predicts the proper trend. However, at the higher altitudes, the differences between the thrust curves are significant at all airspeeds.

Turbofan

The Pratt & Whitney PW4056 is typical of a modern turbofan engine. This engine and other models of the PW4000 series have many applications including the Boeing 747-400, Boeing 767-200/300, McDonnell-Douglas MD-11, Airbus A300-600, and the Airbus A310-300. Some of the characteristics of this engine are presented in Table 6.2.

Figure 6.41 presents the rated takeoff thrust for this engine as a function of ambient temperature for altitudes up to 6000 feet. The engine is seen to be flat-rated below ambient temperatures ranging from approximately 90° to 70°F depending upon the altitude and Mach number. The flat-rated thrust is seen to decrease approximately 17% in going from sea level to an altitude of 6000 ft.

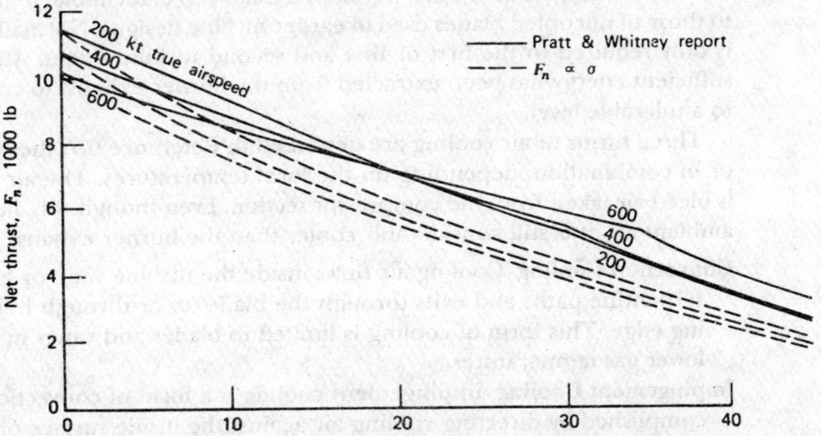

Figure 6.40 Estimated net thrust for the JT4A-3 turbojet at normal and maximum continuous ratings. (Courtesy, Pratt & Whitney)

Table 6.2 Characteristics of the PW4056 Engine

Type—Turbofan	
Sea-Level Standard Day	
Takeoff thrust	55,600 lb
Maximum continuous thrust	48,000 lb
Bypass ratio at takeoff	4.9
Airflow at takeoff	1700 lb/sec
Size	
Engine weight	9213 lb
Engine diameter	96.4 in.
Engine length	148.5 in.
Configuration	**Number of stages**
Fan	1
Low-compressor	4
High-compressor	11
High-turbine	2
Low-turbine	4

Figures 6.42 and 6.43 give the rated maximum climb thrust and cruise thrust, respectively, as a function of Mach number for constant values of altitude up to 45,000 ft. Both figures also include lines of constant TSFC values. At the lower altitudes, the net thrust is seen to decrease rapidly with Mach number. However, at the higher altitudes, the thrust is nearly constant and even increases slightly with Mach number above approximately 35,000 ft.

The range of operating Mach number decreases in the preceding figures at the higher altitudes. This is a reflection of the limitations of the operating envelope presented in Figure 6.44. Such an envelope can result from several limitations including temprature restrictions, stress limits, surge, and compressor stall.

Temperature restrictions are normally associated with the turbine inlet temperature (TIT). High-pressure turbines in the latest high bypass turbofan engines operate with gas temperatures in the 2000 to 2300°F (1094 to 1260°C) range. Various techniques have been developed that keep blade metal temperatures equal to those of uncooled blades used in earlier turbine designs. Normally, blade cooling is only required in the first or first and second turbine stages. After these stages, sufficient energy has been extracted from the burner exhaust to cool the hot gases to a tolerable level.

Three forms of air cooling are described in Reference 6.7; these are used singly or in combination, depending on the local temperatures. The air for this cooling is bleed air taken from the compressor section. Even though this air is warmer than ambient air, it is still considerably cooler than the burner exhaust.

Convection Cooling Cooling air flows inside the turbine vane or blade through serpentine paths and exits through the blade tip or through holes in the trailing edge. This form of cooling is limited to blades and vanes in the area of the lower gas temperatures.

Impingement Cooling Impingement cooling is a form of convection cooling, accomplished by directing cooling air against the inside surface of the airfoil through small, internal, high-velocity air jets. Impingement cooling is concentrated mostly at critical sections such as the leading edges of the vanes and blades.

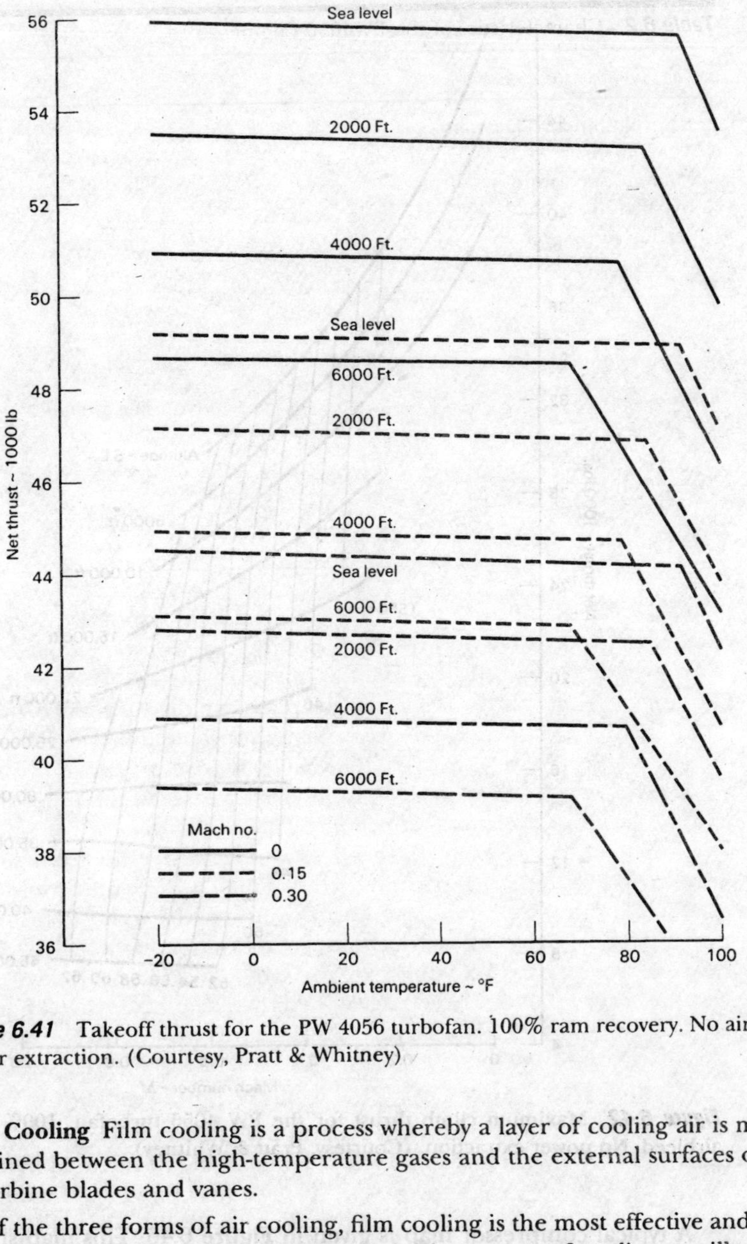

Figure 6.41 Takeoff thrust for the PW 4056 turbofan. 100% ram recovery. No airbleed. No power extraction. (Courtesy, Pratt & Whitney)

Film Cooling Film cooling is a process whereby a layer of cooling air is maintained between the high-temperature gases and the external surfaces of the turbine blades and vanes.

Of the three forms of air cooling, film cooling is the most effective and the least demanding as far as airflow is concerned. These types of cooling are illustrated in Figure 6.45, which shows their application to both stationary and rotating turbine stages.

Surge and compressor stall are related but are not the same thing. Surge refers to oscillations in the rotational speed of the entire engine. This surge is usually related to compressor stall, where the local angles of attack of the rotor blades, for various reasons, achieve sufficiently high values to cause local stalling. Some of these reasons include inlet airflow distortion from gusts, inlet design or uncoordinated maneuvering, rapid power changes, water ingestion, and Reynolds number effects.

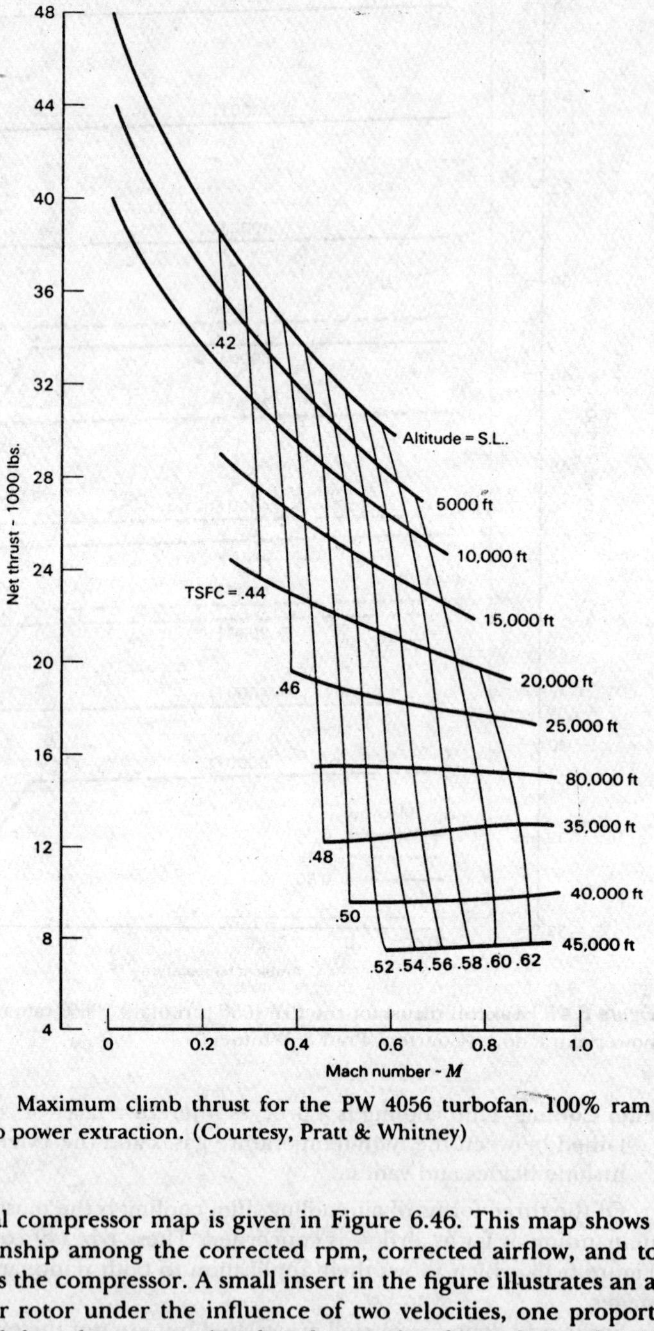

Figure 6.42 Maximum climb thrust for the PW 4056 turbofan. 100% ram recovery. No airbleed. No power extraction. (Courtesy, Pratt & Whitney)

A typical compressor map is given in Figure 6.46. This map shows qualitatively the relationship among the corrected rpm, corrected airflow, and total pressure ratio across the compressor. A small insert in the figure illustrates an airfoil on the compressor rotor under the influence of two velocities, one proportional to the airflow and the other proportional to the rotational speed. At a fixed blade angle, the angle of attack of this section obviously increases as N increases or W_a decreases. This is reflected in the map, which shows one approaching the surge zone as N increases for a constant W_a or as W_a decreases for a constant N.

As the altitude increases, the surge zone drops down, mainly because of Reynolds number effects. At the same time, the steady-state operating line moves up. Thus, compressor stall and surge are more likely to be encountered at the higher altitudes.

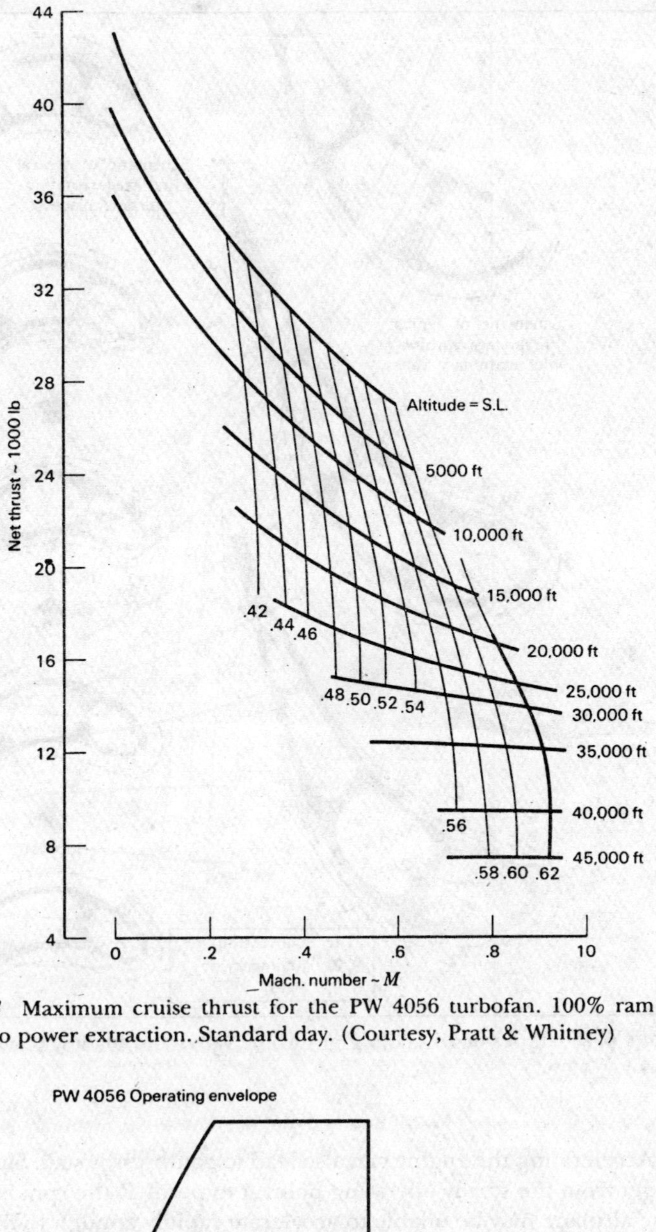

Figure 6.43 Maximum cruise thrust for the PW 4056 turbofan. 100% ram recovery. No airbleed. No power extraction. Standard day. (Courtesy, Pratt & Whitney)

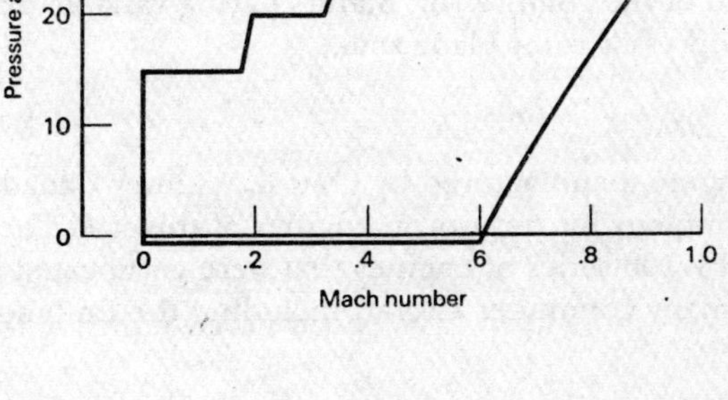

Figure 6.44 PW 4056 turbofan operating envelope. (Courtesy, Pratt & Whitney)

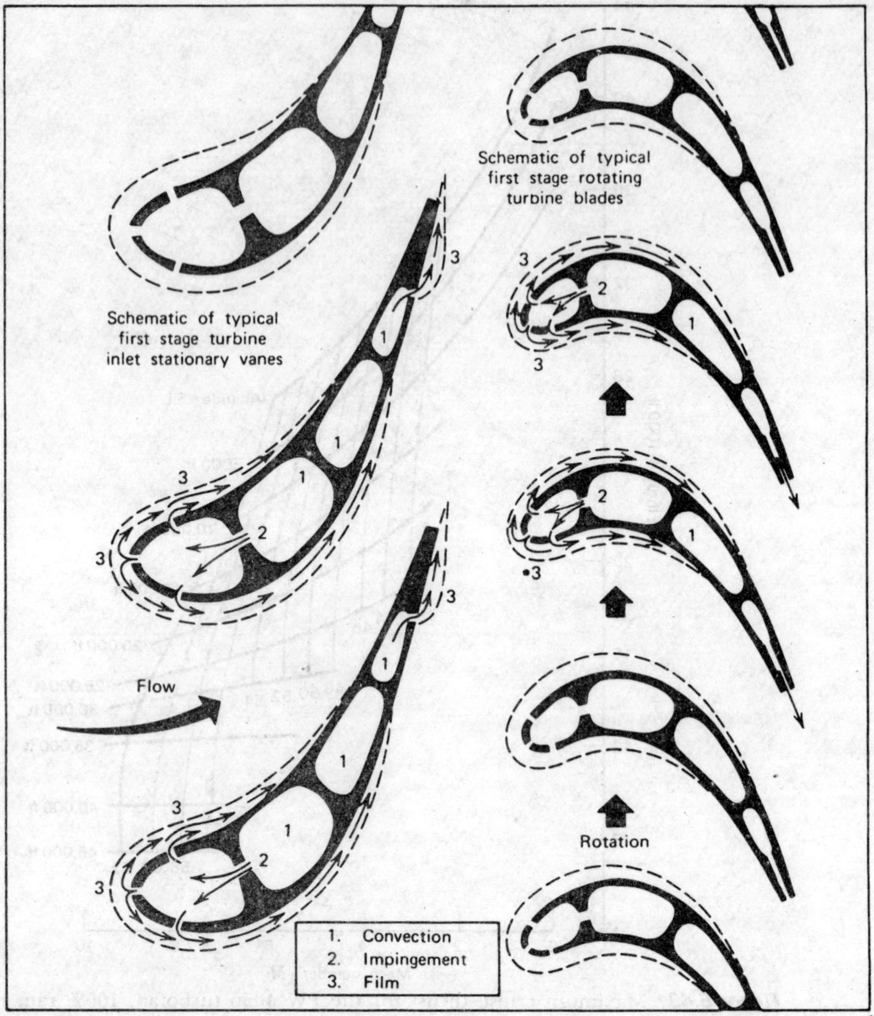

Figure 6.45 Types of air cooling of turbine vanes and blades. (Courtesy, General Electric Co.)

Accelerating the engine can also lead to compressor stall. Suppose, in attempting to get from the steady operating point *A* to point *B*, the rpm is suddenly increased. The airplane may be unable to accelerate rapidly enough to follow the rpm, so the airflow is less than the steady-state value. Surge can be alleviated by unloading the compressor during certain operating conditions. This is accomplished by bleeding air near the middle or end of the compressor. Stators having variable blade angles are also used to delay compressor rotor blade stall.

Turboprop

The PW120 turboprop engine manufactured by Pratt & Whitney Canada is representative of the latest technology for turboprop engines of this size. The PW120 is only one member of the PW100 series of engines that were certificated beginning in 1983 and selected for many commuter aircraft including the Embraer 120, the

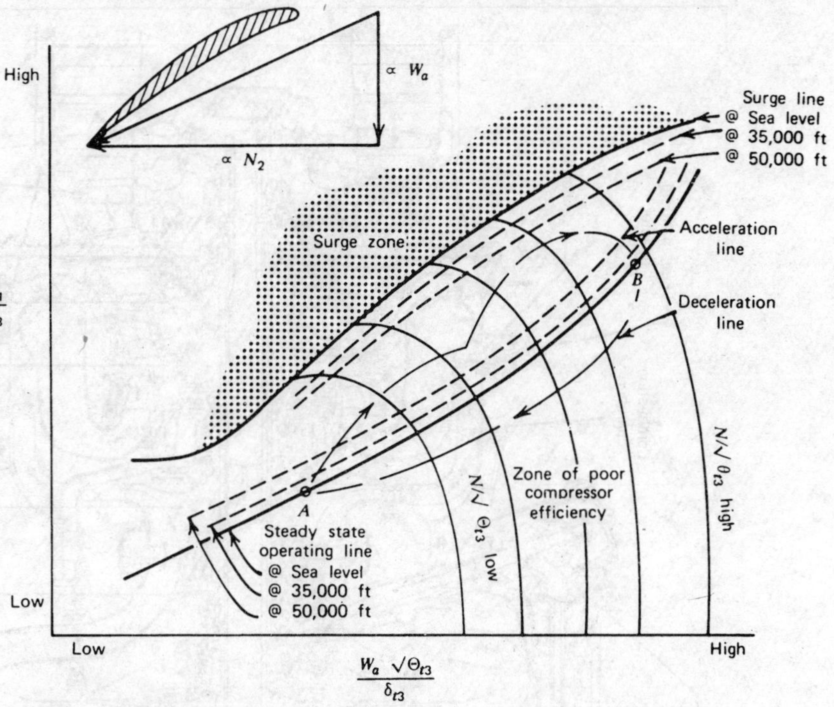

Figure 6.46 Typical compressor map.

De Havilland Dash-8 and the Aerospatiale ATR. The arrangement of the engine is shown in Figure 6.47. The PW120 is known as a free turbine engine since the shaft providing the power is driven by a turbine that is not connected to any compressors and is thus free to turn at its own speed. Proceeding from left to right in Figure 6.47, the inlet air is compressed by two centrifugal impellers, each driven by its own turbine. The air is guided from one stage to another by a series of curved pipes. Following the compression, the flow is turned forward, mixed with fuel, and ignited in the combustion chamber. The hot gases are turned again to the rear and expelled through the two turbines driving the impellers and then through the two-stage, free, power turbine.

The engine is compact with a length of 84 in., a width of 25 in., and a height of 31 in. The dry weight (no fluids) of the engine is only 921 lb. For this size and weight, one obtains the performance shown in Figures 6.48, 6.49, and 6.50.

Figure 6.48 presents the maximum take-off shaft horsepower (SHP) of the PW120 as a function of ambient temperature for altitudes from sea level to 6000 ft. Note that this engine, like the PW4056 turbofan is flat rated, in this case, to 2000 shp for takeoff. Fuel consumption curves for the take-off rating are also included in Figure 6.48. The maximum, continuous climb rating for the PW120 engine is shown in Figure 6.49 for altitudes from sea level to 10,000 ft. Again, fuel consumption curves are included. The maximum continuous cruise rating for the PW120 engine is given in Figure 6.50 along with the fuel consumption for altitudes from 5000 ft to 30,000 ft. These curves, together with those for the PW4056 turbofan will be used in the next chapter to predict the performance of airplanes in which they are installed.

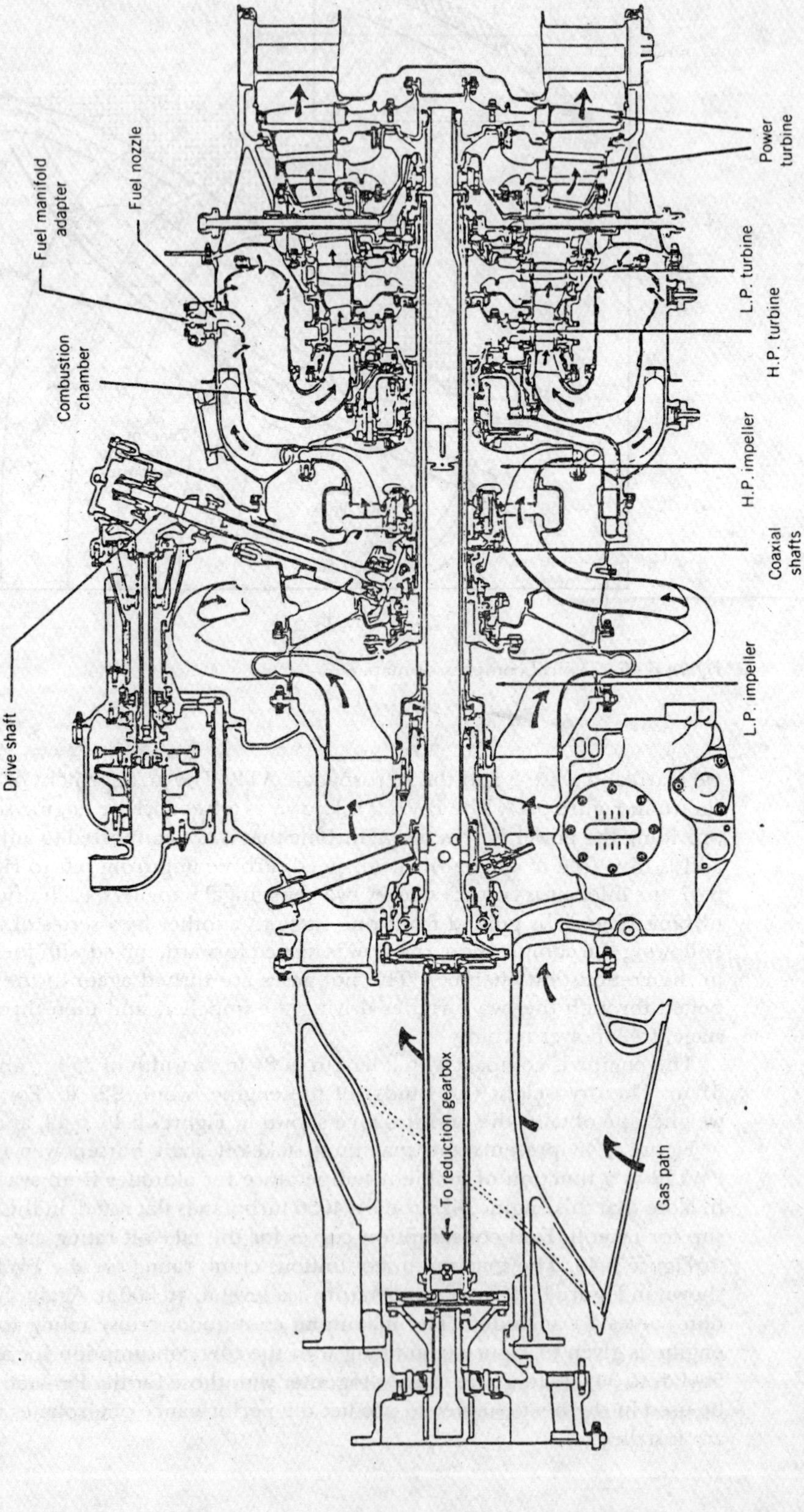

Figure 6.47 Schematic of PW 120 turboprop. (Courtesy, Pratt & Whitney, Canada)

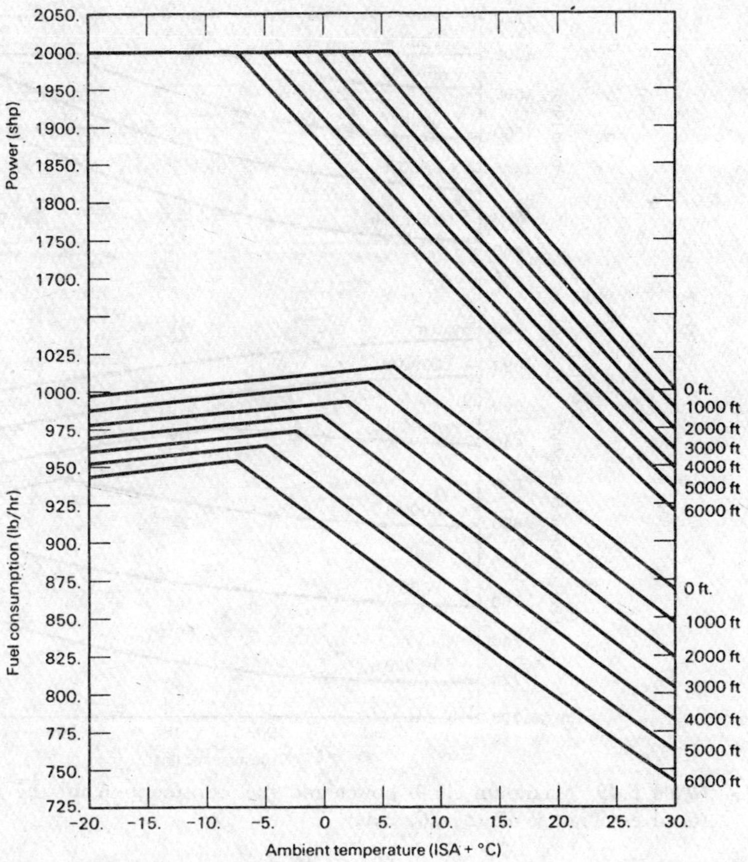

Figure 6.48 Maximum takeoff power and fuel consumption for the PW 120 turboprop. (Courtesy, Pratt & Whitney, Canada)

Installation Losses

The performance curves that have been presented for the JT4A-3 turbojet, the PW4056 turbofan, and the PW120 turboprop are all optimistic since they do not include installation losses. These losses result from

- Total pressure loss in the inlet ducting
- Total pressure loss in the exhaust nozzle
- Bleed air requirements
- Power extraction for accessories
- De-icing requirements

Methods for calculating these losses are not included here because of the extensive information that is required. In practice, an engine manufacturer will supply a database and computer codes to the airframe manufacturer to estimate these losses. Typically, these losses amount to approximately 0.4% for inlet, 5% for anti-icing and 8–22 hp per engine for accessories. To be more specific, for the PW120 in cruise, the power for the accessories amounts to approximately 9.5 shp for a typical commuter installation.

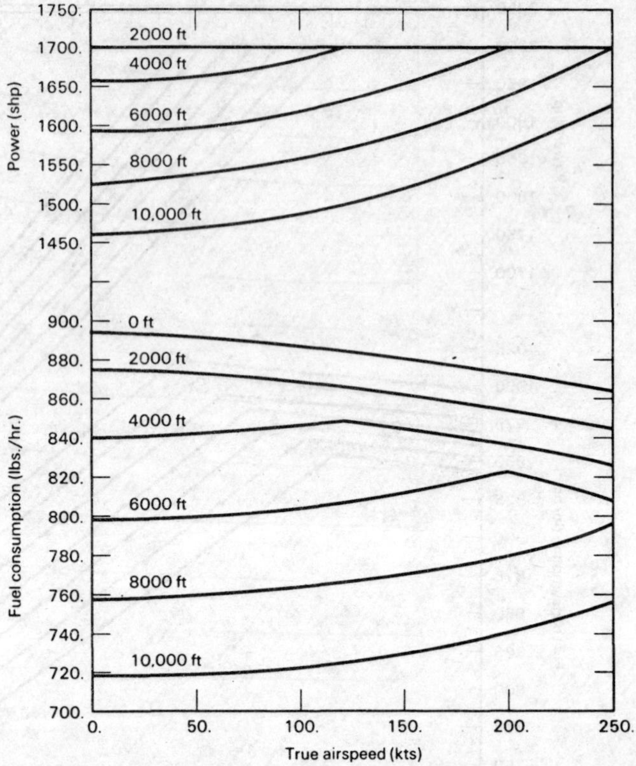

Figure 6.49 Maximum climb power and fuel consumption for the PW 120 turboprop. (Courtesy, Pratt & Whitney, Canada)

TRENDS IN AIRCRAFT PROPULSION

The title of this section was borrowed from an interesting paper (Ref. 6.11) by Rosen. The paper was published in 1971, a few years before the first edition of this book. It addressed the state of the art with regard to propulsion in that period and projected what the future might hold through 1980. It is interesting, in light of the fact that we are, at the time of this writing, 14 years beyond 1980 to compare today's technology with the trends shown by Rosen and to consider future trends. Our considerations will be limited to subsonic airspeeds. To do otherwise is beyond the limitations of this text. It may be beyond the price that society is willing to pay for speed with the emphasis on fuel economy, noise, and the environment. However, despite these concerns, at this time, a considerable effort is being devoted to hypersonic propulsion.

Regarding fuel consumption, Figure 6.51 presents the static TSFC as a function of net thrust for turbojets with afterburners, turbojets, and turbofans. The points represent engines that were operational in 1970. Today, with improved materials allowing for higher combustion temperatures, the TSFC values are somewhat lower than those shown on the graph, as will be discussed shortly. Generally, the trends shown by Figure 6.51 remain unchanged. There is a tendency, as with piston engines, for the specific fuel consumption to improve with size for any given engine type, particularly at the lowest thrust values. However, the important point is the obvious gain to be realized by going to higher bypass ratios (BPR).

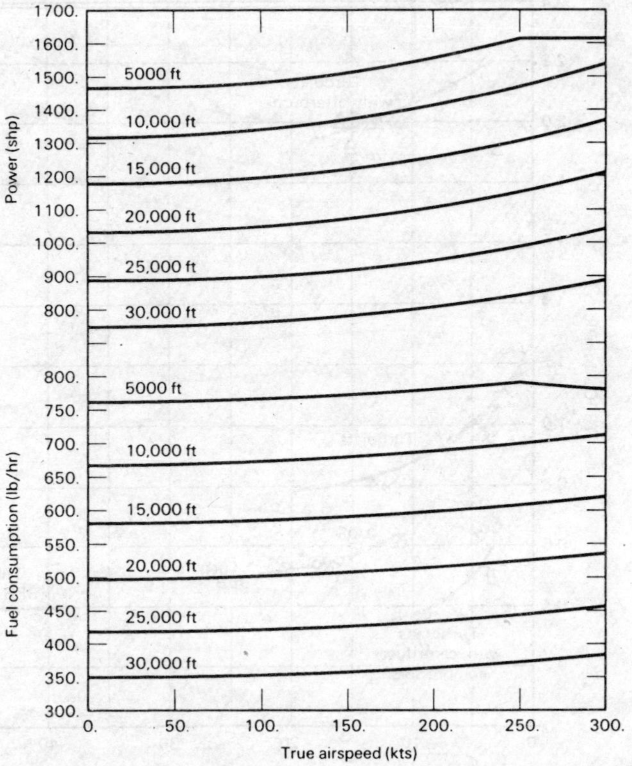

Figure 6.50 Maximum cruise power and fuel consumption for the PW 120 turboprop. (Courtesy, Pratt & Whitney, Canada)

The effect of BPR on performance is emphasized in a slightly different manner in Figure 6.52 based on a figure by Rosen. In this figure, a low BPR is approximately 1:1, whereas a high BPR is 5:1 or higher. It is interesting to note that the PW4056 and PW120 engines of today fall right in line with the trends shown by Rosen over a decade ago.

Turbine engine weight trends are presented in Figure 6.53, again based on Rosen's paper. Again, the PW4056 engine lies fairly well along the trend line shown by the reference. Reflecting on Figures 6.52 and 6.53, it is impressive to note that, since the introduction of the turbojet engine, the fuel required for a pound of thrust per hour for a turbofan engine has decreased by 28% compared to a turbojet engine, whereas the engine weight has more than halved.

It is interesting that, in a sense, the application of gas turbines to commercial aircraft propulsion has nearly completed a cycle. The sudden transition to the turbojet for commercial transportation in the 1950s introduced the air traveler to above-the-weather flying at significantly higher speeds with a power plant that was almost vibrationless. In doing so, the bypass ratio went from a high value, where most of the air goes through the propulsor as compared to the air that goes through the power plant, to a value of zero, where all of the air goes through the power plant. Over the years, the BPR has gradually increased, *but at no sacrifice in comfort or convenience to the passenger.* Indeed, today's high bypass ratio turbofan is quieter, consumes less fuel, and is relatively much lighter than the turbojet.

As noted in Figures 6.52 and 6.53, for a given BPR, there is a gradual improve-

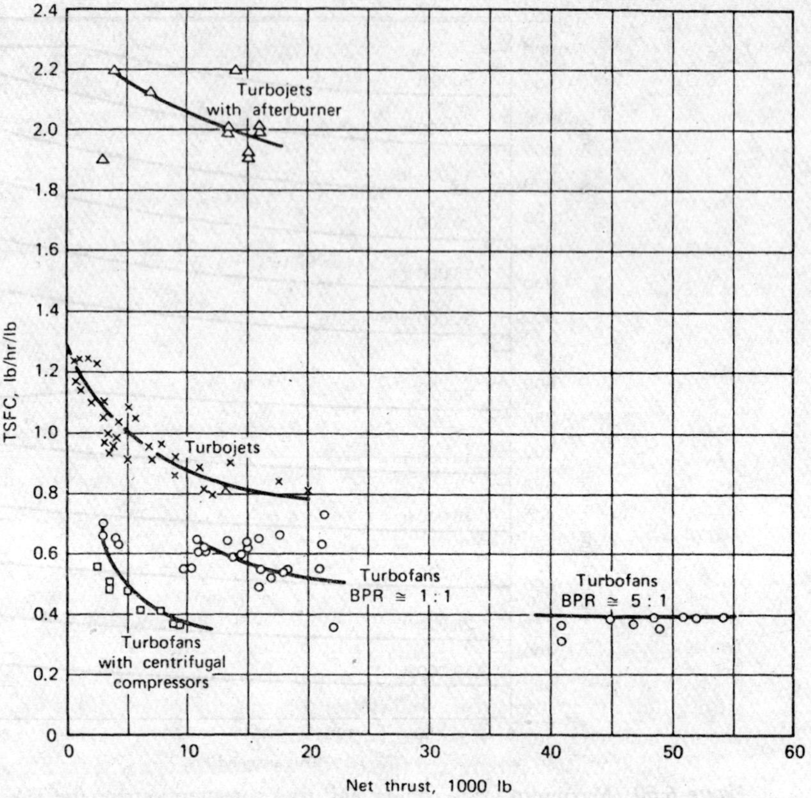

Figure 6.51 Static specific fuel consumption for turbojet and turbofan engines.

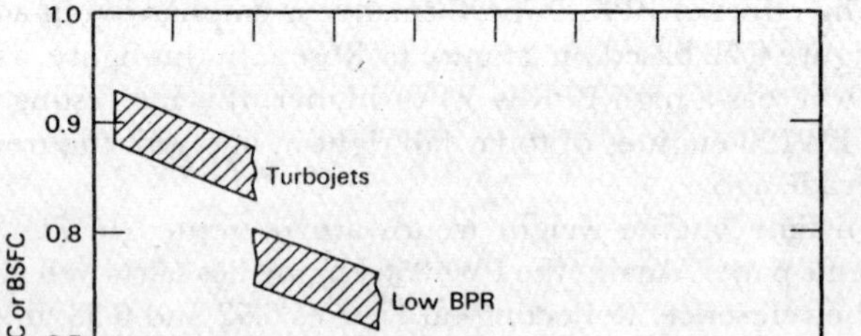

Figure 6.52 Trends in cruise specific fuel consumption.

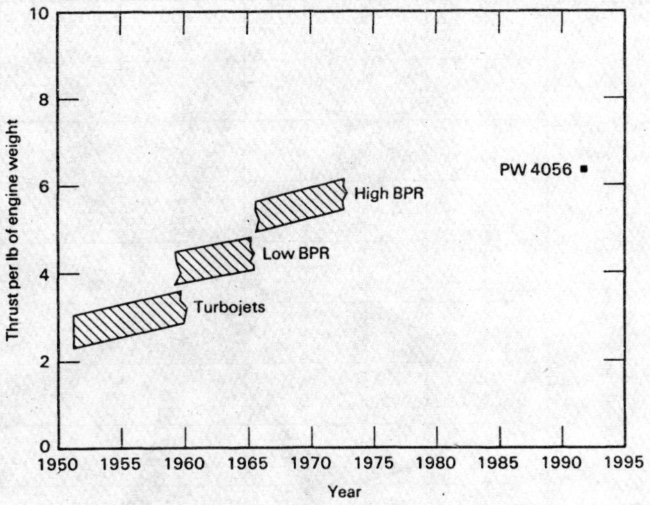

Figure 6.53 Trends in specific thrust.

ment in engine performance with time. This improvement is the result of better materials and cooling techniques, which allow operation at higher pressure ratios and turbine inlet temperatures. It would therefore appear that improved propulsion efficiency in the future will depend on further increases in the pressure ratio, turbine inlet temperatures, and bypass ratios.

During the fuel embargo of the 1970s when fuel prices escalated rapidly, research was begun by NASA and others on advanced turboprop propellers (ATP) for application to high subsonic speed. In effect, a turboprop is simply an extension of the turbofan to a higher bypass ratio. Generally, however, propellers operating at high subsonic Mach numbers suffer a serious loss in efficiency as the result of compressibility effects. Thus, the beneficial effect of the higher BPR is lost. The compressibility losses, together with noise and aeroelastic problems, were the major obstacles that NASA had to overcome in the development of a successful ATP.

Reference 6.12 discusses the design philosophy of these propellers and reports on some early test results. These propellers are multiblades and incorporate thin, transonic airfoil sections. One such propeller that has been tested by NASA's Lewis Research Center is shown in Figure 6.54. The design characteristics for this propeller are

0.8 cruise Mach number
10.7-km (35,000-ft) altitude
8 blades
203 activity factor per blade
301 kW/m^2 (37.5 hp/ft^2) power loading
243.8 m/s (800 fps) tip speed
0.08 integrated design lift coefficient

Test data obtained in Lewis's supersonic wind tunnel (see, e.g., Figure 6.55) shows that the goal of an efficiency of 80% at a Mach number of 0.8 can be met. Subsequent to the wind tunnel testing, ATP configurations were flight tested by Boeing and McDonnell-Douglas. However, currently, there is little activity being devoted to the ATP for several reasons. Fuel prices have declined and stabilized. The noise produced by an ATP is relatively high and presents a problem. Economic

Figure 6.54 Advanced turboprop design for operation at 0.8 Mach number. (Courtesy, NASA Lewis Research Center)

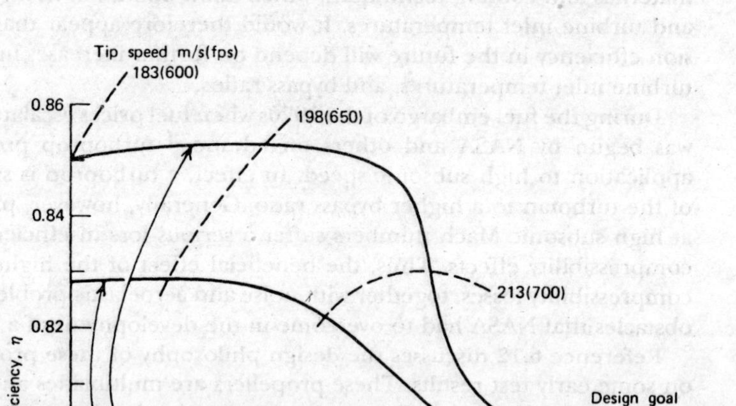

Figure 6.55 Preliminary test data on advanced turboprop from the NASA Lewis Research Center 8- by 6-ft wind tunnel.

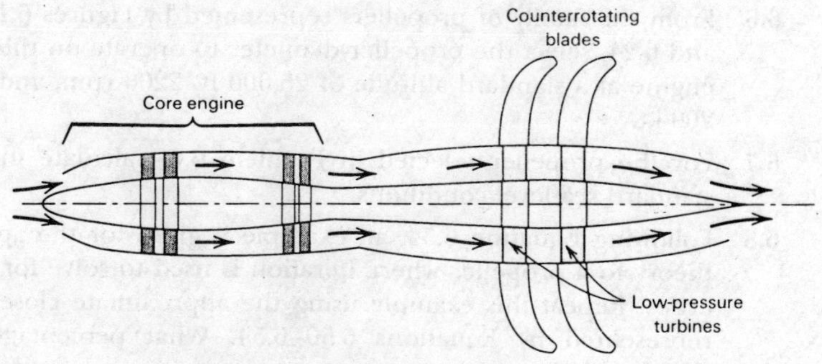

Figure 6.56 A schematic drawing of the UDF concept.

conditions are such that no one is willing to take the financial risk that is required to develop the engine and airplane. Finally, there is the question of whether or not the traveling public will accept the return to a propeller-driven airplane. The last question apparently so bothered the General Electric Co. that they named their entry into this area an "Unducted Fan" known only as UDF.

The UDF is an interesting concept, evolved by GE, that is worthy of note (shown schematically in Figure 6.56). A gas producer, typical of a turbojet engine, is placed ahead of the propeller blades. There are two, counterrotating rows of propeller blades, which are attached directly to low-speed turbines, which are driven by the exhaust from the gas producer. The concept has two attractive features, aside from the effective high BPR from the propeller. First, the direct coupling of the propeller blades to the low-speed turbine negates the need for a gear box. Second, the counterrotating propellers offer an appreciable efficiency gain compared to a single propeller and also apply no torque to the airframe.

If the ATP, in some configuration, can be successfully developed, it offers a reduction of approximately 30% in fuel consumption. However, aerodynamic, structural, noise, and economic problems must be solved before these savings can be realized.

PROBLEMS

6.1 A light airplane powered by the Lycoming 0-360 engine (Figure 6.3) and the propeller of Figure 6.18 is cruising at a pressure altitude of 1500 m. The OAT is 20°C. In addition, rpm = 2400, MAP = 61 cm hg, and IAS = 60 m/s. Calculate the fuel flow and propeller thrust.

6.2 An airplane powered by a single JT4A-3 turbo jet engine has a maximum cruising speed of 500 kt at a standard altitude of 6000 m. What is the airplane's drag in Newtons and at what rate is fuel being expended?

6.3 A JT4A-3 turbojet engine is operating at a pressure altitude of 12,000 m. The ambient temperature is −70°C. At its maximum continuous thrust rating, calculate fuel flow, turbine discharge temperature, and high rotor rpm at a true airspeed of 450 kt.

6.4 Two PW4056 turbofan engines power a transport airplane. At maximum climb power, at what rate is fuel being consumed as the airplane climbs through an altitude of 3500 m at a true airspeed of 300 kt? Assume a standard atmosphere and neglect installation losses.

6.5 Determine the maximum cruise thrust and fuel flow for the PW4056 turbofan engine at a standard altitude of 9 km and an indicated airspeed of 130 m/s.

6.6 From the family of propellers represented by Figures 6.12, 6.13, 6.14, 6.15, and 6.24, select the propeller diameter to operate on the PW120 turboprop engine at a standard altitude of 25,000 ft, 2200 rpm, and a true airspeed of 200 kt.

6.7 For the propeller selected in Problem 6.6, calculate the takeoff thrust at standard sea level conditions.

6.8 Following Equation 6.54, an example is given for the application of vortex theory to a propeller where iteration is used to solve for the induced properties. Repeat this example using the approximate closed-form expressions represented by Equations 6.50–6.54. What percentage error results in $\frac{dC_T}{dx}$ and $\frac{dC_p}{dx}$ from using the small angle approximation?

6.9 Estimate the effective BPR for the PW120 at maximum cruise power at 20,000 ft and 300 kt given a propeller diameter of 7.5 ft and a propeller efficiency of 87%. Assume that the ratio by weight of air flow to fuel flow through the gas producer is 300:1.

6.10 This is an open-ended problem. A single-engine airplane has a gross weight of 4000 lb, a flat plate area of 7 ft^2, Oswald's efficiency factor of 0.7, and a wing loading of 40 psf. The airplane is designed to cruise at 200 mph at 10,000 ft at 2300 rpm. Select the design variables for the propeller for this airplane.

REFERENCES

6.1 Torell, Bruce N., *The Significance of Propulsion in Commercial Aircraft Productivity,* The Seventeenth Sir Charles Kingsford-Smith Memorial Lecture, September 15, 1975, Royal Aeronautical Society, Sydney, Australia.

6.2 Wilkinson, Paul H., *Aircraft Engines of the World,* Published annually by Paul H. Wilkinson, Washington, D.C.

6.3 Hartman, E. P., and Biermann, David, *The Aerodynamic Characteristics of Full-Scale Propellers Having 2, 3, and 4 Blades of Clark Y and R.A.F. 6 Airfoil Sections,* NACA Report 640, November 1937.

6.4 Neumann, G., *Powerplants—Past, Present and Future,* The Fifth William Littlewoval Memorial Lecture, SAE SP-398, 1975.

6.5 Taylor, John W. R., and Munson, Kenneth, *History of Aviation,* Crown, New York, 1972.

6.6 Anonymous, *The Aircraft Gas Turbine Engine and its Operation,* PWA Operating Instruction 200, Pratt & Whitney Aircraft, East Hartford, CT, June 1952 (rewritten August 1970).

6.7 Anonymous, *Aircraft Gas Turbine Guide,* The General Electric Aircraft Engine Group, Cincinnati, Ohio/Lynn, MA, AEG-607, April 1972.

6.8 Anonymous, *JT9D Commercial Turbofan Engine Installation Handbook,* Pratt & Whitney Aircraft, East Hartford, CT, March 1967.

6.9 Hesse, W. J., and Mumford, N. V. S., *Jet Propulsion for Aerospace Applications,* 2nd edition, Pitman, New York, 1964.

6.10 Anonymous, *PT6 Gas Turbine Installation Handbook,* Pratt & Whitney Aircraft of Canada, Longueuil, Quebec, Canada, revised July 1977.

6.11 Rosen, George, *Trends in Aircraft Propulsion,* CASI Paper No 27/10, 12th Anglo-American Aeronautical Conference, July 1971.

6.12 Mikkelson, D. C., Blaha, B. J., Mitchell, G. A., and Wikete, J. E., *Design and Performance of Energy Efficient Propellers for Mach 0.8 Cruise,* NASA TM X-73612, presented at 1977 National Business Aircraft Meeting, Wichita, KS, March 29, 1977 (SAE Paper No. 770458).

7 AIRPLANE PERFORMANCE

INTRODUCTION

When a customer buys an airplane, whether it be a private individual, a corporation, an airline, or the military, the buyer wants to know what the airplane will do. How fast will it fly, how high, and how far? How long a runway is required from which the airplane will operate? How expensive will it be to operate, and what are the operating limitations? How fast will it climb? Will it take a half hour to get up to cruising altitude or only 5 min? This chapter provides methods for answering these questions and others related to the general subject of airplane performance. The groundwork for doing so has been presented in the preceding chapters. With the use of this material, one can calculate lift, drag, and thrust. Aside from weight, these are the principal forces acting on an airplane that determine its performance.

The material in this chapter is presented in the same sequence one encounters in an airplane flight. First the determination of takeoff distance is covered. Next, the rate-of-climb and time-to-climb to a given altitude is explained. This is followed by the calculation of the distance, called the *range*, which can be flown at that altitude for a given amount of fuel. Finally, the descent and landing is considered. These parts of a flight are the topics that one normally relates to airplane performance. However, in addition to these topics, other topics will be covered that relate to airplane design and operations.

TAKEOFF

The takeoff of an airplane certified in the transport category is illustrated in Figure 7.1. Starting from a resting position at the far left, the airplane accelerates under takeoff power. At some point the velocity exceeds the stalling speed, V_s. Beyond this point, the airplane is capable of flying. However, the airplane continues to accelerate on the ground until the *minimum control speed,* V_{mc}, is reached. At this speed, if a critical engine fails, the manufacturer has demonstrated that the airplane is able to maintain straight flight at that speed with zero yaw or with a bank angle of less than 5°. Under these conditions at this speed, the required rudder force may not exceed 180 lb. Continuing to accelerate on the ground, the airplane reaches a calibrated airspeed of V_1, the *critical engine failure speed.* This speed may not be less than V_{mc} and represents the speed at which the *average* pilot could safely continue with the takeoff in the event of a critical engine failure. At a speed that can equal V_1 but that must be 5% higher than V_{mc}, the pilot rotates the airplane. This speed, V_R, is called *takeoff rotation speed.*

Because of tail interference with the ground, the angle of attack at V_R may not be sufficient to lift the airplane. The pilot therefore continues to accelerate up to a speed of V_{mu}, the *minimum unstick speed.* At this speed, the pilot could lift the airplane off the runway and continue the takeoff, even with one engine inoperative, without any hazardous characteristics. However, to provide an additional margin of safety, the airplane continues to accelerate to the *lift-off speed,* V_{LOF}, at which point the airplane becomes airborne. V_{LOF} must be at least 10% higher than V_{mu},

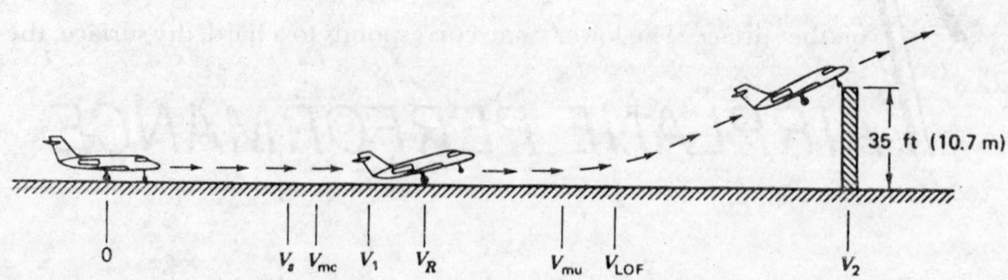

Figure 7.1 FAR Part 25 takeoff.

with all engines operating or 5% higher than V_{mu} with one engine inoperative. After lift-off, the airplane continues to accelerate up to the *takeoff climb speed*, V_2. V_2 is the speed attained at a height of 35 ft (10.7 m) above the ground. V_2 must be greater than 1.2 V_s in the takeoff configuration *and* 1.1 V_{mc}.

This description, applicable to the takeoff of a turbojet or turbofan transport, is in accordance with the definition of these various speeds as presented in the Federal Air Regulations (FAR) Part 25. These regulations govern the airworthiness standards for airplanes in the transport category. Similar regulations for other categories of nonmilitary airplanes can be found in FAR Part 23. The Cherokee 180, which has been used as an example in the preceding chapters, is certified under FAR Part 23. The total horizontal distance, ground roll and airborne, which is required to reach the altitude of 35 ft, starting from rest, is referred to as the FAR takeoff distance.

FAR Part 23 is simpler in specifying the takeoff procedure. For airplanes over 6000 lb (26,700 N), maximum weight in the normal, utility, and acrobatic categories, it is stated simply that the airplane must attain a speed at least 30% greater than the stalling speed with one engine out, V_{S_1}. For an airplane weighing less than 6000 lb, the regulations state simply that the takeoff should not require any exceptional piloting skill. In addition, the elevator power must be sufficient to lift the tail (for a "tail dragger") at 0.8 V_{S_1} or to raise the nose for a nose-wheel configuration at 0.85 V_{S_1}.

Ground Roll

The forces acting on an airplane during the ground roll portion of the takeoff are shown in Figure 7.2. From Newton's second law of motion,

$$T - D - \mu(W - L) = m\dot{V} \tag{7.1}$$

μ is the coefficient of rolling friction, and a dot above the V denotes differentiation with respect to time. μ values can range from approximately 0.02 to 0.1, depending

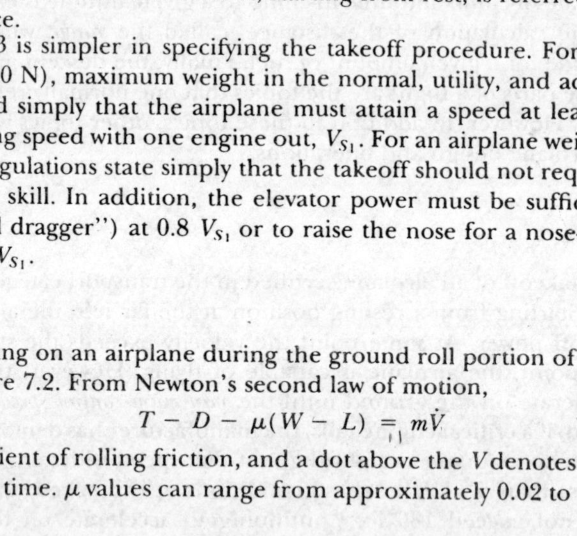

Figure 7.2 Forces on an airplane during the ground roll.

on the surface. The lower value corresponds to a hard, dry surface; the higher value might correspond to moderately tall grass.

Before rotation, the attitude of the airplane on the ground is constant and hence, C_L and C_D are constant. After rotation, C_L and C_D increase, but still remain constant until lift off. The most direct means of solving Equation 7.1 is numerically, using a digital computer. This will require a table lookup or a curve fit for T as a function of D. After V is obtained as a function of time, it can then be numerically integrated to obtain s so that V will be known as a function of s.

A word of caution regarding C_D is in order. Ground effect may reduce the induced drag significantly. Hence, C_D as a function of C_L is less during the ground roll than it is in the air. In view of this, a fairly good approximation is to neglect the induced drag for calculating the total airplane drag during the ground roll, particularly for tricycle landing gears where the wing is nearly level during the ground roll.

For some configurations, such as airplanes designed for STOL operations, neglecting C_{Di} during takeoff may not be satisfactory. In this case, a correction to C_{Di} can be easily obtained by modifying the lifting line model written as an exercise in Chapters 3 and 4 to include ground effect. Consider Figure 7.3, which portrays, in a transverse plane, the two legs of a trailing horseshoe vortex of strength $\gamma(I)$ a distance h above ground. The potential flow for this figure must be found such that the velocity normal to the ground is zero. This can be obtained by placing an *image* vortex system of opposite strength a distance of h below the ground. From symmetry, the ground will then be a streamline, which can be replaced by a solid boundary.

The velocity induced at the point I along the bound vortex by the trailing vortices of the horseshoe vortex centered at the point J and trailing to infinity from the bound vortex can be written from the Biot–Savart law as

$$\Delta w(I) = \frac{\gamma(J)}{4\pi}\left[\left(\frac{1}{Y1 - Y(I)}\right) - \left(\frac{1}{Y2 - Y(I)}\right)\right] \quad (7.2)$$

The velocity induced at I by the image system at J must be resolved in the vertical direction and therefore is slightly different in form from the above.

$$\Delta w(I) = -\frac{\gamma(J)}{4\pi}\left[\left(\frac{Y1 - Y(I)}{R1^2}\right) - \left(\frac{Y2 - Y(I)}{R2^2}\right)\right] \quad (7.3)$$

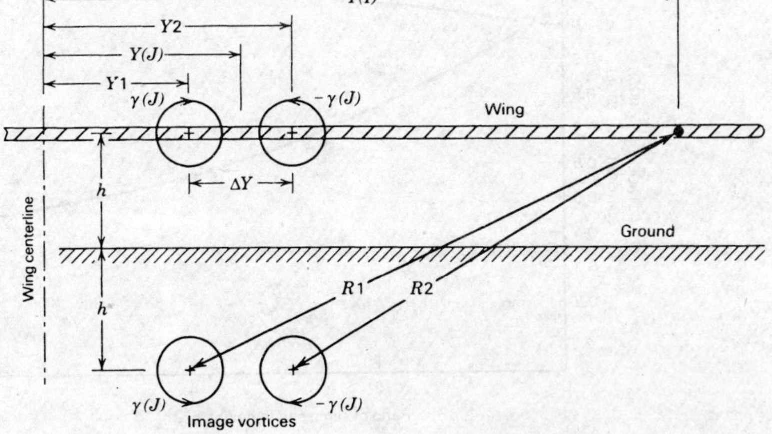

Figure 7.3 Trailing horseshoe vortex and image below the ground.

The sum of Equations 7.2 and 7.3 for a $\gamma(J)$ of unity will therefore be the influence coefficient, $w(I,J)$, for application to a lifting line model, which includes ground effect.

Computer Exercise 7.1 *LIFTLINE (modified for ground effect)*

Modify the lifting line model written as computer exercises in Chapters 3 and 4 to include ground effect. Run the program for an elliptic wing with an aspect ratio of 6 using 100 horseshoe elements across the span and compare your results with Figure 7.3.

The graph of Figure 7.4 is based on the above modification to the lifting line model covered earlier in Chapters 3 and 4. This figure presents the ratio of the induced drag in-ground-effect (IGE) to the induced drag out-of-ground-effect (OGE) at the same lift coefficient as a function of the reciprocal of the ratio of the height of the wing above the ground to the semispan. The ratio of the lift coefficient, IGE, to C_L, OGE at a constant angle of attack is also given in the figure.

The first edition of this text did not rely on computational methods to the extent of this second edition. Therefore, a closed form approximation to Figure 7.4 was obtained simply by determining the effect of the ground on the downwash midway between a pair of vortices representing a completely rolled-up vortex system. Image vortices, similar to those pictured in Figure 7.3, were used to account for the ground. The result of this approximate model is given by

$$\frac{C_{D_i}(IGE)}{C_{D_i}(OGE)} = \frac{16\left(\dfrac{h}{b_v}\right)^2}{1 + 16\left(\dfrac{h}{b_v}\right)^2} \tag{7.4}$$

where b_v is the spanwise distance between the centers of the vortex pair. This approximation agrees almost exactly with Figure 7.4 for heights greater than a

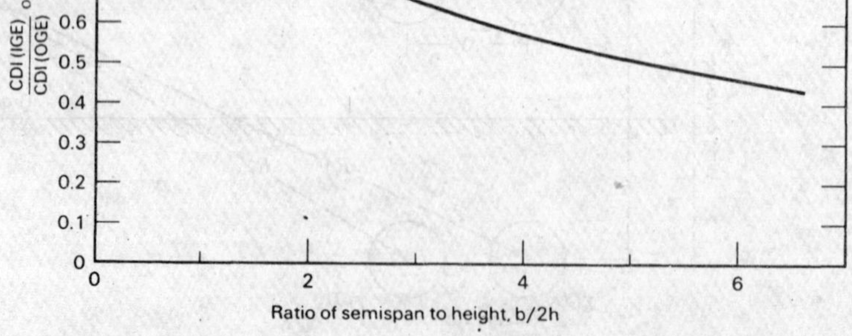

Figure 7.4 Effect of proximity to ground on the lift coefficient and induced drag coefficient.

semispan. However, at a height equal to $\frac{1}{3}$ of a semispan, Equation 7.4 is approximately 20% lower than Figure 7.4. Below a height less than this, Equation 7.4 should be used with caution. Observe from Figure 7.4 for a typical height above the ground of 10% of the span for takeoff that the induced drag is only half of the OGE drag for the same lift coefficient.

During the ground roll an increment to the parasite drag is required for airplanes with retractable landing gear. This increment can be estimated using the material presented in Chapter 4 or on the basis of Figure 7.5. (Reference 7.4.) This figure presents the equivalent flat plate area, f, as a function of gross weight for three different types of landing gear.

The Boeing 767-300ER powered with two PW4056 engines is shown in Figure 7.6 and will be used as an example for the calculation of takeoff distance. It is emphasized that the numbers used in this example are strictly estimates by your author and are not sanctioned by the Boeing Co.

The equation of motion along the runway is numerically integrated using a Taylor series.

$$V(t + \Delta t) = V(t) + a(t)\,\Delta t \tag{7.5}$$

$$s(t + \Delta t) = s(t) + V(t)\,\Delta t + a(t)\,\frac{(\Delta t)^2}{2} \tag{7.6}$$

where t is the time, a is the acceleration, V is the velocity, and s is the ground-roll distance, all in consistent units.

In order to do the numerical integration, the takeoff thrust curve for the PW4056 engine at sea level is fitted by

$$T = 1000\left[55.60 - 4.60\left(\frac{V}{100}\right) + 0.357\left(\frac{V}{100}\right)^2\right]$$

where T is in pounds and V is in f/s.

The following values were used in predicting the ground-roll for the 767 at standard sea level conditions.

Maximum takeoff gross weight, W	387,000 lb
Flat plate area, f	127.5 ft^2
Wing planform area	3084 ft^2
Wing span	156.08 ft
Oswald's efficiency factor	0.7
$\dfrac{C_{D_i}\,(IGE)}{C_{D_i}\,(OGE)}$	0.4
C_L on landing gear with flaps down (IGE)	1.0

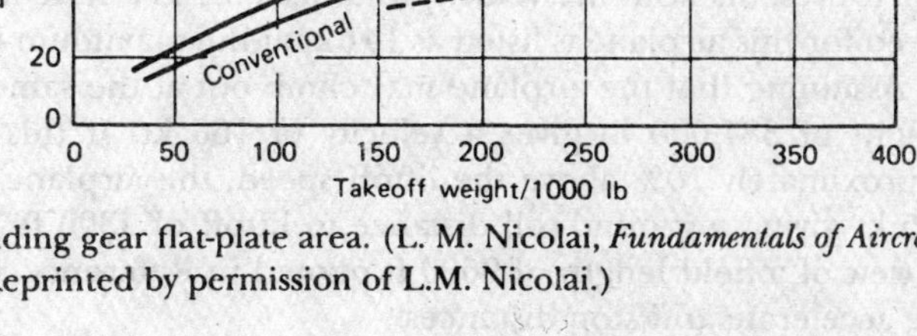

Figure 7.5 Landing gear flat-plate area. (L. M. Nicolai, *Fundamentals of Aircraft Design*, L. M. Nicolai, 1975. Reprinted by permission of L.M. Nicolai.)

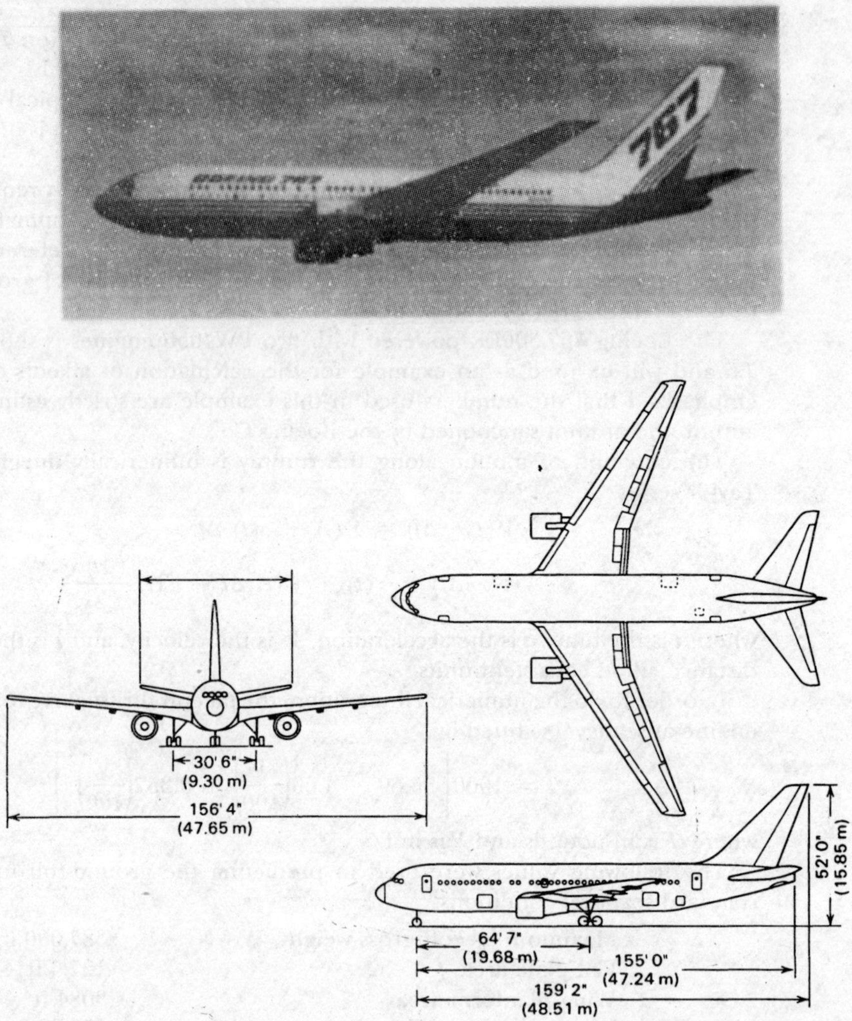

Figure 7.6 Boeing 767-200.

The flat plate area above resulted from assuming a skin friction coefficient, C_f, equal to 0.0035 (see Table 4.2) and estimating the total wetted area at 12,138 ft^2. Added to this is 65 ft^2. for the landing gear and 20 ft^2 for the flaps. The lift coefficient of 1.0 is assumed based on the fact that the takeoff is made with partially deflected flaps. One can estimate the C_L of the wing, flaps up, from the fact that the wing incidence is set to keep the fuselage approximately level in cruise and on the ground.

Integrating Equation 7.1 using the numbers given above results in Figure 7.7, which presents both the velocity and distance as a function of time. The approach speed for this airplane is listed as 141 kt at the maximum landing weight of 300,000 lb. Assuming that the airplane may climb out at the same C_L at the higher takeoff weight of 387,000 lb gives a velocity of 160 kt. If this speed is assumed to be approximately 10% above the liftoff speed, the airplane should lift off at around 145 kt giving a ground-roll distance to liftoff of 4300 ft. This figure is reasonable in view of a field length of 9600 ft quoted in Reference 5.11, which includes both the accelerate and stop distances.

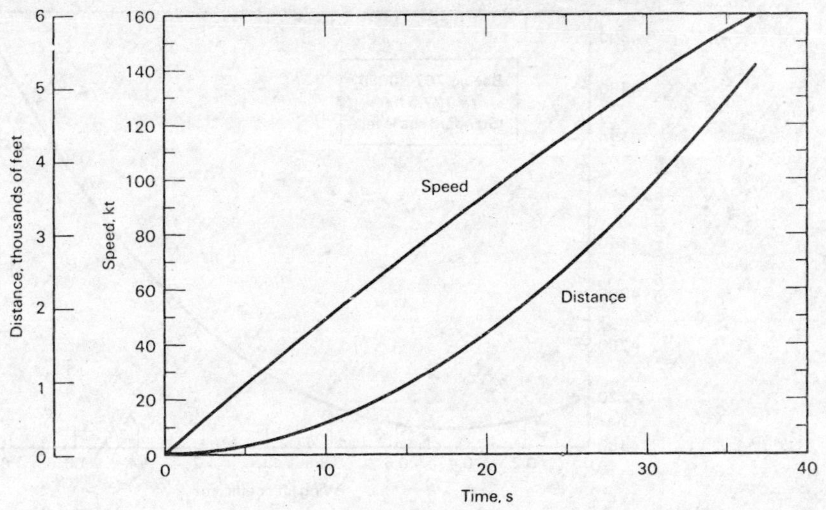

Figure 7.7 Estimated speed and ground roll for Boeing 767-300ER with PW4056 engines.

Since f and C_L are uncertain, Figures 7.8 and 7.9 were prepared to show the sensitivity of the ground-roll distance to these parameters for a constant lift-off speed of 145 kt. The ground-roll distance is seen to increase almost linearly with f, whereas the variation with C_L is highly nonlinear. This latter variation is due to the fact that, at a constant velocity, increasing C_L increases the induced drag but decreases the rolling friction due to the increased lift.

Computer Exercise 7.2 "TAKEOFF"

Write a computer program to predict the ground-roll distance, velocity, and acceleration as a function of time. Input the parameters given for the Boeing 767-300ER and check your results against the appropriate figures.

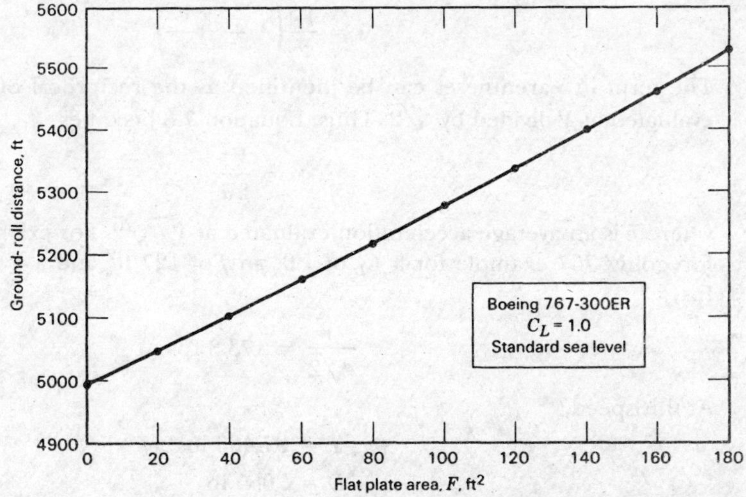

Figure 7.8 Effect of flat-plate area on predicted ground-roll distance to reach a speed of 145 kt.

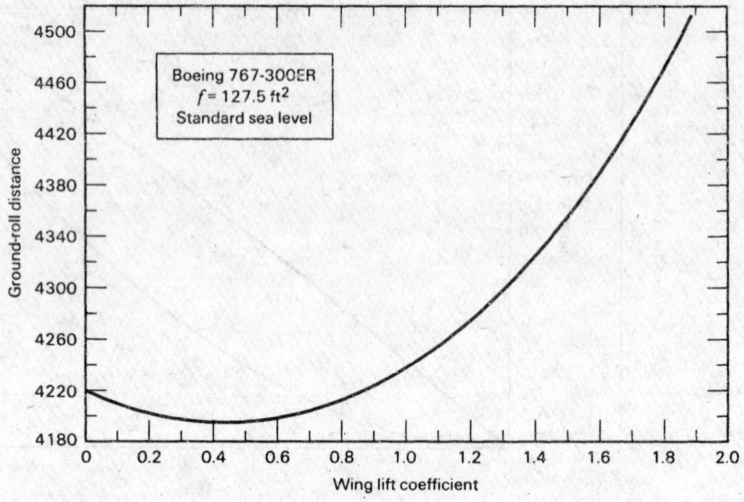

Figure 7.9 Effect of the wing lift coefficient on predicted ground-roll distance to reach a speed of 145 kt.

For preliminary design studies, an approximate method is frequently used to calculate the ground-roll distance. The method is based on assuming that the inverse of the acceleration is a linear function of the square of the velocity. There appears to be no real rational basis for this assumption other than the fact that the results to which it leads are reasonable. To begin, let

$$\frac{dV}{dt} = \frac{dV}{ds}\frac{ds}{dt} = V\frac{dV}{ds} = \frac{1}{2}\frac{dV^2}{ds}$$

Assuming

$$\frac{dV}{dt} = (A + BV^2)^{-1} \tag{7.7}$$

After integrating, the following results

$$s = \frac{V^2}{2}\left(A + B\frac{V^2}{2}\right) \tag{7.8}$$

The term in parentheses can be identified as the reciprocal of the acceleration evaluated at V divided by $\sqrt{2}$. Thus, Equation 7.8 becomes

$$s = \frac{V^2}{2\bar{a}} \tag{7.9}$$

where $\bar{a}$ is an average acceleration evaluated at $V/\sqrt{2}$. For example, consider the foregoing 767 example for a C_L of 1.0, an f of 127 ft^2 and a V of 145 kt. (245.1 fps).

$$\frac{V}{\sqrt{2}} = 173.3 \text{ fps}$$

At this speed,

$$T = 97{,}406 \text{ lb}$$
$$D = 7065 \text{ lb}$$
$$L = 110{,}037 \text{ lb}$$

so that, from Equation 7.3

$$\bar{a} = 7.056 \text{ ft/sec}^2$$

The ground-roll distance is then calculated from Equation 7.9 to be

$$s = \frac{245^2}{2(7.056)} = 4255 \text{ ft}$$

This result, by comparison to Figure 7.7, is seen to be within 0.5% of the more exact value obtained by numerical methods.

Effect of Wind

A headwind will always reduce the ground-roll distance required for an airplane to attain a desired airspeed. Although the headwind increases the drag and decreases the thrust for a given ground speed, it increases the lift and adds directly to the ground speed to increase the airspeed so that the net effect on takeoff distance is favorable.

The effect of the wind is most easily determined by using the approximation of Equation 7.7, keeping in mind that T, D, and L depend on the airspeed and not the ground speed.

If V_G denotes the ground speed and V_w the headwind then, for this case, Equation 7.7 becomes

$$\frac{dV_G}{dt} = [A + B(V_G + V_w)^2]^{-1} \tag{7.10}$$

Integrating gives

$$s = \frac{V_G^2}{2}\left[A + \frac{BV_G^2}{2} + \frac{4BV_wV_G}{3} + BV_w^2\right]$$

Again, if the terms in the brackets are interpreted as equal to the reciprocal of an average acceleration evaluated at some fraction, k, of the *airspeed*, $(V_G + V_w)$, then equating these terms to Equation 7.10 gives

$$k^2(V_G + V_w)^2 = \frac{V_G^2}{2} + \frac{4V_wV_G}{3} + V_w^2$$

This reduces to

$$k = \left[\frac{1}{2}\left(1 + \frac{2x + x^2}{3}\right)\right]^{1/2} \tag{7.11}$$

where

$$x = \frac{V_w}{V_G + V_w}$$

The factor k is presented graphically in Figure 7.10.

Suppose, in the previous example, that the 767 was taking off into a 30-kt (50.7-fps) headwind. Using the same liftoff velocity of 245 fps gives a ratio of headwind to airspeed of 0.207. From Figure 7.10, k = 0.750. Thus the average acceleration should be evaluated at an airspeed of 221.7 fps. At this speed,

$$T = 96{,}702 \text{ lb}$$

$$D = 7{,}957 \text{ lb}$$

$$L = 123{,}360 \text{ lb}$$

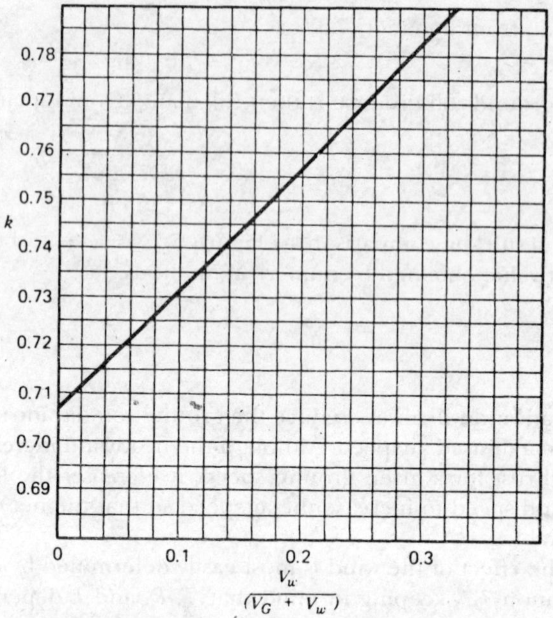

Figure 7.10 Fraction of liftoff airspeed at which to calculate average acceleration as a function of headwind-to-liftoff airspeed ratio.

Therefore, $\bar{a} = 6.95 \text{ ft/sec}^2$.

The ground-roll distance is thus determined to be

$$s = \frac{V_G^2}{2\bar{a}}$$

$$= \frac{(245.0\text{-}50.7)^2}{2(6.95)}$$

$$= 2718 \text{ ft}$$

Compared to the no-wind case, the 30-kt headwind decreases the ground-roll distance by approximately 36%.

It should be noted that FAR Part 25 requires a conservative estimate of headwind effects. Takeoff and landing distances must be calculated on the basis of 50% of reported headwinds and 150% of reported tailwinds. Since the wind can vary from one instant to another, the intent of FAR Part 25 is to use only half of any wind that improves performance but 150% of the opposite.

Airborne Distance

Becoming airborne at the speed V_{LOF}, an airplane continues to accelerate to the speed V_2 over the obstacle height. An airplane that is accelerating both normal to and along its flight path is pictured in Figure 7.11. As shown, $V(d\theta/dt)$ is the acceleration normal to the flight path and dV/dt is the acceleration along the flight path. The equations of motion normal to and along the flight path can be written as

$$T - D - W\sin\theta = \frac{W}{g}\frac{dV}{dt} \tag{7.12}$$

$$L - W\cos\theta = \frac{W}{g}\frac{d\theta}{dt}V \tag{7.13}$$

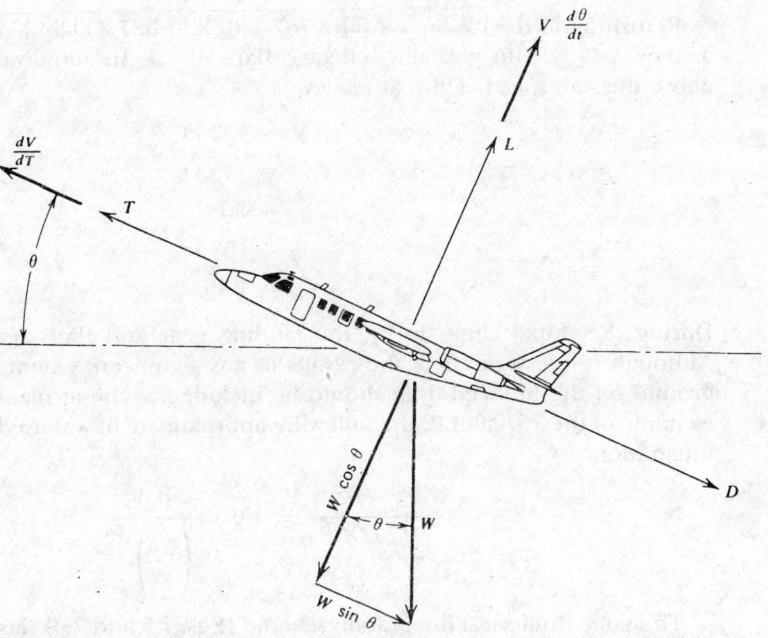

Figure 7.11 The forces on an airplane in an accelerating climb.

The rate of climb, dh/dt, and denoted by V_c in Chapter 1, is found from

$$dh/dt = V \sin \theta \tag{7.14}$$

Rate of climb is also denoted in the literature by R/C. Solving for sin θ from Equation 7.12, Equation 7.14 can be written as

$$\frac{dh}{dt} = V\left[\frac{(T - D)}{W} - \frac{dV/dt}{g}\right] \tag{7.15}$$

This can also be expressed as

$$(T - D)\,V = W\frac{dh}{dt} + \frac{d}{dt}\left(\frac{1}{2}\frac{W}{g}V^2\right) \tag{7.16}$$

TV is the available power and *DV* is the *required power.* Thus $(T - D)\,V$ is the *excess power* that, as Equation 7.16 shows, can be used either to climb or to accelerate. Actually, Equation 7.16 is an energy relationship which states that the excess power equals the sum of the time rates of change of the potential energy and the kinetic energy.

Let us now apply these relationships to the calculation of the horizontal distance required during the takeoff flare to attain a specified height. The actual flight path that is followed during the flare, or transition, segment of the takeoff depends on pilot technique. Referring to Equations 7.12 and 7.13, V and θ are the independent variables, while g, W and T are known, the latter as a function of V. L and D are functions of V and the airplane's angle of attack α. By controlling α and hence, C_L, the pilot can fly a desired trajectory (i.e., the pilot can accelerate or climb or do some of each). During the takeoff, however, in attempting to clear an obstacle, FAR Part 25 limits the operating C_L to approximately $C_{L_{max}}/1.21$ at V_{LOF} and to $C_{L_{max}}/1.44$ at V_2. Therefore, in calculating the flare distance, it will be assumed that C_L varies linearly with V between these limits and is constant for speeds above V_2. Thus, if V_2 is attained before the specified obstacle height is reached, the stall margin on C_L is maintained.

Returning to the example of the 767-300ER at 387,000-lb takeoff weight, let us assume a $C_{L_{max}}$ with partially deflected flaps of 2.1. In addition, let V_{LOF} be 10% above the stall speed. Thus, at sea level,

$$V_s = 132 \text{ kt}$$

$$V_{LOF} = 145 \text{ kt}$$

$$C_{L_{LOF}} = 1.74$$

$$V_2 = 158 \text{ kt}$$

$$C_{l_2} = 1.46$$

During the initial climb to V_2, the landing gear and flaps remain unchanged. Although it does not affect the results to any significant extent, the effect of the ground on the induced drag should be included as the airplane climbs. For this example of the 767-300ER, the following approximate fit was used in the numerical integration.

$$\frac{C_{D_i}\,(IGE)}{C_{D_i}\,(OGE)} = 1.0 - \frac{0.05}{\left(2\dfrac{h}{b}\right)^{1.04}}$$

The same numerical integration scheme (Eqs. 7.5 and 7.6) was used to calculate the initial climb to 35 ft except that Equation 7.14 was included for the altitude, h. The distance, s, in Equation 7.6 is along the flight path. Therefore, to get the horizontal distance along the ground, the increment in s on the right side of Equation 7.6 must be multiplied by $\cos\theta$.

The result of the numerical integration is shown in Figure 7.12 where the horizontal distance and altitude are presented as a function of time. It is seen that from the point of liftoff, it takes the 767-300ER 10 sec to reach an altitude of 35 ft. During this interval, the airplane covers a horizontal distance of 2685 ft. This results in a total time and horizontal distance from the beginning of the takeoff to a 35-ft altitude of 43.3 s and 7008 ft, respectively.

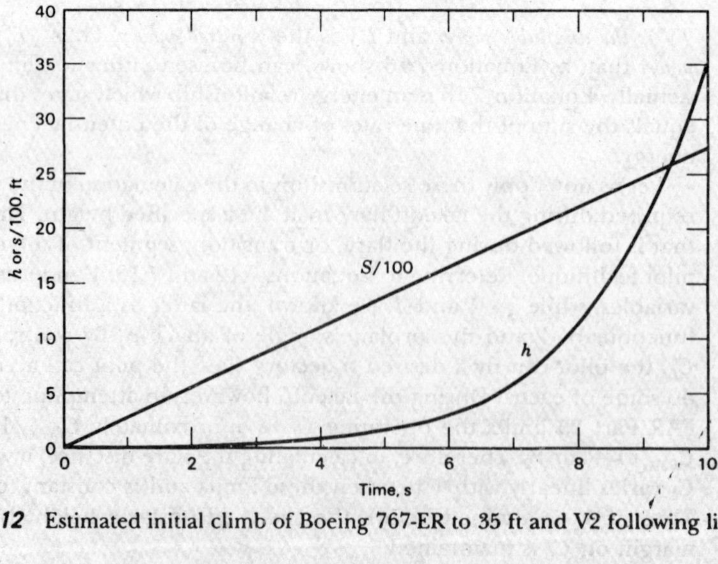

Figure 7.12 Estimated initial climb of Boeing 767-ER to 35 ft and V2 following liftoff.

Balanced Field Length

During the takeoff run the pilot of a jet transport has the option of aborting the takeoff up to the speed V_1. Above this speed, in the event of an engine failure, the takeoff should be continued. Figure 7.11 (from Ref. 7.1) clearly illustrates these options. A balanced field length is defined as one where the distance to continue the takeoff following the recognition of an engine failure at V_1 is equal to the distance required to stop if the takeoff should be aborted. On Figure 7.13, the field length is balanced if the sum of segments *B* and *C* equals the sum of *D* and *E*. FAR Part 25 stipulates the field length to be the greatest of the accelerate-and-go distance, the accelerate-and-stop distance, or 115% of the all-engine-operating distance to a 35-ft height. The stop-and-go portions need not be balanced.

The distance to stop can be found by numerically integrating Equation 7.1. In this case, T will be negative and equal to the reverse thrust. μ is the braking coefficient and, with antiskid systems, can be as high as 0.6 on a dry, hard surface. Detailed considerations relating to the deceleration of an airplane will be deferred until the later section on landing distance.

Computer Exercise 7.3 "CLIMB TO 35 FT and V2"

Add the first climb segment up to 35 ft and V_2 to the program written for Computer Exercise 7.2. Use the same thrust relations and flat-plate area as for the ground-roll program. Run the 767-300ER case and compare with Figure 7.12.

RATE OF CLIMB, TIME TO CLIMB, AND CEILINGS

The rate of climb was given previously by Equation 7.15. For a steady climb this equation becomes

$$\frac{dh}{dt} = \frac{V(T - D)}{W} \tag{7.17}$$

It can be expanded and expressed, as noted previously, as

$$\frac{dh}{dt} = \frac{P_a - P_r}{W} \tag{7.18}$$

where P_a and P_r are the available power and required power, respectively.

$$P_a = TV$$

$$P_r = DV$$

For a gas turbine engine, T is known as a function of altitude and velocity, so that Equation 7.17 is the obvious form to use in this case to determine the rate of climb, R/C. In the case of a propeller-driven airplane,

$$P_a = \eta P_{\text{shaft}}$$

where P_{shaft} is the shaft power delivered to the propeller and η is the propeller efficiency.

As a function of V, the required power is determined by

$$P_r = DV$$

Let us apply Equation 7.17 to the 767-300ER example using the maximum climb thrust given in Figure 6.42. Flaps and gear are assumed to be retracted during the climb so that f reduces to 42.5 ft^2. As an example, take a speed of 350 kt, or 591.5

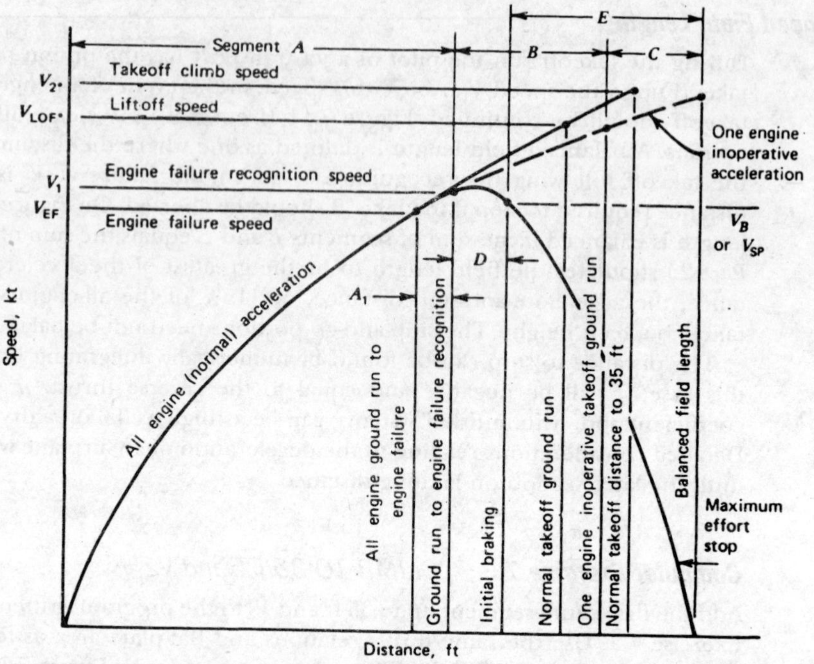

Figure 7.13 Definition of balanced field length.

fps, at an altitude of 20,000 ft. The speed of sound at this altitude equals 1037 fps, giving a Mach number of 0.570. From Figure 6.42,

$$T = 42{,}600 \text{ lb (two engines)}$$

At the given speed and altitude, the dynamic pressure, q, is found to be

$$q = 221.7 \text{ psf}$$

For the gross weight of 387,000 lb and a wing area of 3084 ft^2, the lift coefficient becomes

$$C_L = 0.566$$

The corresponding induced drag coefficient, for an assumed Oswald's efficiency factor of 0.7, is

$$C_{D_i} = 0.01844$$

For this climbing condition the drag breakdown is

$$\text{Parasite drag} = 9422 \text{ lb}$$
$$\text{Induced drag} = 12{,}607 \text{ lb}$$

Therefore, with the above numbers, the calculated rate-of-climb, in feet/minute, from Equation 7.17 becomes

$$R/C = 60\,\frac{591.5(42{,}600 - 22{,}029)}{387{,}000}$$

or

$$R/C = 1886 \; fpm$$

The rate-of-climb, R/C, is expressed in feet per minute (fpm) in accordance with standard American practice. Rate-of-climb meters on American airplanes are calibrated in these units.

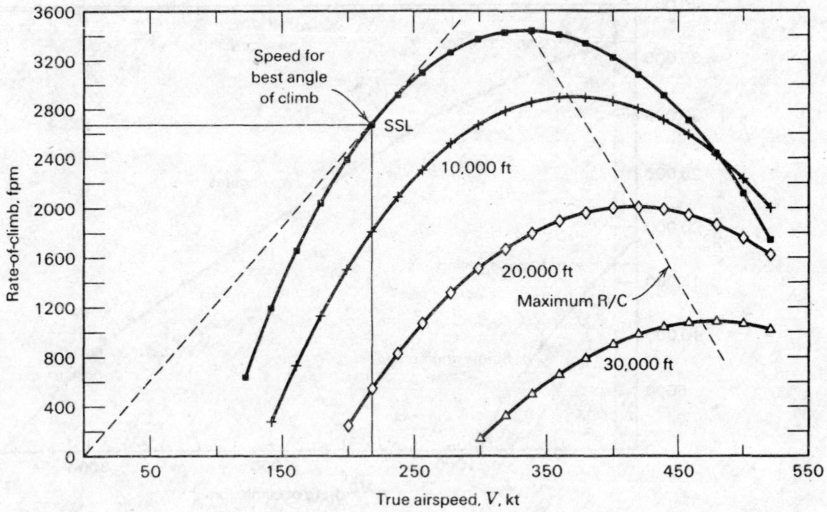

Figure 7.14 Calculated rate-of-climb for Boeing 767-300ER at a weight of 387,000 lb at maximum climb thrust.

In the above manner, the curves presented in Figure 7.14 were prepared using a computer program using the maximum climb thrust. Obviously, the maximum rate of climb and the corresponding true airspeed as a function of altitude is obtained directly from the figure. The maximum angle of climb and the corresponding speed are important for climbing above an obstacle. This angle is given by

$$\theta_c = \tan^{-1} \frac{V_c}{V} \tag{7.18}$$

The above angle will be a maximum at the speed where a straight line from the origin is tangent to the curve of V_c (R/C) versus V as shown on Figure 7.14. Thus, from the figure at SSL conditions,

$$\theta_c(\max) = \tan^{-1} \frac{2675/60}{(220)(1.69)}$$

$$= 6.84°$$

Using the results from Figure 7.14, the curves of Figures 7.15 and 7.16 were prepared. Also, the calculations were repeated using only one engine in order to estimate engine-out performance. Observe that losing an engine more than halves the R/C since the R/C is proportional to the difference between the thrust and drag—not the thrust alone.

The *service ceiling* for an airplane is defined as the altitude at which the R/C equals 100 fpm. From these calculations, the service ceiling is found to equal approximately 17,000 ft with a single engine. This is less than the 21,000 ft quoted in Reference 5.11; however, it is not clear what gross weight corresponds to the 21,000 ft. Figure 7.17 shows the predicted effect of gross weight on the single-engine R/C at 21,000 ft. According to this figure, the gross weight would have reduced to 348,300 lb in order to have an R/C of 100 fpm at 21,000 ft. This is fairly close to an average between the maximum takeoff weight and the maximum landing weight and may explain the difference between the service ceiling predicted here and the one given in Reference 5.11.

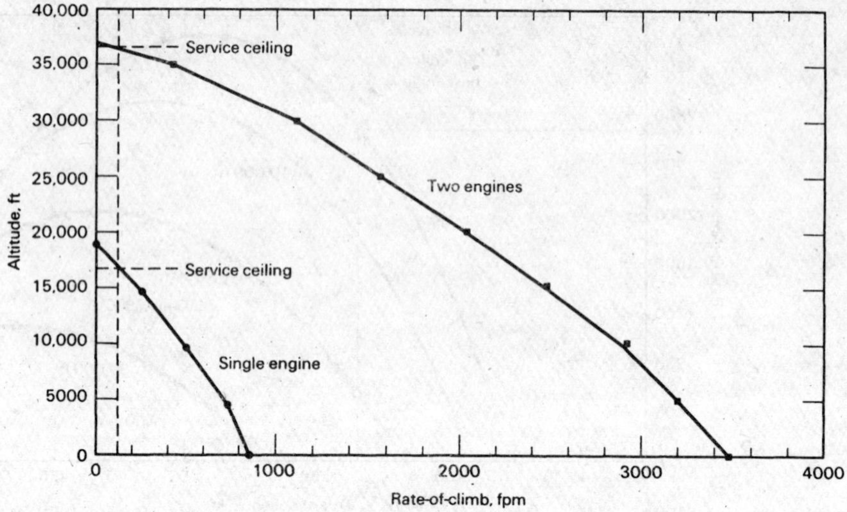

Figure 7.15 Calculated maximum rate-of-climb for the Boeing 767-300ER at 387,000 lb, maximum climb thrust.

The *absolute ceiling* for an airplane is defined as the altitude for which the R/C equals zero. It will be shown later that an airplane can never reach its absolute ceiling. It is obtained experimentally only by extrapolating measurements of R/C at lower altitudes to the altitude for a zero R/C. From Figure 7.15, the absolute ceiling for the 767-300ER at 387,000 lb is estimated to be 37,500 ft.

Consider now the R/C calculation for a propeller-driven airplane. We will use as an example the Piper Cherokee Arrow that is similar to the Cherokee 180 except that the former has retractable landing gear, a 200-bhp (149-kW) piston engine, and a constant speed propeller. The engine is rated at 2700 rpm but, for a contin-

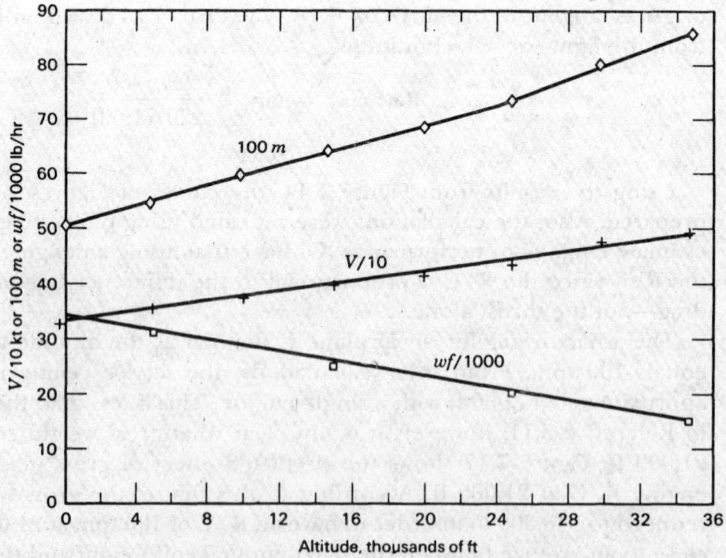

Figure 7.16 True airspeed, fuel flow, and Mach number for maximum rate-of-climb for the Boeing 767-300ER at 387,000 lb.

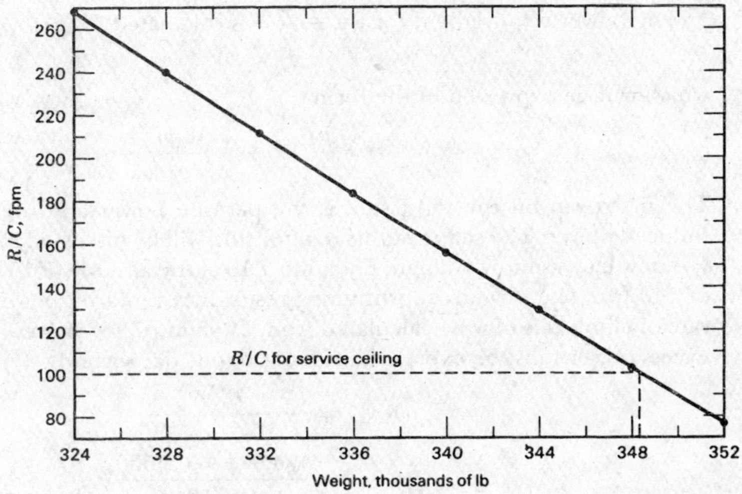

Figure 7.17 Effect of gross weight on single-engine climb for the 767-300ER at 21,000 ft.

uous climb in accordance with recommended practice, an rpm of only 2500 will be used. At standard sea level, the engine develops 185 bhp (138 kW) at this rpm. For higher altitudes, the engine power is estimated on the basis of Figure 6.3. The following values are known or have been estimated for this airplane:

Flat-plate area	f	$= 4.5\ \text{ft}^2\ (0.418\ \text{m}^2)$
Span efficiency	e	$= 0.70$
Propeller diameter	D	$= 6.17$ ft (1.88 m)
Span	b	$= 32.2$ ft (9.81 m)
Wing area	S	$= 169\ \text{ft}^2\ (15.7\ \text{m}^2)$
Aspect ratio	A	$= 6.14$
Weight	W	$= 2650$ lb (11.1 kN)

The propeller performance curves for this airplane were presented in Figures 6.16 and 6.17. These curves, together with the engine power, are used to estimate the available power. As an example, consider an altitude of 10,000 ft (305 m). The engine power at this altitude equals 130 bhp. At a speed of, say, 140 fps, the advance ratio will be

$$\begin{aligned} J &= V/nD \\ &= 140/(2500/60)/6.17 \\ &= 0.545 \end{aligned}$$

The power coefficient equals

$$\begin{aligned} C_P &= \frac{P}{\rho n^3 D^5} \\ &= 130(550)/(0.00176)(41.7)^3(6.17)^5 \\ &= 0.063 \end{aligned}$$

This power coefficient and advance ratio lead to a blade angle of 22° from Figure 6.16. Using this blade angle, together with J, results in a propeller efficiency, η, of 0.73 from Figure 6.17. Thus, the available power at this speed and altitude equals 91 thp where thp stands for "thrust horsepower." The equivalent term in the SI system would be tkW, for "thrust kilowatts."

The power requirement for the Arrow is calculated from

$$P = DV$$

which can be expressed in the form

$$P = \frac{\rho f}{2} V^3 + \frac{2(W/b)^2}{\pi \rho e} \frac{1}{V} \tag{7.19}$$

The first term on the right side is the parasite power and the second term the induced power. The shape of this relationship will be discussed in more detail later. For now let us simply evaluate Equation 7.19 at the altitude of 10,000 ft and a speed of 140 fps. The obvious substitutions result in a required power of 65.3 hp. The rate of climb can now be calculated from Equation 7.18. However, in so doing, the excess power must be expressed in foot-pounds per second.

$$\begin{aligned} R/C &= \frac{P_a - P_r}{W} \\ &= \frac{(94.9 - 65.3)550}{2650} \\ &= 6.14 \text{ fps} \end{aligned}$$

As stated previously, it is current practice in the American aviation industry to express the R/C in feet per minute, so that the above becomes 369 fpm (1.87 m/s).

In this manner, the rate of climb can be calculated over a range of speeds for several altitudes for the Arrow. The maximum rates of climb thus determined are presented in Figure 7.18, where the R/C is seen to decrease almost linearly with

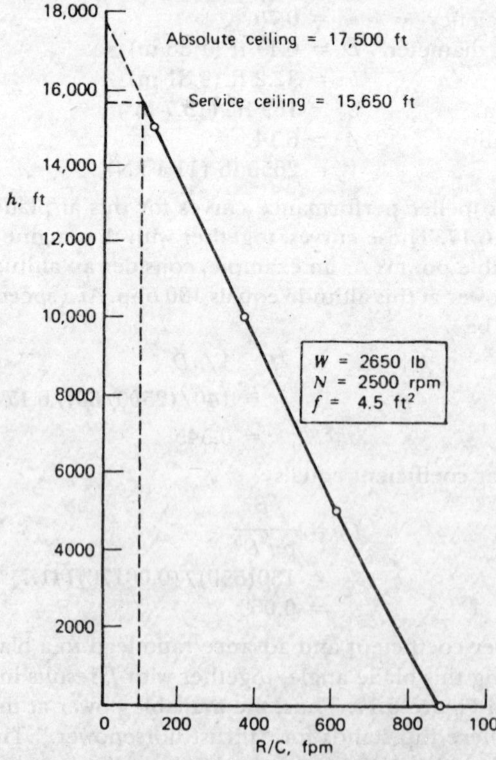

Figure 7.18 Rate-of-climb for the Cherokee Arrow versus altitude.

altitude. In this example, the calculated service ceiling and sea level rate of climb are close to the corresponding values quoted by the manufacturer.

The power-required and power-available curves calculated for this example at sea level are presented in Figure 7.19. Similar to the drag curve, the power-required curve has a minimum value at some speed. Below this speed, it actually requires more power to fly slower. This part of the curve is referred to as the backside of the power curve.

The speed for minimum required power can be found by setting to zero the derivative of Equation 7.19 with respect to V. This leads to

$$V\sqrt{\sigma} = \left[\frac{4(W/b)^2}{3\pi\rho_0^{\,2}\, ef}\right]^{1/4} \tag{7.20}$$

The preceding speed is, of course, not necessarily the speed for maximum rate of climb, since the available power varies with airspeed. In the case of the Arrow, Figure 7.19 shows the optimum speed to be greater than the value given by Equation 7.20. Nevertheless, one might expect the two speeds to be related. Equation 7.20 shows the minimum required power to occur at a constant indicated airspeed. This suggests that the maximum rate of climb will also occur at a constant indicated airspeed. This observation holds closely for piston-engine airplanes but becomes more approximate for turbojets.

Computer Exercise 7.4 "CLIMB"

Write a computer program that loops over speed and altitude to produce the rate-of-climb using maximum climb thrust or power. Program should use a table look-up and interpolation between altitudes and Mach numbers to obtain the thrust or power. Input should be the airplane's flat-plate area, wing area, gross weight, and Oswald's efficiency factor. Output should be the gross weight, altitude, true airspeed, Mach number, fuel flow, and rate-of-climb. Prepare a table of thrust and TSFC from Figure 6.42 and run your program to check Figures 7.14–7.16.

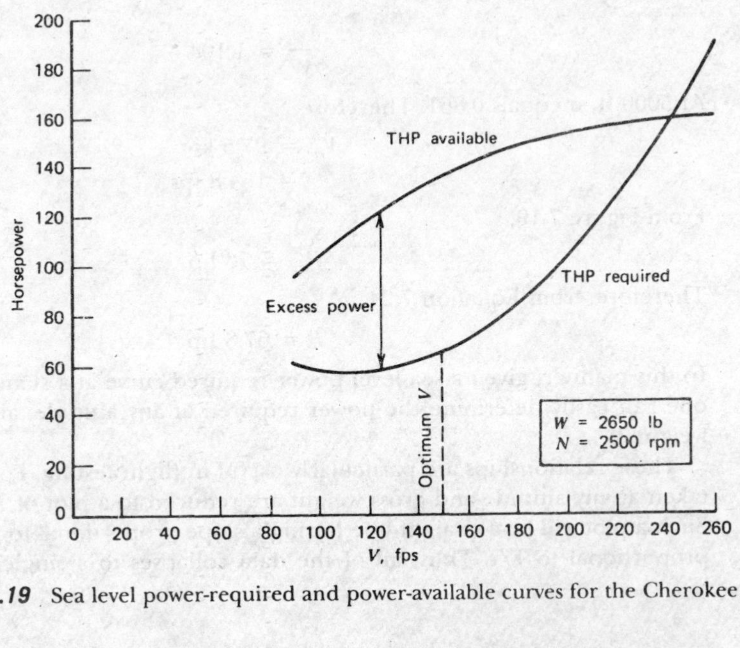

Figure 7.19 Sea level power-required and power-available curves for the Cherokee Arrow.

Generalized Power-Required Curves

In flying, one rarely finds a standard atmosphere. In addition, the gross weight of an airplane is generally different from that used by the manufacturer to quote performance. Thus, if the power required by an airplane is measured at some constant speed, density altitude, and gross weight, how does one determine if it is consistent with the quoted performance or with other measurements? The answer to this question is not as difficult as it might seem.

The drag coefficient of an airplane, neglecting Mach number effects, is a function only of the airplane's lift coefficient. Thus, consider the power measured at a weight of W, a true airspeed of V, and at an altitude for which the density is ρ. At sea level, the velocity, V_{ew} (ew for equivalent weight), giving the *same* C_L as the test point is found from

$$C_L = \frac{W}{\frac{1}{2}\rho V^2 S} = \frac{W_s}{\frac{1}{2}\rho_0 V_{ew}^2 S} \tag{7.21}$$

From the above, V_{ew} becomes

$$V_{ew} = V\sqrt{\sigma}\sqrt{\frac{W_s}{W}} \tag{7.22}$$

For the test point, the power is given by

$$P = \frac{1}{2}\rho_0 V_{ew}^3 S C_D \tag{7.23}$$

Because the C_D values are the same in the expressions for P and P_{ew}, it follows that

$$P_{ew} = P\sqrt{\sigma}\left(\frac{W_s}{W}\right)^{3/2} \tag{7.24}$$

A graph of P_{ew} versus V_{ew} will simply be the sea level power required curve at the standard gross weight. As an example, take the Cherokee Arrow at a gross weight of 2400 lb, a density altitude of 5000 ft and an airspeed of 100 kt. Using Figure 7.19 as the standard,

$$\frac{W_s}{W} = 1.104$$

At 5000 ft, σ equals 0.861. Therefore,

$$\begin{aligned} V_{ew} &= 97.5 \text{ kt} \\ &= 164.8 \text{ fps} \end{aligned}$$

From Figure 7.19,

$$P_{ew} = 73 \text{ hp}$$

Therefore, from Equation 7.24,

$$P = 67.8 \text{ hp}$$

In this manner, given a sea level power required curve at a standard gross weight, one can easily determine the power required at any altitude, airspeed, and gross weight.

These relationships are particularly useful in flight testing. Power-required data taken at any altitude and gross weight are reduced to a plot of $P_{ew} V_{ew}$ versus V_{ew}^4. Such a plot will be a straight line having a slope proportional to f and an intercept proportional to $1/e$. Thus, all of the data collapses to a single, easily fitted line enabling one to determine accurately e and f. The equation of this straight line follows directly from Equation 7.19.

$$P_{ew} V_{ew} = \frac{2(W_s/b)^2}{\pi \rho_0 e} + \frac{\rho_0 f}{2} V_{ew}^4 \tag{7.25}$$

Time to Climb

The time required to climb from one altitude, h, to another, h_2, can be determined by evaluating the integral

$$t = \int_{h_1}^{h_2} \frac{dh}{R/C} \tag{7.26}$$

Knowing the R/C as a function of h, this integral can easily be evaluated numerically. A solution in closed form can be obtained if one assumes that the rate of climb decreases linearly with altitude. A feeling of how valid this assumption is can be gained from Figures 7.15 and 7.18.

Let

$$R/C = (R/C)_0(1 - h/h_{abs}) \tag{7.27}$$

where h_{abs} = absolute ceiling.

Equation 7.26, for $h_1 = 0$, then reduces to

$$t = \frac{h_{abs}}{(R/C)_0} \ln \frac{1}{1 - h/h_{abs}} \tag{7.28}$$

This represents the time required to climb from sea level to the altitude, h. The time required to climb from one altitude to another is obtained directly from Equation 7.28 by subtracting, one from the other, the times required to climb from sea level to each altitude.

Time to climb is presented for the Cherokee Arrow in Figure 7.20. This curve was calculated on the basis of Equation 7.28. For this particular example, it requires approximately 42.5 min to climb to the service ceiling. This number is accurate since the dependence of R/C is linear for the Cherokee Arrow. In the case of a nonlinear curve for R/C, such as Figure 7.15, it is recommended that Equation 7.26 be applied over altitude increments and added to get the cumulative time to

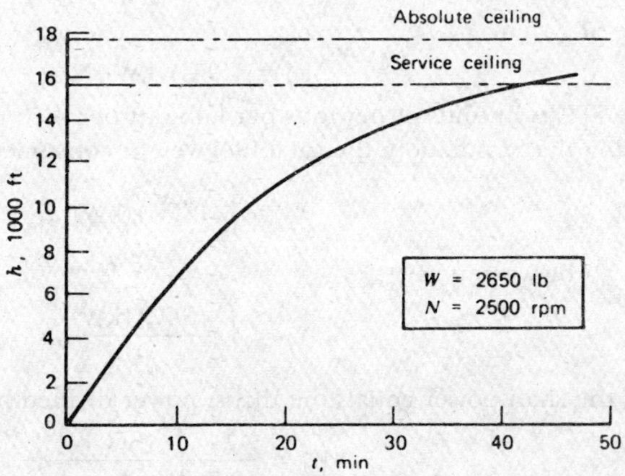

Figure 7.20 Time-to-climb for Cherokee Arrow.

an altitude. In this case, the time-to-climb from altitude 1 to altitude 2 will be given by

$$\Delta t = \frac{h_2 - h_1}{(\mathrm{R/C})_2 - (\mathrm{R/C})_1} \ln \left[\frac{(\mathrm{R/C})_2}{(\mathrm{R/C})_1} \right] \qquad (7.29)$$

Applying Equation 7.29 to Figure 7.15 in increments of 5000 ft results in a time to climb to 35,000 ft of 20.9 min for the two-engine case. Otherwise, Equation 7.28 results in a time of approximately 30 min, an appreciable error compared to the more exact calculation.

In determining the time to climb, it is sometimes tempting to do an iteration on the gross weight since the rate-of-climb will increase as fuel is burned off. In the case of the 767-300ER, from Figure 7.16, fuel is being expended at the rate of approximately 28,000 lb/hr between 0 and 35,000 ft. Thus, if the gross weight at the start of the climb was 387,000 lb, at the end of 20.1 min, the weight would be down approximately 9380 lb. Neglecting the effect of this small change in weight on the excess thrust, it is reasonable to assume that the change in the R/C is inversely proportional to gross weight. Thus, using half of the above loss in weight to give an average gross weight of 382,310-lb results in a 1.2% increase in the R/C or a predicted time-to-climb to 35,000 ft of 20.6 min, a decrease of only 18 sec from the time obtained by assuming the weight to be constant during the climb. Thus, from this example it can be seen that for a typical airplane the change in weight during a climb can be neglected in calculating the R/C or time-to-climb.

RANGE

The *range* of an aircraft is the distance that the aircraft can fly. Range is generally defined subject to other requirements. In the case of military aircraft, one usually works to a mission profile that may specify a climb segment, a cruise segment, a loiter, an enemy engagement, a descent to unload cargo, a climb, a return cruise, a hold, and a descent. In the case of civil aircraft, the range is usually taken to mean the maximum distance that the airplane can fly on a given amount of fuel with allowance to fly to an alternate airport in case of bad weather.

Let us put aside the range profile for the present and consider only the actual distance that an airplane can fly at cruising altitude and airspeed on a given amount of fuel. For a propeller-driven airplane, the rate at which fuel is consumed is

$$w_f = (\mathrm{BSFC})(\mathrm{bhp}) \ \mathrm{lb/hr}$$

In the SI system,

$$w_f = (\mathrm{SFC})(\mathrm{kW}) \ \mathrm{N/s}$$

where SFC is in units of newtons per kilowatt per second.

Using the SI notation, the total fuel weight consumed over a given time will be

$$W_f = \int_0^t (\mathrm{SFC})(\mathrm{kW}) \, dt$$

This can be written as

$$W_f = \int_0^s \frac{(\mathrm{SFC})(\mathrm{kW})}{V} \, ds$$

Since the shaft power equals the thrust power divided by the propeller efficiency,

$$W_f = \frac{1}{1000} \int_0^s \frac{(\mathrm{SFC})\,D}{\eta} \, ds \qquad (7.30)$$

where D is the drag. The constant represents the fact that l kW equals 1000 mN/s.

Given the velocity and weight, Equation 7.30 can be integrated numerically. One of the difficulties in evaluating Equation 7.30 rests with the weight, which is continually decreasing as fuel is burnt off.

A closed-form solution can be obtained for Equation 7.30 by assuming that the SFC and η are constant and that the airplane is flown at a constant C_L. With these assumptions, the fuel flow rate, with respect to distance, becomes

$$\frac{dW_f}{ds} = \frac{(\text{SFC})\,\epsilon}{1000\,\eta} W$$

where ϵ is the drag-to-lift ratio, which is a function of C_L, and W is the airplane weight. dW_f/ds is the negative of dW/ds. Thus,

$$\frac{dW}{W} = -\frac{(\text{SFC})\,\epsilon}{1000\eta} ds$$

Integrating gives

$$\ln \frac{W_I}{W} = \frac{(\text{SFC})\,\epsilon}{1000\,\eta} s$$

where W_I is the initial weight of the airplane. Finally, if W_F is the total fuel weight, the distance or range, R, that the airplane can fly in meters, on this fuel is

$$R = \frac{1000\,\eta}{(\text{SFC})\,\epsilon} \ln\left(1 + \frac{W_F}{W_E}\right) \tag{7.31}$$

W_E denotes "weight empty," meaning "empty of fuel." Normally, weight empty refers to the airplane weight without any fuel or payload. This equation, which holds only for propeller-driven aircraft, is a classical one known as the *Breguet range equation.*

In the case of a turbojet-propelled airplane, the fuel flow becomes

$$w_f = (\text{TSFC})\, D$$

so that

$$\frac{dW}{W} = -\frac{(\text{TSFC})\,\epsilon\, ds}{V}$$

In order to integrate this relationship, we must assume that the airplane operates at a constant ϵ/V and that TSFC is constant. When this is done, the modified Breguet range equation for jet-propelled aircraft is obtained.

$$R = \frac{V}{(\text{TSFC})\,\epsilon} \ln\left(1 + \frac{W_F}{W_E}\right) \tag{7.32}$$

Thus, for maximum range, ϵ should be minimized for propeller-driven airplanes and ϵ/V should be minimized for turbojets. In the case of turbojets, this can lead to the airplane cruising slightly into the drag rise region that results from transonic flow.

Everything being constant except the speed, the drag of an airplane can be written as

$$D = C_1 V^2 + \frac{C_2}{V^2} \tag{7.33}$$

Thus, the drag-to-lift ratio, ϵ, will be a minimum when

$$V_{\text{opt}} = \left(\frac{C_2}{C_1}\right)^{1/4} \quad \text{(propeller-driven airplane)} \tag{7.34}$$

ϵ/V will have a minimum at

$$V_{opt} = \left(\frac{3C_2}{C_1}\right)^{1/4} \quad \text{(turbojet)} \tag{7.35}$$

For propeller-driven airplanes this leads to a minimum ϵ value of

$$\epsilon_{min} = 2\left(\frac{f/S}{\pi eA}\right)^{1/2} \tag{7.36}$$

$$V_{opt} = \left[\frac{4}{\pi eA\rho_0{}^2}\frac{(W/S)^2}{\sigma^2(f/S)}\right]^{1/4} \tag{7.37}$$

For turbojet-propelled airplanes, ϵ/V has a minimum value of

$$\left(\frac{\epsilon}{V}\right)_{min} = (3^{1/4} + 3^{-3/4})\left[\frac{\rho_0{}^2(f/S)^3}{4\pi eA(W/S)^2}\right]^{1/4}\sigma^{1/2} \tag{7.38}$$

The optimum V for the above is equal to that given by Equation 7.37 multiplied by $3^{1/4}$.

Some interesting observations can be made based on Equations 7.36, 7.37, and 7.38. For either propeller or turbojet airplanes, the indicated airspeed for maximum range is constant independent of altitude. However, for the same wing loading, effective aspect ratio, and parasite drag coefficient, the optimum cruising speed for the turbojet airplane is higher than that for the propeller-driven case by a factor of 1.316. The optimum range for a propeller-driven airplane is independent of density ratio and hence altitude. However, with the indicated airspeed being constant, the trip time will be shorter at a higher altitude.

The optimum range for a turbojet is seen to increase with altitude being inversely proportional to the square root of the density ratio. This fact, together with the increase in true airspeed with altitude, results in appreciably higher cruising speeds for jet transports when compared with a propeller-driven airplane. The turbofan, with a high bypass ratio, falls somewhere between the pure turbojet and the propeller. Since the expressions for the optimum cruise velocity for propellers and turbojets differ by only a constant ratio, let us assume, rather arbitrarily, that the optimum cruise velocity for a high BPR turbofan lies halfway between the two. Thus, with this assumption, Equation 7.38 is multiplied by a constant of 1.158 for the turbofan.

We will now examine the 767-300ER with the PW4056 engine in cruise and compare the results using the engine charts with the foregoing closed-form expressions. However, before this can be done, the TSFC of a turbofan engine operating at part-throttle must be examined. Figure 6.43 is limited to the case where the engine is operating at maximum cruise thrust. Generally, this will not be the case since, for steady level flight, the thrust must equal the drag. Thus, at a given cruise speed, the throttle will be retarded resulting in an engine thrust that is below the maximum available at that speed.

When an engine is operating at part throttle, the TSFC is usually higher than the value at maximum thrust. The engine curves for the JT-4A in Chapter 6 show this directly. Unfortunately, similar curves for the PW4056 engine were not available; however, TSFC values at part throttle were found for the JT-9D, a high bypass ratio (BPR) turbofan on which a correction to the TSFC for part throttle operation could be based. This correction is presented in Figure 7.21 where the ratio of the TSFC at part throttle to the TSFC for the maximum rated thrust is given as a function of the ratio of the thrust to the maximum thrust. Data for the JT-9D at two different Mach numbers at 35,000 ft are included on the figure and are seen

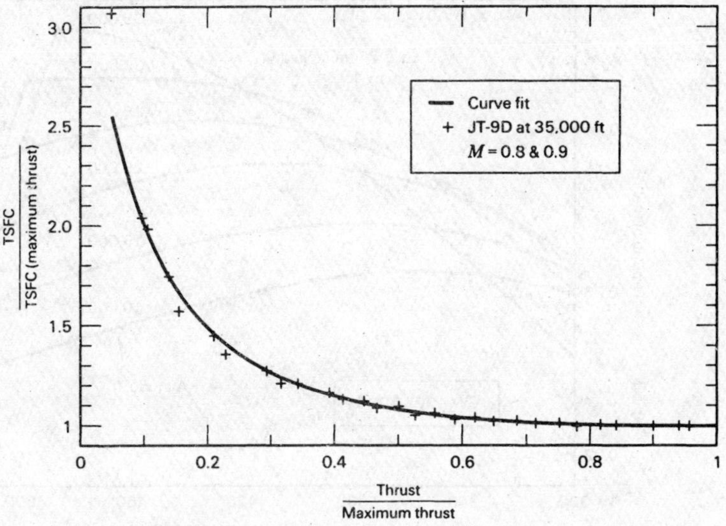

Figure 7.21 Correction to specific fuel consumption for operation at reduced thrust.

to collapse to a single curve when plotted in this manner. An empirical fit to the data is included and is seen to agree closely with the data over the normal range of interest. The equation of this curve is

$$F_R = (0.2452 + 2.9671\,T_R - 4.5139T_R^2 + 3.2937\,T_R^3 + 0.9921\,T_R^4)^{-1} \qquad (7.38)$$

where T_R is the ratio of the thrust to the maximum thrust and F_R is the ratio of the fuel flows, or TSFC values.

The computer program written to calculate the climb performance is easily adapted to perform the cruise performance. For a given true airspeed, altitude, and weight, one calculates the Mach number and airplane drag. The table for the maximum climb thrust and TSFC is replaced by a corresponding table for the maximum cruise thrust and TSFC. The program calls for the maximum thrust and TSFC as a function of altitude and Mach number. It then determines T_R as the ratio of the drag to the maximum cruise thrust. The TSFC can then be found from Equation 7.37.

The distance traveled per unit weight of fuel, miles per pound (meters per kilogram in the SI system), is determined from

$$\frac{\text{mi}}{\text{lb}} = \frac{\text{mi/hr}}{(\text{TSFC})\ T}$$

or

$$\frac{\text{mi}}{\text{lb}} = \frac{V}{(\text{TSFC})\ T} \qquad (7.39)$$

For the predictions to be realistic, the program must recognize that as the Mach number exceeds the drag–divergence Mach, the drag coefficient will increase rapidly. This kind of data is closely held by the manufacturers, so for purposes of illustration it is simply assumed that the drag–divergence Mach for the 767-300ER equals 0.85. Further, based on some of the drag information given in Chapter 5, it is assumed that for M greater than M_{cr}, the drag is increased by the factor

$$1 + \frac{M - M_{cr}}{0.1}$$

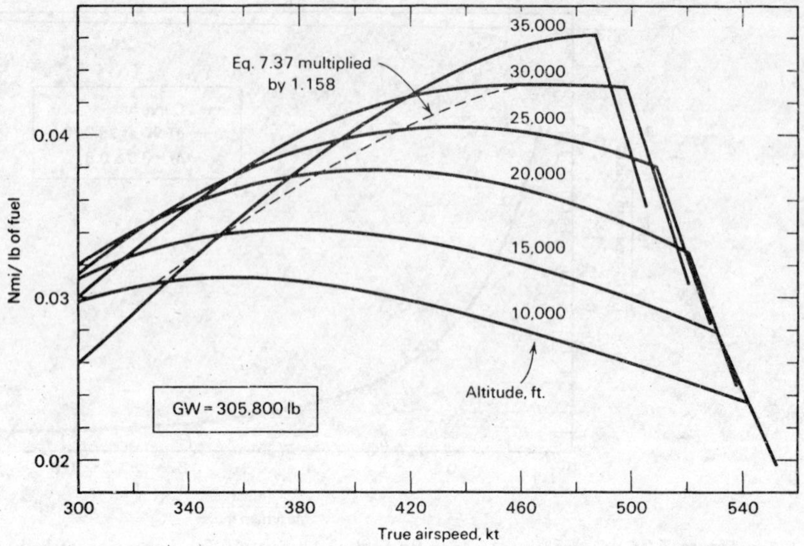

Figure 7.22 Fuel consumption per mile for the Boeing 767-300ER at an average gross weight.

Thus, for example, if the Mach number equals 0.95, the drag is taken to be twice the value, which is calculated without reference to *M*.

As an example, the results presented in Figure 7.22 were determined for the 767-300ER at a gross weight of 305,800 lb. This weight corresponds to the maximum takeoff weight minus half of the fuel weight. Also included in the figure is a plot of the optimum velocity as a function of altitude as predicted by Equation 7.37. Generally, these results, based on a specific airplane and engine, confirm the closed-form predictions concerning the range. The distance covered per pound of fuel and the optimum true airspeed increase appreciably with altitude. The sudden decrease in the miles per lb of fuel at the highest speeds is due, of course, to exceeding the drag–divergence Mach number. These trends with altitude are shown in Figure 7.23.

In the case of propeller-driven airplanes, the optimum cruising velocity given by Equation 7.37 does not reflect practice. To see why, consider the Cherokee Arrow. In this case, at a gross weight of 2650 lb, the optimum velocity is calculated to equal 87.8 kt. This velocity is appreciably slower than the speeds at which the airplane is capable of flying. It is generally true of a piston engine airplane that the installed power needed to provide adequate climb performance is capable of providing an airspeed appreciably higher than the speed for optimum range. Therefore, ranges of such aircraft are quoted at some percentage of rated power, usually 65 or 75%.

The cruising speed at some specified percentage of the rated power can be found from the power curves such as those presented in Figure 7.19 for the Cherokee Arrow. For example, 75% of the rated power corresponds to approximately 81% of the available power shown in Figure 7.19. This increase results from the rating of 200 bhp at 2700 rpm as compared to only 185 bhp output at 2500 rpm for which the figure was prepared. A line that is 81% of the available power crosses the power-required curve at a speed of 223 fps or 132 kt. This speed is therefore estimated to be the cruising speed at 75% of rated power at this particular rpm.

The penalty in the range incurred by cruising at other than the optimum speed can be found approximately from Equations 7.18 and 7.34. The ratio of the drag

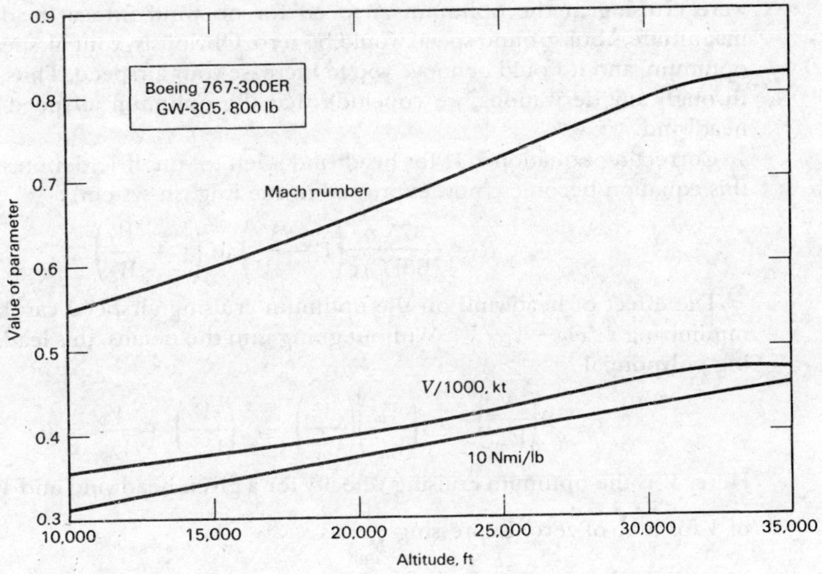

Figure 7.23 Range parameters as a function of altitude.

at any speed to the minimum drag can be expressed as a function of the ratio of the speed to the optimum speed. The result is

$$\frac{D}{D_{\min}} = \frac{1}{2}\left[\left(\frac{V}{V_{opt}}\right)^2 + \left(\frac{V_{opt}}{V}\right)^2\right]$$

This relationship is presented graphically in Figure 7.24. In the preceding example of the Cherokee, this figure shows a loss of approximately 25% in the range by cruising at 75% power instead of the optimum. Of course, the time required to get to your destination is 33% less by cruising at 75% power.

The effect of wind on range is pronounced. To take an extreme, suppose you

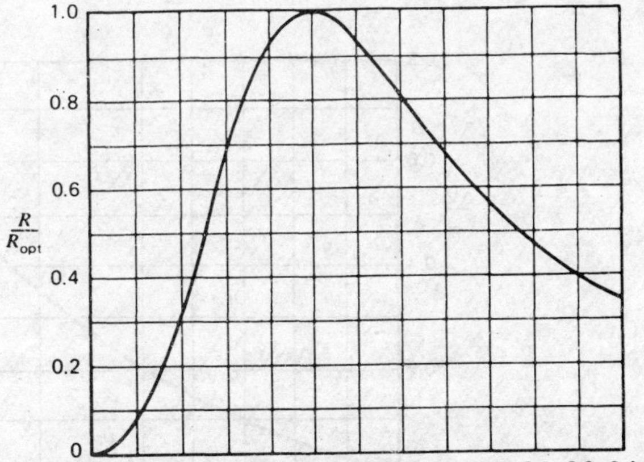

Figure 7.24 Effect of departing from optimum velocity on range for propeller-driven aircraft.

were cruising at the optimum airspeed for no wind into a headwind of equal magnitude. Your ground speed would be zero. Obviously, your airspeed is no longer optimum, and it would behoove you to increase your airspeed. Thus, without going through any derivations, we conclude that the optimum airspeed increases with headwind.

Correcting Equation 7.31 for headwind is left to you. If V_w denotes the headwind, this equation becomes (now expressed in the English system),

$$R = \frac{375\ \eta}{(\text{BSFC})\,\epsilon}\left(1 - \frac{V_w}{V}\right)\ln\left(1 + \frac{W_f}{W_E}\right) \tag{7.40}$$

The effect of headwind on the optimum cruising airspeed can be obtained by minimizing $\epsilon/(1 - V_w/V)$. Without going into the details, this leads to the following polynomial

$$2\left(\frac{V}{V_{\text{opt}}}\right)^5 - 3\left(\frac{V_w}{V_{\text{opt}}}\right)\left(\frac{V}{V_{\text{opt}}}\right)^4 - 2\left(\frac{V}{V_{\text{opt}}}\right) + \frac{V_w}{V_{\text{opt}}} = 0$$

Here, V is the optimum cruising velocity for a given headwind and V_{opt} is the value of V for a V_w of zero. Expressing $\frac{V}{V_{\text{opt}}}$ as

$$\frac{V}{V_{\text{opt}}} = 1 + \Delta V/V_{\text{opt}}$$

$\Delta V/V_{\text{opt}}$ is presented as a function of V_w/V_{opt} in Figure 7.25. This figure shows, for example, that if one has a headwind equal to 50% of the optimum velocity for no wind, he or she should cruise at an airspeed 20% higher than the optimum, no-wind velocity.

Computer Exercise 7.5 "CRUISE"

Modify the "Climb" program written previously to produce a program that will calculate the cruise performance of a piston-engine, turboprop or a pure jet-powered airplane. Use a table look-up to determine thrust or power and correct for part-load operation. Program should loop over altitude, weight, and airspeed to produce distance covered per unit weight of fuel for a given airplane gross weight,

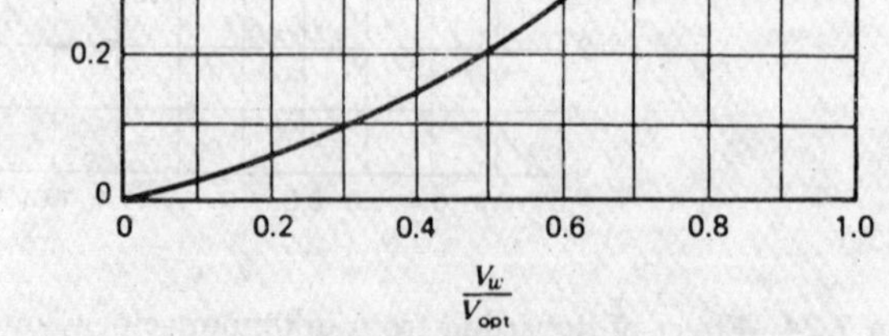

Figure 7.25 Effect of headwind on optimum cruising velocity.

altitude, and true airspeed. Also, a by-product should be the time-rate of fuel consumption for endurance calculations.

MAXIMUM ENDURANCE

Endurance refers to the time that elapses in remaining aloft. Here, one is concerned with the time spent in the air, not the distance covered. A pilot in a holding pattern awaiting clearance for an instrument landing is concerned with endurance. The maximum endurance will be obtained at the airspeed that requires the minimum fuel flow rate. In the case of a turbojet or turbofan engine, the product of TSFC and the drag is minimized.

Assuming a constant TSFC for the jet airplane leads to Equations 7.36 and 7.37, which relate to the maximum range for a propeller-driven airplane. Thus, for a given weight of fuel, W_f, the maximum endurance, t_e, of a turbojet will be

$$t_e = \frac{1}{2(\mathrm{TSFC})}\left(\frac{W_f}{W}\right)\left[\frac{\pi e A}{(f/S)}\right]^{1/2} \tag{7.41}$$

Notice that the endurance is independent of altitude. This follows from the fact that the minimum drag does not vary with altitude. To obtain this endurance, the airplane is flown at the airspeed given by Equation 7.37.

Assuming a constant SFC, a propeller-driven aircraft will have its maximum endurance when flown at the airspeed for minimum required power. This speed has been given previously as Equation 7.20. At the minimum power, the induced power is three times the parasite power. Hence.

$$P_{\min} = 2\rho f V^3 \tag{7.42}$$

$$= \frac{2}{\sqrt{\sigma}}\left[\frac{4(W/b)^2}{3\pi e}\right]^{3/4} \frac{f^{1/4}}{\rho_0{}^{1/2}}$$

The endurance time for a propeller-driven airplane then becomes

$$t_e = \frac{\sqrt{\sigma}}{2\ (\mathrm{SFC})}\left[\frac{3\pi e}{4\ (W/b)^2}\right]^{3/4} \frac{\rho_0^{1/2}\ W_f \eta}{f^{1/4}} \tag{7.43}$$

(SFC) in this equation has units consistent with the other terms; that is, weight per power second where the weight is in newtons or pounds with the power in newton meters per second or foot-pounds per second. For example, if

$$\mathrm{BSFC} = 0.5\ \mathrm{lb/bhp/hr}$$

then

$$\mathrm{SFC} = 2.53 \times 10^{-7}\ \mathrm{lb/(ft\text{-}lb/sec)/sec}$$

Actually SFC used in this basic manner has the units of 1/length. In the English system this becomes ft^{-1} and in the SI system it is m^{-1}. Notice that the endurance of a propeller-driven airplane decreases with altitude. This follows from the fact that the minimum power increases with altitude.

Some of the foregoing equations for range and endurance contain the weight which, of course, varies with time as fuel is burned. Usually, for determining the optimum airspeed or the endurance time, it is sufficiently accurate to assume an average weight equal to the initial weight minus half of the fuel weight. Otherwise, numerical and graphical procedures must be used to determine range and endurance.

DESCENT

The relationships previously developed for a steady climb apply as well to descent. If the available thrust is less than the drag, Equation 7.17 results in a negative R/C. In magnitude this equals the rate of descent, R/D. The angle of descent, θ_D, in radians, is given by

$$\theta_D = \frac{D - T}{W} \tag{7.44}$$

Civil aircraft rarely descend at angles greater than 10°. The glide slope for an ILS (instrument landing system) approach is only 3°. Steeper slopes for noise abatement purposes are being considered, but only up to 6°.

The minimum θ_D value in the event of an engine failure is of interest. From Equation 7.44 we see that this angle is given by

$$\theta_{D_{\min}} = \epsilon_{\min} \text{ rad} \tag{7.45}$$

Thus, the best glide angle is obtained at the C_L giving the lowest drag-to-lift ratio. This angle is independent of gross weight. However, the greater the weight, the higher the optimum airspeed will be. The minimum ϵ and corresponding airspeed have been given previously as Equations 7.36 and 7.37. Of course, in the event of an engine failure, one must account for the increase in f caused by the stopped or windmilling propeller, or by the stopped turbojet.

LANDING

The landing phase of an airplane's operation consists of three segments; the approach, the flare, and the ground roll. FAR Part 25 specifies the total landing distance to include that required to clear a 50-ft (15.2-m) obstacle. A sketch of the landing flight path for this type of approach is shown in Figure 7.26. The ground roll is not shown, since it is simply a continuous deceleration along the runway. FAR Part 25 specifies the following, taken verbatim:

§ 25.125 Landing.

(a) The horizontal distance necessary to land and to come to a complete stop (or to a speed of approximately 3 knots for water landings) from a point 50 feet above the landing surface must be determined

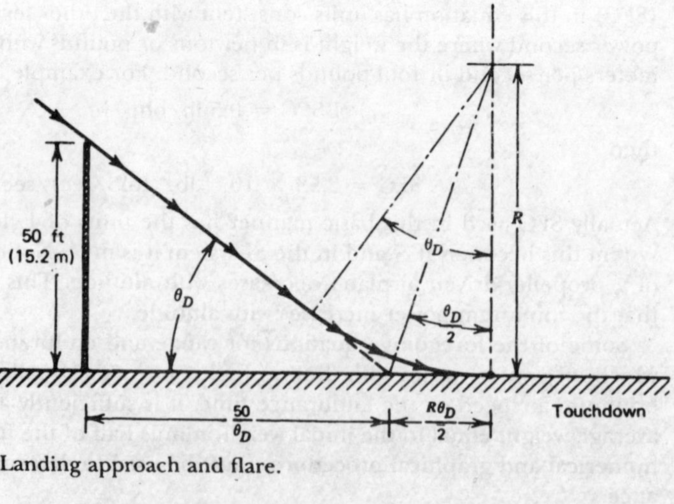

Figure 7.26 Landing approach and flare.

(for standard temperatures, at each weight, altitude, and wind within the operational limits established by the applicant for the airplane) as follows:

(1) The airplane must be in the landing configuration.

(2) A steady gliding approach, with a calibrated airspeed of not less than 1.3 V_s, must be maintained down to the 50-ft height.

(3) Changes in configuration, power or thrust, and speed, must be made in accordance with the established procedures for service operation.

(4) The landing must be made without excessive vertical acceleration, tendency to bounce nose over, ground loop, porpoise, or water loop.

(5) The landings may not require exceptional piloting skill or alertness.

(b) For landplanes and amphibians, the landing distance on land must be determined on a level, smooth, dry, hard-surfaced runway. In addition—

(1) The pressures on the wheel braking systems may not exceed those specified by the brake manufacturer;

(2) The brakes may not be used so as to cause excessive wear of brakes or tires; and

(3) Means other than wheel brakes may be used if that means—

(i) Is safe and reliable;

(ii) Is used so that consistent results can be expected in service; and

(iii) Is such that exceptional skill is not required to control the airplane.

(c) For seaplanes and amphibians, the landing distance on water must be determined on smooth water.

(d) For skiplanes, the landing distance on snow must be determined on smooth, dry snow.

(e) The landing distance data must include correction factors for not more than 50 percent of the nominal wind components along the landing path opposite to the direction of landing, and not less than 150 percent of the nominal wind components along the landing path in the direction of landing.

(f) If any device is used that depends on the operation of any engine, and if the landing distance would be noticeably increased when a landing is made with that engine inoperative, the landing distance must be determined with that engine inoperative unless the use of compensating means will result in a landing distance not more than that with each engine operating.

The total distance thus calculated must be increased by a factor of 1.667.

FAR Part 23 is somewhat simpler in defining the landing for airplanes certified in the normal, utility, or acrobatic categories. It states the following:

§23.75 Landing.

(a) For airplanes of more than 6,000 pounds maximum weight (except skiplanes for which landplane landing data have been determined under this paragraph and furnished in the Airplane Flight Manual), the horizontal distance required to land and come to a complete stop (or to a speed of approximately three miles per hour for seaplanes and amphibians) from a point 50 feet above the landing surface must be determined as follows:

(1) A steady gliding approach with a calibrated airspeed of at least 1.5 V_{s_1} must be maintained down to the 50 foot height.

(2) The landing may not require exceptional piloting skill or exceptionally favorable conditions.

(3) The landing must be made without excessive vertical acceleration or tendency to bounce, nose over, ground loop, porpoise, or water loop.

(b) Airplanes of 6,000 pounds or less maximum weight must be able to be landed safely and come to a stop without exceptional piloting skill and without excessive vertical acceleration or tendency to bounce, nose over, ground loop, porpoise, or water loop.

Airborne Distance

From Figure 7.26, since θ_D is a small angle, the total airborne distance, s_A, is given in meters by

$$s_A = \frac{15.2}{\theta_D} + \frac{R\theta_D}{2} \tag{7.46}$$

This assumes the flare to be a circular arc having a radius of R. If V_A is the approach velocity, this velocity is assumed to remain constant throughout the flare. The acceleration toward the center of curvature, a_n, will therefore be

$$a_n = \frac{V_A^2}{R}$$

However,

$$L - W = \frac{W}{g} a_n$$

Thus,

$$R = \frac{V_A^2}{g(L/W - 1)}$$

If C_{LA} denotes the lift coefficient during the steady approach then, during the flare,

$$\frac{L}{W} = \frac{C_L}{C_{LA}}$$

The flare radius can therefore be expressed as

$$R = \frac{V_A^2}{g[(C_L/C_{LA}) - 1]} \tag{7.47}$$

FAR Part 25 requires that V_A exceed the stalling speed in the landing configuration by 30%. Thus,

$$C_{LA} \leq \frac{C_{L\max}}{1.69} \tag{7.48}$$

The ratio C_L / C_{LA}, thus, can vary anywhere from just above 1 to 1.69 or higher. A typical value of this ratio for jet transports is 1.2. Using this value, but keeping in mind that it can be higher, the total airborne distance, in feet, becomes

$$s_A = \frac{50}{\theta_D} + \frac{V_A^2 \theta_D}{0.4 g} \tag{7.49}$$

The 767-300ER approaches at a speed of 141 kt at an angle of 3°. Thus,

$$\begin{aligned} s_A &\simeq 955 + 231 \\ &= 1186 \text{ ft } (362 \text{ m}) \end{aligned}$$

After touchdown, an approximately 2-sec delay is allowed while the pilot changes from the landing to the braking configuration. During this period the airplane continues to roll at the speed V_A. Actually, practice has shown that the speed decreases during the flare by approximately 5 kt typically. Denoting this portion by a subscript "tran" for transition,

$$\begin{aligned} s_{\text{tran}} &\simeq 2(V_A) \\ &\simeq 460 \text{ ft } (140 \text{ m}) \end{aligned}$$

Ground Roll

The calculation of the ground roll in landing follows along the same lines used for a takeoff ground roll, but with different parameters and initial conditions. The braking coefficient of friction varies from approximately 0.4 to 0.6 on a hard, dry surface to 0.2 on wet grass or 0.1 on snow. With spoilers, the lift is essentially zero. Also, with flaps and spoilers, the parasite drag coefficient may be higher.

Beginning with an initial value of V_A, the equations of motion can be numerically integrated, accounting for the variation with V of any reverse thrust, drag, and possibly lift. One can also use the approximate relationship derived previously, Equation 7.9. In the case of landing, this becomes

$$s = \frac{V_A{}^2}{2\bar{a}} \tag{7.50}$$

where $\bar{a}$ is the magnitude of the deceleration evaluated at $V_A/\sqrt{2}$.

As an example, consider the 767-300ER at its maximum landing weight of 300,000 lb at SSL conditions. Let us assume a μ of 0.4, an f of 150 ft^2 and V_A equal to 141 kt. The velocity for calculating the average deceleration will be $141/\sqrt{2}$ or 99.7 kt. At this speed, the drag will equal 5063 lb. The frictional retarding force will be equal to μW, or 120,000 lb, assuming the lift is zero. This is reasonable for two reasons: (1) on the wheels, the C_L is small and (2) spoilers are deployed by the pilot shortly after touchdown to cancel the lift. These forces result in an average deceleration of

$$\bar{a} = 13.4 \text{ ft/s}^2$$

or 0.42 g.

Therefore, the total ground-roll distance is estimated from Equation 7.50 to equal 2115 ft. Added to the airborne distance, this gives a total estimated FAR landing distance, including the factor of 1.667, of 6270 ft (1912 m).

Lighter aircraft, except on an instrument approach, tend to descend at an angle steeper than 3°. With their lower wing loadings, light aircraft also touch down at much lower speeds. Hence, their landing distances are significantly less than those for a jet transport. A Cherokee Arrow, for example, touching down at approximately 65 kt can be stopped with moderate braking within 300 m (1000 ft).

RANGE PAYLOAD

Specifications of military aircraft and many larger civil aircraft include range-payload curves. This is a graph that for a particular mission profile, presents the effect of trading off payload for fuel on the range of an airplane. In determining such a curve, one must consider the operational phases that have been treated thus far in this chapter.

Airplanes are designed and certified to operate at a gross weight not to exceed some maximum weight. In many designs, the volume of the fuel tanks is sufficient so that, with full payload, the certified gross weight will be exceeded if the tanks are filled. Thus, if the payload is sufficiently high, the fuel tanks can only be partially filled resulting in a lower range. Specifications for many military and civil aircraft include a *range-payload* curve. This is a graph that, for a particular mission profile, presents the effect of trading off fuel for payload on the range of the airplane.

As an example in calculating a range-payload curve, consider a commuter-type airplane, the DHC-8, referred to as the Dash 8. This airplane is illustrated in Figure 7.27a and b. The pertinent dimensions, weights, and performance values are listed as follows:

Wing span, b	85 ft
Wing area	585 ft^2
Maximum ramp (parking area) weight	34,700 lb
Maximum takeoff gross weight (MTOGW)	34,500 lb
Maximum landing weight	33,900 lb
Maximum fuel	5678 lb

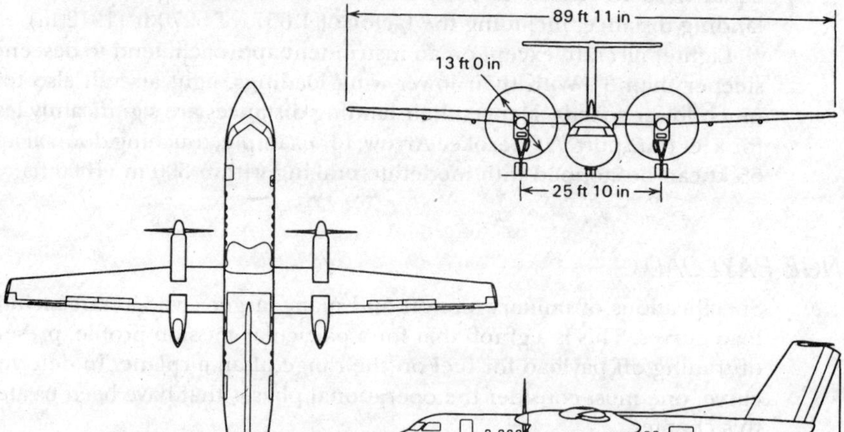

Figure 7.27 The de Havilland DHC-8 Commuter Airplane.

Maximum payload (passengers)	8400 lb
Maximum payload (cargo)	9349 lb
Empty weight	22,600 lb
Cruising speed, 15,000 ft, 34,500 lb	265 kt
Maximum R/C at SSL, 34,500 lb	1560 fpm
Power plant (Figures 6.47–6.50)	PW120

The total wetted area is estimated to equal 3232 ft^2 and the skin friction coefficient to be 0.0046 resulting in an equivalent flat plate area of 14.9 ft^2.

The range will be determined according to the following mission definition:

- Taxi and takeoff clearance
- Climb from sea level to 15,000 ft
- Cruise at 80% of maximum SSL cruise power
- Descend to sea level
- Land with 45-min reserve

The reserve fuel is calculated on the basis of holding at 15,000 ft.

To construct the range-payload curve, rate-of-climb and cruise curves must first be determined similar to those that were done for the 767-300ER. These curves are presented in Figures 7.28–7.30 and were obtained by applying the climb and cruise programs previously developed in this chapter. Given these figures, the process begins by assuming a payload, maybe zero initially. The initial fuel weight is then calculated as the maximum ramp weight minus the sum of the empty weight and the payload. If the fuel weight determined in this manner exceeds the maximum allowable fuel weight, then the fuel weight is equated to this maximum. The takeoff gross weight is then found as the sum of the empty weight, payload, and fuel weight minus an allowance for warm-up, clearance, and taxi. For this example, this amount is assumed to equal 200 lb, which is the difference between the maximum ramp weight and the maximum takeoff gross weight.

It is assumed initially that the takeoff gross weight is constant during the climb to 15,000 ft. Also, since the maximum climb power for the PW120 turbojet is given only to 10,000 ft in Figure 6.49, the R/C curve is assumed to be linear above an altitude of approximately 4000 ft. Therefore, using Figure 7.29, the time-to-climb to 15,000 ft is found from Equation 7.29 to be nearly constant over a range of gross weights and equal to approximately 11.8 min at an assumed true airspeed of 120 kt. An iteration on the climb produces a reduction in the time of only a few seconds. Therefore, the fuel used and the forward distance gained is assumed to be the same for all of the gross weights. During the climb to 15,000 ft, it is estimated that 305 lb of fuel are burned and the airplane travels a distance of 23 N mi.

Now consider Figure 7.30. Notice for this turboprop airplane that, unlike the

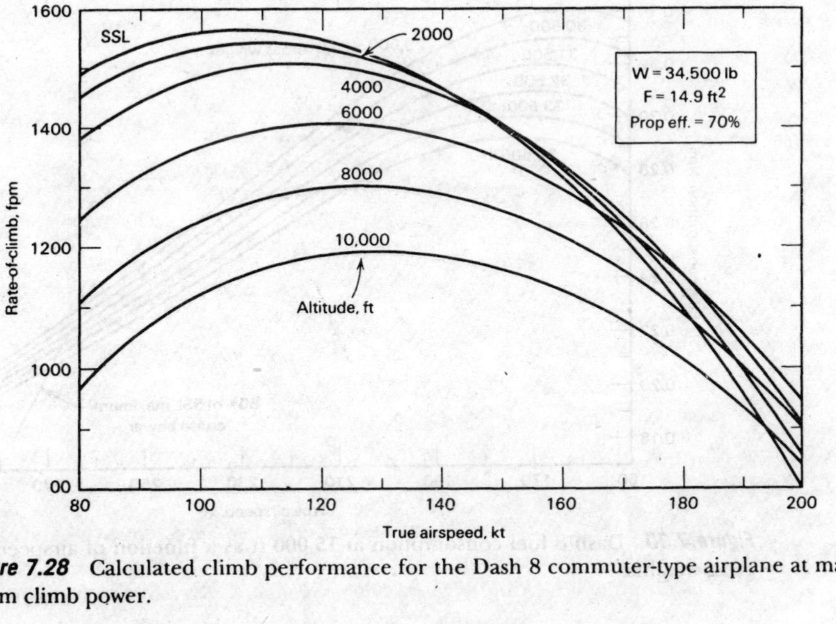

Figure 7.28 Calculated climb performance for the Dash 8 commuter-type airplane at maximum climb power.

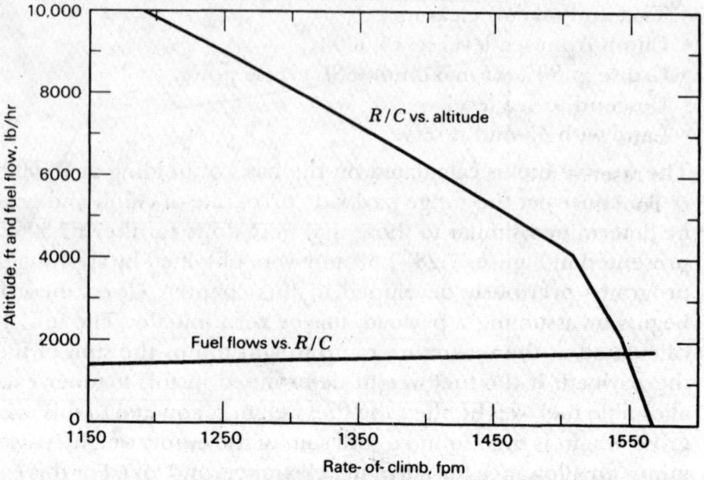

Figure 7.29 Calculated maximum rate-of-climb and corresponding fuel flow for the Dash 8 airplane at maximum climb power.

jet transport, the fuel burned per mile peaks at a speed well below that corresponding to the maximum cruise thrust. Thus, although the range can be extended considerably by flying at a power less than the specified power of 80% maximum, the time required to do so would be almost doubled. Since "time is money," the higher speed is chosen. A line representing 80% of the maximum SSL cruise power is shown on the figure. The intersection of this line with the altitude curves is to be used for determining the range.

The minimum rate at which fuel is burned is found from the cruise program to be a true airspeed slightly less than 150 kt. The curve is fairly flat with speed so that the values for lb/hr of fuel burned while holding were chosen at 150 kt. The reader

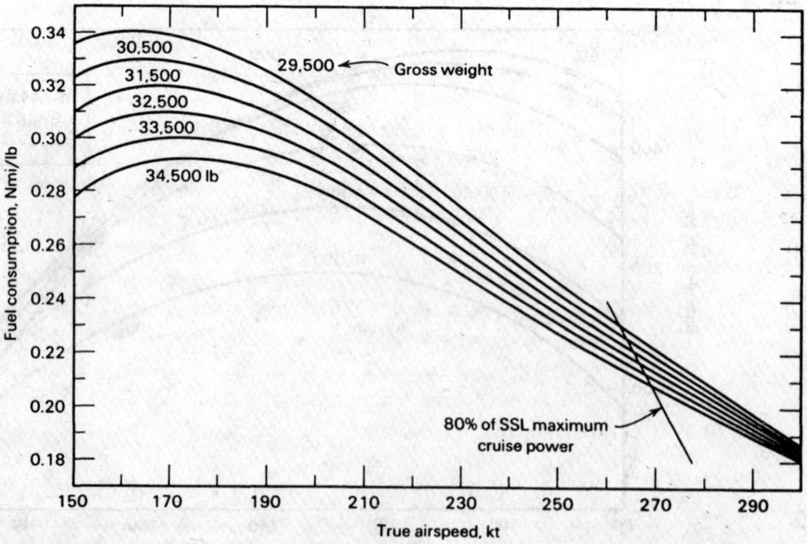

Figure 7.30 Dash 8 fuel consumption at 15,000 ft as a function of airspeed for a range of gross weights.

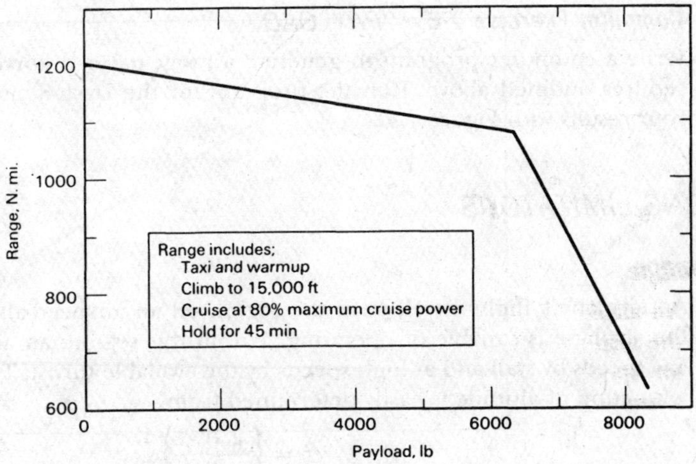

Figure 7.31 Range-payload curve for Dash 8.

can obtain these from Figure 7.30 with a small effort, and it will be found that the fuel flow, w_f, varies linearly with gross weight according to

$$w_f = 0.019W\text{-}114.5 \text{ lb/hr} \tag{7.51}$$

Thus, for a hold of 45 min, the total amount of fuel expended will be equal to

$$W_f = 0.01425W\text{-}85.9 \text{ lb} \tag{7.52}$$

The descent to sea level will not figure into the range calculation done here. No credit will be taken for the distance covered during the descent nor will a penalty be taken for the fuel burned during the descent. This is not an unreasonable procedure since the hold is normally performed over a beacon near the destination airport and, once cleared, the descent is made at a reduced power near idle.

The process of calculating the range-payload curve begins by starting with the airplane at 15,000 ft, having expended 505 lb of fuel and traveled 23 N mi. A small amount of fuel is then burned and the distance calculated, which can be traversed with this fuel decrement. In so doing, Figure 7.30 is entered along the 80% line at an average gross weight equal to the beginning gross weight minus half of the fuel burned. The total distance covered at this point is calculated as the sum of the distance covered at 15,000 ft and the 23 N mi traveled during the climb. The fuel to hold for 45 min is then calculated at a gross weight equal to the gross weight before the fuel burn minus the decrement of fuel burned. All of the fuel burned to this point is then compared to the fuel that was loaded into the plane on the ramp. If the fuel burned is equal to or greater than the initial fuel, then the range for the payload is equal to the total distance covered to this point.

The process described above is repeated over a range of payloads from zero up to the maximum allowable value. The procedure lends itself nicely to a computer program, particularly since the inputs for fuel to cruise and hold can be fitted with linear functions of the gross weight. Using such a program, the range-payload curve presented in Figure 7.31 was produced. The break in the slope of the curve at a payload of 6422 lb is the maximum payload above which the fuel load must be reduced. Again, as with the 767-300ER, it is emphasized that this estimated performance was obtained by your author and is not endorsed or confirmed by the manufacturer.

Computer Exercise 7.6 "PAYLOAD"

Write a computer program to generate a range-payload curve following the procedures outlined above. Run the program for the Dash-8 airplane and compare your results with Figure 7.31.

OPERATING LIMITATONS

Flight Envelope

An airplane's flight envelope is the region on an airspeed-altitude plot in which the airplane is capable of operating. Within this region, an airplane is limited at low speeds by stall and at high speeds by the available thrust. The stall boundary as a function of altitude is easily determined from

$$V^2 = \left(\frac{2\ W/S}{\rho_0 C_{L_{\max}}}\right)\frac{1}{\sigma} \tag{7.53}$$

The high-speed boundary is determined from power-available, power-required curves such as those presented in Figure 7.18. As an example, let us again consider the Cherokee Arrow. Figure 7.32 was prepared using a gross weight of 11.8 kN, a $C_{L_{\max}}$ of 1.6, a constant propeller efficiency of 0.85, and a sea level engine power of 149 kW. The Cherokee is capable of level flight within the region bounded by the two curves labeled "stall" and "maximum" power.

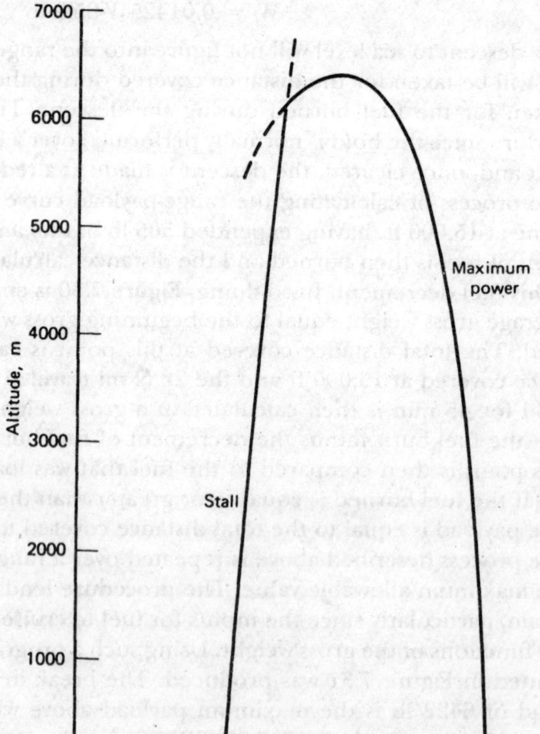

Figure 7.32 Flight envelope for the Cherokee Arrow.

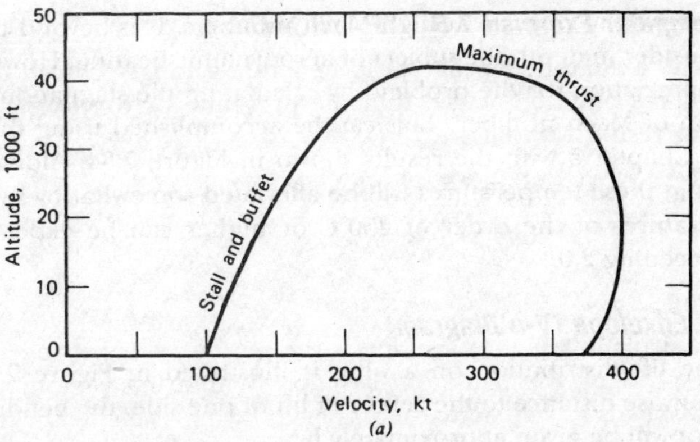

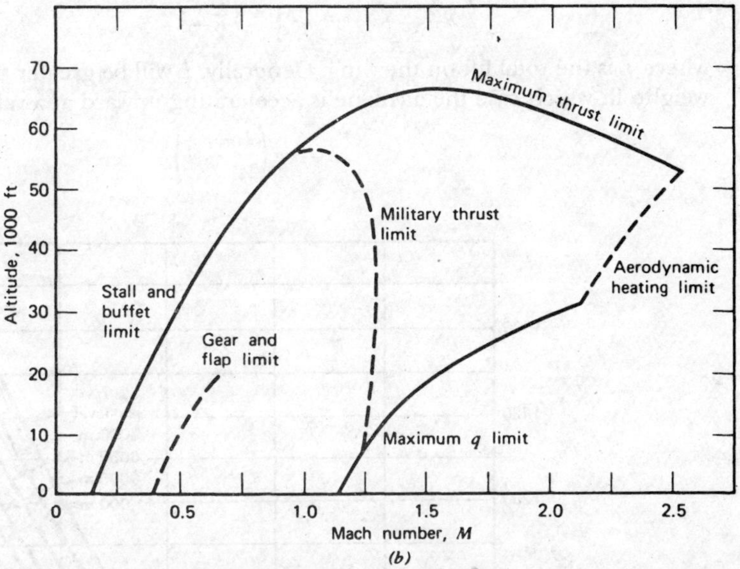

Figure 7.33 Typical aircraft flight envelopes. (*a*) Subsonic aircraft. (*b*) Supersonic aircraft. (L. M. Nicolai, *Fundamentals of Aircraft Design*, L. M. Nicolai, 1975. Reprinted by permission of L. M. Nicolai.)

A typical flight envelope for a supersonic aircraft is given in Figure 7.33. At high subsonic Mach numbers, a phenomenon known as *buffet* can limit flight to speeds higher than the stalling speeds. This type of buffeting is caused by an instability in the position of the shock waves near the trailing edge of the upper and lower wing surfaces. As the stall is approached, these waves begin to move fore and aft out of phase with each other, producing a periodic flow behind the wing that resembles a Karman vortex street. The tail, in proximity to this unsteady flow, can produce a severe shaking of the airplane.

Also shown in Figure 7.33 is a limit on the maximum dynamic pressure that can be tolerated. This boundary arises from structural considerations and involves items such as flutter, torsional divergence, and static pressure within an engine inlet diffuser.

An aerodynamic heating limit as shown in Figure 7.33 also exists for airplanes designed to operate at high Mach numbers. It is beyond the scope of this text to consider in depth the subject of aerodynamic heating. However, one can gain some appreciation for the problem by calculating the stagnation temperature as a function of Mach number. This can be accomplished using the relationships covered in Chapter 5 with the results shown in Figure 7.34. Along the leading edge of a wing, these temperatures will be alleviated somewhat by sweep. Nevertheless, temperatures of the order of 250°C or higher can be expected for Mach numbers exceeding 2.0.

Maneuvering Envelope (V-n Diagram)

The lift distribution on a wing is illustrated in Figure 7.35. If y represents the spanwise distance to the center of lift of one side, the bending moment at the wing root will be given approximately by

$$M \simeq y\frac{L}{2} \tag{7.54}$$

where L is the total lift on the wing. Generally, L will be greater than the airplane's weight, in which case the airplane is accelerating upward at a value equal to

$$a = g\left(\frac{L}{W} - 1\right)$$

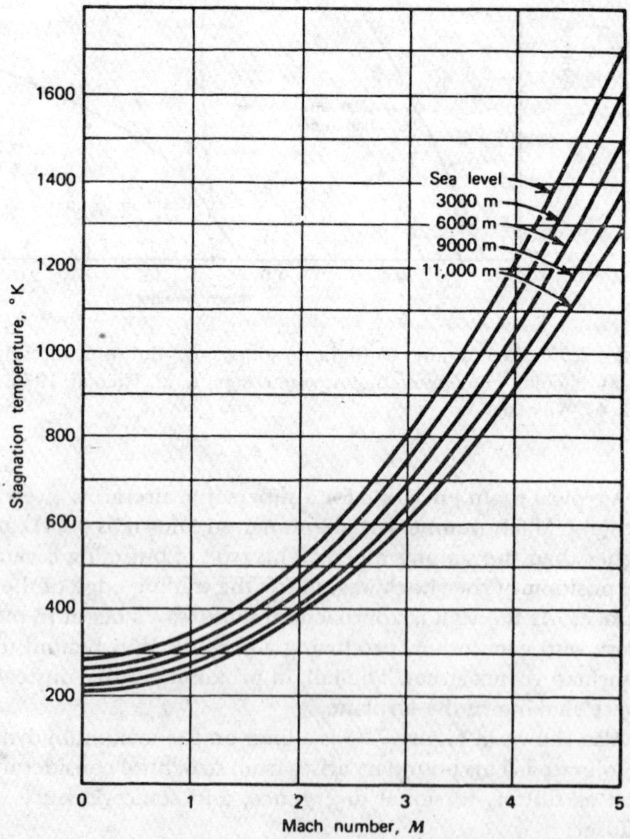

Figure 7.34 Stagnation temperature as a function of altitude and Mach number.

In terms of acceleration,

$$L = W\left(1 + \frac{a}{g}\right) \tag{7.55}$$

so that the bending moment becomes

$$M = y\frac{W}{2}\left(1 + \frac{a}{g}\right)$$

The term $(1 + a/g)$ is known as the *load factor, n.*

$$n = 1 + \frac{a}{g} \tag{7.56}$$

In this example, the wing bending moment in steady flight is seen to increase by the factor n. Similarly, n is a measure generally of the increase in the loads on any member of the airplane resulting from accelerations. In steady, level flight, n is equal to 1. As a result of maneuvering or gusts, n can increase in magnitude to high values and can be positive or negative.

The value of n that can be achieved by maneuvering can be obtained from

$$n = \frac{L}{W}$$

$$= \frac{\frac{1}{2}\rho V^2 S C_{L_{\max}}}{W}$$

But $2W/\rho S C_{L_{\max}}$ equals the stalling speed, V_s. Therefore,

$$n = \left(\frac{V}{V_s}\right)^2 \tag{7.57}$$

Since V can be appreciably greater than the square of the stalling speed, v_s, it is not practical to design an airplane's structure to withstand the highest possible load factors that it could produce. Instead, based on experience, airplanes are certified to withstand different limit load factors, depending on the airplane's intended use. A limit load is one that can be supported by a structure without yielding. In addition to designing to the limit loads, FAR Parts 23 and 25 require factors of safety of 1.5 to be applied to the sizing of the structure. Since the ultimate

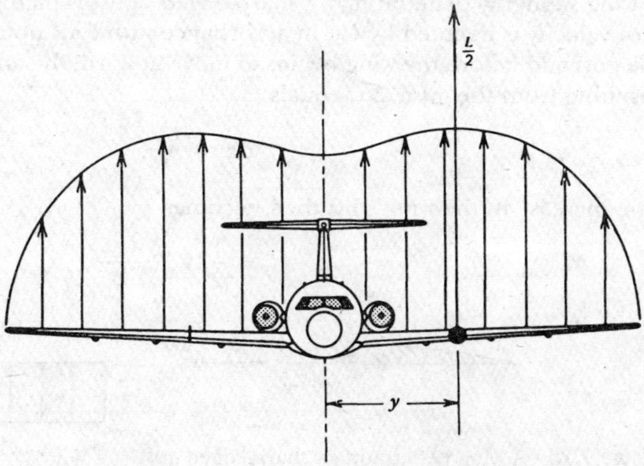

Figure 7.35 Wing bending moment due to spanwise lift distribution.

allowable stress of aluminum alloys is approximately 50% greater than the yield stress, a factor of safety of 1.5 applied to the limit loads is approximately equivalent to designing to ultimate load factors with no factor of safety.

Civil airplanes are designed in the normal, utility, acrobatic, and transport categories. For the first three categories, FAR Part 23 states:

§ 23.337 Limit maneuvering load factors.

(a) The positive limit maneuvering load factor n may not be less than

[(1) $2.1 + \dfrac{24{,}000}{W + 10{,}000}$ *for normal category airplanes, except that n need not be more than 3.8;]*

(2) 4.4 for utility category airplanes; or

(3) 6.0 for acrobatic category airplanes.

(b) The negative limit maneuvering load factor may not be less than—

(1) 0.4 times the positive load factor for the normal and utility categories; or

(2) 0.5 times the positive load factor for the acrobatic category.

(c) Maneuvering load factors lower than those specified in this section may be used if the airplane has design features that make it impossible to exceed these values in flight.

For the transport category, FAR Part 25 states:

§ 25.337 Limit maneuvering load factors.

(a) Except where limited by maximum (static) lift coefficients, the airplane is assumed to be subjected to symmetrical maneuvers resulting in the limit maneuvering load factors prescribed in this section. Pitching velocities appropriate to the corresponding pull-up and steady turn maneuvers must be taken into account.

(b) The positive limit maneuvering load factor n for any speed up to V_D *may not be less than 2.5.*

(c) The negative limit maneuvering load factor—

(1) May not be less than − 1.0 at speeds up to V_C*; and*

(2) Must vary linearly with speed from the value at V_C *to zero at* V_D*.*

(d) Maneuvering load factors lower than those specified in this section may be used if the airplane has design features that make it impossible to exceed these values in flight.

V_c, known as the design cruising speed, need not exceed V_H, the maximum speed in level flight at maximum continuous power. Otherwise it may not be less than $33\sqrt{W/S}$ knots for the normal category or V_B plus 43 knots for transports. V_B is the lowest speed for a load factor of 2.5. V_D, the design dive speed, need not be greater than V_H for a transport or $1.4V_C$ for the normal category.

Gust Load Factors

A wing suddenly penetrating a "sharp-edged" gust is pictured in Figure 7.36. The gust velocity is denoted by U_{de} in accordance with FAR notation. After penetrating the gust and before the wing begins to move upward, the angle-of-attack increment resulting from the gust, $\Delta\alpha$, equals

$$\Delta\alpha = \frac{U_{de}}{V} \tag{7.58}$$

The increase in the wing's lift then becomes

$$\Delta L = \frac{1}{2}\rho V^2 S a \frac{U_{de}}{V}$$

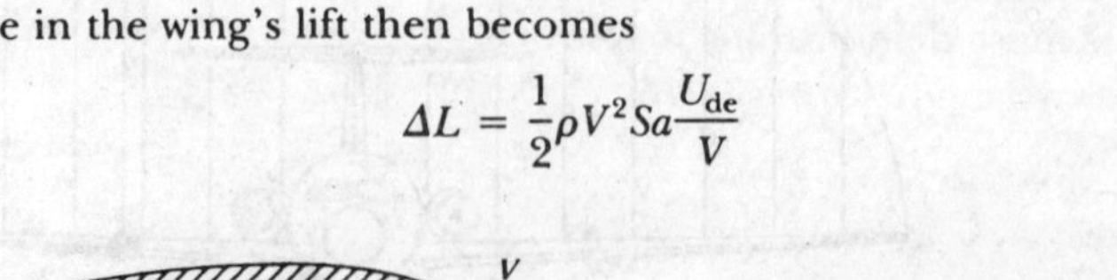

Figure 7.36 A wing penetrating a sharp-edged gust.

Before encountering the gust, in level flight,

$$W = \frac{1}{2}\rho V^2 Sa\alpha$$

The load factor, n, resulting from the gust encounter therefore becomes

$$\begin{aligned} n &= \frac{L}{W} \\ &= \frac{W + \Delta L}{W} \\ &= 1 + \frac{U_{de}}{V\alpha} \end{aligned}$$

or

$$n = 1 + \frac{\rho V U_{de} a}{2(W/S)} \tag{7.59}$$

In practice, one never encounters a truly sharp-edged gust. Therefore, U_{de} is multiplied by an alleviation factor less than unity, again based on experience, which lessens the acceleration due to the gust. The final result, given in FAR Part 23, for the load factor resulting from a gust is expressed as follows:

$$n = 1 + \frac{K_g U_{de} Va}{498\ (W/S)} \tag{7.60}$$

where

$$K_g = \frac{0.88\ \mu}{5.3 + \mu} = \text{gust alleviation factor}$$

$$\mu = \frac{2(W/S)}{\rho \bar{c} a g} = \text{airplane mass ratio}$$

U_{de} = "derived" gust velocity, fps

$\bar{c}$ = mean geometric chord, S/b

V = equivalent airspeed in knots

a = slope of the airplane normal force coefficient $C_N/\,rad$

Positive and negative values of U_{de} up to 50 fps must be considered at v_c at altitudes between sea level and 20,000 ft. The gust velocity may be reduced linearly from 50 fps at 20,000 ft to 25 fps at 50,000 ft. Positive and negative gusts of 25 fps at V_D must be considered at altitudes between sea level and 20,000 ft. This velocity can be reduced linearly to 12.5 fps at 50,000 ft.

For FAR Part 23, the foregoing criteria for the maneuvering and gust loads results in the type of *V-n* diagram pictured in Figure 7.37. In certifying an airplane, one must demonstrate the structural integrity of the airplane subjected to the aerodynamic loadings that can exist throughout the *V-n* diagram.

ENERGY METHODS FOR OPTIMAL TRAJECTORIES

The problem to be considered briefly in this section concerns the altitude-velocity (or Mach number) schedule, which should be flown to minimize the time or fuel required to go from one speed and altitude to another speed and altitude. As pointed out in Reference 7.4, this problem can be solved by the application of

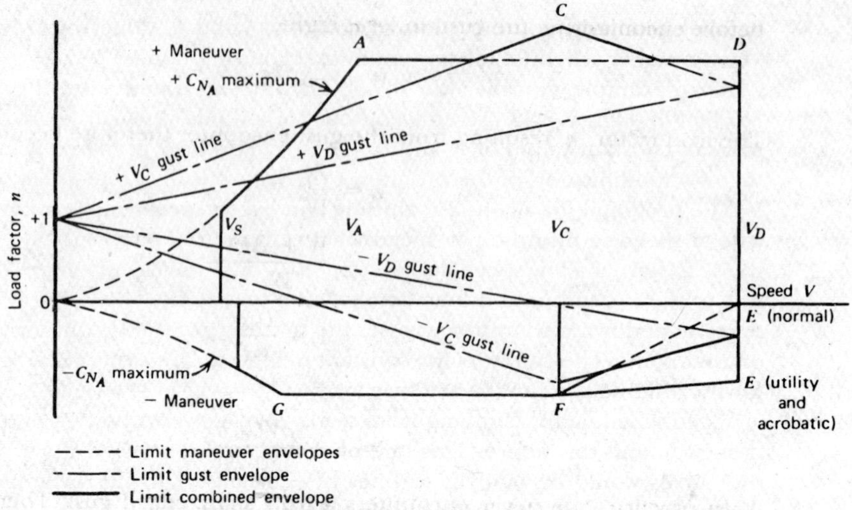

Figure 7.37 *V-n* diagram for airplanes in the normal, utility, and acrobatic categories.

variational calculus. However, the result is a formidable computer program. As an alternate method, which is approximate but close to the more exact solution, one can obtain a graphical solution by considering the energy state of the airplane.

As noted previously, Equation 7.16 is an energy relationship for the rate-of-climb. If we let h_e denote the total energy, kinetic and potential, per unit weight of the airplane, Equation 7.16 can be written in terms of this specific energy as

$$\frac{dh_e}{dt} = \frac{V(T - D)}{W} \tag{7.61}$$

where

$$h_e = h + \frac{V^2}{2g} \tag{7.62}$$

dh_e/dt will be denoted by P_s and is called the *excess specific power.*

The rate of change of h_e with respect to fuel weight, W_f, will be denoted by f_s and can be written as

$$\begin{aligned} f_s &= \frac{dh_e}{dW_f} \\ &= \frac{dh_e/dt}{dW_f/dt} \\ &= \frac{dh_e/dt}{T(\text{SFC})} \text{ (ft/lb or m/N)} \end{aligned} \tag{7.63}$$

The time required to go from one energy level to another will be given by

$$\Delta t = \int_{h_1}^{h_2} \frac{1}{P_s}\, dh_e \tag{7.64}$$

The path to minimize Δt at any altitude and airspeed will be the one that gives the maximum rate of change of h_e for a given P_s value. Therefore, if contours of constant h_e and constant P_s values are plotted as a function of altitude and Mach number, the path for minimum time will be the locus of points for which the contours are parallel. Similarly, contour plots of constant f_s and h_e values provide an altitude-Mach number schedule for minimum fuel consumption.

As an example, consider the 767-300ER at its maximum takeoff gross weight of 387,000 lb. Figure 7.38 presents contours of altitude versus true airspeed for contours of constant values of P_s and h_e. These curves were prepared using a program that is a modification of the program written to calculate rates-of-climb.

The program, for a constant altitude and excess specific power, iterates over the true airspeed to find the two speeds that result in the given P_s. One way of doing this is to start at a low speed and increment V until the calculated p_s crosses the left side of the P_s contour and then converge on that point. Then start at a high V and decrement downward until P_s crosses the right side of the P_s contour and converge on that point. The contours for constant height, h_e, are easily generated by solving for the altitude as a function of true airspeed for a given, constant h_e. The altitude-airspeed schedule for climbing from sea level to a given altitude is indicated by the dashed line in this figure. This line passes through points on the P_s curves where the curves would be tangent to lines of constant h_e. In this example, where the thrust and drag are well behaved, the result is about as one would expect.

The results are substantially different, however, for an airplane designed to operate through Mach 1, particularly if the thrust is marginal in the transonic region. Such a case is presented in Figure 7.39 (taken from Ref. 7.4). As indicated by the dashed line, in this case the optimum trajectory consists of a subsonic climb at a nearly constant Mach number to 33,000 ft followed by a descent through the transonic drag rise region to 20,000 ft and a Mach number of 1.25. A climb to 39,000 ft at increasing Mach numbers then ensues up to 39,000 ft and Mach 2.1. The remainder of the climb up to 50,000 ft is accomplished at a nearly constant Mach number, as shown.

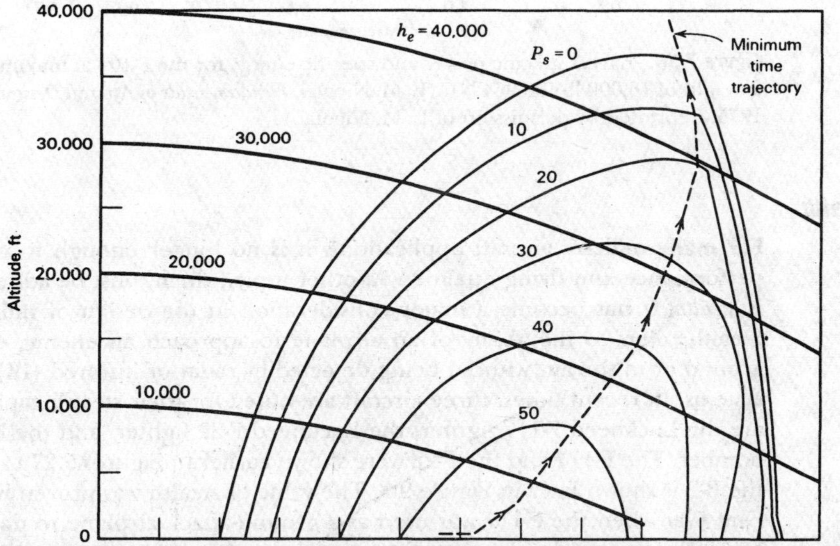

Figure 7.38 Estimated specific power and specific energy for the Boeing 767-ER at MTOGW and climb thrust.

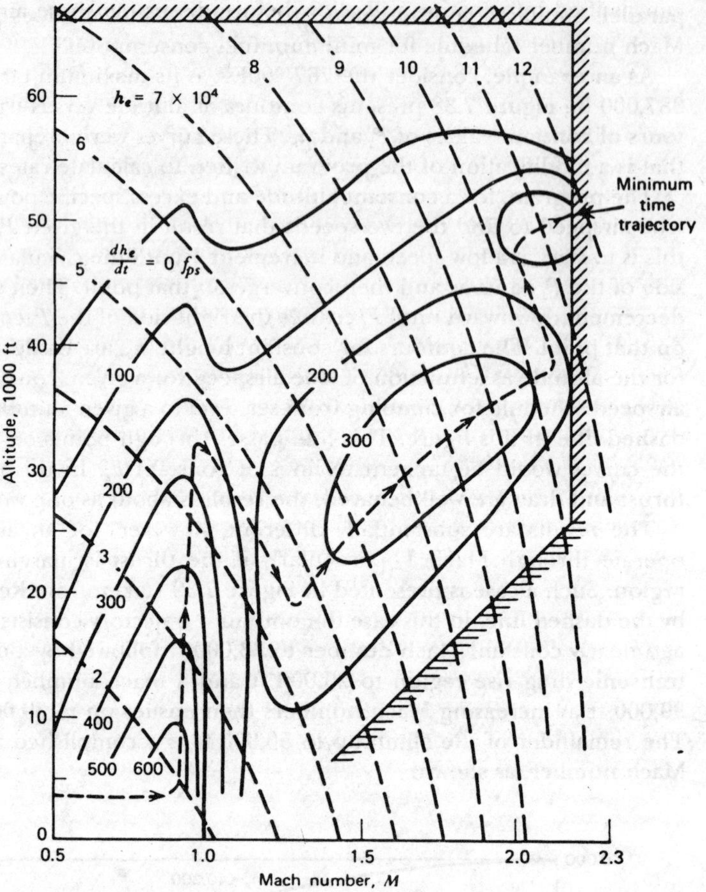

Figure 7.39 Excess specific power and specific energy for the F-104 at maximum power and a weight of 18,000 lb (80,064 N). (L. M. Nicolai, *Fundamentals of Aircraft Design*, L. M. Nicolai, 1975. Reprinted by permission of L. M. Nicolai.)

Stealth

For many military aircraft applications, it is no longer enough to consider only performance and flying qualities. Another topic, which must be added to this list, is *stealth.* It has become a major consideration in the design of military aircraft. Stealth refers to the ability of an airplane to approach an enemy, either on the ground or in the air, without being detected by radar or infrared (IR) surveillance systems. In recent times, three aircraft are noted for their stealth capability. These are the Lockheed F-117 fighter, the Lockheed F-22 fighter, and the Northrop B-2 bomber. The F-117 and the F-22 were shown earlier in Figures 5.27*a* and 5.27*b* and the B-2 is shown later in Figure 9.3. The value of stealth was proven in the Persian Gulf War when the F-117 was used as a ground-attack airplane to destroy ground control centers and radar sites. Early warning radar and ground-to-air missiles were ineffective against the F-117 with the result that approximately 20 of the stealth airplanes were able to complete their missions with relative impunity.

A brief introduction to radar and IR detection will be given to the extent that unclassified information will allow before discussing these airplanes. Most of this material is based on References 7.5 and 7.6. The presentation is qualitative with specifics such as operating frequencies and the like being left to the electronic engineers. The title of Reference 7.5 is somewhat misleading as it contains more than simply a description of the F-117. It provides a good introduction, particularly for the uninformed aerodynamicist, into radar principles and the history of stealth technology.

There are several means of diminishing the effectiveness of an enemy's surveillance system. One method is to jam the signal with an electronic countermeasure (ECM). This is not always effective since it depends, in part, on knowing the operating characteristics of the enemy system, particularly the operating frequencies. Another means is to deploy chaff from the target airplane to scatter the radar beam. To counter IR detection, a heat-source, such as flares, can be released to decoy a heat-seeking missile. However, these require that you carry the added weight and, for a given sortie, are limited in the number of times they can be used. Probably the best way to counter detection from radar or an IR system is to design the airplane to have small radar and IR *signatures,* meaning *low observables* (LO).

The factors affecting the ability of a radar system to detect a passive target include the power of the radar, the pattern of the radar beam, the area of the target, the reflectivity of the target, the directivity of the reflected signal, and the ability of the antenna to receive the reflected energy. The size of the target, its reflectivity, and directivity are lumped into one parameter known as its *radar cross section* (RCS). An airplane may be very large, and it may reflect 100% of the electromagnetic energy striking it from the radar. However, if all of this reflected energy is directed away from the radar's antenna, the RCS of the airplane will be zero. To a varying degree, stealth airplanes attempt to reduce all three factors, that is, size, reflectivity, and directivity, in order to achieve a low RCS.

IR systems detect the radiation from heat. Therefore, to provide a low IR signature, the temperature of an airplane's engine exhaust should be kept low and the heat internal to the engine should be shielded. The engine exhaust can be lowered by mixing and diffusing the hot gases with cold air entrained from the free-stream. Shielding is accomplished by not having a line-of-sight (LOS) engine installation. As the name implies, if one looks into the inlet or tailpipe, he or she should not be able to see the gas-producer or hot-turbine sections.

The RCS is defined such that its value, multiplied by the power density of the radar beam at the target, equals the power density reflected back in the direction of the radar. RCS is usually expressed in units of m^2. Since RCS depends upon reflectivity, the aspect of the target to the radar is important. Reference 7.5 quotes some typical values from the side, which are considerably higher than the geometric area. For example,

Airplane	Side RCS
F-15	400 m^2 (estimated)
B-52	1076 ft^2
B-1A	108 ft^2
B-1B	11 ft^2

The above value for the F-15 is approximately 16 times the geometric side area of that airplane.

Some indication of the relationship of RCS to detection range is found in a graph presented in Reference 7.5. Five points are noted on the graph:

RCS—m^2	Detection Range—miles
0.01	16
0.1	32
1.0	55
5.0	84
10.0	100

Little can be done about the geometric size in reducing RCS. This tends to be fixed by the mission requirements with respect to payload and range. This statement may not be quite true for the flying wing configuration of the B-2, which tends to minimize the geometric area for a given payload.

The design of a stealth airplane appears unique because it is intended to alter directivity, that is, the direction in which the radar waves are reflected from the airplane's surface. This is particularly true of the F-117, which has been described as a first-generation stealth fighter. According to Reference 7.5, a study by Lockheed concluded that almost all radar viewing of airplanes occurred within an angle of 30° above or below the plane of the aircraft. Therefore, except for the wings, the exterior of the F-117 is made of flat panels at angles greater than 30° so that most radar rays will not be reflected back to the radar antenna. The sharp breaks along the edges of the panels are not desirable from the standpoint of drag, which probably explains in part why the F-117 is a subsonic airplane.

The second-generation stealth airplanes, the F-22 and the B-2, have avoided the use of flat panels by carefully contouring the geometry and blending the fuselage into the lifting surfaces. Particularly noticeable on the B-2 is the sawtooth appearance of the wing's trailing edge. You will also note that none of the leading or trailing edges on the F-22 are perpendicular to the direction of flight. Truncated, diamond-shaped planforms are used for the wing and tail surfaces. This airplane is capable of supersonic speeds ($M < 2$) so that the stealth requirements on the planform shapes are compatible with aerodynamic requirements.

All three of the stealth airplanes utilize *radar absorbent material* (RAM), which is not apparent from a photograph or drawing. The development of such material was started by the Germans during World War II in an attempt to hide the periscopes of submarines from radar. Later, they attempted to apply it to an airplane but without much success. Many RAM types have been studied including rubberized paint with particulates, polyurethane foams, and sheet materials with rubber, silicones, and polyurethane bases. The type of RAM used on the stealth airplanes that are currently operational is not known.

In addition to the flat panels, the F-117 has a unique exhaust system to lower the exhaust temperature. According to Reference 7.5:

> *Exhaust gases are mixed with relatively cool ambient air in a plenum just aft of the engine compartment. The cooling air arrives via ducting that brings it from slots located in front of and below the intakes. Once mixed, the exhaust then is passed through a horizontal slot-type nozzle assembly that is some six feet wide and approximately six inches deep. This slot is divided into twelve separate ports which serve to channel the exhaust gases into an extended lower lip which is actually the flattened empennage of the aircraft. There the exhaust gases are again mixed rapidly with ambient air. By the time they enter the aircraft slipstream, temperature levels have been lowered significantly and the exhaust plume presents a minimal infrared target.*

The F-22 probably has a comparable system to cool the exhaust gases, but it is not mentioned in Reference 7.6. However, there are several key aerodynamic features of the F-22 that are noted in the reference and repeated here.

1. Blended wing-body with internal weapon bays. (As noted above, the blending is probably done as much to reduce reflectivity as for aerodynamic reasons.)
2. Modified diamond wing with ailerons, flaperons (flaps that can move differentially) and full leading flaps.
3. Twin, canted vertical tails.
4. All moveable clipped diamond horizontal tails.
5. Two-dimensional, convergent–divergent exhaust nozzles with independent throat and exit area actuation and pitch axis thrust vectoring.
6. Free-stream fixed geometry supersonic inlets with swept cowl lips, boundary layer bleed and overboard bypass systems, and a relatively long subsonic diffuser having 100% line-of-sight RF blockage.
7. All exterior edge angles aligned with either the wing leading or trailing edge angles.

Available data on the three stealth airplanes are presented in Appendix F. It is emphasized that some of the performance data are estimated.

THE ART OF ESTIMATING AND SCALING

Sometimes one must give a reasonable estimate of an airplane's characteristics or performance without having all the facts at hand. It is therefore a good idea to commit to memory a few principles and numbers. The "square-cube" scaling law is a good one to remember. For two geometrically similar airplanes designed to the same stress levels and using the same materials, one would expect their areas to be proportional to the characteristic length squared and their volumes, and hence weight, proportional to the length cubed.

$$S \propto l^2$$

$$W \propto l^3$$

It follows that $l \propto W^{1/3}$. One would therefore expect the wing loading of aircraft to vary as

$$\frac{W}{S} \propto W^{1/3} \tag{7.65}$$

Figure 7.40 was prepared with Equation 7.65 in mind. This figure indicates that while the square-cube law can be helpful in estimating the gross weight of an airplane, other factors must also be considered. For performance reasons, wing loadings are sometimes made purposefully higher or lower than the average. Generally, the aircraft with higher cruising speeds lie on the high side of the shaded portion in Figure 7.33. This upper boundary is given by

$$\begin{aligned}\frac{W}{S} &= 2.94\,(W^{1/3} - 6)\text{ psf}\\ &= 85.5\,(W^{1/3} - 9.9)\text{ N/m}^2\end{aligned} \tag{7.66}$$

The lower boundary is approximated by

$$\begin{aligned}\frac{W}{S} &= 1.54\,(W^{1/3} - 6)\text{ psf}\\ &= 44.8\,(W^{1/3} - 9.9)\text{ N/m}^2\end{aligned} \tag{7.67}$$

The foregoing must be qualified somewhat. Scaling, such as this, is valid only if pertinent factors other than size remain constant, for example, the structural effi-

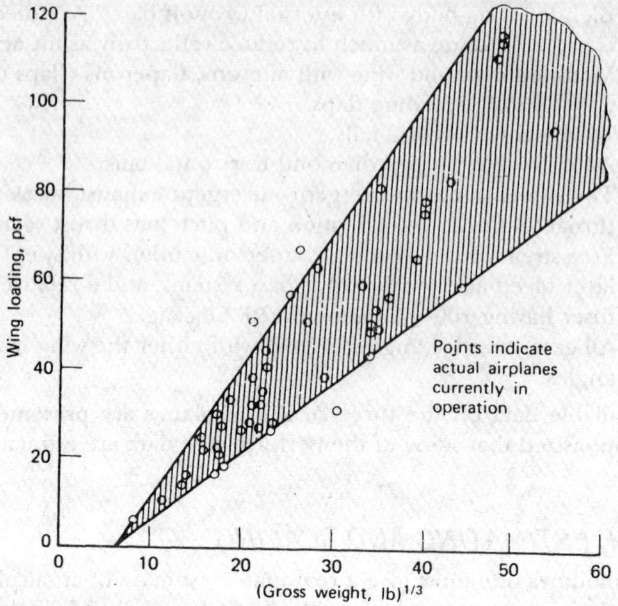

Figure 7.40 Square-cube law.

ciency of materials. Also, for purposes of their mission, aircraft are designed for different load factors.

In sizing an aircraft, it is also of value to note that the empty weights of aircraft average close to 50 or 60% of the design gross weights as shown in Figure 7.41.

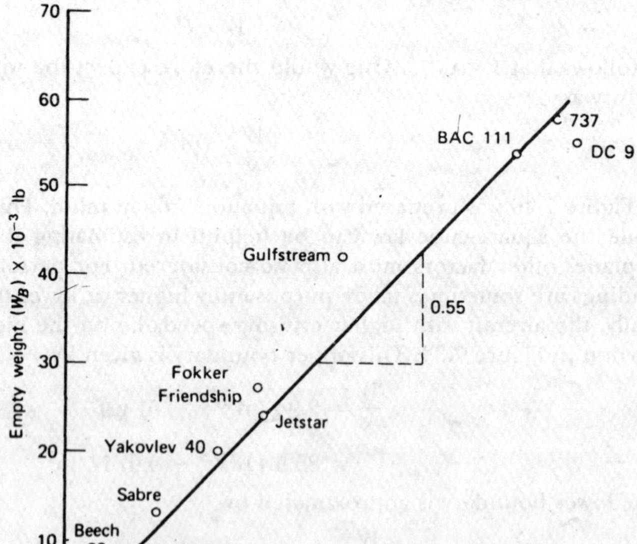

Figure 7.41 Relationship between gross and empty weight.

Table 7.1 Summary of Approximate Relationships

$\dfrac{\text{Weight empty}}{\text{Gross weight}} \simeq 0.55$	
$W/S = 2.94\ (W^{1/3} - 6)$ psf	High performance (W in lb)
$W/S = 85.5\ (W^{1/3} - 9.9)\ \text{N/m}^2$	High performance (W in N)
$W/S = 1.54\ (W^{1/3} - 6)$ psf	Low performance (W in lb)
$W/S = 44.8\ [W^{1/3} - 9.9]\ \text{N/m}^2$	Low performance (W in N)
$C_f \simeq 0.0065$	Light aircraft, fixed gear
$\simeq 0.0045$	World War II prop-driven fighters
$\simeq 0.0035$	Turbojet aircraft
$C_{D_0} \simeq 0.037$	Light aircraft, fixed gear
$\simeq 0.025$	Light aircraft, retractable gear
$e \simeq 0.6$	Low-wing aircraft
$\simeq 0.8$	High-wing aircraft
$C_{L_{max}} \simeq 1.3$	No flaps
$\simeq 1.8$	Plain flaps
$\simeq 2.5$	Double-slotted flaps
BSFC $\simeq$ 0.5 lb/bhp-hr	Piston engines
$\simeq$ 3.0 N/kW-hr	
TSC $\simeq$ 0.35 1/hr	High-bypass turbofan
$\simeq$ 0.6 1/hr	Moderate-bypass turbofan
$\simeq$ 0.8 1/hr	Turbojet
$\eta \simeq 0.85$	Propeller in cruise
$\simeq 0.70$	Propeller in climb

Note: Output from turbojet engines, approximately proportional to density ratio, σ. For piston engines, $P \simeq P_0\ (\sigma - 0.1)/0.9$.

Thus, knowing the payload and fuel and having some idea of the aerodynamic "cleanliness" of the aircraft, one can undertake a preliminary estimate of its weight and performance.

For example, suppose we are designing a four-place, light aircraft with fixed gear. Let us arbitrarily decide on 300 lb of fuel. The gross weight will be approximately

$$\text{Gross weight} = \text{Weight empty} + \text{Payload} + \text{Fuel weight}$$

$$W = W_E + W_{PL} + W_F \tag{7.68}$$

Assuming each passenger and baggage to weigh 200 lb gives

$$W = W_E + 800 + 300$$

But $W_E \simeq 0.55\ W$, so that

$$W \simeq 2400 \text{ lb}$$

From Figure 7.39 for this weight, the wing loading should be approximately

$$W/S = 14.7 \text{ psf}$$

This results in a wing area of $S = 163\ \text{ft}^2$. If we decide on a low-wing airplane, then an e of 0.6 is reasonable. Also, for a fixed-gear aircraft, a parasite C_D of 0.037 was recommended in Chapter 4. We are now in a position to construct power-required curves for various altitudes. The next step would be to select a power plant so that performance estimates can be made. In so doing, we would make use of a typical BSFC of 0.5 lb/bhp-hr for piston engines (unless performance curves on the particular engine selected were available). The value of developing a "feeling" for reasonable values for airplane parameters should be obvious. Table 7.1 summarizes some of these.

PROBLEMS

7.1 All questions refer to the same airplane. In answering a given question, use any assumptions given in preceding questions.

(a) An "ultralight" aircraft (such as a BD-5) is powered with a sea level-rated 60-hp piston engine driving a 3.7-ft diameter propeller. It carries 200 lb of payload and 10 gal of gasoline. What would be a reasonable estimate of its empty weight?

(b) Assuming the gross weight of the airplane to be 600 lb, what would be a reasonable estimate of its wing loading?

(c) Assuming a wing area of 65 ft^2 and an aspect ratio of 6 with no flaps, at 20% above its stalling speed, what would you estimate for its landing speed?

(d) Assuming a takeoff speed of 55 mph, what would the ground-roll distance be for the airplane?

(e) Assuming retractable gear and a low-wing configuration, construct power required curves for sea level and 8000 ft. What is the minimum power required at these altitudes?

(f) Assuming a propeller efficiency during climb of only 70%, calculate the rate-of-climb at sea level and 8000 ft at the speed for minimum power required.

(g) Cruising at 75% of sea level-rated power, calculate the cruising velocities and ranges (no reserve and neglect climb) at sea level and 8000 ft. Assume an η of 85%.

7.2 For a propeller-driven airplane, show that the minimum required power varies inversely with the square root of the density ratio. If the available power is assumed proportional to $(\sigma - 0.1)/0.9$, show that the density ratio for the absolute ceiling is governed by the relationship

$$\sigma^{3/2} - 0.1\sigma^{1/2} = 0.9\left[1 - \frac{W(R/C)_0}{P_0}\right]$$

where P_0 and $(R/C)_0$ are the sea level values for the available power and rate-of-climb.

7.3 An airplane has a constant thrust to weight ratio of 0.25 and a braking friction coefficient, μ, of 0.5. While taking off on an 850-m strip and not yet airborne, the engine fails. At that instant, the plane had attained a speed of 69 kt. The pilot immediately applies the brakes. Will the plane stop before the end of the runway (neglect aerodynamic drag)? Justify your answer.

7.4 A fairing is to be added to a Cherokee Arrow. This fairing will reduce f by 0.25 ft^2, but weighs 12 lb. At its maximum gross weight this will require decreasing the fuel by 2 gal. From a range standpoint, will the fairing be beneficial?

7.5 An advanced medium STOL transport has an approach speed of 85 kt with a descent angle of 6.6°. It weighs 670 kN and has a wing loading of 4300 N/m^2. On touching down it has a constant deceleration of 0.25 g. What is its total landing distance over a 15-m obstacle?

7.6 An airplane has an R/C of 1000 fpm at sea level and 500 fpm at 7000 ft. How much time will it require to climb from an altitude of 2000 ft up to 10,000 ft?

7.7 Calculate the available thrust for the Cherokee Arrow as a function of airspeed up to 45 m/s at SSL conditions. Assume 2700 rpm and 149 kW. How do your results compare with Figures 6.21 and 6.22? For the standard gross weight of 11.8 kN, calculate the ground-roll distance to attain a velocity of 36 m/s.

7.8 For a normally aspirated piston engine airplane, show that the absolute ceiling is a function of the sea level rate-of-climb. Derive an equation relating the two performance items.

7.9 This is an open-ended problem. What kind of a performance penalty do you think the F-117 is paying for its stealth performance?

REFERENCES

7.1 Anonymous, *Jet Transport Performance Methods,* The Boeing Co., Commercial Airplane Group, Seattle, WA, Boeing Document No. D6-1420, 6th Edition, May 1969.

7.2 Anonymous, *Cessna/Citation I Specification and Description,* Cessna Aircraft Co., Commercial Jet Marketing Division, Wichita, KS, January 1977.

7.3 Anonymous, *Cessna/Citation I Flight Planning Guide,* Cessna Aircraft Co., Commercial Jet Marketing Division, Wichita, KS, February 1977.

7.4 Nicolai, L. M., *Fundamentals of Aircraft Design,* distributed by School of Engineering, University of Dayton, Dayton, OH, 1975.

7.5 Miller, Jay, *Lockheeed F-117 Stealth Fighter,* Aerofax, Inc., Specialty Press, Stillwater, MN, 1991.

7.6 Mullin, Sherman N., *The Evolution of the F-22 Advanced Tactical Fighter,* 1992 Wright Brothers Lecture, AIAA Aircraft Design Systems Meeting, August 24, 1992, Hilton Head, SC.

8 HELICOPTERS AND V/STOL AIRCRAFT

This chapter treats the aerodynamics of helicopters and aircraft designed to take off in a very short distance, or even vertically. Such aircraft are known as *V/STOL aircraft*, an acronym denoting vertical or short takeoff and landing. Perturbations on this acronym include STOL, VTOL, STOVL, and ASTOVL. The meaning of the first two are obvious. The third acronym means "short takeoff and vertical landing," whereas the "A" on the last one refers to "advanced" and usually means a STOVL aircraft with supersonic capability. Whether or not an aircraft can be considered to have STOL capability is a relative matter. A Piper Cub, which might take off in under 1000 ft because of its light wing loading, is not a STOL aircraft. On the other hand, a turbojet aircraft, which normally requires a 10,000-ft-long runway, may be considered to have STOL capability if the aircraft can get off in 3000 ft.

V/STOL capability is the result of some special design feature that produces a vertical force at low speeds. This force may be either lift in the usual sense or it may be a vertical force produced by some type of thruster.

HELICOPTERS

A helicopter is an aircraft that can take off and land vertically. However, as will be seen later, the performance of a helicopter in forward flight is penalized because of the lift and thrust being produced by the rotor instead of a fixed wing and separate thruster. Therefore, the helicopter is not usually referred to as a VTOL aircraft. Despite its poor cruising performance, the helicopter is the obvious choice for missions where the emphasis is on vertical flight. Although the helicopter has a reputation, mainly due to the Vietnam war, as a weapon of destruction, the helicopter has proven to be a life-saving machine, which has rescued thousands from the ravages of storms and fires. Its ability to maneuver in and out of restricted areas and to hover efficiently for long periods of time have made it invaluable for emergency medical service (EMS), police activities, logging operations, and support of offshore oil rigs.

A Brief History of the Helicopter

The helicopter evolved and was refined over a period of approximately 35 years, culminating in the development of the main rotor-tail rotor configuration by Igor Sikorsky in late 1941. In September 1907, the Brequet-Richet Gyroplane first flew in Douai, France. However, lifted by four rotors, the machine was restrained by ropes; it lifted only a few feet off the ground, but without a control system, it was far from being a practical helicopter.

In 1907 a French mechanic, Paul Cornu, was the first person in history to rise vertically, completely unrestrained. A photograph of his helicopter is shown in Figure 8.1. It was a tandem rotor configuration powered by a 24-hp engine, which turned two rotors approximately 90 rpm. Cornu was possibly the first experimenter

Figure 8.1 The Carnu Helicopter; the first helicopter to rise unconstrained on November 13, 1907. Courtesy, National Air & Space Museum.

to be concerned with helicopter control. Just below each rotor was a set of vanes, which deflected the downwash for purposes of maneuvering and providing forward thrust. Although Cornu achieved a historic first, the performance and control of his helicopter was marginal and it never developed into a practical machine.

Juan de la Cierva, who paved the way for the development of the successful helicopter, never built a helicopter himself. Cierva is credited with the development of the autogiro, which resembles the helicopter, but utilizes an unpowered rotor. The rotor turns, or *autorotates,* as the autogiro is pulled through the air by a separate propeller. In January 1923, Cierva successfully flew his C.4 autogiro, which incorporated rotor blades free to flap up and down in response to the unsteady aerodynamic forces that arise in forward flight. This revolutionary approach to alleviating undesirable blade hub stresses and moments was conceived by Cierva after observing that one of his models with flexible blades was stable in forward flight, unlike his earlier full-scale machines. This concept of an *articulated* rotor was the technical breakthrough that led others to develop the successful helicopter. Cierva might have done so himself, except that he met an untimely death at age 42 in the crash of a Dutch airliner departing from London's Croydon Aerodrome in December 1936.

In October 1930, an Italian, Corradino D'Ascanio, established the first recognized helicopter record when he flew for one-half mile at an altitude of 59 ft for 8 min 45 sec. It is seen in Figure 8.2 that two coaxial, contrarotating rotors, fragile in appearance, were controlled by flaps supported on booms trailing from each blade near the tip. This type of control system is used today by the Kaman Co. in some of their designs.

Germany, just prior to and during World War II, made rapid strides in helicopter development. The FA-61 designed by Heinrich Focke was flown for the first time on June 26, 1936. Later, on May 10, 1937, as a propaganda stunt for the Nazi regime, the FA-61 was flown inside Berlin's Deutschlandholle sports arena by the renowned female pilot Hanna Reitsch. Another German helicopter, the FL-282 (Figure 8.3) designed by Anton Flettner, became operational with the German Navy. Over 1000 of these helicopters were produced. Named the Kolibri, the helicopter had a forward speed of 90 mph and could operate at 13,000 ft with a payload of 800 lb.

On December 8, 1941, Igor Sikorsky flew his final version of the VS-300. Unlike previous helicopter designs, the VS-300 employed a tail rotor to counteract the

Figure 8.2 The D'Ascanio Helicopter; the first helicopter to establish recognized records when, in October 1930, it flew one-half mile at an altitude of 59 ft. Courtesy, National Air & Space Museum.

torque of the main rotor. This represented a major accomplishment, which has been copied by the majority of successive helicopter designs. The VS-300 is pictured in Figure 8.4 with Sikorsky at the controls.

The second successful helicopter to fly in the United States was a single-rotor machine designed by a small team led by Frank Piasecki. This work led to the founding of the Piasecki Helicopter Co. and the development of the first successful tandem-rotor helicopter, the XHRP-X in Figure 8.5.

Figure 8.3 The Flettner FL-282 Kolibri; the first production helicopter. Over 1000 were built and operational with the German Navy in World War II. Courtesy, National Air & Space Museum.

Figure 8.4 Sikorsky's VS-300; the first U.S. helicopter and the world's first employing a main-rotor, tail-rotor configuration. Courtesy, National Air & Space Museum.

Helicopter Technology

An approximate expression for the thrust and ideal power for a rotor system in forward flight can be obtained from *Glauert's hypothesis.* This hypothesis applies the momentum theorem for a lifting system to the flow through a circular area, which circumscribes the tips of the system and is normal to the resultant velocity vector through the system. To understand the basis for this hypothesis, consider the wing and the propeller as two extremes of a "lifting" system. For the wing, the circular area will equal $\pi b^2/4$.

Assuming the induced velocity to be small, the resultant velocity is equal approximately to the free-stream velocity, V. Thus, the mass flow through the system will be

$$m = \rho\pi \frac{b^2}{4} V \tag{8.1}$$

The downwash far downstream of the wing is taken to be twice the value of w at the wing. Thus, from momentum principles, the induced velocity and the lift are related by

$$L = 2w\rho\left(\pi\frac{b^2}{4}\right)V \tag{8.2}$$

The downwash, w, is then obtained as

$$\frac{w}{V} = \frac{2L}{\pi\rho V^2 b^2}$$

or, for an aspect ratio, A,

$$\frac{w}{V} = \frac{C_L}{\pi A} \tag{8.3}$$

Figure 8.5 Piasecki's XHRP-X; the second U.S. helicopter and the world's first successful tandem-rotor machine. Courtesy, National Air & Space Museum.

This is identical to the result that was obtained from the lifting line model for the elliptic wing.

The other extreme is the propeller where the thrust is aligned with the velocity. In this case, the application of Glauert's hypothesis is identical to the classical momentum theory of propellers.

Since Glauert's hypothesis leads to reasonable results for the two extreme cases where the direction of the force vectors differs by 90° relative to the resultant velocity, it is assumed to hold for a helicopter where the direction of the resultant velocity can lie anywhere between the vertical and horizontal directions. A side view of a rotor in forward flight is shown in Figure 8.6 with the thrust vector taken normal to the plane of the rotor. From Glauert's hypothesis, the induced velocity and rotor thrust are related for a disc area of A by

$$T = 2\rho A \bar{V} w \tag{8.4}$$

where

$$\bar{V} = \sqrt{(w - V\sin\alpha)^2 + (V\cos\alpha)^2}$$

α is the angle-of-attack of the rotor plane and is defined positively nose up even though the thrust vector is normally inclined forward in forward flight.

The ideal power is equal to the product of the thrust and the component of the resultant velocity normal to the rotor and opposite in direction to the thrust.

$$P = T(w - V\sin\alpha) \tag{8.5a}$$

In steady, level flight, the sum of the forces acting on the helicopter in the direction of flight leads to

$$D + T\sin\alpha = 0 \tag{8.5b}$$

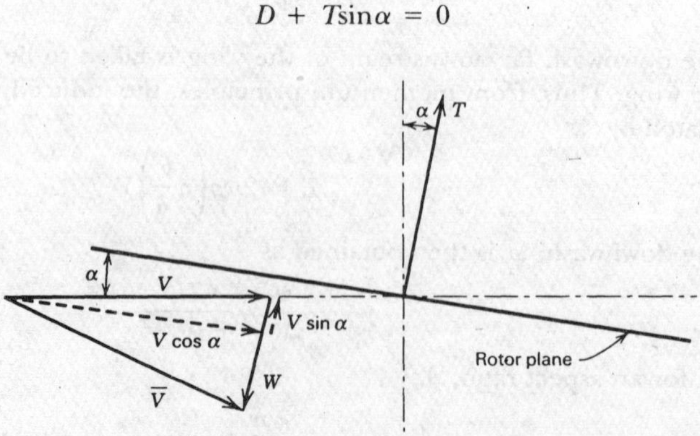

Figure 8.6 Side view of rotor.

It follows that

$$P = DV + Tw \tag{8.6}$$

Thus, it is found that the ideal power required by a helicopter, similar to a fixed-wing airplane, is composed of two parts: the *parasite power, DV*, and the *induced power, Tw*. There is an additional power required by the helicopter that must be added to Equation 8.6, which is called the *profile power.* The profile power is the power required to overcome the profile drag of the rotor blades as they rotate.

Figure 8.7 illustrates an element of a rotor blade under the influence of the forward velocity and the velocity due to rotation. The blade is pictured at an instantaneous angular position known as the *azimuth angle,* ψ, which is measured positively from the downstream position in the direction of rotation. For some unknown reason, the rotors on U.S.-built helicopters rotate counterclockwise when viewed from above, whereas most European-built helicopters rotate oppositely. Thus, on U.S. helicopters the blades on the right side advance into the flow, whereas the blade retreats from the flow on the left side. One speaks of the *advancing* blade or the *retreating* blade.

The velocity normal to the blade element in Figure 8.7 is equal to $\omega r + V\sin\psi$. Thus, the instantaneous torque on the rotor due to the profile drag of the blades is given by

$$Q(\psi) = B\frac{\rho}{2}\int_0^R r(\omega r + V\sin\psi)^2 cC_d dr \tag{8.7}$$

where B is the number of blades and ω is the angular velocity of the rotor.

The power will equal the work done in one revolution divided by the time required for the revolution.

$$P = \frac{\omega \int_0^{2\pi} Q(\omega)\, d\psi}{2\pi} \tag{8.8}$$

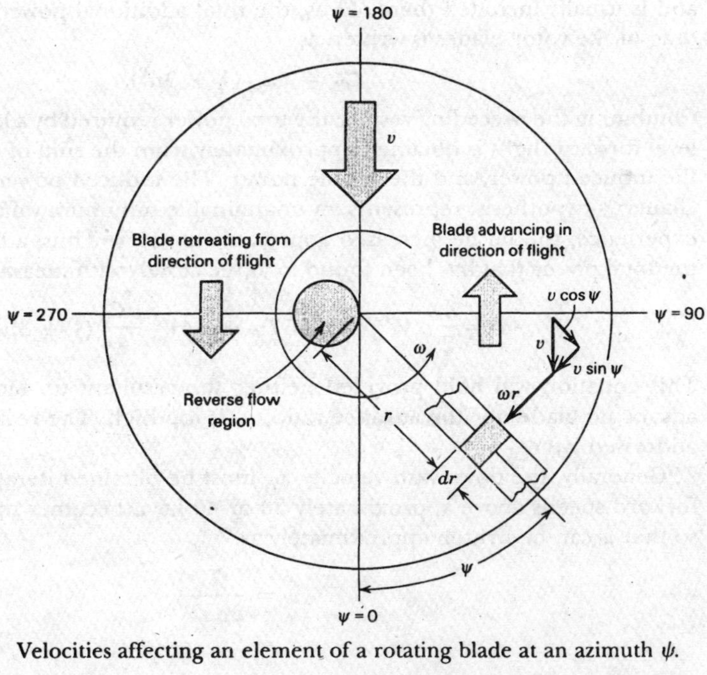

Figure 8.7 Velocities affecting an element of a rotating blade at an azimuth ψ.

The thrust and power coefficients for a helicopter are defined in terms of the tip speed, $V_T = \omega R$, and disc area, A, as

$$C_T = \frac{T}{\rho A V_T^2} \tag{8.9a}$$

$$C_P = \frac{P}{\rho A V_T^3} \tag{8.9b}$$

It is left to the reader to show that if C_d is assumed constant, the profile power coefficient in forward flight can be expressed as

$$C_{P_p} = C_{P_{po}} (1 + \mu^2) \tag{8.10}$$

μ is the ratio of the forward speed to the rotor tip speed, V_T, or $V/\omega R$. $C_{P_{po}}$ is the profile power coefficient in hover and is given by

$$C_{P_{po}} = \frac{\sigma C_d}{8} \tag{8.11}$$

σ is the rotor solidity and is equal to the ratio of the blade planform area to the rotor disc area. For B blades and a chord of c, σ is defined by

$$\sigma = \frac{Bc}{\pi R} \tag{8.12}$$

Because of the nonlinear dependence of the drag on the velocity, the net profile drag on a blade in one revolution is not zero. Therefore the parasite power must be increased to compensate for the profile drag. In one revolution, this average drag will equal the impulse divided by the time.

$$\overline{D} = \frac{\omega}{2\pi} \int_0^{2\pi/\omega} \frac{1}{2} \rho\, (\omega r + V \sin\psi)^2\, c C_d \sin\psi\, dt$$

Since $\omega dt = d\psi$, it follows that, for B blades, in coefficient form, the increment in parasite power is equal to $2C_{P_{po}}\mu^2$. This is of the same form as part of Equation 8.10 and is usually included there. Thus, the total additional power due to the profile drag of the rotor blades is written as

$$C_{P_p} = C_{P_{po}} (1 + 3\mu^2) \tag{8.13}$$

Combining the preceding results, the total power required by a helicopter in steady, level forward flight is obtained approximately from the sum of the parasite power, the induced power, and the profile power. The induced power, as obtained from Glauert's hypothesis, represents an unattainable minimum value, which, based on experience, should be increased approximately 15%. Thus, a final expression for the total power that has been found to agree closely with measurements is given by

$$P = \frac{1}{2} \rho\, V^3 f + 1.15 T w + \rho A V_T^3 \frac{\sigma C_d}{8} (1 + 3\mu^2) \tag{8.14}$$

This equation will hold provided neither the resultant tip Mach number of the advancing blade nor the advance ratio, μ, is too high. The reasons for this will be addressed later.

Generally, the downwash velocity, w, must be obtained iteratively. However, for forward speeds above approximately 30 or 40 kt, w becomes small compared to V so that w can be written approximately as

$$w = \frac{T}{2\rho A V} \tag{8.15}$$

The average blade profile drag coefficient, $\bar{C}_d$, can be related to an average blade lift coefficient, $\bar{C}_l$. The instantaneous thrust can be written as

$$T = \frac{B}{2}\rho\int_0^R (\omega r + V\sin\psi)^2 \, cC_l dr$$

Averaging the thrust over one revolution, the average blade lift coefficient can be expressed in terms of the rotor thrust coefficient, solidity, and advance ratio as

$$\bar{C}_l = \frac{6\,C_T}{\sigma\left(1+\dfrac{3\mu^2}{2}\right)} \tag{8.16}$$

Thus, for given operating conditions, one can calculate an average lift coefficient and, knowing the airfoil characteristics, determine a corresponding value for the average drag coefficient.

As an example in the use of the foregoing, consider the attack helicopter, the Bell AH-1J, at a true airspeed of 30 kt at a standard altitude of 1000 ft. This helicopter has a gross weight of approximately 9500 lb, a rotor diameter of 44 ft, a flat plate area of approximately 22 ft^2 with external weapons, and a rotor tip speed of 738 fps. The two-bladed rotor has a constant chord of 2.25 ft, and it will be assumed that the average C_l and average C_d are related by

$$\overline{C_d} = 0.008 + 0.008\bar{C}_l^2 \tag{8.17}$$

At 30 kt, from Equation 4.35, the parasite drag of the helicopter will be 65.27 lb resulting in an angle-of-attack of $-0.39°$ from Equation 8.5*b*. In order to iterate for the downwash, it will be assumed initially that w is equal to the value for hovering, w_0. This can be obtained from Equation 8.4 setting V equal to zero.

$$w_0 = \sqrt{\frac{T}{2\rho A}} \tag{8.18}$$

Thus, the first iteration on w gives $w = 36.79$ fps. This value is then substituted into Equation 8.4 to give a second value of 21.53. The iteration converges rapidly to a value of $w = 24.05$. Once w is obtained, the induced power can be determined. For this particular operating condition the following values are calculated:

$V = 30$ kt	altitude $= 1000$ ft
HP_i	477.8 hp
HP_p	204.4 hp
HP_{par}	6.0 hp
Total HP	688.2

In this manner, the curves of Figure 8.8 presenting the power breakdown for the AH-1J were determined. Observe that the total required power decreases rapidly as the speed increases from zero reaching a minimum at around 70 kt. This decrease results from the drop in the induced power. As the speed increases, the parasite power increases so that a minimum value is reached above which the increase in the parasite power predominates. Note that the profile power amounts to approximately 21% of the total power in hover and increases gradually with forward speed. At the minimum power, this percentage increases to 41%. Also note the required power does not rise above the power to hover until a speed of 139 kt is exceeded.

Computer Exercise 8.1 HELICOPTER POWER

Write a computer program to calculate the components of power required by a helicopter in forward flight. The program should loop on the true airspeed at a

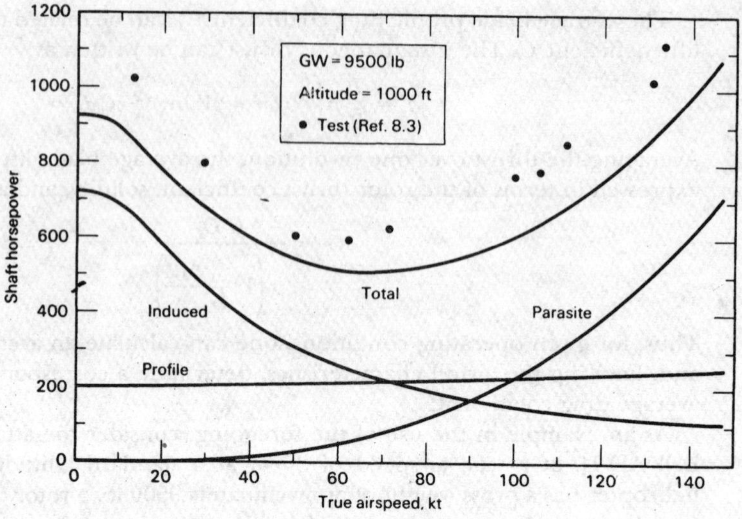

Figure 8.8 Power required for a Bell AH-1J helicopter.

fixed altitude. Input will be the gross weight, rotor diameter, rotor tip speed, number of blades, blade chord, and altitude. Run your program for the AH-1J and compare with Figure 8.8.

Experimental measurements of the total power required for the AH-1J are included in Figure 8.8 as taken from Reference 8.1. It can be seen that the power relationships developed thus far are optimistic notwithstanding the empirical increase in the induced power. There are several additional sources of power, which must be included in a more sophisticated analysis. First, one must consider nonaerodynamic requirements such as power losses in the gear boxes, which are necessary to transmit the engine power to the main rotor. This power is typically of the order of 1% per gear mesh. Power is also needed for accessories such as a cooling fan or avionics.

Aerodynamic power requirements not yet considered include, for the single rotor-tail rotor configuration, the power to drive the tail rotor. In the case of a tandem rotor configuration where each rotor cancels the torque of the other, an additional induced power is required by the rear rotor acting in the downwash of the forward rotor. The tandem configuration is discussed in Reference 8.2. If the resultant tip speed of the advancing blade is too high, compressibility losses will occur requiring additional profile power. Also, if the resultant tip speed of the retreating blade is too low, retreating blade stall can occur resulting again in an increase in the profile power. In addition, in hover or at low forward speeds, the rotor downwash produces a vertical drag downward on the fuselage and appendages that adds to the thrust required by the main rotor. This *download*, which typically amounts to 5% of the gross weight for a helicopter, will be covered in more detail later.

In order to examine compressibility and retreating blade stall, it is necessary to examine the local angles of attack along the blade as it traverses the azimuth. To do this, the dynamic motion of a rotor blade must be examined since the aerodynamic forces and dynamic motion of a blade are coupled. This presents a real challenge to the rotary-wing aerodynamicist.

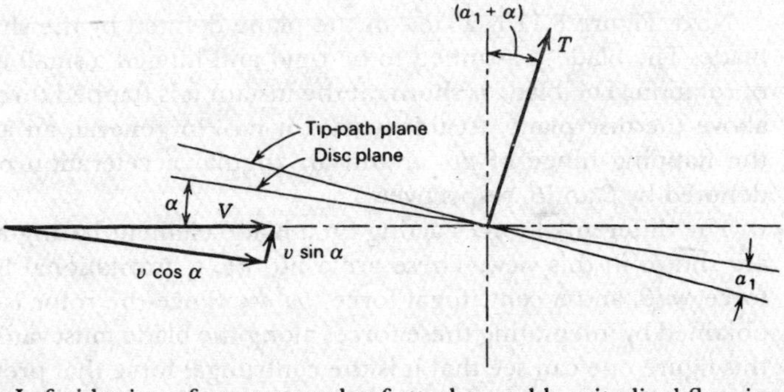

Figure 8.9 Left-side view of rotor at angle-of-attack α and longitudinal flapping of a_1.

Rotor Dynamics; Blade Flapping

It was mentioned earlier that de la Cierva achieved a breakthrough in rotor technology with the introduction of the articulated rotor. This section will examine the consequences of allowing a blade to flap in and out of its plane of rotation. To begin, consider Figure 8.9, which is a side view of an articulated rotor. Instead of having simply a rotor plane, we will now define two planes: the disc plane, which is normal to the rotor shaft axis, and the tip path plane, which is the plane described by the blades as they flap relative to the disc plane. The angle-of-attack of the rotor is defined relative to the disc plane, while the motion of the blades follows the tip path plane, which is nosed up from the disc plane through an additional angle, a_1, referred to as the *longitudinal flapping*. In this view, the free-stream velocity is resolved into two components as shown; one parallel to the disc plane and the other one directed upward and normal to the disc plane.

Figure 8.10 is a planview of the disc plane showing a blade at an azimuth angle of ψ. From this view, the velocity normal to the blade section is obtained as equal to

$$V_N = \omega r + V \cos\alpha \sin\psi \tag{8.19}$$

From the above it is obvious that something is required to compensate for the difference in the resultant velocity on the advancing and retreating blades. This is done by allowing the blades to flap.

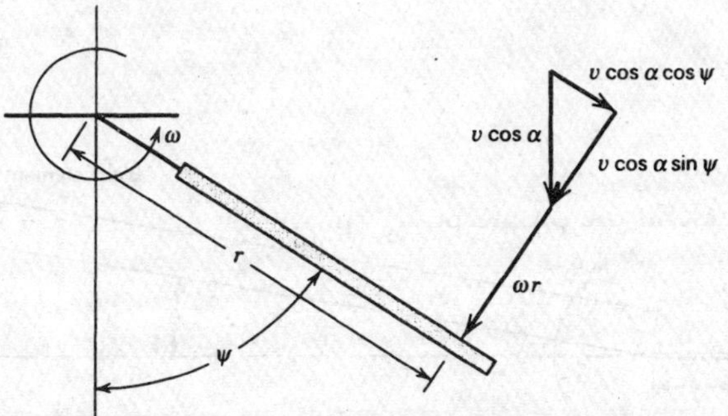

Figure 8.10 Disc plane showing blade at an azimuth with corresponding velocity components.

Next, Figure 8.11 is a view in the plane defined by the shaft axis and the rotor blade. The blade is assumed to be rigid and hinged a small distance from the axis of rotation. The blade is shown at the instant it is flapped through an angle of $\beta(\psi)$ above the disc plane. At this instant it has, in general, an angular velocity about the flapping hinge of $d\beta/dt$ and an angular acceleration of $d^2\beta/dt^2$, which are denoted by $\dot{\beta}$ and $\ddot{\beta}$, respectively.

The differential forces acting on a blade element having an elemental mass dm are shown in this view. These are a lift, dL, a gravitational force $g\,dm$, an inertial force, $ml\ddot{\beta}$, and a centrifugal force, $r\omega^2 dm$. Since the rotor is hinged, the moment obtained by integrating these forces along the blade must vanish at the hinge. From this figure one can see that it is the centrifugal force that prevents the rotor blades from folding up when they are hinged.

Before examining in detail the forces on the blade element, it is informative to consider the behavior of the integrated moments along the blade. The problem is made more tractable without any significant loss in accuracy if the hinge offset is taken to be zero. Therefore, for the following, the distance of the hinge from the shaft axis, e, will be set to zero. The blade motion, $\beta(\psi)$, can be expressed as a Fourier series in ψ.

$$\beta(\psi) = \beta_0 - \sum_{1}^{N} (a_n \cos\psi + b_n \sin\psi) \tag{8.20}$$

The constant term, β_0, is referred to as the *coning*. The higher harmonics are important to the prediction of vibrations and noise. However, for performance and control considerations, the blade motion will be limited to first harmonic flapping.

The moments of the lift and weight about the flapping hinge are designated by M_L and M_W, respectively. The integrated moment resulting from the inertial force is given by

$$M_I = \int_0^R - r(r\ddot{\beta}\, dm) \tag{8.21}$$

and the moment from the centrifugal force by

$$M_C = -\int_0^R r\beta\,(r\omega^2\, dm) \tag{8.22}$$

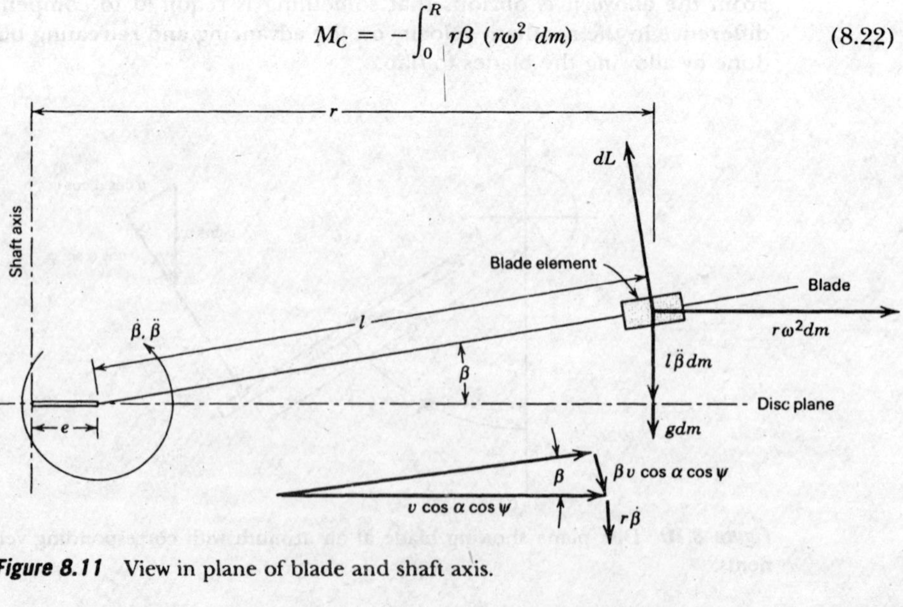

Figure 8.11 View in plane of blade and shaft axis.

The β terms can be taken out from under the integral signs and the remaining integral of $r^2\,dm$ is recognized as I_F, the mass moment of inertia of the blade about the flapping hinge. Thus, the sum of the moments about the flapping hinge leads to

$$M_L + M_W = I_F\ddot{\beta} + I_F\omega^2 I_F\beta \tag{8.23}$$

The derivative of β with respect to time can be changed to a derivative with respect to ψ.

$$\begin{aligned}\frac{d\beta}{dt} &= \frac{d\beta}{d\psi}\frac{d\psi}{dt} \\ &= \omega\frac{d\beta}{d\psi}\end{aligned} \tag{8.24}$$

Thus, substituting Equation 8.24 into Equation 8.23, the surprising result—when seen for the first time—is obtained, which states that the moment due to the lift is a constant, independent of ψ, for first harmonic flapping.

$$\begin{aligned}M_L &= -M_W + \omega^2\beta_0 I_F \\ &= \text{constant}\end{aligned} \tag{8.25}$$

The aerodynamic forces acting on a blade element to satisfy Equation 8.25 are determined by reference to Figure 8.12. This is a view looking along the blade toward the hub and shows the blade section and the various velocity components that determine its lift and drag. The angle, $\theta(\psi)$, is the blade pitch angle relative to the disc plane. As indicated, this angle can be varied with ψ in order to provide *cyclic pitch control.* The angle of attack of the section is given by

$$\alpha(\psi) = \theta(\psi) + \tan^{-1}\frac{V\sin\psi - w - r\dot{\beta} - \beta V\cos\alpha\cos\psi}{\omega r + V\cos\alpha\sin\psi} \tag{8.26}$$

The preceding relationships are nondimensionalized in terms of the advance ratio, μ, and an inflow ratio, λ. λ is the net flow up and normal to the disc plane divided by the tip speed. It does not include any contribution resulting from the blade motion.

$$\begin{aligned}\lambda &= \frac{V\sin\alpha - w}{\omega R} \\ &= \mu\sin\alpha - \frac{w}{V_T}\end{aligned} \tag{8.27}$$

The cyclic pitch control, $\theta(\psi)$, is expressed as

$$\theta(\psi) = \theta_0 + \theta_1\cos\psi + \theta_2\sin\psi + \theta_T x \tag{8.28}$$

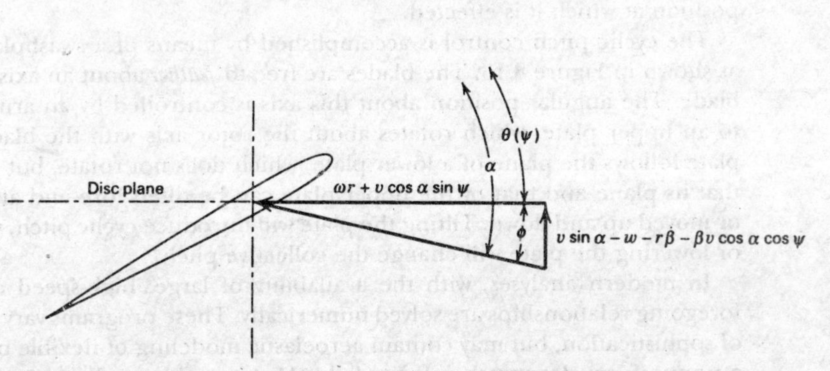

Figure 8.12 View looking in toward hub of showing blade element.

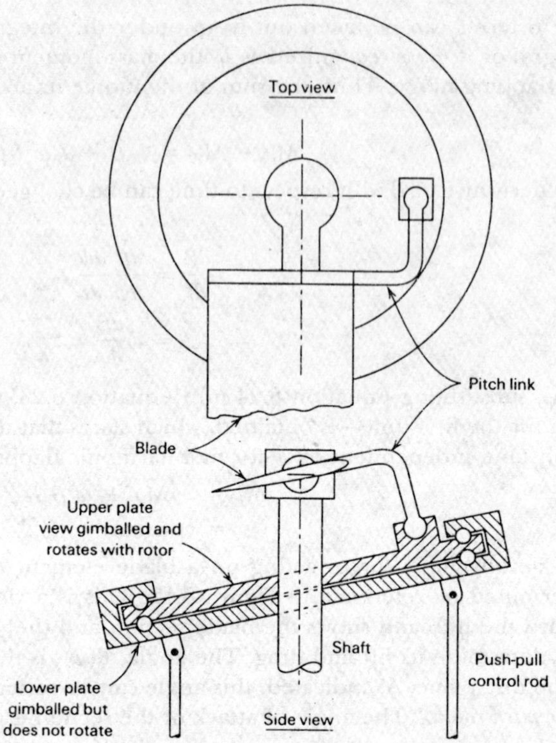

Figure 8.13 Schematic of swashplate mechanism for providing cyclic pitch.

The first term, θ_0, is called the *collective pitch*; the second term, θ_1, the *lateral cyclic pitch*; and the third term, θ_2, the *longitudinal cyclic pitch.* θ_T is the total twist of the blade from root to tip and is usually negative. Note that the term θ_2, which is called the longitudinal cyclic pitch, has its maximum input at $\psi = 90°$ and 270°. The reason for this relates to the behavior of a second order system with damping. When forced at its undamped natural frequency, the response of such a system lags the input by 90°. Since an articulated rotor blade approximates a damped, second order system with an undamped natural frequency equal to the rotational velocity, a cyclic pitch at a ψ of 90° results in a displacement 90° later. Thus, the cyclic pitch is denoted according to the flapping, which is produced and not to the azimuth position at which it is effected.

The cyclic pitch control is accomplished by means of a swashplate mechanism as shown in Figure 8.13. The blades are free to *feather* about an axis parallel to the blade. The angular position about this axis is controlled by an arm that is linked to an upper plate, which rotates about the rotor axis with the blade. This upper plate follows the plane of a lower plate, which does not rotate, but is gimballed so that its plane and that of the upper plate can be tilted fore and aft to either side or moved up and down. Tilting the plate will introduce cyclic pitch, whereas raising or lowering the plate will change the collective pitch.

In modern analyses, with the availability of large, high-speed computers, the foregoing relationships are solved numerically. These programs vary in their degree of sophistication, but may contain aeroelastic modeling of flexible blades as well as a nonuniform downwash velocity field. Many modern helicopter designs employ composite flexures or elastomeric bearings in place of mechanical fixtures. As in any design option, there are trade-offs. The advocates of the hingeless rotor claim reduced maintenance and increased control power since a hub moment is transferred at the blade root.

The classical solution for the rigid, articulated rotor, which was obtained prior to the computer, will be outlined here. Not only does it provide an insight into the dynamic and aerodynamic behavior of a rotor, but the results are sufficiently accurate for many purposes. Indeed, in the middle 1950s, the helicopter had been flying for some time, but the computer was still on the runway waiting for takeoff. The mainframe computers of that time would be no match for the PCs of today.

It is assumed that the downwash is uniform and given by Equation 8.2. Also, it is assumed that the angles θ and ϕ in Figure 8.12 are small. To a small angle approximation, the angle-of-attack will be given by

$$\alpha = \frac{\lambda - x\dfrac{d\beta}{d\psi} - \beta\mu\cos\psi}{x + \mu\sin\psi} \tag{8.29}$$

The differential lift on a blade section can then be written as

$$dL = \frac{1}{2}\rho V_T^2 (x + \mu\sin\psi)^2\, a\left(\theta + \frac{\lambda - x\dfrac{d\beta}{d\psi} - \beta\mu\cos\psi}{x + \mu\sin\psi}\right) cdr \tag{8.30}$$

where $a = \dfrac{dC_l}{d\alpha}$.

Equation 8.30 multiplied by the radius, r, will give an expression for dM_l/dr. When the equations for $\beta(\psi)$ and $\theta(\psi)$ are substituted into dL/dr and dM_l/dr, expressions are obtained containing constants and terms multiplying powers of $\sin\psi$ and $\cos\psi$. The average lift, which will be the thrust, is obtained by multiplying Equation 8.30 by the number of blades, dividing by 2π, and integrating over ψ and r from 0 to 2π and from 0 to BR, respectively. Here, BR is an effective radius, with B equal to approximately 0.97, which is used to account for the loss in lift at the tips of the blades. B, as used here, is not the number of blades.

The instantaneous hub moment is obtained by integrating the differential moment from 0 to BR. Again, the result will contain constant terms and terms multiplying powers of $\sin\psi$ and $\cos\psi$. The result is simplified by retaining only constant and first harmonic terms. In so doing, it must be remembered that higher powers of sin and cos terms can contain constants. For example,

$$\sin^2\psi = \frac{1}{2} - \frac{1}{2}\cos(2\psi)$$

In order for the result to satisfy Equation 8.25 for all values of ψ, it follows that the sets of constant terms, coefficients of the sin terms, and coefficients of the cos terms must each satisfy the equation. The algebra is laborious and will not be given here; instead, the reader is referred to the original Reference 8.3 for the details.

The results of the algebraic reduction can be expressed (Ref. 8.2) by a series of equations for the thrust coefficient, blade coning, longitudinal flapping, and lateral flapping. The quantities are generally a function of the advance ratio, μ, the inflow ratio, λ, and the control angles θ_0, θ_1 and θ_2.

$$C_T = \frac{a\sigma}{2}\left[\lambda T_1 + \theta_0 T_2 + \theta_T T_3 + \theta_2 T_4\right] \tag{8.31}$$

where

$$T_1 = \frac{1}{2}\left(B^2 + \frac{\mu^2}{2}\right) \qquad T_3 = \frac{1}{4}B^2(B^2 + \mu^2)$$

$$T_2 = \frac{1}{3}B^3 + \frac{1}{2}\mu^2 B \qquad T_4 = \frac{1}{2}\mu\left(B^2 + \frac{1}{4}\mu^2\right)$$

$$\beta = \frac{\gamma}{2}[\lambda F_1 + \theta_0 F_2 + \theta_T F_3 + \theta_2 F_4] - \tau \tag{8.32}$$

where

$$F_1 = \frac{B^3}{3} \qquad F_3 = B^3\left(\frac{1}{5}B^2 + \frac{1}{6}\mu^2\right)$$

$$F_2 = \frac{1}{4}B^2(B^2 + \mu^2) \qquad F_4 = \frac{1}{3}\mu B^3$$

$$\tau = \frac{M_w}{I_f\omega^2} \qquad \gamma = \frac{c\rho a R^4}{I_F}$$

In the above, B is the effective radius fraction, a is the section lift curve slope, and c is the blade chord. γ is known as the *Lock number* and, in effect, is the ratio of the aerodynamic moment on a blade to the moment resulting from the inertial forces.

$$a_1 = \lambda A_{11} + \theta_0 A_{12} + \theta_T A_{13} + \theta_2 A_{14} \tag{8.33}$$

where

$$A_{11} = \frac{4\left(\dfrac{\mu B^2}{2} - \dfrac{\mu^3}{8}\right)}{B^2\left(B^2 - \dfrac{1}{2}\mu^2\right)} \qquad A_{12} = \frac{8\mu B}{3\left(B^2 - \dfrac{1}{2}\mu^2\right)}$$

$$A_{14} = \frac{b^2 + \dfrac{3}{2}\mu^2}{b^2 - \dfrac{1}{2}\mu^2} \qquad A_{13} = \frac{2\mu B^2}{B^2 - \dfrac{1}{2}\mu^2}$$

$$b_1 = \beta_0 B_{11} + \theta_1 \tag{8.34}$$

where

$$B_{11} = \frac{4\mu B}{3\left(B^2 + \dfrac{1}{2}\mu^2\right)}$$

Some articulated rotors have coupling between flapping and pitching, which is accomplished by effectively rotating the flapping hinge through an angle δ_3. This *delta three effect*, as it is called, can be expressed as

$$K_\beta = \frac{\partial\theta}{\partial\beta} \tag{8.35}$$

This effect can be included in the above equations by increasing θ_0 by $K_\beta\beta_0$, decreasing θ_1 by $K_\beta a_1$, and decreasing θ_2 by $K_\beta b_1$ in the above equations.

The assumption that the section angle-of-attack, α, is small does not hold for a region on the retreating side of the rotor. An examination of Equation 8.29 shows that the expression for α can become infinite on the retreating side as x approaches zero. If the denominator of that equation is equated to zero, then

$$x = \mu\sin\psi \tag{8.36}$$

Within this region the resultant velocity normal to a blade section is approaching to the trailing edge instead of the leading. For obvious reasons, the area defined by Equation 8.36 is called the *reverse flow region*. This region will lie within a circle, which is centered on the $\psi = 270°$ line and passes through the rotor axis and the point $x = \mu$. This region, shown on Figure 8.7, is excluded in the integration of the differential lift to obtain the thrust and instantaneous hub moment. In other words, it is assumed that the lift on the blade vanishes in the reverse flow region.

Helicopter Trim and Control Angles

Figure 8.14 is a side view of a helicopter showing the pertinent angles, forces, and dimensions needed to calculate the longitudinal trim and control angles for a helicopter in forward flight. The cg is located a distance of x ahead of the rotor axis and a distance of h below the hub. The aerodynamic center of the horizontal tail is located a distance of l_t behind the cg. The moments produced by the thrust and the horizontal tail about the cg are obvious. Not as obvious is the hub moment labeled M_H in Figure 8.14. This moment arises as the result of the hinge offset when the blades flap away from the disc plane. This is illustrated in Figure 8.15 where a force couple is shown at the instant the blades of a two-bladed rotor are positioned fore and aft. The longitudinal flapping angle, a_1, is small so that the magnitude of the force along each blade is equal approximately to the total centrifugal force on the blade. At this instant, the nose-up pitching moment is equal to

$$M = 2e\,(CF)\,a_1$$

At any other azimuth angle, a_1 can be replaced by β and the moment resolved about the pitching axis. The resolved moment is then integrated with respect to ψ from 0 to 2π and divided by 2π. In this manner, the hub pitching moment for a B-bladed rotor will be found to equal

$$M_H = -\frac{Be\,(CF)\,a_1}{2} \tag{8.37}$$

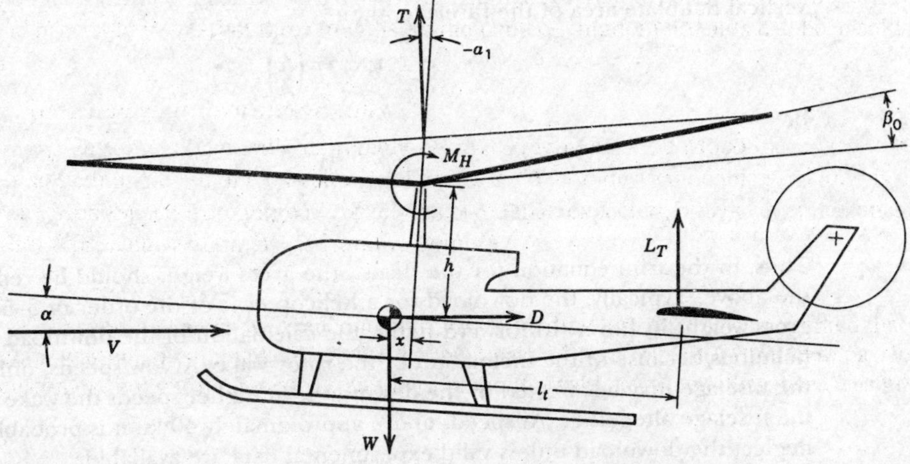

Figure 8.14 Schematic of helicopter showing forces, dimensions, and angles.

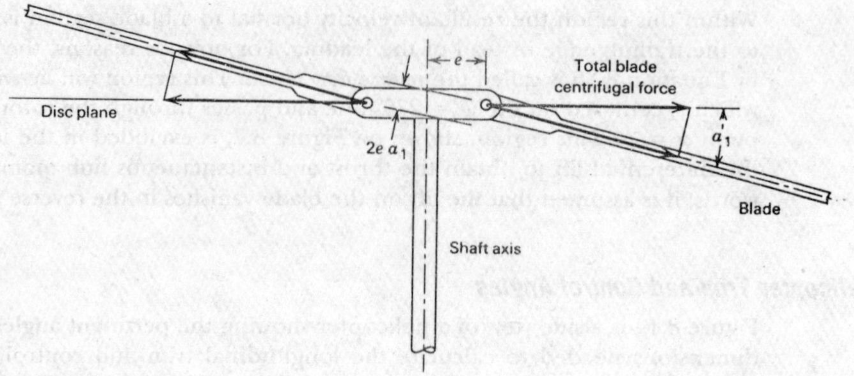

Figure 8.15 Side view of articulated rotor showing origin of hub moment resulting from longitudinal flapping.

In steady, level flight, the sum of the forces and moments on the helicopter must equal zero. Thus, assuming a_1 and α to be small angles,

$$\sum F_x = 0 = -T(\alpha + a_1) - D \tag{8.38a}$$

$$\sum F_z = 0 = W - T - L_t \tag{8.38b}$$

$$\sum M = 0 = M_H - L_t l_t + T a_1 h - Tx \tag{8.38c}$$

A more complete analysis will also include the aerodynamic forces and moments on the fuselage and thrust from the turboshaft engine. Also, the drag can be treated more exactly by placing the drag components at their corresponding locations. Additional lifting surfaces, such as stub wings, can also be considered in the trim analysis (Ref. 8.4). For the purposes of this text, the total drag of the helicopter is simply placed at the cg. Also, the horizontal tail is assumed to be fixed.

Another refinement to the above analysis is the *download*. The download is the vertical, downward drag on the fuselage as the result of the downwash from the rotor. For equilibrium, the rotor thrust must be increased by the download. In hover, the calculation of the download is straightforward. It is assumed that the dynamic pressure experienced by the fuselage is equal to that far below the rotor. From momentum theory, this is equal to the disc loading T/A. If f_v is the equivalent vertical flat-plate area of the fuselage, then

$$T = W + f_v\left(\frac{T}{A}\right)$$

or

$$T = \frac{W}{1 - \dfrac{f_v}{A}} \tag{8.39}$$

Thus, in the trim equation for the thrust, the gross weight should be replaced by the above. Typically, the download for a helicopter is of the order of 3–5% of the gross weight in hover. In forward flight, the calculation of the download is rather nebulous because of the inclination of the rotor wake. At low speeds, only part of the fuselage may be affected by the downwash. At higher speeds the wake can miss the fuselage altogether. At speeds above approximately 40 kt, it is probably best to neglect the download unless valid experimental data are available.

For some helicopters, the incidence angle of the tail is linked to the longitudinal cyclic pitch in order to provide a suitable stick position–airspeed relationship. This aspect of static stability and control is discussed in the next chapter. Also, in calculating the lift on the horizontal tail, the downwash from the main rotor must be taken into account. If w_T is the downwash estimated at the horizontal tail, the angle-of-attack at the tail must be reduced by w_T/V since the resultant flow at the tail will be inclined downward by this angle. Again, the inclination of the rotor wake as the forward speed increases makes this calculation difficult.

One can resort to an iterative technique for complicated stems in order to solve for the trim angle of attack and the corresponding pitch control angles. However, for illustrative purposes, a simple helicopter configuration that does not have a horizontal tail will be considered here. In this case, the longitudinal flapping is obtained immediately from Equation 8.38*c*. Without any other mechanism present to introduce a moment about the cg, the longitudinal flapping must be such as to direct the thrust vector through the cg. Therefore,

$$a_1 = \frac{x}{h} \tag{8.40}$$

It then follows from Equation 8.38*a* that

$$\alpha = -\left(\frac{D}{W} + a_1\right) \tag{8.41}$$

As an example in the use of the equations for the trim and control angles, consider again the AH-1J helicopter (without a horizontal tail). The rotor axis is located at station 200, and the cg will be taken at the 197 station. In accordance with practice, these station designations are in inches. The rotor for the AH-1J is a two-bladed teetering type with no hinge offset, a blade Lock number of 5.047, a tip speed of 738 fps, and a total blade twist of $-10.0°$. The dimensionless blade weight moment, τ, is equal to 0.00195. The teetering hinge is approximately 7.57 ft above the cg. A true airspeed of 30 kt will be assumed at a standard altitude of 1000 ft.

Using the above relations'.ips, the following values are determined:

$T_1 = 0.4717$	$T_2 = 0.3063$	$T_3 = 0.2222$	$T_4 = 0.03237$
$F_1 = 0.3042$	$F_2 = 0.2222$	$F_3 = 0.1724$	$F_4 = 0.02088$
$A_{11} = 0.1460$	$A_{12} = 0.1896$	$A_{13} = 0.1376$	$A_{14} = 1.0101$
$\mu = 0.0687$	$\alpha = -2.286°$	$1 = 1.892°$	$\theta_0 = 14.97°$
$\theta_2 = 0.716°$	$\beta_0 = 2.433°$	$w = 23.8$ fps	$\overline{V} = 56.86$ fps

Computer Exercise 8.2 "HELICOPTER TRIM"

Modify the program written in Computer Exercise 8.1 to include the calculation of the control angles, flapping angles, and angle-of-attack. Compare your results with the numbers from the AH-1J example.

Figures 8.16 and 8.17 present calculated values of the trim and control angles for the AH-1J helicopter from hover to 160 kt. Also, test results are included for the trim angle-of-attack as taken from Reference 8.4. The predictions for α are seen to agree within 1 or 2° even though the refinements mentioned previously are not included. The measurements of α are tending above the predictions as the speed increases, which is probably the result of the horizontal tail incidence being linked to the longitudinal cyclic pitch. Indeed, the calculated results do not include the effect of the horizontal tail. For a more precise treatment of this helicopter, the reader is referred to References 8.2, 8.5, 8.6, and 8.7.

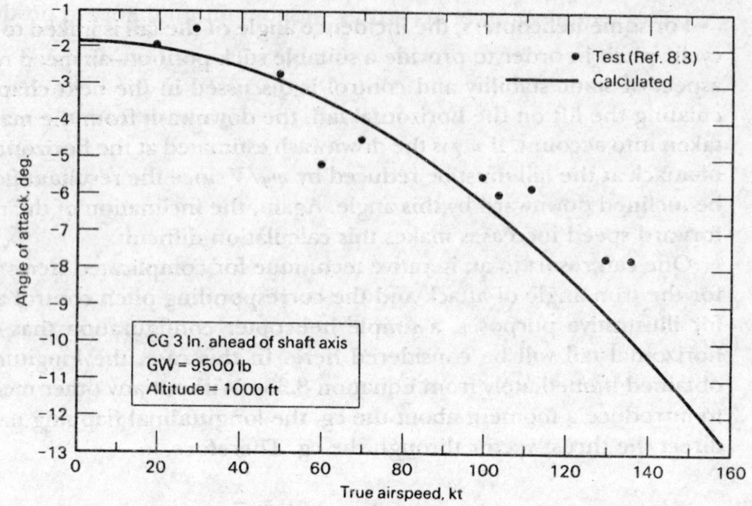

Figure 8.16 Calculated and measured trim angles-of-attack for the Bell AH-1J helicopter.

HELICOPTER PERFORMANCE

Hover

A hovering helicopter rotor is akin to a statically thrusting propeller. Thus, the relationships developed earlier for a propeller in Chapter 6 can be applied here. One cannot, however, speak of an efficiency for a hovering rotor since, by definition, for the static case the efficiency is identically zero. Instead, hovering rotors are measured against the ideal rotor based on momentum theory by means of a *figure of merit, M.*

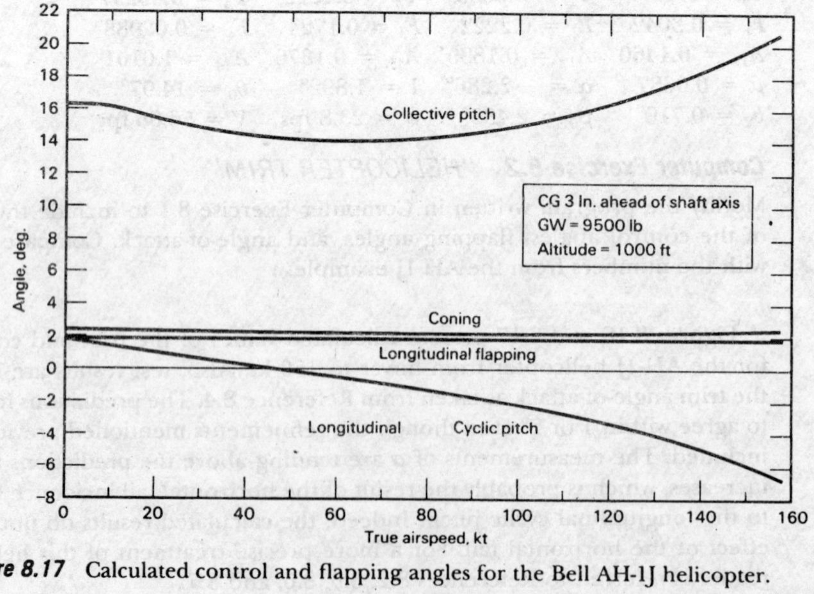

Figure 8.17 Calculated control and flapping angles for the Bell AH-1J helicopter.

The required power according to momentum theory to produce a given thrust, T, statically by a rotor having a disc area, A, was given earlier by Equation 6.17 as

$$P_i = \frac{T^{3/2}}{\sqrt{2\rho A}} \tag{8.42}$$

The figure of merit is then defined as the ratio of the ideal power given above to the actual power required by the rotor.

$$M = \frac{P_i}{P} \tag{8.43}$$

Use the figure of merit with caution because it is generally higher for the rotor with the highest disc loading. This can be shown by writing M approximately as

$$M = \frac{P_i}{1.12P_i + P_{pi}}$$

The factor of 1.12 is included to account for the fact that the actual induced power, statically, is approximately 12% higher than that predicted from momentum theory. With some algebraic reduction, using Equations 8.11 and 8.16, M becomes

$$M = \frac{1}{1.12 + \dfrac{3\epsilon}{4}\sqrt{\dfrac{2}{C_T}}} \tag{8.44}$$

ϵ is the average drag-to-lift ratio for the blade. This equation is only approximate, and a more precise calculation of the actual power required to hover can be performed using the numerical integrations developed in Chapter 6.

Typically, consider a disc loading of 5 psf, a tip speed of 700 fps, and an average C_L of 0.5 at sea level. The drag-to-lift ratio for this C_L should be approximately 0.024 resulting in a figure of merit from Equation 8.44 of 0.663. Now consider a rotor having a disc loading of 10 psf operating at the same average C_L and tip speed. ϵ will remain the same, but the figure of merit increases to 0.717. Thus, the one rotor shows a higher figure of merit than the other even though the drag-to-lift ratio and the margin below stall are the same for both rotors.

Forward Flight

In forward flight (above approximately 40 kt) and for small angles of climb below approximately 5°, the performance of a helicopter follows closely the same procedure used for a propeller-driven, fixed-wing airplane. One calculates the required power for level flight and subtracts it from the available power to give an excess power for climbing. However, for vertical flight or for large angles of climb, the power relationships change sufficiently from the level flight case so that a more precise approach to calculating the rate of climb is required. A novel approach is taken in Reference 8.4, which allows one to make use of any performance program written for level flight.

The reference notes that in a steady climb, the gross weight can be resolved into two components, normal to and along the flight path. The normal component is then taken as an equivalent merit weight in level flight, whereas the component in the drag direction results in a modified equivalent flat-plate area, which depends on the true airspeed and climb angle. Thus, the same program used to calculate the required power in level flight can be used for climbing if the following expressions are used for the gross weight and equivalent flat-plate area.

$$W' = W\cos\theta_c \tag{8.45}$$

$$f' = f + \frac{W \sin\theta_c}{\frac{1}{2}\rho V^2} \tag{8.46}$$

where θ_c is the angle of climb. V is the resultant velocity along the flight path and includes the rate-of-climb. Given V and the angle of climb, the rate-of-climb will be

$$R/C = V \sin\theta_c \tag{8.47}$$

Thus, for a fixed true airspeed, one can increase the angle of climb until the calculated required power is equal to the available power. The maximum rate-of-climb for that airspeed follows from Equation 8.47.

As the angle of climb approaches 90°, or vertical, the change of f with α should be considered. Note that for θ_c equal to 90°, the equivalent drag from Equation 8.46 replaces the weight, whereas the equivalent weight vanishes. Thus, an additional vertical drag should be taken into account. However, this drag for typical maximum vertical rates of climb is only 2 or 3% of the gross weight and will not affect the calculated vertical rate of climb significantly.

The power required by a tail rotor can be determined using the same relationships as those developed for the main rotor. The tail rotor thrust is equal to the torque of the main rotor divided by the distance between the axes of the two rotors. Using this thrust, the tail rotor geometry, and tip speed, the total power of the tail rotor can be determined as the sum of the induced power and profile power. The parasite power of the tail rotor will be zero because the thrust does not have a forward component.

The whirling blades of a tail rotor present a hazard during ground operations and can sometimes become an unintended "weed eater." Increasingly, companies are opting for a different means of countering the main rotor torque. These include the NOTAR (no tail rotor) configuration developed by the McDonnell-Douglas Helicopter Co. and the ducted fan configuration, called the Fenestron by the Aerospatiale Co.

A comparison of the three types of tail thrusters is shown schematically in Figure 8.18. The NOTAR configuration incorporates a fan, mounted in the tail boom, which compresses the air and forces some of it out of a long slot along the right side of the boom. The slot directs the thin sheet of air downward and tangential to the rounded surface of the boom causing a circulation around the boom. This circulation in combination with the downwash from the rotor produces a force on the boom directed to the right. The remainder of the compressed air exits to the

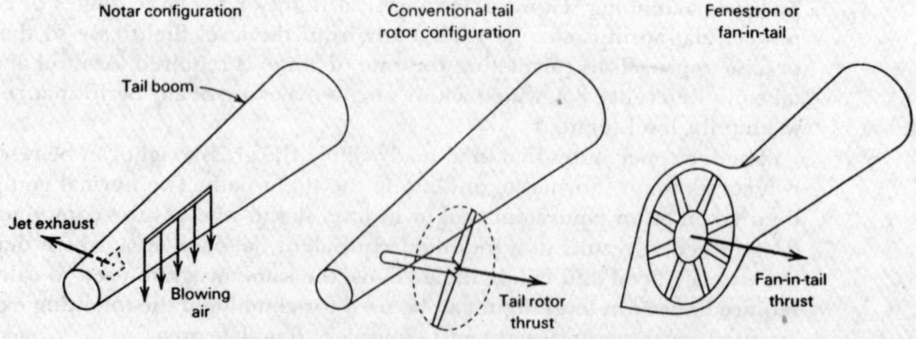

Figure 8.18 Schematic drawings of various types of tail thrusters mounted on the end of single-rotor tail booms.

left through a nozzle at the end of the boom, which is throttled by a moveable sleeve. The circulatory force and the nozzle thrust provide the antitorque, whereas the throttling of the nozzle provides yaw control.

Ground Effect

Close to the ground, the downwash induced by a rotor is diminished. Thus, the induced power decreases for a given thrust. Conversely, for a given power, the available thrust increases. At a high gross weight or at a high altitude, a helicopter may be able to hover in ground effect (IGE) but not be able to do so out of ground effect (OGE). Sometimes a helicopter, which cannot hover OGE, will lift off IGE and then climb after reaching a forward speed for which the required power is less. Figure 8.19 presents a graph that can be used to estimate the reduction in required power, or the increase in available thrust, which results from ground effect. As can be seen, ground effect essentially disappears if the rotor is higher than one diameter above the ground.

Design Variables

Table 8.1 lists some characteristics for a number of current, operational helicopters as obtained from Reference 8.5 and other sources. Figure 8.20 shows the dependence of disc loading on the gross weight. The reader is directed to Reference 8.5 for further data and analysis on helicopter performance.

The trend shown in Figure 8.20 is a result of the square-cube law discussed earlier at the end of Chapter 7. If the weight varies as the cube, and the area as the square, of a characteristic length, then the disc loading, W/A, must vary as the cube root of the weight. This is shown in Figure 8.20 where a curve has been faired through the data points.

The rotor solidity, tip speed, and disc loading are selected for a helicopter to satisfy several criteria. First, if the tip speed is too high, compressibility effects will be encountered at relatively low forward speeds. Also, a high tip speed results in a high noise level. Conversely, if the tip speed is too low, retreating blade stall will incur at a relative low forward speed. For a given tip speed and disc loading, the rotor solidity is chosen to give an average blade lift coefficient, according to Equa-

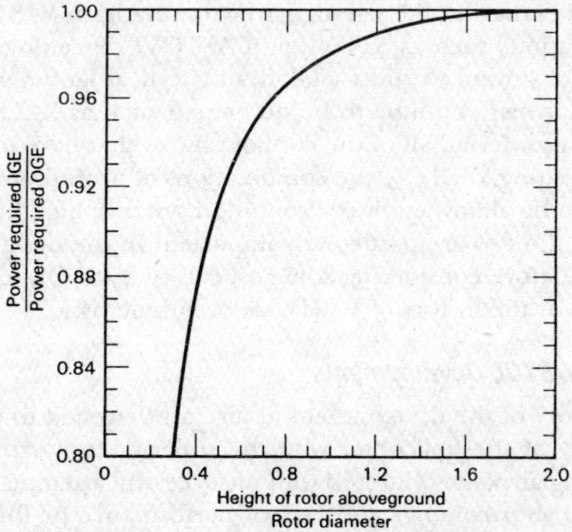

Figure 8.19 Effect of ground proximity on the power required by a helicopter to hover.

Table 8.1 Helicopter Characteristics

Parameter	Aerospatiale AS350L Ecuriel	Messerschmidt-Bolkow-Blohm BK117	Bell UH-1H Huey Iroquois	McDonnell-Douglas AH-64a Apache	Boeing Ch-47C Chinook (Tandem rotors)	Sikorsky CH-53E Super Stallion
Max gross weight, lb	4299.00	6283.00	9500.00	20500.00	50000.00	73500.00
Max. continuous hp	592.00	410.00	1250.00	1536.00	3000.00	2696.00
Main rotor Diam, ft	35.07	36.08	48.00	48.00	60.00	79.00
Main rotor chord	1.15	1.04	1.75	1.75	2.67	2.50
Number of blades	3.00	4.00	2.00	4.00	3.00	6.00
Disc loading, psf	4.45	6.15	5.25	11.33	8.84	14.99
Tip speed, fps	712.00	725.00	780.00	726.00	707.00	732.00
Rotor solidity	0.06	0.07	0.05	0.09	0.08	0.12
Hover ceiling OGE, ft	7380.00	8200.00	6004.00	12400.00	5900.00	9500.00
Service ceiling, ft	15600.00	14760.00	12700.00	20500.00	8500.00	18500.00
R/C, fpm	1555.00	1950.00	1700.00	2880.00	1485.00	2750.00
Never-exceed V, kt	147.00	150.00	128.00	197.00	160.00	170.00
Max. cruise V, kt	125.00	136.00	110.00	145.00	158.00	150.00
Thrust coef. SSL	0.00369	0.00492	0.00363	0.00904	0.00744	0.01177
Avg. blade lift coef	0.35	0.40	0.47	0.58	0.53	0.58

tion 8.16 in hover, of approximately 0.5. Since $C_{l_{max}}$, for unflapped airfoils, is of the order of 1.4, the C_l value of 0.5 allows a margin for maneuvering before the rotor blade stalls. If a rotor is hovering at a condition where the blade is close to stalling, a sudden increase in the collective pitch can result in a loss in thrust rather than the desired increase.

This introductory material to the helicopter will allow the reader to perform preliminary analyses relating to helicopter performance and design. However, reference books and textbooks devoted entirely to the helicopter are available and should be consulted for more advanced treatment of problems relating to the aerodynamics and dynamics of helicopters (References 8.5–8.7).

V/STOL

As noted earlier, in this age of acronyms, the term V/STOL has been extended to configurations such as STOVL and ASTOVL, meaning short takeoff and vertical landing or advanced short takeoff and vertical landing. New configurations (and new acronyms) continue to be developed such as RALS for *Remote Augmented Lift System.* Considering all of the configurations that have been studied in the past for accomplishing V/STOL, the combinations of propulsor, rotors, and lifting surfaces appear to be almost endless. Propulsion systems, high lift systems, and interactions between the two are particularly important to the design and analysis of V/STOL aircraft. Before considering some aspects of V/STOL technology, a brief look will be taken at the history of V/STOL developments.

A Brief History of V/STOL Developments

The history of the development of aircraft designed to combine the vertical flight capability of the helicopter with the superior forward flight performance of the fixed-wing airplane is littered with unsuccessful attempts. While many of these flew, technical shortcomings and lack of performance or financing prohibited further development. A brief, concise chronological history of these aircraft follows.

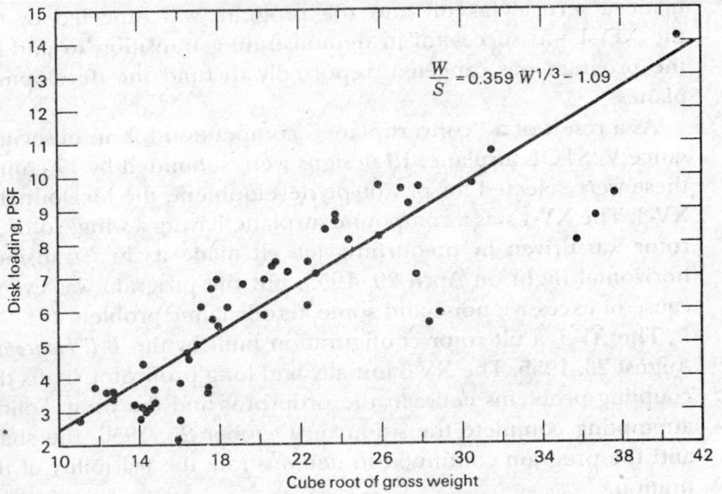

Figure 8.20 The trend of disc loading with gross weight for helicopters.

The 1930s

The *Herrick Convertiplane* was a biplane configuration that could take off in a conventional manner. The upper wing would then rotate so that it could land vertically.

The *Baynes Heliplane* was far ahead of the available technology of its time. It was a tilt rotor configuration that was patented, but never built. The rotors were to be powered by a gas generator providing gas out to turbines in each nacelle.

The 1950s

In the 1940s *The Transcendental Aircraft Co.* designed and built the Model 1-G. This single place, research aircraft, shown in Figure 8.21, has a gross weight of 1750 lb and two, 17-ft-diameter, tilting rotors. It experienced dynamic problems with coupling between the wing and rotors and crashed in 1951. It was rebuilt and flew again in 1956.

The *POGO VTOL* competition resulted in two "tail sitters"; aircraft that stood on their tails with the nose pointed vertically upward. The Convair XFY-1 had a delta wing, weighed 15,000 lb, and was powered by an Allison T-40 engine driving two coaxial, 16-ft-diameter propellers. On November 2, 1954, the XFY-1 made its first transition from vertical to horizontal flight and back. The *Lockheed XFV-1* was similar to the XFY-1 except that its wing was straight. This airplane was equipped with a temporary conventional landing gear and flew horizontally. However, it never

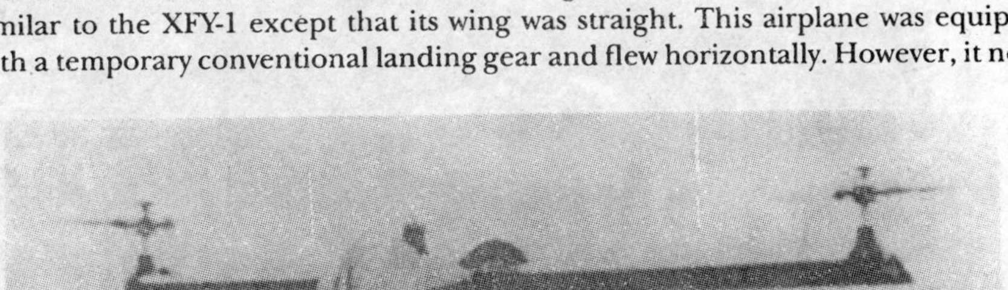

Figure 8.21 The Transcendental 1-G; the first tilt-rotor VTOL aircraft.

made a vertical takeoff and the program was canceled by the Navy. Although the XFY-1 was successful in demonstrating transition to and from vertical flight, the program was canceled, reportedly to fund the development of turbojet airplanes.

As a result of a "convertiplane" competition for an observation and reconnaissance V/STOL airplane, 19 designs were submitted by 17 companies. Only two of these were selected for prototype development, the McDonnell XV-1 and the Bell XV-3. The XV-1 was a compound airplane having a wing, rotor, and propeller. The rotor was driven by tip-burning jets. It made its first transition from vertical to horizontal flight on April 29, 1955, but the program was eventually canceled because of excessive noise and some aerodynamic problems.

The *XV-3*, a tilt rotor configuration built by the *Bell Helicopter Co.*, first flew on August 23, 1955. The XV-3 initially had long proprotor shafts that caused dynamic coupling problems between the proprotor and the wing. Following a crash while attempting complete transition on October 25, 1956, the shafts were shortened and the program continued to demonstrate the feasibility of the tilt-rotor configuration.

In 1957, the *Bell X-14A*, a single-place, experimental airplane was developed under a U.S. Air Force contract and first hovered on February 19, 1957. This aircraft, shown in Figure 8.22, subsequently made many successful transitions from the hovering mode to forward flight and back. This airplane achieved its VTOL capability from swivelling cascade nozzles that redirected the exhaust from two Armstrong-Siddeley Viper engines. It was successful as a research vehicle and was converted to a variable stability vehicle for NASA in 1968. The configuration could probably have been developed into an operational vehicle.

Also in the late 1950s, the *Ryan X-13 Vertijet* was flown successfully. This turbojet-powered airplane was remindful of the earlier tail sitters except that, instead of sitting in an upright position on landing gear, it was supported upright by a hook at the nose that engaged a cable attached to a mooring structure. The first complete transition from hover to forward flight and back was made on April 11, 1957. Although its performance was promising, the configuration was not developed further because it had no STOL capability and the pilots objected to the vertical attitude for takeoff and landing.

In the mid-1950s the U.S. Army funded a series of technology demonstrators, or flying test beds. Four of them were flown. The *Vertol VZ-2* was a tilt-wing airplane, which was designed and rolled out in one year. Completed in 1957, it successfully

Figure 8.22 The Bell X-14A VTOL aircraft utilizing vectored jet thrust. Courtesy, National Air & Space Museum.

went through transition and was tested over an eight-year period at NASA Langley and Edwards AFB. A photograph of the VZ-2 is shown in Figure 8.23. Your author has a warm spot in his heart for this particular airplane because he worked on the design and analyzed its transition from the hover mode to forward flight.

The *Ryan VZ-3* was a deflected slipstream configuration employing extremely large flaps mounted behind twin propellers. This rather bizarre-looking aircraft is pictured in Figure 8.24. It demonstrated exceptional STOL performance, but was not able to take off or land vertically.

The *Doak VZ-4,* first tested in February 1958, utilized tilting, ducted propellers mounted on each wing tip to achieve VTOL performance. After flying successfully, the design was purchased by the Douglas Aircraft Co., but after losing to the Bell XV-15 in the tri-service competition, it was never developed further.

The *Fairchild VZ-5* was another deflected-slipstream configuration, but with four propellers in front of a large flap system. It suffered from some of the same problems as the VZ-3.

There were two other tilt-wing designs that followed the VZ-2. These were the *Kaman K-16* funded by the U.S. Navy and the *Hiller X-18* funded by the Air Force. Both of these were unsuccessful. The K-16 was built, but never flown, and the X-18 took off and landed vertically, but never achieved transition to the airplane mode. The program was canceled after only 19 flights.

In the 1950s the development of an engine and control system intended for V/STOL application was begun, namely, the Pegasus engine. A Frenchman, Michael Wibault, sold the idea of the rotating nozzles to the Bristol Co. of England after some persuasion by an American Colonel, John O'Driscoll. Bristol revised the engine design to an axial flow configuration and, with most of the funding supplied by the United States, successfully developed the *Pegasus engine.* In 1959–60, the British Ministry of Aviation ordered six prototypes of the P.1127 Kestrel aircraft employing the Pegasus engine. The aircraft, built by the British Hawker Aircraft, Ltd., began hovering tests on October 21, 1960, only 17 months after its construction began. It was from this aircraft that the Harrier aircraft, to be discussed later, evolved.

The 1960s

The U.S. Tri-Service Competition in the early 1960s resulted in three aircraft being built and flown. The *XC-142A,* a four-engine, tilt-wing transport, monitored by the

Figure 8.23 The Vertol VZ-2; the first tilt-wing VTOL aircraft.

Figure 8.24 The Ryan VZ-3; a deflected slipstream STOL aircraft. Courtesy, National Air & Space Museum.

Navy, was built by a consortium of *Hiller, Vought, and Ryan.* This consortium won the competition over Vertol despite the success of the VZ-2 compared to the Hiller and Ryan tilt-wing airplanes. Five XC-142 airplanes were built. Only one survived after a number of accidents and can be found today at the Air Force museum in Dayton, Ohio. The performance, both in hover and forward flight, of the XC-142 was marginal. Thus, what was to have been the first production V/STOL airplane in the United States never materialized.

Another airplane in the Tri-Service competition was the Bell X-22A. This airplane, supported in part by the U.S. Navy and U.S. Air Force, employed four ducted rotors mounted at the tips of tandem wings. One of a kind, it was intended purely for research. It proved to be reliable and valuable over the years as a variable-stability airplane to study stability and control problems relating to V/STOL flight.

The program begun in the 1950s by the Army continued into the 1960s with the *Lockheed XV-4* and the *Ryan XV-5*. The XV-4, the Hummingbird, employed thrust augmentation, whereby the primary jet from the engine was ducted vertically downward along the centerline of the fuselage. This jet, mixing with the surrounding air, entrained a secondary flow so that the net thrust was greater than the initial momentum of the engine's jet. Unfortunately, a thrust augmentation ratio of only 1.2 was achieved compared to the predicted value of 1.4. A second aircraft was converted to a configuration employing six J-85 engines for direct lift. After this aircraft crashed, the program was terminated.

The Ryan XV-5, shown in Figure 8.25, was a unique configuration referred to as a "fan in wing." Two large fans, covering much of the planform, were mounted on each side of the airplane in the wing. These fans were driven by turbine blades mounted directly to a ring around the fan blade tips. For vertical takeoff, panels on the wing surface were opened and the engine exhaust was diverted to drive the tip turbines. After transitioning to forward flight, the panels were closed to form a solid wing surface and the engine exhaust directed rearward. The XV-5 first went

Figure 8.25 The Ryan XV-5; a fan-in-wing VTOL aircraft. Courtesy, National Air & Space Museum.

through transition from vertical flight and back on November 5, 1964. It subsequently made approximately 100 flights totaling 42 hr up to speeds of 450 mph. Unfortunately, the XV-5 suffered two crashes that resulted in the program being canceled. Neither crash was connected with the VTOL system. If support had been given to the fan-in-wing program, as was done with the Harrier, the U.S. military might well have another operational V/STOL airplane in its inventory today. With the lower disk loading of the fan as compared to a lift engine, the fan-in-wing configuration would appear to offer, for the same cruise performance, a better hover performance than a deflected jet VTOL system.

There were also many V/STOL developments outside of the United States in the 1960s worthy of note. These include the following:

Dassault Balzac This French airplane employed separate lift engines. It was canceled after control problems and two crashes.

Dassault Mirage III-V This supersonic French VTOL airplane was powered by a SNECMA TF-106 turbofan with 19,820 lb of thrust. Vertical lift was provided by eight Rolls Royce RB.162-1 turbojets with a thrust of 4400 lb each. It first hovered on February 12, 1965, and subsequently went through successful transitions. However, the program was canceled in 1966 following a crash.

Canadair CL-84 This was a tilt-wing airplane of which three were built for the Canadian military. It was first flown on May 7, 1965, but, although the performance was impressive, the design never went into production.

Dornier-Do-31 This West German design employed two Pegasus engines plus 10 lift engines in pods mounted at the wing tips. The first transition from hover to forward flight was made on December 16, 1967. By mid-1969, over 200 takeoffs and transitions had been made. Unfortunately, excessive noise discouraged any future development.

Yakolev Freehand This Russian design appeared to use an engine and control system similar to the Pegasus system in the Harrier. It was first photographed in July 1967 at an air show in Domodedovo where it took off vertically, went through transition, and then landed vertically.

The 1970s and 1980s

The *Bell XV-15,* a tilt-rotor airplane that first hovered on May 3, 1977, has proven to be a durable and reliable V/STOL airplane. The first full transition to the airplane mode was made on July 24, 1979. Based on experience gained from the XV-3, the XV-15 has demonstrated many types of operational capabilities including nap-of-the-earth, ship on-board exercises, air-to-air evasive maneuvers, in-flight refueling and weapons delivery. The XV-15 employs two 25-ft-diameter proprotors driven by two 1550 shp turboshaft engines. At a gross weight of 13,000 lb it can hover OGE at an altitude of 8650 ft.

The *McDonnell-Douglas/British Aerospace Harrier II*, the *AV-8B*, is a growth version of the British *Hawker Siddeley AV-8A Harrier*. The performance of the B model is improved over the A model because of the extensive use of composites and refinements in the propulsion system and in the aerodynamic design. Currently, the Harrier, pictured in Figure 8.26, is the only operational VTOL airplane in the Western world. The Rockwell *XFV-12A* represented another attempt to employ flow augmentation to increase the thrust available from the engine. The wings were designed to open to form primary and secondary nozzles. So-called hypermixing nozzles were utilized to induce mixing of the primary jet, the diverted engine exhaust, with the secondary flow. Unfortunately, the augmentation ratio that was achieved fell considerably below that which was predicted. As a result, the airplane never flew. The low augmentation ratio may have been the result of losses incurred in the turning and ducting of the engine exhaust to the nozzles.

The *X-Wing* aircraft, currently in limbo, employs a unique rotor known as a circulation-control rotor (CCR). The airfoil sections of such a rotor are relatively thick with an elliptical shape. Air is blown over the upper surface near the leading edge and/or the trailing edge depending upon whether the rotor blade is in the advancing or retreating position. At Mach numbers below 0.25, the blowing results in high lift coefficients and low drag coefficients despite the thick section. Hence, the rotor is structurally capable of being stopped in an *X* configuration in forward flight to act as a fixed wing. A CCR rotor was installed on the Rotor Systems Research Aircraft (RSRA) at NASA Ames Research Center; however, the program has been mothballed because of excessive costs and technical problems. The pneumatic system required to cyclically control the air to each blade is complex with very demanding requirements.

The *Bell-Boeing V-22 Osprey* in Figure 8.27 is a tilt-rotor airplane currently undergoing development. The airplane has a maximum gross weight of approximately 60,500 lb and is powered by two Allison T406-AD-400 engines, each rated at 6150 shp and driving a three-bladed, 38-ft-diameter, gimballed rotor. Additional details of this airplane will be covered later when discussing the aerodynamics of propellers and rotors at high angles-of-attack.

Figure 8.26 The Harrier AV-8B.

Figure 8.27 The Bell-Boeing V-22 Osprey; a tilt-rotor VTOL aircraft currently under development.

Technical Milestones and Other Notes

To the interested engineer, the history of V/STOL flight has been one of disillusionment. Over the last 35 years, research and development studies, prototype evaluations, and mission studies have pointed toward a production V/STOL aircraft. Competitions have been undertaken—Tri-Service, AMST, the Navy Type A, and Type B—which have had the industry preparing for a V/STOL aircraft. It is amazing to reflect on all of the research and the millions of dollars that have been spent on prototype V/STOL aircraft and then to realize that, outside of Russia, there is only one production V/STOL aircraft in existence; namely, the Harrier.

In addition to prototype developments, let us take a brief look at a few of the research efforts and other happenings relating to the development (or nondevelopment) of V/STOL over the last several decades. In 1933, Galen B. Shubauer investigated the jet flap experimentally. This early work on a high lift device was reported in NACA Technical Note 442 in January 1933. It was over 20 years later that an analytical solution to the jet flap was obtained by D. A. Spence in a paper appearing in the Proceedings of the Royal Society.

Around 1949, NACA at the Langley Research Center initiated an extensive experimental program to gather data on many types of high lift devices including blown flaps, propellers at high angles-of-attack, wings and propellers in combination, and multiple flap configurations. Two names prominent in this work are Richard Kuhn and John Campbell. In the early 1950s John Anttinello demonstrated that a choked supersonic jet flap would perform as well as a subsonic flap at the same blowing coefficient. In the mid-1950s Kenneth Razak and others at the University of Wichita (now Wichita State University) performed the first systematic series testing of jet flaps.

In the February 1966 issue of *Aeronautics and Astronautics* an article appeared entitled "V/STOL—Still Just Around the Corner." The article stated ". . . everyone at the Pentagon, NASA, and Transportation Department agrees that V/STOL aircraft are going to do great things someday but not today." That same year, the Air Force Chief of Staff said, "We just have not gotten far enough in the development stage to know how to build a pure V/STOL aircraft that will carry any payload a reasonable distance."

In the September 1968 issue of *Aeronautics and Astronautics*, an article entitled "V/STOL—Its Day Has Come," a proposed V/STOL service is described in the Northeast corridor. Eastern Airlines was anxious to start, but was thwarted by manufacturers' inability to develop an economically feasible aircraft coupled with failure of municipal authorities to provide the landing facilities. They announced that they would fly the Breguet 941, licensed by McDonnell-Douglas as the Model 188, on shuttle routes. This airplane was powered by four 1500 shp turboshaft engines. It could take off in 1050 ft over a 35-ft height and could land in 820 ft over a 50-ft obstacle. Its maximum cruising speed at 10,000 ft was 259 mph with a range of 620 miles with maximum payload. Eastern operated the airplane over an extended period; however, the airplane did not prove to be economical and the need for STOL performance was never shown.

The Custer Channel Wing Aircraft is shown in Figure 8.28, not because of its impact on V/STOL developments, but because many people are not familiar with the configuration and your author has had some personal contact with the machine pictured in the figure. In order to achieve high lift at low speeds, the propeller on each wing is mounted in a channel formed by the wing. The increase in the velocity over the upper surface of the wing produced by the propeller results in an increment in the lift on the wing above that caused by the forward motion. Even statically, a lift is produced on the wing due to the propeller slipstream. The concept has been tested in wind tunnels by both the NACA and the Air Force and as a high-lift device using power, it compares favorably with other high lift devices. Your author viewed critically the slow speed performance of the aircraft shown in Figure 8.28 and was impressed by its STOL performance. However, with the propeller blades rotating in and out of the channel, the noise was somewhat objectionable and one wonders about the fatigue stresses on the propeller blades as they sweep in and out of the channel.

In the May 1971 issue of *A & A*, a report was given on hearings for a proposed STOLport on New York City's West Side between 24th and 34th streets. The report stated that the honor of scaling the peak of absurdity fell to Percival Goodman, Professor of Urban Design at Columbia University, who compared the STOLport to Nazi cremation chambers and the Mylai massacre. Like most of the others (witnesses), he offered no facts in evidence. "Figures lie, and liars figure," he said, "and who needs facts and figures when, on prima facie and common sense grounds, the STOLport is not a good idea."

In the December 12, 1977, issue of *Aviation Week* magazine, it was stated that the Army and the Navy are lending increased support to the Air Force's plan to acquire the Advanced Medium STOL Transport (AMST). One month later, in the January 16, 1978, issue, it was stated that "all AMST funding has been deleted from the

Figure 8.28 Custer channel wing aircraft. (Courtesy, National Air & Space Museum)

fiscal 1979 budget." By then, a considerable amount of money had been spent in developing the McDonnell-Douglas YC-15 employing externally blown flaps (EBF) and the Boeing YC-14 with upper surface blowing (USB), both of which demonstrated outstanding STOL performance and maneuverability. A sketch of the Boeing YC-14 is shown in Figure 8.29. The engines of this aircraft are mounted on the upper surface of the wing. When the flaps are deflected the engine exhausts follow the upper surface of the flaps resulting in a high lift.

In this same period, the Navy was pushing for V/STOL development and defined three types of V/STOL aircraft: **Type A:** Large, subsonic, fleet support transport with ASW capability; **Type B:** Tactical, attack, or fighter with supersonic capability; **Type C:** Smaller, multipurpose, subsonic aircraft for destroyers and frigates. But in the January 2, 1978, issue of *Aviation Week,* it was reported that "the Carter administration has reduced the Navy's budget by 1.9 billion which means that V/STOL R & D will be reduced by 50%." And thus, it has gone over the years for V/STOL development. It is hoped that the V-22 Osprey program will meet with a better fate. Possibly, the competition from a European design, similar to the V-22, known as EUROFAR, will stimulate more interest in the Osprey.

Nearly all of the V/STOL programs to date have been funded from federal sources. A tilt-wing airplane being funded entirely as a commercial venture by the Ishida Corporation of Fort Worth, Texas, is shown in Figure 8.30. Unfortunately,

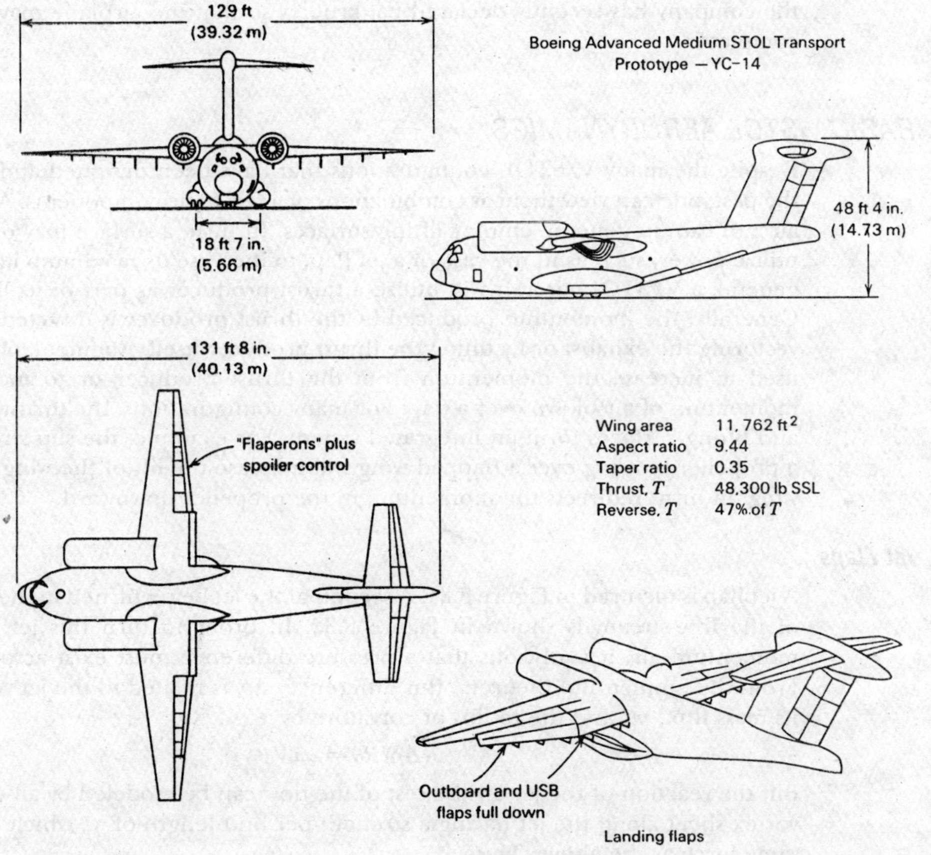

Figure 8.29 The Boeing YC-14 advanced medium STOL transport (AMST) with upper-surface blowing (USB).

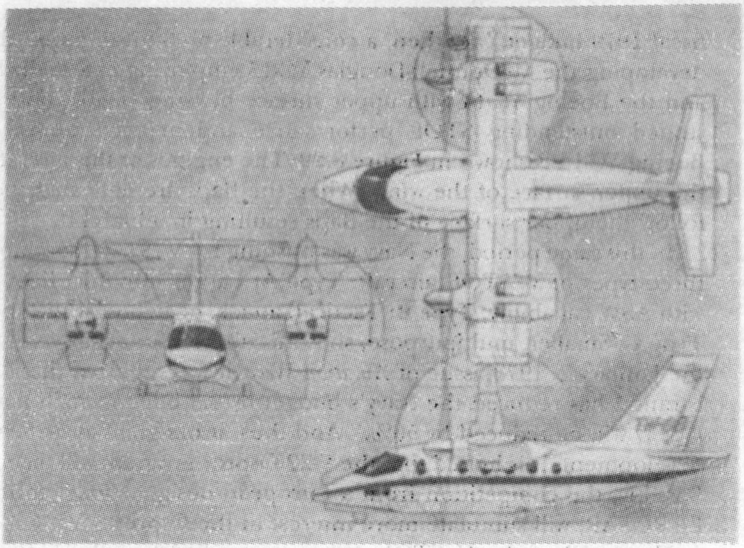

Figure 8.30 Ishida TW-68 tilt-wing VTOL aircraft.

the company has recently declared bankruptcy so that the airplane may never be completed.

BASIC V/STOL AERODYNAMICS

Despite the many V/STOL configurations that have been designed and tested in the past, one can view them as combinations of a few basic components. A V/STOL aircraft can, in general, employ lifting surfaces. In turn, a surface may or may not utilize power, such as in the case of a jet flap, to increase its maximum lift. Also, in general, a V/STOL aircraft will utilize a thrust producer as part of its lift system. Generally, the momentum produced by the thrust producer is directed either by vectoring the exhaust or by tilting the thrust producer itself. Augmentation may be used to increase the momentum from the thrust producer or to increase the momentum of air blown over a flap. For many configurations, the thrust producer and lifting surfaces form an integrated system. For example, the slip stream from a propeller blowing over a flapped wing will increase the lift of the wing while the wing, in turn, redirects the momentum in the propeller downward.

Jet Flaps

A jet flap is pictured in Figure 8.31. A section of the jet being turned in the direction of the free-stream is shown in Figure 8.32. In order to turn this jet of higher momentum air, it is obvious that a pressure difference must exist across the jet. From the momentum theorem, this difference, Δp, is related to the jet velocity, V_j, its mass flux, m_j, and the radius of curvature by

$$(\Delta p) R\theta = m_j V_j \theta \tag{8.48}$$

But the reaction of the jet on the rest of the flow can be modeled by an equivalent vortex sheet along the jet having a strength per unit length of γ_j, which exerts the same force as the above. Thus,

$$m_j V_j \theta = \rho V \gamma_j \tag{8.49}$$

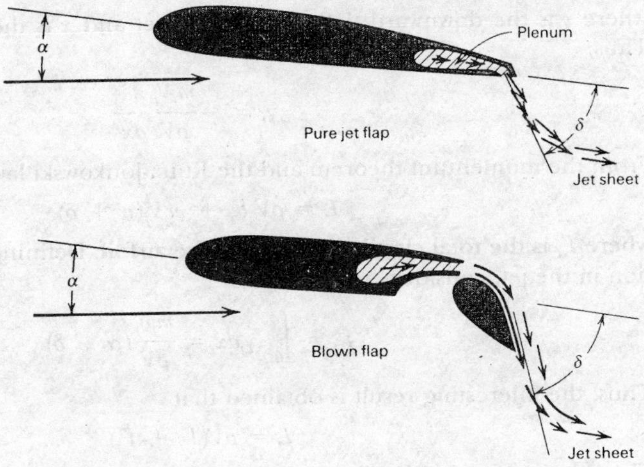

Figure 8.31 An airfoil with a jet flap.

or

$$\Delta p = \frac{m_j V_j}{R} \tag{8.50}$$

and

$$\gamma_j = \frac{\Delta p}{\rho V} \tag{8.51}$$

For small deflections of the jet,

$$R = -\left(\frac{d^2 y}{dx^2}\right)^{-1} \tag{8.52}$$

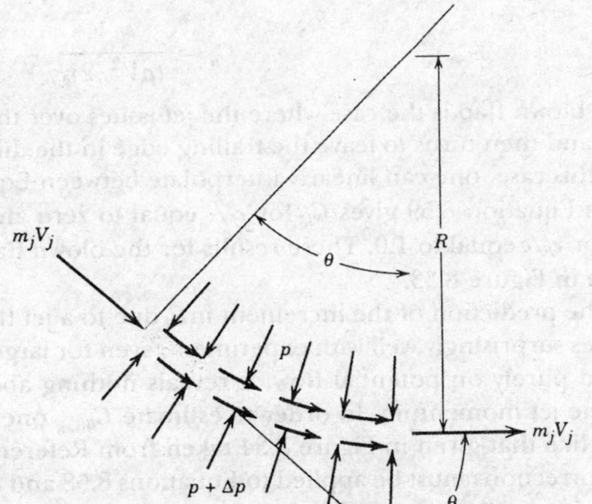

Figure 8.32 Section of a high-momentum jet being turned by the pressure difference across the jet.

where y is the downward deflection of the jet and x is the downstream distance. Thus,

$$\gamma_j = -\frac{m_j V_j}{\rho V}\frac{d^2 y}{dx^2} \tag{8.53}$$

From the momentum theorem and the Kutta-Joukowski law,

$$L = \rho V \Gamma_c + m_j V_j(\alpha + \delta) \tag{8.54}$$

where Γ_c is the total circulation around the airfoil. Defining Γ_j as the total circulation in the jet, consider

$$\Gamma_j = \int_0^\infty \gamma_j dx = \frac{m_j v_j}{\rho V}(\alpha + \delta) \tag{8.55}$$

Thus, the interesting result is obtained that

$$L = \rho V(\Gamma_c + \Gamma_j) \tag{8.56}$$

The solution for the distribution of circulation along the airfoil and jet sheet was first obtained by Spence, Reference 8.8. The solution was obtained by expressing both distributions in series form and then satisfying that the resultant velocity is everywhere tangent to the surfaces of the airfoil and jet vortex sheet. In addition, the relationship between the shape of the vortex sheet and the momentum in the jet given by Equation 8.53 was satisfied.

A numerical fit to Spence's results for a pure jet flap, where c_F/c, equals zero, is given by

$$C_l = C_{l_\alpha}\alpha + C_{l_\delta}\delta \tag{8.57}$$

where

$$C_{l_\alpha} = 2\pi(1 + 0.151\sqrt{C_\mu} + 0.219C_\mu) \tag{8.58}$$

$$C_{l_\delta} = [4\pi C_\mu(1 + 0.151\sqrt{C_\mu} + 0.139C_\mu)^{1/2} \tag{8.59}$$

C_μ is the jet momentum coefficient and can be interpreted as the ratio of the jet thrust to the product of the free-stream dynamic pressure and the airfoil area.

$$C_\mu = \frac{m_j V_j}{(\rho V^2/2)\,c} \tag{8.60}$$

A blown flap is the case where the jet issues over the upper surface of a physical flap and then turns to leave the trailing edge in the direction of the flap deflection. For this case, one can linearly interpolate between Equations 8.58 and 8.59 for C_{l_δ}, since Equation 8.59 gives C_{l_δ} for c_F/c equal to zero and Equation 8.58 is, in effect, C_{l_δ} for c_F/c equal to 1.0. These results for the blown flap are presented in graphical form in Figure 8.33.

The prediction of the increment in C_l due to a jet flap given by the linear theory agrees surprisingly well with experiment, even for large flap angles. However, being based purely on potential flow, it reveals nothing about the dependency of $C_{l_{max}}$ on the jet momentum. In order to estimate $C_{l_{max}}$ one must resort to experimental data like that given in Figure 8.34 taken from Reference 8.9.

Corrections must be applied to Equations 8.58 and 8.59 for application to a wing of finite aspect ratio, A. The correction factor, F, can be found in Reference 8.2 and is given as follows. Using this factor, C_{l_α} and C_{l_δ} become

$$C_{L_\alpha} = FC_{l_\alpha} \tag{8.61}$$

$$C_{L_\delta} = FC_{l_\delta} \tag{8.62}$$

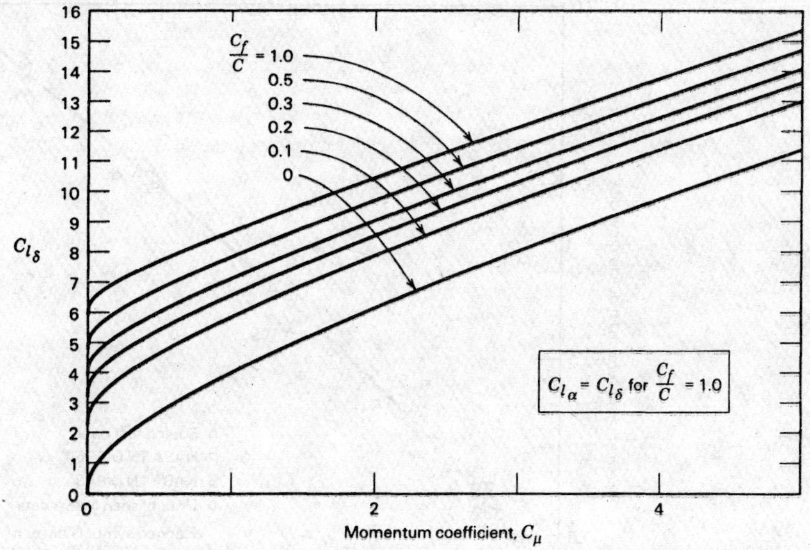

Figure 8.33 The effectiveness of a blown flap.

where

$$F = \frac{A + 2C_\mu/\pi}{A + 2 + 0{,}604\sqrt{C_\mu} + 0.876C_\mu} \tag{8.63}$$

The classical expression for the induced drag coefficient is also modified to account for the jet flap and is given by

$$C_{D_i} = \frac{C_L^2}{2C_\mu + \pi A} \tag{8.64}$$

For partial span flaps, Reference 8.2 suggests simply correcting Equations 8.61 and 8.62 by the ratio of wing area, which is jet flapped to the total wing area.

For a finite wing, there is a limit on the amount of circulatory lift that can be produced. As the circulation is increased, the downwash velocity becomes large by comparison to the free-stream velocity. As shown in Figure 8.35, this causes the resultant force vector ($\mathbf{F} = \rho \mathbf{V} \times \boldsymbol{\Gamma}$) to increase and tilt backward. Despite the fact that the circulation, Γ, continues to increase, the vertical component of the resultant force, namely the lift, reaches a limiting value. In coefficient form, this limit is given approximately by

$$C_{L_{\text{limit}}} = 0.86A \tag{8.65}$$

There are three high lift configurations that have been tested on prototype airplanes and are closely related to the jet flap. These are *upper surface blowing* (USB), *externally blown flaps* (EBF), and the *augmenter wing*. The former two have been mentioned previously in connection with the YC-14 and YC-15 airplanes. The augmenter wing was applied to a modified de Havilland Buffalo aircraft and, like the other two configurations, was tested extensively. All of these configurations generate high lift by deflecting a stream of high-momentum air downward from the trailing edge of the wing. Sketches of these configurations and some experimental data are presented in Figure 8.36.

For the USB configuration, the turbojet exhaust is fanned downward and along the upper surface of the wing immediately ahead of the flap. The high-momentum

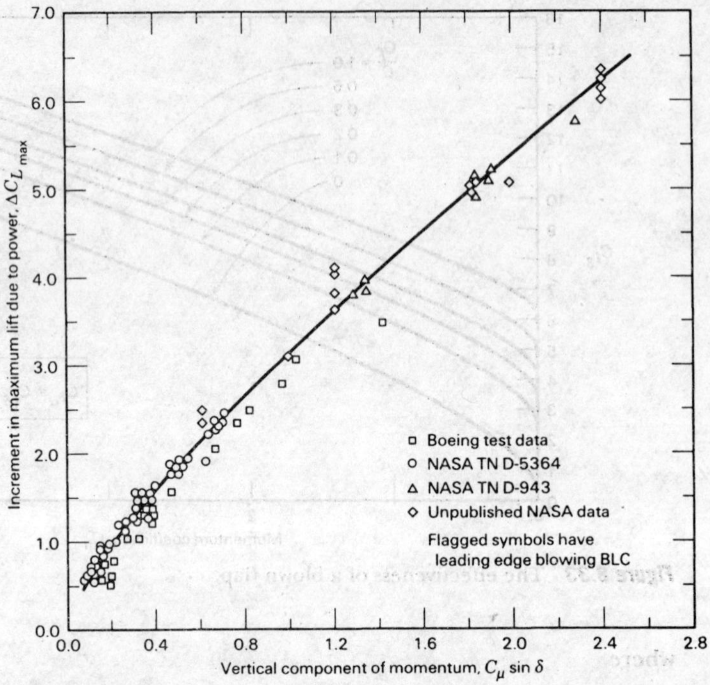

Figure 8.34 Increment in $C_{l_{max}}$ for a jet-flapped airfoil. (Courtesy, Iowa State University Press)

exhaust is turned by the *Coanda effect* and leaves in the direction of the flap. The Coanda effect is that phenomenon whereby a flow tends to adhere to, and follow, the contour of a continuous surface. (The author is sure the reader has had the experience of pouring from a cup and having the liquid follow, unwantedly, around the lip.) For the EBF configuration, the turbojet exhaust enters a multislotted flap below the wing and is turned by the resulting aerodynamic forces on the flap. The Augmenter Wing is essentially a blown-flap configuration except, when lowered, the flap divides into two parts to form a secondary nozzle surrounding the jet being blown from the forward part of the wing. As described in the next section, this results in an entrained flow through the nozzle formed by the flaps thus increasing the momentum in the jet.

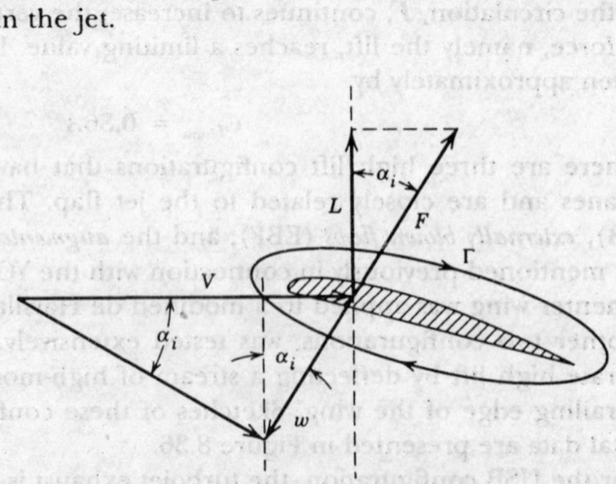

Figure 8.35 The induced velocity and force on an airfoil with high circulation.

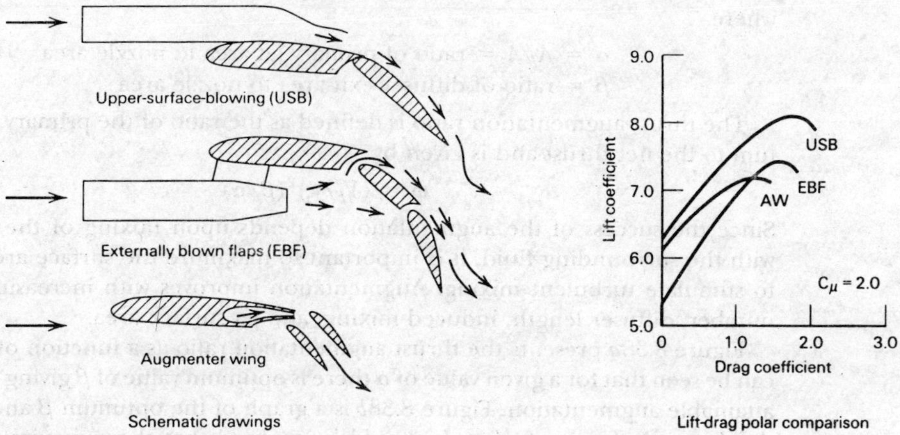

Figure 8.36 Upper-surface blowing (USB), externally blown flaps (EBF), and augmenter wing concepts.

Augmentation

An augmenter is composed of a primary jet issuing into a nozzle as shown in Figure 8.37. The viscous shear along the sides of the jet entrains a secondary flow so that the total mass flow through the nozzle is greater than that issuing from the primary jet. The thrust will be increased because of the additional mass flow. A further increase in the thrust can be achieved by diffusing the mixed flow as shown. An approximate solution for the augmentation is presented in Reference 8.2 and can be obtained by the following steps:

1. Apply Bernoulli's equation to the secondary flow from the free-stream to the plane of the primary jet.
2. Apply the momentum theorem from the plane of the primary jet to the beginning of the diffuser, assuming the mixing to be complete at that point.
3. Apply Bernoulli's equation through the diffuser.
4. Apply continuity throughout the flow.

The foregoing is similar to Von Karman's original solution except that he did not include the diffuser. Using the above assumptions, the following quadratic equation is obtained for the velocity at the diffuser exit.

$$(V_E/V_j)^2 + (V_E/V_j)\frac{2\alpha\beta(1-2\alpha)}{1-2\alpha+\alpha^2(1+\beta^2)} - \frac{(2\alpha-3\alpha^2)}{(1-2\alpha+\alpha^2(1+\beta^2))} = 0.0 \quad (8.66)$$

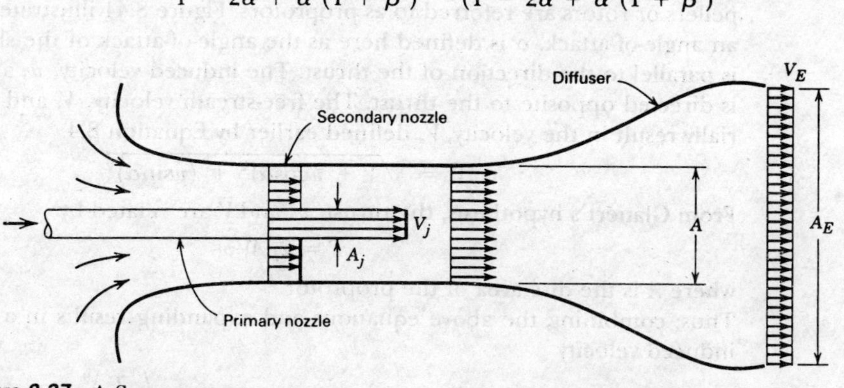

Figure 8.37 A flow augmenter.

where

$\alpha = A_j/A$ = ratio of primary jet area to nozzle area
β = ratio of diffuser exit area to nozzle area

The thrust augmentation ratio is defined as the ratio of the primary jet momentum to the net thrust and is given by

$$\phi = (V_E/V_j)^2(\beta/\alpha) \tag{8.67}$$

Since the success of the augmentation depends upon mixing of the primary jet with the surrounding fluid, it is important to maximize the surface area of the jet to stimulate turbulent mixing. Augmentation improves with increasing Reynolds number, diffuser length, induced mixing, and jet surface area.

Figure 8.38*a* presents the thrust augmentation ratio as a function of α and β. It can be seen that for a given value of α there is optimum value of β giving a maximum attainable augmentation. Figure 8.38*b* is a graph of the optimum β and maximum ϕ values as a function of α. It should be realized that these maximum ϕ values represent upper limits on ϕ, which are probably not attainable since complete mixing and uniform flow throughout the augmenter is assumed in the theory.

The entrainment of the secondary flow in an augmenter depends upon the mixing in the shear layer between the primary and secondary flows. This mixing is enhanced for a given flow rate if the surface of the shear layer is made as large as possible. Figure 8.39 depicts various nozzle geometries that have been tested with this in mind. Experimental data is included in the figure and are seen to agree fairly well with the predictions. Unfortunately, these nozzle configurations are difficult to package within the confines of an airframe. More recently, *hypermixing nozzles* have been studied, which promise similar results, but in a smaller space. This type of nozzle consists of a row of primary jets, the angular directions of which alternate to produce a vortex downstream of each jet boundary. These vortices increase considerably the mixing between the primary and secondary flows as compared to simply issuing the primary jets all in the same direction.

Some additional data are presented in Figure 8.40 to illustrate the strong effect of Reynolds number on the performance of an augmenter. In this case, the primary jet issues around the inside periphery of the nozzle to maximize the surface area of the jet as shown in the sketch.

The Tilt-Rotor Configuration; Propellers and Rotors at High Angles-of-Attack

Glauert's hypothesis was discussed earlier and is the basis for analyzing propellers or rotors at high angles-of-attack. For application to a tilt-rotor airplane, the propellers or rotors are referred to as proprotors. Figure 8.41 illustrates a propeller at an angle-of-attack. α is defined here as the angle-of-attack of the shaft axis, which is parallel to the direction of the thrust. The induced velocity, w, at the proprotor is directed opposite to the thrust. The free-stream velocity, V, and w added vectorially result in the velocity, $\overline{V}$, defined earlier by Equation 8.4.

$$\overline{V} = \sqrt{V + w\cos\alpha)^2 + (w\sin\alpha)^2} \tag{8.68}$$

From Glauert's hypothesis, the thrust, w, and $\overline{V}$ are related by

$$T = 2\rho A\overline{V}w \tag{8.69}$$

where A is the disc area of the proprotor.

Thus, combining the above equations and expanding results in a quartic for the induced velocity

$$w^4 + 2V(\cos\alpha)w^3 + V^2w^2 = \left(\frac{T}{2\rho A}\right)^2 \tag{8.70}$$

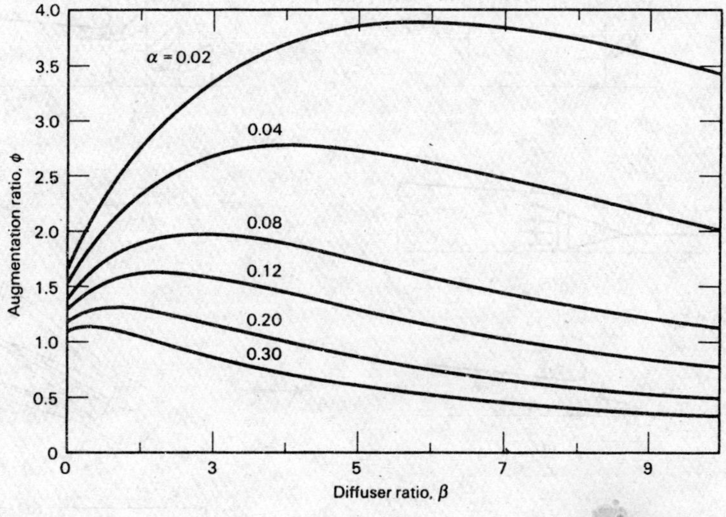

(a)

Figure 8.38a Predicted augmentation ratio as a function of diffuser ratio and nozzle area ratio.

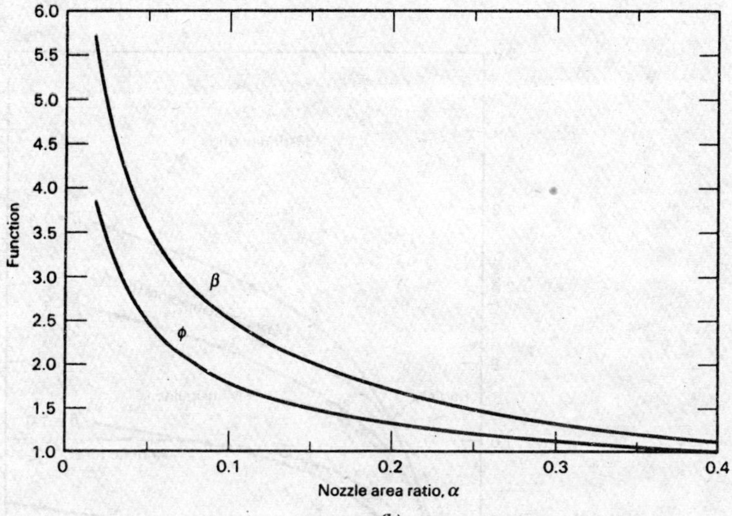

(b)

Figure 8.38b Maximum thrust augmentation ratio and corresponding diffuser ratio as a function of the nozzle area ratio.

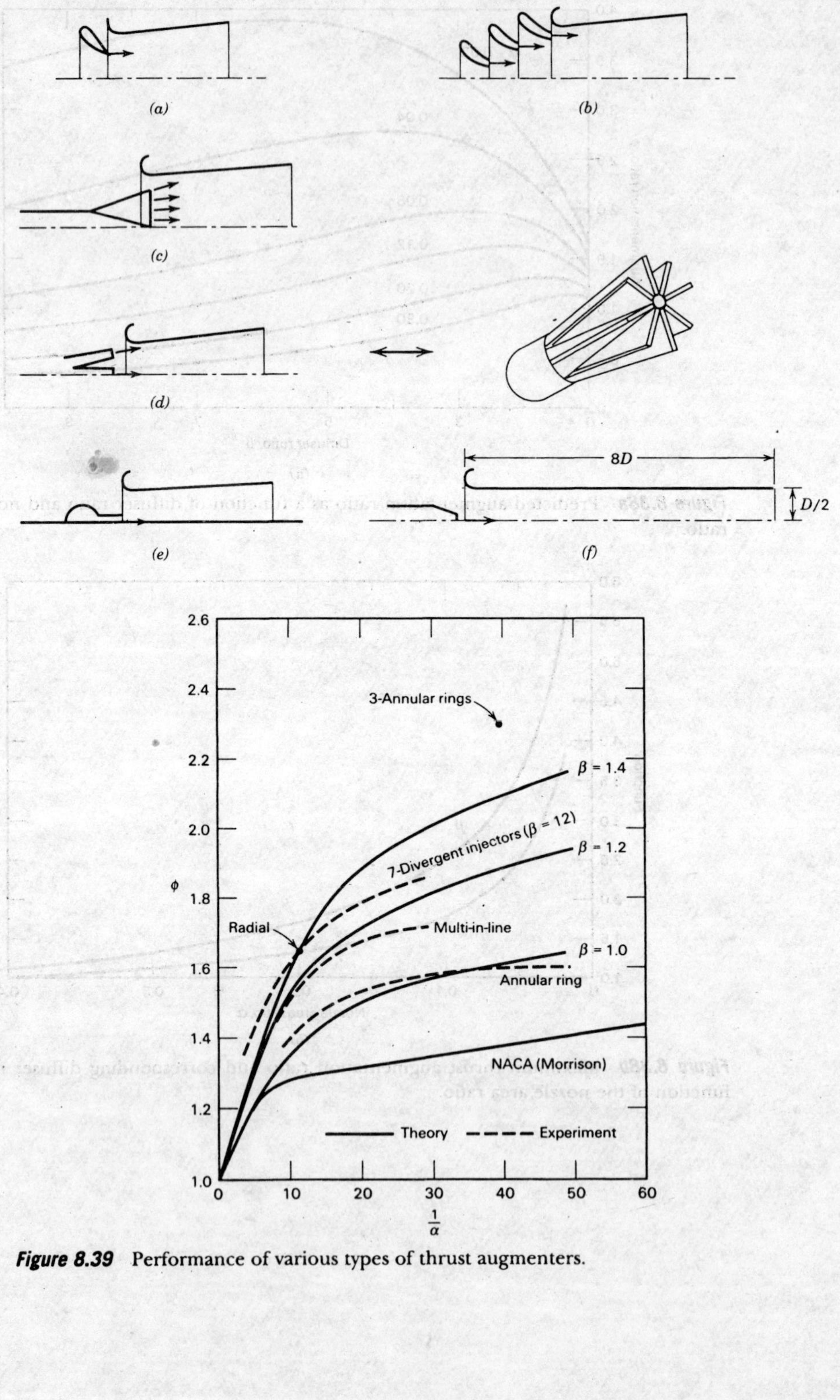

Figure 8.39 Performance of various types of thrust augmenters.

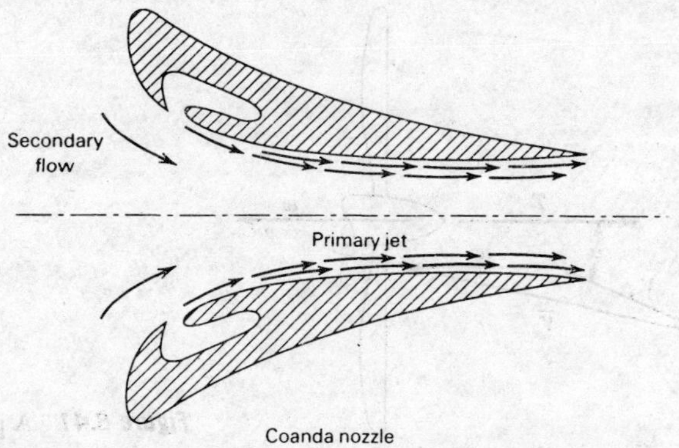

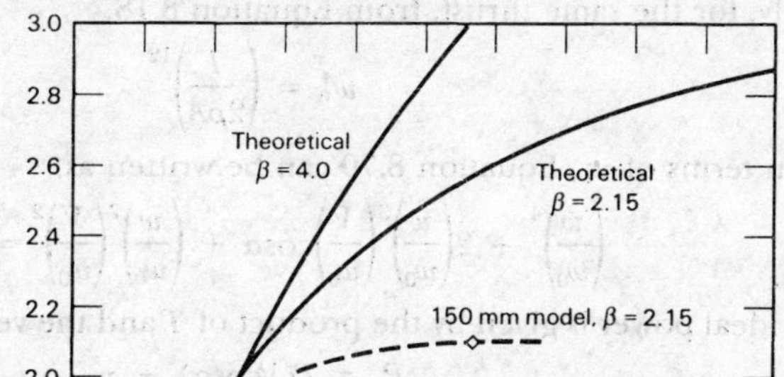

Figure 8.40 Effect of model size on thrust augmentation.

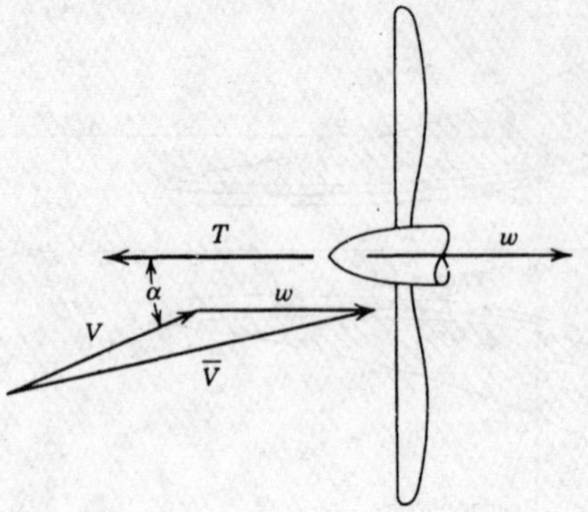

Figure 8.41 A propeller at an angle of attack.

Statically, for the same thrust, from Equation 8.18,

$$w_0^4 = \left(\frac{T}{2\rho A}\right)^{12}$$

Thus, in terms of w_0, Equation 8.70 can be written as,

$$\left(\frac{w}{w_0}\right)^4 + 2\left(\frac{w}{w_0}\right)^3\left(\frac{V}{w_0}\right)\cos\alpha + \left(\frac{w}{w_0}\right)^2\left(\frac{V}{w_0}\right)^2 = \left(\frac{T}{T_0}\right)^2 \tag{8.71}$$

The ideal power is given by the product of T and the velocity normal to the disk.

$$P_i = T(V\cos\alpha) + w \tag{8.72}$$

or

$$\frac{P_i}{P_0} = \frac{T}{T_0}\left[\frac{V}{w_0}\cos\alpha + \frac{w}{w_0}\right] \tag{8.73}$$

where

$$P_0 = Tw_0 \tag{8.74}$$

Setting T to a constant in one case, or P to a constant in the other case, leads to the functions presented graphically in Figures 8.42 *a, b, c,* and *d*. In practice, the induced part, Tw, of the ideal power is found to be approximately 12–15% higher than the value from momentum theory. However, this correction is not included in Figure 8.42.

The profile power required for the profile drag of the blades must be added to the ideal power to obtain the total power required. The profile power can be obtained from Equation 8.13 by replacing μ with $\overline{\mu}$ where

$$\overline{\mu} = \mu\sin\alpha$$

where $\mu = \dfrac{V}{V_T}$ and C_{po} is given by Equation 8.11.

$$C_p = C_{po}(1 + 3\overline{\mu}^2) \tag{8.75}$$

The static performance of propellers or rotors as a function of disk loading, T/A, was given earlier in Figure 6.22. Typical disk loadings for various classes of

Figure 8.42 Ideal propeller performance at an angle-of-attack (*a*) Variation of induced velocity with speed for constant thrust (*b*) Variation of required ideal power for constant thrust (*c*) Variation of induced velocity with speed for constant ideal power (*d*) Variation of available thrust with speed for constant ideal power.

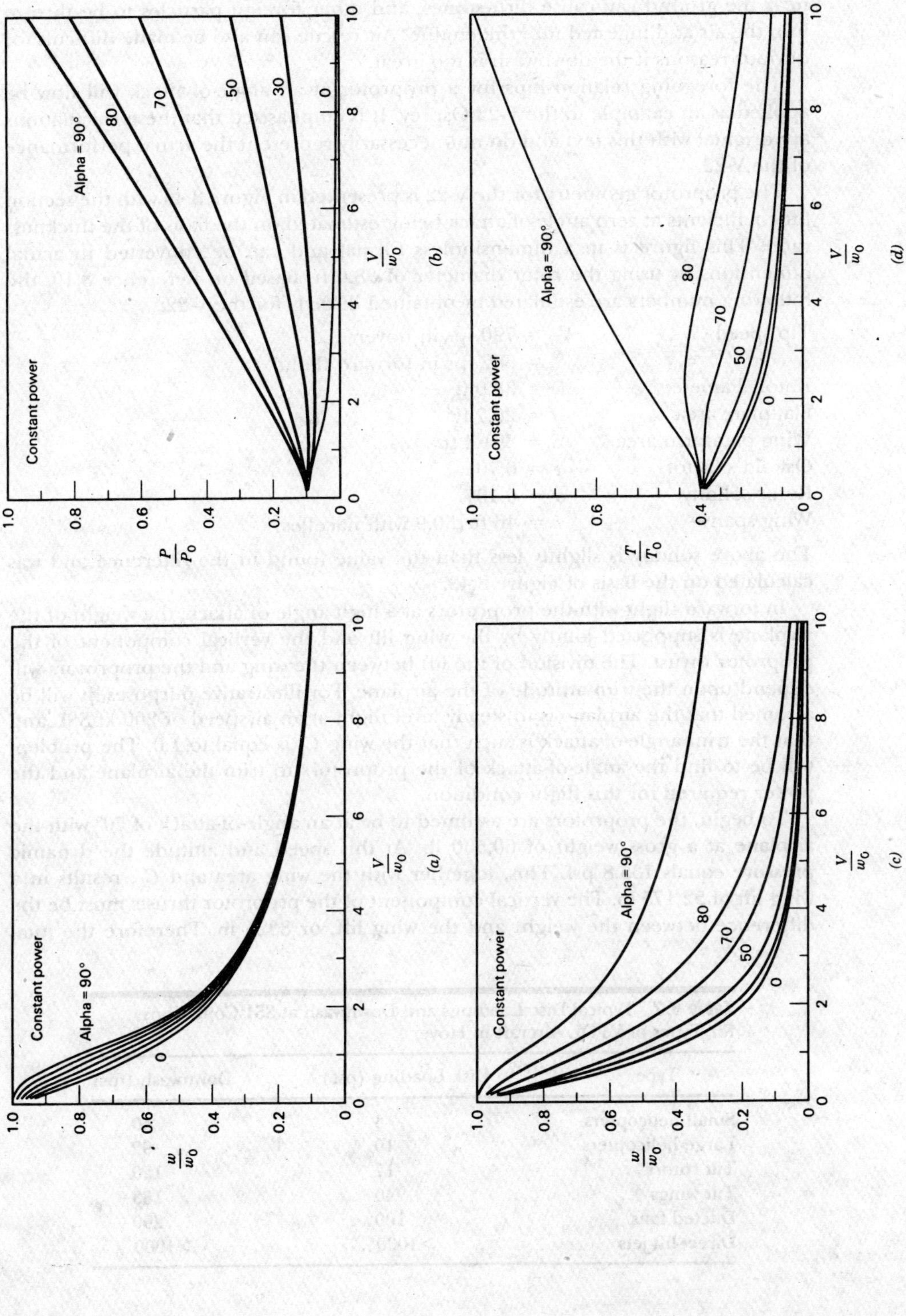
Constant power
Alpha = 90°
80
70
50
30
0
$\frac{P}{P_0}$
$\frac{V}{w_0}$
(b)
Constant power
Alpha = 90°
80
70
50
0
$\frac{T}{T_0}$
$\frac{V}{w_0}$
(d)
Constant power
Alpha = 90°
0
$\frac{w}{w_0}$
$\frac{V}{w_0}$
(a)
Constant power
Alpha = 90°
80
70
50
0
$\frac{w}{w_0}$
$\frac{V}{w_0}$
(c)

VTOL aircraft are presented in Table 8.2 together with the downwash velocity, which depends upon the disc loading. The relationship of downwash to disk loading can be as important as the relationship of power to disc loading. A high downwash near the ground can cause dirt, stones, and other foreign particles to be thrown into the air and ingested into the engine. Air rescue can also be made difficult for obvious reasons if the downwash is too great.

The foregoing relationships for a proprotor at an angle-of-attack will now be applied as an example to the V-22 Osprey. It is emphasized that these calculations are original with this text and do not necessarily represent the actual performance of the V-22.

The proprotor geometry for the V-22 is presented in Figure 8.43 with the section lift coefficients at zero angle-of-attack being estimated on the basis of the thickness ratio. This figure is in a dimensionless format and can be converted to actual dimensions by using the rotor diameter of 38.0 ft. Based on Reference 8.10, the following numbers are estimated or obtained directly for the V-22:

Tip speed	$V_T =$	790 fps in hover
	$=$	662 fps in forward flight
Rotor diameter	$D =$	38.0 ft
Flat-plate area	$f =$	26.7 ft^2
Wing planform area	$S =$	381.4 ft^2
Oswald's factor	$\epsilon =$	0.70
Rotor solidity	$\sigma =$	0.107
Wing span	$b =$	46 ft (50.9 with nacelles)

The above solidity is slightly less than the value found in the reference and was calculated on the basis of Figure 8.43.

In forward flight with the proprotors at a high angle of attack, the weight of the airplane is supported jointly by the wing lift and the vertical component of the proprotor thrust. The division of the lift between the wing and the proprotors will depend upon the trim attitude of the airplane. For illustrative purposes, it will be assumed that the airplane is in steady level flight at an airspeed of 200 kt SSL and that the trim angle-of-attack is such that the wing C_L is equal to 1.0. The problem will be to find the angle-of-attack of the proprotors to trim the airplane and the power required for this flight condition.

To begin, the proprotors are assumed to be at an angle-of-attack of 70° with the airplane at a gross weight of 60,500 lb. At this speed and altitude the dynamic pressure equals 135.8 psf. This, together with the wing area and C_L, results in a wing lift of 52,175 lb. The vertical component of the proprotor thrusts must be the difference between the weight and the wing lift, or 8325 lb. Therefore the total

Table 8.2 Typical Disc Loadings and Downwash at SS1 Conditions for Types of VTOL Aircraft in Hover

Type	Disk Loading (psf)	Downwash (fps)
Small helicopters	3	50
Large helicopters	10	92
Tilt rotors	17	120
Tilt wings	40	183
Ducted fans	100	290
Direct-lift jets	>1000	>1000

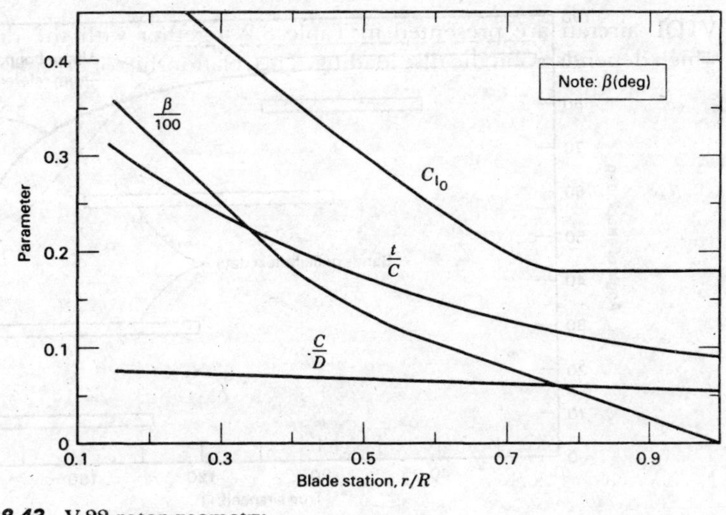

Figure 8.43 V-22 rotor geometry.

thrust of the proprotors is equal to this value divided by the sine of 70°, or 8859 lb. Using the flat-plate area of 26.7 ft^2 and the ϵ of 0.7 results in a drag of 7971 lb. However, for steady flight, the horizontal component of the total thrust must equal the drag. In this case the horizontal component equals the total thrust multiplied by the cosine of 70°, or 3030 lb. Since this number is less than the drag, the airplane would be decelerating. Therefore, the assumed angle-of-attack of the proprotor must be decreased until a value is found, which trims both the vertical and horizontal forces. A small computer program can be written to do the iteration whereby the trim angle-of-attack will equal 46.2°. The total proprotor thrust at this angle required for trim will equal 11,526 lb. The total disk area is 2268 ft^2 resulting in a disk loading for each proprotor of 5.08 psf.

The power required can be found by first calculating the hover requirements for the same thrust. For this disk loading the static downwash, w_0, will equal 32.68 fps. Thus V/w_0 equals 10.34 and the ideal power, which is induced power for hover, is 684.8 hp. For the trim angle-of-attack and speed, the ratio of the ideal power to the ideal hover power for the same thrust, P_i/P_{i0}, equals 7.25. Thus, the ideal power required in forward flight at the trim conditions is calculated to be 4965 hp.

The above numbers can be improved upon by calculating the profile power from vortex theory for the same thrust and adding it to the ideal power to obtain the total power. This will be considerably more accurate than the result obtained from Glauert's hypothesis alone. This has been done using the program developed in Chapter 6 and the V-22 geometry in Figure 8.43. In helicopter terminology, for hover, an induced power coefficient of 0.000252 and a total power coefficient of 0.000348 is found for a thrust coefficient of 0.00342. Thus, the result is a profile power coefficient of 0.96×10^{-4} giving an additional power of 464 hp. In forward flight, these numbers translate to a total power at 200 kt of 5429 hp.

Observe that this power is approximately half the power available. Thus, at maximum power for the conditions considered in the example, the airplane would be accelerating. In transitioning from the helicopter mode to the airplane mode, the proprotors rotate downward continuously and the airplane accelerates to a speed above which the weight is supported entirely by the wing.

The program mentioned above was used to generate the predicted curve of

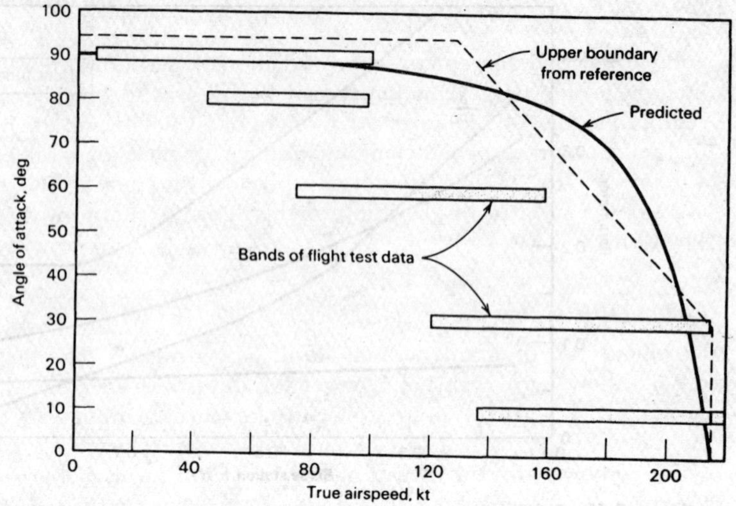

Figure 8.44 Predicted trim angles for the V-22.

proprotor trim angle of attack for the Osprey presented in Figure 8.44. The predictions appear reasonable in view of the test data and boundaries relating to the transition corridor as presented in Reference 8.11.

The download was not considered in this example since the hover condition is a fictitious reference for calculating the power in forward flight. However, with essentially a flat plate below the rotors, the download for a tilt-rotor airplane is appreciably higher than for a helicopter. One of the main reasons for installing a flap on the V-22 is to reduce the download. With the flap up, in hover the download is approximately 16.5% of the total thrust and decreases nearly linearly down to approximately 9% for a flap angle of 65°.

The figure of merit of an isolated, model V-22 rotor is presented in Figure 8.45

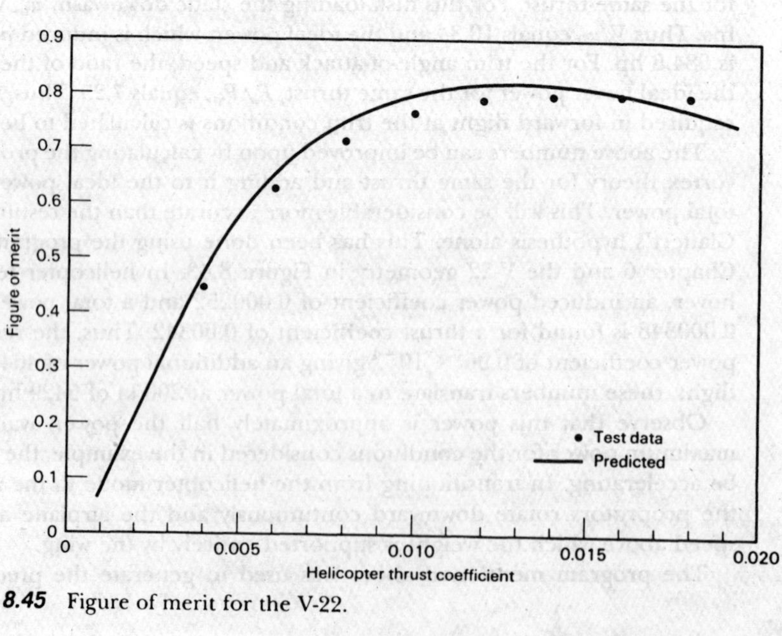

Figure 8.45 Figure of merit for the V-22.

and shows good agreement with predictions based on vortex theory modified to include profile drag. This figure can be used as to quickly estimate the power required by the V-22 to hover. At a weight of 60,500 lb. and the hovering tip speed, the helicopter thrust coefficient, including a 9% download, will equal 0.0196 giving a figure of merit of 0.8. The downwash velocity is equal to 78.2 fps resulting in an ideal power of 9376 hp. Dividing by the figure of merit gives an actual power of 11,720 hp.

Computer Exercise 8.3 *"TILTPROP AND TILTTRIM"*

Formulate two programs to analyze tiltrotor airplanes. The first program should predict the relative thrust and power variations with speed as a function of angle-of-attack for constant power or constant thrust. Input will be the proprotor geometry, speed, angle-of-attack, and either thrust or power. Output will be the required power (for the thrust) or the thrust (for the available power). The second program is to predict the trim angle-of-attack of the proprotors for a tilt-rotor airplane given the aerodynamic characteristics, geometry, and weight of the airplane.

Compare the results of the program with the predictions made for the V-22 and with the curves of Figure 8.42.

The Tilt–Wing Configuration

A wing-propeller combination is pictured in Figure 8.46*a*. Unlike a tilt-rotor, the wing is always at a positive angle-of-attack to the propeller slipstream. Also, the disc loading for a tilt-wing airplane is relatively high so that, even at high angles-of-attack, the wing is at a much lower angle-of-attack relative to the resultant velocity because of the propeller's slipstream. A considerable amount of wind tunnel testing was performed on propeller-wing combinations in the 1950s and 1960s and forms the aerodynamic base for the Ishida development, which will be described later.

There is no completely satisfactory theory in existence that adequately predicts the performance of a wing-propeller combination at high angles of attack. For conventional wings and propellers, the theory of Smelt and Davies (Ref. 8.12) satisfactorily predicts the increase in wing lift coefficient due to a propeller slipstream, but it is not applicable to the high angles-of-attack associated with the tilt wing. In Reference 8.2 an approximate, semiempirical theory is developed, applicable to the tilt wing, which gives fairly good results. If the wing is completely submerged in the slipstream, then the situation shown in Figure 8.46 is comparable to a jet flap, where a stream of higher momentum air leaves the wing trailing edge and is turned in the direction of the free-stream. Admittedly, in this case the jet is very thick.

For N propellers, the equivalent momentum coefficient will be given by

$$C_\mu = \frac{NT}{qS} \tag{8.76}$$

From Figure 8.46*b*,

$$V_R \sin\alpha_s = 2w\sin\alpha_p \tag{8.77}$$

The vertical component of the thrust is

$$T\sin\alpha_p = mV_R\sin\alpha_p \tag{8.78}$$

While the vertical component of the deflected thrust, that is, the slipstream lift, is given by

$$L_{ss} = mV_R\sin(\alpha_s + \theta) \tag{8.79}$$

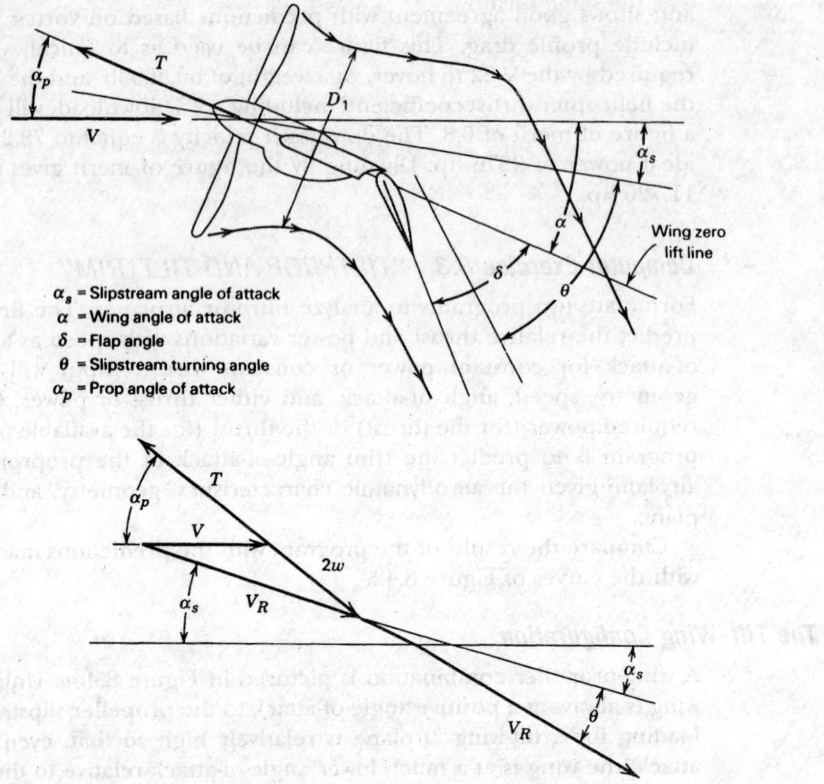

Figure 8.46 Wing–propeller combination.

Thus, it is assumed that the lift of the wing–propeller combination can be given as the sum of three parts:

$$C_L = C_{LT=0} + C_{L\Gamma} + C_\mu \frac{\sin\alpha_p}{\sin\alpha_s}\sin(\alpha_s + \theta) \qquad (8.80)$$

$C_{LT=0}$ is the lift that the wing will produce relative to the angle-of-attack of the incoming slipstream velocity. The second term in Equation 8.80 results from the additional circulation produced by the slipstream acting as a jet flap, and the last term is the vertical component of the slipstream momentum given by Equation 8.78.

The drag of the wing in the slipstream causes a loss of momentum, which must be considered.

$$D = \frac{1}{2}\rho V_R^2 S C_D = m\Delta V_R \qquad (8.81)$$

or

$$\frac{\Delta V_R}{V_R} = \frac{D\sin\alpha_s}{T\sin\alpha_p} \qquad (8.82)$$

V_R is the resultant velocity in the propeller slipstream as shown in Figure 8.46*b*. The magnitude of this velocity is assumed to remain constant as the velocity vector is turned by the wing. Hence,

$$C_L = C_{LT=0} + C_{L\Gamma} + C_\mu \sin(\alpha_s + \theta)\left(\frac{\sin\alpha_p}{\sin\alpha_s} - \frac{D}{T}\right) \qquad (8.83)$$

For the same turning,

$$\frac{D}{T} = \frac{\frac{1}{2}\rho V_R^2 C_d S}{\rho A \acute{V}(2w)} \tag{8.84}$$

where A is the disc area of the propeller. But, for static conditions,

$$\left(\frac{D}{T}\right)_{V=0} = \frac{\frac{1}{2}\rho(2w_o)^2 C_d S}{2\rho A w_o^2} = \frac{C_d S}{A} \tag{8.85}$$

Therefore, for the same turning,

$$\frac{D}{T} = \frac{V_r^2}{4\acute{V}w}\left(\frac{D}{T}\right)_{V=0} \tag{8.86}$$

where

$$\acute{V} = \sqrt{(V + w\cos\alpha_p)^2 + (w\sin\alpha_p)^2}$$

Defining the quantity, $1 - (D/T_{V=0}$, as the thrust recovery factor, F/T, the equation for C_L becomes

$$C_L = C_{L_{T=0}} + C_{L_\Gamma} + C_\mu \sin(\alpha_s + \theta)\left[\frac{\sin\alpha_p}{\sin\alpha_s} - \left(1 - \frac{F}{T}\right)\frac{V_R^2}{4\acute{V}w}\right] \tag{8.87}$$

The above set of equations can be solved with the use of Figures 8.47 and 8.48. Figure 8.47 presents data on the turning performance of various types of flaps, while the thrust loss factor, F/T, is obtained from Figure 8.48. The term, C_{L_Γ}, can be obtained from Equation 8.57. However, in so doing, $C_\mu(\alpha + \delta)$ must be subtracted since Equation 8.57 already takes into account the vertical momentum of the slipstream as it leaves the wing.

The net forward thrust coefficient, C_X, defined in terms of q and s can be derived in a similar manner and will not be given here (see Ref. 8.2 for details).

$$C_X = C_\mu\left\{\cos(\alpha_s + \theta)\left[\frac{\sin\alpha_p}{\sin\alpha_s} - \left(1 - \frac{F}{T}\right)\frac{V_R^2}{4\acute{V}w}\right] - \frac{V\sin\alpha_p}{V_R\sin\alpha_s}\right\} \tag{8.88}$$

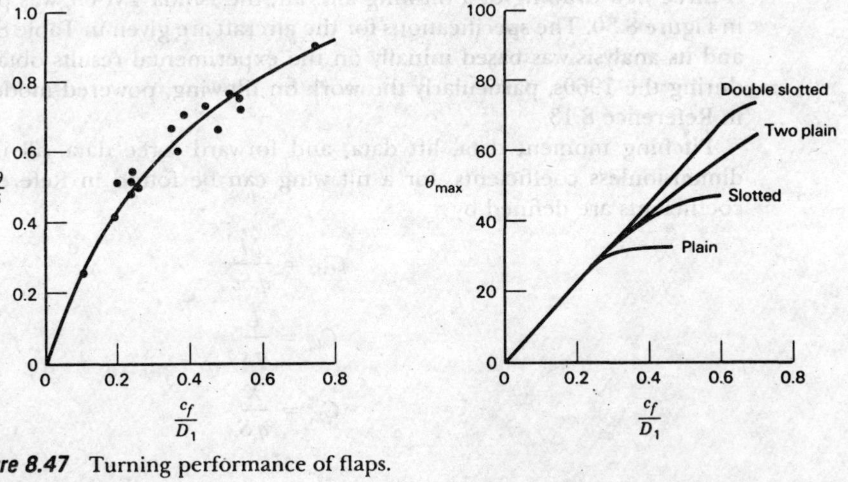

Figure 8.47 Turning performance of flaps.

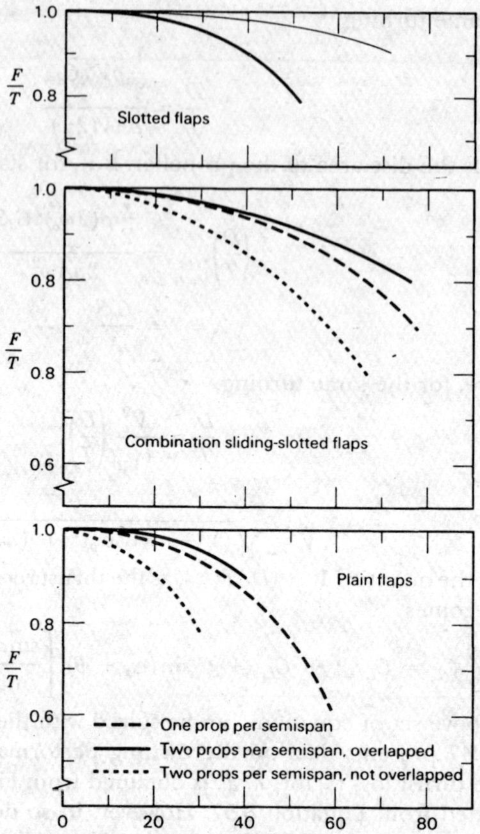

Figure 8.48 Thrust loss factor due to turning of slipstream.

A computer program can be written to accomplish the above procedure by curve-fitting Figures 8.47 and 8.48. Predictions from such a program are compared with experimental results in the next section.

The Ishida TW-68

A three-view drawing of a tilt-wing aircraft, the Ishida TW-68, was pictured earlier in Figure 8.30. The specifications for the aircraft are given in Table 8.3. This design and its analysis was based initially on the experimental results obtained by NASA during the 1960s, particularly the work on tilt-wing, powered models as reported in Reference 8.13.

Pitching moment data, lift data, and forward force data, all in the form of dimensionless coefficients, for a tilt wing can be found in Reference 8.13. The coefficients are defined by

$$C_{M_s} = \frac{M}{q_s Sc} \tag{8.89}$$

$$C_{L_s} = \frac{L}{q_s S} \tag{8.90}$$

$$C_{X_s} = \frac{X}{q_s S} \tag{8.91}$$

Table 8.3 Specifications for the Ishida TW-68 Tilt-Wing VTOL Turboprop

Power plant
Four Pratt & Whitney Canada PT6A-67

Propeller
Six blades
16.7 ft. diameter
860 RPM VTOL
731 RPM Cruise

Exterior Dimensions

Length	45.5 ft.
Height	17.0 ft.
Wing Span	38.0 ft.
Total Span (includes propellers)	41.4 ft.
Wing Area	269 ft²
Wing Chord	7.1 ft.

Interior Dimensions

Cabin Length	18.3 ft.
Cabin Width	6.1 ft.
Cabin Height	5.8 ft.
Cabin Volume	558 ft³
Baggage Volume	120 ft³

Seating Capacity

Corporate Configuration	1-2 Pilots	9 Passengers
Commuter Configuration	2 Pilots	14 Passengers

Weights	VTOL	STOL
Maximum takeoff, lb	14,000	16,500
Empty, lb*	8,960	8,960
Useful Load, lb**	5,040	7,540
Fuel Capacity, lb	3,685	3,685
Maximum Fuel Payload, lb***	1,355	3,855

Performance (Estimated)

Maximum Speed	350 kt
Maximum Cruise Speed	297 kt
Maximum Range Speed	187 kt
Maximum Range/14 Passenger (STOL)	1180 sm
Maximum Range/14 Passenger (VTOL)	630 sm

*Estimated equipped empty weight
**Takeoff weight less operating weight
***Takeoff weight less full fuel weight and operating weight

where q_s is the propeller slipstream dynamic pressure defined as the sum of the free-stream q and the disc loading, T/A. T is the propeller thrust and A is the disc area, $\pi D^2/4$. To obtain coefficients based on the free-stream q, one divides the "s" coefficients by the quantity, $1 - C_{T_s}$, where

$$C_{T_s} = \frac{T}{q_s A} \tag{8.92}$$

Figure 8.49 presents a comparison of the approximate theory developed above with some of the lift data from Reference 8.13. The theoretical method agrees closely for the unflapped wing at the high equivalent C_μ value, but the agreement is not as good for the other cases.

One specific data point from Reference 8.13 will be evaluated as an example. Expressing all dimensions in terms of the chord, the wing–propeller combination tested by the reference has the following geometry:

- Rectangular planform
- Chord = 1.0
- Single slotted, 40% chord flap
- Propeller diameter = 2.0
- Distance from propeller to quarter-chord = 0.7
- One propeller on each side
- Angle-of-attack of wing zero-lift line = 6.5° relative to propeller axis
- Wing span = 4.0

The actual span on which the measured coefficients are based includes the width of the fuselage, which increases the above span by approximately 1.0. The fuselage is neglected in estimating the lift of the wing–propeller combination, but the span of the fuselage is included in the reference area for determining the coefficients.

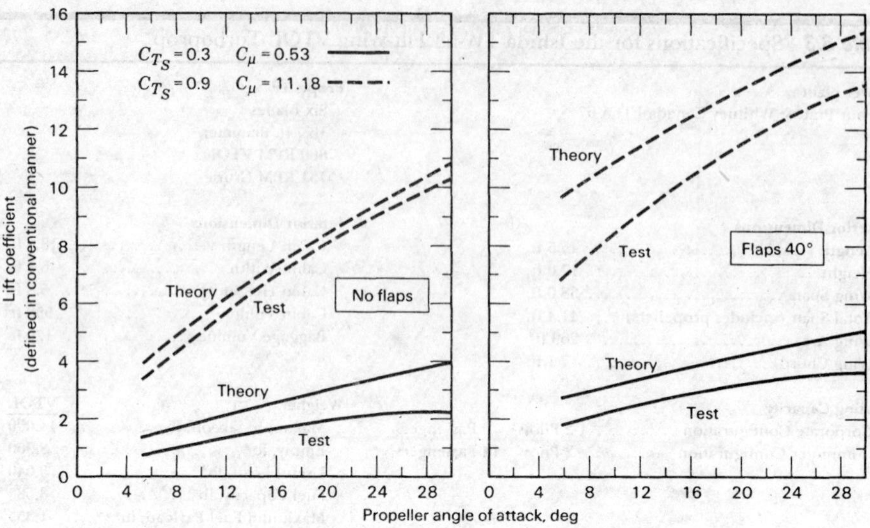

Figure 8.49 Aerodynamic data for a wing–propeller combination (Ref. 8.11).

As an example, let us take an angle-of-attack of the propeller of 25°, a flap angle of 40°, and a thrust coefficient, C_{T_s}, of 0.9 based on the sum of the free-stream dynamic pressure and the disk loading. In terms of the free-stream dynamic pressure only, the thrust coefficient, C_T, equals 0.9 divided by $1 - C_{T_s}$ or 9.0. The momentum coefficient, C_μ, is therefore equal to 11.2. All velocities can be expressed in terms of w_0, the static induced velocity. Thus,

$$V = 2\sqrt{\frac{1 - C_{T_s}}{C_{T_s}}}$$

or, $V = 0.667$. The rate of turning, θ/δ, because of the wing is estimated to equal 0.743 and for the flap, 0.366. Thus, the angle of the slipstream is found to equal 15.9° resulting in a total turning of 26.2°. The ratio of thrust loss to thrust is estimated to equal 0.932. The slope of the wing lift curve (less the momentum reaction) is found to equal 6.94/rad and the increase in C_L per radian of flap deflection is 10.12. Thus, $C_{l_{T=0}} = 0.94$, $C_{L_T} = 2.35$, and the C_L from the vertical component of the momentum equals 11.18 resulting in a total estimated C_L of 14.23.

One of the concerns of the tilt wing is that of wing buffet during transition or descent. From Figure 8.46*b*, this limit must obviously depend on the propeller induced velocity, the required wing C_L, and the angle-of-attack. These factors are reflected in Figure 8.50 taken from Reference 8.14.

The reference shows the importance for twin-engine, tilt-wing airplanes of the direction of propeller rotation. If the propellers are rotated up in the center, the allowable flight path angle increases appreciably because of the upwash induced at the center of the wing by the propellers. For example, for a flap deflection of 20° and a C_{T_s} of 0.9, the allowable descent angle increases from 22° to a positive climb angle of 10°.

The vertical placement of the propellers relative to the wing is also important. The propellers should be placed well below the wing so that, at a high angle-of-attack, the slipstream is carried back over the wing. Some estimate of the limitations of buffet due to stall can be obtained by examining the maximum C_L predicted by

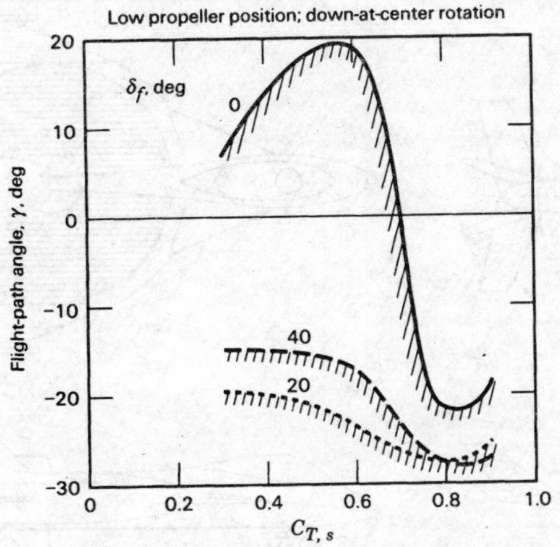

Figure 8.50 Operating boundaries for a tilt-wing aircraft.

the approximate theory for the wing–propeller combination as just presented. The approximate theory, of course, does not consider refinements such as direction of rotation or placement of the propeller.

High-Speed Rotorcraft

Aircraft such as the tilt-wing and tilt-rotor configurations are referred to as high-speed rotorcraft since they derive their VTOL capability from a rotor, but are capable of higher speeds than a conventional helicopter. There are perturbations on these configurations that have been evaluated over the last three decades and that continue to receive attention.

In the early 1990s, the NASA Ames Research Center supported studies of high-speed rotorcraft configurations by four major helicopter manufacturers: Bell, Boeing, McDonnell-Douglas, and Sikorsky. The goal of these studies was to delineate the critical technologies for the future advancement of high-speed rotorcraft. A summary of these results can be found in References 8.15 through 8.18. Cruise speeds in the vicinity of 450 kt and ranges of the order of 1000 n.mi were considered encompassing military transport, civil transport, ASW, and scout attack missions.

From these studies, several configurations were considered promising. In addition to the standard tilt wing and tilt rotor, the following configurations were considered:

1. *Folding tilt-rotor system (TFS):* Here, the rotor tilts to the rear and then folds to trail behind, thereby reducing the drag. The propulsive thrust is provided by a *convertible* engine, which can serve as either a turboshaft or turbofan engine.
2. *Variable diameter tilt rotor (VDTR):* Here the rotor blades telescope to produce a smaller diameter in the airplane mode. Otherwise, the configuration is similar to a tilt-rotor airplane.
3. *Rotor Wing:* Somewhat similar to the X-Wing, a three-bladed rotor with a large blade area (high solidity) is stopped in forward flight with two of the blades forming a forward-swept wing. The third blade lies on top of the fuselage, pointing downstream.
4. *Shrouded Rotor:* This configuration is similar to the Ryan XV-5, the fan-in-wing technology demonstrator, shown in Figure 8.25.

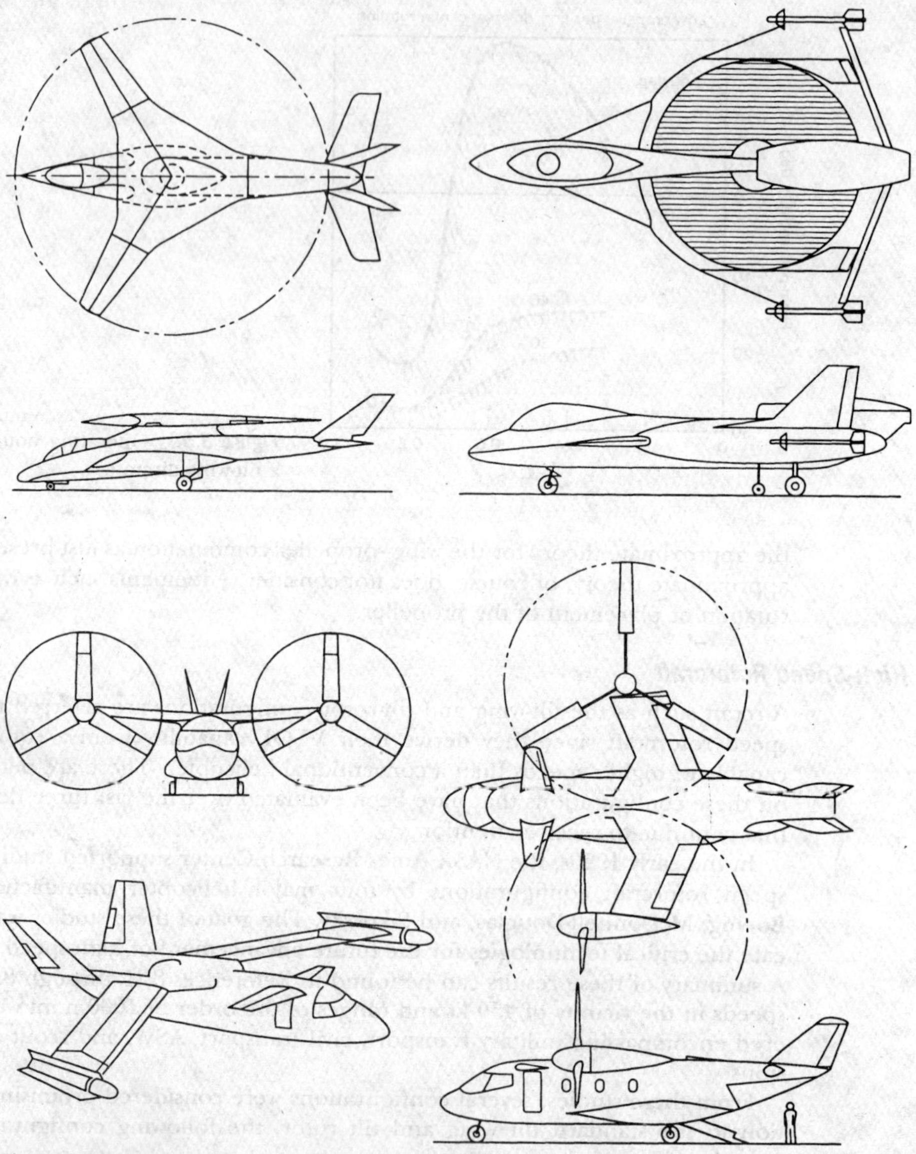

Figure 8.51 Sketches of high-speed VTOL rotorcraft.

The results are too voluminous to be given here, but sketches of the various configurations that were studied are shown in Figure 8.51. Some of the critical technologies discussed in the references include rotor aerodynamics, weight, aeroelasticity, materials and manufacturing, convertible engines, internal and external noise, drive train, and flight controls. That just about covers everything!

VTOL Performance from Turbojet Engines

The Harrier, pictured earlier in Figure 8.26, is an example of a VTOL airplane where the vertical force required to take off vertically comes entirely from the engine's exhaust. A sketch of the Rolls-Royce Pegasus, vectored-thrust turbofan is shown in Figure 8.52. The configuration is a bypass engine where the air through the fan is directed through the forward set of nozzles, while the turbine exhaust flows through the rearward set of nozzles. After leaving the engine, each air flow is turned twice. First, the air is turned 90° and directed outward. Then, the flow is turned another 90° by a nozzle that can swivel to vector the exhaust momentum over a range of directions from slightly forward of vertical to the rear. The latest version of this engine is rated at a maximum static lift thrust of 105.07 kN, or 23,620 lb. This compares with a quoted maximum VTO gross weight of 20,595 lb for the Harrier. This weight can be increased to 31,000 lb if the STOL capability of the airplane is used.

The vectored-thrust concept used by the Harrier was also employed by other designs including the Bell X-14 and the Yakovlev 36 Freehand. The alternative to vectored thrust is the use of separate engines to provide the lift and forward thrust. The Shorts SC1, in the UK in 1960, was the first VTOL aircraft to demonstrate the separate lift-engine concept. In 1971, the VFW-Fokker VAK191B was flown in Germany, which also used direct lift engines, but in combination with a Rolls-Royce engine similar to the Pegasus.

The performance of these types of airplanes is closely connected to the engine performance. The analysis of the lift system essentially reduces to an analysis of the engine. However, there are some nonengine aspects of the design and operation that are worthy of note. First, for the vectored-thrust configuration, as opposed to a direct-lift engine, which is always pointed vertically, the exhaust must be turned by a vectorable nozzle. This is accomplished by means of a cascade of airfoils that must be carefully designed to provide the desired turning without excessive losses. This may have been the reason for the poor performance of the Rockwell XFV-12, as mentioned earlier. The reader is referred to Reference 8.2 for an introduction to the subject of aerodynamics of cascades and to Reference 8.19 for more depth.

There are two phenomena relating to the performance of VTOL airplanes employing lifting jets that are important: *suckdown* and *fountain effect.* Both of these are very configuration dependent so that a general, quantitative analysis is not feasible. Instead, one must resort to testing to determine their magnitudes.

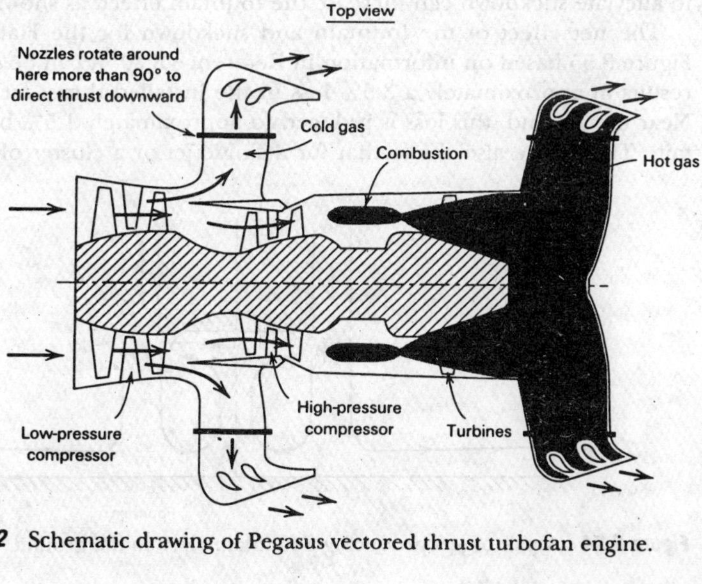

Figure 8.52 Schematic drawing of Pegasus vectored thrust turbofan engine.

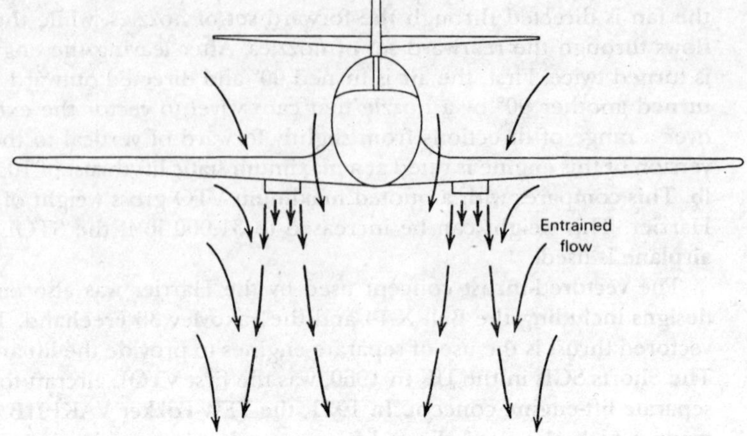

Figure 8.53 Vertical jets entraining a secondary flow, which causes suckdown.

Suckdown can be explained by reference to Figure 8.53. Somewhat similar to an augmenter, at altitude a vertical jet can entrain a secondary flow downward around the airplane. If this flow follows around the bottom of the fuselage, it can produce a low pressure that results in a downward force. The obvious way to avoid this adverse force is to avoid a curved contour on the bottom of the fuselage or to install strakes along the bottom of the sides to cause the flow to separate. In addition to the reduced pressure on the bottom of the fuselage, the downward flow can produce a download on other parts of the airplane. Collectively, the sum of these forces is called *suckdown.*

Fountain effect is normally a favorable effect that occurs near the ground. As shown in Figure 8.54, a jet impinging on the ground will spread radially outward. However, if another jet is present, the jets will interact in the region between them causing a "fountain" to be directed upward. This upward flow increases the pressure underneath the fuselage resulting in increased lift. The same strakes that help to alleviate suckdown can increase the fountain effect, as shown in Figure 8.55.

The net effect of the fountain and suckdown for the Harrier is found from Figure 8.55 based on information in Reference 8.20. It can be seen that suckdown results in approximately a 3.5% loss in the installed thrust far above the ground. Near the ground, this loss is reduced to approximately 1.5% because of the fountain. This figure also shows that for a single jet or a cluster of jets, the suckdown

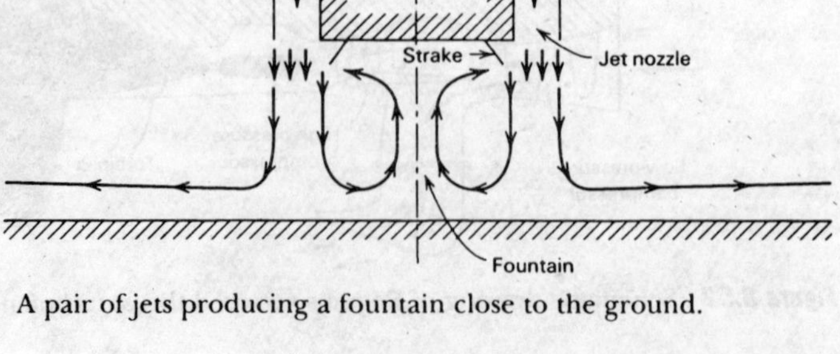

Figure 8.54 A pair of jets producing a fountain close to the ground.

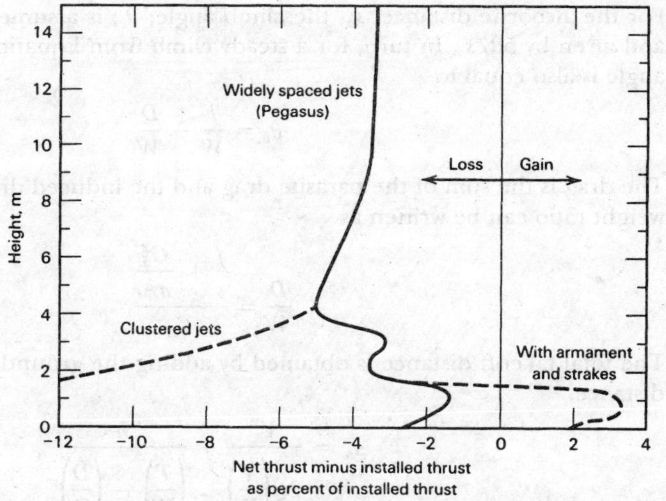

Figure 8.55 Effect of fountain and suckdown on the AV-8B Harrier.

increases near the ground as the radially spreading flow reduces the pressure under the airplane.

Another problem that applies to almost all types of aircraft hovering near the ground is that of *re-ingestion.* This refers to the same air being repeatedly drawn into the engine as a result of the circular flow patterns caused by ground interference. This can cause abrasive particles or foreign objects to be ingested into the rotor or compressor resulting in mechanical wear or failure. Also, if the warm exhaust is ingested, a significant power loss can occur. The designer of a VTOL airplane must be aware of these problems and take measures to avoid them.

STOL Performance

Methods for predicting takeoff and landing distances were presented in Chapter 7, and these are applicable to STOL aircraft. The difference, of course, between conventional and STOL aircraft is the ability of the latter to produce high-lift coefficients. However, a simple analysis will show that C_L is not the whole story. To take advantage of the high C_L on takeoff, the airplane must also have a high thrust-to-weight ratio. For a given thrust-to-weight ratio, there is an optimum C_L value that results in a minimum total takeoff distance over a specified obstacle.

Instead of the more exact methods presented in Chapter 7, for the purpose of disclosing the essential behavior of an STOL airplane, we will assume the following simplified, but reasonably exact, model. During the ground roll, the thrust is the only force assumed to be acting on the airplane and this force is taken to constant during the entire takeoff. As derived in Chapter 7, this assumption will result in fairly accurate results if the thrust is determined at an airspeed equal to the liftoff airspeed divided by $\sqrt{2}$. The transition between ground roll and climb is neglected. It is assumed that upon reaching a velocity equal to that which is 10% above the stall, the airplane instantaneously rotates and climbs at this constant velocity over the prescribed obstacle height of 50 ft. Thus, for the ground-roll distance, s_{gr},

$$\left(\frac{T}{W}\right)s_{gr} = \frac{V^2}{2g}$$

For the airborne distance, s_a, the climb angle, θ_c, is assumed a priori to be small and given by $50/s_a$. In turn, for a steady climb from Equations 7.12 and 7.13, this angle is also equal to

$$\theta_c = \frac{T}{W} - \frac{D}{W} \tag{8.93}$$

The drag is the sum of the parasite drag and the induced drag so that the drag-to-weight ratio can be written as

$$\frac{D}{W} = \frac{\frac{f}{s} + \frac{C_L^2}{\pi A e}}{C_L} \tag{8.94}$$

The total takeoff distance is obtained by adding the ground roll and the airborne distance.

$$s_{to} = \frac{V^2}{2g\left(\frac{T}{W}\right)} + \frac{h}{\left(\frac{T}{W}\right) - \left(\frac{D}{W}\right)} \tag{8.95}$$

If the airplane is operating at 20% above the stalling speed, then

$$C_L = \frac{C_{L_{max}}}{1.44}$$

As an example, assume that $A = 8.0$, $e = 0.7$, and $f/S = 0.03$.

Substituting these numbers into Equations 8.89–8.91 produces the results shown in Figure 8.56. This figure illustrates the fact, because of the induced drag during the climb phase, increasing the $C_{L_{max}}$ value beyond a certain value can result in an increase in the total takeoff distance. It also shows that for moderately high thrust-to-weight ratios, very little is to be gained in the takeoff distance by increasing

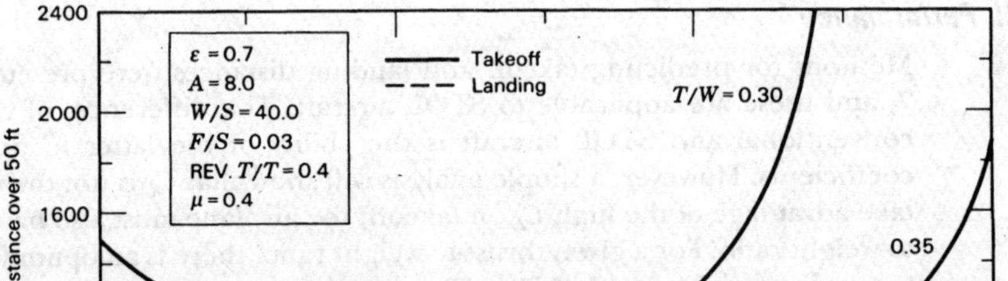

Figure 8.56 Effect of maximum lift coefficient and thrust-to-weight ratio on the total takeoff and landing distance of an STOL airplane.

$C_{L_{max}}$. Thus, for good STOL performance, the installed thrust appears to be as important as the high lift system.

Calculated landing distances are also included in Figure 8.56 using the same parameters as for the takeoff distance calculations. In addition, it was assumed that the ratio of reverse thrust to forward thrust was 0.4 and that the coefficient of braking friction, μ, was 0.4. In this case the descent angle is given by D/W with the retarding force equal to $\mu W + (T_{rev}/T)\, T$. Thus, the expression for the total landing distance becomes

$$S_{ldg} = \frac{V^2}{2g\left(\frac{T}{W}\right)} + \frac{h}{\left(\frac{T}{W}\right) - \left(\frac{D}{W}\right)} \tag{8.96}$$

The landing distances are seen to be fairly insensitive to the thrust-to-weight ratio and depend primarily upon $C_{L_{max}}$ for the fixed-wing loading. In designing an STOL airplane, one would like to have approximately the same takeoff and landing distances. Thus, in this example, for a thrust-to-weight ratio of 0.4, a $C_{L_{max}}$ of approximately 3.5 would be appropriate.

Actually, to achieve a short landing, the important parameter is the approach speed, which depends on both $C_{L_{max}}$ and the wing loading. This was pointed out many years ago by Reference 8.21, which considered two possible limiting STOL landing scenarios called "routine" and "maximum performance." The routine landing assumed a rate-of-descent of 500 fpm, a circular arc transition with a normal acceleration of 0.1 g, and a stopping deceleration of 0.3 g. The maximum performance assumed a rate-of-descent of 1000 fpm with no transition and a deceleration of 0.8 g. As I state in Reference 8.2, the landing performance in actual practice will probably lie somewhere between these two extremes. This is shown in Figure 8.57, where Kuhn's limits are presented together with landing distances for eight STOL airplanes, as quoted by Reference 1.6.

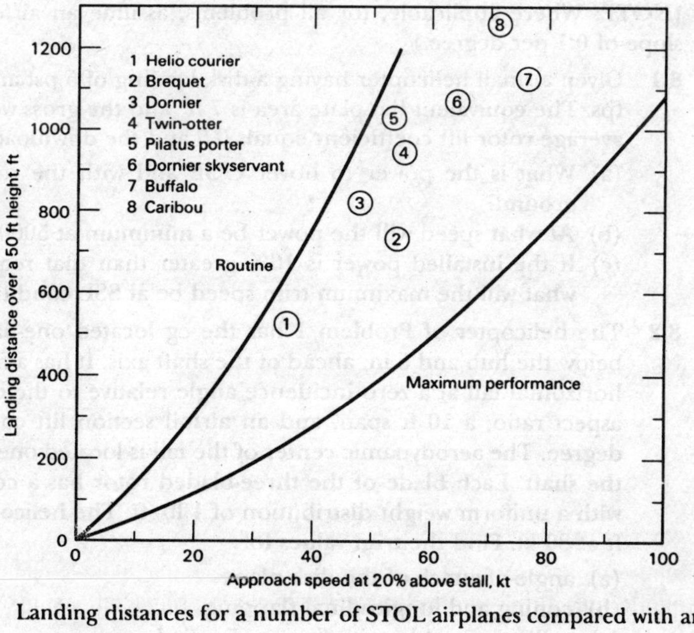

Figure 8.57 Landing distances for a number of STOL airplanes compared with analysis.

SUMMARY

This chapter has presented some of the technology concerned with the design and analysis of aircraft capable of V/STOL flight. Generally, the type of aircraft will be dictated by the aircraft's mission. If it is required to hover efficiently, then the disc loading should be low. If cruising performance is not too important, then the helicopter is probably the choice. However, if efficient hovering together with a fairly high cruising speed and range is needed, then a high speed rotorcraft will be required. Finally, one may need the VTOL capability only to take off and land with the primary mission of the airplane being to carry a payload rapidly for a fairly long distance. In this case, a turbofan-powered configuration with an extremely high disc loading will be the choice.

In the case of the ASTOVL airplane, the thrust required for VTOL capability is compatible with that required for supersonic flight. Thus, there is currently considerable interest in providing the additional controls and thrust vectoring for supersonic aircraft to give them VTOL or STOL capability. Recently, the USAF/McDonnell-Douglas F-15 STOL/Maneuver Technology Demonstrator (S/MTD) showed that "today's fighters can take off and land on an undamaged 1500-ft section of a bomb-cratered runway if the aircraft are modified with integrated flight and propulsion control systems" (*Aviation Week & Space Technology,* September 16, 1991). The S/MTD F-15b testbed is fitted with a set of vanes placed behind the engine, which swivel up or down to vector the thrust vertically. In addition, a set of canard control surfaces are installed on the airplane and integrated with the nozzles.

For STOL performance one must strive for a low stalling speed compatible with the thrust-to-weight ratio. The thrust-to-weight ratio and stalling speed are primary factors in determining the takeoff distance. It is only the stalling speed that is the primary factor in determining landing distance.

PROBLEMS

(NOTE: Where applicable, for all problems, assume an airfoil section lift curve slope of 0.1 per degree.)

8.1 Given a small helicopter having a disk loading of 5 psf and a tip speed of 700 fps. The equivalent flat-plate area is 7 ft^2 and the gross weight is 1100 lb. The average rotor lift coefficient equals 0.5 and the download is 5%.

(a) What is the power to hover OGE and with the rotor 10 ft above the ground?

(b) At what speed will the power be a minimum at 5000-ft altitude?

(c) If the installed power is 10% greater than that required to hover SSL, what will the maximum trim speed be at SSL conditions?

8.2 The helicopter of Problem 1 has the cg located one-half of a rotor radius below the hub and 6 in. ahead of the shaft axis. It has a fixed, constant chord horizontal tail at a zero incidence angle relative to the disk plane with a 5.0 aspect ratio, a 10 ft span, and an airfoil section lift curve slope of 0.1 per degree. The aerodynamic center of the tail is located one rotor radius behind the shaft. Each blade of the three-bladed rotor has a constant cross section with a uniform weight distribution of 1 lb/ft. The helicopter is flying at 6000 ft at 80 kt. Find the trim values for:

(a) angle-of-attack of the disk plane

(b) coning and longitudinal flapping

(c) collective and longitudinal cyclic pitch

8.3 Derive Equations 8.10 and 8.13.

8.4 Derive an expression for the undamped, natural flapping frequency of a blade with a hinge offset of $\epsilon = e/R$ and a uniform weight distribution rotating at a speed of ω rad/sec.

8.5 A horizontal jet having a cross-sectional area of 0.5 ft^2 expands isentropically from a reservoir pressure of 26 psia and a temperature of 80°F into SSL conditions. A small wing is completely immersed in the jet and is producing a lift of 20 lb causing the jet to deflect downward from the horizontal at a constant angle behind the wing. Find this angle, which you can assume is small.

8.6 A wing with a full-span, pure-jet flap is at an angle-of-attack of 5° and the flap is deflected 30°. The jet expands isentropically from the same reservoir conditions as Problem 8.5. The planform is rectangular with an aspect ratio of 7. The ratio of the jet thickness to the chord is 0.04. The wing is moving at a speed of 150 mph, SSl conditions. Calculate the lift coefficient. What is the limiting C_L that can be obtained from the jet reaction and circulation?

8.7 The jet from the wing of Problem 8.6 is surrounded with a secondary nozzle having a cross-sectional area that is 10 times greater than the primary jet area. Assuming the augmented flow is completely mixed with no losses and then blown over the flap, calculate the lift coefficient.

8.8 Using the numbers given for the V-22,

(a) Estimate its hover ceiling OGE assuming that the power available decreases with altitude in the same proportion as the power for the PW-120 turboshaft engine (see Chapter 6).

(b) Find the trim speed in m/s at an altitude of 3000 m for proprotor angles-of-attack of zero and 85°.

8.9 This is an open-ended problem. Do you think the Ishida tilt-wing airplane will stall during transition from the hover mode to the airplane mode? Assume a slow, steady transition so that inertia forces can be neglected.

8.10 Check the example given for the tilt wing at an angle-of-attack of 25° and a flap angle of 40° and a C_{T_s} of 0.9. Repeat the exercise for C_{T_s} of 0.3.

REFERENCES

1. Hennis, R. P., and McCormick, B. W., *A Computer Model for Determining Weapon Release Parameters for a Helicopter in Non-Accelerated Flight,* Naval Weapons Surface Center, NSWC/TR-3823, October 1978.
2. McCormick, B. W., *Aerodynamics of V/STOL Flight,* Academic Press, New York, 1967.
3. Wheatly, J. B., *An Aerodynamic Analysis of the Autogyro Rotor with a Comparison Between Calculated and Experimental Results,* NACA TR 487, 1934.
4. Hennis, R. P., and McCormick, B. W., *Computer Model for Predicting Dynamic Behavior of a Helicopter for Application to Weapons Delivery and Subsequent Safe Escape,* Naval Weapons Surface Center, NSWC TR 85-285, September 1986.
5. Prouty, Raymond W., *Helicopter Performance, Stability and Control,* PWS Engineering, Boston, 1986.
6. Gessow, Alfred, and Myers, G. C., *Aerodynamics of the Helicopter,* 1952, republished by Frederick Ungar Publishing Co., New York, 1967.
7. Johnson, Wayne, *Helicopter Theory,* Princeton University Press, 1980.
8. Spence, D. A., *The Lift of a Thin, Jet-Flapped Wing,* Proc. Roy. Soc., A238, 46–68, 1956.
9. Kohlman, David L., *Introduction to V/STOL Airplanes,* Iowa State University Press, 1981.
10. Rosenstein, Harold, *Aerodynamic Development of the V-22 Tilt Rotor,* Twelfth European Rotorcraft Forum, September 22–25, 1986, Garmisch-Partenkirchen, Germany.

11. Dunford, P. J., Lunn, Ken, Magnuson, R. A., and Marr, R. L., *V-22 Tiltrotor Flight Test Development,* 48th Annual Forum, American Helicopter Society, Washington, DC, June 3–5, 1992.

12. Smelt, R., and Davies, H., *Estimation of Increase in Lift Due to Slipstream,* ARC R & M 1788, 1937.

13. Fink, M. P., *Aerodynamic Data on Large Semispan Tilting Wing With 0.5-Diameter Chord, Single-Slotted Flap, and Single Propeller 0.08 Chord Below Wing,* NASA TN D-4030, July 1967.

14. Hassell, K., Kirby, R. H., *Descent Capability of Two-Propeller Tilt-Wing Configurations,* Conference on V/STOL and STOL Aircraft NASA Sp-116, April 1966.

15. Rutherford, J., O'Rourke, M., Lovenguth, M., Mitchell, C., *Conceptual Assessment of Two High-Speed Rotorcraft,* NASA/Industry High Speed Rotorcraft Meeting, May 8, 1991, Phoenix, AZ.

16. Conway, Scott M., *Conclusions from High-Speed Rotorcraft Studies,* NASA/Industry High Speed Rotorcraft Meeting, May 8, 1991, Phoenix, AZ.

17. Wilkerson, Joseph, and Schneider, John, *Technology Needs for High-Speed Rotorcraft-Conceptual Designs and Performance,* NASA/Industry High Speed Rotorcraft Meeting, May 8, 1991, Phoenix, AZ.

18. Scott, Mark, *Summary of Technology Needs for High-Speed Rotorcraft Study,* NASA/Industry High Speed Rotorcraft Meeting, May 8, 1991, Phoenix, AZ.

19. Gostellow, J. P., *Cascade Aerodynamics,* Pergamon Press, New York, 1984.

20. Fozard, J. W., *Harri-ing Sir Isacc Vertically: Secrets of Engineering Elegance in Jumping Jets,* Proceeding of the Royal Institution, Volume 58.

21. Kuhn, R. E., "Take-Off and Landing Distance and Power Requirements of Propeller-Driven STOL Airplanes," *IAS Preprint 690,* presented at Twenty-fifth Annual Meeting, New York, January 28–31, 1957.

9 STATIC STABILITY AND CONTROL

INTRODUCTION

FAR Part 23 (REf. 9.1) states that, "the airplane must be safely controllable and maneuverable during—(1) take off; (2) climb; (3) level flight; (4) dive; and (5) landing (power on and off) (with the wing flaps extended and retracted)." Part 23 also says that, "The airplane must be longitudinally, directionally, and laterally stable." This particular FAR, which defines the airworthiness standards for normal, utility, and acrobatic airplanes, then goes into detail to state what requirements an airplane must meet in order to satisfy the foregoing general statements on controllability and stability.

Stability of an airplane refers to its movement in returning, or the tendency to return, to a given state of equilibrium, frequently referred to as trim. More specifically, static stability refers to the tendency of an airplane under steady conditions to return to a trimmed condition when disturbed rather than any actual motion it may undergo following the disturbance. The forces and moments are examined to determine if they are in the direction to force the airplane back to its equilibrium flight conditions. If so, the airplane is statically stable.

Dynamic stability encompasses the unsteady behavior of an airplane responding to time-dependent aerodynamic forces and moments produced by the airplane's motion. Following a disturbance of the airplane from a trimmed condition, its movement under the influence of the unsteady forces and moments is examined to determine whether or not the airplane ultimately returns to its trimmed condition. To the uninitiated, it comes as a surprise to learn that many airplanes exhibit a dynamic instability known as spiral divergence. Unless counteracted by control input, many airplanes will gradually drop one wing and go into an ever tightening spiral dive. Fortunately, this spiral divergence initially is so gradual that the pilot will compensate for it without even realizing it. However, if a noninstrument-rated pilot is suddenly without a reference to the horizon, the spiral divergence instability can prove disastrous.

An airplane that is statically stable will not necessarily be dynamically stable. However, one that is statically unstable will be dynamically unstable; that is, static stability is necessary but not sufficient for dynamic stability. In this chapter, we will consider static stability and control; in other words, we will consider the forces and moments acting on an airplane undergoing steady motion.

COORDINATE SYSTEM: FORCES, MOMENTS, AND VELOCITIES

The right-handed set of body axes that will be used in the material to follow has been discussed earlier in Chapter 1 and is illustrated in Figure 1.10. These axes are fixed at the center of gravity relative to the body and, hence, move with the body. The system is easily remembered. x is directed forward, y to the right, and z downward. X, Y, and Z are the aerodynamic force components along their respective axes; u, v, and w the velocity components of the center of gravity along these axes; L, M, and N are the moments about these axes; and P,Q, and R are the angular velocities about the x-, y-, and z-axes, respectively. L, M, and N are defined as the rolling, pitching, and yawing moments, respectively. Similarly, P, Q, and R are called the roll, pitch, and yaw rates.

We will see later that for many problems concerning stability and control of an airplane, its motion in the plane of symmetry (x-z plane) can be uncoupled from the motion of the plane of symmetry. The former is referred to as longitudinal motion and treats linear motion along the x- and z-axes, and rotation about the y-axis. Motion of the plane of symmetry, known as lateral or lateral-directional motion, deals with linear motion along the y-axis and angular rotations about the x- and z-axes.

LONGITUDINAL STATIC STABILITY

Because this material is an introduction to the subject of airplane stability and control, a certain amount of elegance will be foresaken for the sake of simplicity and clarity. Most of the derivations to follow will relate to a "basic airplane" consisting simply of a wing and a tail. Analysis of this simple configuration will disclose the basic principles involved in determining the motion of the complete airplane. Following these fundamental developments, the effects of the fuselage and propulsion system will then be considered.

An airplane is in trim if all of the forces, aerodynamic and gravitational, and aerodynamic moments acting on the airplane about the cg are in balance. These forces and moments will be examined in detail later. For now, consider the graph shown in Figure 9.1. Here, the pitching moment about the cg of an airplane is graphed for two possible cases; first, where the aerodynamic moment increases as the angle of attack is increased and, second, where the moment decreases with increasing α.

The airplane is in trim at point A, where the total pitching moment about the cg, M, is zero. At this point α, the angle of attack of the wing's zero lift line must be positive in order to produce a lift. Now consider what happens if the angle of attack is increased from the value at point A to that at point $\acute{A}$. M will vary approximately linearly with α and can either increase or decrease with increasing α as depicted by the two lines shown in the figure. Suppose that M increases to point B as α is increased. At this point, M is now positive, which means a nose-up moment. Thus for this case the change in the moment tends to increase α even more as indicated by the arrow at B. Obviously, this is an unstable situation. Conversely, if

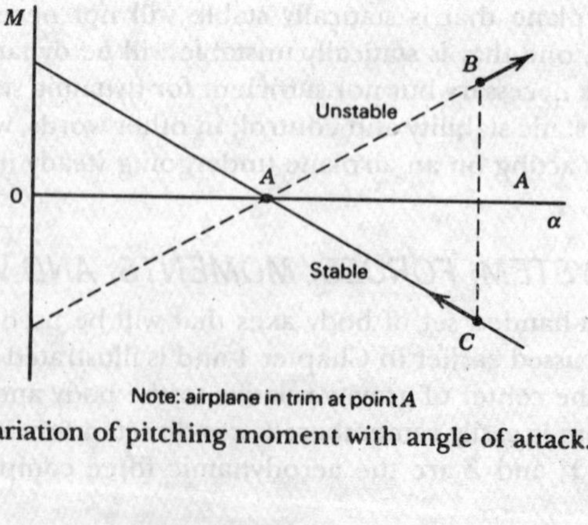

Figure 9.1 Possible variation of pitching moment with angle of attack.

the moment decreases to point C as α is increased, the result is a nose-down pitching moment tending to return α to its trimmed value. Therefore, for an airplane to be *statically stable in pitch, and in trim,* the following two conditions must hold:

$$\frac{dM}{d\alpha} < 0$$

$$M > 0 \quad \text{for} \quad \alpha = 0$$

In dimensionless form, these become

$$C_{M\alpha} < 0 \tag{9.1}$$

$$C_{M\alpha=0} > 0 \tag{9.2}$$

Before becoming too involved in the formulation of the equations governing the above conditions, consider qualitatively the implications of these two conditions. First, consider an airplane with a wing and tail as illustrated in Figure 9.2 with the wing at a zero angle of attack. In this two-dimensional portrayal, the wing and tail are represented simply by airfoils having lift and moment characteristics analogous to the three-dimensional surfaces. The representative airfoil section has a chord equal to the *mean aerodynamic chord* (MAC) discussed in Chapter 3.

Unless otherwise stated, the angle of attack in this chapter will always be referenced to the zero lift line of the wing or tail. For the situation shown in Figure 9.2 the moment about the cg will equal

$$M = M_{ac} - l_t L_t \tag{9.3}$$

The moment about the wing's aerodynamic center is normally small by comparison to the moment provided by the tail. In any case, for a positively cambered wing, M_{ac} is negative. Thus, in order to satisfy Equation 9.2 it follows from Equation 9.3 that the lift on the tail must be negative and the zero lift line for the tail must be nose down relative to the wing's zero lift line. The angle, i_t, is known as the tail incidence angle and is defined positive as shown, contrary to the right-hand rule applied to other angles.

At a nonzero angle of attack, the moment about the cg will be given by Equation 9.3 plus the increment caused by the added wing lift.

$$M = M_{ac} + l_w L_w - l_t L_t \tag{9.4}$$

Therefore, Equation 9.1, specifying that the moment have a negative rate of change with the angle of attack, will be satisfied, if

$$l_w \frac{dL_w}{d\alpha} < l_t \frac{dL_t}{d\alpha} \tag{9.5}$$

This can be satisfied by placing the cg sufficiently close to or ahead of the wing's aerodynamic center, or by making either the tail planform area or the tail length sufficiently large.

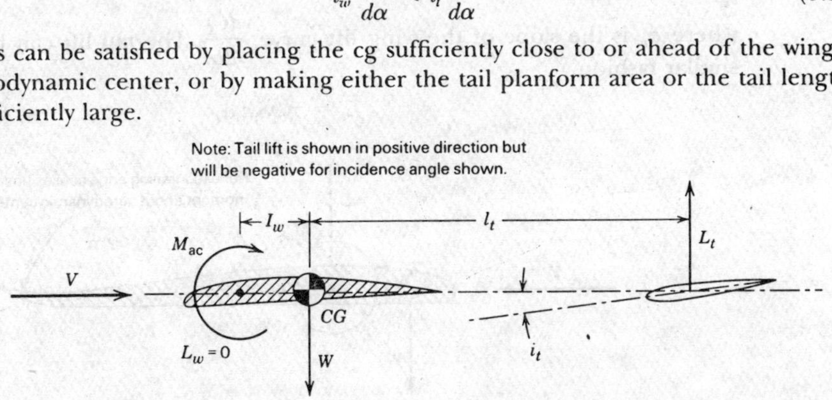

Figure 9.2 Forces on a wing–tail combination at zero angle of attack.

Figure 9.3 The B-2, a Stealth Bomber.

Consider how a tailless airplane, that is, a flying wing like the B-2 shown in Figure 9.3, can satisfy Equations 9.1 and 9.2. Figure 9.4 illustrates this configuration. Here, an airfoil is shown with a reflexed trailing edge. With the use of this geometry, or also negative camber, M_{ac} can be made positive. Thus, the moment will be positive for zero lift, and, with the cg ahead of the M_{ac}, the moment about the cg will become negative as the angle of attack increases.

Total Airplane Lift

In order to assure that an airplane satisfies the above criteria for static stability and trim, one must examine in detail the aerodynamic forces and moments on the airplane. To do this, we will again consider the wing and tail, which were discussed previously in Figure 9.2 except that now, as shown in Figure 9.5, we will place the wing at an angle of attack and include the interaction between the two lifting surfaces.

As stated previously, the wing is represented by its *mean aerodynamic chord* (MAC), $\bar{c}$, with the lift and moment acting at the aerodynamic center. The tail is represented in a similar manner.

The wing lift can be obtained from the slope of the lift curve, the angle of attack, and the free-stream dynamic pressure.

$$L_w = qSa_w\alpha \tag{9.6}$$

where a_w is the slope of the wing lift curve, $\dfrac{dC_L}{d\alpha}$. The tail lift can be written in a similar fashion.

$$L_t = \eta_t q S_t a_t \alpha_t \tag{9.7}$$

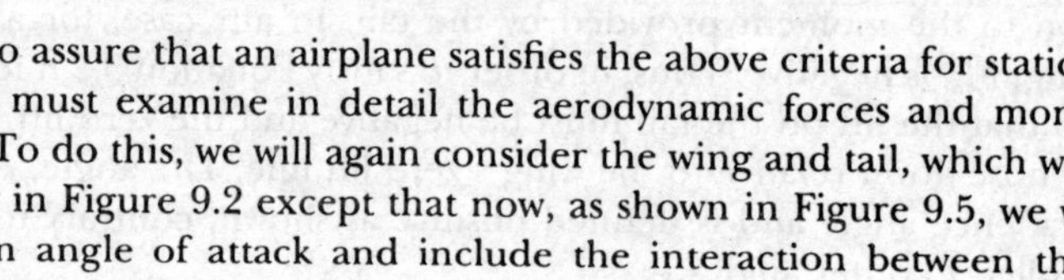

Figure 9.4 Forces and moments on a flying wing (tailless airplane).

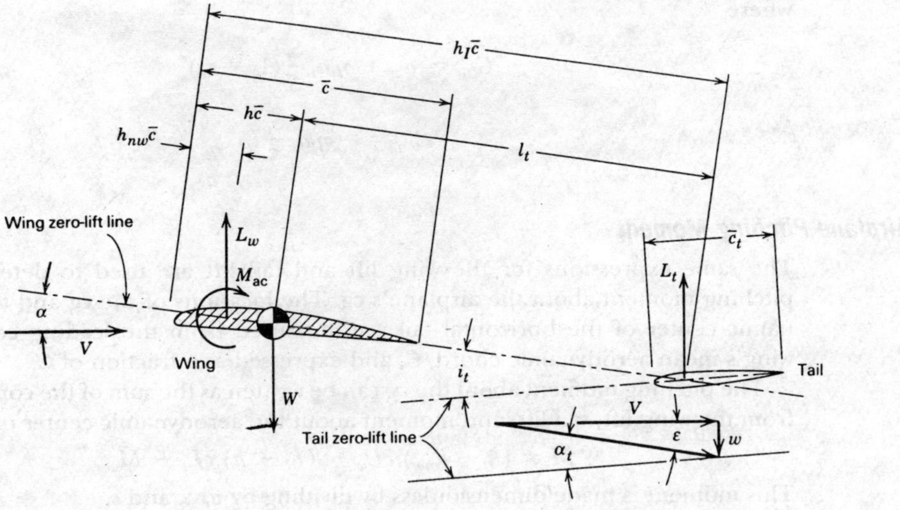

Figure 9.5 A wing-horizontal tail in combination showing forces, velocities, angles, and dimensions.

In this text, it will normally be assumed that the dynamic pressure, q, at the tail is the same as the free-stream value. (η_t set equal to 1). On the average, over the span of the tail, q may be greater or less than the free-stream value because of the boundary layer over the fuselage or the propeller slipstream. However, the distribution of q is difficult to determine and usually the dynamic pressure over the tail is close to the free-stream value anyway.

An important effect that cannot be neglected at the tail is the downwash, w, which is induced there by the wing. This parameter also varies along the tail span; however, in order to gain an understanding of the basic principles, a value will be used, which varies with the wing angle of attack, but which represents an average value affecting the tail. The effect of w is to cause a deflection of the flow downward at the tail thus reducing its angle of attack. This induced angle, denoted by ϵ, varies linearly with α_w and can be written as

$$\epsilon = \frac{w}{V}$$

$$= \frac{d\epsilon}{d\alpha} d\alpha \tag{9.8}$$

$$= \epsilon_\alpha \alpha$$

The angle of attack of the tail will therefore equal the angle of attack of the wing minus the incidence angle of the tail minus the downwash angle.

$$\alpha_t = \alpha_w - i_t - \epsilon_\alpha \alpha \tag{9.9}$$

The total lift on the airplane will be the sum of the wing and tail lifts, or

$$L = qS\left[\dot{a}_w + a_t \frac{S_t}{S}(1 - \epsilon_\alpha)\right]\alpha - \eta_t q a_t S_t i_t \tag{9.10}$$

Defining the airplane lift coefficient in terms of q and S, Equation 9.10 becomes

$$C_L = C_{L\alpha}\alpha + C_{L0} \tag{9.11}$$

where

$$C_{L\alpha} = a_w + \eta_t a_t \frac{S_t}{S}(1 - \epsilon_\alpha) \tag{9.12}$$

$$C_{L0} = -\eta_t a_t \frac{S_t}{S} i_t \tag{9.13}$$

Airplane Pitching Moment

The same expressions for the wing lift and tail lift are used to determine the pitching moment about the airplane's cg. The locations of the cg and the aerodynamic center of the horizontal tail are measured from the leading edge of the wing's mean aerodynamic chord, $\overline{C}$, and expressed as a fraction of $\overline{C}$.

The pitching moment about the cg can be written as the sum of the contributions from the wing lift, tail lift, and moment about the aerodynamic center of the wing.

$$M = (h - h_{nw})\,\bar{c}L_w - (h_l - h)\,\bar{c}L_t + M_{ac} \tag{9.14}$$

This moment is made dimensionless by dividing by q, S, and $\bar{c}$.

$$C_M = (h - h_{nw})\,C_{L_w} - \eta_t\,(h_l - h)\,\frac{S_t}{S}\,C_{L_t} + C_{M_{ac}} \tag{9.15}$$

This equation is often found in the literature in terms of the tail length, l_t, as

$$C_M = (h - h_{nw})\,C_{L_w} - \eta_t \frac{S_t}{S}\frac{l_t}{\bar{c}}\,C_{L_t} + C_{M_{ac}} \tag{9.16}$$

The quantity multiplying the tail lift coefficient is called the *horizontal tail volume*, V_H.

$$V_H = \frac{S_t}{S}\frac{l_t}{\bar{c}} \tag{9.17}$$

Thus, Equation 9.16 becomes

$$C_M = (h - h_{nw})\,C_{L_w} - \eta_t\,V_H C_{Lt} + C_{M_{ac}} \tag{9.18}$$

The problem with using Equation 9.18 in place of Equation 9.15 is that one tends to forget that V_H is not a constant and must vary as the cg location is changed.

Let us express Equation 9.15 in terms of the angle of attack by the use of Equation 9.9.

$$C_M = (h - h_{nw})\,a_w\alpha - (h_l - h)\,\eta_t \frac{S_t}{S}\,a_t\,[\alpha\,(1 - \epsilon_\alpha) - i_t] + C_{M_{ac}} \tag{9.19}$$

Similar to a cambered airfoil, we see that the moment coefficient about the cg of an airplane can be expressed in the following form:

$$C_m = C_{M_\alpha}\,\alpha + C_{m0} \tag{9.20}$$

where

$$C_{M\alpha} = a_w\,(h - h_{nw}) - (h_l - h)\,\eta_t \frac{S_t}{S}\,a_t\,(1 - \epsilon_\alpha) \tag{9.21}$$

$$C_{M0} = (h_l - h)\,\eta_t \frac{S_t}{S}\,a_t i_t + C_{M_{ac}} \tag{9.22}$$

Neutral Point

Similar to the aerodynamic center for an airfoil, a cg location exists for an airplane about which the moment is constant, independent of the angle of attack. To determine this point, designated in dimensionless form as h_n, Equation 9.21 is equated to zero.

$$C_{M\alpha} = 0 = (h_n - h_{nw})\, a_w - (h_t - h_n)\, \eta_t \frac{S_t}{S} a_t (1 - \epsilon_\alpha) \tag{9.23}$$

Note that h has now been replaced by h_n. This can be solved for h_n to give

$$h_n = \frac{h_{nw} + h_t \eta_t \dfrac{S_t}{S} \dfrac{a_t}{a_w} (1 - \epsilon_\alpha)}{1 + \eta_t \dfrac{S_t}{S} \dfrac{a_t}{a_w} (1 - \epsilon_\alpha)} \tag{9.24a}$$

Additional moments, such as those due to propellers or the fuselage, can be included easily in Equation 9.24a if it is assumed that they do not depend upon the cg location. For example, if $C_{M_{\alpha b}}$ is the slope of the moment coefficient for a body based on S and $\bar{c}$, then the above equation becomes

$$h_n = \frac{h_{nw} + h_t \eta_t \dfrac{S_t}{S} \dfrac{a_t}{a_w} (1 - \epsilon_\alpha) + \dfrac{C_{M\alpha b}}{a_w}}{1 + \eta_t \dfrac{S_t}{S} \dfrac{a_t}{a_w} (1 - \epsilon_\alpha)} \tag{9.24b}$$

The denominator in Equation 9.24a is seen to equal $C_{L\alpha} / a_w$ from Equation 9.12. If the above is substituted into Equation 9.21, the following is obtained:

$$C_{M\alpha} = (h - h_n)\, C_{L\alpha} \tag{9.25}$$

This deceptively simple result is an important one. It states that, in dimensionless form, the slope of the moment curve, $C_{M\alpha}$, is equal to the slope of the lift curve, $C_{L\alpha}$, multiplied by the dimensionless distance of the cg behind the neutral point, $h - h_n$. But for static stability $C_{M\alpha}$ must be negative. Therefore, it follows *for an airplane to be statically stable, the cg must be ahead of the neutral point.*

The quantity $(h_n - h)$ is known as the *static margin.* It represents the distance that the center of gravity is ahead of the neutral point expressed as a fraction of the mean aerodynamic chord. A static margin of at least 5% is recommended to assure adequate static longitudinal stability.

There are configurations other than the conventional wing–tail combination that are trimmable and longitudinally statically stable. Since the preceding relationships regarding longitudinal static stability place no restrictions on the relative sizes of the wing and tail, it is easily seen that the smaller surface can be placed ahead of the larger. This is called a Canard configuration and all of the above relationships will hold if the forward lifting surface is treated as the wing. In this case one will find that the angle of attack of the Canard surface must be greater than that of the wing to trim. Also, the cg must be placed toward the rear, just ahead of the wing. The original Wright Flyer (Figure 1.1) was a Canard configuration. However, the placement of the cg was not forward enough to provide inherent stability and the airplane was difficult to control.

Downwash Angle

The estimation of the downwash angle resulting from the wing's vortex system is a difficult task. First, the system, in the real world, is difficult to define, particularly in the presence of the fuselage and engines. The trailing vortex sheet, as it leaves the wing, rolls up into a pair of trailing vortices. This roll-up is influenced significantly by viscosity and other nonlinear effects, which are difficult to model precisely. Secondly, even if the wing's vortex system is known, one is then faced with the problem of defining an average downwash at a point on the tail that will represent the effect on the tail. Thus, to use a simple correction to the tail angle of attack leaves something to be desired. With these reservations in mind, a model will now be presented, amenable to a relatively easy numerical solution, which gives results in fair agreement with downwash angle presented by others.

Referring to Figure 9.6, a wing is replaced by a single bound vortex with a vortex trailing from each tip of the bound vortex. The span of the bound vortex is less than the wing span because of the fact that, as a vortex sheet shed from a wing rolls up, the edge of the sheet moves inboard. This process approaches asymptotically a span of b' as the sheet completely rolls up into a pair of vortices, each having a circulation equal to the midspan value of the bound circulation of the wing, Γ_0. Figure 9.6 shows the system closed far behind the wing by a lifting line having a constant circulation of Γ_0 opposite in direction to the bound circulation on the wing. Viewed as a closed system, the net lift on the system must equal zero. Thus, it follows that the total vertical force on the downstream lifting line must equal in magnitude the lift on the wing.

$$\rho V \Gamma_0 b' = \frac{1}{2} \rho V^2 S C_L$$

For an elliptic wing, Γ_0 is given by (see Chapter 3)

$$\Gamma_0 = \frac{1}{2} C_0 C_L V$$

The wing area, S, for an elliptic wing equals $(\pi/4)\, bC_0$. Combining the above relationships results in

$$\frac{b'}{b} = \frac{\pi}{4} \tag{9.26}$$

The bound vortex system is placed along the quarter-chord line of the wing. The representative point on the tail at which the downwash angle is determined is

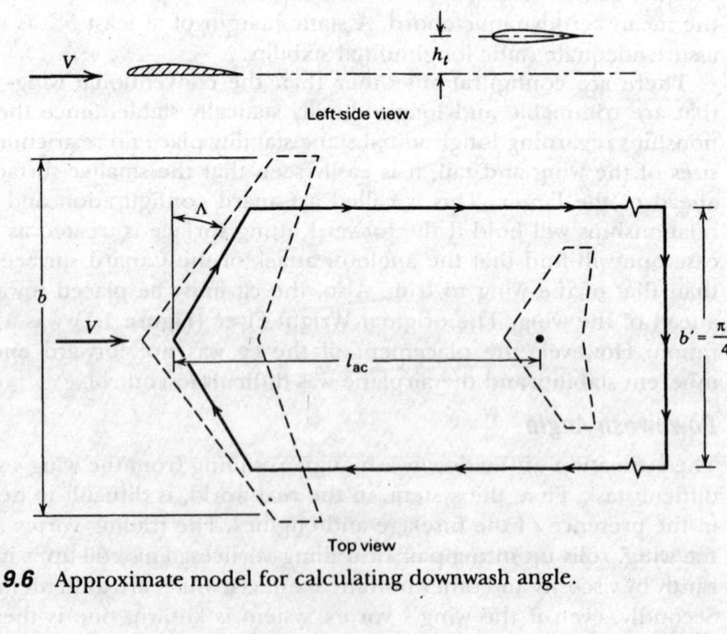

Figure 9.6 Approximate model for calculating downwash angle.

taken as the aerodynamic center of the tail. To determine the rate of increases of ϵ with α, ϵ_α, given the geometry in Figure 9.6, one first calculates the downwash at the distance l_{ac} back from the apex and a height of h_t above the plane of the vortex system for a vortex strength of 1.0. This can be easily accomplished using the program written as Computer Exercise No. 2.2 or manually, using the Biot–Savart law for a straight line as given by Equation 2.65. If the result of this calculation is denoted as w_1, then ϵ_α follows immediately from

$$\epsilon_\alpha = w_1 \frac{d\Gamma_0}{d\alpha}$$

or

$$\begin{aligned}\epsilon_\alpha &= \frac{w_1}{V}\Gamma_0 \\ &= \frac{2S}{\pi b} a_w w_1\end{aligned} \tag{9.27}$$

In this manner, the results presented in Figures 9.7 and 9.8 were generated. h_t is the distance of the tail above the plane of the trailing vortex system. Although this distance is determined by the geometry of the airplane, the downward deflection of the vortex system, and the angle of attack of the airplane, this degree of sophistication is generally not justified in view of the approximate model used to determine ϵ_α. Instead, it is suggested that h_t be taken as simply the height of the tail chord above the midspan chord of the wing. Also note that the possible difference between the velocity at the tail and the free-stream velocity is neglected in determining the downwash angle. Figure 9.7 applies only to the case of zero sweepback. Figure 9.8 is a correction for sweepback for a zero tail height. However, on a relative basis, this correction does not depend significantly on the tail height, so that Figure 9.8 can be used in general.

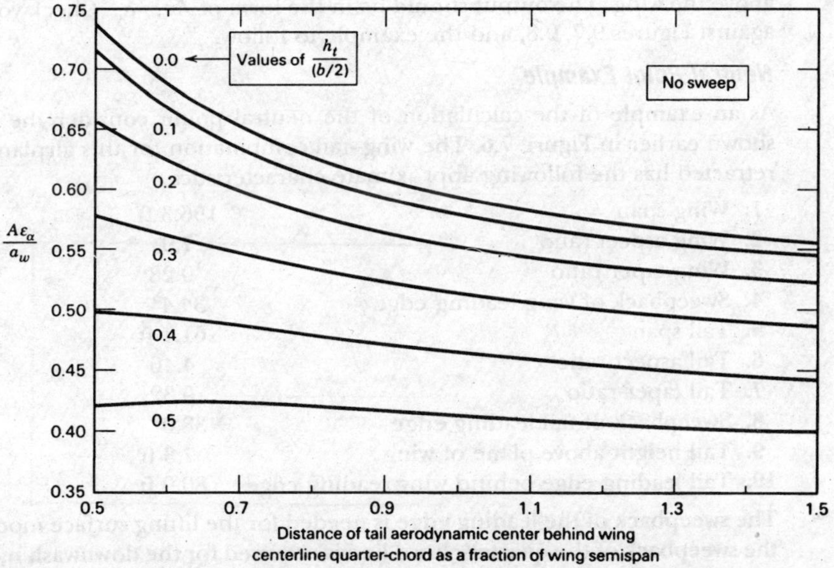

Figure 9.7 Downwash angle as a function of the distance of the tail behind and above the wing.

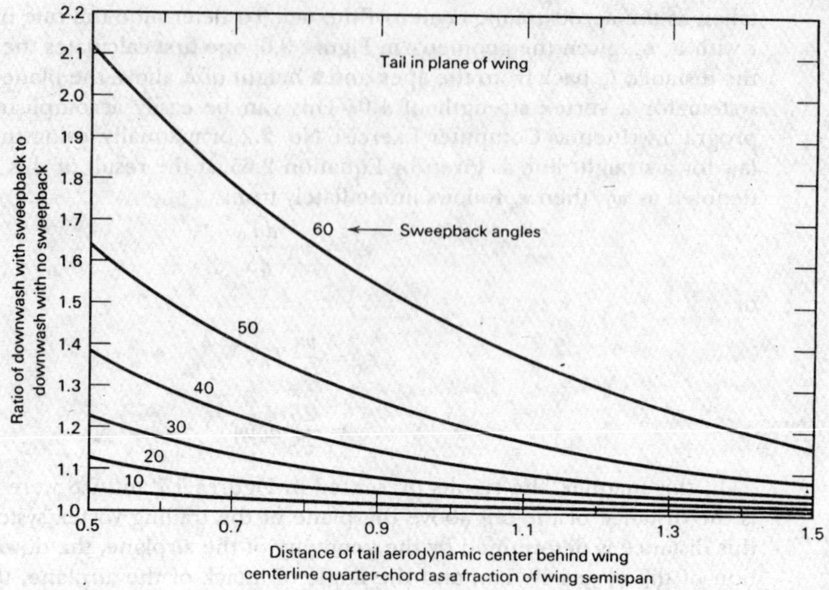

Figure 9.8 Effect of sweepback on downwash angle.

Computer Exercise 9.1 "EPSILON"

Formulate a computer program to determine ϵ_α, the rate of change of the downwash angle at the tail with the angle of attack of the wing. Use the approximate model described above. Input will be the wing span, the distance of the tail ac behind the centerline quarter-chord point of the wing, and the height of the tail above the wing. The output should be in the form of $A\epsilon_\alpha / a_w$. Check your program against Figures 9.7, 9.8, and the example to follow.

Neutral Point Example

As an example of the calculation of the neutral point, consider the Boeing 767 shown earlier in Figure 7.6. The wing–tail combination for this airplane with flaps retracted has the following approximate characteristics.

1. Wing span	156.3 ft
2. Wing aspect ratio	7.9
3. Wing taper ratio	0.28
4. Sweepback of wing leading edge	34.4°
5. Tail span	61.1 ft
6. Tail aspect ratio	4.46
7. Tail taper ratio	0.32
8. Sweepback of tail leading edge	38.0°
9. Tail height above plane of wing	7.3 ft
10. Tail leading edge behind wing leading edge	80.9 ft

The sweepback of the leading edge is needed for the lifting surface model, whereas the sweepback of the quarter-chord line is required for the downwash model. These two sweep angles are related by

$$\tan\Lambda_{1/4} = \tan\Lambda_{LE} - \frac{1 - \lambda}{A(1 + \lambda)}$$

The mean aerodynamic chord of the wing is calculated as 21.7 ft from Equation 3.79*b*. Using the vortex lattice program in Chapter 3, the wing can be modeled with no twist or camber and 1° angle of attack. The results give a calculated lift coefficient, which will equal a_w, of 0.0760 and a moment coefficient about the centerline leading edge of -0.0458. It follows, using the value of $\bar{c}$ and Equation 3.81, that the aerodynamic center of the wing is located a distance of 13.07 ft behind the centerline leading edge. Similarly, for the tail, C_M about the leading edge equals -0.0273, the C_L, or a_t, equals 0.0641 and $\bar{c}$ is 14.9. Thus, the aerodynamic center of the tail is calculated to lie 6.3 ft behind its centerline leading edge. It follows that the distance of the aerodynamic center of the tail behind the centerline quarter-chord point of the wing is equal to 79.6 ft. The distance between aerodynamic centers is 74.2 ft.

For clarity, the longitudinal placement of these elements is shown in Figure 9.9. The MAC of the wing is placed such that its quarter-chord point lies on the ac of the wing. It then follows that $h_{n_w} = 0.25$ and $h_l = 3.67$. From Figures 9.7 and 9.8, the quantity, $A\epsilon_\alpha / a_w$, is found to equal 0.611, resulting in a value of ϵ_α equal to 0.264. Finally, using Equation 9.24, the dimensionless neutral point distance, h_m, is found to be 0.745. This results in the neutral point being 23.78 ft behind the centerline leading edge.

Computer Exercise 9.2 "NP"

Formulate a computer program that determines the dimensional distance of the neutral point of an airplane relative to the centerline leading edge of the wing. The input should be a data file containing the same kind of data, which was listed for the 767 example. Run the lifting surface program developed in Computer Exercise 3.3 twice; first for the wing and then for the tail to determine a_w, a_t, and the corresponding moment coefficients about the centerline leading edges. The equation for the MAC should be added to the lifting surface program as a reference length for $C_{M_{LE}}$, and the program should be run with no twist or camber in order to determine the location of the aerodynamic center. Having determined the location of the tail ac, the program for the downwash, Computer Exercise 9.1, can be used to determine ϵ_α. However, we can improve on the approximate model in this program for ϵ_α during the calculation of the wing characteristics. After deter-

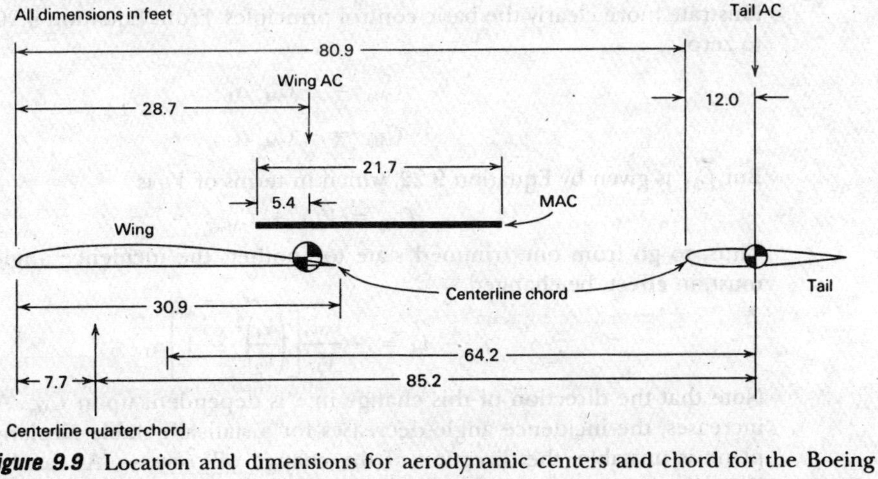

Figure 9.9 Location and dimensions for aerodynamic centers and chord for the Boeing 767.

mining the strengths of the vortex lattices for the wing, one can apply the Biot–Savart law again usng the location of the aerodynamic center of the horizontal tail, the wing geometry, and the strengths of the vortex lattices to determine the downwash at the tail. From either method, having ϵ_α and knowing all of the distances shown in Figure 9.9, the neutral point can then be determined from Equation 9.24.

LONGITUDINAL CONTROL

The Wright Brothers approached the design of their airplane from a perspective, which was significantly different from others. Their predecessors and their contemporaries felt that an airplane had to possess inherent stability. The Wright Brothers, however, believed that a person was capable of controlling an airplane. Therefore, they gave considerable thought and effort to providing control about all three airplane axes. They were successful in their efforts, even though their airplane was unstable, because they had provided for adequate control.

Like the Wright Brothers' Flyer, today's modern combat aircraft are inherently unstable, but are controllable because of feedback provided by computer-controlled servo systems. Obviously, the degree to which an airplane can be controlled, the *control power,* is as important as its degree of stability. However, one should guard against the philosophy that "black boxes" can fix any stability and control problems of an airplane. The aerodynamic design of the airplane determines the magnitude of the aerodynamic forces and moments that are available to an automatic control system. If these are not adequate, no amount of digital wizardry will help. Therefore, it is important to have an understanding of the aerodynamic aspects of control.

To begin, consider the simple question of longitudinal control required to change an airplane from one trimmed, straight, and level state to another. Let V_1 be the speed for the initial state and let V_2 be the speed for the new state. In either state, the lift must equal the weight and the moment must be zero. If α_1 is the initial trimmed angle of attack, then in order to maintain the same lift,

$$\alpha_2 = \alpha_1\left(\frac{V_1}{V_2}\right)^2 \tag{9.28}$$

This relationship neglects the contribution of the tail to the total lift in order to illustrate more clearly the basic control principles. From Equation 9.20 with C_M set to zero,

$$C_{M0_1} = -C_{M\alpha}\,\alpha_1 \tag{9.29}$$

$$C_{M0_2} = -C_{M\alpha}\,\alpha_2 \tag{9.30}$$

But C_{M0} is given by Equation 9.22, which in terms of V_H is

$$C_{M0} = V_H i_t + C_{M_{ac}} \tag{9.31}$$

Thus, to go from one trimmed state to another, the incidence angle of the tail must, in effect, be changed.

$$i_{t2} - i_{t1} = -\frac{C_{M\alpha}}{V_H}\left[\left(\frac{V_1}{V_2}\right)^2 - 1\right]\alpha_1 \tag{9.32}$$

Note that the direction of this change in i_t is dependent upon $C_{M\alpha}$. As the speed increases, the incidence angle decreases for a statically stable airplane. If the airplane is unstable, the direction of the change will reverse. As can be seen from Equation 9.25, this control reversal will occur if the airplane is loaded so that the cg is behind the neutral point. Obviously, this control reversal is a dangerous situation.

An airplane's flight manual will contain a loading diagram, which prescribes the aft cg limit as a function of the airplane's gross weight based on maintaining adequate longitudinal stability. The diagram will also specify a forward cg limit, which is usually dictated by the tail power required to lift the nose on takeoff.

The above analysis is now repeated in more detail, including the contribution of the tail to the total lift.

Tail Incidence to Trim as a Function of Lift Coefficient

One tends to think of the tail incidence angle as a fixed angle built into the airplane. However, let us consider the incidence angle as variable in which case both the airplane lift coefficient and pitching moment coefficient about the cg can be written as linear functions of α and i_t.

$$C_M = C_{M_{ac}} + C_{M_\alpha}\alpha + C_{M_i}i_t \tag{9.33}$$

$$C_L = C_{L_\alpha}\alpha + C_{L_i}i_t \tag{9.34}$$

where

$$C_{L_i} = -\eta_t a_t \frac{S_t}{S}$$

$$C_{M_i} = \eta_t a_t V_H$$

Solving for α from Equation 9.34, substituting the result into Equation 9.33 and setting C_M to zero gives an expression for i_t, which satisfies the trim conditions

$$i_t = -\frac{C_{M_{ac}}C_{L_\alpha} + C_{M_\alpha}C_L}{C_{L_\alpha}C_{M_i} - C_{M_\alpha}C_{L_i}} \tag{9.35}$$

Stabilizer–Elevator

Equation 9.35 applies to either a stabilator or an elevator if i_t is taken to mean the incidence angle of the tail's zero lift line. To pursue this further, we know from Chapter 3 that the lift coefficient for a flapped airfoil can be expressed as

$$C_l = a(\alpha + \tau\delta)$$

Here, τ is the value of the flap effectiveness factor obtained from Figure 3.35 multiplied by η from Figure 3.36. Thus, referring to the tail configuration of Figure 9.10*b*, if the elevator is deflected downward through the angle δ_e, the incidence angle of the zero lift line is decreased by the amount $\tau\delta_e$. If i_{hs} denotes the incidence angle of the fixed horizontal stabilizer, then the incidence angle of the tail zero lift line will be

$$i_t = i_{hs} - \tau\delta_e \tag{9.37}$$

The coefficients $C_{M\delta}$ and $C_{l\delta}$ can be obtained from the corresponding derivatives with respect to i_t.

$$\begin{aligned} C_{L\delta} &= -\tau C_{L_i} \\ &= \tau\eta_t a_t \frac{S_t}{S} \end{aligned} \tag{9.38}$$

$$\begin{aligned} C_{M\delta} &= -\tau C_{M_i} \\ &= -\tau\eta_t V_H a_t \end{aligned} \tag{9.39}$$

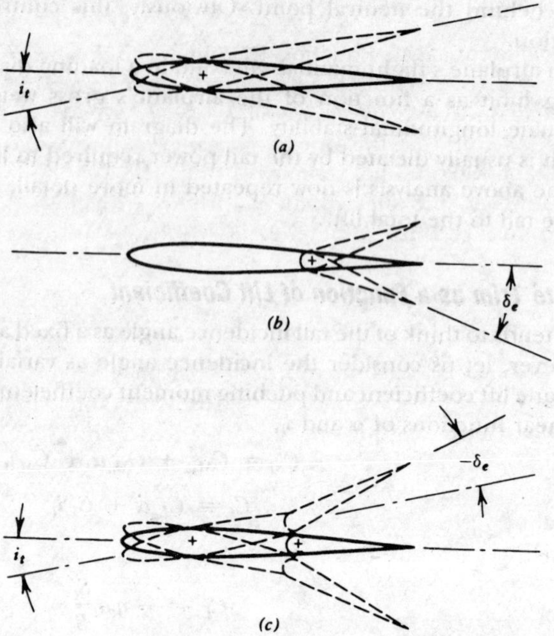

Figure 9.10 Horizontal tail configuration. (a) All movable. (b) Horizontal stabilizer and elevator. (c) All-movable tail with linked tab (stabilator).

Using Equations 9.38 and 9.39, the elevator angle to trim becomes

$$\delta_e = -\frac{C_{M_\alpha}C_L + C_{L_\alpha}C_{M_{ac}}}{C_{M\delta}C_{L\alpha} - C_{M_\alpha}C_{L\delta}} + \frac{i_{hs}}{\tau} \tag{9.40}$$

Thus, as a function of C_L, δ_e has the same form as i_t, except that the corresponding rates are the negative of each other since the positive rotation of the elevator is opposite that of the incidence angle.

In order to simplify the algebra, which will follow later, Equation 9.40 will be written as

$$\delta_e = A_\delta C_L + B_\delta + \frac{i_{hs}}{\tau} \tag{9.41}$$

Stabilator

The configuration shown in Figure 9.10c is referred to as a stabilator and is employed, for example, on the light airplane shown in Figure 3.59. Let i_{h_s}, which is now variable, and δ_e be related by

$$\delta_e = k_e i_{h_s} + \delta_0 \tag{9.42}$$

We will see shortly that the constant term, δ_0, is added above in order to allow for trim. Using Equation 9.37, i_t now becomes

$$i_t = i_{hs}(1 - \tau k_e) - \tau\delta_0 \tag{9.43}$$

Thus, from Equation 9.35,

$$i_{h_s} = -\frac{C_{M_\alpha}C_L + C_{L_\alpha}C_{M_{ac}}}{C_{M_i}C_{L_\alpha} - C_{M_\alpha}C_{L_i}}\frac{1}{(1 - \tau k_e)} + \frac{\tau\delta_0}{1 - \tau k_e} \tag{9.44}$$

This is identical to Equation 9.35 in form except for the factor $1/(1 + \tau k_e)$. By varying k_e, one is able to alter the rate of change of the tail incidence angle with C_L. Usually a negative value of k_e is used to make the tail more effective when rotated. Note that this does not change a_t for a fixed position of the tail. Hence, the static longitudinal stability of the airplane, defined by $C_{M\alpha}$, is not changed by k_e. However, the stick gradient with speed will change with k_e, thereby giving a different feel to the pilot and can effectively satisfy FAR Part 23 (Ref. 9.1) requirements regarding longitudinal static stability.

Control Forces

For obvious reasons, in the design of an airplane's control system, the stick or control wheel forces must lie within acceptable limits throughout the operating envelope (*V-n* diagram) of the airplane. In addition, the variation of these forces with airspeed about any trim point should be such as to give a proper "feel" to the pilot. Generally, this means that a push forward on the longitudinal control should be required to increase the airspeed, and a pull should be required to fly slower.

With regard to the longitudinal control forces, FAR Part 23 allows a maximum of 60 lb for a temporary application to a stick and 75 lb for a wheel. A prolonged application is not allowed to exceed 10 lb for either type of control.

Gearing

The control force, P, is directly proportional to the control surface hinge moment, H.

$$P = GH \tag{9.45}$$

G is referred to as the gearing. To determine G, refer to the sketch in Figure 9.11. Here a schematic of a control linkage is shown between a stick and an elevator. If the system is in equilibrium,

$$PL = \left(\frac{H}{l_e}\right) l$$

where the stick force, P, and the hinge moment, H, are shown in their positive directions. Now allow the stick at the point of application of P to move a distance

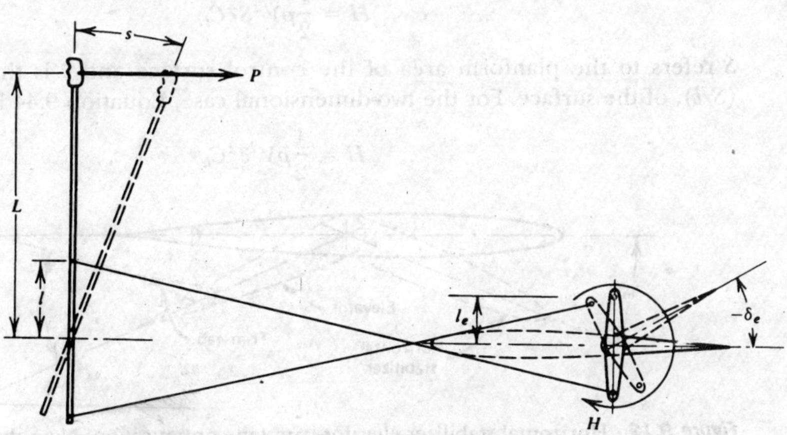

Figure 9.11 Schematic of a longitudinal control system.

s in the direction of P. In so doing the elevator will rotate through an angle δ_e, given by

$$\delta_e = -\frac{s}{L}\frac{l}{l_e}$$

Thus,

$$P = \left(-\frac{\delta_e}{s}\right)H \tag{9.46}$$

Comparing Equation 9.46 to Equation 9.45 gives

$$G = -\frac{\delta_e}{s} \tag{9.47}$$

This result is general and independent of the details of the linkage. Since δ_e is negative for a positive stick displacement, it follows that G is positive. It can also be obtained simply by noting that if the system is in equilibrium throughout its displacement, the net work done on the system is zero. Hence,

$$Ps + H\delta_e = 0$$

from which Equation 9.47 follows.

Stick Force for a Horizontal Stabilizer–Elevator Combination

The configuration shown in Figure 9.12 is probably the most common configuration used for the horizontal tails of airplanes. The forward-most part of the tail is the horizontal stabilizer. Usually this is fixed to the fuselage, but on some airplanes it is pivoted about its trailing edge so that the incidence angle can be changed for trim. The portion of the tail behind the horizontal stabilizer is the elevator, the angular position of which is controlled by the movement of the stick as shown in Figure 9.11.

In the case where the horizontal stabilizer is fixed, trim is provided by an additional movable surface at the rear of the elevator, which is known as the trim tab. As shown, the incidence angle of the horizontal stabilizer, i_{h_s}, is measured relative to the airframe. The elevator angle, δ_e, is measured relative to the horizontal stabilizer and the tab angle, δ_t, is measured relative to the elevator. i_{h_s} is defined positive nose down, but the other two are defined positive tail down.

The moment about the hinge line of any control surface is written in terms of a hinge moment coefficient, C_H.

$$H = \frac{1}{2}\rho V^2 S\bar{c} C_H \tag{9.46}$$

S refers to the planform area of the control surface and $\bar{c}$ is the mean chord, (S/b), of the surface. For the two-dimensional case, Equation 9.44 becomes

$$H = \frac{1}{2}\rho V^2 \bar{c}^2 C_h \tag{9.49}$$

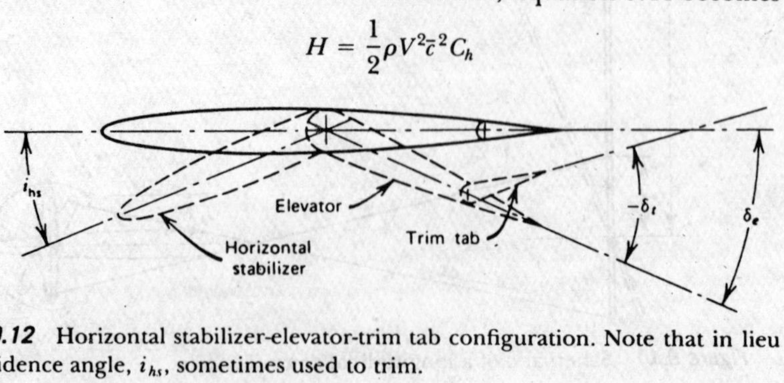

Figure 9.12 Horizontal stabilizer-elevator-trim tab configuration. Note that in lieu of trim tab, incidence angle, i_{h_s}, sometimes used to trim.

Using the same convenient notation as Reference 9.2 and assuming a symmetrical section, the elevator hinge moment can be written as

$$C_H = b_1\alpha_{ht} + b_2\delta_e + b_3\delta_t \qquad (9.50)$$

To avoid confusion, *ht* has been used here to denote the entire horizontal tail and *t* to denote the tab. The constants, b_1, b_2, *and* b_3, are the rates of change of C_H with the angles α_{ht}, δ_e, and δ_t, respectively.

Thus the stick force for the stabilizer–elevator configuration can be written as

$$\frac{P}{GS_e\bar{c}_e} = \frac{1}{2}\rho V^2(b_1\alpha_{ht} + b_2\delta_e + b_3\delta_t)$$

But

$$\alpha_{ht} = \alpha(1 - \epsilon_\alpha) - i_{hs}$$

$$\frac{1}{2}\rho V^2 = \frac{W}{SC_{L_\alpha}\alpha}$$

and δ_e is given by Equation 9.40. Thus,

$$\frac{P}{GS_e\bar{c}_e} = \frac{b_1(1 - \epsilon_\alpha)}{C_{L\alpha}}\left(\frac{W}{S}\right) + \frac{1}{2}\rho V^2\left[i_{h_s}\left(\frac{b_2}{\tau} - b_1\right) + b_2(A_\delta C_L + B_\delta) + b_3\delta_t\right]$$

Substituting for C_L in terms of the wing loading and V, the above reduces finally to

$$\frac{P}{GS_e\bar{c}_e} = \left(\frac{W}{S}\right)\left[\frac{b_1(1 - \epsilon_\alpha)}{C_{L\alpha}} + b_2A_\delta\right] + \left[i_{h_s}\left(\frac{b_2}{\tau} - b_1\right) + b_2B_\delta + b_3\delta_t\right]\frac{1}{2}V^2 \qquad (9.51)$$

To save some writing, this is expressed as

$$\bar{P} = A + BV^2 \qquad (9.52)$$

Equation 9.52 can be used to estimate the stick position as a function of speed, but it is more illuminating to reduce this equation further in terms of a trim speed.

Suppose that the trim tab is adjusted so that the stick force is zero at a trim speed of V_{TR}. It then follows that

$$B = -\frac{A}{V_{TR}^2}$$

Thus, Equation 9.52 becomes

$$\bar{P} = A\left[1 - \left(\frac{V}{V_{TR}}\right)^2\right] \qquad (9.53)$$

where the A coefficient is given by

$$A = \frac{W}{S}\left[\frac{b_1(1 - \epsilon_\alpha)}{C_{L\alpha}} - \frac{b_2C_{M_\alpha}}{C_{M\delta}C_{L\alpha} - C_{M_\alpha}C_{L\delta}}\right] \qquad (9.54)$$

and the B coefficient is

$$B = i_{hs}\left(\frac{b_2}{\tau} - b_1\right) - \frac{b_2C_{L_\alpha}C_{M_{tr}}}{C_{M\delta}C_{L\alpha} - C_{M_\alpha}C_{L\delta}} + b_3\delta_t \qquad (9.55)$$

The behavior of the above relationships will be discussed later by means of an example.

Stick Force for a Stabilator

In the case of an all-movable tail with a linked elevator, the hinge point is somewhere around the aerodynamic center of the tail. The hinge moment, in this case, is comparable to the pitching moment about some arbitrary chordwise position.

Thus, the moment coefficient, C_M, will be used in place of C_H with the MAC of the tail as the reference chord.

$$C_M = b_1\alpha_{ht} + b_2\delta_e \tag{9.56}$$

where

$$b_1 = \frac{\partial C_H}{\partial \alpha} \text{ and } b_2 = \frac{\partial C_H}{\partial \delta} \tag{9.57}$$

The angle of attack, α_{HT}, will be given by

$$\alpha_{HT} = \alpha(1 - \epsilon_a) - i_{hs} \tag{9.58}$$

with i_{hs} given by Equation 9.42.

Following the same algebraic procedure as for the stabilizer–elevator configuration, equations are obtained for the stick force, which are identical in form to Equations 9.52 and 9.53. However, for the stabilator configuration

$$\bar{P} = \frac{P}{GS_T\bar{c}_t} \tag{9.59}$$

$$A = \frac{W}{S}\left[\frac{b_1(1 - \epsilon_a)}{C_{L_\alpha}} + \frac{C_{M_\alpha}(b_1 - k_e b_2)}{(C_{L_\alpha}C_{M_i} - C_{M_\alpha}C_{L_i})(1 - \tau k_e)}\right] \tag{9.60}$$

$$B = \frac{(b_1 - k_e b_2)}{(1 - \tau k_e)}\left(\frac{C_{M_{ac}}C_{L_\alpha}}{C_{L_\alpha}C_{M_i} - C_{M_\alpha}C_{L_i}} + \tau\delta_0\right) + b_2\delta_0 \tag{9.61}$$

Hinge Moment Coefficients

To repeat, C_H for a stabilizer–elevator configuration is based on the moment about the hinge line of the elevator and the chord of the elevator. For a stabilator, C_H is replaced by C_M about an arbitrary chordwise location ahead of the tab hinge line. C_M will not be covered here because material on this can be found in Chapter 3. If the dimensionless chordwise station, x, is known and x_{ac} and $C_{M_{ac}}$ found from Chapter 3, possibly using Computer Exercise 3.1, then C_M at x can be determined from

$$C_M = C_{M_{ac}} + (x - x_{ac})C_L \tag{9.62}$$

The constants b_1 and b_2 are obtained by determining the rate at which C_M changes with α and δ.

Reference 9.3, similar to Reference 5.5, is a comprehensive collection of data relating to aerodynamics, structures, and performance. The reader should refer to data such as this in order to determine effects such as a gap between the stabilizer and elevator or aerodynamic balance where part of the elevator is ahead of the hinge line. These effects are illustrated in Figure 9.13. For the purposes of this text, the elevator will be taken as a full-span, plain flap. Thus, the hinge moment can be determined on a 2-D basis and the correction for aspect ratio applied as given by Equation 3.70.

Figures 9.14*a* and *b* present 2-D results for C_h obtained using the airfoil program developed in Computer Exercise 3.1. The numerical results should agree closely with an exact solution because 300 vortex elements were used in modeling the airfoil. Figure 9.14*a* shows the effect of elevator chord on the change of C_h with angle of attack and elevator angle. Figure 9.14*b* gives the change in C_h with tab angle as a function of elevator chord for two ratios of tab-to-elevator chord.

The empennage on some airplanes consists of twin vertical tails mounted on each tip of the horizontal tail. These "end plates" inhibit the loss of lift toward the tips of the horizontal tail thereby effectively increasing its aspect ratio. A correction

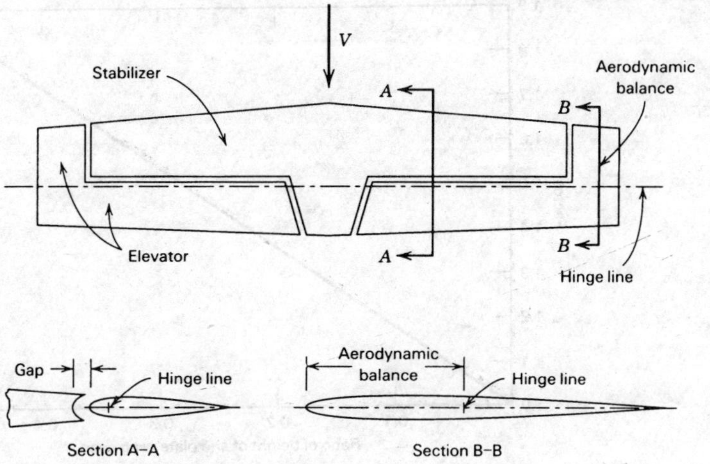

Figure 9.13 Horizontal tail geometry.

to the aspect ratio for end plates is given in Figure 9.15 and should be applied to both C_{L_α} and C_{H_α}.

At this point, we have already encountered a number of derivatives of coefficients taken with respect to an angular displacement. Such derivatives are called *stability derivatives* or *control derivatives*. Many more of these will be defined in the remainder of this chapter and the next.

Computer Exercise 9.3 "HINGE"

Modify the 2-D airfoil program written for Computer Exercise 3.1 to predict the hinge moment coefficient derivatives with elevator and tab angles for a plain flap. Input should be c_e/c and c_t/c_e. Output should be C_{h_α}, $C_{h_{\delta_e}}$, and $C_{h_{\delta_t}}$. Check your results against Figure 9.14.

Example Calculation of Stick Force

As an example in the use of the foregoing relationships, consider a fictitious executive jet transport that is geometrically similar to the Boeing 767 and operating at a 10,000-ft altitude. All linear dimensions will be taken as equal to those for the

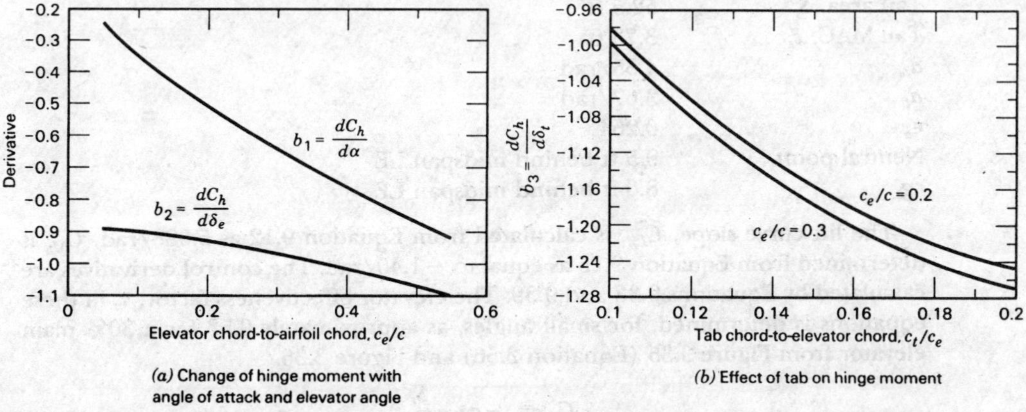

Figure 9.14 Hinge moment derivatives (per radian).

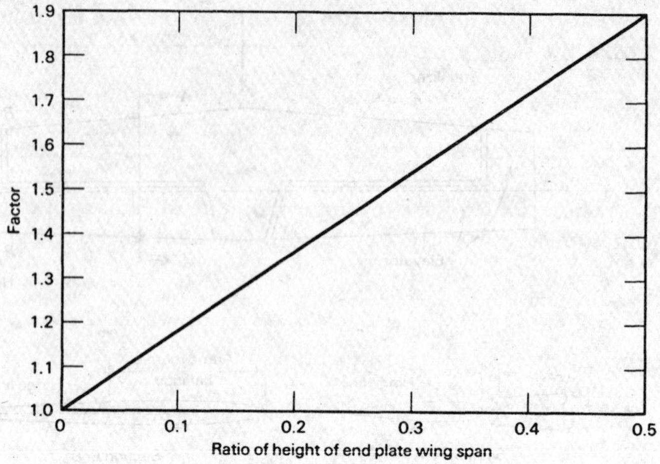

Figure 9.15 Correction factor to aspect ratio for effect of end plates (multiply geometric aspect ratio by factor to get effective aspect ratio).

767 reduced by a factor of 1/4. Thus, the numbers used for the neutral point example, suitably reduced, will apply to this example. However, the wing loading, *W/S*, will be reduced to 50 psf. and, unlike the 767, the control system will be a mechanical one with the gearing, *G*, equal to 0.4 rad/ft. Also, it is assumed that $c_e/c = 0.3$ and $c_t/c_e = 0.1$. The airplane is loaded so that the cg lies 4.0 ft behind the leading edge of the centerline chord. From the neutral point example, the neutral point lies 5.95 ft behind the leading edge. Thus, the airplane has an 8.3% static margin based on an MAC of 5.425 ft.

The pertinent parameters for this airplane are tabulated as follows. These are simply scaled from the neutral point example for the 767 or taken directly in the case of the dimensionless coefficients.

Wing span, b	39.1 ft
Wing aspect ratio, A	7.9
Wing area, S	193.5 ft^2
Wing MAC, $\bar{c}$	5.425
Tail span, b_t	15.3
Tail aspect ratio, A_t	4.46
Tail area, S_t	52.5 ft^2
Tail MAC, $\bar{c}_t$	3.73 ft
a_w	4.35/rad
a_t	3.67/rad
ϵ_α	0.264
Neutral point	9.5 ft behind midspan LE
cg	8.0 ft behind midspan LE

The lift curve slope, C_{L_α}, is calculated from Equation 9.12 as 5.088/rad. C_{M_α} is determined from Equation 9.25 as equal to -1.40/rad. The control derivatives are calculated by Equations 9.38 and 9.39. The elevator effectiveness factor, τ, in these equations is determined, for small angles, as approximately 0.53 for a 30% plain elevator from Figure 3.35 (Equation 3.56) and Figure 3.36.

$$C_{l_\alpha} = -\tau\eta_t a_t \frac{S_t}{S}$$
$$= -0.528/\text{rad}$$

For the given elevator and tab chord ratios, the following values are determined from the same program that was used to produce Figure 9.14.

$$b_1 = C_{h_\alpha} = -0.627/\text{rad}$$
$$b_2 = C_{h_{\delta_e}} = -0.968$$
$$b_3 = C_{h_{\delta_t}} = -1.273$$

These values for the hinge moment rates are for the 2-D case and must be multiplied by a correction factor to account for finite aspect ratio. This correction factor, obtained from Equation 3.7, using the aspect ratio of the tail, is determined to be 0.630.

An absolute value for the elevator angle and stick force cannot be determined because we have no information on the $C_{M_{ac}}$. However, for illustrative purposes the assumption is made that $C_{M_{ac}}$ is zero. This is a fairly good assumption since the contribution of $C_{M_{ac}}$ to the elevator trim angle and stick force is normally small except with flaps down at a low speed.

We can calculate the change in δ_e and in the stick force when the airplane departs from a trimmed condition even without knowing $C_{M_{ac}}$ or i_{hs}. This is done by using Equations 9.40 and 9.53. Note that the constant, A, in Equation 3.53 does not involve the altitude or speed. Using the derivatives given above, this constant equals 3.42.

Figure 9.16 presents, for this example, the calculated elevator angles to trim for cg positions of 8.0, 9.0, and 10.0 ft behind the leading edge of the midspan chord. The up-elevator angle is seen to increase with decreasing speed if the cg is ahead of the neutral point. However, with the cg behind the neutral point, the pilot must move the stick back to fly faster. This latter case, from the control standpoint, is an unnatural and unstable situation, which can be dangerous.

The variation of the stick force gradient with trim speed for a cg position well forward of the neutral point is illustrated in Figure 9.17. In this case, the static margin is 28%. As expected, and as required by FAR Part 23, at all trim speeds a pull is required to decrease the airspeed and a push to increase the speed.

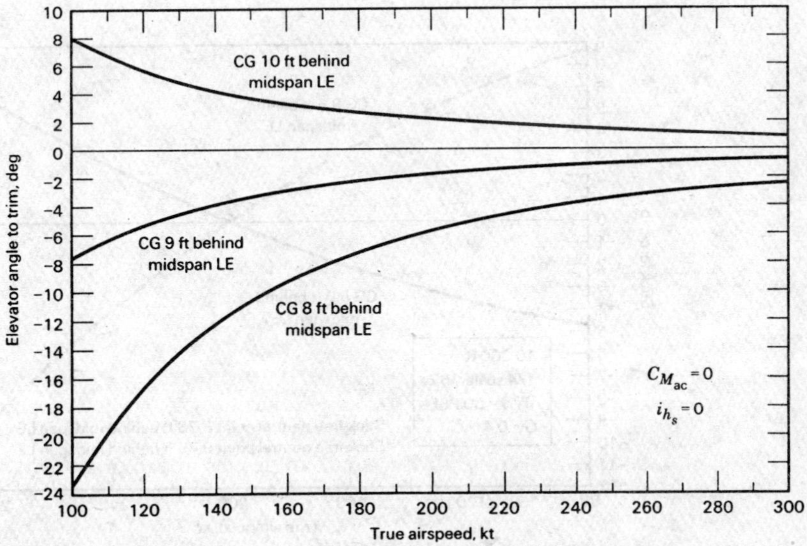

Figure 9.16 Calculated elevator to trim executive jet at 10,000 ft for three different cg positions.

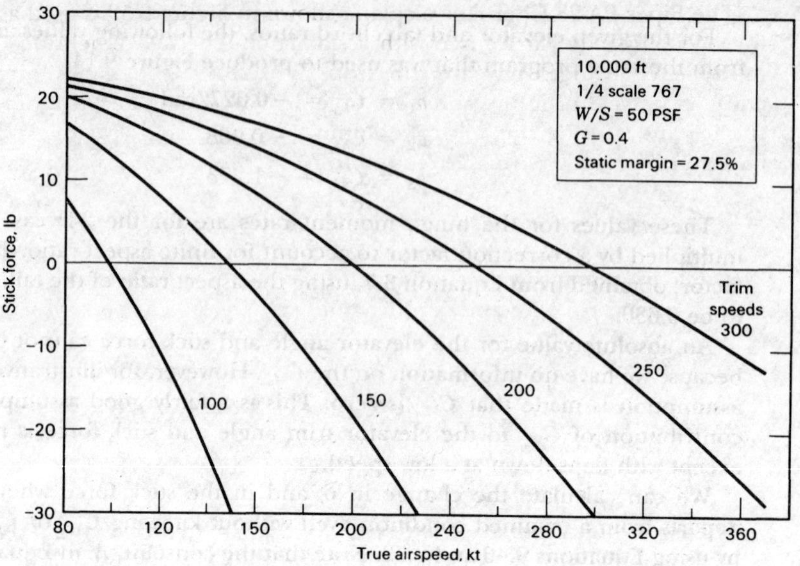

Figure 9.17 Calculated stick force for executive jet with a forward cg (cg 4.0 ft behind midspan leading edge).

The results presented in Figure 9.18 are not expected when seen for the first time. Here, the stick force gradient is shown about a trim speed of 200 kt for aft cg positions of 8.5 and 9.0 ft behind the leading edge of the midspan chord. These correspond to static margins of 18.4% and 9.2%, respectively. Even though the static margin is positive for both cases, the stick force gradient is seen to be unstable for the cg location of 9.0 ft. This is definitely a dangerous situation, which goes counter to the "feel" experienced by a pilot. The reason for this unexpected behavior of the stick force gradient lies with a parameter known as the *stick-free neutral point,* which will be covered shortly.

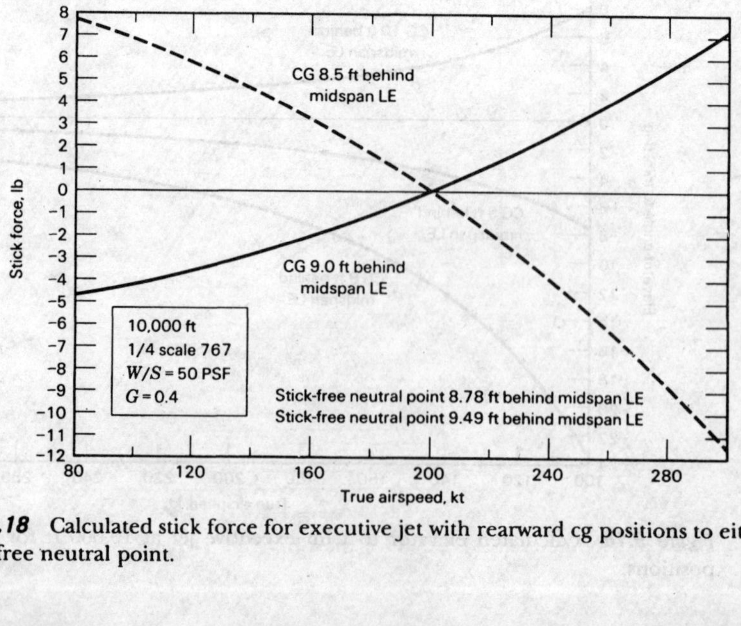

Figure 9.18 Calculated stick force for executive jet with rearward cg positions to either side of stick-free neutral point.

The Piper PA-28 Cherokee airplane shown in Figure 3.59 is used as an example for calculating the stick force with a stabilator. The tail dimensions and angular movements, as measured, are shown in Figure 9.19. The angles for the stabilizer and elevator are arbitrary and were obtained by placing an inclinometer on each surface and moving the control column. From the slopes of the two curves, the linkage factor and gearing are found to be

$$k_e = 1.50$$

$$G = 0.5 \text{ rad/ft}$$

The following quantities are estimated from the tail geometry assuming the airfoil to be symmetrical with the aerodynamic center at $0.25c$ and a section lift curve slope equal to $0.106C_l/\text{deg}$.

$$a_t = 0.0642 C_L/\text{deg}$$

$$\tau = 0.44$$

$$\frac{\Delta C_M}{\Delta C_l} = -0.185 \qquad \text{(Figure 3.34)}$$

For the symmetrical airfoil with $\delta_e = 0$, $C_{M_{ac}} = 0$, so C_M at the pivot line, which is 1.7% of the chord behind the $C/4$ point, will be

$$C_M = 0.017 a_t \alpha_t$$

Thus,

$$b_1 = 0.0625/\text{rad}$$

The increment in C_{L_t}, due to the elevator deflection, will be

$$\Delta C_{L_t} = a_t \tau \eta \, \delta_e$$

$$= 1.616 \, \delta_e$$

with δ_e in radians.

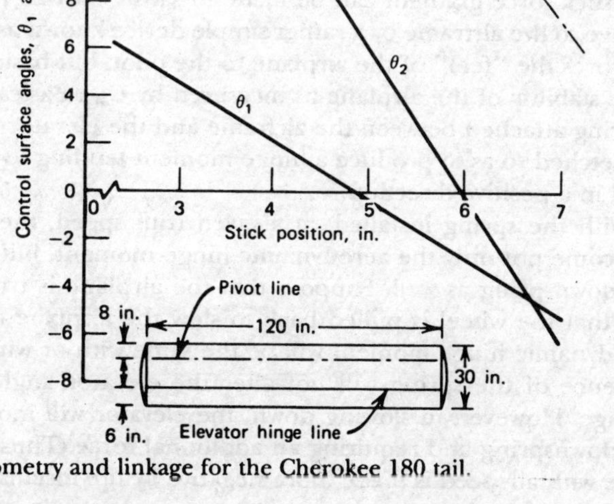

Figure 9.19 Geometry and linkage for the Cherokee 180 tail.

The increment in C_M about the pivot line produced by δ_e will equal the sum of the increment in $C_{M_{ac}}$ and the moment produced by the increment in C_L acting ahead of the pivot.

$$\begin{aligned}\Delta C_M &= -0.185(1.616)\ \delta_e + 1.616(0.017)\delta_e \\ &= 1.616\ \delta_e(-0.185 + 0.017)\end{aligned}$$

Thus, it follows that

$$b_2 = -0.271$$

A gross weight of 2255 lb is typical of a Cherokee 180 with four passengers, full-fuel, but no baggage. The cg position, shown in Figure 3.59, of 13 in. behind the leading edge is equivalent to h = 0.206. ϵ_a has been given earlier as equal to 0.447 with h_n equal to 0.423. Thus,

$$C_{M_\alpha} = -1.10$$

Also, from Equations 9.34*a* and *b*,

$$C_{M_i} = 1.438$$

$$C_{L_i} = -0.559$$

With the above numbers, the constant, *A*, in Equation 9.60 becomes

$$A = 0.586$$

The stick force, from Equation 9.53, is obtained as

$$\begin{aligned}P &= GS_t\bar{c}_t A\left[1 - \left(\frac{V}{V_{TR}}\right)^2\right] \\ &= (0.5)(25)(2.5)(0.586)\left[1 - \left(\frac{V}{V_{TR}}\right)^2\right] \\ &= 18.3\left[1 - \left(\frac{V}{V_{TR}}\right)^2\right]\end{aligned}$$

Stick force gradients, as calculated for the Cherokee equipped with a stabilator, are shown in Figure 9.20. Here, the gradients are seen to be stable for all trim speeds. Although not shown for this case, it is interesting to note that the stick force gradient will be stable for this airplane for any positive static margin. Again, this behavior, contrary to that found for the stabilizer–elevator configuration, is explained by the stick-free neutral point.

Downsprings

The stick force gradient can be made to satisfy Part 23 requirements without any change to the airframe by a rather simple device known as a *downspring*. This device improves the "feel" of the airplane to the pilot, but basically does not change the static stability of the airplane as measured by C_{M_α}. Basically, the device consists of a spring attached between the airframe and the elevator control horn. The spring is stretched so as to produce a hinge moment tending to rotate the elevator downward in a positive direction.

With the spring installed, at a given trim speed, the trim tab must be set to overcome not only the aerodynamic hinge moment, but the moment imposed by the downspring as well. Suppose that the airplane is trimmed at a certain speed and that the wheel is pulled back to slow the airplane down. The change in the aerodynamic hinge moment will be the same with or without the spring since the presence of the spring will not alter the elevator angle required for the speed change. However, in slowing down, the elevator will move up, thereby stretching the downspring and requiring an additional force. Thus, the gradient of the stick force with airspeed is made more negative by the installation of the downspring.

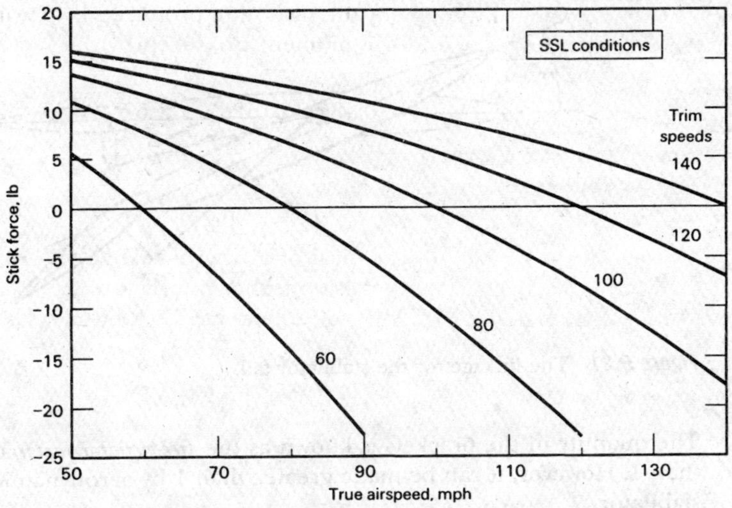

Figure 9.20 Calculated stick force for the Cherokee with a stabilator configuration.

The Piper Cheyenne II, a light twin-turboprop airplane, utilizes an ingenious modification of the downspring known as the *variable downspring.* For most airplanes, as the flaps and gear are lowered and the speed is reduced for landing, the stick gradient becomes very small, possibly even changing sign. A downspring can remedy this problem at the low speed, but then may prove unsatisfactory at higher speeds. To accommodate a wide speed range, the Cheyenne II employs an angle of attack sensor that controls a servo connected to the downspring. When the angle of attack increases beyond a prescribed angle, the servo is activated and stretches the downspring.

Stick-Free Neutral Point

Consider the tail with a stabilizer and elevator shown in Figure 9.21*a*. Without any knowledge of aerodynamics, one's intuition tells you that the aerodynamic forces on the elevator are such as to force it upward. Although intuition can sometimes be a dangerous thing when it comes to aerodynamics, in this case it is correct. Therefore, if the elevator is released so that it is free to *float,* the elevator will move up to some position as shown in Figure 9.21*b* where the hinge moment will be zero. If the angle of attack of the tail is increased further, the elevator will continue to move up.

The relationship between the floating elevator angle and the angle of attack can be found by setting the expression for the hinge moment to zero.

$$C_h = 0 = b_1\alpha + b_2\delta_e + b_3\delta_t$$

or

$$\delta_e = -\frac{b_1}{b_2}\alpha - \frac{b_3}{b_2}\delta_t \tag{9.63}$$

Assuming that the trim tab has a negligible contribution to the tail lift, the lift coefficient of the tail with a floating elevator is

$$\begin{aligned} C_{L_t} &= a_t(\alpha + \tau\delta_e) \\ &= a_t\alpha\left(1 - \tau\frac{b_1}{b_2}\right) \end{aligned} \tag{9.64}$$

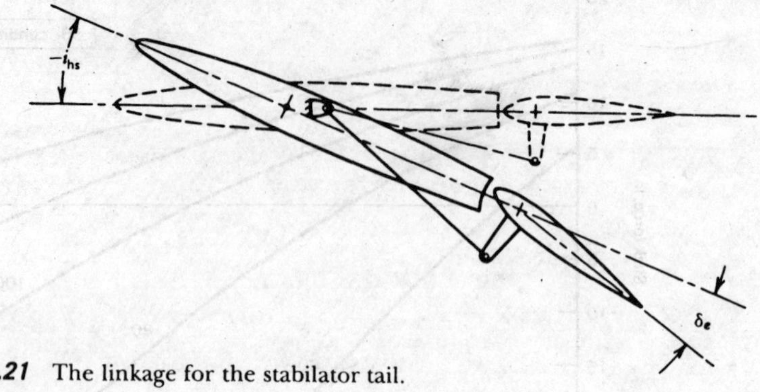

Figure 9.21 The linkage for the stabilator tail.

The quantity in the brackets is known as the *free-elevator factor* and is normally less than 1. However, it can be made greater than 1 by aerodynamic balance or with a stabilator.

$$F_e = 1 - \frac{b_1}{b_2} \tag{9.65}$$

It is seen that the free-elevator factor effectively decreases the slope of the tail lift curve. Using this reduced value of a_t in Equation 9.24 for the neutral point results in another similar point known as the *stick-free neutral point,* h'_n.

The elevator angle to trim an airplane is, of course, not dependent on the free-elevator factor. The same wing lift and tail lift are required for trim whether the elevator is floating or not. Indeed, in trimming the airplane by means of a trim tab, the pilot is really making the floating angle equal to that required for trim.

Suppose the airplane is in trim and the pilot decides to slow down without adjusting the trim tab. He or she will have to reduce the power and pull back on the wheel, thereby increasing the angle of attack. With the increased angle of attack for the new trim condition, the elevator will try to float upward in the direction that it must move. If this change in the floating angle is less than the change required to go from trim speed to a lower one, which is normally the case, then the pilot will have to exert a pull on the wheel to provide the *additional* elevator travel. Thus, the force required of the pilot is proportional to the difference between the change in the trim δ_e and the floating δ_e rather than to the change in the trim δ_e alone.

The equation for the stick-free neutral point is

$$h'_n = h_{nw} + V_H\frac{a_t}{a_w}(1 - \epsilon_\alpha)F_e$$

In terms of the stick-fixed neutral, the above becomes

$$h'_n - h = h_n - h - V_H\frac{a_t}{a_w}(1 - \epsilon_\alpha)(1 - F_e)$$

Substituting for $h_n - h$ in terms of C_{M_α} and C_{L_α}, the above reduces to

$$h'_n - h = \frac{C_{M_\alpha}}{C_{L_\alpha}} - V_H\frac{a_t}{a_w}(1 - \epsilon_\alpha)\tau\frac{b_1}{b_2}$$

But

$$C_{M_\delta} = -V_H\tau a_t$$

Thus,

$$h'_n - h = -\frac{C_{M\alpha}}{C_{L\alpha}} + \frac{C_{M\delta}(1 - \epsilon_\alpha)\, b_1}{a_w b_2}$$

If the contribution of the tail to the lift is neglected so that $C_{L_\alpha} \approx a_w$, then it follows that the constant, A, in Equation 9.51 is given approximately by

$$A = \frac{W}{S}\frac{b_2}{C_{M\delta}}(h'_n - h) \tag{9.66}$$

The above clearly shows that it is the stick-free neutral point, and not the stick-fixed neutral point, that determines whether or not a stick-force gradient is stable. If the cg is ahead of the stick-free neutral point, the coefficient, A, will be positive. From Equation 9.53, this will require a pull on the stick to reduce the trim airspeed, which is the stable condition.

Equation 9.66 is an approximate expression presented only to show the relationship between the stick-force gradient and the stick-free neutral. The expression given by Equation 9.54 for the coefficient, A, is more exact and should be used in any calculations. For example, for the executive jet results (Figure 9.17), the use of Equation 9.66 results in stick forces that are 14% lower than those obtained from using Equation 9.54.

Stabilator Configuration

The stabilator trim tab is mechanically linked to the fuselage, as shown schematically in Figure 9.21. The tail is trimmed by moving the attachment point of the link to the fuselage. With the stick free and the hinge moment trimmed to zero, the entire tail is floating around the pivot line. If the fuselage angle of attack is disturbed, the tail will tend to maintain a constant angle of attack. However, the tab, being linked to the fuselage, will deflect, thereby changing the aerodynamic moment about the pivot line. This, in turn, will cause the entire tail to float to a different position at which the moment is again zero. Thus, to determine the free-elevator factor for a stabilator, the linkage to the fuselage must be taken into account. The tail cannot be treated as an entity, as was just done for the horizontal stabilizer-elevator configuration. We begin by setting Equation 9.56 equal to zero and substituting Equation 9.42 for δ_e. This leads to an expression for the *floating* incidence angle, i_{hs}.

$$i_{hs} = \frac{b_1(1 - \epsilon_\alpha)}{b_1 - b_2 k_e}\alpha + \frac{b_2}{b_1 - b_2 k_e}\delta_0 \tag{9.67}$$

The tail lift coefficient can be written as

$$C_{L_t} = a_t[\alpha(1 - \epsilon_\alpha) - i_{hs} + \tau(k_e i_{hs} + \delta_0)]$$

Substituting Equation 9.67 for i_{hs}:

$$C_{L_t} = a_t\left[\alpha(1 - \epsilon_\alpha) - \frac{(1 - \tau k_e)\, b_1(1 - \epsilon_\alpha)}{b_1 - b_2 k_e}\alpha + \left[\tau - \frac{b_2}{b_1 - b_2 k_e}\right)\delta_0\right] \tag{9.68}$$

Thus the rate of change of tail lift coefficient with α will be

$$\frac{dC_{L_t}}{d\alpha} = a_t(1 - \epsilon_\alpha)\left[1 - \frac{(1 - \tau k_e)\, b_1}{b_1 - b_2 k_e}\right] \tag{9.69}$$

For the stick-fixed case, where the incidence angle is fixed, the tail lift curve slope in the downwash of the wing is simply

$$\frac{dC_{L_t}}{d\alpha} = a_t(1 - \epsilon_\alpha)$$

By comparison to Equation 9.69, it follows that for the stabilator configuration, the free-elevator factor is given by

$$F_e = 1 - \frac{(1 - \tau k_e)\, b_1}{b_1 - b_2 k_e} \tag{9.70}$$

Using the numbers previously estimated for the Cherokee 180, the value for F_e is estimated to be

$$F_e = 1.30$$

Thus, the effectiveness of the stabilator in providing longitudinal static stability is not degraded by freeing the stick; indeed, it is actually improved!

Steady Maneuvering

A wing–tail combination is pictured in Figure 9.22 undergoing a steady pull-up maneuver. For clarity, the angle $\Delta\alpha_t$ is shown much larger than it actually is; that is, the ratio of l_t to the radius of the flight path, R, is much less than one. The acceleration, a_n, of the airplane toward the center of curvature requires a lift in excess of the weight given by

$$L = W\left(1 + \frac{a_n}{g}\right) \tag{9.71}$$

The lift-to-weight ratio is known as the load factor, n.

$$n = 1 + \frac{a_n}{g} \tag{9.72}$$

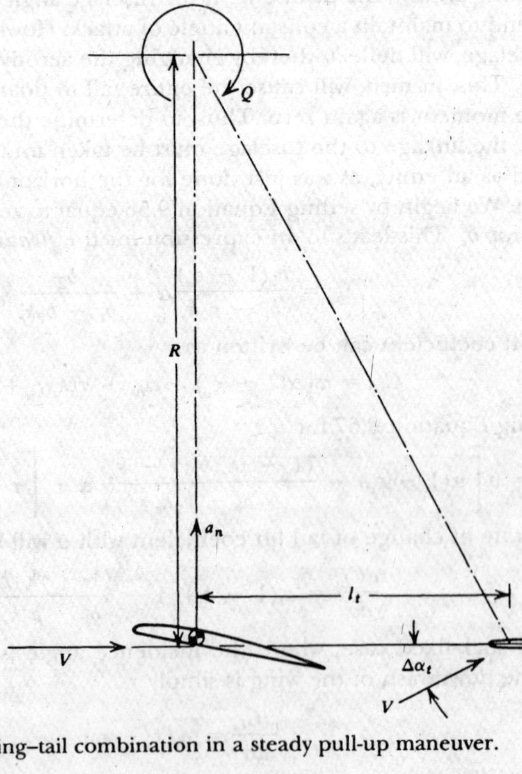

Figure 9.22 Wing–tail combination in a steady pull-up maneuver.

The relationships derived thus far that relate C_M to C_L (or α) hold for the maneuvering case. However, an additional increment must be added to C_M as a result of the increment in α_t due to the angular velocity, Q.

$$\Delta M = -\tfrac{1}{2}\rho V^2 \eta_t S_t a_t \,\Delta\alpha_t l_t$$

But

$$\Delta\alpha_t = \frac{l_t}{R} = \frac{Ql_t}{V} \tag{9.73}$$

Thus, in coefficient form,

$$\Delta C_M = \frac{\Delta M}{\tfrac{1}{2}\rho V^2 S\bar{c}}$$

or

$$\Delta C_M = -\eta_t a_t V_H \frac{l_t Q}{V} \tag{9.74}$$

Notice that we can also visualize the airplane as translating with a linear velocity of V while rotating about its center of gravity with an angular velocity of Q. The tail is then moving down with a velocity of $l_t Q$, which leads to Equation 9.73.

Horizontal Stabilizer-Elevator Configuration: Elevator Angle per g

Including the effect of Q, we can now write Equation 9.33*a* as

$$C_M = C_{M0} + C_{M_\alpha}\alpha + C_{M\delta}\,\delta_e + C_{MQ}Q \tag{9.75}$$

However, neither Q nor C_{MQ} is dimensionless (although their product is). Thus, it is sometimes desirable to write Equation 9.75 as

$$C_M = C_{M0} + C_{M_\alpha}\alpha + C_{M\delta}\,\delta_e + C_{Mq}\bar{q} \tag{9.76}$$

where $\bar{q}$ is a dimensionless pitch rate defined by

$$\bar{q} = \frac{Q\bar{c}}{2V} \tag{9.77}$$

The choice of $\bar{c}/2$ as a reference length is simply according to convention and has no physical significance. Note that the stability derivation C_{M_q} is dimensionless, which means that the derivative is taken with respect to the dimensionless pitch rate, $\bar{q}$.

From Equation 9.74, C_{M_q} is seen to equal approximately

$$C_{M_q} = -\eta_t a_t V_H \frac{2l_t}{\bar{c}} \tag{9.78}$$

C_{M_q} is referred to as "pitch damping," since it represents a negative moment proportional to the rotational velocity about the pitch axis.

In a steady pull-up maneuver, there is no angular acceleration, so C_M must be zero as it is in trimmed level flight. Assuming C_{M_0} to remain constant, we can therefore equate the increment C_M resulting from the steady maneuver to zero. Since C_M is assumed linear with α, δ, and $\bar{q}$, we can write

$$0 = C_{M\alpha}\,\Delta\alpha + C_{M\delta}\,\Delta\delta_e + C_{Mq}\bar{q}$$

or

$$\Delta\,\delta_e = -\left(\frac{C_{M\alpha}\,\Delta\alpha + C_{Mq}\bar{q}}{C_{M\delta}}\right) \tag{9.79}$$

Remember that $\bar{q}$ now represents the dimensionless pitch rate and *not* the dynamic pressure. Since

$$V = QR$$

and

$$a_n = \frac{V^2}{R}$$

it follows from Equation 9.77 that

$$\bar{q} = \frac{g\bar{c}}{2V^2}\left(\frac{a_n}{g}\right) \tag{9.80}$$

The quantity (a_n/g) is the acceleration expressed as the ratio to the acceleration due to gravity. When, for example, we speak about "pulling 4g's," we mean that $a_n/g = 4.0$.

$\Delta\alpha$ can be obtained from Equations 9.71 and 9.72.

$$\frac{L}{W} = 1 + \frac{a_n}{g}$$

or

$$\frac{C_L + \Delta C_L}{C_L} = 1 + \frac{a_n}{g}$$

Thus,

$$\begin{aligned}\Delta C_L &= C_L \frac{a_n}{g} \\ &= C_{L\alpha}\,\Delta\alpha + C_{L\delta}\,\Delta\delta_e\end{aligned}$$

so that

$$\Delta\alpha = \frac{C_L}{C_{L\alpha}}\frac{a_n}{g} - \frac{C_{L\delta}}{C_{L\alpha}}\,\Delta\,\delta_e \tag{9.81}$$

where C_L is calculated for 1 g flight. Combining Equations 9.79, 9.80, and 9.81 results in an expression for the elevator angle per g of acceleration.

$$\frac{\Delta\,\delta_e}{(a_n/g)} = -\frac{C_{M_\alpha}C_L + (g\bar{c}/2\,V^2)\,C_{M_q}C_{L\alpha}}{C_{M\delta}C_{L\alpha} - C_{M_\alpha}C_{L\delta}} \tag{9.82}$$

Substituting Equation 9.25 for $C_{M\alpha}$, Equation 9.82 becomes

$$\frac{\Delta\,\delta}{(a_n/g)} = -\frac{C_{L\alpha}C_L}{C_{M\delta}C_{L\alpha} - C_{M\alpha}\,C_{L\delta}}\left(h - h_n + \frac{\rho\bar{c}\,S}{4\,m}\,C_{M_q}\right)$$

where m is the mass of the airplane. We now define a relative mass parameter, μ.

$$\mu = \frac{2\,m}{\rho\bar{c}\,S} \tag{9.83}$$

$\bar{c}/2$ is the same reference length used in the definition of $\bar{q}$ (Equation 9.77). This length multiplied by the wing area gives a reference volume. The parameter μ is then the ratio of the airplane mass to the air mass within this volume. Using the definition of μ, Equation 9.82 becomes

$$\frac{\Delta\,\delta_e}{(a_n/g)} = \frac{C_{L\alpha}C_L}{C_{M\delta}C_{L\alpha} - C_{M\alpha}\,C_{L\delta}}\left(h_n - h - \frac{1}{2\mu}\,C_{M\mu}\right) \tag{9.84}$$

Similar to the neutral point, there is a particular value of h, known as the stick-fixed maneuver point, h_m, for which no elevator deflection is required to produce an acceleration. Indeed, if the center of gravity moves aft of this point, the effect of the elevator is reversed. From Equation 9.84,

$$h_m = h_n - \frac{1}{2\mu} C_{M_q} \tag{9.85}$$

Since C_{M_q} is negative, h_m lies aft of the neutral point.

In terms of the neutral point,

$$\frac{\Delta \delta_e}{(a_n/g)} = \frac{C_{L\alpha} C_L}{C_{M\delta} C_{L\alpha} - C_{M\alpha} C_{L\delta}} (h_m - h) \tag{9.86}$$

The quantity $(h_m - h)$ is known as the stick-fixed maneuver margin. Again, it is emphasized that C_L is calculated for straight and level flight. It is not the total airplane C_L required during the steady pull-up maneuver.

Stabilator Angle per g

The preceding developments will apply to the stabilator configuration if $\Delta \delta_e$ is replaced by Δi_{hs}, $C_{M\delta}$ by C_{M_i}, and $C_{L\delta}$ by C_{L_i}. A separate analysis of the stabilator with regard to a steady maneuver is therefore unnecessary.

Stick Force per g

Stabilizer-Elevator Configuration In order to calculate the stick force required per g of normal acceleration, we again use Equations 9.45 and 9.50.

$$C_H = b_1 \alpha_t + b_2 \delta_e + b_3 \delta_t$$

But, in the steady pull-up,

$$\alpha_t = \alpha(1 - \epsilon_\alpha) - i_{hs} + 2\frac{l_t}{\bar{c}}\bar{q} \tag{9.87}$$

Thus,

$$\eta\tau \frac{P}{GqS_e\bar{c}_e} = b_1\left[\alpha(1 - \epsilon_\alpha) - i_{hs}\, 2\frac{l_t}{\bar{c}}\bar{q}\right] + b_2 \delta_e + b_3 \delta_t$$

Now let δ_t be adjusted to trim P to zero for unaccelerated flight. For this condition let $\alpha = \alpha_0$ and $\delta_e = \delta_{e0}$, such that $\alpha = \alpha_0 + \Delta\alpha$, $\delta_e = \delta_{e0} + \delta_e$. It then follows that

$$\frac{P}{G\eta_t qS_e\bar{c}_e} = b_1\left[\Delta\alpha(1 - \epsilon_\alpha) + 2\frac{l_t}{\bar{c}}\bar{q}\right] + b_2 \Delta \delta_e$$

$\Delta\alpha$ is obtained from Equation 9.81 and $\Delta \delta_e$ from Equation 9.86. Also, from Equation 9.80

$$2\frac{l_t}{\bar{q}} = g\frac{l_t}{V^2}\left(\frac{a_n}{g}\right)$$

Thus, the stick force per g becomes

$$\frac{P}{a_n/g} = G\eta_t S_e\bar{c}_e\left(\frac{W}{S}\right)\left[b_1\left(\frac{(1 - \epsilon_\alpha)}{C_{L\alpha}} + \frac{1}{\mu}\frac{l_t}{\bar{c}}\right) + \left(\frac{b_2 C_{L\alpha} - b_1 C_{L\delta}(1 - \epsilon_\alpha)}{C_{M\delta}C_{L\alpha} - C_{M\alpha}C_{L\delta}}\right)(h_m - h)\right] \tag{9.88}$$

The position of h for which P equals zero is known as the stick-free maneuvering point and is denoted by h'_m. In terms of h'_m, the stick force per g becomes

$$\frac{P}{a_n/g} = G\eta_t S_e\bar{c}_e\left(\frac{W}{S}\right)\left[\frac{b_2 C_{L\alpha} - b_1 C_{L\delta}(1 - \epsilon_\alpha)}{C_{M\delta}C_{L\alpha} - C_{M\alpha}C_{L\delta}}\right](h'_m - h) \tag{9.89}$$

where

$$h'_m = h_m + b_1\left[\frac{(1-\epsilon_\alpha)}{C_{L\alpha}} + \frac{1}{\mu}\frac{l_t}{\bar{c}}\right]\left(\frac{C_{M\delta}C_{L\alpha} - C_{M\alpha}C_{L\delta}}{b_2C_{L\alpha} - b_1C_{L\delta}(1-\epsilon_\alpha)}\right) \tag{9.90}$$

Stabilator The stick force per g for the stabilator is obtained in a manner similar to that which was followed for the stabilizer-elevator configuration. The moment coefficient about the pivot line is once again given by Equation 9.56. Also, in the steady pull-up, Equation 9.87 again applies and δ is related to i_{hs} by Equation 9.42. Thus, for the stabilator,

$$\frac{P}{\eta_t GqS_t\bar{c}_t} = b_1\left[\alpha(1-\epsilon_\alpha) - i_{hs} + 2\frac{l_t}{\bar{c}}\bar{q}\right] + b_2(k_e i_{hs} + \delta_0)$$

Now let δ_0 be adjusted so that $P = 0$ for $\bar{q} = 0$. For this trim condition we again let $\alpha = \alpha_0$ and $i_{hs} = i_{ho}$, so that, generally, $\alpha = \alpha_0 + \Delta\alpha$ and $i_{hs} = i_{ho} + \Delta i_{hs}$. Thus,

$$\frac{P}{G\eta_t qS_t\bar{c}_t} = b_1\left[\Delta\alpha(1-\epsilon_\alpha) - \Delta i_{hs} + g\frac{l_t}{V^2}\left(\frac{a_n}{g}\right)\right] + b_2k_e\Delta i_{hs}$$

or

$$\frac{P}{a_n/g} = G\eta_t S_t\bar{c}_t\left(\frac{W}{S}\right)\left[b_1\left(\frac{1-\epsilon_\alpha}{C_{L\alpha}} + \frac{1}{\mu}\frac{l_t}{\bar{c}}\right) + \left(\frac{b_2C_{L\alpha}k_e - b_1C_{L_i}(1-\epsilon_\alpha)}{C_{M_i}C_{L\alpha} - C_{M\alpha}C_{L_i}}\right)(h_m - h)\right] \tag{9.91}$$

The similarity of this relationship to the corresponding equation for the stabilizer-elevator tail configuration is obvious. Again, we can express the stick force per g in terms of the stick-free maneuvering point.

$$\frac{P}{a_n/g} = G\eta_t S_t\bar{c}_t\left(\frac{W}{S}\right)\left[\left(\frac{b_2C_{L\alpha}k_e - b_1C_{L_i}(1-\epsilon_\alpha)}{C_{M_i}C_{L\alpha} - C_{M\alpha}C_{L_i}}\right)(h'_m - h)\right] \tag{9.92}$$

where

$$h'_m = h_m + b_1\left(\frac{1-\epsilon_\alpha}{C_{L\alpha}} + \frac{1}{\mu}\frac{l_t}{\bar{c}}\right)\left[\frac{C_{M_i}C_{L\alpha} - C_{M\alpha}C_{L_i}}{b_2C_{L\alpha}k_e - b_1C_{L_i}(1-\epsilon_\alpha)}\right] \tag{9.93}$$

An airplane loaded so that the center of gravity is close to the stick-free maneuvering point presents a dangerous situation. The pilot can impose extreme inertia loads on the airplane with the application of little or no control force. Such a situation, however, is rarely encountered if an adequate static margin is maintained, since the stick-fixed and stick-free maneuver points are aft of the corresponding neutral points.

Effect of Fuselage and Nacelles

A body, such as a fuselage or nacelle, is not a very efficient producer of lift by comparison to a wing or tail surface. Thus, except for missilelike configurations, where the body is relatively large, the contribution of the fuselage or nacelles to the total airplane lift can be neglected. However, in the case of the moment, fuselage and nacelle contributions can be fairly significant and result in a measurable shift of the neutral point. Generally, the increments to $C_{M\alpha}$ are positive, resulting in a decrease in h_n. Depending on the incidence of the wing relative to the fuselage and nacelles, these components can also contribute to C_{M0}. Thus, including the fuselage and/or nacelles, Equation 9.20 becomes

$$C_M = C_{M0} + \Delta C_{M0} + (C_{M_e} + \Delta C_{M_e})\alpha \tag{9.94}$$

The increments ΔC_{M0} and $\Delta C_{M\alpha}$ result from the addition of the fuselage and/or nacelles to the basic wing-tail combination. Thus, the relationships derived thus far are unchanged if the total values for C_{M0} and $\Delta C_{M\alpha}$ are used. An increment to $C_{L\alpha}$ can also be included.

It is somewhat futile to present either data or theoretical results for predicting ΔC_{M0} and $\Delta C_{M\alpha}$ in view of the many possible fuselage-nacelle-wingtail combinations. Real fluid effects and lifting surface-body interference effects severely limit any application of theoretical solutions of the body alone.

For approximate purposes, the lift coefficient of a fuselage or nacelle can be estimated by curve-fitting existing data. This type of effort can be found in Reference 5.6. The best solution, of course, is to conduct reliable wind tunnel tests of models. For the purpose of this text, based on Reference 5.6, the rate of change of a body lift coefficient with angle of attack will be calculated from

$$\frac{dC_{L_b}}{d\alpha} = 2\left[1 - 1.76\left(\frac{l_b}{d_b}\right)^{-3/2}\right] /\text{rad} \tag{9.95}$$

This body lift coefficient, C_{L_b}, is based on the maximum cross-sectional area of the body, S_b. The length of the body is denoted by l_b, whereas d_b is the diameter of a circle having the same area as the body's maximum cross-sectional area, S_b. For purposes of calculating the moment, the body lift can be assumed to act approximately at a point midway between the nose and the station where the body's cross-sectional area is a maximum. This is shown in Figure 9.23 with a plot of Equation 9.95. Again it is emphasized that there is no accepted way to calculate the lift and moment of a fuselage, and the above is strictly a "guesstimate," which is better than a guess, but does not replace a good wind tunnel test result.

Just as the contribution of the tail involves the ratio of S_t/S, the fuselage lift coefficient obtained from Equation 9.95 must be multiplied by S_b/S when adding it to the total lift coefficient of the airplane. The fuselage contribution to the airplane moment coefficient is calculated by multiplying C_{L_b} by S_b/S and by the distance of the cg behind the center of pressure of the fuselage divided by $\bar{c}$.

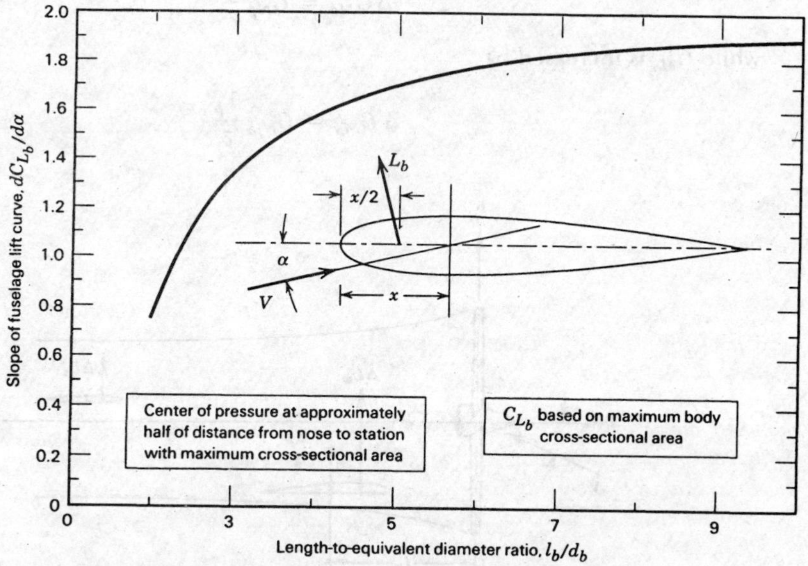

Figure 9.23 Approximate lift curve slope for fuselages and nacelles.

Computer Exercise 9.4 "STKFRC"

Formulate a program that determines the stick-fixed and stick-free neutral points and the stick force as a function of airspeed for a given trim airspeed. The program can be limited to the stabilizer–elevator configuration. The input should consist of the wing and horizontal geometry, ϵ_α, hinge moment coefficients b_1 b_2, A b_3, and gearing, G. Also input $C_{M\alpha}$ and $C_{L\alpha}$ for the fuselage. Output should include the neutral points, stick force, elevator trim angles, and the following stability and control derivatives: $C_{M\alpha}$, $C_{L\alpha}$, $C_{L\delta_e}$, and $C_{M\delta_e}$. Check results against the numbers given in the example for the executive jet, which resembles a small Boeing 767.

Effects of Propulsion System

Propellers

Both propellers and jets can affect longitudinal static stability. Consider first the case of a propeller-driven airplane, shown schematically in Figure 9.24. As illustrated in the figure, a propeller develops not only a thrust, T, directed along its axis of rotation, but also a force, P_N, directed normal to its axis. It will be shown that P is proportional to the propeller thrust, α, and the advance ratio, J, and is directed upward for a positive angle of attack.

Since the thrust is directed along the propeller axis and rotates with the rest of the airplane, its contribution to the moment about the center of gravity does not change with α. Hence $C_{M\alpha}$ is not affected by T. On the other hand, C_{M0} is affected by T since, generally, the thrust line will not pass through the center of gravity. If the thrust line lies a distance of Z_p above the center of gravity, then C_{M0}, given by Equation 8.5, is decreased by the amount

$$\Delta C_{M0} = -\left(\frac{T}{qS}\right)\left(\frac{Z_p}{\bar{c}}\right) \tag{9.96}$$

The normal force, P_N, affects both the trim, C_{M0}, and the stability, $C_{M\alpha}$. C_{M0} is increased by the amount

$$\Delta C_{M0} = C_{Np}\frac{l_p}{\bar{c}} \tag{9.97}$$

while $C_{M\alpha}$ is increased by

$$\Delta C_{M\alpha} = C_{Np\alpha}\frac{l_p}{\bar{c}} \tag{9.98}$$

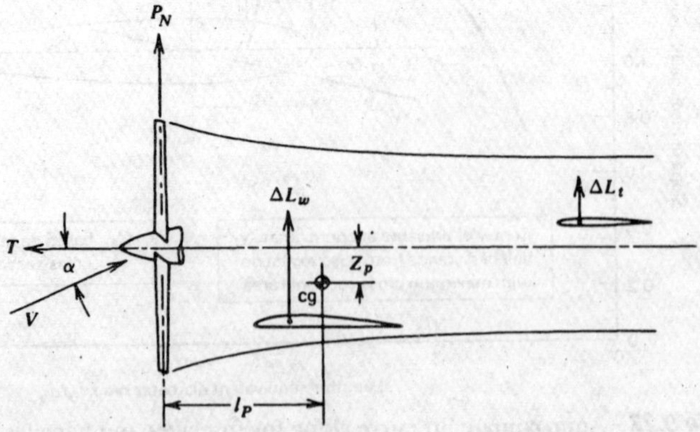

Figure 9.24 Effect of propeller forces and slipstream on longitudinal static stability.

where

$$C_{N_p} = \frac{P_N}{qS}$$

$$C_{N_{p\alpha}} = \frac{\partial C_{N_p}}{\partial \alpha}$$

In addition to contributing to the forces and moments affecting longitudinal motion, a propeller at an angle of attack produces a yawing moment that couples with the lateral-directional behavior of the airplane. Let us now examine the origin of both the normal force and yawing moment. Figure 9.25*a* shows a side view of a propeller at an angle of attack α. It is seen that a component of the free-stream velocity, $V\alpha$, lies in the plane of the propeller and is directed upward. Figure 9.25*b* is a view in the plane of rotation looking in the direction of flight. Measuring the displacement of a blade from the vertical bottom line, it is seen that a blade section experiences a velocity normal to the blade given by the sum of ωr and $\alpha V \sin \psi$. In Figure 9.25*c* we view a section of the blade looking in toward the hub. The section lift and drag combine vectorially to produce a thrust per unit blade length of

$$\frac{dT}{dr} = \frac{dL}{dr}\cos\phi - \frac{dD}{dr}\sin\phi$$

Similarly, a differential force, dF/dr, lying in the plane of rotation is generated given by

$$\frac{dF}{dr} = \frac{dL}{dr}\sin\phi + \frac{dD}{dr}\cos\phi$$

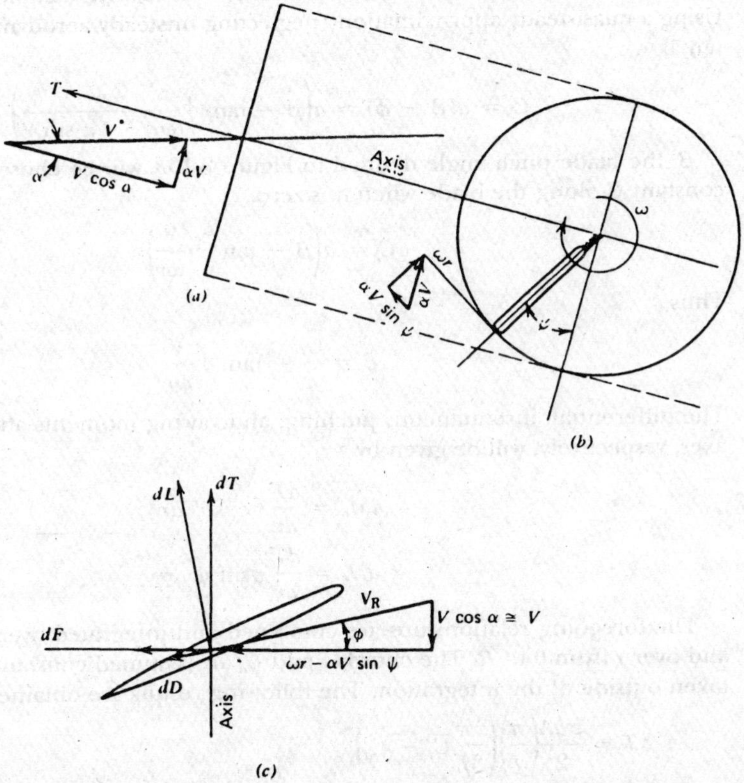

Figure 9.25 Forces on a propeller at an angle of attack.

dF/dr will have components of dP_N/dr directed upward and dY_p/dr directed to the right and given by

$$\frac{dP_N}{dr} = -\frac{dF}{dr}\sin\psi$$

$$\frac{dY_p}{dr} = \frac{dF}{dr}\cos\psi$$

Since all three of these force components vary with the blade position, ψ, average values are obtained by integrating each with respect to ψ from 0 to 2π and then dividing by 2π. For example,

$$T = \frac{B}{2\pi}\int_0^{2\pi}\int_0^R \frac{dT}{dr}\,dr\,d\psi \tag{9.99}$$

where B is the number of blades and R is the propeller radius.

In order to cast the equations in a tractable form, it is assumed that the angle ϕ is small so that

$$V_R \simeq \omega r - V\alpha \sin\psi$$

With this assumption dL/dr and dD/dr become

$$\frac{dL}{dr} = \tfrac{1}{2}\rho(\omega r - V\alpha\sin\psi)^2 cC_l \tag{9.100}$$

$$\frac{dD}{dr} = \tfrac{1}{2}\rho(\omega r - V\alpha\sin\psi)\, cC_d \tag{9.101}$$

C_l, as well as the local velocity, varies with ψ, whereas C_d is assumed to be constant. Using a quasisteady approximation (neglecting unsteady aerodynamics), C_l is written as

$$C_l = a(\beta - \phi) \simeq a\left(\beta - \tan^{-1}\frac{V}{\omega r - V\alpha\sin\psi}\right)$$

β, the blade pitch angle defined in Figure 6.10*b*, will be chosen so as to give a constant C_l along the blade when α is zero.

$$C_l = a\left(\beta - \tan^{-1}\frac{V}{\omega r}\right) \tag{9.102}$$

Thus,

$$\beta = \frac{\overline{C}_l}{a} + \tan^{-1}\frac{V}{\omega r}$$

The differential, instantaneous pitching, and yawing moments about the y- and z-axes, respectively, will be given by

$$dM_p = \frac{dT}{dr}\, r\cos\psi\, dr$$

$$dN_p = \frac{dT}{dr}\, r\sin\psi\, dr$$

The foregoing relationships are combined and integrated over ψ from 0 to 2π and over r from 0 to R. The chords c and C_d are assumed constant, so they can be taken outside of the integration. The following results are obtained:

$$T = \frac{\sigma q A}{2}\left(\frac{\pi}{J}\right)\left(\frac{2\pi}{3J}\overline{C}_l - C_d\right) \tag{9.103}$$

$$P_N = \frac{\sigma q A}{2}\left\{\overline{C}_l + \frac{aJ}{2\pi}\ln\left[1 + \left(\frac{\pi}{J}\right)^2\right] + \frac{\pi}{J}C_d\right\}\alpha \tag{9.104}$$

$$N_p = \frac{-\sigma q A R}{2}\left\{\frac{2\pi}{3J}\overline{C}_l + \frac{a}{2}\left[1 - \left(\frac{J}{\pi}\right)^2 \ln\left(1 + \left[\frac{\pi}{J}\right]^2\right)\right] - \frac{\pi}{J}C_d\right\}\alpha \tag{9.105}$$

$$M_p = 0$$
$$Y_p = 0$$
$$\sigma = \text{blade area/disc area}$$

It is convenient to express P_N and N_p in terms of T and the product TR, respectively. Since the same assumptions are involved in calculating T, P_N, and N_p, the ratios of P_N and N_p to T are probably more accurate than the predicted absolute values of these quantities. Also, (σqA) cancels out, so that one needs only to estimate the average lift coefficient of the blades in order to obtain the force and moment ratios.

In steady, level flight, the thrust is equal to the airplane drag. Thus, knowing the drag, propeller geometry, advance ratio, and q, one can calculate the $\overline{C}_l$ from Equation 9.103. The result is relatively insensitive to C_d, so that a typical value of 0.01 can be assumed for this quantity.

Figures 9.26 and 9.27 present the derivatives of P_N and N_p with respect to α in ratio form as a function of advance ratio for constant values of $\overline{C}_l$. If the propeller geometry is not known, a reasonable value of $\overline{C}_l$ would probably be around 0.8.

The direction of P_N is independent of the direction of the propeller rotation. However, the sign of the moment, N_p, is reversed if the propeller rotates opposite to that shown in Figure 9.25. As derived, all rotations and moments follow the right-handed coordinate system of Figure 1.10.

The method presented here is original with this text. However, preliminary tests show that the method is fairly accurate.

An alternate method for predicting P_N and N_p is offered in Reference 9.12. Graphs based on this reference can be found in other sources, but they seem to be specific to particular propellers, and their general applicability appears questionable. The conclusions of Reference 9.12 may be applied, however, without these

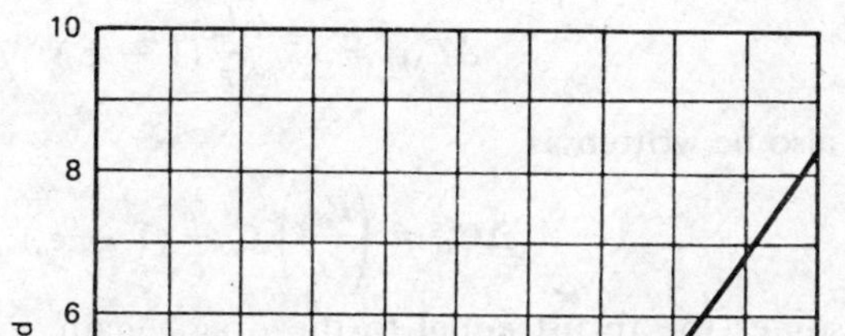

Figure 9.26 Ratio of propeller normal force derivative to thrust.

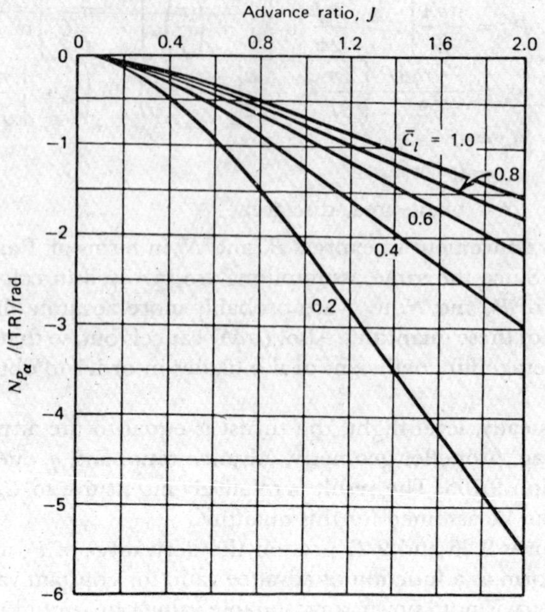

Figure 9.27 Ratio of propeller moment derivative to product of thrust and radius.

graphs. They state that a propeller behaves like a fin having the same area as the average projected side area of the propeller with an effective aspect ratio of approximately 8. The moment produced by the propeller is approximately equal to the lift on the fin multiplied by the propeller radius. "Average" side area means an average area projected in one revolution. This is given approximately by one-half the number of blades times the projected side area of one stationary blade.

In calculating the propeller normal force, one must account for the fact that there is an upwash ahead of the wing that effectively increases the angle of attack. Thus, Equation 9.98 becomes

$$\Delta C_{M\alpha} = C_{N_{p\alpha}} \frac{l_p}{\bar{c}} (1 - \epsilon_\alpha) \tag{9.106}$$

This can also be written as

$$\Delta C_M = \left(\frac{P_{N\alpha}}{T}\right) C_D \frac{l_p}{\bar{c}} (1 - \epsilon_\alpha) \tag{9.107}$$

which assumes the thrust equal to the drag. Again, ϵ_α is the rate of change of *downwash* with α. Ahead of the wing, ϵ_α is negative so that $(1 - \epsilon_\alpha)$ is greater than unity. This factor is presented in Figure 9.28 using the vortex system shown in Figure 9.6. Again, in the final analysis, one should resort to wind tunnel testing of a powered model to determine these effects accurately.

The effect of the propeller slipstream on the wing and tail will not be treated here in any detail. At higher speeds, these effects are small and can usually be neglected. At the lower speeds, however, the increased q in the propeller slipstream can increase both the wing and tail lift. In addition, particularly for V/STOL (vertical/short takeoff and landing) applications, the propeller-induced velocity can become appreciable relative to the free-stream velocity, resulting in significant changes in the section angles of attack of both the wing and tail.

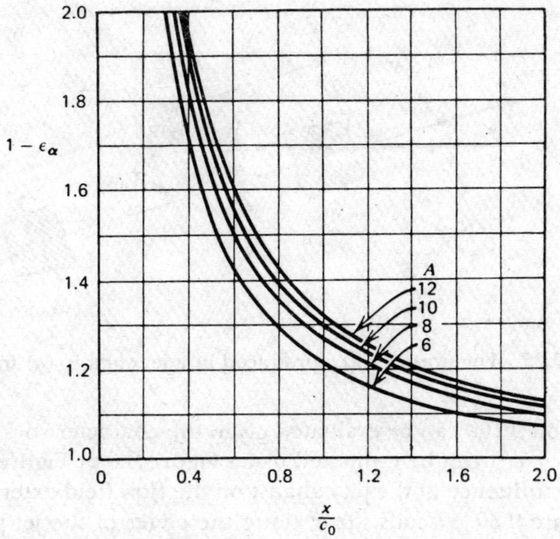

Figure 9.28 Correction for upwash ahead of a wing. Model of Figure 9.6 assumed using an unswept elliptic wing. c_o = root chord, x = distance ahead of quarter-chord line.

Jets

The two principal effects of a jet propulsion system entail the forces developed internally on the jet ducts and the influence of the jet exhaust on the flow field of the tail. The following is based on material in Reference 9.2.

Figure 9.29 depicts schematically the flow through a jet propulsion system. The notation is similar to that of the reference. First, it is obvious that an increment to the trim moment results from the thrust, T, which is independent of α. This increment can be written as

$$\Delta C_M = \left(\frac{T}{qS}\right)\frac{Z_j}{\bar{c}} \tag{9.108}$$

Second, as in the case of the propeller, a normal force is developed by the jet. This force, N_j, is directed normal to the exhaust velocity V_j and acts at the intersection of lines parallel to V_i and V_j through the centers of the inlet and exhaust. This is usually close to the inlet. Applying momentum principles in the system, the magnitude of N_j is found to be

$$N_j = m_j V_i \theta$$

where m_j is the mass flow rate through the engine. The angle θ equals the angle of attack of the thrust line α minus the downwash angle ϵ_i at the inlet. If the inlet is ahead of the wing, ϵ_i will generally be negative.

$$\theta = \alpha - \epsilon_i$$

From this, it follows that the normal force, N_j, contributes an increment to $C_{M\alpha}$, given by

$$\Delta C_{M\alpha} = \left(\frac{m_j V_i}{qS}\right)\frac{X_j}{c}(1 - \epsilon_{i\alpha} \tag{9.109}$$

Since

$$m_j = \rho_i A_i V_i$$

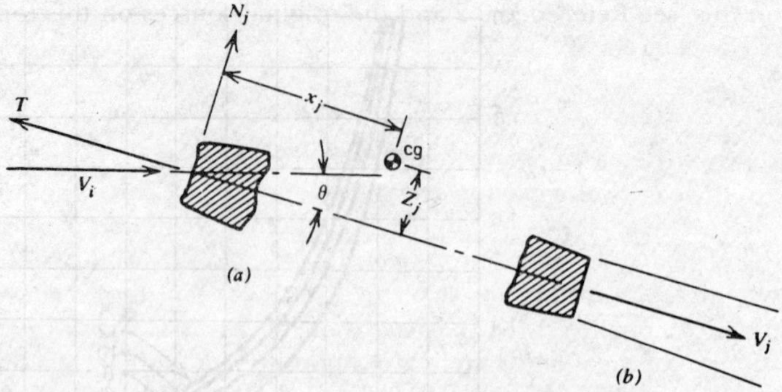

Figure 9.29 The normal force produced by a jet engine. (*a*) Intake. (*b*) Tailpipe.

Equation 9.109 can be evaluated given the characteristics of the engine. The quantity $(1 - \epsilon_{i_\alpha})$ can be estimated using Figure 9.6 or Figure 9.7.

The influence of the jet exhaust on the flow field external to the jet is illustrated in Figure 9.30. Viscous shear along the edges of the jet produces an entrainment of the external flow into the jet. To the external flow, the jet appears somewhat like a distributed line of sinks. As shown in Figure 9.30, this jet-induced flow can produce a change in the local flow direction for a lifting surface in proximity to the jet.

Again, the calculation of flow inclination and the effect on a lifting surface is a difficult task and is best determined from wind tunnel tests. The effect can be appreciable and generally should not be neglected. According to Reference 9.2, it may vary with α enough to reduce the stability significantly.

From material to be found in the appendix of Reference 9.2, an approximate estimate of the inclination angle due to jet entrainment can be calculated from

$$\epsilon = \frac{C_D}{n\left(\dfrac{x}{b}\right)\left(\dfrac{r}{b}\right)A} \tag{9.110}$$

where

C_D = airplane drag coefficient
n = number of engines
x = distance aft of exhaust exit plus 2.3 diameters of the jet at its exit
r = radial distance from jet centerline
b = wingspan
A = aspect ratio

Although α does not appear in Equation 9.110, the inclination of the jet itself varies with α, which causes the distance, r, to change. For more detailed information

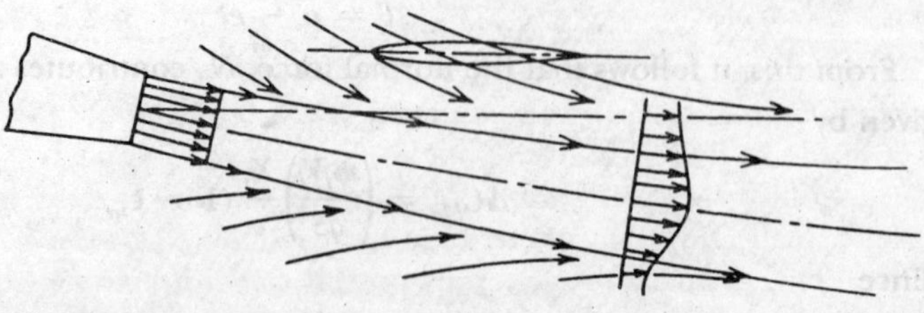

Figure 9.30 Flow inclination resulting from flow entrainment.

on this, see Reference 9.2 and the original sources on this topic, which are noted in the reference.

Ground Effect

Downwash velocities are decreased when the airplane is operating close to the ground. Since the ground is a solid boundary to which the normal component of velocity must vanish, its presence alters the streamline pattern that exists around the airplane out-of-ground effect. In order to determine ground effect, an image system representing the airplane's vortex system is placed below the airplane a distance equal to twice the height of the airplane above the ground. The vortex strength of the image system is of opposite sign to the original vortex system. Thus, midway between the two systems, their induced velocities in the vertical direction will cancel, satisfying the boundary condition along the ground.

An additional graph to estimate the effect of the ground on downwash is not needed in light of this information. Instead, one can again make use of Figures 9.7 and 9.8. For example, suppose an airplane with no sweep is operating a height above the ground, h, equal to one-quarter of its span. Also, let the distance $l_{t_{ac}}$ equal $b/2$ and $h_t/b = 0.05$. Relative to the image system, the tail is located at an $\dfrac{h}{b/2}$ equal to 1.05. Thus, from Figure 9.7

$$\frac{A\epsilon_\alpha}{C_{L\alpha}} = 0.40$$

Out-of-ground effect, the above value from Figure 9.7, is equal to 0.61. The velocity induced by the image vortices subtract from that induced OGE. Thus, for this IGE condition, ϵ_α is reduced to 34% of its OGE value.

LATERAL–DIRECTIONAL STATIC STABILITY AND CONTROL

We now turn our attention to the static forces and moments that tend to rotate the airplane about its x- and z-axes or translate it along its y-axis. First consider directional stability and control about the z-axis.

Directional Static Stability

Figure 9.31 illustrates an airplane undergoing a positive sideslip. In this case, the component of its velocity vector along the y-axis is nonzero and positive. The airplane is "slipping" to the right. This results in a positive sideslip angle, β, being defined as shown. As a result of the velocity vector no longer lying in the plane of symmetry, a yawing moment, N, is produced by the fuselage and by the side force on the vertical tail. The airplane will possess positive static directional stability (sometimes called weathercock stability, after the weathervane) if

$$\frac{\partial N}{\partial \beta} = N_\beta > 0 \tag{9.111}$$

The criterion for longitudinal static stability required the slope of the moment curve to be negative. In this case, the sign is opposite because the yawing moment is opposite in direction to the angle β.

Generally, the yawing moment from the fuselage is destabilizing. However, it is usually small and easily overridden by the stabilizing moment contributed by the vertical tail, so airplanes generally possess static directional stability. However, the vertical tail is not normally sized by any considerations of static directional stability.

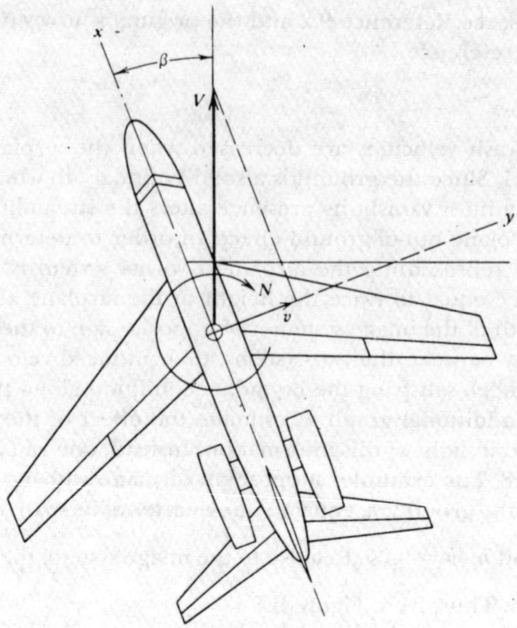

Figure 9.31 An airplane having a positive sideslip.

Instead, the minimum tail size is determined by controllability requirements in the event of an asymmetric engine failure or flying qualities requirements related to dynamic motion.

The yawing moment is expressed in coefficient form by

$$N = C_N q S b \tag{9.112}$$

The contribution to C_N from the fuselage can be estimated on the basis of Equation 9.95. However, in calculating $C_{N_{\beta H}}$ corresponding to $C_{M_{\alpha B}}$, the difference in signs and reference lengths must be remembered. $C_{y\beta H}$ will also be of opposite sign to $C_{L\alpha B}$.

The contribution of the vertical tail to the static directional stability is formulated in a manner similar to that followed in determining the horizontal tail's effect on longitudinal stability. If the aerodynamic center of the vertical tail is located a distance of l_v behind the center of gravity, then

$$N_v = \eta_t q S_v l_v a_v [\beta(1 - \epsilon_\beta)]$$

or

$$C_{N_v\beta} = \eta_t \frac{S_v}{S}\frac{l_v}{b} a_v (1 - \epsilon_\beta) \tag{9.113}$$

The vertical tail volume, V_v, is defined by

$$V_v = \frac{S_v}{S}\frac{l_v}{b}$$

so that Equation 9.113 becomes

$$C_{N_v\beta} = \eta_t V_v a_v (1 - \epsilon_\beta) \tag{9.114}$$

a_v is the slope of the lift curve for the vertical tail. An estimate of this quantity is made more difficult by the presence of the fuselage and horizontal tail. If the vertical tail is completely above the horizontal tail, then it is recommended that an effective aspect ratio be calculated for the vertical tail equal to the geometric aspect ratio multiplied by a factor of 1.6. If the horizontal tail is mounted across the top of the vertical tail (the so-called T-tail configuration), this factor should be increased to approximately 1.9 to allow for the end-plate effect of the fuselage on the bottom of the vertical tail and the horizontal tail on the top. This latter factor of 1.9 is only typical of what one might expect for an average ratio of fuselage to tail size. For a more precise estimate of this factor, one should resort to wind tunnel tests or Reference 9.3 or 5.6.

The sidewash factor, ϵ_β, is extremely difficult to estimate with any precision. For preliminary estimates, it can be taken to be zero. However, one should be aware of its possible effects on N_v.

Directional Control

Control of the yawing moment about the z-axis is provided by means of the rudder. The rudder is a movable surface that is hinged to a fixed vertical stabilizer. The rudder is the vertical counterpart to the elevator, and its effectiveness is determined in the same way. Referring to Figure 9.32, the vertical tail is shown at zero sideslip angle with the rudder deflected positively through the angle δ_r. The rudder produces an increment in the vertical tail lift (or side force) that results in an increment in the yawing moment, given by

$$\Delta N = -l_v \, \Delta L_v$$

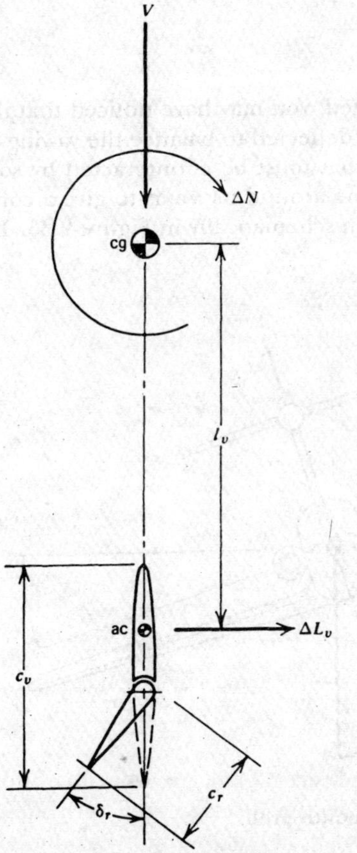

Figure 9.32 Rudder control.

The increment in the tail lift is given by

$$\Delta L_v = \eta_t q S_v a_v \tau\, \delta_r$$

so that in coefficient form, the rate of change of C_N with respect to the rudder angle becomes

$$C_{N\delta} = -\eta_t V_v a_v \tau \qquad (9.115)$$

The subscript r is dropped on δ, since it should be obvious that δ refers to the rudder when considering C_N. τ is the effective change in the angle of the zero lift line of the vertical tail per unit angular rotation of the rudder and is estimated in the same manner in which τ was obtained for the elevator.

As an example in the use of the rudder, suppose the airplane pictured in Figure 9.31 lost power on its right engine. If each engine is located a distance of Y_e from the fuselage centerline, the resulting asymmetric thrust would produce a yawing moment about the center of gravity equal to TY_e. In steady trimmed flight the thrust, T, must equal the drag and N must be zero. Therefore,

$$DY_e + qSbC_{N\delta}\,\delta_r = 0$$

or

$$\delta_r = \frac{C_D Y_e}{\eta_t V_v a_v \tau b}$$

Given the airplane and rudder geometry and C_d as a function of V, one can then calculate the vertical tail volume necessary to keep δ_r within prescribed limits, usually a linear operating range.

Lateral Control

In the engine-out example just presented, you may have noticed that the airplane was really not in trim. With the rudder deflected to balance the yawing moment of the engine, a side force is produced that must be counteracted by some means. This can be done by rolling the airplane around its x-axis to give a component of the weight along its y-axis. This is shown schematically in Figure 9.33. To carry the

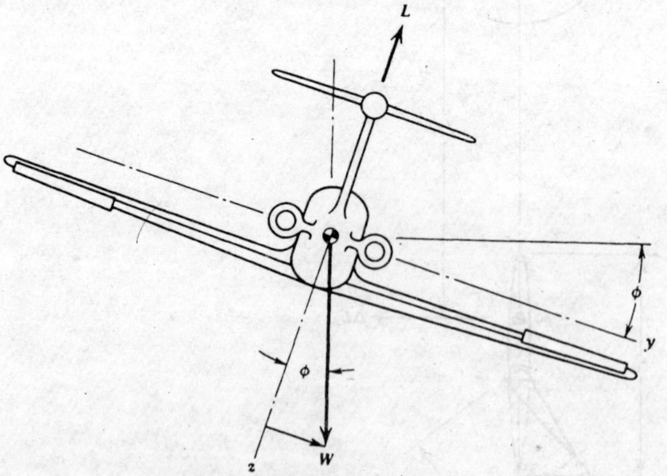

Figure 9.33 Rear view of an airplane with positive roll.

preceding engine-out example further, the sum of the forces along the y-axis must be zero for trim.

Thus,

$$\Delta L_v + W\phi = 0$$

or

$$\phi = \frac{-\eta_t q S_r a_r \tau \delta_r}{W}$$

Substituting for δr gives

$$\phi = -\frac{C_D}{C_L}\frac{Y_e}{l_v} \tag{9.116}$$

Thus, to trim the engine-out situation completely, the pilot must have the capability to roll the airplane. This is accomplished by movable control surfaces that are hinged on the outer rear portion of each side of the wing, as shown in Figure 9.34. These surfaces, known as ailerons, move differentially: as one moves up, the other moves downward. Since the ailerons do not necessarily move the same amount, the aileron deflection is defined as the total angle between the two ailerons. For example, if the right aileron moves down 10° and the left aileron moves up 14°, the aileron deflection, δ_v, is equal to 24°. δ_a is defined to be positive when the right aileron rotates in a positive direction.

It was the incorporation of roll control that distinguished the efforts of the Wright Brothers from those of the other aviation pioneers of their day. Instead of ailerons, however, the Wright Brothers warped their box wing to provide a differential angle of attack from one wing tip to the other. This was modified by Glenn Curtiss, who developed the hinged aileron, which was a considerable improvement over having to warp the wing. Since they had first recognized the need for complete lateral control and had devised the means to do so, the Wright Brothers brought a legal suit against Glenn Curtiss, claiming that his ailerons infringed on their patent rights. After a lengthy and bitter legal fight, the courts ruled in favor of the Wright Brothers. I have only the greatest respect and admiration for the Wright Brothers, but I question this particular legal decision, since it seems to me that Curtiss' development of the separate movable aileron control surface was, indeed, a control system distinct from the wing-warping system developed by the Wrights.

Leaving history and returning again to the technical aspect of roll control, observe that, when an airplane is rolled through an angle, ϕ, there is no mechanism

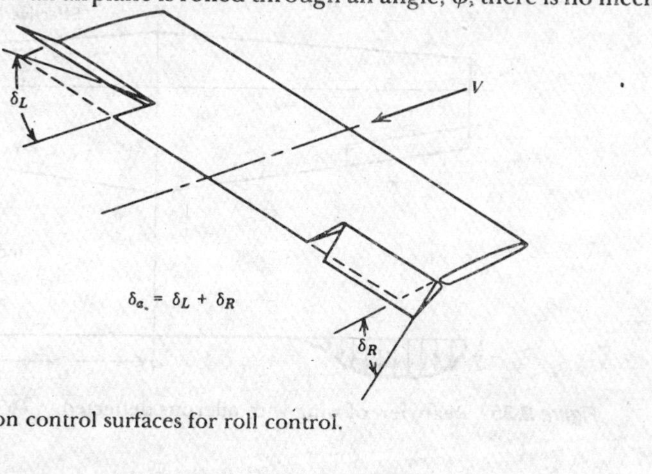

Figure 9.34 Aileron control surfaces for roll control.

present to generate an aerodynamic restoring moment. In the preceding cases involving angle of attack and sideslip angle, we encountered the derivatives $C_{M\alpha}$ and $C_{N\beta}$, respectively, which had nonzero values. In the case of the roll angle, however, the corresponding derivative $C_{l\phi}$ is always zero. Because of this, the concept of roll stability, in the static sense, does not exist. At the most, one could say that an airplane possesses neutral static stability in roll.

We define the rolling moment coefficient by reference to the wing area and wingspan, as with the yawing moment coefficient.

$$L = qSbC_l \tag{9.117}$$

Here we have a real problem with notation, since one tends to think of lift when the letters L or l are used. You will have to be careful and aware of the application of the coefficients in order to make a distinction. Generally, if the lowercase l is being used with reference to the entire airplane, it has reference to the rolling moment.

When the aileron on one side is deflected through an angle and the aileron on the other side is deflected oppositely, the section lift coefficients will change along the entire span of the wing. However, the effect is most pronounced over the span of the ailerons. The rolling moment is best determined by using the lifting surface model written in Chapter 3. In lieu of this, an approximate method can be used, which produces fairly accurate results. The method is based on first predicting the rolling moment without accounting for any induced effects. Then the results are corrected for three-dimensional effects by visualizing the portion of the wing with the aileron as a low-aspect ratio wing.

When deflected, an outboard aileron will produce an incremental lift distribution something like that sketched in Figure 9.35. Thus, on either end of the aileron, an incremental vortex system is shed in a manner similar to that shed at the tips of a wing. This incremental vortex system will induce an additional downwash over the aileron, thereby reducing its incremental lift. It is assumed that this reduction is in proportion to the reduction in lift that one would expect from a finite wing having an aspect ratio equal to the aspect ratio of that portion of the wing spanned by the aileron. If this ratio is denoted by A_a, the correction factor F_a becomes

$$F_a = \frac{A_a}{A_a + [2(A_a + 4)]/(A_a + 2)}$$

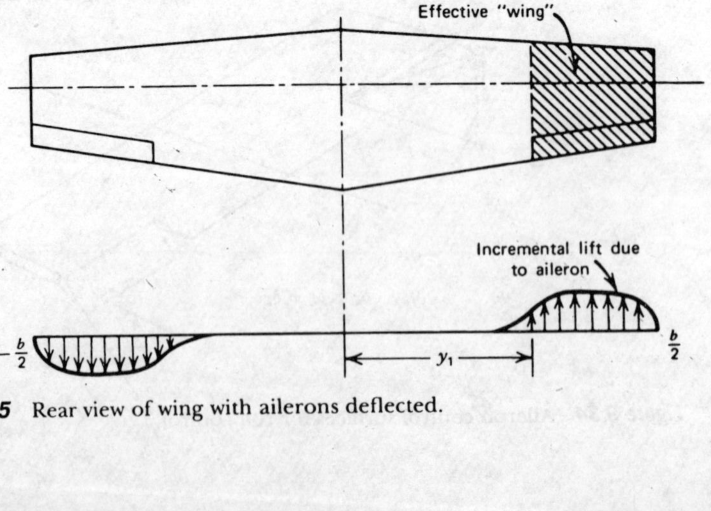

Figure 9.35 Rear view of wing with ailerons deflected.

For the case where $x_2 = 1.0$, A_a, A, x_1, and λ are related by

$$A_a = \left[\frac{(1 - x_1)(1 + \lambda)}{1 - x_1 + \lambda(1 + x_1)}\right]\frac{A}{2}$$

Neglecting induced effects, the rolling moment produced by the ailerons can be estimated by reference to Figure 9.36. As shown, the inboard end of the aileron is located a distance of y_1 from the wing centerline. The aileron extends from there out to a distance of y_2. As a result of the aileron deflection, an increment in the section lift per unit span is produced on the right side and is given by

$$d(\Delta L) = qca_0\tau\,\delta_{aR}\,dy$$

This incremental lift results in a differential rolling moment.

$$dL = -\,yqca_0\tau\,\delta_{aR}\,dy$$

Integrating this over the spanwise extent of the right aileron gives

$$L = -\,qa_0\tau\,\delta_{aR}\int_{y1}^{y2} cy\,dy$$

Since the wing is symmetrical, it is easily shown that a similar contribution to the rolling moment is obtained from the left side, so the total rolling moment is given by

$$L = -\,qa_0\tau(\delta_{aR} + \delta_{aL})\int_{y1}^{y2} cy\,dy$$

$$= -\,qa_0\tau\,\delta_a\int_{y1}^{y2} cy\,dy \tag{9.118}$$

In dimensionless form, C_l becomes

$$C_l = -\tfrac{1}{4}\,a_0\tau\,\delta_a A\int_{x1}^{x2}\left(\frac{c}{b}\right)x\,dx \tag{9.119}$$

where $x = y/(b/2)$.

For a linearly tapered wing this can be integrated to give

$$C_l = -\,a_0\tau\,\delta_a\,\frac{[3({x_2}^2 - {x_1}^2) - 2(1 - \lambda)({x_2}^3 - {x_1}^3)]}{12(1 + \lambda)}F_a \tag{9.120}$$

The induced effects, accounted for by the factor, F_a, are significant and can be confirmed by comparison to the experimental data presented in Reference 9.8.

Figure 9.37 presents $C_{l\delta}$ on the basis of the foregoing relationships. In preparing this figure, it was assumed that

$a_0 \simeq 0.106C_l/\text{deg}$

$\eta \simeq 0.80$ (correction to τ—see Equation 3.54 and Figure 3.36)

$x_2 = 1.0$

$\dfrac{c_a}{c} = 0.25$

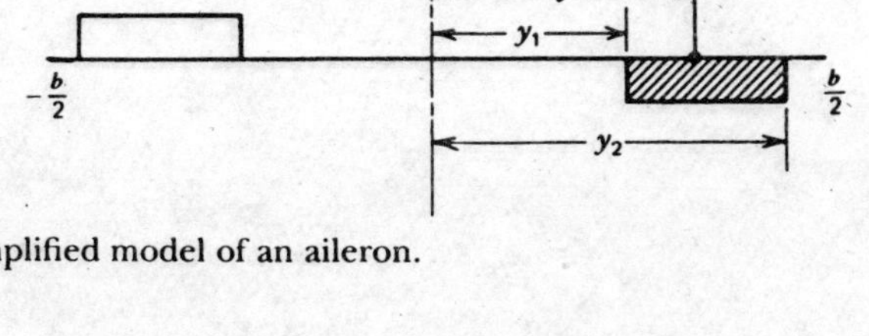

Figure 9.36 Simplified model of an aileron.

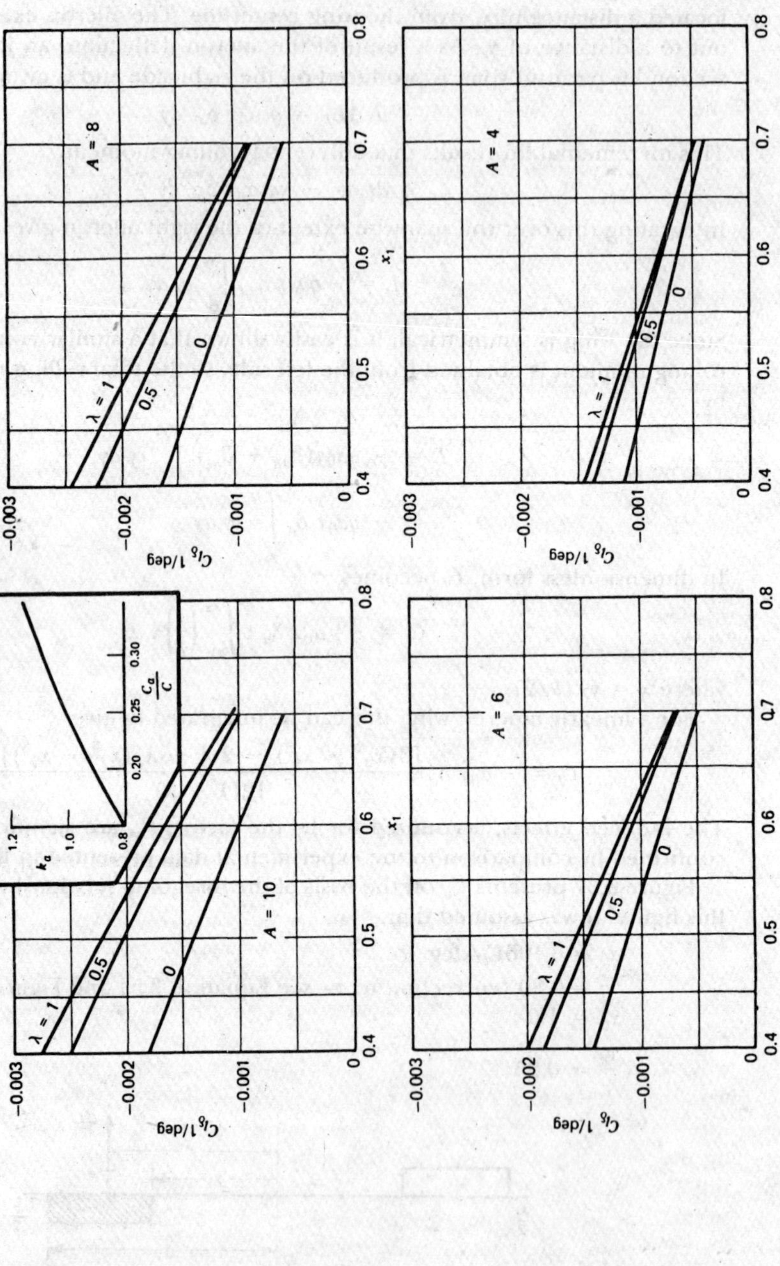

Figure 9.37 Aileron control as a function of wing and aileron geometry $\Lambda = 0$, $x_2 = 1.0$.

For other aileron-wing chord ratios, this figure must be corrected for the effect on τ. This correction is provided as an insert in Figure 9.37. In order to determine $C_{l\delta}$ for a value of c_a/c different from 0.25, the value of $C_{l\delta}$ read from the figure is multiplied by the factor k_a. It is interesting to note that Figure 3.36 for η results in extreme nonlinearities in τ for flap angles beyond about 15 or 20°. Yet aileron data appear linear with δ_a up to angles of approximately 30°. This is further evidence that, because of induced effects, the section angles of attack are reduced over the ailerons. Indeed, according to Reference 9.8, as the aspect ratio of the portion of the wing covered by the aileron decreases, the range of δ_a over which C_l is linear with δ_a increases.

If x_2 is less than unity, superposition can be used to calculate $C_{l\delta}$. For example, suppose $x_1 = 0.5$ and $x_2 = 0.9$. Then $C_{l\delta}$ can be calculated from

$$C_{l\delta}\begin{pmatrix} x_1 = 0.5 \\ x_2 = 0.9 \end{pmatrix} = C_{l\delta}\begin{pmatrix} x_1 = 0.5 \\ x_2 = 1.0 \end{pmatrix} - C_{l\delta}\begin{pmatrix} x_1 = 0.9 \\ x_2 = 1.0 \end{pmatrix}$$

Again, the subscript a is dropped on the δ in writing $C_{l\delta}$, since δ_a is the primary control for producing a rolling moment.

Minimum Control Speed, V_{MC}

The material on the trimming of an airplane with an engine-out condition pertains to Federal Air Regulations (FAR) Part 23 concerning the minimum control speed, V_{MC}. Specifically, FAR Part 23 states that "V_{MC} is the calibrated airspeed, at which, when the critical engine is suddenly made inoperative, it is possible to recover control of the airplane with that engine still inoperative, and maintain straight flight either with zero yaw or, at the option of the applicant, with an angle of bank of not more than five degrees. The method used to simulate critical engine failure must represent the most critical mode of powerplant failure with respect to controllability expected in service."

V_{MC} must be demonstrated at the most unfavorable conditions including, flaps in the takeoff position, maximum power on the good engines, windmilling propeller, airplane trimmed for takeoff, gear up, OGE, maximum takeoff weight, and with the cg in the most unfavorable position. This last requirement normally will mean the most allowable aft cg position at maximum TOGW. Under these restrictions, V_{MC} may not exceed 1.2 V_{S_1}, where V_{S_1} is determined at maximum TOGW and equals the stalling speed at idling power in the configuration for demonstrating V_{MC}.

Asymmetric thrust on a twin-engine airplane at high power due to an engine or propeller failure can present a serious situation to the pilot, particularly where the engines are mounted on the wing. When thrust is lost on one side only, the airplane will experience a sudden yawing moment because of the thrust on the one side and the excess drag of the inoperative propeller on the other. Concurrent with this yawing moment will be a severe rolling moment caused by an unbalanced lift on the wing. This asymmetric lift results from the flow over the wing on the good side being increased by the propeller slipstream and reduced on the bad side due to the drag of the inoperative propeller.

If the power loss occurs below V_{MC}, the pilot has insufficient rudder and aileron control to overcome the yawing and/or rolling moments. If the failure occurs near the ground during takeoff or landing, a crash is almost inevitable. If the failure occurs at altitude, say while practicing stall recovery, a crash may still ensue if the pilot does not react before the airplane enters a spin. Thus, in designing a multiengine airplane, the engine-out requirements must be considered in the analysis of the control system.

Adverse Yaw

A pilot initiates a turn from straight flight by rolling the airplane in the desired direction of the turn while at the same time applying a little rudder in order to start the airplane yawing in the desired direction. If done properly, a so-called coordinated turn is achieved where the resultant of the gravitational and inertial forces remains perpendicular to the wing. The pilot may feel a little heavier in a coordinated turn but, at least, forces tending to throw him or her to one side or the other are not experienced. A full pitcher of beer will not slosh in a coordinated turn. (Note that FAA regulations forbid the consumption of alcohol by a pilot eight hours prior to flying.)

A pair of aileron surfaces, which are simply plain flaps that travel asymmetrically when deflected, will usually produce a motion known as adverse yaw. At an angle of attack, a flap deflected downward will produce a drag increment greater than that produced by the same deflection upward. Thus, for example, suppose the right aileron moves up and the left one moves down to initiate a turn to the right. The higher drag on the left aileron will produce a yawing moment that tends to yaw the airplane to the left, opposite to what was desired. This is known as adverse yaw due to ailerons, a characteristic that can make an airplane uncomfortable to fly.

To alleviate adverse yaw, the mechanical linkages are sometimes designed so that, for a given control movement, the upward movement on one aileron is greater than the downward movement on the opposite aileron. For example, on the Cherokee 180, the aileron moves up 30° but down only 15°.

Another means of alleviating the adverse yaw is in the design of the aileron. Figure 9.38 illustrates a possible way of accomplishing this. The configuration shown in Figure 9.38, known as a Frise aileron, has the hinge point below the aileron surface. As the aileron is raised, the nose projects down into the flow, thereby increasing the drag on the up aileron.

Roll Control by the Use of Spoilers

A spoiler is a device on the upper surface of an airfoil that, when extended into the flow, causes the flow to separate, resulting in a loss of lift. The effect is illustrated in Figure 9.39, which also illustrates several types of spoilers.

Roll control can be accomplished by the use of spoilers installed on the outboard extent of each wing in place of conventional ailerons. If a roll to the right is desired, the spoiler on that side is raised, reducing the lift on the right wing and causing a positive rolling moment. The drag will also be increased on the right wing, producing a favorable yawing moment.

Spoilers for roll control are currently used on several military airplanes and most jet transports to augment the primary aileron controls. They are also used on at least one general aviation airplane, the twin-turboprop Mitsubishi MU-2, as the sole means of roll control. In this case, the use of spoilers only for roll control permits the use of full-span Fowler flaps with a higher wing loading. According to Reference

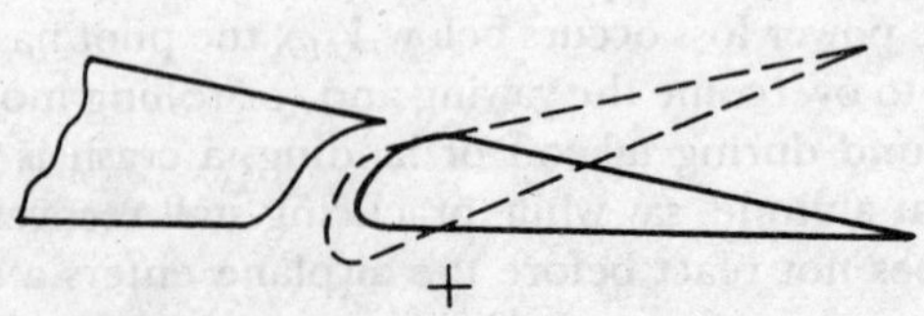

Figure 9.38 The Frise aileron designed to eliminate adverse yaw.

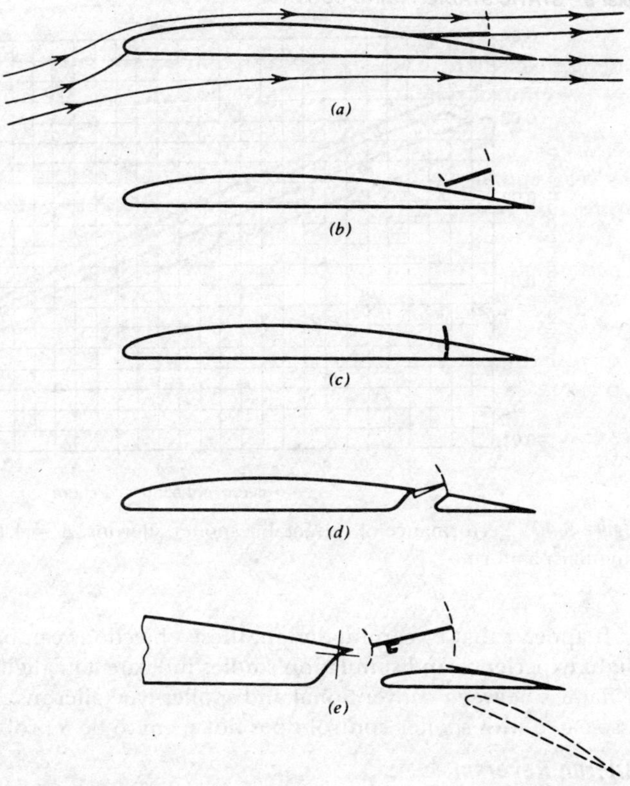

Figure 9.39 Types of spoiler ailerons. (*a*) Retractable aileron. (*b*) Gapped retractable aileron. (*c*) Plug aileron. (*d*) Slot-lip aileron. (*e*) Modified slot-lip aileron with gap and Fowler flap.

9.10, "As the requirements increase for higher wing loading to improve ride and performance, better handling qualities, and improved high lift devices, spoilers will probably be applied more extensively to light aircraft in the future."

Figure 9.40 (taken from Ref. 9.8) presents some rolling moment data typical of the retractable spoiler aileron of the type pictured in Figure 9.39. Note that for aileron projections in excess of 2% of the chord, the variation of C_l with the projection is quite linear. However, for projections less than this amount, the graphs are highly nonlinear. In fact, little or no rolling moment is produced until the projection exceeds 1% of the chord. This "dead" region would not produce a very satisfactory feel to the pilot.

The configurations of Figure 9.39*b* and 9.39*e* are intended to remove the dead region at low spoiler deflections. As soon as the gap is opened, its influence is felt in producing separated flow. The configuration of Figure 9.39*e* is the type used on the MU-2.

In the past, spoilers have not been used as the sole primary source of roll control because of their nonlinear characteristics. At high angles of attack their effectiveness can be reduced or even reversed. This nonlinear behavior is also reflected in the control forces that result from nonlinear hinge moments. In addition, a lag can occur in reestablishing the attached flow pattern when the spoiler is retracted. Objections to spoiler ailerons have also been based on the fact that the rolling moment is produced by the loss of lift on one side of the wing. This results in an accompanying loss in altitude.

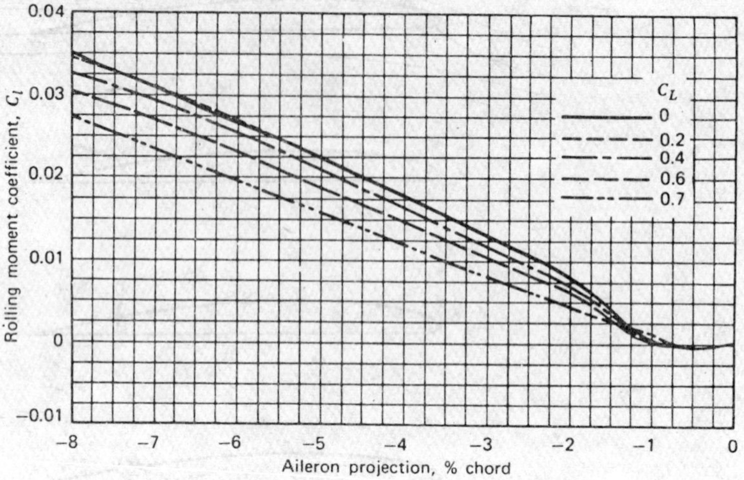

Figure 9.40 Performance of retractable spoiler ailerons. $A = 4.13$, $x_1 = 0.4$, $x_2 = 1.0$, rectangular planform.

It appears that by careful design, these objections can be overcome. Also, actual flight experience and simulation studies indicate a negligible difference in the roll dynamics between conventional and spoiler-type ailerons. Thus the loss in altitude associated with spoiler control does not seem to be a problem.

Aileron Reversal

Before leaving the subject of ailerons, mention should be made of a behavior known as aileron reversal. As the name suggests, it is a behavior whereby the action of an aileron, as related to its deflection, is opposite to what one would normally expect. This anomalous behavior is primarily associated with airplanes that have thin, flexible, sweptback wings that operate at high speeds. As a result of the aerodynamic moment, such a wing will tend to twist as the aileron is deflected. For example, an up aileron, which is intended to drop the wing, may instead twist the wing nose up sufficiently to cause a net increase in the wing lift.

To pursue this behavior, consider Figure 9.41, which illustrates the situation with a simplified model. A symmetrical airfoil section is shown mounted on a torsional spring with a spring constant of k_θ. When the flap is deflected through a positive angle δ, an aerodynamic moment is produced that causes the airfoil to rotate through an angle α (shown positively). This results in an opposing moment generated by the spring that, at some α, balances the aerodynamic moment. Thus,

$$k_\theta \alpha = qc^2 C_{m_\delta} \delta$$

and

$$C_l = (C_{l_\alpha} \alpha + C_{l\delta} \delta)$$

Thus,

$$C_l = \left(\frac{C_{l\theta} C_{M\delta} q c^2}{k_\theta} + C_{l\delta} \right) \delta \tag{9.21}$$

Since $C_{M\delta}$ is negative, it is obvious that, for a given k_θ, there is a q above which the flap deflection will produce a decrease in the lift.

Aileron reversal can be avoided by the obvious means of limiting the operational

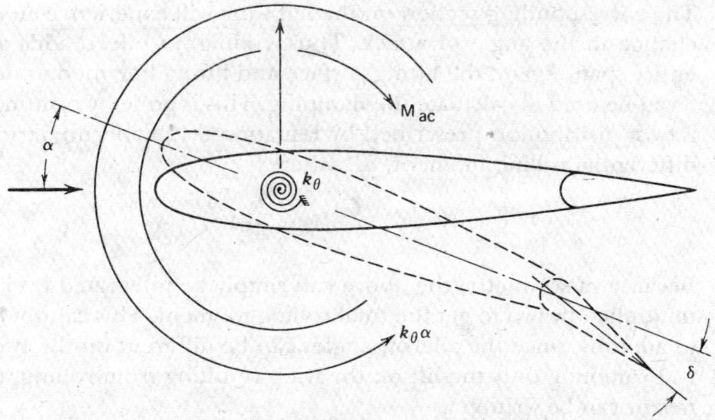

Figure 9.41 A wing mounted on a torsional spring to illustrate aileron reversal.

q or by stiffening the wing; either method can impose operational penalties on the airplane. A less obvious method is used to lock out conventional outboard ailerons at high speeds and employ spoiler control instead.

Steady Rolling Motion

A measure of aileron control power is afforded by the steady roll rate, *P*, produced by the aileron. This rate is such that the rolling moment from the ailerons equals the damping moment resulting from *P*. The origin of this damping moment can be seen from Figure 9.42.

A section of the right wing located a distance *y* from the centerline will experi ence an increment in its angle of attack because of *P*, given by

$$\Delta\alpha = \frac{Py}{V} \tag{9.122}$$

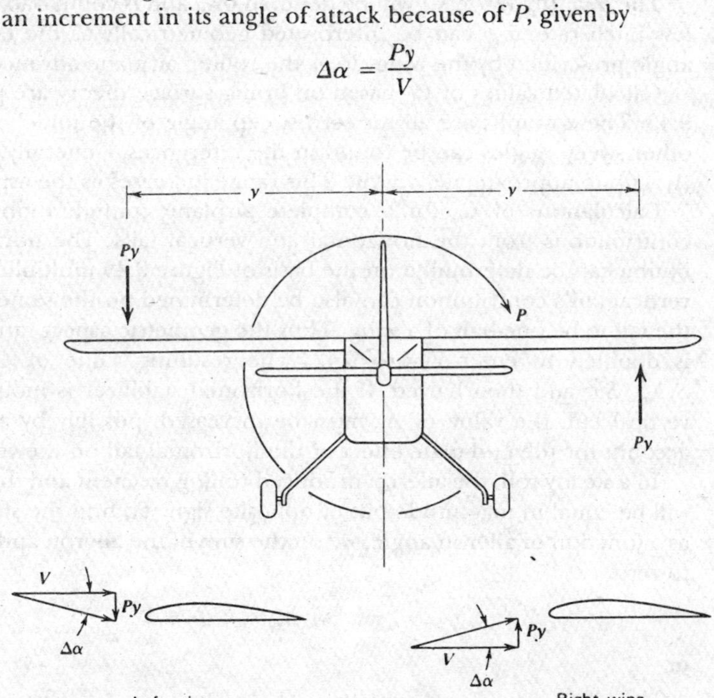

Figure 9.42 Origin of damping moment resulting from roll rate, *P*.

The corresponding section on the left wing will experience an equal, but opposite, change in the angle of attack. Thus, rolling, in effect, adds a linear twist to the entire span. Again, the lifting surface and lifting line models described in Chapter 3 can be used to calculate the damping. This is done by running those models with a twist distribution prescribed by Equation 9.122 and numerically integrating the differential rolling moment, dL, where

$$dL = \frac{1}{2}\rho V^2 c C_l y dy \tag{9.123}$$

Because of symmetry, the above can simply be integrated over the right wing and multiplied by two to get the total rolling moment. This cannot be done, in the case of ailerons, since the aileron angles can be different on the two sides.

Examining only the lift on the wing resulting from rolling, the section lift coefficient can be written as

$$C_l = a\left(\frac{Py}{V} - \alpha_i\right)$$

If α_i is neglected, the above can be integrated over the span and expressed as a rolling moment coefficient.

$$C_l = -\frac{aA}{2}\left(\frac{Pb}{2V}\right)\int_0^1\left(\frac{c}{b}\right)x^2\,dx \tag{9.124}$$

One is cautioned against substituting a wing geometry into this equation. The integrated result will be in serious error because of neglecting the induced effects. The principal value of Equation 9.124 lies in revealing that the coefficient of roll damping is proportional to the quantity $Pb/2V$.

The quantity $(Pb/2V)$ will be denoted by $\bar{p}$ and is comparable to the dimensionless pitch rate, $\bar{q}$. $\bar{p}$ can be interpreted geometrically as the tangent of the helix angle prescribed by the wing tip as the rolling airplane advances through the air.

Calculated values of C_{l_p} based on lifting surface theory are presented in Figure 9.43. These graphs are all for zero sweep angle of the midchord line. Results for other sweep angles can be found in the references. Generally, C_{l_p} is insensitive to $\Lambda_{1/2}$ up to approximately $\pm 20°$. This range increases as the aspect ratio decreases.

Calculations of C_{l_p} for a complete airplane configuration must also include contributions from the horizontal and vertical tails. The horizontal tail's contribution can be determined on the basis of Figure 9.43 multiplied by $S_t b_t^2/Sb^2$. The vertical tail's contribution can also be determined on the same basis by visualizing the tail to be one-half of a wing. Thus the geometric aspect ratio of the vertical tail is doubled to enter Figure 9.43. The resulting value of C_{l_p} is multiplied by $S_v b_v^2/Sb^2$ and then halved. If the horizontal stabilizer is mounted on top of the vertical tail, the value of A_v must be increased, possibly by as much as 20%, to account for the end-plate effect of the horizontal tail on the vertical tail.

In a steady roll, the aileron-produced rolling moment and the damping moment will be equal in magnitude but of opposite sign. To find the steady rolling velocity as a function of aileron angle, we set the sum of the aileron and damping moments to zero.

$$C_{l\delta}\,\delta_a + C_{l_p}\bar{p} = 0$$

or

$$\frac{Pb}{2V} = -\frac{C_{l\delta}}{C_{l_p}}\delta_a \tag{9.125}$$

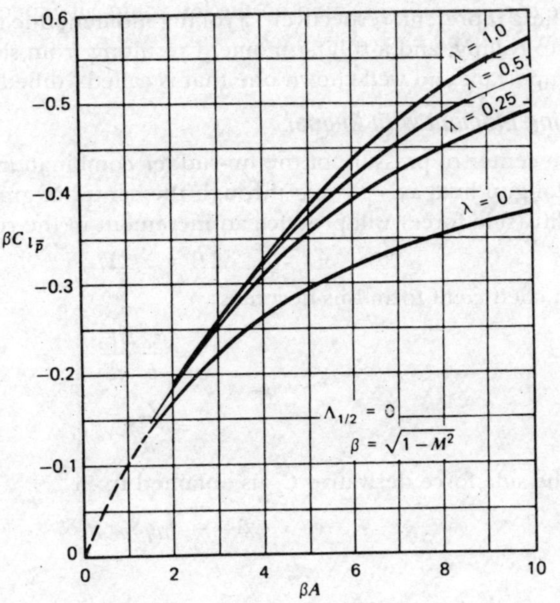

Figure 9.43 Roll damping coefficient derivative. Note that C_{l_p} should be corrected by ratio of section lift curve to 2π.

Thus, the dimensionless roll rate is directly proportional to the aileron deflection angle. $Pb/2V$ will typically range from approximately 0.05 to 0.10, depending on the maneuverability requirements imposed on the airplane.

Past practice was to specify values of $\bar{p}$ but both civil and military requirements (Refs. 9.1 and 9.12) now state specific requirements on P or, to be more specific, a displacement of ϕ in a given time. For example, FAR Part 23 requires that an airplane weighing less than 6000 lb on the approach must have sufficient lateral control to roll it from 30° in one direction to 30° in the other in 4 sec. In landing at 20% above the stalling speed, this number is increased to 5 sec. Mil-F-8785 B requires an air-to-air fighter to be rolled 360° in 2.8 sec, an interceptor aircraft 90° in 1.3 sec, a transport or heavy bomber 30° in 1.5 sec, and a light utility aircraft 60° in 1.4 sec.

Coupling Effects

There are several interactions or couplings involving lateral-directional controls and motions that are important considerations in providing satisfactory handling or riding qualities for an airplane. We have already seen one of these, a coupling between yaw and aileron displacement. Similarly, the rudder deflection can produce a rolling moment, since the center of pressure of the rudder-vertical fin combination generally lies above the center of gravity. These two control couplings are denoted in coefficient form by:

$$C_{N_{\delta_a}}$$

$$C_{l_{\delta_R}}$$

Other couplings include:

$$C_{l_r}$$

$$C_{N_p}$$

$$C_{l_\beta}$$

These represent, respectively, a rolling moment due to yawing, a yawing moment due to rolling, and a rolling moment resulting from sideslip. This last coupling is an important and well-known one that is called "dihedral effect."

Rolling Moment with Rudder

If the center of pressure of the fin-rudder combination lies a distance of z_v above the longitudinal axis passing through the center of gravity, an increment, ΔL_v, in the lift (side force) will produce an increment in the rolling moment, given by

$$\Delta L = z_v \, \Delta L_v$$

In coefficient form this becomes

$$\Delta C_l = \frac{z_v}{b} \frac{\Delta L_t}{qS}$$

$$= \frac{Z_v}{b} C_{Y\delta} \, \delta_r \tag{9.126}$$

The side force derivative $C_{Y\delta}$ is obtained from

$$\Delta Y = \eta_t q S_v a_v \tau \, \delta_r$$

or

$$C_{Y\delta} = \eta_t \frac{S_v}{S} a_v \tau \tag{9.127}$$

Rolling Moment with Yaw Rate

Figure 9.44 illustrates a wing having a tapered planform that is yawing at the rate of R rad/sec. The wing is operating at a lift coefficient of C_L. It will be assumed that the section lift coefficients are constant and equal to C_L. Because of the rotational velocity R, a section on the right side located at y experiences a local velocity equal to $V - Ry$ while the corresponding section on the left wing has a velocity of $V + Ry$. Thus, a differential rolling moment is produced equal to

$$dL = y \tfrac{1}{2} \rho c C_L [(V + Ry)^2 - (V - Ry)^2] \, dy$$

If the chord is constant, expanding and integrating from 0 to $b/2$ results in

$$L = \frac{\rho c C_L R V b^3}{12}$$

$$= \tfrac{1}{3} q S b \bar{r} C_L$$

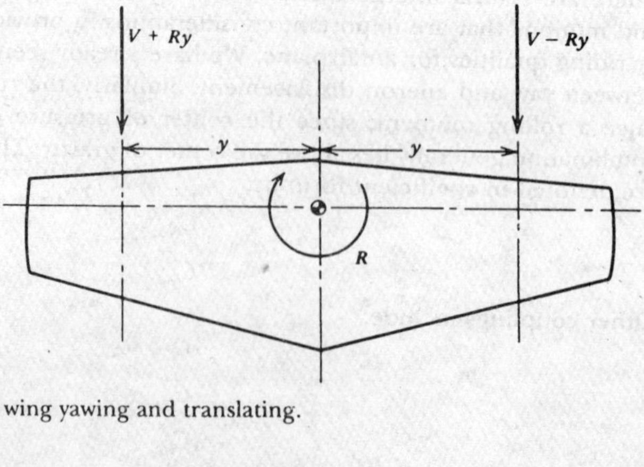

Figure 9.44 A wing yawing and translating.

Therefore,

$$C_{l_r} = \frac{C_L}{3}$$

Here, r denotes differentiation with respect to a dimensionless yaw rate comparable to $\bar{p}$ and defined by

$$\bar{r} = \frac{Rb}{zv}$$

As an exercise, show that, for a linearly tapered wing,

$$C_{l_r} = \frac{C_L}{6}\left(\frac{1 + 3\lambda}{1 + \lambda}\right) \tag{9.128}$$

The vertical tail can also contribute to C_{l_r}. Because of the yaw rate, its angle of sideslip is decreased by Rl_v/V. Thus, an incremental side force on the tail is generated, given by

$$\Delta Y = \eta_t q S_v a_v \frac{l_v R}{V}$$

In coefficient form,

$$C_Y = 2\eta_t a_v V_v \bar{r} \tag{9.129}$$

This incremental side force acting above the center of gravity produces an increment in the rolling moment, given by

$$\Delta L = \eta_t q S_v a_v \frac{l_v R}{V} Z_v$$

or, in coefficient form,

$$c_{l_r} = C_{y_r} \frac{Z_v}{b}$$

$$= 2\eta_t a_v V_v \left(\frac{Z_v}{b}\right) \tag{9.130}$$

The total C_{l_r} equals

$$C_{l_r} = \frac{C_L}{6}\left(\frac{1 + 3\lambda}{1 + \lambda}\right) + C_{y_r}\frac{Z_v}{b} \tag{9.131}$$

Yawing Moment with Roll Rate

Referring again to Figure 9.42, it is seen that on the right wing, which is moving down, the lift vector is inclined forward through the angle Py/V. On the left wing, the inclination is to the rear. The components of these tilted lift vectors in the x direction give rise to a differential increment in the yawing moment equal to

$$dN = -2yqcC_L \frac{Py}{V} dy$$

Integrating from 0 to $b/2$, the total yawing moment increment in coefficient form for a linearly tapered wing becomes

$$C_N = -\frac{C_L\bar{p}}{12}\left(\frac{1 + 3\lambda}{1 + \lambda}\right) \tag{9.132}$$

Rolling Moment with Sideslip Angle—Dihedral Effect

The rate of change of rolling moment with sideslip angle, $C_{l\beta}$, is important to the handling qualities of an airplane. Generally, a small negative value of C_{l_β} is desirable,

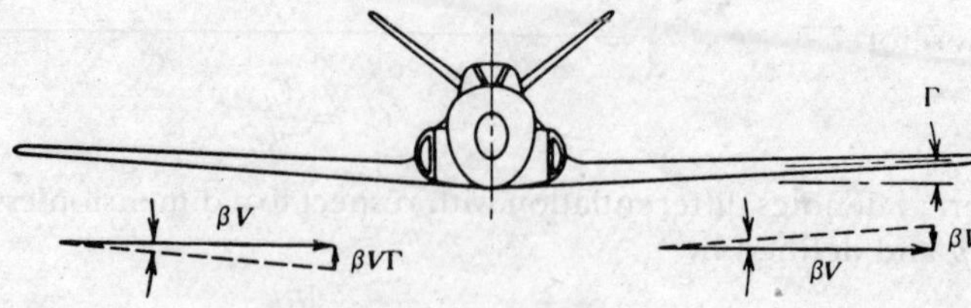

Figure 9.45 Definition of dihedral angle.

but too much dihederal effect makes an airplane uncomfortable to fly. The principal factors affecting $C_{l\beta}$ are sweepback, placement of the wing on the fuselage, and the dihedral angle of the wing, Γ.

The primary control over $C_{l\beta}$ is exercised through the dihedral angle, Γ, shown in Figure 9.45. Although the sweepback angle also affects $C_{l\beta}$ significantly, Λ is normally determined by considerations other than $C_{l\beta}$. From Figure 9.45, it can be seen that a positive sideslip results in an upward velocity component along the right wing and a downward component along the left wing. Added to the free-stream velocity, this results in an increase in the angle of attack over the right wing equal to

$$\Delta\alpha = \beta\Gamma$$

Over the left wing, an opposite change in α occurs. This differential increment in α results in a differential rolling moment, given by

$$dl = -2qca\beta\Gamma y\, dy$$

or

$$C_l = -2a\beta\Gamma \frac{\int_0^{b/2} cy\, dy}{bS}$$

For a linearly tapered wing, this reduces to

$$C_l = -\frac{a}{6}\left(\frac{1+2\lambda}{1+\lambda}\right)\beta\Gamma$$

or

$$C_{l\beta} = -\frac{a}{6}\left(\frac{1+2\lambda}{1+\lambda}\right)\Gamma \tag{9.133}$$

The derivation of Equation 9.133 has neglected induced effects that are appreciable for low aspect ratios. Its primary value lies in disclosing the linear relationship between Γ and $C_{l\beta}$.

It is recommended that the contribution to $C_{l\beta}$ from Γ be estimated using Figure 9.46. This figure represents the departure of $C_{l\beta}$ from the following normal case and is based on graphs presented in Reference 5.5.

$$\lambda = 0.5$$

$$A = 6.0$$

$$\Lambda_{1/2} = 0$$

$$C_{l\beta} = -0.00021\ \Gamma/\text{deg}(\Gamma \text{ in deg})$$

Thus,

$$C_{l\beta} = -0.00021 k_\lambda k_A k_\Lambda \Gamma \tag{9.134}$$

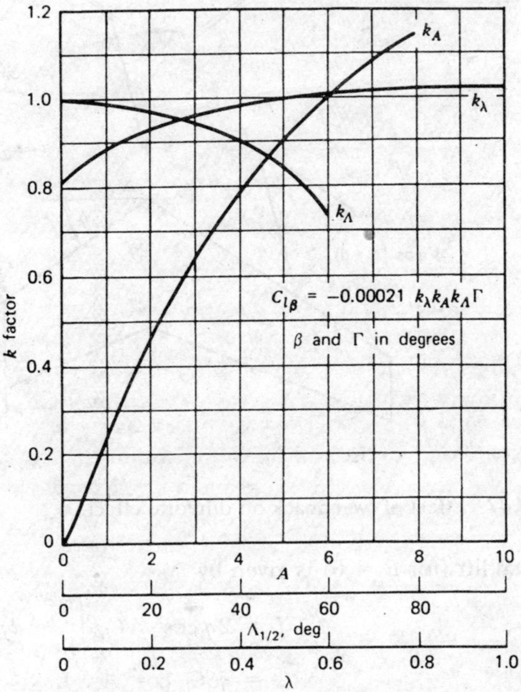

Figure 9.46 Effect of wing dihedral angle on C_{l_β}. Note that factors represent relative effects of varying each parameter independently from the normal case $\lambda = 0.5$, $A = 6.0$, $\Lambda_{1/2} = 0.0$.

where the functions k_λ, k_A, and k_Λ are obtained from Figure 9.46 as functions of λ, A, and $\Lambda_{1/2}$, respectively.

The effect of sweepback on $C_{l\beta}$ is determined with the help of Figure 9.47. A swept wing is shown operating at a positive sideslip angle β. From the geometry, the velocity component normal to the leading edge of the right wing is given by

$$V \cos (\Lambda - \beta)$$

The corresponding velocity on the left wing is

$$V \cos (\Lambda + \beta)$$

If C_{l_n} is the section lift coefficient based on the normal velocity and "normal chord," then the differential lift on the right and left wings will be

$$dL_R = q \cos^2 (\Lambda - \beta) c \cos \Lambda C_{l_n} ds$$

$$dL_L = q \cos^2 (\Lambda + \beta) c \cos \Lambda C_{l_n} ds$$

The differential rolling moment is therefore

$$dl = qcC_{l_n} y[\cos^2 (\Lambda + \beta) - \cos^2 (\Lambda - \beta)] \cos \Lambda\, ds$$

But

$$y = s \cos \Lambda$$
$$dy = ds \cos \Lambda$$

Thus,

$$l = qC_{l_n}[\cos^2 (\Lambda + \beta) - \cos^2 (\Lambda - \beta)] \int_0^{b/2} cy\, dy$$

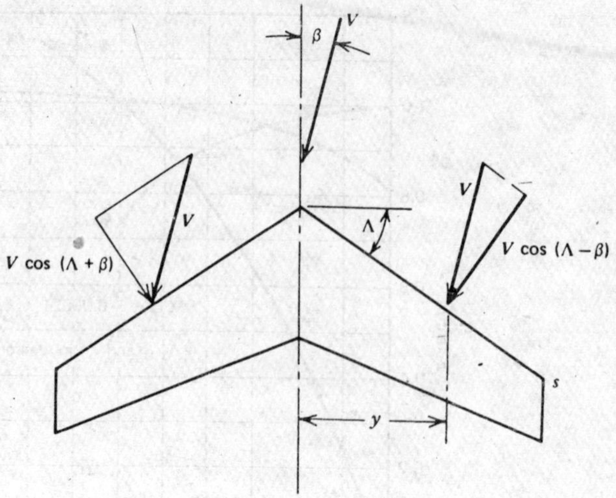

Figure 9.47 Effect of sweepback on dihedral effect.

The total lift (for $\beta = 0$) is given by

$$L = 2q \cos^2 \Lambda\, C_{l_n} \int_0^{b/2} c\, dy$$

$$= qSC_{l_n} \cos^2 \Lambda$$

Thus, the wing C_L and the normal section C_{l_n} are related by

$$C_L = C_{l_n} \cos^2 \Lambda$$

and the rolling moment coefficient becomes

$$C_l = \frac{C_L}{\cos^2 \Lambda} [\cos^2 (\Lambda + \beta) - \cos^2 (\Lambda - \beta)] \frac{\int_0^{b/2} cy\, dy}{Sb}$$

If this equation is differentiated with respect to β and evaluated at $\beta = 0$, the following results:

$$C_{l_\beta} = -4 C_L \tan \Lambda \left[\frac{\int_0^{b/2} cy\, dy}{Sb} \right] \tag{9.135}$$

For a linearly tapered wing, Equation 9.135 reduces to

$$C_{l_\beta} = -\frac{1 + 2\lambda}{3(1 + \lambda)} C_L \tan \Lambda \tag{9.136}$$

Again, this result is only qualitatively correct. Generally,

$$C_{l_\beta} = -f(A, \lambda) C_L \tan \Lambda \tag{9.137}$$

Figure 9.48 (based on Ref. 5.6) presents C_{l_β}/C_L as a function of Λ for a range of aspect ratios. The variation with tan Λ is seen to hold only for the higher aspect ratios. This figure can be used with Equation 9.136 to estimate C_{l_β} for other taper ratios.

Observe that wing sweep can contribute significantly to dihedral effect. In order to avoid an excessive dihedral effect on aircraft with highly swept wings, it is frequently necessary to employ a negative dihedral angle on the wing, particularly if the wing is mounted high on the fuseLage.

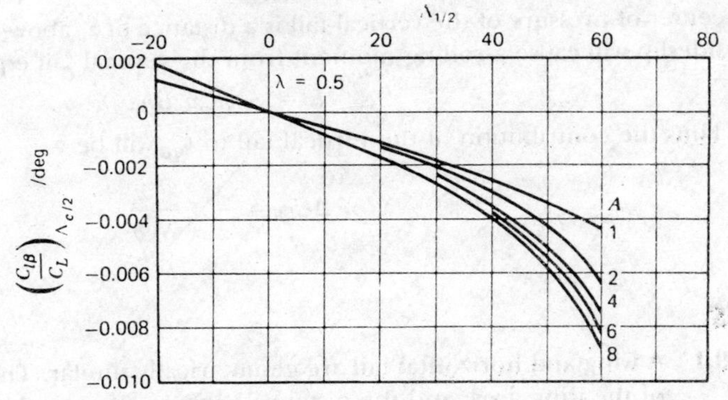

Figure 9.48 Effect of wing sweep on $C_{l\beta}$.

The effect that the wing placement on the fuselage has on $C_{l\beta}$ is seen by reference to Figure 9.49. In a plane across the top of the fuselage, the cross-flow around the fuselage is seen to go up on the right side and down on the left. Thus, for a high wing, this flow increases the angle of attack of the right wing while decreasing α on the left wing. This results in a negative rolling moment comparable to a positive dihedral effect. For a low wing, the effect is just the opposite. This is the reason, as you may have observed, that the dihedral angle for unswept low wings is generally greater than for high wings. Many high-wing airplanes do not have any dihedral angle at all.

As a rule, it is recommended that the following be added to $c_{l\beta}$ to account for the fuselage cross-flow (Ref. 9.6).

$$\text{High-wing } \Delta C_{l\beta} = -0.00016/\text{deg}$$

$$\text{Midwing } \Delta C_{l\beta} = 0$$

$$\text{Low-wing } \Delta C_{l\beta} = 0.00016/\text{deg}$$

The horizontal tail can also contribute to $C_{l\beta}$. Its contribution is calculated in the same manner as for the wing. On many airplanes the horizontal tail has a pronounced dihedral. On others, such as the McDonnell-Douglas F-4 Phantom, the horizontal tail has a pronounced negative dihedral known as *anhedral.*

There can also be a significant contribution to $C_{l\beta}$ from the vertical tail. If the

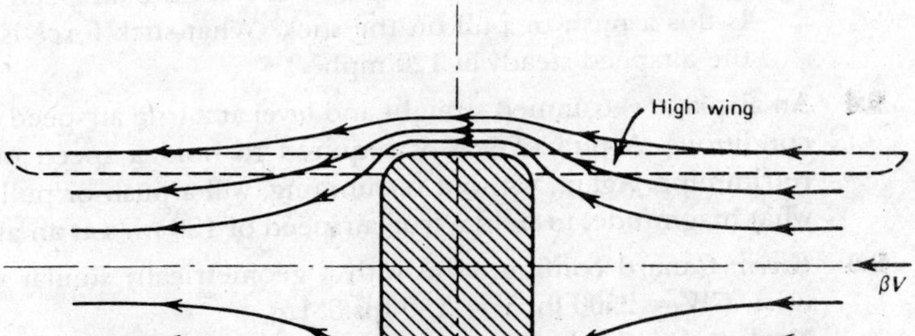

Figure 9.49 Rear view showing the cross-flow around a fuselage.

center of pressure of the vertical tail is a distance of z_v above the cg, then a positive sideslip will cause a rolling moment from the vertical tail equal to

$$L = -qS_v a_v \beta h_v$$

Thus the contribution of the vertical tail to $C_{l\beta}$ will be

$$\Delta C_{l\beta} = -a_v \frac{S_v}{S}\frac{z_v}{b} \quad (9.137)$$

PROBLEMS

9.1 A wing and horizontal tail are geometrically similar. The tail area is one-third of the wing area, and the distance between the aerodynamic centers is equal to the semispan of the wing.

(a) If $\epsilon_\alpha = 0.3$, what is the maximum distance that the cg can be behind the ac of the wing without the wing–tail combination being statically unstable?

(b) What is the angle of incidence of the tail to trim for a 5% static margin if $C_L = 1$, $a_w = 0.08$/deg, $\bar{c} = \frac{b}{6}$, $b = 50$ ft, and $C_{M_{ac}} = 0$?

9.2 Sketch a graph of C_{l_α} versus α for various flap angles on a airfoil and use this to explain graphically why the slope of the lift curve of a horizontal tail is reduced when the elevator is free to rotate.

9.3 An airplane has a stick-fixed static margin of 10%. It is flying at 109 mph SSL. The wing loading is 30 psf and the zero lift line of the entire airplane is at an angle of attack of 10°. The wing has an aspect ratio of 6.0. The slope of the wing lift curve is 0.075/deg, and ϵ_α is the same as that in the plane of a wing having an elliptic lift distribution. The wing and tail are geometrically similar and both employ symmetrical airfoils. In answering each question below, use results from the previous parts as needed.

(a) What is the planform area of the horizontal tail compared to the wing area if the incidence angle of the tail is zero?

(b) If the cg is a distance of $\bar{c}$ behind the aerodynamic center of the wing, how far is the tail aerodynamic center back from the cg in terms of $\bar{c}$?

(c) What tail incidence angle is required to trim?

(d) If τ equals 0.6, what δ_e is required to trim at 90 mph?

(e) A stick force of 10 lb is required to hold the airspeed steady at 90 mph. Is this a push or pull on the stick? What stick force is required to hold the airspeed steady at 120 mph?

9.4 An airplane is trimmed straight and level at a true airspeed of 100 m/s at SSL conditions. A pull of 5 N is required to hold a speed of 90 m/s without retrimming. Again, without retrimming, will a push or pull be required, and what magnitude, to hold a true airspeed of 120 m/s at an altitude of 3000 m?

9.5 *Given:* Canard configuration with a geometrically similar wing and tail. No twist. $GW = 2500$ lb. $V = 150$ fps SSL
Neglect downwash
Wing $\bar{c} = 4.5$ ft. $C_{M_{ac}} = -0.1$ $a_w = 0.07$/deg
Distance from one ac to the other is 30 ft
Wing planform area $= 150$ ft^2
Canard planform area $= 50$ ft^2
Find: Distance of neutral point behind the canard ac

9.6 *Given:* Same configuration as Problem 9.5. *Assume* the neutral point lies 20 ft behind the tail ac.

Find: Canard incidence angle to trim at 150 mph and 8000 ft for a static margin of 5%. In this case, take i_t positive nose up.

9.7 A small wing is fastened on one end of a rod 5 ft long. The rod is rotating about the other end at 16 rpm, causing the wing to move downward. The aspect ratio of the wing is 7.0, a = 0.08, and b = 2.0 in. V = 100 fps SSL. Find the wing lift coefficient at the instant the wing is at a 10° angle to V as shown.

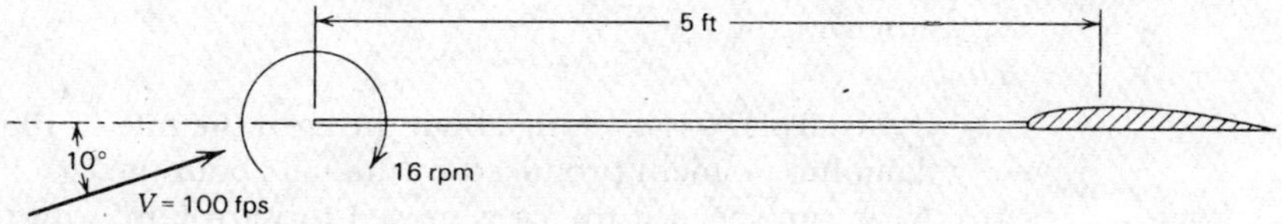

9.8 *Given:* A 10% thick symmetrical airfoil with a plain flap. The flap is hinged to the forward part of the airfoil through a torsional spring.

(a) Derive an equation for the airfoil lift curve, $dC_l/d\alpha$ as a function of an equivalent spring constant, k_θ ft-lb/rad, the flap area and chord, b_1, b_2, τ, q, and $\partial C_l/\partial\alpha$ with no flap (denoted by a).

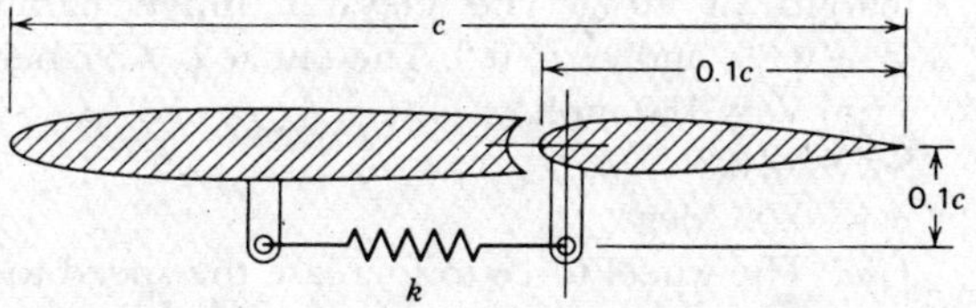

(b) If the airfoil chord equals 10 ft and the flap is a plain flap with a 30% chord and no aerodynamic balance, calculate $dC_l/d\alpha$ at a speed of 100 kt at an altitude of 10,000 ft for k_θ = 0 and for k_θ = 1.6 ft-lb/deg. Use the theoretical value of $\partial C_l/\partial\alpha$ and τ for thin airfoils with no empirical corrections. Use program "Hinge" or Figure 9.14 to determine b_1 and b_2.

9.9 This is an open-ended problem. The design of the empennage for an airplane must satisfy control requirements as well as stability requirements. For example, the area of the horizontal tail for a T-tail configuration can be reduced since it is away from the wing's downwash and, hence, more effective in providing stability. However, the maximum force, which the tail can provide, is reduced in proportion to the area. Thus, a smaller T-tail may have problems in lifting the nose on takeoff at a forward cg or in providing sufficiently high pitch rates for maneuvering. Considering these factors, and possibly others, design a horizontal tail for the Cherokee (vertical placement, area, aspect ra[illegible]), which will provide adequate static stability and control.

9.10 *Given:* A flying wing airplane. The airfoil of the wing is symmetrical and the wing is untwisted with a leading edge sweep of 15°, a taper ratio of 0.6, and an aspect ratio of 9.0. The wing is equipped with 25% full-span elevons (flaps that can be deflected differentially to serve as elevators or ailerons). The wing loading is 60 psf and the gross weight is 20,000 lb. The trim speed is 200 kt at 20,000-ft altitude.

(a) If the airplane is statically stable, will the elevons be up or down and why?
(b) Find the location of the neutral point from the centerline leading edge in feet (use program "Lattice").
(c) If the airplane is trimmed and has a 5% static margin, what is the deflection of the elevons? (use program "Lattice").

9.11 *Given:* A "simple" airplane composed of only a wing and a tail with the following: $S = 150$ ft^2, $C_{M_{ac}} = 0$, $\bar{c} = 6$ ft, $a_w = 0.075$/deg $S_t = 50$ ft^2, $a_t = 0.065$/deg, $\epsilon_\alpha = 0.4$, and $W = 3000$ lb
Distance from the wing ac to the tail ac = 20 ft
Airplane trimmed at $V = 100$ kt SSL
cg at the wing ac
Find:

(a) If the airplane is pitching about the cg at the rate of 100°/sec, what is the damping moment produced by the tail about the cg.
(b) New trim speed if the cg is moved forward 6 in. with the tail incidence angle and elevator angle held fixed.

9.12 *Given:* An airplane with a horizontal stabilizer–elevator configuration. The tail span is 35 ft and the chord (stabilizer + elevator) equals 10 ft. The elevator is 10% of the chord. The total wheel movement from full forward to full aft is 14 in. and causes the elevator to move from 29° down to 20° up. The airplane has a wing loading of 50 psf, a total lift curve slope of 0.08/deg, and a static margin of 10%. The elevator hinge parameters are $b_1 = -0.55$, $b_2 = -0.89$, and $\epsilon_\alpha = 0.3$. The tail ac is 3.5 $\bar{c}$ behind the cg.
Trim V = 150 mph
$S_T/S = 0.3$
$a_t = 0.05$/deg
Find: The wheel force to increase the speed to 180 mph without retrimming.

9.13 If the Cherokee 180 (Figure 3.62) ailerons are held hard over (30° up, 15° down), what will the steady roll rate be at an airspeed of 100 mph, SSL conditions?

9.14 A Cherokee 180 is making a crosswind landing under SSL conditions. The wind is directly from the right at 10 mph. The true airspeed for landing is 90 mph at a weight of 2400 lb. The pilot slips and banks the airplane into the wind so as to keep the fuselage aligned with the runway with no sideward drift. Calculate the rudder and aileron control needed to make the crosswind landing.

9.15 Given: A wing–tail combination as shown with $i_t = 2°$ and $C_{M_{ac}} = 0$. Using only closed-form expressions and graphs found in the text, calculate:

1. a
2. a_t
3. b'
4. ϵ_α
5. V_H
6. C_{L_α}
7. h_t
8. h_n
9. τ
10. $C_{M\delta}$
11. $C_{L\delta}$
12. b_1

13. b_2
14. b_3
15. C_{M0}
16. F_e

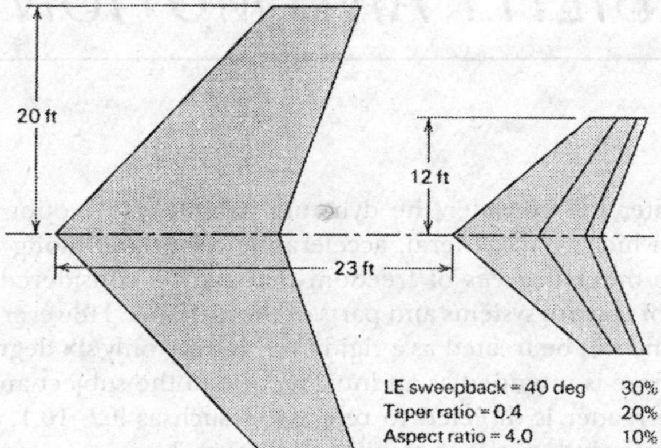

Wing and tail planform geometrically similar

9.16 Given: The same geometry as Problem 9.15 except that the *entire* tail is fixed at a zero incidence angle relative to the fuselage reference line. Longitudinal control is accomplished by rotating the wing about its aerodynamic center. What angle of the wing's zero-lift line relative to the fuselage reference line is required to trim the airplane at a speed of 100 mph for a weight of 3000 lb and SSL conditions?

REFERENCES

1. Anonymous, *Federal Air Regulations Part 23-Airworthiness Standards: Normal, Utility and Acrobatic Category Airplanes,* continuously revised.
2. Etkin, Bernard, *Dynamics of Flight,* John Wiley, New York and London, 1959.
3. Anonymous, *Royal Aeronautical Society Data Sheets,* continuously updated.
4. Ames, M. B., and Sears, R. I., *Determination of Control Surface Characteristics from NACA Plain-Flap and Tab Data,* NACA R 721, 1941.
5. Roskam, J., *Flight Dynamics of Rigid and Elastic Airplanes,* published by author, Lawrence, KS, 1973.
6. Perkins, C. D., and Hage, R. E., *Airplane Performance, Stability and Control,* John Wiley, New York and London, 1949.
7. Campbell, J. P., and McKinney, M. O., *Summary of Methods for Calculating Dynamic Lateral Stability and Response and for Estimating Lateral Stability Derivatives,* NACA R 1098, 1952.
8. Fischel, Jack, Rodger, L. N., Hagerman, J. R., and O'Hare, W. M., *Effect of Aspect Ratio on the Low-Speed Lateral Control Characteristics of Untapered Low-Aspect-Ratio Wings Equipped with Flap and with Retractable Ailerons,* NACA R 1091, 1952.
9. Fink, M. P., Freeman, D. C., Jr., and Greer, H. D., *Full-Scale Wind-tunnel Investigation of the Static Longitudinal and Lateral Characteristics of a Light Single-Engine Airplane,* NASA TN D-5700, March 1970.
10. Roskam, J., and Kohlman, D. L., *Spoilers for Roll Control of Light Airplanes,* AIAA Paper No. 74-861, AIAA Mechanics and Control of Flight Conference, Anaheim, CA, August 5–9, 1974.
11. Anonymous, *Mil-F-8785 B Military Specification-Flying Qualities of Piloted Aircraft,* August 1969.
12. Ribner, H. A., *Propellers in Yaw,* NACA R 820, 1945.

10 OPEN-LOOP DYNAMIC STABILITY AND MOTION

INTRODUCTION

This chapter will consider the dynamic stability and motion of an uncontrolled airplane, which is, in general, accelerating about and along all three of its axes. There are other degrees of freedom that can be considered, such as the elastic behavior of control systems and parts of the airframe. However, in this presentation the airplane will be treated as a rigid body having only six degrees of freedom. The material here is intended as an introduction to the subject and for more in-depth study, the reader is directed to references such as 9.2, 10.1, and 11.1, which are devoted entirely to airplane stability and control.

Newton's well-known second law of motion

$$\bar{F} = m\bar{a} \tag{10.1}$$

is a vector relationship that states that the acceleration, $\bar{a}$, of a particle of mass, m, in a given direction is equal to the total force acting on the mass in that direction divided by the mass. The acceleration, $\bar{a}$, is relative to a fixed inertial system, which, for airplanes, is taken to be Earth-fixed axes. For spacecraft, it's a different matter.

There are several difficulties with applying Equation 10.1 to an airplane. First, the airplane is not a point mass, but rather an assemblage of mass elements. Second, the aerodynamic forces, in general, acting on the airplane are dependent on the motion of the airplane relative to the air and not on the motion of the airplane relative to Earth axes. For example, the yawing moment produced by the vertical tail depends upon the sideslip angle and not on whether the airplane is heading North, South, East, or West.

Although the equations of motion could be written with respect to Earth axes, the transformation of the aerodynamic forces and moments to these axes at any instant is very unwieldy. It is therefore expedient to adapt Equation 10.1 to a set of noninertial, moving axes, which are fixed relative to the airplane.

Equations of Motion

Although these equations can be formulated in an elegant manner using vector calculus, the classical approach taken by Reference 10.2 offers more physical insight into the development and will be followed here. Figure 10.1 shows a coordinate system that is fixed relative to the airplane with the origin placed at its center of gravity. The origin of the system is moving relative to Earth axes with instantaneous linear velocity components of U, V, and W along its own x-, y-, and z-axes, respectively. In addition to these linear velocities, the system is rotating relative to Earth axes at instantaneous angular velocity components of P, Q, R, about its own axes. Thus, at the instant, a mass element, Δm, shown in Figure 10.1, has linear velocity components relative to inertial axes of

$$\dot{x} = U + Qz - Ry \tag{10.1a}$$

$$\dot{y} = V + Rx - Pz \tag{10.1b}$$

$$\dot{z} = W + Py - Qx \tag{10.1c}$$

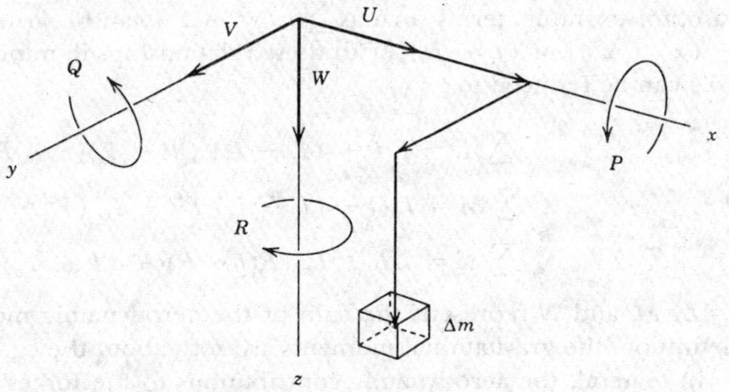

Figure 10.1 Element of mass in a moving coordinate system.

The corresponding accelerations can be obtained by differentiating the above with respect to time.

$$\ddot{x} = \dot{U} + Q\dot{z} + z\dot{Q} - R\dot{y} - y\dot{R} \tag{10.2a}$$

$$\ddot{y} = \dot{V} + R\dot{x} + x\dot{R} - P\dot{z} - z\dot{P} \tag{10.2b}$$

$$\ddot{z} = \dot{W} + P\dot{y} + y\dot{P} - Q\dot{x} - x\dot{Q} \tag{10.2c}$$

The accelerations, multiplied by the elemental mass, Δm must equal the elemental force on the mass element. However, any forces internal to the airplane acting on the mass elements will cancel one another leaving only the instantaneous external forces. Thus, when the set of Equations 10.1 are then substituted into the above, summed over the entire mass of the airplane, and equated to the total external force components $\sum X$, $\sum Y$, and $\sum Z$, the following equations are found:

$$\sum X = m(\dot{U} + QW - RV) \tag{10.3a}$$

$$\sum Y = m(\dot{V} + RU - PW) \tag{10.3b}$$

$$\sum Z = m(\dot{W} + PV - QU) \tag{10.3c}$$

In order to obtain the above, it must be recognized that, by the definition of the center of gravity,

$$\sum x\Delta m = \sum y\Delta m = \sum z\Delta m = 0$$

where $\sum$ indicates a sum over the entire mass of the airplane.

The angular moment of the mass element about an axis will equal the product of the normal distance from the mass to the axis (moment arm), the mass and the linear velocity normal to the plane of the arm and axis. Thus, the instantaneous components of the total angular momentum about the x-, y-, and z-axes will be

$$L = \sum m(y\ddot{z} - z\ddot{y}) \tag{10.4a}$$

$$M = \sum m(z\ddot{x} - x\ddot{z}) \tag{10.4b}$$

$$N = \sum m(x\ddot{y} - y\ddot{x}) \tag{10.4c}$$

In reducing these equations it should be recognized that

$$I_x = \sum m(y^2 + z^2) \quad Iy = \sum m(x^2 + z^2) \quad I_z = \sum m(x^2 + y^2)$$

$$I_{xy} = \sum mxy \qquad I_{xz} = \sum mxz \qquad I_{yz} = \sum myz$$

and because of symmetry,

$$I_{xy} - I_{yz} - 0$$

Also, for example, terms such as $\sum m(y^2 - z^2)$ can be written as $\sum m[(x^2 + y^2) - (x^2 + z^2)]$ or $(I_z - I_y)$. With these relationships in mind, the set of Equations 10.4 can be reduced to

$$\sum L = I_x\dot{P} + (I_z - I_y)\,QR - I_{xz}(\dot{R} + PQ) \tag{10.5a}$$

$$\sum M = I_y\dot{Q} - (I_z - I_x)\,PR + I_{xz}(P^2 - R^2) \tag{10.5b}$$

$$\sum N = I_z\dot{R} + I_{xz}(RQ - \dot{P}) + (I_y - I_x)\,PQ \tag{10.5c}$$

$\sum L$, M, and N represent the sum of the aerodynamic moments only since, by definition, the gravitational moments are zero about the cg.

In general, the aerodynamic contributions to the forces and moments on the left side of the above equations will depend upon the control positions, the linear velocities, the angular velocities, the angle of attack, and the sideslip angle. The gravitational force resolved along the airplane axes will depend upon the attitude of the moving axes with respect to the vertical. If the forces and moments can be obtained knowing the parameters on which they depend, then the above equations can be numerically integrated to give the resulting motion.

For our purposes, it is sufficiently instructive and accurate for most cases, to linearize Equations 10.5*a*, *b*, and *c*. This will be done by linearizing all forces, moments, and velocities about a trimmed condition. The ***aerodynamic*** forces and moments are expanded about the trimmed condition to read

$$X = X_0 + \Delta X \tag{10.6a}$$

$$Y = Y_0 + \Delta Y \tag{10.6b}$$

$$Z = Z_0 + \Delta Z \tag{10.6c}$$

$$L = \Delta L \tag{10.6d}$$

$$M = \Delta M \tag{10.6e}$$

$$N + \Delta N \tag{10.6f}$$

where, for example, ΔX is a small perturbation in the X force about the trimmed value, X_0. The trimmed values for the moments in the above are all zero because these moments are taken about the cg. The trimmed values of the forces are not zero because they must balance the gravitational forces.

Similarly, the velocity components are expressed as a steady state value plus a perturbation. In doing do, the coordinate system will be chosen initially so that the x-axis is aligned with the free-stream velocity with no side slip. As such, the axes are referred to as *stability axes*. In the stability axis system, the trim values for V and W will be zero and the velocity components become

$$U = U_0 + u \tag{10.7a}$$

$$V = v \tag{10.7b}$$

$$W = w \tag{10.7c}$$

Similarly, the angular rates become

$$P = p \tag{10.7d}$$

$$Q = q \tag{10.7e}$$

$$R = r \tag{10.7f}$$

The perturbation forces and moments will replace the forces and moments on the left side of the set of Equations 10.3 and 10.5 since, in trim, the sum of the aerodynamic and gravitational forces must vanish. On the right side of the equations, only first-order terms in the perturbation velocities will be retained. Thus, these equations, when linearized, become

$$\Delta X_G + \Delta X = m\dot{u} \tag{10.8a}$$

$$\Delta Y_G + \Delta Y = m(\dot{v} + U_0 r) \tag{10.8b}$$

$$\Delta Z_G + \Delta Z = m(\dot{w} - U_0 q) \tag{10.8c}$$

$$\Delta L = I_x\dot{p} - I_{xz}\dot{r} \tag{10.8d}$$

$$\Delta M = I_y\dot{q} \tag{10.8e}$$

$$\Delta N = I_z\dot{r} - I_{xz}\dot{p} \tag{10.8f}$$

ΔX_G, ΔY_G, and ΔZ_G are the perturbation gravitational force components.

Note that Equations 10.8*a*, 10.8*c*, and 10.8*e* contain only terms involving forces, moments, and motion in the *x-z* plane. Such motion *in the plane of symmetry* is called *longitudinal motion*. Equations 10.8*b*, 10.8*d*, and 10.8*f* govern only motion *of the plane of symmetry*, which is known as *lateral-directional* motion. From the nature of the above set of equations, the longitudinal motion can be solved independent of the lateral-directional motion. Thus, we can separate the six-degree-of-freedom motion into two uncoupled three-degree-of-freedom motions.

Euler Angles

The transformation from the moving, noninertial body axis system to an inertial, or earth-fixed, system can be written in terms of the Euler angles ϕ, θ, and ψ. Both coordinate systems are right-handed with x directed forward, y to the right, and z downward. All rotations are defined positive according to the right-hand rule.

In order, the body is first rotated about its *z*-axis through an angle ψ, then rotated about its *y*-axis through an angle θ, and finally rotated about its *x*-axis through an angle ϕ. These rotations are not commutative. The final orientation of the airplane will depend upon the order in which the rotations are performed. To illustrate this, hold a small model in front of you, heading directly away from you with wings level. Now rotate the model 90° about its *x*-axis, then 90° about its *y*-axis, and then 90° about its *z*-axis, all directions positive in accordance with the right-hand rule. The model will now be pointing nose down with its top toward you. Now reverse the order. Rotate it first about its *z*-axis, then its *y*-axis, and then the *x*-axis. In this case, the model's final orientation will be nose-up with its top toward you.

The angles ϕ, θ, and ψ can be related to the angular rates P, Q, and R by

$$\phi = \phi_0 + \int_0^t (P + Q\sin\phi\tan\theta + R\cos\phi\tan\theta)\, dt \tag{10.9a}$$

$$\theta = \theta_0 + \int_0^t (Q\cos\phi - R\sin\phi)\, dt \tag{10.9b}$$

$$\psi = \psi_0 + \int_0^t \sec\theta\,(Q\sin\phi + R\cos\phi)\, dt \tag{10.9c}$$

The initial values of the Euler angles, ϕ_0, θ_0, and ψ_0 will be zero if the inertial and body axes are initially aligned. However, it is convenient to take the *x-y* plane of the earth-fixed axes to be horizontal. Thus, if the airplane is generally in a steady climbing turn, the initial values of ϕ and θ are not zero. Instead, the initial value of θ_0 will equal the angle of climb and ϕ_0 will be the bank angle. ψ_0 in this case will still be taken as zero with all calculated headings then being relative to the initial heading.

Denoting the earth-fixed coordinates by a prime, the position of the airplane in this coordinate system can be found by the following:

$$x' = \int_0^t [U\cos\phi\cos\psi + V(\sin\phi\sin\theta\cos\psi - \cos\phi\sin\psi) + W(\cos\phi\sin\theta\cos\psi + \sin\phi\sin\psi)]\,dt \tag{10.10a}$$

$$y' = \int_0^t [U\cos\theta\sin\psi + V(\sin\phi\sin\theta\sin\psi + \cos\phi\cos\psi) + W(\cos\phi\sin\theta\sin\psi - \sin\phi\cos\psi)]\,dt \tag{10.10b}$$

$$z' = \int_0^t [-U\sin\theta + V\sin\phi\cos\theta + W\cos\phi\cos\theta]\,dt \tag{10.10c}$$

All of the above positions are, of course, relative to the initial position of the airplane.

If the *x-y* plane of the earth-fixed axes is horizontal, the components of the gravitational force along the body axes can be written in terms of the Euler angles as

$$X_G = -mg\sin\theta\cos\phi \tag{10.11a}$$

$$Y_G = mg\cos\theta\sin\phi \tag{10.11b}$$

$$Z_G = mg\cos\theta\cos\phi \tag{10.11c}$$

The perturbation gravitational forces are therefore

$$\Delta X_G = -mg(\cos\theta_0\cos\phi_0\theta - \sin\theta_0\sin\phi_0\phi) \tag{10.11d}$$

$$\Delta Y_G = mg(-\sin\theta_0\sin\phi_0\theta + \cos\theta_0\cos\phi_0\phi) \tag{10.11e}$$

$$\Delta Z_G = mg(-\sin\theta_0\cos\phi_0\theta - \cos\theta_0\sin\phi_0\phi) \tag{10.11f}$$

For many cases of interest, the trim roll is zero and the pitch angle is small so that Equations 10.9, 10.10, and 10.11 are greatly simplified and

$$P \approx \dot{\phi} \tag{10.12a}$$

$$Q \approx \dot{\theta} \tag{10.12b}$$

$$R \approx \dot{\psi} \tag{10.12c}$$

$$x' = \int_0^t U\cos\psi\,dt \tag{10.12d}$$

$$y' = \int_0^t (U\sin\psi + V\cos\psi)\,dt \tag{10.12e}$$

$$z' = \int_0^t (-U\theta + w)\,dt \tag{10.12f}$$

$$\Delta X_G = -mg\cos\theta_0\theta \tag{10.12g}$$

$$\Delta Y_G = mg\cos\theta_0\phi \tag{10.12h}$$

$$\Delta Z_G = 0 \tag{10.12i}$$

Longitudinal Motion

A change of variable will be made from the perturbation velocity along the *z*-axis to a perturbation angle of attack. Assuming small angles, w and α are related by

$$\alpha = \frac{w}{U_0 + u} \approx \frac{w}{U_0} \tag{10.13a}$$

It is now assumed that any aerodynamic force or moment behaves linearly with any perturbation quantity. Thus, for example,

$$\Delta X = \frac{\partial X}{\partial u}u + \frac{\partial X}{\partial v}v + \frac{\partial X}{\partial \alpha}\alpha + \frac{\partial X}{\partial \dot{\alpha}}\dot{\alpha} + \frac{\partial X}{\partial p}p + \frac{\partial X}{\partial q}q$$
$$+ \frac{\partial X}{\partial r}r + \frac{\partial X}{\partial \delta}\delta + \frac{\partial X}{\partial \phi}\phi + \frac{\partial X}{\partial \theta}\theta + \frac{\partial X}{\partial \psi}\psi \quad (10.13b)$$

All of the derivatives in the above are evaluated at the trimmed condition. δ signifies one of the control angles, and the terms involving ϕ, θ, and ψ relate to possible gravitational forces. The terms $(dZ/d\dot{\alpha})\dot{\alpha}$ and $(dM/d\dot{\alpha})\dot{\alpha}$ are related to unsteady aerodynamics, which will be discussed later.

The derivatives in Equation 10.13*b*, such as dX/du, will now be replaced by so-called stability derivatives. In the case of forces, the stability derivatives are obtained by dividing the derivatives by the mass. The moment derivatives are divided by the appropriate mass moment of inertia. The resulting stability derivatives are denoted by a subscripted notation. Thus, for example,

$$X_u = \frac{1}{m}\frac{\partial X}{\partial u}$$

$$M_\alpha = \frac{1}{I_y}\frac{\partial M}{\partial \alpha}$$

Equation 10.13*b* is written very generally, and many of the derivatives will equal zero. Table 10.1 is a matrix listing the possible stability and control derivatives with an indication of their significance. With this table in mind, the longitudinal equations of motion become

$$X_u u + X_\alpha \alpha - g\cos\theta_0 \theta = \dot{u} \quad (10.14a)$$

$$Z_u u + Z_\alpha \alpha + Z_{\dot{\alpha}}\dot{\alpha} + Z_q \dot{\theta} - g\sin\theta_0 \theta + Z_\delta \delta_e = U_0(\dot{\alpha} - \dot{\theta}) \quad (10.14b)$$

$$M_\alpha \alpha + M_{\dot{\alpha}}\dot{\alpha} + M_q \dot{\theta} + M_\delta \delta_e = \ddot{\theta} \quad (10.14c)$$

Table 10.1 Aerodynamic Stability and Control Derivatives

Independent Variable	Dependent Variable					
	X	Y	Z	L	M	N
u	+	−	+	−	−	−
v	−	+	−	+	−	+
α	+	−	+	−	+	−
$\dot{\alpha}$	−	−	+	−	+	−
p	−	−	−	+	−	+
q	−	−	+	−	+	−
r	−	+	−	+	−	+
δ_e	−	−	+	−	+	−
δ_r	−	+	−	+	−	+
δ_a	−	−	−	+	−	−
ϕ	−	+	−	−	−	−
θ	+	−	−	−	−	−
ψ	−	−	−	−	−	−

− value is normally zero or small.
\+ value can be significant.

Note in the above that q has been replaced by $\dot{\theta}$ resulting in three linear, simultaneous, partial-differential equations with constant coefficients in u, α, and θ.

Before proceeding with the solution of these equations, let us examine the derivatives. Again, it is emphasized that these derivatives are partial derivatives and evaluated at the trim condition.

X Derivatives

Referring to Figure 10.2, an airplane is shown in a steady climb. The airplane is rotated through the initial climb angle θ_0 about its y-axis resulting in a component of the gravitational force in the x-direction given by Equation 10.11a. From this figure the net force in the x-direction will be

$$\sum X = T - D - W \sin \theta \cos \phi \tag{10.15}$$

Thus,

$$X_u = \frac{1}{m}\left(\frac{\partial T}{\partial u} - \frac{\partial D}{\partial U}\right) \tag{10.16}$$

$$X_\alpha = \frac{1}{m}\left(\frac{\partial T}{\partial \alpha} - \frac{\partial D}{\partial \alpha}\right) \tag{10.17}$$

For the trimmed condition, $\sum X = 0$ and

$$T_0 = D_0 + W \sin \theta_0$$

The drag can be written as $D = \frac{1}{2}\rho (U_0 + u)^2 C_D S$. Differentiating this with respect to u, with α constant, and evaluating the derivative at the initial condition gives

$$D_u = \rho U_0 C_D S/m \tag{10.18}$$

Since the gravitational term does not depend upon u, it contributes nothing to the X_u derivative. The thrust, however, can vary with speed and must be considered.

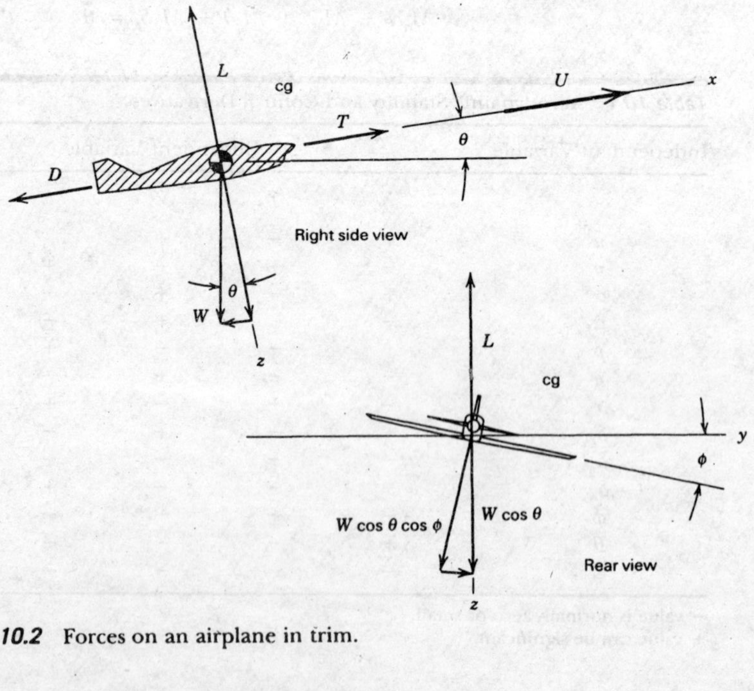

Figure 10.2 Forces on an airplane in trim.

Following the lead of Reference 9.2, it is observed that the thrust of a jet remains fairly constant over a small speed range. Thus, for this type of powerplant, $T_u \approx 0$. The same is true, of course, for gliding flight since the thrust itself is zero. For propeller-driven airplanes, over a small speed range, it can be assumed that the power and efficiency remain constant. This means that in increasing the speed from U to $U + \Delta U$ with the thrust changing from T to $T + \Delta T$, the following must hold

$$TU = (T + \Delta T)\ (U + \Delta U)$$

Expanding the above and dropping higher-order terms results in the derivative of the thrust with respect to airspeed, evaluated at the trim condition being

$$\frac{\partial T}{\partial u} = -\frac{T_0}{U_0}$$

Combining the above relationships, the derivative of the X force with respect to the forward speed is obtained as

$$X_u = -\frac{1}{m}\left(\frac{T_0}{U_0} + \rho\, U_0\, C_{D_0}\, S\right) \quad \text{(propeller-driven airplanes)} \tag{10.19a}$$

$$X_u = -\rho\, U_0\, C_{D_0}\, S/m \quad \text{(jet-driven and gliding airplanes)} \tag{10.19b}$$

From the material in Chapter 4, the drag of an airplane can be expressed as

$$D = \frac{1}{2}\rho\,(U_0 + u)^2\left(f + \frac{C_L^2}{\pi A e}\right)$$

Thus, differentiating this with respect to α and evaluating the result at the trim condition gives

$$\frac{\partial D}{\partial \alpha} = \frac{\rho\, U_0^2\, S C_{L_0}\, C_{L\alpha}}{\pi A e}$$

Since the thrust and the x component of the weight do not depend upon α, the above is equal to the negative of the $\partial X/\partial \alpha$. Thus,

$$X_\alpha = \frac{g}{\cos(\theta)} - \frac{1}{m}\,\frac{\rho\, U_0^2\, S C_{L_0}\, C_{L\alpha}}{\pi A e} \tag{10.20}$$

The X_q derivative is assumed to be zero. This is not quite true since the pitching will affect to some extent the lift on both the wing and the tail, and hence, the drag of both surfaces. However, this is a secondary effect, which will be neglected, but which should be included in a more sophisticated analysis.

Z Derivatives

From Figure 10.2, the force along the airplane's z-axis is given by

$$Z = W \cos\theta \cos\phi - \frac{1}{2}\rho\,(U_0 + u)^2\, S C_L \tag{10.21}$$

Differentiating this equation in turn with respect to θ, u, and w results in the following stability derivatives evaluated at the trim condition.

$$Z_u = -\frac{1}{m}\rho\, U_0\, S C_{L_0} \tag{10.22}$$

$$Z_\alpha = -\frac{1}{m}\frac{1}{2}\rho\, U_0^2\, S C_{L\alpha} \tag{10.23}$$

In the last chapter, it was reasoned that the angle of attack at the tail increases because of pitching. This increase is given by

$$\Delta\alpha = \frac{l_t Q}{U_0}$$

Thus, the change in the Z force, because of the increase in α, will change by

$$\Delta Z = -\frac{1}{2}\rho\, U_0^2\, a_t S_t \frac{l_t Q}{U_0}$$

Thus the Z derivative, Z_q, is given by

$$Z_q = -\eta_t \frac{1}{m}\frac{1}{2}\rho\, U_0 a_t S_t l_t \tag{10.24}$$

M Derivatives

The moment about the cg can be written as

$$M = \frac{1}{2}\rho\,(U_0 + u)^2\, S\bar{c}\, C_M$$

Since the derivatives are evaluated at the trim condition, and since C_M is zero for this condition, it follows that $M_u = 0$. M_α is obtained immediately from the above by differentiating with respect to α.

$$M_\alpha = \frac{1}{I_y}\frac{1}{2}\rho\, U_0^2\, S\bar{c}\, C_{M\alpha} \tag{10.25}$$

The derivative M_q is obtained directly from Z_q and is equal to

$$M_q = \frac{m}{I_y} l_t Z_q \tag{10.26}$$

Control Derivatives

The control derivatives in coefficient form, $C_{l_{\delta a}}$, $C_{M_{\delta e}}$, and $C_{N_{\delta r}}$, which provide moments about the x-, y-, and z-axes, were covered earlier in Chapter 9. (See Equations 9.39, 9.120, and 9.115.) The control derivative for longitudinal motion in physical units will simply be $C_{M_{\delta e}}$ multiplied by the dimensional quantities by which it is defined.

$$M_\delta = \frac{1}{I_y}\frac{1}{2}\rho\, U_0^2\, S\bar{c}\, C_{M_{\delta r}} \tag{10.27}$$

In addition to providing a pitching, the deflection of the elevator produces an increased lift, or a decrease in the Z force. Thus, we have another derivative of δ_e.

$$Z_\delta = -\eta_t \frac{1}{m}\frac{1}{2}\rho\, U_0^2\, S_t a_t \tau \tag{10.28}$$

Effects of Compressibility and Unsteady Aerodynamics

To this point, the stability derivatives have all been determined on the basis of steady aerodynamics, which neglects the rate of change of angle of attack on the lift of a wing. This effect is very important to the subject of aeroelasticity where the frequency at which a surface oscillates can be very high. The period of oscillation for a wing that is fluttering can be of the same order or less as the time it takes the flow to travel one chord length. The typical frequency for rigid-body motion of an entire airplane, however, is relatively much slower. We will be examining the normal modes of motion for a typical airplane and will find that, typically, the shortest period of a mode, and it is a heavily damped mode, is of the order of 1 or 2 sec. During this time the flow will travel 50 or so chord lengths. Thus, unlike an aeroelastic phenomena, at any instant one can safely apply a quasisteady approximation in determining the lift. However, there is one unsteady effect concerning the downwash at the tail, which can influence airplane motion.

If the lift on a wing varies with time, a sheet of spanwise vorticity is shed downstream. If the rate at which the lift changes is sufficiently high, this shed vorticity can significantly change the lift, as a function of α, from the steady-state value.

One will find a fairly comprehensive discussion of stability derivatives in Reference 9.2 and 10.3. Following the earlier reference, an estimate of the unsteady effect of the wing on the tail can be made based on the fact that it takes a finite time for the wing wake to be convected downstream to the tail. Thus, $\epsilon(t)$ at the tail is produced by the wing α at t-Δt. If α is increasing at a constant rate of $\dot{\alpha}$,

$$\epsilon(t) = \epsilon_\alpha - \epsilon_\alpha \dot{\alpha}\, \Delta t$$

where $\epsilon_\alpha \alpha$ would be the steady-state downwash. Δt is the time required for the wake to be transported from the wing to the tail. This is given approximately by $\Delta t = l_t / U_0$. Thus,

$$\epsilon(t) = \epsilon_\alpha \alpha - \epsilon_\alpha \dot{\alpha} \frac{l_t}{U_0}$$

or

$$\epsilon_{\dot{\alpha}} = -\epsilon_\alpha \frac{l_t}{U_0} \tag{10.29}$$

The product $\epsilon_\alpha \alpha$ represents an increase in the tail angle of attack. Hence, both Z and M will vary with $\dot{\alpha}$, so terms $Z_{\dot{w}}$ and $M_{\dot{w}}$ should be added to the left-hand sides of Equations 10.13*b* and *c*, respectively. Specifically, these derivatives will be

$$Z_{\dot{\alpha}} = -\eta_t \frac{1}{2m} \rho U_0 a_t S_t l_t \epsilon_\alpha \tag{10.30}$$

$$M_{\dot{\alpha}} = \left(\frac{m}{I_y}\right) l_t Z_{\dot{\alpha}} \tag{10.31}$$

Since this effect is normally small and has the largest influence on the pitching moment, only Equation 10.31 will be included in the material to follow.

It may be necessary to include additional terms involving other stability derivatives in the set of Equations 10.13, depending on the airplane configuration and its operating regime. For example, an airplane operating at transonic speeds will be sensitive to speed changes because of Mach number effects. Hence, a term will have to be added to X_u to account for the increase in C_D with Mach number.

Solution of Equations

The equations of longitudinal motion represent a set of three, simultaneous, linear differential equations. The classical solution of these types of equations is discussed in Appendix D. In general, the solution for any dependent variable will be composed of the sum of a particular, or steady, solution and a homogeneous, or transient, solution. It is the homogeneous solution that determines whether or not the airplane will be stable. This solution is obtained by setting all control input to zero and solving the resultant set of equations. In general, the solution for this case for any one dependent variable, using α as an example, will take the form of

$$\alpha = \alpha_0 e^{\sigma t}$$

where the amplitude, α_0, and the root, σ, are complex.

For an arbitrary control input it is convenient to solve the set of equations of motion by the methods of Laplace transforms. The Laplace transform can be applied to the set of Equations 10.14 and the result written in matrix form as

$$\begin{bmatrix} g\cos\theta_0 & (s - X_u) & -X_\alpha \\ (U_0 + Z_q)\ s\text{-}g\sin\theta_0 & Z_u & (-U_0 s + Z_\alpha) \\ s(s - M_q) & 0 & -(M_{\dot\alpha} s + M_\alpha) \end{bmatrix} \begin{bmatrix} \theta \\ u \\ \alpha \end{bmatrix} = \begin{bmatrix} 0 \\ mZ_\delta \\ I_y M_\delta \end{bmatrix} \bar{\delta}_e \tag{10.32}$$

where s is the independent transform variable and $\bar{\delta}_e$ is the transform of the elevator angle.

Before determining the response of the system to an elevator input, the stability of the system and the normal modes can be examined by solving the *characteristic equation* obtained by equating the 3 × 3 matrix on the left side to zero, replacing s by σ, and solving it as a determinant in σ. This is directly comparable to what one obtains by assuming the exponential form for the dependent variables as given above. The resulting values of σ are the *eigenvalues* for the system and are the only values for which the homogeneous solution will exist. Expanding this determinant, a quartic is obtained of the form

$$A\sigma^4 + B\sigma^3 + C\sigma^2 + D\sigma + E = 0 \tag{10.33}$$

where

$$A = -U_0$$
$$B = U_0(X_u + M_Q) + Z_\alpha + (U_0 + Z_q)\ M_{\dot\alpha}$$
$$C = X_\alpha Z_u - X_u Z_\alpha - M_q(U_0 X_u + Z_\alpha) - (U_0 + Z_q)\ (X_u M_{\dot\alpha} - M_\alpha) - gM_{\dot\alpha}\sin(\theta_0)$$
$$D = M_q(X_u Z_\alpha - X_\alpha Z_u) - (U_0 + Z_q) X_u M_\alpha - gZ_u M_{\dot\alpha}\cos(\theta_0) + g(X_u M_{\dot\alpha} - M_\alpha)\sin(\theta_0)$$
$$E = gM_\alpha(X_u \sin\theta_0 - Z_u \cos\theta_0)$$

To determine whether or not a particular airplane is longitudinally, dynamically stable at a particular operating condition, the stability derivatives are first calculated. Then the above quartic is solved for its roots. In general, these roots will be complex of the form

$$\sigma = \eta \pm i\omega$$

where η and ω are real numbers. These roots characterize the normal modes of the longitudinal motion and determine if the motion is stable. If η is negative, the motion will be periodic, displaying a damped oscillation as shown in Figure 10.3*a*. If η is positive, the motion will again be periodic but will be a diverging oscillation as shown in Figure 10.3*b*. If ω is zero, the root is real and the motion will be aperiodic. It will converge if η is negative or diverge if η is positive. These two cases are shown in Figure 10.3*c*.

A dynamic system is stable, if, when perturbed from a state of equilibrium, the system returns to its original state. Obviously, the condition that the normal modes of motions shown in Figures 10.3*a*, 10.3*b*, and 10.3*c* be stable is that *any real, or real part, of a root be negative.* A value of s in the complex plane equal to a root is referred to as a *pole* and represents a value of s for which the characteristic equation will be zero. In line with the previous statement regarding the real part of a root, a system will be stable if all of its poles lie in the left half of the complex plane.

Complex roots of the characteristic equation will always appear in conjugate pairs. Thus, the motion corresponding to a pair of complex roots can be written in the form

$$x = \bar{x}e^{\eta}(e^{i\omega t} + e^{-i\omega t})$$

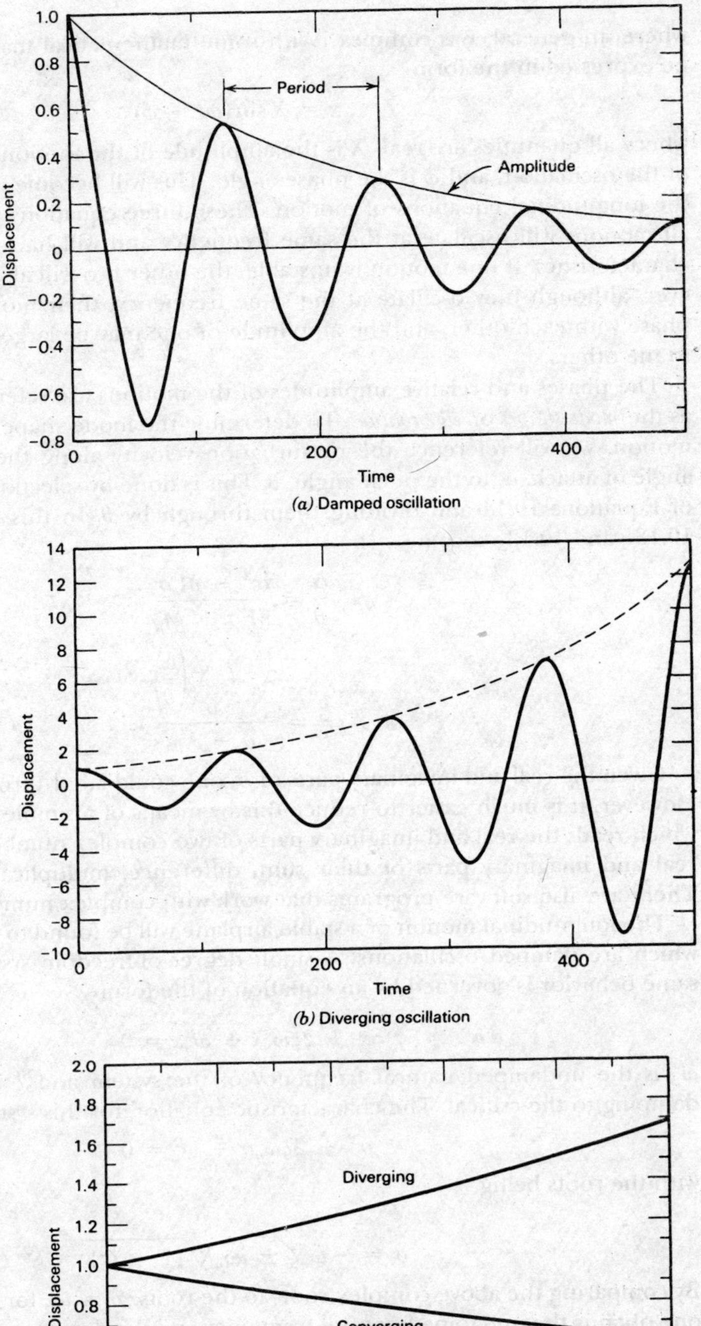

Figure 10.3 Possible types of unforced linear motion.

where, in general, $\bar{x}$ is complex. With some mathematical manipulation, this can be expressed in the form

$$x = X \sin(\omega t + \phi)$$

where all quantities are real. X is the amplitude of the motion, ω is the frequency of the oscillation, and ϕ is the phase angle. This will become clear by considering the longitudinal equations of motion. These three equations are coupled so that all motions will oscillate at the same frequency and will have the same damping characteristics. If one motion is unstable, the other two will also be unstable. However, although they oscillate at the same frequency, their motions will not be in phase with each other, and the amplitude of one may be large or small compared to the others.

The phases and relative amplitudes of the motions are referred to, collectively, as the *mode shapes* or *eigenvalues*. To determine the mode shape of the longitudinal motion, we will reference the perturbation velocity along the x-axis, u, and the angle of attack, α, to the pitch angle, θ. This is done by selecting any two of the set of Equations 10.14 and dividing them through by θ. In this manner, Equations 10.13a and 10.13c become

$$\frac{\alpha}{\theta} = \frac{\sigma^2 - M_q\sigma}{M_{\dot{\alpha}}\sigma + M_\alpha} \tag{10.34}$$

$$\frac{u}{\theta} = \frac{X_\theta + X_\alpha\left(\dfrac{\alpha}{\theta}\right)}{\sigma - X_u} \tag{10.35}$$

Given the real and imaginary parts of σ, one could labor through the reduction. However, it is much easier to reduce this by means of a simple computer routine, which reads the real and imaginary parts of two complex numbers and returns the real and imaginary parts of their sum, difference, multiplication, and division. There are also software programs that work with complex numbers directly.

The longitudinal motion of a stable airplane will be found to exhibit two modes, which are damped oscillations. A single-degree-of-freedom system displaying the same behavior is governed by an equation of the form

$$\ddot{x} + 2\zeta\omega_n\dot{x} + \omega_n^2 x = 0 \tag{10.36}$$

ω_n is the undamped natural frequency of the system and ζ is the ratio of the damping to the critical. The characteristic equation for this system is

$$\sigma^2 + 2\zeta\omega_n\sigma + \omega_n^2 = 0$$

with the roots being

$$\sigma = -\omega_n\zeta \pm i\omega_n\sqrt{(1 - \zeta^2)}$$

By comparing the above complex roots to the roots, $\eta \pm i\omega$, for a particular mode, one obtains the undamped natural frequency and the damping ratio for the mode.

$$\omega_n = \sqrt{\omega^2 + \eta^2} \tag{10.37a}$$

$$\zeta = -\frac{\eta}{\omega_n} \tag{10.37b}$$

For oscillatory motion, the period, T, is defined as the time required to go from one peak displacement to the next. This time is found from

$$\omega T = 2\pi \tag{10.38a}$$

The time for the amplitude of a stable oscillation to damp to half-amplitude, $T_{1/2}$, is determined by the value of η.

$$\frac{e^{\eta(t+T_{1/2})}}{e^{\eta t}} = \frac{1}{2}$$

This reduces to

$$T_{1/2} = -\frac{\log_e 2}{\eta} \tag{10.38b}$$

If η is positive, the oscillation is unstable and the amplitude of the oscillation will diverge. The time for the oscillation to double, T_2, is equal to Equation 35*b* without the minus sign.

$$T_2 = \frac{\log_e 2}{\eta} \tag{10.38c}$$

Equations 10.38*b* and *c* apply also to aperiodic motion, that is, to a nonoscillatory convergent or divergent motion. As will be seen later, these parameters are important in specifying the handling qualities of an airplane.

Example Calculation of the Longitudinal Modes

The Piper Cherokee, pictured in Figure 3.59, will be used to exemplify the determination of the longitudinal modes of an airplane. This airplane will be used extensively for the remainder of this text for purposes of illustration. It is therefore emphasized, as noted on Figure 3.59, that dimensions and parameters for this airplane can be found in the appendix. These parameters are those that have been estimated or calculated and are used in many of the examples or computer exercises. For convenience, some of the numbers are given here.

Operating conditions

$$U_0 = 100 \text{ mph SSL} \quad \theta_0 = 0 \quad \phi_0 = 0$$

Weight and inertia

$$W = 2400 \text{ lb} \quad I_y = 1249 \text{ slug ft}^2 \quad \text{cg is 1.0 ft behind wing LE}$$

Wing

$$b = 30 \text{ ft} \quad c_0 = 5.25 \text{ ft} \quad \lambda = 1.0 \quad \Lambda = 0$$

Tail

$$b = 10 \text{ ft} \quad c_0 = 2.5 \text{ ft} \quad \lambda = 1.0 \quad \Lambda = 0$$

Distance from wing LE to tail LE = 13.96 ft
Height of tail above wing = 1.0 ft

Fuselage

$$l_n = 5.0 \qquad l_b = 23.0 \text{ ft}$$

Distance of nose from wing LE = 5.0 ft
Length of nose section = 5.0 ft

Drag parameters

$$W = 2400 \text{ lb} \quad f = 5 \text{ ft}^2 \quad e = 0.6 \quad l_b = 23.0 \text{ ft}$$

With this input to the lifting surface programs undertaken in Chapter 3 and the static stability programs described in Chapter 9, the following parameters and coefficients can be determined:

$$\epsilon_\alpha = 0.413 \quad a_w = 4.20/\text{rad} \quad a_t = 3.66/\text{rad}$$

$$\bar{c} = 5.25 \text{ ft} \quad C_{L\alpha} = 4.50/\text{rad} \quad h_n = 0.428$$

These numbers do not include the effect of the fuselage or propellers.

Adding the fuselage results in the following increments.

$$\Delta C_{L_\alpha} = 0.14/\text{rad} \qquad \Delta C_{M_\alpha} = 0.093$$

The propeller adds a small contribution to C_{L_α} and C_{M_α}, which can be determined approximately, not knowing the propeller geometry, by assuming reasonable values for C_l and C_d in Equations 9.104 and 9.105. The propeller solidity is then determined which would result in these values while producing T_0. For this example, C_l of 1.0 and a C_d of 0.01 were assumed together with a propeller rpm of 2400 and a diameter of 6.17 ft. In this way, the following increments are found for the propeller.

$$\Delta C_{L_\alpha} = 0.01/\text{rad} \qquad \Delta C_{M_\alpha} = 0.012$$

Adding the fuselage and propeller to the wing and tail results in the following:

$$C_{L_\alpha} = 4.75/\text{rad} \qquad h_n = 0.40650$$

Thus, the effect of adding the fuselage and propeller is to move the stick-fixed neutral point forward 2.3% of the mean chord, or 1.4 in.

With the parameters given above, the following values can be determined for the force and moment derivatives:

$$X_\alpha = -27.52 \text{ ft/sec/sec/rad} \qquad X_u = -0.0715/\text{sec}$$
$$Z_\alpha = -256.9 \text{ ft/sec/sec/rad} \qquad Z_u = -0.439/\text{sec}$$
$$Z_Q = -2.655 \text{ ft/sec/rad} \qquad M_\alpha = -17.34/\text{sec/sec}$$
$$M_{\dot\alpha} = -0.7149/\text{sec} \qquad M_q = -1.927/\text{sec}$$

Since these stability derivatives represent forces and moments divided by mass and mass moments of inertia, respectively, they have only the units of feet, seconds, and radians.

After dividing through to set the first coefficient to unity, the coefficients of the characteristic equation become

$$A = 1.0 \quad B = 4.452 \quad C = 20.64 \quad D = 1.369 \quad E = 1.671$$

The roots of this equation are a pair of complex root.

$$\sigma_1 = -0.02477 \pm 0.2856i \qquad \sigma_2 = -2.201 \pm 3.936i$$

The first root gives a period of 22.0 sec and a time-to-damp to half-amplitude of 28.0 sec. The second root has a period of 1.60 sec and a time-to-damp to half-amplitude of only 0.31 sec.

These two modes are characteristic of most stable airplanes. The first mode, with light damping and a relatively long period, is called the *long-period* or *phugoid mode.* The second, heavily damped mode is referred to simply as the *short-period mode.*

As the cg moves aft toward the neutral point, the longitudinal motion will become unstable. To investigate this, one generates a *root locus plot.* This is a plot of the roots in the complex plane as a parameter is varied, in this case the cg position. Root locus plots for the example being considered here are presented in Figures 10.4 and 10.5. Any roots that lie in the left-half plane of these figures are stable and any that are in the right-half plane correspond to unstable motion. If the root is complex, the motion is oscillatory; otherwise, the motion is aperiodic. Beginning with a cg position of 1.0 ft, which has been used in the example thus far, it can be seen from this figure that the phugoid mode of the Cherokee remains stable and oscillatory until the cg is close to the stick-fixed neutral point. Beyond this, for a short distance, the pair of complex roots become two real roots, one positive, or unstable, and the other negative. As the cg continues to move back, the

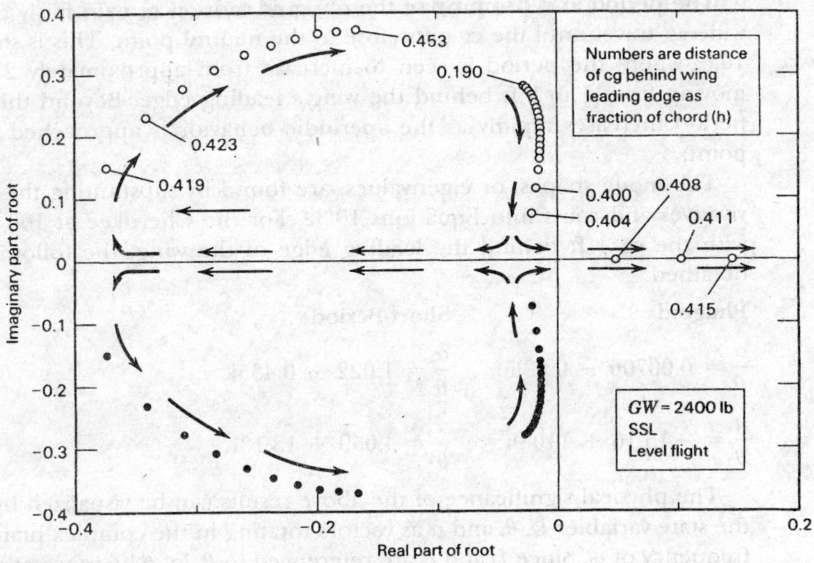

Figure 10.4 Root locus plot of phugoid motion for the Cherokee.

stable root splits again into a pair of stable, complex roots, while the other real root merges with a root of the short-period mode.

As the cg moves back, Figure 10.5 shows that the roots of the short period mode behave initially, like those of the phugoid mode. However, this mode does not become unstable until the cg is slightly behind the neutral point at a static margin of approximately −1.3%. This is interesting, but does not change the fact that the airplane is unstable at this condition because of the behavior of the phugoid mode. If the cg is behind the neutral point, the system is statically unstable and, as such, will be unstable. *A system can be statically stable but dynamically unstable. However, it cannot be statically unstable, but dynamically stable.*

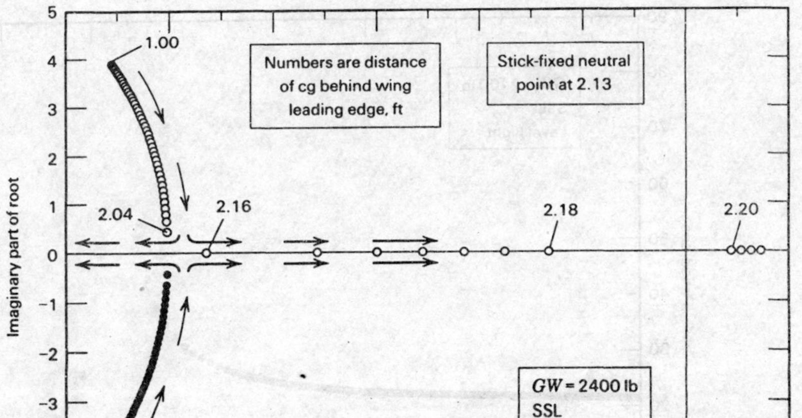

Figure 10.5 Root locus plot of short-period motion for the Cherokee.

The period and damping of the phugoid motion remain fairly constant over a wide cg travel until the cg gets close to the neutral point. This is shown in Figure 10.6, where the period is seen to increase from approximately 21 to 31 sec in moving from 1 to 2 ft behind the wing's leading edge. Beyond this location, the period increases rapidly as the aperiodic behavior is approached at the neutral point.

The mode shapes, or eigenvalues, are found by substituting the calculated derivatives and roots into Equations 10.32. For the Cherokee at 100 mph and SSL with the cg 1 ft behind the leading edge of the wing, the following results are obtained.

Phugoid	**Short-Period**
$\dfrac{\alpha}{\theta} = 0.00706 - 0.031i$	$\dfrac{\alpha}{\theta} = 1.022 + 0.435i$
$\dfrac{u}{\theta} = -15.16 + 110.9i$	$\dfrac{u}{\theta} = 4.059 + 13.13i$

The physical significance of the above results can be visualized by representing the state variables U, θ, and α as vectors rotating in the complex plane at a circular frequency of ω. Since U and α are referenced to θ, let θ be represented by a vector lying instantaneously along the positive real axis having a magnitude of unity. Any of the above quantities plotted in the same plane will then show their relationship to θ. This is done to scale in Figure 10.7 for α/θ. It can be seen that the magnitude of α is greater than the magnitude of θ by a factor of 1.11, whereas the displacement of α leads that of θ by 23.1°. In general, if the ratio of x to y is given by $a + ib$, the ratio of the magnitude of x to y will be $\sqrt{a^2 + b^2}$ and x will lead y by the angle $\tan^{-1} b/a$. Therefore, we can write the above relationships between u, α, and θ as

Phugoid	**Short-Period**
$\alpha = 0.0318\theta_0 e^{-\eta t}\cos(\omega t - \angle 77.2°)$	$\alpha = 1.11\theta_0 e^{-\eta t}\cos(\omega t + \angle 23.1°)$
$U = 111.9\theta_0 e^{-\eta t}\cos(\omega t + \angle 97.8°)$	$U = 13.7\theta_0 e^{-\eta t}\cos(\omega t + \angle 72.8°)$

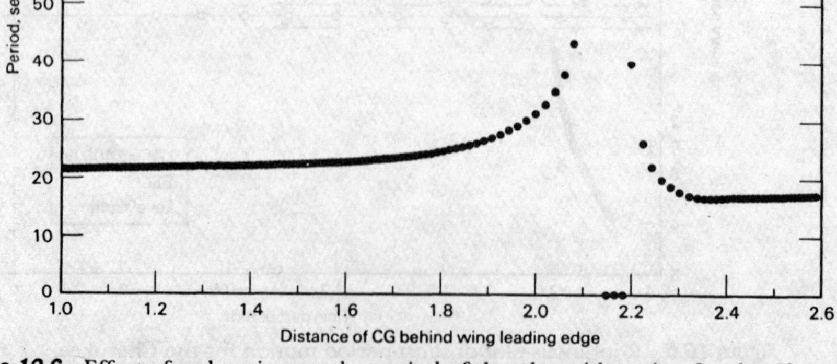

Figure 10.6 Effect of cg location on the period of the phugoid motion for the Cherokee.

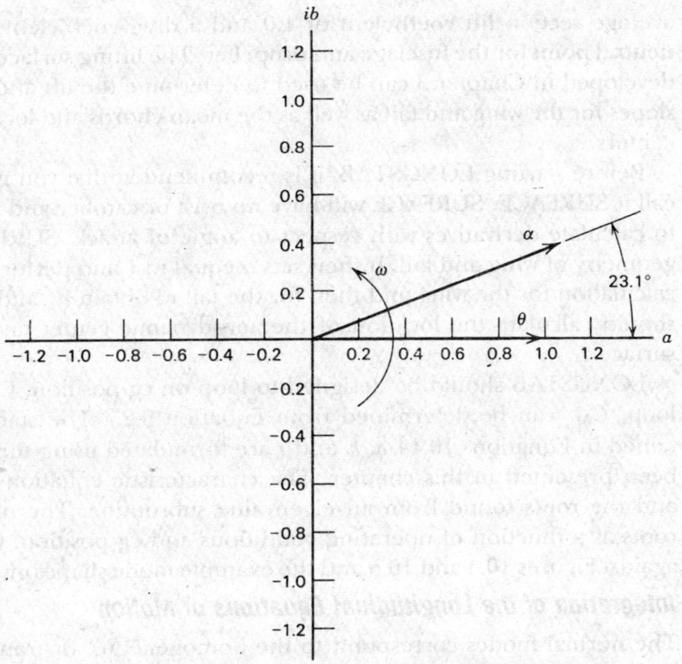

Figure 10.7 Representation of state variables in complex plane.

Approximate Mode Shapes

From an examination of the above results, it can be seen that the phugoid mode corresponds to motion at nearly a constant angle of attack, whereas the airspeed is nearly constant for the short-period mode. Thus, if Equations 10.14*a* and *b* are used with α and $\dot{\alpha}$ equal to zero, a simple quadratic equation is obtained for the roots of the phugoid mode. If Equations 10.14*b* and *c* are used with u and $\dot{u}$ set to zero, another simple quadratic equation is obtained for the short-period mode.

Phugoid Approximation

$$\sigma^2 - X_u\,\sigma - \left(\frac{gZu}{U_0 + Z_q}\right) = 0 \qquad (10.39)$$

Short-Period Approximation

$$\sigma^2 - \left(\frac{U_0 M_q + Z_\alpha + M_{\dot{\alpha}}(Z_q + U_0)}{U_0}\right)\sigma + \left(\frac{M_q Z_\alpha - M_\alpha(U_0 + Z_q)}{U_0}\right) = 0 \qquad (10.40)$$

For the example of the Cherokee at 100 mph SSL, the approximations result in periods that are 8.3% and 4.6% lower than the exact values for the phugoid and short period, respectively.

Computer Exercise 10.1 "LONGSTAB"

Write a computer program to predict the roots and mode shapes for the longitudinal equations of motion. The input, provided by a data file, should contain the geometry of the wing, horizontal tail, fuselage, and propeller. The output from program NP (Computer Exercise 9.2) can be used to provide the downwash and neutral point for the wing–tail combination. For treating the propeller, assume an average section lift coefficient of 1.0 and a drag coefficient of 0.01. Correct the neutral point for the fuselage and propeller. The lifting surface program, LATTICE, developed in Chapter 3 can be used to determine the lift and moment coefficient slopes for the wing and tail as well as the mean chords and location of aerodynamic centers.

Before starting LONGSTAB, it is recommended that you modify LATTICE and call it SURFACE. SURFACE will have no twist or camber and will be designed only to calculate derivatives with respect to angle of attack. SURFACE should read in geometry of wing and tail. It then sets α equal to 1 and performs the lifting surface calculation for the wing and then for the tail to obtain a_w and a_t. In the process, it should calculate the location of the aerodynamic center and the MAC for each surface.

LONGSTAB should be designed to loop on cg position. Each time around the loop, C_{M_α} can be determined from Equation 9.25. The stability derivatives contained in Equations 10.14 *a, b,* and *c* are formulated using the equations that have been presented in this chapter. The characteristic equation is then determined, and the roots found from an eigenvalue subroutine. The output should be the roots as a function of operating conditions and cg position. Check your program against Figures 10.4 and 10.5 and the example mode shapes given for the Cherokee.

Integration of the Longitudinal Equations of Motion

The normal modes correspond to the homogeneous, or transient solution of the simultaneous differential equations of motion. The particular solution, also known as the forced response or steady-state solution, is found by numerically integrating the equations. This can be done using almost any acceptable numerical integration scheme, but for our purposes, a fourth-order Runge–Kutta method will be used. Referring to Equations 10.13 *a, b,* and *c*, in order to deal with first-order derivatives only, a supplementary equation will be defined.

$$q = \dot{\theta} \tag{10.41a}$$

The original set of equations can then be written as

$$\dot{u} = X_u u + X_\alpha \alpha + g\theta \tag{10.41b}$$

$$(U_0 - Z_{\dot{\alpha}})\,\dot{\alpha} = Z_u u + Z_\alpha \alpha + Z_\theta \theta + U_0 q + Z_\delta \delta_e \tag{10.41c}$$

$$\dot{q} = M_\alpha \alpha + M_{\dot{\alpha}} \dot{\alpha} + M_q q + M_\delta \delta_e \tag{10.41d}$$

One additional equation is added to the above in order to obtain the altitude, namely Equation 10.10*c* with U replaced by $U_0 + u$, V, and ϕ set to zero and W replaced by $U_0\alpha$.

Thus, we have five first-order, simultaneous equations in the form required by the Runge–Kutta scheme. However, since $\dot{\alpha}$ appears on the right-hand side of Equation 10.41*d*, this equation should be evaluated before Equation 10.41*c*, so that $\dot{\alpha}$ for the previous time-step is being used in 10.41*d*. The reader can refer to any one of the many references available on numerical methods, so details will not be presented here concerning the numerical integration.

Returning to the example of the Cherokee at 100 mph, SSL, the set of Equations 10.41 have been integrated for three particular cases. The first case models a flight testing experiment, which your author has flown many times. This experiment is a good means of identifying the phugoid mode of an airplane and can be done with very simple instrumentation, namely, a voice recorder and a stopwatch.

The experiment begins by trimming the airplane straight and level at 100 mph. A pull is then exerted on the wheel and held until the airplane is steady at 80 mph. The airplane will now have a slight rate of climb. On the count of zero, the wheel is quickly pushed forward and held at the original trimmed position. The airplane will then nose over, lose altitude, and accelerate. However, at some speed above the trim speed it will nose up and begin to climb and slow down. The cycle will be repeated until the motion damps out to the original trim speed.

Observers can continuously read into a recorder the airspeed and altitude since the period of the phugoid is sufficiently long to allow this with a fair degree of accuracy. The data is then taken from the audio tapes by playing them back repeatedly and timing the readings with a stopwatch. The method is not very elegant, but it is surprisingly effective. Of course, commercial and government flight testing facilities utilize very accurate (and expensive) recording gyroscopes with the data frequently being telemetered to the ground and reduced in real time.

The initial conditions for this case are found by setting all the derivatives to zero and u to its initial value, in this case, -20 mph (converted to fps). In this manner, the initial values are

$$u = 29.3 \text{ fps} \quad \delta_e = -7.9^0 \quad \alpha = 2.2^0 \quad \theta = 0.7^0$$

Using these initial conditions, the longitudinal equations of motion were integrated to obtain the results shown in Figures 10.8*a* and 10.8*b*. In Figure 10.8*a*, the velocity and pitch angle are plotted as a function of time. The results can be seen to confirm the predictions based on the roots and mode shapes with regard to period and time to damp. You can turn that statement around and say that the numerical integration is confirmed by the roots. From this plot, one can only see the phugoid motion. However, if the time and *y*-scales are expanded, the heavily damped, short-period mode is found as shown for α in Figure 10.8*b*.

Computer Exercise 10.2 "MOTION"

Formulate a computer program to integrate the longitudinal equations using a fourth-order Runge–Kutta scheme. Input will be the stability derivatives determined by Program LONGSTAB, the mass properties of the airplane, and the initial state (U, Q, θ, α, altitude) of the airplane. Output should be the airplane state as a function of time for the initial conditions. Check your program by comparing it with Figures 10.8*a* and *b*.

Nondimensional Equations of Motion

We have worked with the equations of motion in dimensional form to this point since this seems to be the current practice in industry. However, in the past, much of the stability and control analyses were done on an entirely nondimensional basis. Some companies today use large computer codes where the dimensional input is made nondimensional, the calculations performed, and the nondimensional results then converted to dimensional units for output.

It is convenient to work in physical units; however, there are some advantages to working with nondimensional derivatives as they tend to have the same values from one airplane to another. As a simple example, consider Z_α and $C_{z\alpha}$. The latter is approximately equal to the negative of the lift curve, $C_{L\alpha}$, which is around 6.0/rad. The dimensional derivative, however, can have almost any value, depending on the size of the airplane, its airspeed, and its altitude.

In view of the above, the equations of motion in a dimensionless form will be considered briefly. Beginning with the longitudinal equations of motions, all force derivatives are nondimensionalized by dividing by the dynamic pressure, q, and the wing area, S. All moment derivatives are divided by $qS\bar{c}$. A reference time, t^*, is defined by

$$t^* = \frac{\bar{c}}{2U_0} \tag{10.42}$$

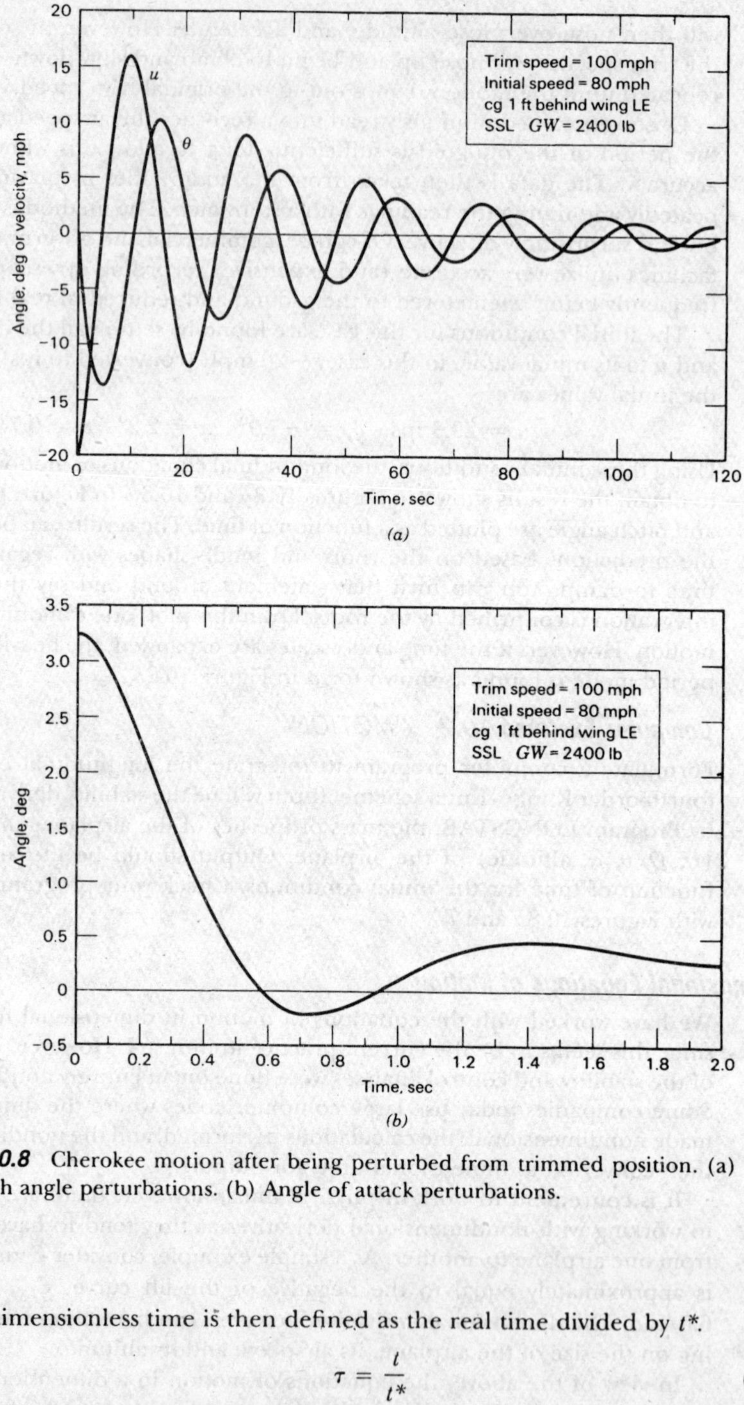

Figure 10.8 Cherokee motion after being perturbed from trimmed position. (a) Velocity and pitch angle perturbations. (b) Angle of attack perturbations.

and a dimensionless time is then defined as the real time divided by t^*.

$$\tau = \frac{t}{t^*} \tag{10.43}$$

τ has no units, but it is frequently measured in terms of air seconds. All velocities are referenced with respect to U_0 and again, the z-component of velocity is transformed to the angle of attack. The mass of the airplane is nondimensionalized and denoted by μ, where

$$\mu = \frac{m}{\rho S \frac{\bar{c}}{2}} \tag{10.44}$$

The dimensionless mass moment of inertia is defined by

$$i_y = \frac{I_y}{\rho S \left(\frac{\bar{c}}{2}\right)^3} \tag{10.45}$$

Details of the algebraic reduction will not be presented because they are fairly straightforward although somewhat laborious. A few examples of the process should suffice to show the process that is followed to arrive at the final equations. Consider, for example, the term $M_q \dot{\theta}$. First, any derivative with respect to time is changed to a derivative with respect to τ. For example,

$$\begin{aligned}\frac{\partial \theta}{\partial t} &= \frac{\partial \theta}{\partial t}\frac{d\tau}{dt} \\ &= \frac{2U_0}{\bar{c}}\frac{\partial \theta}{\partial \tau}\end{aligned}$$

Therefore, a term like $M_q \dot{\theta}$ can be written as

$$\frac{1}{2}\rho U_0^2 S \bar{c}\, C_{M_{\bar{q}}} \bar{q}$$

where it is now understood that the partial derivative of any dimensionless coefficient with respect to an independent variable implies that the independent variable is, itself, dimensionless. Thus, in the above,

$$C_{M_q} \equiv \frac{\partial C_M}{\partial \bar{q}}$$

where $\bar{q}$ is a dimensionless rate.

$$\bar{q} = \frac{Q\bar{c}}{2\,U_0}$$

Making these substitutions into Equations 13 *a*, *b*, and *c*, one arrives at a corresponding set of nondimensional equations, which govern the dimensionless motion of an airplane with respect to dimensionless time.

$$C_{x_u} u + C_{x_\alpha}\alpha - C_{L_0}\cos\Theta_0\theta = 2\mu\dot{u} \tag{10.46a}$$

$$2C_{L_0} u + 2\mu\dot{\alpha} - C_{Z_\alpha}\alpha - (2\mu + C_{Z_q})\,\dot{\theta} + C_{L_0}\sin\Theta_0\theta = C_{Z_\delta}\delta_e \tag{10.46b}$$

$$-\,C_{M_\alpha}\alpha + i_y\ddot{\theta} - C_{M_q}\dot{\theta} = C_{M_\delta}\delta_e \tag{10.46c}$$

It is emphasized that *all* quantities in the above equations are nondimensional.

Nondimensional Stability Derivatives

The stability derivatives in dimensionless form are obtained by first dividing the forces by qS and the moments by $qS\bar{c}$. However, care must be taken in differentiating the coefficients with respect to the dimensionless velocity component, u, and the dimensionless rates, q and $\dot{\alpha}$. As an example, two derivatives will be developed in detail.

C_{X_u}

The aerodynamic portion of the X force can be written as

$$X = \frac{1}{2}\rho\,[(U_0 + u)^2 + v^2 + w^2]\,SC_X$$

Solving the above for C_X and dropping higher-order terms gives

$$C_X = \frac{X}{\frac{1}{2}\rho U_0^2\,(1 + 2\bar{u})\,S}$$

where $\bar{u}$ is now dimensionless with respect to U_0. Differentiating this with respect to $\bar{u}$ and evaluating the result for the trimmed condition gives

$$C_{X_u} = \frac{2X_u}{\rho U_0 S} - 2C_{X0} \tag{10.47}$$

From Equation 10.15, the X coefficient at the trimmed condition is obtained as

$$C_{X0} = C_{L0}\sin\theta_0\cos\phi_0 \tag{10.48}$$

LATERAL-DIRECTIONAL MOTION

Lateral-directional motion, motion of the plane of symmetry, is governed by Equations 10.8*b*, 10.8*d*, and 10.8*f*. As with the longitudinal motion, the aerodynamic force and moments affecting the motion of the plane of symmetry are linearized about a trimmed condition.

$$\Delta Y = \frac{dy}{d\beta}\beta + \frac{dY}{dp}p + \frac{dY}{dr}r + \frac{dY}{d\delta_r}\delta_1 \tag{10.49a}$$

$$\Delta L = \frac{dl}{d\beta}\beta + \frac{dL}{dr}r + \frac{dL}{dp}p + \frac{dL}{d\delta_r}\delta_r + \frac{dL}{d\delta_a}\delta \tag{10.49b}$$

$$\Delta N = \frac{dN}{d\beta}\beta + \frac{dN}{dr}r + \frac{dN}{dp}p + \frac{dN}{d\delta_r}\delta_1 \tag{10.49c}$$

The perturbed gravitational force in the Y direction is obtained from Equation 10.11*b* as

$$\Delta Y_G = mg\cos\theta_0\cos\phi_0\phi \tag{10.49d}$$

Thus, for an airplane in a wings-level, steady climb, the linearized lateral-directional equations of motion become

$$Y_\beta\beta + Y_r r + Y_p\phi + Y_{\delta_r}\delta_r + g\cos\theta_0\phi = U_0\,(\dot{\beta} + r) \tag{10.50a}$$

$$L_\beta\beta + L_r r + L_p\dot{\phi} + L_{\delta r}\delta_r + L_{\delta_a}\delta_a = \ddot{\phi} - \frac{I_{XZ}}{I_x}\dot{r} \tag{10.50b}$$

$$N_\beta\beta + N_r r + N_p\dot{\phi} + N_{\delta_r}\delta_r = \dot{r} - \frac{I_{XZ}}{I_z}\ddot{\phi} \tag{10.50c}$$

Similar to the longitudinal motion, the stability derivative Y_β is equal to $dY/d\beta$ divided by the mass. L_β is equal to $dL/d\beta$ divided by I_x, whereas N_β is obtained by dividing $dN/d\beta$ by I_z. The other stability derivatives are defined in a similar manner.

Side Force Y Derivatives

The Y forces on an airplane arise mainly from the fuselage, vertical tail, and propeller as the result of sideslip and yawing. Figure 10.9 is the top view of an airplane

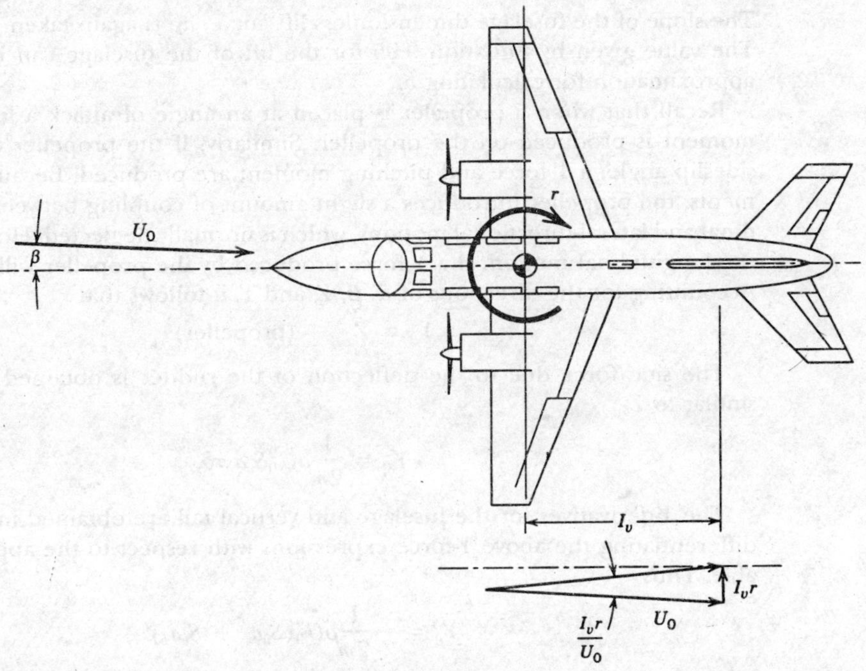

Figure 10.9 An airplane that is yawing and slipping.

that is both yawing and slipping. If the aerodynamic center of the vertical tail is located a distance of l_v behind the cg, then the sideslip angle at the tail produced by yawing will be given by

$$\Delta\beta = -\frac{rl_v}{U_0} \tag{10.51}$$

Adding the above to the sideslip angle at the cg results in a total vertical tail Y force given by

$$\Delta Y = -\frac{1}{2}\rho U_0^2 S_v a_v \left(\beta - \frac{rl_v}{U_0}\right) \qquad \text{(vertical tail)} \tag{10.52}$$

Be careful of the positive direction for Y and β. In this equation, a_v is the slope of the dimensionless tail lift curve, $dC_L/d\beta$, and is positive in the usual sense of lift and angle of attack.

A small additional amount of side force can be produced by the vertical tail because of rolling. If the aerodynamic center of the tail is placed a distance of h_v above the cg, then rolling about the x-axis will produce a change in the sideslip at the vertical tail given by

$$\Delta\beta = \frac{ph_v}{U_0} \tag{10.53}$$

This can be added to Equation 10.52, but it is normally small and frequently neglected.

Similar to longitudinal motion, the Y force produced by the fuselage is written as

$$\Delta Y = -\frac{1}{2}\rho U_0^2 a_b S_b \beta \qquad \text{(fuselage)} \tag{10.54}$$

The slope of the fuselage dimensionless lift curve, a_b, is again taken to be positive. The value given by Equation 9.95 for the lift of the fuselage can be used as an approximation for calculating a_b.

Recall that when a propeller is placed at an angle of attack a lift and yawing moment is produced on the propeller. Similarly, if the propeller is placed at a sideslip angle, a Y force and pitching moment are produced. Because of the moments, the propeller introduces a slight amount of coupling between the longitudinal and lateral-directional motions, which is normally neglected. However, as with the longitudinal motion, the Y force produced by the propeller will be included. Accounting for the directions of α, β, Z, and Y, it follows that

$$Y_\beta = Z_\alpha \qquad \text{(propeller)} \tag{10.55}$$

The side force due to the deflection of the rudder is obtained in a manner similar to Z_{δ_e}:

$$Y_{\delta_r} = \frac{1}{2m}\rho U_0^2 S_v a_v \tau \delta_r \tag{10.56}$$

The Y-derivatives for the fuselage and vertical tail are obtained immediately by differentiating the above Y-force expressions with respect to the appropriate variable. Thus

$$Y_\beta = -\frac{1}{2m}\rho U_0^2(S_v a_v + S_b a_b) \tag{10.57}$$

$$Y_r = \frac{1}{2m}\rho U_0 S_v a_v l_v \tag{10.58}$$

$$Y_p = -\frac{1}{2m}\rho U_0 S_v a_v h_v \tag{10.59}$$

The contribution from the propeller must be added to Equation 10.57 to obtain the total derivative with respect to β.

As an added refinement, one can also include a yaw damping term due to the propeller since yawing results in an effective sideslip at the propeller. If the propeller is located a distance of l_p ahead of the cg, then, because of yawing, a sideslip angle will be produced at the propeller given by $\Delta\beta = l_p r/U_0$. Combining this with Equation 10.55 results in

$$\Delta Y_r = \Delta Y_\beta \frac{l_p}{U_0} \qquad \text{(propeller)} \tag{10.60}$$

Rolling Moment L Derivatives

Fuselage–wing interactions, the wing dihedral angle, and the vertical tail all contribute to the rate of change of the rolling moment with sideslip angle. L_β, the dihedral effect, has been discussed earlier in terms of the coefficient, C_{l_β} (Equations 9.133–9.137). Knowing this coefficient, L_β can be calculated, by definition, from

$$L_\beta = \frac{1}{2I_x}\rho U_0^2 S b C_{l_\beta} \tag{10.61}$$

The derivative, L_p, can be obtained from material presented previously in Chapter 9. From the definition of the rolling moment coefficient,

$$L_p = \frac{\rho U_0 S b^2 C_{l_p}}{4I_x} \tag{10.62}$$

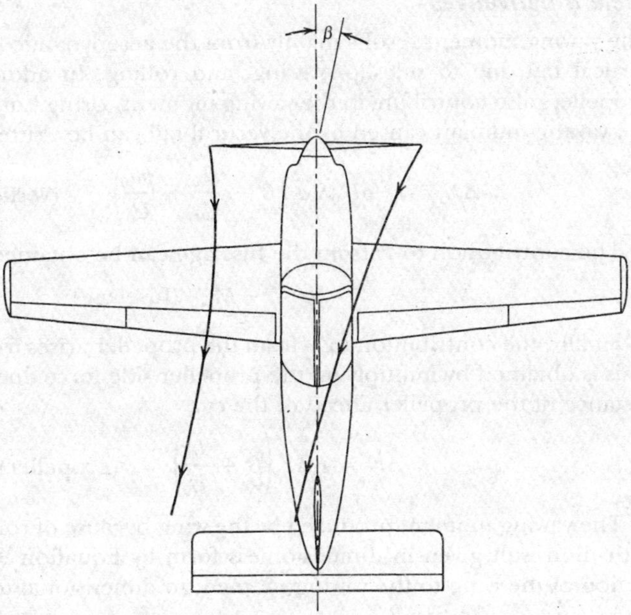

Figure 10.10 Effect of slipstream on dihedral effect.

An effect that was not discussed earlier, but that can be important, particularly at low speeds with flaps down, is illustrated in Figure 10.10. As a propeller-driven airplane slips to one side, the slipstream trails to the other side causing an increased lift on that side. This can result in a significant increase in C_{l_β}. Unfortunately, the interaction is difficult to predict in a general way and is best obtained by means of wind tunnel testing with a powered model.

The coefficient C_{l_p} can be obtained from Figure 9.43 or calculated using the lifting surface model developed in Chapter 3. In determining C_{l_p} for the entire airplane, one should include the contributions from the tail surfaces as well as the wing.

A rolling moment can be produced on a wing by yawing because of the difference in the resultant velocities on either side of the wing. Also, the vertical tail can produce a rolling moment because of yawing. The yawing produces a sideslip angle at the tail, resulting in an added side force on the tail. Acting above the cg, this force will cause a rolling moment. These contributions from the wing and tail have been discussed previously in Chapter 9 in dimensionless form and the results presented by Equations 9.128 and 9.130. The dimensional derivative, L_r, can be written in terms of the dimensionless coefficient as

$$L_r = \frac{\rho U_0 S b^2 C_{l_r}}{4 I_x} \tag{10.63}$$

Roll control, as presented in Chapter 9, is provided by the ailerons. The methods presented there, summarized in Figure 9.37, can be used to obtain the coefficient $C_{l_{\delta_a}}$ from which the actual rolling moment can be obtained.

$$\Delta L = \frac{1}{2}\rho U_0^2 S b C_{l_{\delta_a}} \delta_a \tag{10.64}$$

Yawing Moment N Derivatives

The yawing moment results mainly from the aerodynamic forces produced on the vertical tail due to sideslip, yawing, and rolling. In addition, the fuselage and propellers also contribute to the yawing moment. Using Equations 10.49 and 10.50, the yawing moment caused by the vertical tail can be written as

$$\Delta N = l_v \frac{1}{2}\rho U_0^2 S_v a_v \left(\beta - \frac{r l_v}{U_0} + \frac{p h_v}{U_0}\right) \quad \text{(vertical tail)} \tag{10.65}$$

The contribution to N from the fuselage can be obtained approximately as

$$N_\beta = -M_\alpha \quad \text{(fuselage)} \tag{10.66}$$

Finally, the contribution to N from the propeller arises from sideslip and yawing. This is obtained by multiplying the propeller side force due to these factors by the distance of the propeller ahead of the cg.

$$\Delta N = l_p \Delta Y_\beta \left(\beta + \frac{l_p r}{U_0}\right) \quad \text{(propeller)} \tag{10.67}$$

The yawing moment produced by the wing because of rolling was derived earlier with the result given in dimensionless form by Equation 9.132. Thus, the contribution of the wing to the yawing moment in dimensional form becomes

$$\Delta N = \frac{\rho U_0 b^2 S}{4} C_{N_p} \quad \text{(wing)} \tag{10.68}$$

Yaw control is provided by the rudder. Again, this was covered in Chapter 9, Equation 9.115. The actual yawing moment can be obtained from the control derivative by

$$\Delta N = \frac{1}{2}\rho U_0^2 S b C_{N_{\delta_r}} \delta_r \quad \text{(rudder)} \tag{10.69}$$

Solution of Equations

The transform of the set of Equations 10.50 can be written in matrix form as

$$\begin{bmatrix} (U_0 s - Y_\beta) & (U_0 - Y_r) & -(Y_p s + g\cos\theta_0) \\ -L_\beta & -(\frac{I_{xz}}{I_x}s + L_r) & (s^2 - L_p s) \\ -N_\beta & (s - N_r) & -(\frac{I_{xz}}{I_z}s^2 + N_p s) \end{bmatrix} \begin{bmatrix} \beta \\ r \\ \phi \end{bmatrix} = \begin{bmatrix} 0 \\ L_{\delta_r} \\ N_{\delta_r} \end{bmatrix} \delta_r + \begin{bmatrix} 0 \\ L_{\delta_r} \\ 0 \end{bmatrix} \delta_a \tag{10.70}$$

Similar to the longitudinal case, the characteristic equation for lateral-directional case can be written as

$$As^4 + Bs^3 + Cs^2 + Ds + E = 0 \tag{10.71}$$

where

$$A = -U_0 \left(1 - \frac{I_{xz}^2}{I_x I_z}\right)$$

$$B = U_0 I_{xz}\left(\frac{N_p}{I_x} + \frac{L_r}{I_z}\right) + Y_\beta\left(1 - \frac{I_{xz}^2}{I_x I_z}\right) + U_0 (L_p + N_r)$$

$$C = U_0 (L_r N_p - L_p N_r) + \frac{I_{xz}}{I_x}(Y_p N_\beta - Y_\beta N_p) - \frac{I_{xz}}{I_z}[Y_\beta L_r + L_\beta (U_0 - Y_r)]$$
$$- N_\beta (U_0 - Y_r) + L_\beta Y_p - Y_\beta (L_p + N_r)$$

$$D = Y_\beta(N_r L_p - L_r N_p) + L_\beta(Y_r N_p - N_r Y_p) + N_\beta(L_r Y_p - L_p Y_r)$$
$$+ U_0\,(N_\beta L_p - N_p L_\beta) + g\cos\theta_0\left(L_\beta + N_\beta \frac{I_{xz}}{I_x}\right)$$
$$E = g\cos\theta_0 (N_\beta L_r - L_\beta N_r)$$

Example Calculation of the Lateral-Directional Modes

Again the Piper Cherokee will be used as an example to illustrate the normal modes of the lateral-directional motion. In addition to the numbers given earlier for the longitudinal motion, the following additional data apply for this example.

Weights and Inertia

$$I_x = 1070 \text{ slugft}^2 \quad I_z = 2312 \text{ slugft}^2 \quad I_{xz} = 0$$

Vertical Tail

$$b = 4.47 \text{ ft} \quad c_0 = 3.8 \text{ ft} \quad \lambda = 0.50 \quad \Lambda = 35.0°$$

One can use the graphs presented in Chapter Nine to obtain dimensionless stability derivatives such as $C_{l\beta}$ or C_{N_p} for the wing and tails. However, the lifting surface program, suitably modified, was used instead to determine the contribution of the wing and tails to the stability derivatives. This modification will be discussed later. In this way, the following dimensionless derivatives can be determined.

Wing

$$C_{l\beta} = -0.09882 \quad C_{l_p} = -0.4632 \quad C_{N_p} = -0.08329 \quad C_{l_r} = 0.1666$$

Horizontal Tail

$$C_{l\beta} = 0.0 \quad C_{l_p} \; -0.3655$$

Vertical Tail

$$C_{l_p} = -0.1427 \qquad a_v = 0.0562$$

In the above, all rates are dimensionless and all angles are in degrees. Specifically, $q = Qb/2b$ and $r = Rb/2b$. In addition, the values of C_{N_p} and C_{N_r} are for a unit angle of attack of 1° and must be multiplied by α corresponding to the trim C_L in degrees to obtain the actual values.

To treat the vertical tail with the lifting surface program, the planform is mirrored about the root chord doubling the span and area. The rolling moment and lift coefficient rates are then obtained for this fictitious tail. The lift and rolling moment are then determined in the usual manner using the fictitious tail and then halving the result. The derivatives given above for the vertical tail apply to the fictitious tail.

In dimensional form for this example, the derivatives become:

Propeller

$$Y_\beta = -0.637 \text{ ft/sec/sec/rad} \qquad Y_r = -0.0239 \text{ ft/sec/rad}$$
$$N_\beta = -0.1129\text{/sec/sec} \qquad N_r = -0.00423\text{/sec}$$

Fuselage

$$Y_\beta = -0.600 \text{ ft/sec/sec/rad} \qquad N_\beta = -0.8486\text{/sec/sec}$$

Total Configuration

$$Y_\beta = -15.71 \text{ ft/sec/sec/rad} \qquad Y_p = -0.189 \text{ ft/sec/rad}$$
$$Y_r = 1.227 \text{ ft/sec/rad} \qquad L_\beta = -12.06\text{/sec/sec}$$
$$L_p = -5.436\text{/sec} \qquad L_r = 1.923\text{/sec}$$
$$N_\beta = 4.952\text{/sec/sec} \qquad N_p = -0.3694\text{/sec}$$
$$N_r = -0.492\text{/sec}$$

When the above numbers are substituted into Equation 10.67, dividing through by the first coefficient, the coefficients for the characteristic equation become

$$A = 1.0 \quad B = 6.036 \quad C = 8.915 \quad D = 34.13 \quad E = -0.7886$$

This equation is found to have a pair of complex roots and two real roots.

$$\sigma_1 = -0.2579 + 2.476i \quad \sigma_2 = -0.2579 - 2.476i$$
$$\sigma_3 = 0.02399 \qquad \sigma_4 = -5.543$$

The pair of complex roots represents a damped oscillation and is referred to as the *Dutch-roll mode.* The positive real root defines the *spiral mode,* in this case referred to as *spiral divergence.* The spiral mode can also be stable, in which case it is referred to as *spiral convergence.* It may be surprising to the uninitiated to learn that most airplanes exhibit spiral divergence; that is, most airplanes actually have an unstable mode. The negative real root corresponds to the *roll mode.* The shapes of these modes will now be discussed.

Using Equations 10.50*b* and *c*, one can formulate two equations defining ϕ/β and r/β. Then, knowing r/β, the change in heading, ψ/β, can be found since

$$\psi = \frac{r}{\sigma}$$

In terms of a root, σ, Equations 10.47*b* and *c* can be written as

$$\frac{\left(\frac{I_{xz}}{i_x}\sigma + L_r\right)}{L_\beta}\left(\frac{r}{\beta}\right) - \frac{(\sigma - L_p)\,\sigma}{L_\beta}\left(\frac{\phi}{\beta}\right) = -1 \qquad (10.72a)$$

$$-\frac{(\sigma - N_r)}{N_\beta}\left(\frac{r}{\beta}\right) + \frac{\left(\frac{I_{xz}}{I_z}\sigma + N_p\right)\sigma}{N_\beta}\left(\frac{\phi}{\beta}\right) = -1 \qquad (10.72b)$$

The coefficients of each term in the above are, in general, complex, and so these two equations can be written in the form of

$$z_1\left(\frac{r}{\beta}\right) + z_2\left(\frac{\phi}{\beta}\right) = -1$$
$$z_3\left(\frac{r}{\beta}\right) + z_4\left(\frac{\phi}{\beta}\right) = -1$$

Thus,

$$\frac{r}{\beta} = \frac{z_2 - z_4}{z_1 z_4 - z_2 z_3} = \frac{z_r}{z_D}$$
$$\frac{\phi}{\beta} = \frac{z_3 - z_1}{z_1 z_4 - z_2 z_3} = \frac{z_\phi}{z_D}$$

As with the longitudinal mode shapes, one can work his or her way through the above equations with a simple program, which takes the real and imaginary parts of two complex numbers and returns the real and imaginary parts of the sum, difference, product, and quotient. The final quotients of z_r/z_D and z_ϕ/z_D will give the real and imaginary parts of r and ϕ relative to β for each root.

When this is done for the oscillatory Dutch roll mode defined by the pair of complex roots, the following is predicted for the Cherokee at 100 mph SSL.

$$RP\left(\frac{r}{\beta}\right) = 0.440$$

$$IP\left(\frac{r}{\beta}\right) = -2.207$$

$$RP\left(\frac{\phi}{\beta}\right) = 0.0440$$

$$IP\left(\frac{\phi}{\beta}\right) = 0.807$$

$$RP\left(\frac{\psi}{\beta}\right) = -0.890$$

$$IP\left(\frac{\psi}{\beta}\right) = -0.186$$

Thus, the amplitude of r/β equals 2.250 and r lags β by 78.7°. The amplitude of ϕ/β equals 0.808 and ϕ leads β by 86.9°. The amplitude of ψ/β is 0.908 and ψ lags β by 168.2°. Note that ψ, β, and ϕ are approximately of the same magnitude. The motion of the airplane for this mode can be described as follows. Imagine that the airplane begins to yaw to the right. As it does so, it slips to the left, so that its path remains nearly as a straight line. As it yaws to the right, it begins to roll in that direction. While still rolled to the right, the airplane begins to yaw to the left and slip to the right. This turning and rolling motion is somewhat mindful of the weaving and twisting that an ice skater undergoes in skating along the ice, Hence, the mode has come to be called the *Dutch roll* after the country well-known for its ice skating. For this example, the Dutch roll mode has a period of 2.54 sec and is well damped, requiring only 2.69 sec to halve amplitude. In most airplanes, the Dutch roll mode can be excited by a step input to the rudder, but is, typically, barely noticeable by the pilot. However, your author has flown one airplane with a lightly damped Dutch roll mode and experienced the uncomfortable feeling that the rear end was trying to pass the front end.

The root of the spiral mode is real and positive, with r, ψ, and ϕ related to β by

$$\frac{r}{\beta} = 8.94 \quad \frac{\psi}{\beta} = 389.0 \quad \frac{\phi}{\beta} = 40.9$$

All of these angles are, of course, in phase with β. This is a spiral motion with a time to double amplitude, for this example, of 29.5 sec. This is a relatively long time compared to a pilot's reaction time. As a result, even though the airplane is unstable, the pilot is oblivious to the instability and instinctively provides the necessary control to keep the wings level. This can be done with the rudder pedals or with the ailerons since roll and sideslip are coupled by L_β. Spiral divergence can be demonstrated by trimming an airplane straight and level and then releasing all controls. A slight disturbance will start the airplane turning to the left or right in a slight bank. As time progresses, the bank angle and turn rate will increase. If no corrective action is taken by the pilot, the load factor will increase to a point where structural failure can occur. Although spiral divergence cannot be described as unsafe, it can result in extreme attitudes if the pilot should be studying a chart and forgets to fly the airplane for a few moments. It can prove catastrophic for the noninstrument-rated pilot who finds herself or himself in instrument conditions.

The root for the spiral mode is normally small and can be approximated by neglecting all terms in σ higher than the first. Thus, from the expansion of the characteristic equation,

$$\sigma \approx -\frac{E}{D}$$
$$= 0.0235$$

which is close to the exact value.

The coefficient, D, in the characteristic equation is normally positive so that the sign of the coefficient, E, predominately determines the sign of the root for the spiral mode. Since most of the contribution to L_r results from the wing, this derivative is not too easily adjusted. Varying the vertical tail size will change N_β and N_r in approximately the same proportion. Also, the vertical tail size is normally fixed by other considerations. Hence, the primary control on the spiral mode is exercised through L_β, the dihedral. Increasing the dihedral effect will tend to make the spiral mode more stable. However, as stated previously, too much dihedral leads to an unpleasant feel to the airplane.

For the roll mode, the root is real and negative. Thus, the mode is a stable, aperiodic motion.

$$\frac{\psi}{\beta} = -0.569 \quad \frac{r}{\beta} = 3.152 \quad \frac{\phi}{\beta} = -10.2$$

Typically, the roll mode is heavily damped. For the Cherokee, the predicted time to damp to half amplitude is only 0.13 sec. For this mode, both β and ψ are small compared to ϕ. Thus, this mode is primarily a damped rolling motion. Indeed, if one neglects all but the ϕ terms in Equation 10.47*b*, the root for this mode becomes

$$\sigma = \frac{L_p}{I_x}$$

This gives a value for the root equal to -5.44, which is within 0.5% of the exact value.

Integration of the Lateral-Directional Equations of Motion

The set of Equations 10.47*a*, *b*, and *c* can be numerically integrated to obtain the displacements and accelerations as a function of time for given control inputs. This has been done using a fourth-order Runge-Kutta scheme with the results presented in Figures 10.11*a*, *b*, and *c*. Two supplemental equations are used; namely,

$$\dot{\phi} = p$$
$$\dot{\psi} = r$$

The above equations are used to solve for ϕ and ψ. $\dot{\beta}$ is obtained from Equation 10.47*a*, $\dot{p}$ from Equation 10.47*b*, and $\dot{r}$ from Equation 10.47*c*.

The Dutch Roll mode can be seen by pulsing the rudder and examining the output for a few seconds after the pulse is stopped. Such a case is presented in Figure 10.11*a* using the Cherokee as an example. Referring to the rudder deflection as a function of time, which is given on the figure, it can be seen that

$$\int \delta_r \, dt = 1$$

A pulse of this magnitude is referred to as a *unit impulse*, and the response is called the *impulsive admittance.* The significance of the impulsive admittance will be discussed in the next chapter. From Figure 10.11*a*, it can be seen, as predicted by the

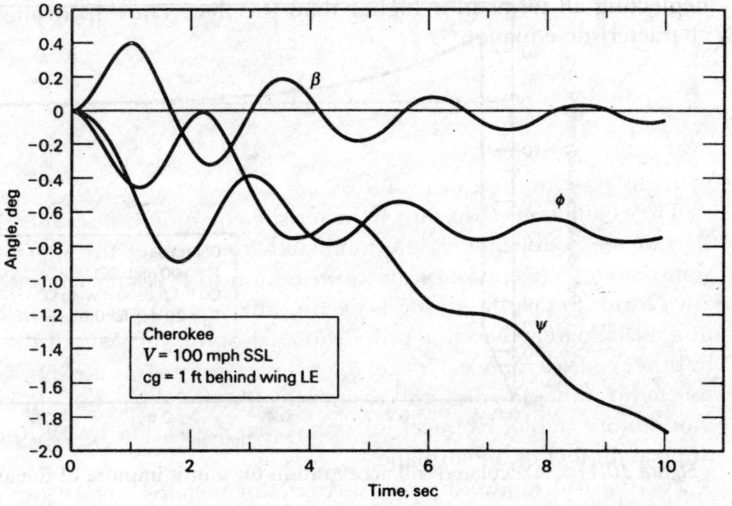

(a)

Figure 10.11a Short-term behavior of the sideslip, roll, and heading produced by a rudder displacement of 1° for 1 sec. (rudder unit impulse).

roots, that β, ϕ, and ψ are all of the same magnitude with the period and time-to-damp to one-half equal to approximately 2.7 and 2.3 sec, respectively.

Expanding the time scale for the response to a rudder pulse results in Figure 10.11*b*. Here, after approximately 20 sec, the Dutch Roll mode has damped out and one is left with only the diverging spiral mode. The time-to-double is seen to equal approximately 30 sec in agreement with the prediction from the roots.

The roll mode is a heavily damped, aperiodic mode, which is best seen by pulsing

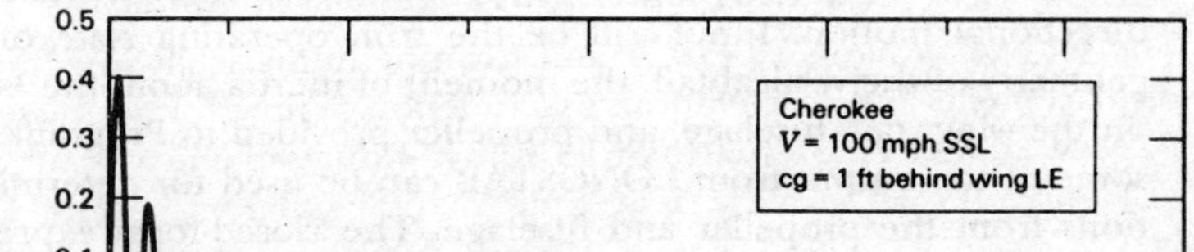

(b)

Figure 10.11b Long-term behavior of the sideslip, roll, and heading produced by a rudder displacement of 1° for 1 sec. (rudder unit impulse).

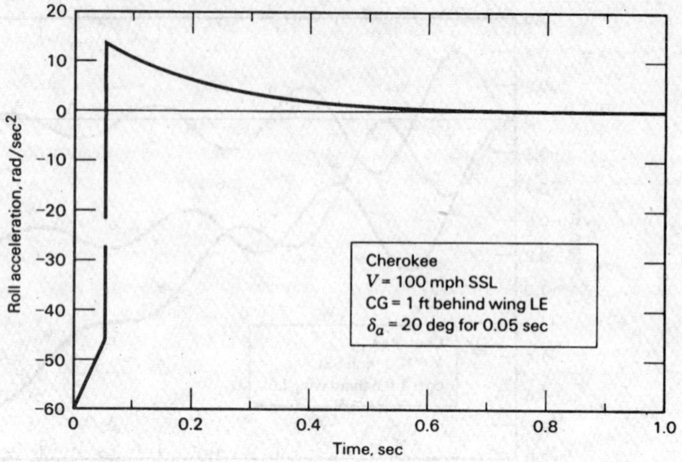

Figure 10.11c Calculated roll acceleration for a unit impulse of the ailerons.

the ailerons and examining the rolling acceleration, $\ddot{\phi}$ Such a result is presented in Figure 10.11*c* for the Cherokee where, at time zero, the ailerons are deflected to 20° and then returned to zero at a time of 0.05 sec. Again, the input to the ailerons is a unit impulse so that Figure 10.11*c* represents the impulsive admittance of the roll mode. During the impulse, the rolling acceleration is seen to be negative because of the negative rolling moment provided by the positive aileron deflection. When the ailerons are returned to zero, the damping moment resulting from the rolling velocity, which was produced by the impulse, then causes a positive rolling acceleration, which damps rapidly to zero. In accordance with the root for this mode, the time-to-damp to one-half is equal to 0.13 sec.

Computer Exercise 10.3 "LATSTAB"

Formulate a computer program to predict the roots and mode shapes for lateral-directional motion. Input will be the trim operating state of the airplane, the geometry of the vertical tail, the moment of inertia about the *x*-axis, and the input on the wing, tail, fuselage, and propeller provided to Program LONGSTAB. Also, some of the output from LONGSTAB can be used for determining the contributions from the propeller and fuselage. The closed-form expressions and figures given in the chapter can be used to obtain rolling and yawing moment coefficients for the wing and tails, or the program SURFACE (see Computer Exercise 10.1) can be modified to determine these coefficients. To do this, one modifies the local angles of attack over the surface to account for the increments resulting from ϕ, Γ, and β. Also, since $\boldsymbol{F} = \rho \boldsymbol{V} \times \boldsymbol{\Gamma}$, additional local lift increments must be considered because of *Y*-velocity components from β. Check your program against the Cherokee example given in the text. You may also wish to expand the program "Motion" to include the lateral-direction equations of motion.

Nondimensional Equations for Lateral-Directional Motion

As is the case with longitudinal motion, the equations defining lateral-directional motion are frequently used in a dimensionless form given by

$$C_{Y_\beta}\beta + C_{Y_p}\dot{\phi} + C_{L_0}\phi + C_{Y_r}r + C_{Y_{\delta_r}}\delta_r = 2\mu\,(\dot{\beta} + r) \qquad (10.73a)$$

$$C_{l\beta}\beta + C_{l_p}\phi + C_{l_r}r + C_{l\delta_a}\delta_a + C_{l\delta_r}\delta_r = i_x\ddot{\phi} - i_{xz}\dot{r} \qquad (10.73b)$$

$$C_{N\beta}\beta + C_{N_p}\phi + C_{N_r}r + C_{N\delta_a}\delta_a + C_{N\delta_r}\delta_r = i_z\dot{r} - i_{xz}\ddot{\phi} \qquad (10.73c)$$

In the above, *all* quantities are dimensionless. The forces are divided by qS and the moments by qSb to obtain the coefficients. The dimensionless time, τ, is defined as the real time divided by a reference time, t^*, where

$$t^* = \frac{b}{2U_0} \qquad (10.74)$$

The rates, p and r, which are now dimensionless, are defined by

$$\bar{p} = \frac{pb}{2U_0} \qquad (10.75a)$$

$$\bar{r} = \frac{Rb}{2U_0} \qquad (10.75b)$$

The mass is made dimensionless by dividing it by the mass density multiplied by a reference volume.

$$\mu = \frac{m}{\rho S \dfrac{b}{2}}$$

The mass moments of inertia are nondimensionalized by dividing by the product of the reference mass and the square of a reference length, $b/2$. For example,

$$i_x = \frac{I_x}{\rho S\left(\dfrac{b}{2}\right)^3} \qquad (10.76)$$

FLYING QUALITIES

Longitudinal Flying Qualities

The evaluation of an airplane's flying quality is subjective. It is difficult to quantify how an airplane feels to a pilot. An airplane may even have an unstable mode (as we will see in the next chapter) and yet feel fine to the pilot if the time to double amplitude is sufficiently long.

Figure 10.12 (taken from Ref. 10.4) presents the so-called Cooper-Harper Scale for evaluating airplane flying qualities. Obviously, with adjectives such as "excellent," "good," "fair," "moderate," "considerable," and "extensive," different pilots will give the same airplane different ratings. Nevertheless, this system does provide a rational and somewhat objective base for measuring an airplane's flying quality. Because of the subjective nature of the Cooper-Harper Scale, Reference 9.1 mainly emphasizes three levels within the scale:

Level 1: Cooper-Harper Scale = 1 − 3.5
Level 2: Cooper-Harper Scale = 3.5 − 6.5
Level 3: Cooper-Harper Scale = 6.5 − 9+

In order to assure that an airplane lies within one of these levels (level 3 is really undesirable but flyable), Reference 10.4 specifies definite dynamic characteristics that the airplane should possess.

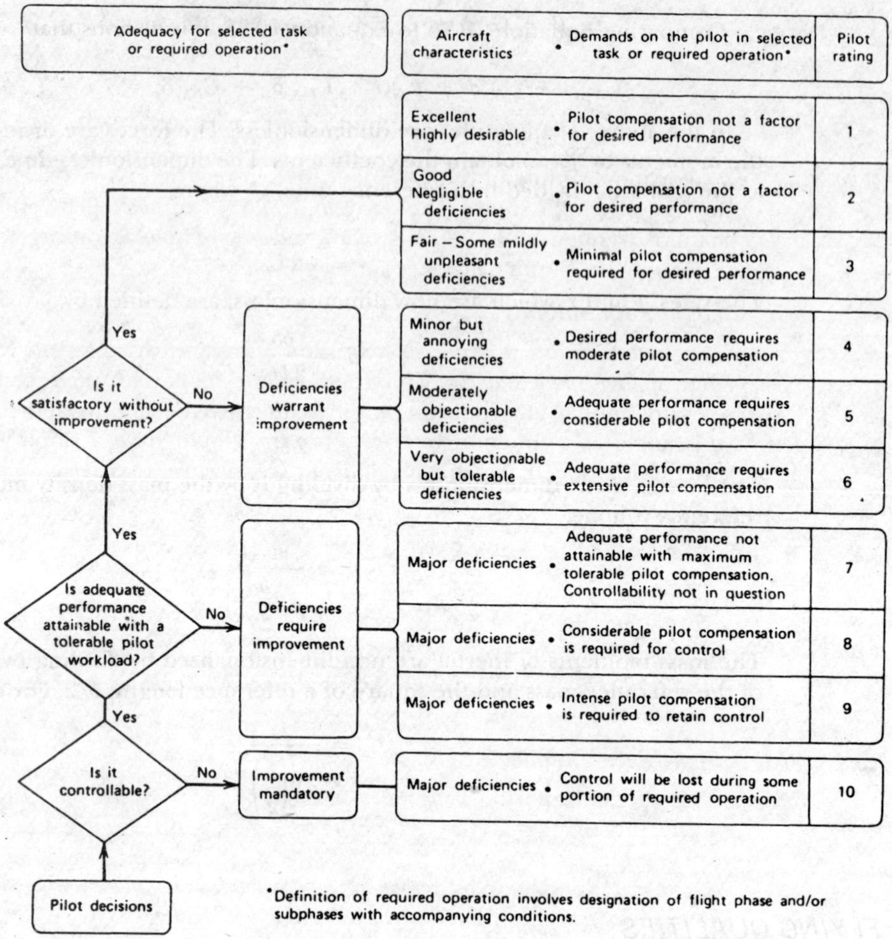

Figure 10.12 Cooper-Harper Scale for rating airplane handling qualities.

Phugoid Mode

First, with regard to the *phugoid mode.*

$$\text{Level 1: } \zeta > 0.04 \tag{10.77a}$$

$$\text{Level 2: } \zeta > 0 \tag{10.77b}$$

$$\text{Level 3: } T_2 > 55 \text{ sec} \tag{10.77c}$$

ζ is the damping ratio, and T_2 is the time to double amplitude.

For emphasis, material given earlier is repeated here. The damping ratio, ζ, for a particular mode, is related to the roots of the characteristic equation as follows. Let the roots defining an oscillatory mode be given by

$$\sigma_{1,2} = -a \pm i\omega$$

Now consider the product

$$(\sigma - \sigma_1)(\sigma - \sigma_2) = 0$$

Expanded, it is

$$\sigma^2 + 2a\sigma + a^2 + \omega^2 = 0 \tag{10.78}$$

Comparing Equation 10.76 to Equation 10.33, it is obvious that

$$\omega_n{}^2 = a^2 + \omega^2 \tag{10.79a}$$

$$\zeta = \frac{a}{\omega_n} \tag{10.79b}$$

For the example of the Cherokee 180, in real time, $a = 0.0248$ and $\omega = 0.286$ for the phugoid. Hence, the damping ratio, ζ, equals 0.0864 and the undamped natural frequency, ω_n, equals 0.287 rad/sec. Thus, according to the criteria of Equation 10.77, the Cherokee falls within level 1.

Flight Path Stability

Another criterion relating to longitudinal motion discussed in Reference 10.4 is that of *flight path stability*. This refers to the flight-path-angle change effected by elevator control at a constant power setting. For the landing approach flight phase, this angle as a function of true airspeed should have a slope at the minimum operational speed, which is negative or less positive than:

$$\text{Level 1: } 0.06°/\text{kt} \tag{10.80a}$$

$$\text{Level 2: } 0.15°/\text{kt} \tag{10.80b}$$

$$\text{Level 3: } 0.24°/\text{kt} \tag{10.80c}$$

In effect, these are tantamount to stating that one should not design an airplane to operate too far into the backside of the power-required curve. In this region, the flight-path-angle (positive for climb) will increase as the speed is increased for a constant power setting.

Short-Period Mode

Both the short-period frequency and damping ratio are important to achieving satisfactory flying qualities. When the damping is too low, the short-period response can produce an annoying oscillation. When the damping is too high, the response to control input can be sluggish. Therefore, upper and lower limits to short-period damping are recommended in Reference 10.4 in order to achieve a given Cooper-Harper level. These limits are given in Table 10.2 according to the following flight categories.

Category A These are nonterminal flight phases that require rapid maneuvering, precision tracking, or precise flight path control such as air-to-air combat, in-flight refueling (receiver), terrain-following, and close formation flying.

Category B These are nonterminal flight phases that are normally accomplished using gradual maneuvers and no precision tracking, although accurate flight path control may be required. This category includes climb, cruise, and descent, in-flight refueling (tanker), and aerial delivery.

Category C These are terminal flight phases requiring gradual maneuvers but precise flight path control. These include takeoff, catapult takeoff, approach, wave-off/go-around, and landing.

Specifying the damping ratios alone, as in Table 10.2, is not necessarily sufficient to assure adequate flying qualities. As mentioned earlier, the short-period frequency is also important. Indeed, it appears as if the values of frequency and damping ratio for satisfying flying qualities are interdependent.

Reference 10.4 and other sources present a number of graphs similar to Figure 10.13. These are sometimes referred to as target plots. Since the establishment of these boundaries is subjective in nature, the contours defining the Cooper-Harper levels should not be taken as hard and fast. However, an airplane that falls to the left of level 3 in Figure 10.13 can be dangerous to fly.

For the Cherokee 180, the roots of the characteristic equation for the short-period mode were calculated previously. In real time these were

$$\sigma_{1,2} = -2.201 \pm 3.936i$$

Thus, for this mode

$$\begin{aligned} \zeta &= 0.489 \\ \omega_n &= 4.50 \text{ rad/sec} \\ &= 0.72 \text{ Hz} \end{aligned}$$

These values are seen to lie well within the level 1 region of Figure 10.13.

Lateral-Directional Flying Qualities

Requirements on the three lateral-directional modes to assure a desired level of flying quality can be found in Reference 10.4. Although specific to military aircraft, these criteria can obviously prove of use in the evaluation or design of civil aircraft. Tables 10.3, 10.4, and 10.5 present criteria for the three modes as a function of the Cooper-Harper level, flight phase category, and class of airplane. Airplane class is defined according to:

Class I—Small, light airplanes.
Class II—Medium weight, low-to-medium maneuverability.
Class III—Large, heavy, low-to-medium maneuverability.
Class IV—High maneuverability.

Roll Mode

Table 10.3 presents the maximum roll mode time constants that should not be exceeded in order to achieve a given flying quality level. In order to interpret this table, one needs to understand the meaning of the term "time constant."

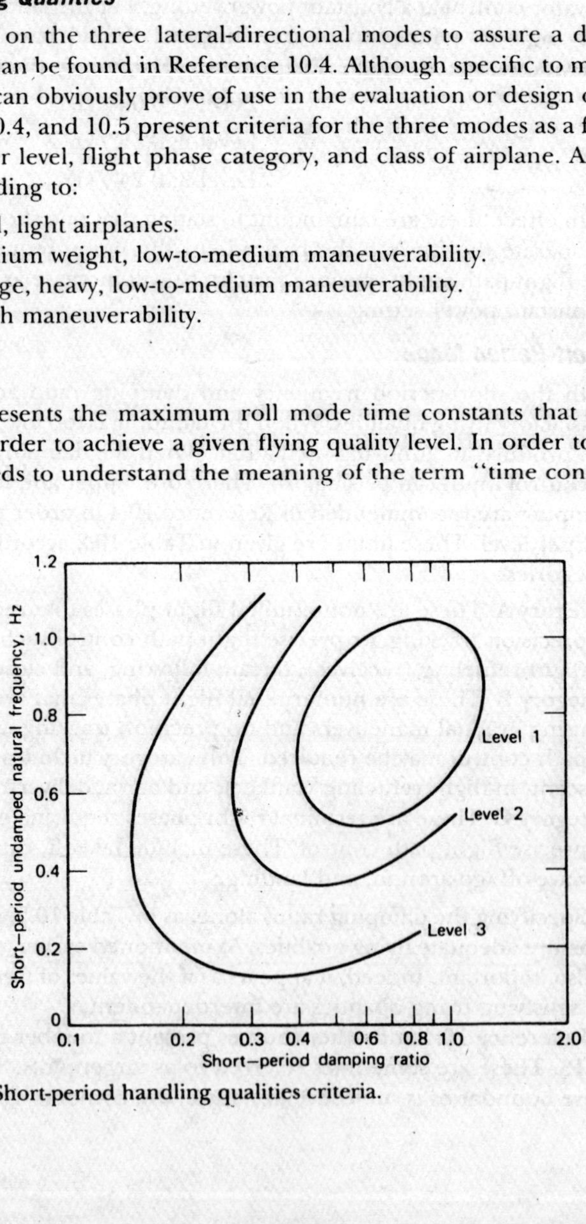

Figure 10.13 Short-period handling qualities criteria.

Table 10.2 Short-Period Damping Ratio Limits

	Categories A and C		Category B	
Level	Minimum	Maximum	Minimum	Maximum
1	0.35	1.30	0.3	2.00
2	0.25	2.00	0.2	2.00
3	0.15	—	0.15	—

Table 10.3 Maximum Roll Mode Constant

Flight Phase Category	Class[a]	Level 1	Level 2	Level 3
A	I, IV	1.0	1.4	
	II, III	1.4	3.0	
B	All	1.4	3.0	10
C	I, II–C, IV	1.0	1.4	
	II–L, III	1.4	3.0	

[a] C and L refer to carrier and land operations.

Table 10.4 Minimum Dutch Roll Frequency and Damping

Level	Flight Phase Category	Class	Min ζ	Min $\zeta\omega_n$, rad/sec	Min ω_n, rad/sec
	A	I, IV	0.19	0.35	1.0
		II, III	0.19	0.35	0.4
1	B	All	0.08	0.15	0.4
	C	I, II–C			
		IV	0.08	0.15	1.0
		II–L, III	0.08	0.15	0.4
2	All	All	0.02	0.05	0.4
3	All	All	0.02	—	0.4

Table 10.5 Spiral Stability—Minimum Time-to-Double Amplitude

Class	Flight Phase Category	Level 1, sec	Level 2, sec	Level 3, sec
I and IV	A	12	12	4
	B and C	20	12	4
II and III	All	20	12	4

A first-order, linear system $x(t)$ obeys a differential equation, which can be written as

$$\dot{x} + \frac{1}{\tau} x = f(t) \tag{10.81}$$

where τ is defined as the time constant. For the homogeneous solution, x will be of the form

$$x = x_0 e^{\sigma t}$$

so Equation 10.36 becomes

$$\left(\sigma + \frac{1}{\tau}\right) x = 0$$

or

$$\sigma = -\frac{1}{\tau} \tag{10.82}$$

The homogeneous solution for x can therefore be written as

$$x = x_0 e^{-t/\tau} \tag{10.83}$$

Thus, the time constant, τ, is a measure of the damping in a first-order system.

The more heavily damped a system, the smaller will be its time constant. Given an initial displacement and released, the system will damp to $1/e$ or 0.368 of its initial displacement in a time equal to the time constant. The time to halve amplitude and the time constant, τ, are related by

$$T_{1/2} = 0.693\tau \tag{10.84}$$

For the Cherokee 180 example, for the roll mode, $\sigma = -5.543$. Thus, in real time,

$$\tau = 0.180 \text{ sec}$$

This value is well within the level 1 criteria for the class I airplane for all flight phases.

Dutch Roll Mode

Table 10.4 presents criteria for the frequency and damping ratio for the Dutch roll mode. Note that minimum values are specified for ζ, ω_n, and for the product $\zeta\omega_n$. The minimum value for ω_n is determined from the column labeled ω_n. However, the governing damping requirement equals the largest value of ζ obtained from either of the two columns labeled ζ and $\zeta\omega_n$.

For the Cherokee 180 example, in real time,

$$\sigma = -0.300 \pm 2.37i$$

Thus, from Equations 10.79a and b.

$$\zeta = 0.126$$
$$\omega_n = 2.39 \text{ rad/sec}$$

With reference to Table 10.4, the Cherokee's parameters in this mode are seen to be greater than the minimum values prescribed for a flying quality level of 1 in the flight phase category B and C. Thus, one would not expect to encounter any problems from the Dutch roll mode with this airplane.

Spiral Stability

The use of Table 10.5 should be obvious and needs no explanation. For the Cherokee example, the "time to double" of 29.5 sec is seen to exceed all of the minimum times specified in this table.

SPINNING

When stalled, an airplane has a tendency to "drop off" into a spin, particularly if the stall is asymmetrical or entered with power. A pilot can purposefully initiate a spin by kicking the rudder hard to one direction at the top of the stall.

In a spin, an airplane rotates around a vertical spin axis as it is descending rapidly in a nose-down attitude. The path of its center of gravity prescribes a helix around the spin axis. The airplane's motion and attitude result in a high angle of attack on the order of 45° or more.

The stall/spin is one of the major causes of light plane accidents today. In my opinion, the blame for this can be placed on pilot training instead of current airplane design. On a landing approach, too many pilots fly too slow. Rarely is there a valid reason for dragging a light airplane in under power at a speed just above the stalling speed. Under such a condition, an unexpected gust or a maneuver on the part of the pilot can produce an asymmetrical stall. If this occurs on the approach, the altitude may be insufficient to recover from either the stall or ensuing spin, regardless of how well the airplane is designed.

FAR Part 23 requires, for the normal category, that a single-engine airplane be able to recover from a one-turn spin in not more than one additional turn, with the controls applied in the manner normally used for recovery. Normally, to recover from a spin, one uses forward stick and rudder opposite to the direction of the spin. Ailerons are generally ineffective for spin recovery, since the wing is fully stalled. FAR Part 23 also requires, for the normal category, that the positive limit maneuvering load factor and applicable airspeed limit not be exceeded in a spin. In addition, there may be no excessive back pressure (on the stick) during the spin or recovery, and it must be impossible to obtain uncontrollable spins with any use of the controls.

The analysis of spinning, in order to design for good spin-recovery characteristics, is difficult because of the nonlinear nature of the problem. However, one can understand the principal factors influencing spin recovery by reference to Figures 10.14*a*, *b*, and *c*. A side view of the spinning airplane is shown in Figure 10.14*a*. As the airplane descends, the aerodynamic forces on the aft fuselage and tail, F_F and F_T, tend to nose the airplane downward. However, in a nose-down attitude, because of the rotation, centrifugal forces are developed on the airplane's mass to either side of the center of gravity. These create a pitching moment about the center of gravity that opposes the nose-down aerodynamic moment.

The wing is completely stalled in a spin so that its resultant force, F_w, is approximately normal to the wing. In a nose-down attitude, the horizontal component of F_w points in toward the spin axis and balances the centrifugal force while the vertical component of F_w balances the weight. If θ is the angle of the nose up from the vertical, as shown in Figure 10.14*a*, then the angular velocity about the spin axis, Ω, the spin radius, R_s and θ are related by:

$$R_s = g/\Omega^2 \tan\theta \tag{10.85}$$

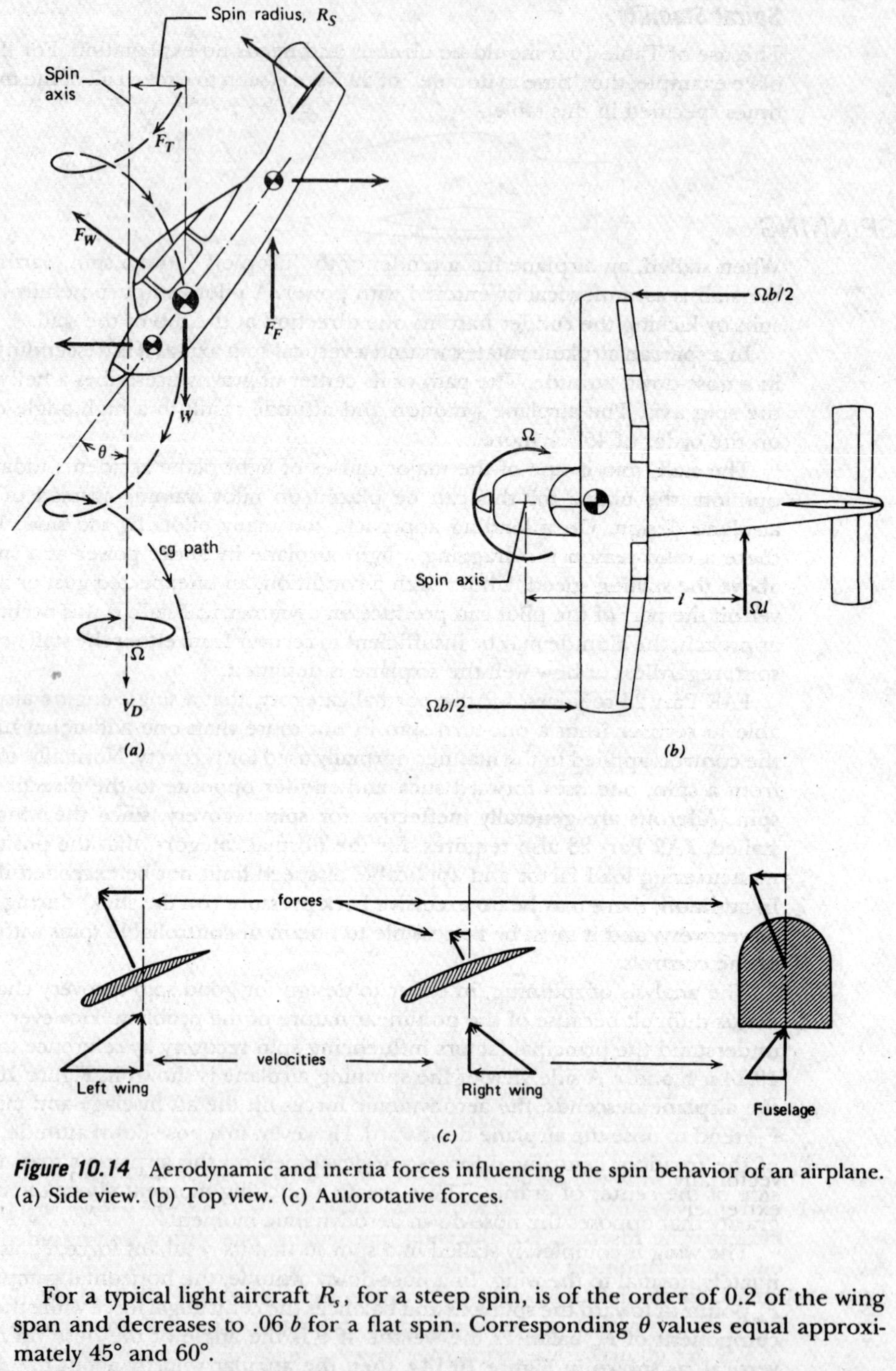

Figure 10.14 Aerodynamic and inertia forces influencing the spin behavior of an airplane. (a) Side view. (b) Top view. (c) Autorotative forces.

For a typical light aircraft R_s, for a steep spin, is of the order of 0.2 of the wing span and decreases to .06 b for a flat spin. Corresponding θ values equal approximately 45° and 60°.

Normally, the angle of roll, ϕ, is small in a spin. Using Equations 10.8*d*, *e*, and *f*, it can be shown that the inertia moments about the spin axis and about the airplane's y-axis are proportional to Ω^2 and $I_z - I_y$). Thus, the spinning behavior

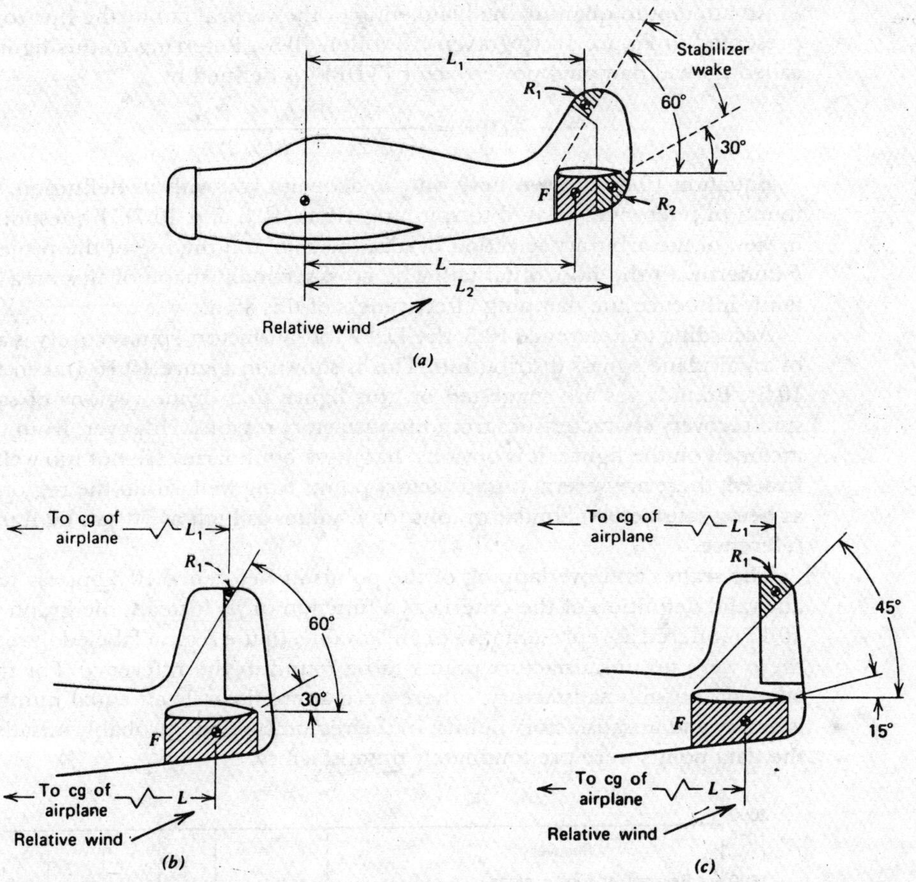

Figure 10.15 Definition of tail damping power factor. (a) Full-length rudder: α assumed to be 45°. (b) Partial length rudder; α assumed to be 45°. (c) Partial-length rudder; α assumed to be 30°.

of an airplane is determined as much, possibly more so, by its mass distribution as by its aerodynamic shape.

The pro-spin, autorotative forces can be produced by both the wing and the fuselage. Referring to Figure 10.14*b*, consider the resultant velocities in a spin at each wing tip and at a typical fuselage section aft of the cg. Assume R_s to be small as in the case of a flat spin. The velocities at each of these locations combine vectorially with the descent velocity, V_D, as shown in Figure 10.14*c* to produce extremely high angles of attack for the wing sections and a nearly vertical flow from beneath the fuselage. For the wing, particularly one with a well-rounded leading edge, a net moment in the direction of rotation results from the higher angle of attack on the left side compared to the right side. The aft fuselage, depending upon its geometry, can develop a force also in the direction of rotation.

Rudder control is the principal means of recovering from a spin. Therefore, in designing the empennage, the placement of the horizontal tail relative to the rudder is important. If the horizontal tail is too far forward, in a spin its wake will blanket the rudder, making it ineffective.

An attempt to quantify the blanketing of the vertical tail by the horizontal tail is presented in Figure 10.15 (taken from Ref. 10.5). Referring to this figure, a term called the tail damping power factor (TDPF) is defined by

$$\text{TDPF} = \frac{FL^2}{S(b/2)^2}\,\frac{R_1 L_1 + R_2 L_2}{S(b/2)} \tag{10.86}$$

Equation 10.86 is given here only to acquaint you with its definition, since it is found in references related to spinning (Refs. 10.6 and 10.7). I question its value in view of the arbitrary selection of α in the spin and the use of the projected area F underneath the horizontal tail. The cross-sectional shape of this area must certainly influence the damping effectiveness of this area.

According to Reference 10.5, the TDPF for satisfactory spin recovery is a function of an airplane's mass distribution. This is shown in Figure 10.16 (taken from Ref. 10.5). Boundaries are suggested on this figure that divide regions of satisfactory spin recovery characteristics from unsatisfactory regions. However, from the points included on the figure, it is obvious that these boundaries are not too well defined. Indeed, there are several unsatisfactory points lying well within the region denoted as being satisfactory. Similar graphs for μ values as high as 70 can be found in the reference.

The scatter and overlapping of the points in Reference 10.5 appear to rule out any valid definition of the criteria as a function of μ. Instead, the graph of Figure 10.17 is offered as representative of any μ value. In the region labeled "satisfactory," there were no unsatisfactory points to be found in the reference. For the region labeled "possibly satisfactory," there were approximately an equal number of satisfactory and unsatisfactory points. In the region labeled "probably unsatisfactory," the data points were predominately unsatisfactory.

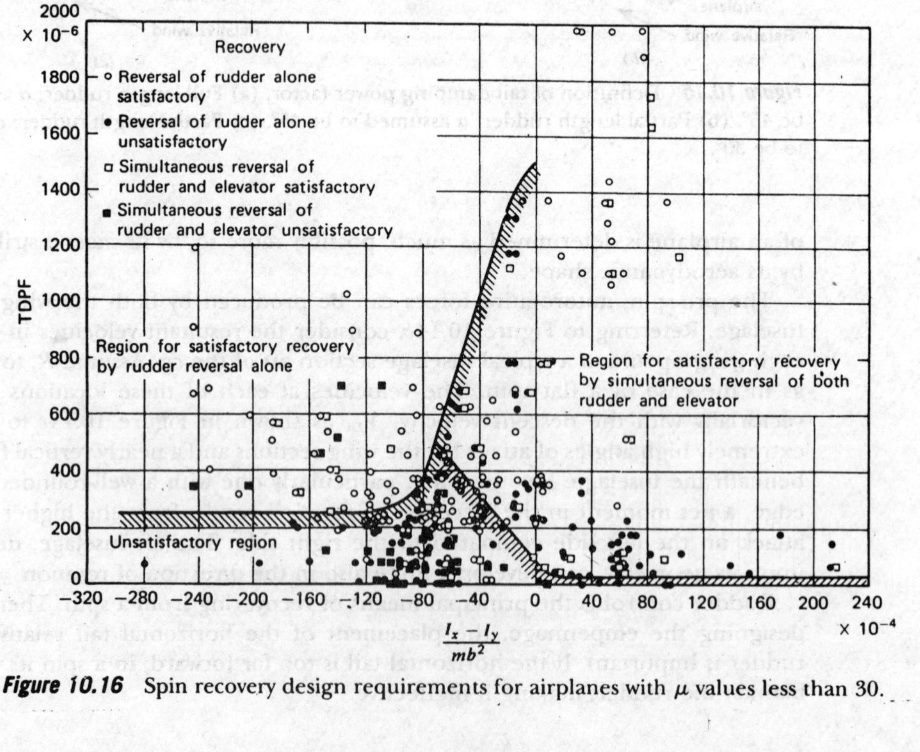

Figure 10.16 Spin recovery design requirements for airplanes with μ values less than 30.

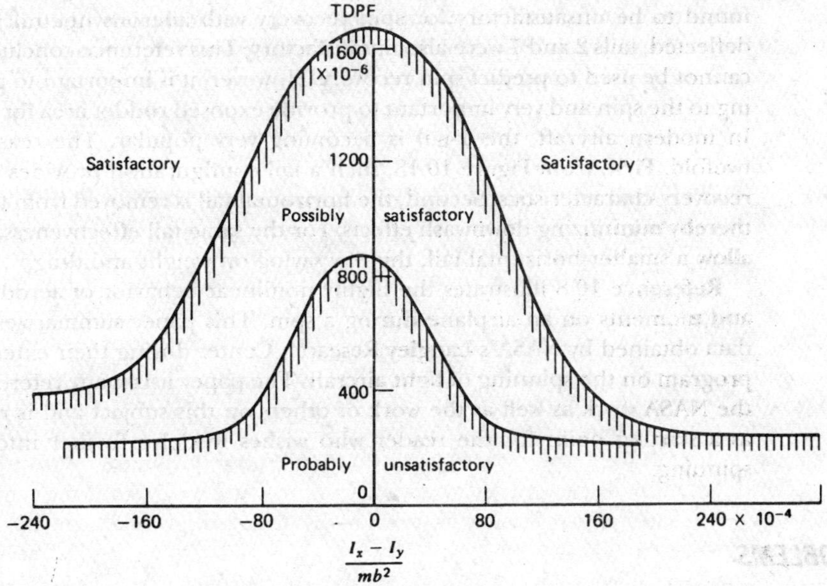

Figure 10.17 Criteria for spin recovery.

Despite the uncertainty associated with this figure, one point is obvious. For satisfactory spin recovery characteristics, the moments of inertia about the pitching and rolling axes should be significantly different.

Figure 10.18 is taken from Reference 10.7. Obviously, neither the shape of the curve dividing the satisfactory region from the unsatisfactory region nor the values of TDPF for satisfactory recovery agree with Figure 10.17. Tails 1, 3, and 4 were

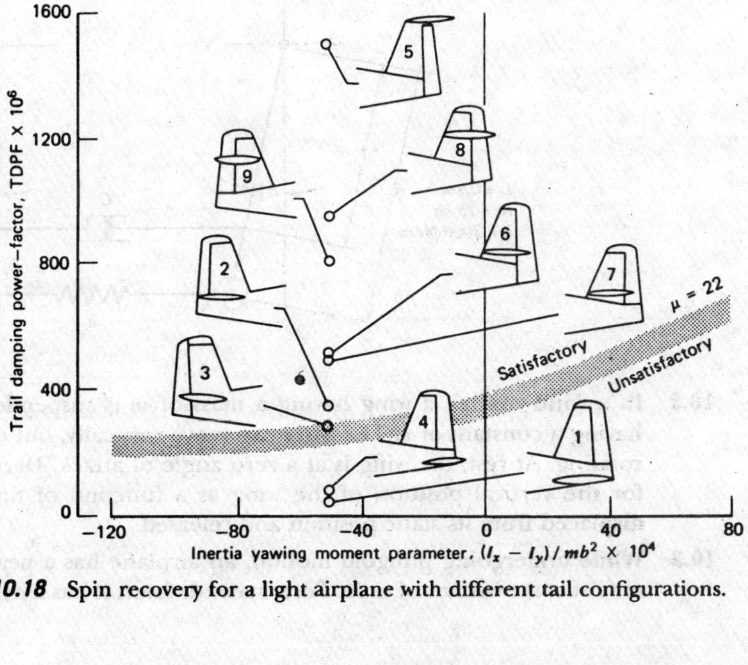

Figure 10.18 Spin recovery for a light airplane with different tail configurations.

found to be unsatisfactory for spin recovery with ailerons neutral. With ailerons deflected, tails 2 and 7 were also unsatisfactory. This reference concludes that TDPF cannot be used to predict spin recovery. However, it is important to provide damping to the spin and very important to provide exposed rudder area for spin recovery. In modern aircraft, the T-tail is becoming very popular. The reason for this is twofold. First, from Figure 10.18, such a tail configuration provides excellent spin recovery characteristics. Second, the horizontal tail is removed from the wing wake, thereby minimizing downwash effects. For the same tail effectiveness, the T-tail will allow a smaller horizontal tail, thereby saving on weight and drag.

Reference 10.8 illustrates the highly nonlinear behavior of aerodynamic forces and moments on an airplane during a spin. This paper summarizes much of the data obtained by NASA's Langley Research Center during their extensive research program on the spinning of light aircraft. The paper lists many references covering the NASA work as well as the work of others on this subject and is recommended as a starting point for the reader who wishes to delve further into the topic of spinning.

PROBLEMS

10.1 Two pendulums are coupled as shown in the figure. Write the equations governing the motion of the system and solve them for the roots and the mode shapes. In addition to defining the modes mathematically, describe the mode shapes qualitatively.

(a) The frequency of oscillation for one mode will be higher than for the other. If this higher frequency is equal to 5 hz and if the time-to-damp to half-amplitude is 0.5 sec, what is the frequency and $T_{1/2}$ for the other mode?

(b) What is the frequency of each mode if the dashpots are removed?

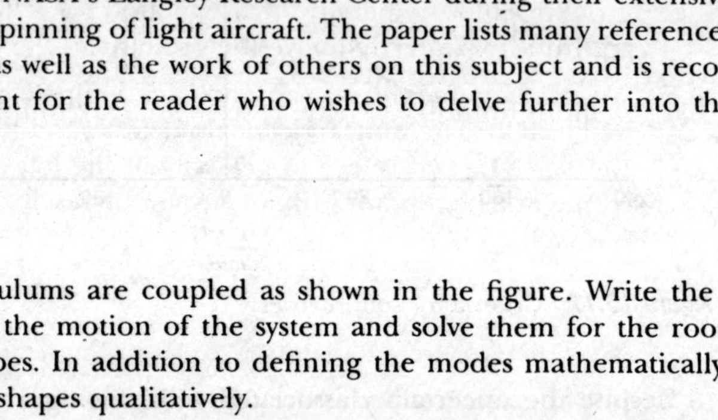

10.2 In a wind tunnel, a wing having a mass of m is suspended from a spring having a constant of k. The wing can move vertically, but is restrained from rotating. At rest, the wing is at a zero angle of attack. Derive an expression for the vertical position of the wing as a function of time if the wing is displaced from its static position and released.

10.3 While undergoing phugoid motion, an airplane has a peak velocity of 150 m/s at some instant of time. Forty seconds later, it has a peak velocity of 134 m/s. As time progresses, the velocity approaches a steady value of 120 m/s. What is the undamped natural frequency and damping ratio for this motion?

10.4 Calculate the flight path stability for the Cherokee 180 at airspeeds of 50, 60, and 70 kt at SSL conditions and a gross weight of 10.7 kN. Compare the results to the criteria specified by Equations 10.78*a*, *b*, and *c* and state what Cooper–Harper level the airplane satisfies for each airspeed.

10.5 An air-cushion vehicle operates on a "cushion" of air in a plenum chamber at its base, which is pressurized by a powered fan. Assuming that the lift provided by the air cushion varies inversely with the gap between the ground and the bottom of the vehicle, derive a linearized equation for the vertical motion of an ACV when perturbed from some initial height above the ground. Find the period for the motion for a 2000-lb machine hovering in trim at a 1.0-ft height above the ground.

10.6 A wing is mounted on the end of a long weightless rod. The wing itself weighs 2 lb. The wing area equals 11.5 ft^2, and its lift-curve slope equals 0.08 C_l/deg. The rod is free to pivot about the left end, but is constrained by a spring at its midpoint with a constant of 100 lb/ft. The airspeed equals 100 mph SSL.

(a) Formulate the equation of motion for the wing when displaced from its static position.
(b) Calculate time-to-damp to half-amplitude and period.

10.7 In dimensionless form, the phugoid motion is approximated by two equations:

$$2\mu\dot{u} - C_{X_u} u + C_{L_0}\,\theta = 0$$
$$2\,C_{L_0}\,u - 2\mu\dot{\theta} = 0$$

Derive the relationship between the coefficients for the mode to be critically damped.

10.8 A system is being forced by $F_0 e^{i\omega t}$, and the response is found to be

$$x = \frac{F_0}{2}(1 + i)\,e^{i\omega t}.$$

(a) Does the response lead or lag the force, and by what angle?
(b) What is the amplitude of the response?

10.9 Given a system with two degrees of freedom, $x(t)$ and $y(t)$: The equations governing the system are:

$$\ddot{x} + 2x - y = 0$$
$$10x + \ddot{y} - 4y = e^{i\omega t}$$

(a) Write the characteristic equation for the system.
(b) Write the particular solution.
(c) Calculate the time to halve-or double amplitude.
(d) Formulate x/y for the two modes and determine the mode shapes.

10.10 A ball is balanced on a spreading jet of air. The speed of the jet decreases linearly with height given by $V = V_0 - V_1 h$, where V_0 and V_1 are positive constants. The air mass density in the jet equals ρ, and C_D of the ball is based on its projected area, A. The ball has a mass of m. When released, the ball positions itself at a height where the weight and drag are in balance. Write an equation governing the linearized vertical motion of the ball when perturbed from its equilibrium height and solve for the period of the motion assuming the damping is less than critical.

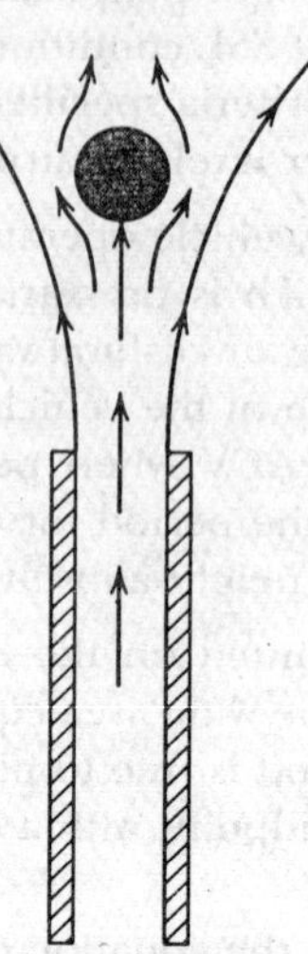

10.11 This is an open-ended problem. Many modern, high-performance fighter airplanes are purposefully designed to be unstable with controls fixed (open loop). They are then equipped with automatic control systems (closed loop) to provide satisfactory handling qualities. The stated reason for doing this is that an airplane, which is unstable and open-loop, is more agile; that is, it can go from one maneuvering state to another more quickly than it can if it were stable and open-loop. Do some type of analysis or investigation to see if you can explain this behavior.

10.12 A point mass on an airplane is located at $x = 0$, $y = 2$, $z = 0$. All dimensions in feet and x, y, and z are body axes through the cg. At a particular instant, V, W, Q, R, ϕ, θ, and ψ are zero, but P equals 40°/sec. If the airplane at this instant is flying at 70 mph and the point mass equals 1 slug, what are the x, y, and z components of the inertial force on the mass at that instant?

10.13 *Given:* An airplane is flying at 10,000 ft.

True airspeed = 400 kt $\quad a_w = 0.08$/deg $\quad a_t = 0.055$/deg

$S_t = 75$ ft² $\quad S = 400$ ft² $\quad I_x = 60{,}000$ slug-ft²

$I_y = 95{,}000$ slug-ft² $\quad I_z = 150{,}000$ slug-ft² $\quad W = 25{,}000$ lb $\quad \epsilon_\alpha = 0.3$

$l_t = 35$ ft $\quad$ cg is 3 ft behind wing ac

The airplane is in trim when it suddenly encounters an upward vertical gust having a velocity of 60 fps.

Find: (a) Instantaneous load factor

(b) Instantaneous pitch acceleration

10.14 A system is being forced by $F = e^{i\omega t}$, and the response is given by $x = \dfrac{1 + 2i}{3 - i} e^{i\omega t}$. What is the amplitude of the response and its phase angle?

10.15 Given a propeller-driven airplane weighing 3000 lb, which requires 150 hp to fly at a speed of 120 mph. The propeller efficiency equals 80%. Find the contribution of the propeller to the stability derivative X_u.

$$\left(\text{Remember: } prop\ \eta = \frac{TV}{\text{engine shaft power}}\right)$$

10.16 *Given:*

$$X_u = -0.04 \quad X_\alpha = -18.0 \quad Z_u = -0.3 \quad Z_q = -2.7$$
$$Z_\alpha = -200.0 \quad M_\alpha = -7.0 \quad \theta_0 = 0 \quad U_0 = 200 \text{ fps}$$

Using the two equations approximating the phugoid (Equations 10.14a and b with $\alpha = \dot{\alpha} = 0$), find the root, solve for u/θ, and find the phase angle between u and θ.

10.17 *Given:* Stability derivatives in Problem 10.16 and $M_\delta = -3.5$, $Z_\delta = -16.0$. The airplane is trimmed in level flight at 200 fps SSL. The elevator angle required for trim equals $-5°$. What elevator angle will be required to trim the airplane at a speed of 180 fps? (Hint: $180 = 200 + u$, where $u = -20$.)

10.18 *Given:*

- True airspeed = 150 mph at SSL
- Distance from cg to vertical tail ac = 15 ft
- Neglect aerodynamic effects from all but the vertical tail
- I_z = 1000 slug-ft^2
- Wing span = 40 ft
- Vertical tail span = 6 ft

A lateral gust of 30 fps from the right causes a yawing angular acceleration of 100 rad/sec/sec. What will the y force be on the vertical tail at 200 mph SSL for a lateral gust of 20 fps from the right.

10.19 A symmetrical airfoil at an angle of attack of 10° produces a lift coefficient of 1.06. The same airfoil at an angle of attack of 2° and with a flap down 30° produces a lift coefficient of 1.35. What is the flap effectiveness factor, τ?

10.20 *Given:* An airplane is modeled by simply a wing and a horizontal tail. The wing and tail are geometrically similar, but the planform area of the wing is three times that of the tail.
Also:

$$W/S = 30 \text{ psf} \quad V = 200 \text{ fps} \quad a = 0.08/\text{deg}$$
$$C_{M_{ac}} = -0.07 \quad A = 7.0 \quad \lambda = 1.0 \quad b = 30 \text{ ft}$$

tail ac is 20 ft behind wing ac
cg is 3 ft behind wing ac

Find: (a) Static margin
(b) Tail incidence angle to trim

10.21 Dynamically scaled, radio-controlled models are frequently used for spin testing. Examine the scaling relationships involving size, mass, and mass distribution required for dynamic simulation of a full-scale airplane with a model.

10.22 Based on Figures 10.15–10.18, do you think the Cherokee 180 will have satisfactory spin recovery characteristics? (Note: This airplane is placarded against spins.)

REFERENCES

1. Nelson, Robert C., *Flight Stability and Automatic Control,* McGraw-Hill, New York, 1989.
2. Seckel, Edward, *Stability and Control of Airplanes and Helicopters,* Academic Press, New York, London, 1964.
3. Roskam, J., *Methods for Estimating Stability and Control Derivatives of Conventional Subsonic Airplanes,* published by author, Lawrence, KS, 1971.
4. Chalk, C. R., Neal, T. P., Harris, T. M., and Pritchard, F. E., "*Military Specification-Flying Qualities of Piloted Airplanes,*" *Background Information and User Guide for MIL-F-8785B (ASG),* AFFDL-TR-69-72, August 1969.
5. Neihouse, A. I., Lichtenstein, J. H., and Pepoon, P. W., *Tail-Design Requirements for Satisfactory Spin Recovery,* NACA TN 1045, April 1946.
6. Bowman, J. S., "Aircraft Spinning," *Aeronautics and Astronautics,* 4(3), March 1966.
7. Bowman, J. S., and Burk, S. M., *Comparison to Results from Spin Tunnel and Radio Control Model Spin Tests with Full-Scale Airplane Results,* presented at General Aviation Technologyfest, AIAA, Wichita, KS, NASA TP 1009, November 18–19, 1977.
8. McCormick, B. W., "Equilibrium Spinning of a Typical Single-Engine Low-Wing Light Aircraft," *AIAA J. of Aircraft,* 18(3), March 1981.

2. Seckel, Edward, *Stability and Control of Airplanes and Helicopters*, Academic Press, New York, London, 1964.
3. Roskam, J., *Methods for Estimating Stability and Control Derivatives of Conventional Subsonic Airplanes*, published by author, Lawrence, KS, 1971.
4. Chalk, C. R., Neal, T. P., Harris, T. M., and Pritchard, F. E., *"Military Specification-Flying Qualities of Piloted Airplanes," Background Information and User Guide for MIL-F-8785B (ASG)*, AFFDL-TR-69-72, August 1969.
5. Neihouse, A. I., Lichtenstein, J. H., and Pepoon, P. W., *Tail-Design Requirements for Satisfactory Spin Recovery*, NACA TN 1045, April 1946.
6. Bowman, J. S., "Aircraft Spinning," *Aeronautics and Astronautics*, 4(3), March 1966.
7. Bowman, J. S., and Burk, S. M., *Comparison to Results from Spin Tunnel and Radio Control Model Spin Tests with Full-Scale Airplane Results*, presented at General Aviation Technologyfest, AIAA, Wichita, KS, NASA TP 1009, November 18–19, 1977.
8. McCormick, B. W., "Equilibrium Spinning of a Typical Single-Engine Low-Wing Light Aircraft," *AIAA J. of Aircraft*, 18(3), March 1981.

11 CONTROLLED MOTION AND AUTOMATIC STABILITY

In the last chapter we were concerned with the uncontrolled motion and stability of an airplane. We will now study the dynamic behavior of the airplane with control input. Further, we will consider the process by which the stability of an airplane is augmented automatically. This is done by sensing the motion of the airplane and feeding an appropriate signal into a controller to move the control surfaces accordingly. It will be seen that this *control feedback* can change completely the dynamic behavior of the airplane response to a pilot-control input. The *open-loop* system, namely, the airplane with no control input, can be unstable, whereas the *closed-loop* system, the airplane with control feedback, can be stable.

RESPONSE TO A CONTROL INPUT

It was shown in Chapter 10 that the equations of motion can be numerically integrated to obtain the motion as a function of time. For the examples that were given, the airplane was initially perturbed by deflecting the controls from, and then returning them to, their trimmed positions. It should be obvious that one can also specify the control positions as a function of time and numerically integrate the equations to obtain the resulting response. For example, one can oscillate the elevator sinusoidally at different frequencies, thereby obtaining the frequency response of the airplane for the longitudinal motion. However, application of Laplace transform methods can prove to be more efficient and informative, particularly with regard to the design of automatic control systems such as autopilots or *stability augmentation systems* (SAS). A short review of Laplace transforms with a table of some commonly used transforms is presented in Appendix E.

TRANSFER FUNCTIONS

The transfer function of a system is defined as the ratio of the transformed output to the transformed input. As noted, the input and output can both have multiple values. For example, for longitudinal motion, repeating Equation 10.32 for $\theta_0 = 0$, the transformed longitudinal equations of motion can be written as

$$\begin{bmatrix} g & (s - X_u) & -X_\alpha \\ -(U_0 + Z_q)\, s & -Z_u & (U_0 s - Z_\alpha) \\ (s^2 - M_q s) & 0 & -(M_{\dot{\alpha}} s + M_\alpha) \end{bmatrix} \begin{bmatrix} \theta \\ u \\ \alpha \end{bmatrix} = \begin{bmatrix} o \\ Z_\delta \\ M_\delta \end{bmatrix} \bar{\delta}_e \qquad (11.1)$$

If δ_e is viewed as a single input, a multiple output will result, namely, u, α, and θ. However, we can, in this case, easily solve for the transfer function, which relates each controlled variable to the single input. For example, using Cramer's rule, we can solve the above equation for the ratio of θ to δ_e. The result will be a transfer function relating the input, δ_e, to the output, θ, and will be denoted by $G_{\theta\delta}(s)$

$$G_{\theta\delta}(s) = \frac{N(s)}{D(s)} \qquad (11.2)$$

where

$$N(s) = \begin{bmatrix} 0 & (s - X_u) & -X_\alpha \\ Z_\delta & -Z_u & (U_0 s - Z_\alpha) \\ M_\delta & 0 & -(M_{\dot\alpha} s + M_\alpha) \end{bmatrix}$$

$$D(s) = \begin{bmatrix} g & (s - X_u) & -X_\alpha \\ -(U_0 + Z_q)s & -Z_u & (U_0 s - Z_\alpha) \\ (s^2 - M_q s) & 0 & -(M_{\dot\alpha} s + M_\alpha) \end{bmatrix}$$

The other transfer functions, $G_{u\delta}(s)$ and $G_{\alpha\delta}(s)$, can be found in a similar manner.

IMPULSIVE ADMITTANCE

The transfer function, $G_{\theta\delta}(s)$, will now be used to determine θ for a given elevator deflection. In this case a special $\delta_e(t)$ will be used, namely, a *unit impulse.* The response to this control input is known as the *impulsive admittance.* A unit impulse of the elevator, δ_e, is defined such that δ_e as a function of time satisfies the following:

$$\lim_{(\delta t \to 0)} \int_0^{\delta t} \delta_e(t)\, dt = 1 \tag{11.3}$$

As long as $\delta_e(t)$ is applied for a short time compared to the period of any one of the normal modes, the shape of the function $\delta_e(t)$ will not affect the response to any extent. Thus, what we are about to calculate theoretically can be obtained experimentally by rapidly displacing the elevator and returning it to its original position. The response is then divided by the integral of δ_e with respect to time to give the impulsive admittance.

The following expression will be used for the elevator input and is seen to satisfy Equation 11.3.

$$\begin{aligned} \delta_e(t) &= \underset{(\Delta t \to 0)}{limit} \frac{1}{\Delta t} & 0 < t < \Delta t \\ &= 0 & t > \Delta t \end{aligned} \tag{11.4}$$

The Laplace transform of the unit impulse is simply equal to unity. Since

$$\theta(s) = G_{\theta\delta}(s)\, \delta_e(s) \tag{11.5}$$

It follows that the impulsive admittance of θ is simply equal to the inverse transform of $G_{\theta\delta}(s)$. This inverse can be obtained by the method of partial fractions, but is most readily determined by the application of Heaviside's expansion, recognizing that the denominator in Equation 11.2 is the characteristic equation for longitudinal motion. Thus, denoting the impulsive admittance for θ by $h_\theta(t)$,

$$h_\theta(t) = \sum_{i\text{-}1}^{4} \left\{ \frac{(S - \sigma_i)\, N(s)}{D(s)} \right\}_{S = \sigma_i} e^{\sigma_i t} \tag{11.6}$$

where σ_i denotes one of the four roots (complex or real) of the longitudinal equation of motion.

From the formulation of the characteristic equation for longitudinal motion as written in Chapter 10 (Equation 10.30), $D(s)$ can be written as

$$D(s) = As^4 + Bs^3 + Cs^2 + D\sigma + E \tag{11.7}$$

Since, in terms of the roots of the polynomial, $D(s)$ can be written as

$$D(s) = A(s - \sigma_1)(s - \sigma_2)(s - \sigma_3)(s - \sigma_4) \tag{11.8}$$

the factor $(s - \sigma_i)$ cancels in the numerator and denominator when $s = \sigma_i$. Thus, for example, the first term in the sum for $h_{\delta\theta}(t)$ becomes

$$\frac{N(\sigma_1)}{A(\sigma_1 - \sigma_2)(\sigma_1 - \sigma_3)(\sigma_1 - \sigma_4)} e^{\sigma_1 t}$$

The notation $N(\sigma_1)$, of course, means that $N(s)$ is to be evaluated with s replaced by σ_1.

Expanding $N(s)$ gives the following quadratic:

$$N(s) = as^2 + bs + c \tag{11.9}$$

where

$$a = M_\delta U_0 + Z_\delta M_{\dot\alpha}$$
$$b = Z_\delta (M_\alpha - X_u M_{\dot\alpha}) - M_\delta (X_u U_0 + Z_\alpha)$$
$$c = M_\delta X_u Z_\alpha - M_\delta Z_u X_\alpha - Z_\delta X_u M_\alpha$$

To obtain the impulsive admittance, one first determines the stability derivatives followed by the roots of the longitudinal equations of motion. These are then substituted into Equations 11.6, 11.8, and 11.9. When the sum given by Equation 11.6 is evaluated, terms of the form $(x + iy)\, e^{\eta + i\omega}$ result. But $x + iy$ can be written as $x + iy = re^{i\phi}$. Thus, the i_{th} term of the sum can be written as,

$$re^{\eta t}[\cos(\omega t + \phi) + i\sin(\omega t + \phi)]$$

We are interested only in the real part of the above. This i_{th} contribution to the total impulsive admittance represents the part contributed by the normal mode corresponding to the i_{th} root.

The above procedure was programmed and a sample case run for the example of the Cherokee used previously. The result is presented in Figure 11.1. Also included in the figure is the motion calculated by numerically integrating the equations of motion. In this case, the elevator was initially deflected 10° and then returned to zero at one-tenth of a second giving an impulse of 1°/sec. Observe that, as it should be, the impulsive admittance obtained from the transfer function is in close agreement with the numerical integration.

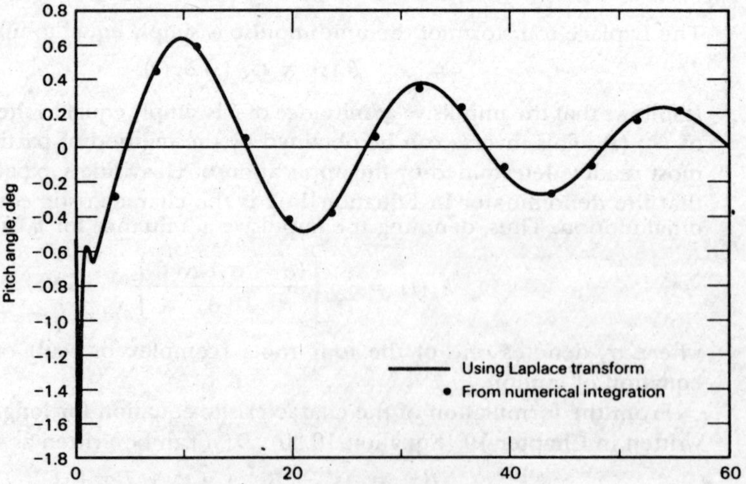

Figure 11.1 Impulsive admittance for the Cherokee pitch angle. One degree-second of the elevator at 100 mph SSL, 2400 lb cg 1 ft behind leading edge.

Computer Exercise 11.1 *"IMPULSE"*

Write a program to calculate the impulsive admittance for the pitch angle using the transfer function. Check your program against the example of a Cherokee at 2400 lb, 100 mph, SSL with the cg 1 ft behind the leading edge. As a check for your work, the following intermediate numbers should be obtained.

$A = 146.7$ (coefficient for σ^4 in characteristic equation)
$a = -1490.2$ (coefficient for s^2 term in numerator of $G_{\delta\theta}$.)
$b = -2489.0$ (coefficient for s term in numerator of $G_{\delta\theta}$.)
$c = -46.85$ (constant term in numerator $G_{\delta\theta}$.)

In the sum for $h_\theta(t)$, each of the four terms can be expressed as $re^{i\phi}$ as noted above. Given below are $r(I)$ and $\phi(I)$ for each term.

$$r(1) = 0.4156 \qquad \phi(1) = -172.4^0$$
$$r(2) = 0.4156 \qquad \phi(2) = 172.4^0$$
$$r(3) = 1.143 \qquad \phi(3) = 68.9^0$$
$$r(4) = 1.143 \qquad \phi(4) = -68.9^0$$

INDICIAL ADMITTANCE

The indicial admittance, $A(t)$, of a system is its response to a unit step input. For example, in the case of δ_e and θ, the indicial admittance of θ would be its response for

$$\delta_e = 1 \qquad t > 0$$

It can be easily shown that the impulsive admittance is equal to the derivative of the indicial admittance. Since the system is linear, the response from two-step inputs applied at time zero and Δt can be added to give the total response. Thus, in the limit as $\Delta t \to 0$, the indicial admittance applied at $t = 0$ minus the indicial admittance applied at $t = \Delta t$ multiplied by $1/\Delta t$ is equivalent to the impulsive admittance. But, by definition, this is the derivative of the indicial admittance.

$$h(t) = \lim_{\Delta t \to 0} \frac{A(t) - A(t - \Delta t)}{\Delta t}$$

or

$$h(t) = \frac{dA(t)}{dt} \tag{11.10}$$

CONVOLUTION INTEGRAL (DUHAMEL'S INTEGRAL)

One of the uses of $h(t)$ is for application to the convolution integral. It will be shown that, knowing $h(t)$, one can calculate the response for any arbitrary elevator input as a function of time. This is done by means of the convolution, or Duhamel's, integral, which utilizes the fact that, for a linear system, responses to different inputs can be superimposed to give the total response to the total input. This is illustrated in Figure 11.2.

The elevator deflection as a function of time, $\delta(t)$, is shown in the upper part of Figure 11.2. This function is approximated by a series of impulses as shown, each with a width of Δt and an amplitude of $\delta(t)$. Consider the response at time t to the impulse at time τ. A time of $(t - \tau)$ has elapsed since the impulse acted upon the

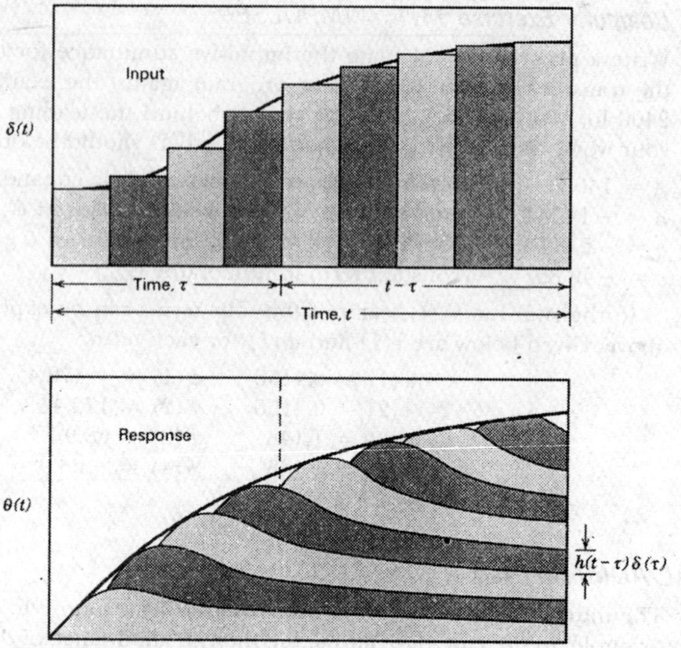

Figure 11.2 Superposition of impulsive inputs to obtain the total response by superposition of the impulsive admittances.

system, so that its contribution to the total displacement, θ, will equal the impulsive admittance of the system evaluated at time $(t - \tau)$ multiplied by the value of the impulse, or $\delta(\tau)\,\Delta\tau$. The total response of the system at time τ will equal the sum of the contributions from all of the impulses up to that time.

$$\theta(t) = \sum_{o}^{t} \delta(\tau)\, h(t - \tau)\, \Delta\tau \tag{11.11}$$

This relationship can be evaluated numerically, or in the limit as $\Delta\tau$ approaches zero, the above becomes

$$\theta(t) = \int_0^t \delta(\tau)\; h\,(t - \tau)\; d\tau \tag{11.12}$$

The convolution integral can also be expressed in terms of the indicial response.

$$\theta(t) = A(t)\,\delta(0) + \int_0^t \frac{d\delta}{d\tau} A(t - \tau)\; d\tau \tag{11.13}$$

The above integral can be easily obtained by superimposing, beginning at time τ, the indicial response, multiplied by an incremental δ, to the total response up to the time τ.

FREQUENCY RESPONSE

The frequency response of a system is defined as the steady-state amplitude and phase angle of the response when forced continuously at a discrete frequency by a sinusoidal force having a unit amplitude. This response can be obtained by the

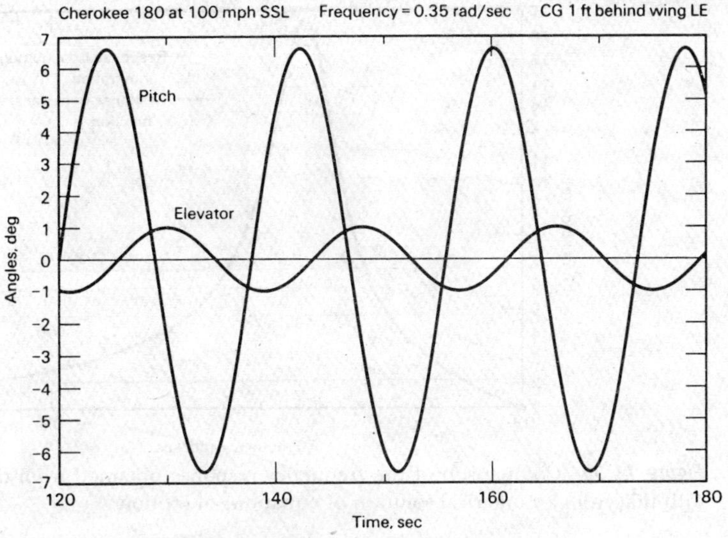

Figure 11.3a Numerical solution of response to elevator oscillation.

"brute force" method of oscillating the elevator at a discrete frequency and numerically integrating the equations of motion. This has been done for the longitudinal motion using the program MOTION with the results shown in Figure 11.3*a*. In this case, the solution was allowed to run for 2 minutes to allow the transient solution to damp. The output was then filed, giving essentially the steady-state solution.

The amplitude of the elevator for Figure 11.3*a* is 1°, so that the amplitude of the output, in this case the pitch angle, θ, will equal the ratio of the amplitude of the output to the amplitude of the input. This ratio for the frequency of 0.35 rad/sec is seen to equal approximately 6.6. It is also seen from this figure that the pitch response is not in phase with the elevator angle. The peak response in θ is seen to lag the peak in δ_e by approximately 6 sec. At a frequency of 0.35 rad/sec, the lag in the time corresponds to a lag in the phase angle of 120°. Of course, to do the numerical integration at many frequencies is tedious and costly; fortunately, there is a better way to determine the frequency response.

The system must be stable in order to define a frequency response and can be obtained immediately from the transfer function for the system. To demonstrate this, we will again use θ as the response and δ_e as the input. Taking the elevator motion to be

$$\delta_e = e^{i\omega t}$$

the transformed frequency response can be written as

$$\theta(s) = \frac{G(s)}{s - i\omega} \tag{11.14}$$

Using the Heaviside expansion, the inverse transform for the above becomes

$$\theta(t) = G(i\omega)\, e^{i\omega t} + \sum_{1}^{4} \left\{ \frac{(s - \sigma_i)\, N(s)}{(s - i\omega)\, D(s)} \right\}_{s = \sigma_i} e^{\sigma_i t}$$

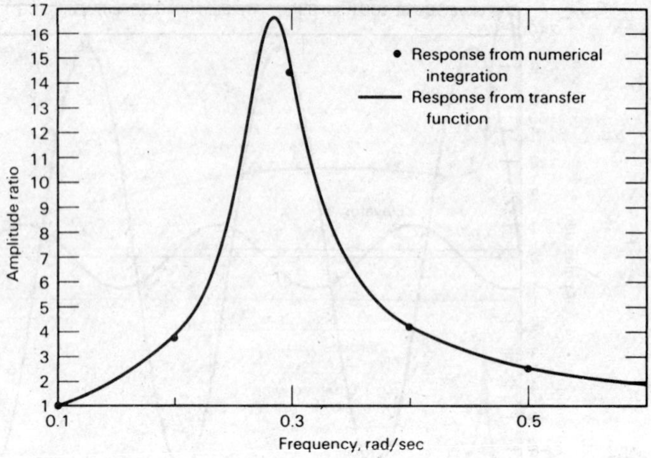

Figure 11.3b Comparison of the frequency response obtained from the transfer function with that from a numerical solution of equations of motion.

Since the system is stable, the real parts of the roots, σ_i, are negative so that in the limit, as $t \rightarrow \infty$, the sum vanishes and the interesting result is obtained that

$$\text{Frequency response} = G(i\omega) \tag{11.15}$$

$G(i\omega)$ is complex and can be written as

$$G(i\omega) = \frac{A_o}{A_i} e^{i\phi}$$

A_o is the amplitude of the output, or response; A_i is the amplitude of the input, or forcing function; and ϕ is the phase angle by which the output leads the input.

The frequency response for the Cherokee 180 example has been determined over a wide range of frequencies by programming the solutions of Equations 11.2, 11.8, 11.9, and 11.15 with the Laplace transform variable equal to $i\omega$. The results are shown in Figures 11.3*b* and 11.3*c* where the ratio of the amplitude of θ to δ_e and the phase angle, ϕ, are presented as a function of the frequency, ω, in radians per second. Such a graph is known as a *Bode* diagram. Normally, such a diagram is plotted using logarithmic scales and covers a wide spectrum of frequencies. In this instance, however, linear scales are used to emphasize agreement with the previously determined eigenvalues for the phugoid and short-period modes. Observe that an amplitude spike, or *resonance,* is seen at the frequency of approximately 0.28 rad/s corresponding to the phugoid mode. A corresponding rapid change in the phase angle with frequency is also found at the phugoid mode. The same observations are not true near the frequency of 4.54 rad/s corresponding to the short-period mode. This behavior is attributable to the relative damping of the two modes.

Figure 11.3*d* is presented in order to emphasize the difference in the dynamic pitching behavior of the airplane near the two modes. This figure presents the numerical solution of the longitudinal motion of the Cherokee when forced at a frequency of 4.0 rad/s. The time scale is expanded to only a fraction of a cycle in order to show precisely the amplitude and phase. In this case the amplitude ratio,

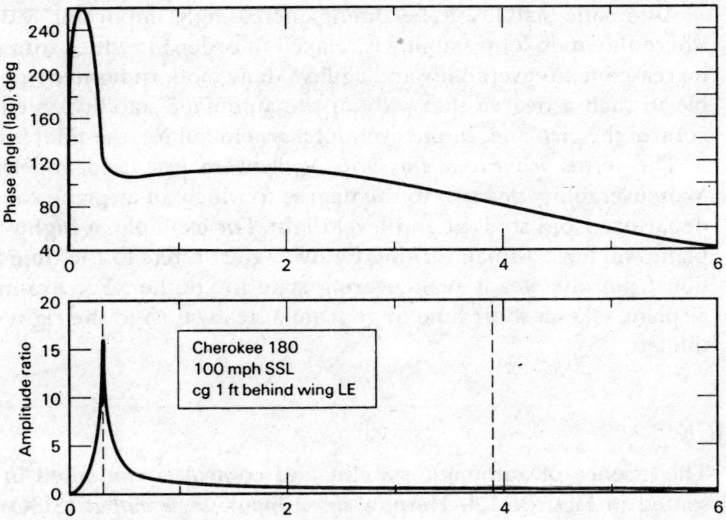

Figure 11.3c Frequency response of longitudinal motion to elevator input over frequency range encompassing phugoid and short-period modes.

even though it is close to the natural frequency of the short period mode, is reduced to 0.6. At this frequency, the phase lag is only 0.36 sec corresponding to 82.5°.

AUTOMATIC STABILITY

Automatic stability is a broad and complex subject; only a brief introduction to the subject will be attempted here. For further study on the subject, one is referred to the many books and papers on the subject as, for example, Reference 10.1 or References 11.1, 11.2, and 11.3.

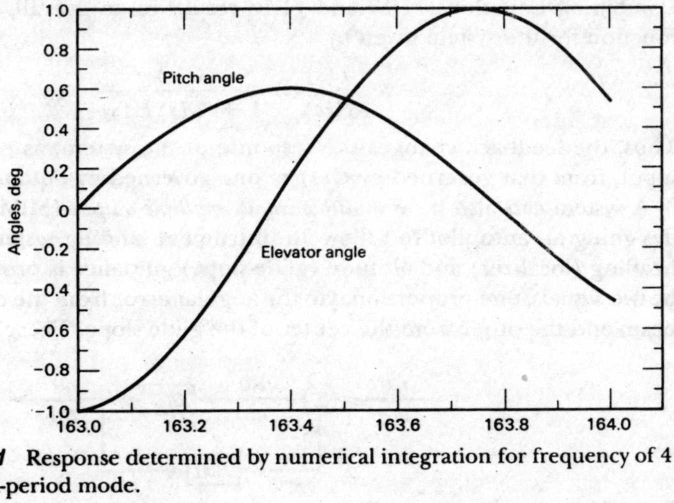

Figure 11.3d Response determined by numerical integration for frequency of 4 rad/s close to the short-period mode.

Automatic stability is becoming increasingly important with modern aircraft where the open-loop stability is relaxed in order to reduce trim drag or to provide increased maneuverability and agility. Many modern fighter-type aircraft are unstable to such a degree that without the automatic stability system the pilot cannot control the airplane. In the event of a system failure, the pilot is instructed to eject.

The terms *maneuverability* and *agility* were just used and should be defined. Maneuverability pertains to the degree to which an airplane can maintain a steady departure from straight and level flight. For example, a highly maneuverable airplane will have a small turning radius. Agility refers to the time required to transition from one steady maneuvering state to another. For example, a highly agile airplane takes a short time to go from a steady turn to the right to a steady turn to the left.

FEEDBACK

The essence of automatic stability and control is embodied in the diagram presented in Figure 11.4. Here, a *single input–single output* (SISO) system is shown. Following the notation of Reference 11.3, the system is represented in both real and Laplace-transformed variables. Beginning at the left of the diagram, the control input in real time is denoted by $r(t)$ or, in terms of s, by $R(s)$. The input, or *command*, is fed into a summing junction where it is combined with a feedback signal to give an error signal $e(t)$ or $E(s)$. The error is then fed into a block representing the open-loop response of the system where $G(s)$ is the *open-loop transfer function.* $G(s)$, in general, will encompass not only the airframe response, but also the behavior of other components such as control actuators and their electronic controls. The output from $G(s)$, or the *controlled variable,* is denoted by $y(t)$ or $Y(s)$ and is equal to

$$Y(s) = G(s)E(s) \tag{11.17}$$

A signal proportional to the output is then fed back through the *feedback transfer function,* $H(s)$, resulting in an output from this block equal to $H(s)Y(s)$. Thus, the error signal will be

$$E(s) = R(s) - H(s)Y(s) \tag{11.18}$$

Combining Equations 11.17 and 11.18 results in an overall, closed-loop, transfer function for the system given by

$$\frac{Y(s)}{R(s)} = \frac{G(s)}{1 + G(s)H(s)} \tag{11.19}$$

Thus, the feedback changes the response of the system, as related to the control input, from that governed by $G(s)$ to one governed by Equation 11.19.

A system can also have *multiple input, multiple output* (MIMO). For example, in designing an autopilot to follow an instrument landing system (ILS), where both heading (localizer) and altitude (glide-slope) guidance is provided, the inputs will be two signals, one proportional to the angular error from the center of the localizer beam and the other from the center of the glide slope. These two reference inputs

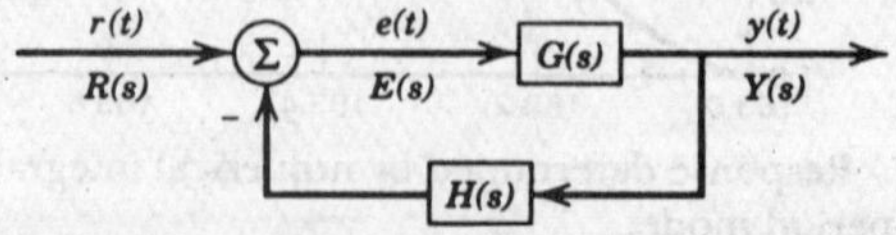

Figure 11.4 A control system incorporating displacement feedback.

result in multiple outputs; namely, the altitude and heading of the airplane. In between the input and output there are of course a myriad of subsystems consisting of radio receivers, amplifiers, actuators, sensors (gyroscopes, altimeter, airspeed) and the open-loop response of the airplane. On a smaller scale, viewing only the airplane as the system, the multiple input would be the airplane controls: aileron, rudder, elevator, and throttle. The multiple output would be the airplane state variables: pitch, yaw, roll, and velocity vector.

LATERAL MOTION WITH FEEDBACK

As an example of automatic control, we will consider a simple device known as a *wing leveler.* This relatively inexpansive device is found on many small airplanes and is, in a sense, a single-axis autopilot commanding a roll angle, which is usually zero. For single-pilot operation on instrument flight rules (IFR) it enhances safety considerably by stabilizing the spiral mode. Because of dihedral effect, this simple device also provides a certain degree of heading control.

The system will appear as shown in Figure 11.5. The command input in this case is some desired roll angle, $\phi_c(s)$, with the output being the actual roll angle, $\phi(s)$. This angle is fed back to the summing junction to give an error signal, $\epsilon(s)$. The error signal is then fed into a controller, $C(s)$, which drives the aileron angle. The aileron angle then commands the airframe through the transfer function, $G_{\phi\delta}(s)$.

The diagram shown in Figure 11.5 is equivalent to that of Figure 11.4. Since

$$\begin{aligned}\phi(s) &= \delta_e(s)\,G(s)\\ &= C(s)\,G(s)\,[\phi_c(s) - \phi(s)]\end{aligned}$$

it follows that

$$\frac{\phi}{\phi_c} = \frac{C(s)\,G(s)}{1 + C(s)\,G(s)} \tag{11.20}$$

Thus, the feedback transfer function, $H(s)$, and the controller, $C(s)$, have the same relative effect on the system stability. If $H(s)$ is equal to $C(s)$, the poles of the denominator will be the same for either closed-loop system.

For the transfer function of the controller, suppose $C(s)$ is of the form

$$C(s) = \frac{k_1}{s} + k_2 + k_3 s \tag{11.21}$$

The first term can be thought of as being responsive to the past displacement and represents the integral of ϕ with respect to time. The second term reacts to the present value of ϕ, whereas the last term, being the derivative of ϕ with respect to time, is looking ahead to the future value of ϕ. The second term, k_2, corresponds to a spring rate, whereas the third term, $k_3 s$, is a damping term. Although the system can be stabilized with sufficient damping, this alone will not assure that ϕ will equal ϕ_c.

In order to keep the example simple, let us use a controller equivalent only to a displacement feedback, namely,

$$\delta_a = k\phi \tag{11.22}$$

Figure 11.5 A single-axis autopilot to command a roll angle.

or

$$C(s) = -k$$

In this case, Equation 11.20 can be written as

$$\frac{\phi}{\phi_c} = -\frac{N(s)}{D(s) + kN(s)}$$

where $N(s)$ and $D(s)$ are the numerator and denominator of $G(s)$. As before, replacing $N(s)$ and $D(s)$ by

$$N(s) = as^2 + bs + c \qquad D(s) = As^4 + Bs^3 + Cs^2 + Ds + E$$

the above becomes

$$\frac{\phi}{\phi_c} = \frac{-N(s)}{As^4 + Bs^3 + (C - ka)s^2 + (D - kb)s + E - kc} \tag{11.23}$$

Thus, closing the loop and adding the controller is seen to modify the characteristic equation.

From Equation 10.67, the coefficients a, b, and c in the expression for the transfer function, $G_{\phi\delta}$, relating the roll angle to the aileron input are found to be

$$a = -L_{\delta_a}U_0$$
$$b = L_{\delta_a}(Y_\beta + U_0N_r)$$
$$c = L_{\delta_a}(N_\beta Y_r - Y_\beta N_r - N_\beta U_0)$$

The coefficients A, B, C, D, and E for the denominator have been given previously immediately below Equation 10.68. In effect, closing the loop is seen to change the coefficients C, D, and E. Therefore, the computer program "LATSTAB" is easily modified to incorporate this change. This has been done for the lateral-directional example presented in Chapter 10 with the results presented in Figure 11.6. This figure presents the root of the spiral (which is entirely real) and the time to double or halve amplitude as a function of the parameter, or gain, k. It is seen that a gain of approximately 0.02 is needed before the closed-loop system is stable for this mode. At the value of k for which the root is zero, the mode is neutrally stable with the times to double or halve approaching infinity.

The numerical integration of the equations of motion can be easily modified to

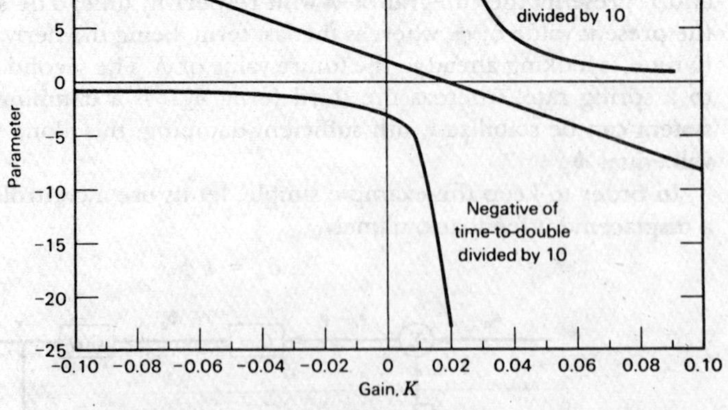

Figure 11.6 Effect of feedback gain on the root of the spiral mode and the time-to-double or halve.

accommodate most feedback systems since, in many cases, the feedback represents, in effect, a modification of a stability derivative or the addition of an effective derivative. For this example, consider Equation 10.47*b*, governing the relationship between the rolling moments and the rolling acceleration. This equation, repeated here, is

$$L_\beta \beta + L_r r + L_p \phi + L_{\delta_a} \delta_a = \ddot{\phi} - \frac{I_{xz}}{I_x} \dot{r}$$

But δ_a is linked to ϕ for this example by Equation 11.22. Thus, substituting this equation with Equation 10.47*b* results in

$$L_\beta \beta + L_r r + L_p \phi + kL_{\delta_a} \phi + L_{\delta_a} \delta_{a_c} = \ddot{\phi} - \frac{I_{xz}}{I_x} \dot{r} \tag{11.24}$$

Here, the control input represented by δ is retained by the term, $L_{\delta_a} \delta_{a_c}$, but the driving of δ_a by the roll angle gives rise to a pseudostability derivative, L_ϕ, where

$$L_\phi = kL_{\delta_a}$$

Suppose, in this same equation for the purpose of performing a coordinated turn, that the aileron angle is linked to the sideslip angle, β, in the same form as Equation 11.22. For this case, the first term in Equation 10.47*b* would then become

$$(L_\beta + kL_{\delta_a})\, \beta$$

Thus, in this case, the feedback has resulted in no additional terms, but, instead, a modification to the stability derivative, L_β.

Program "MOTION" was modified to incorporate the feedback to the ailerons, Equation 11.23, and then run with no gain and with $k = 0.06$. To perturb the system, an initial unit impulse was given to the ailerons. The ensuing rolling motions are compared in Figure 11.7, where the "wing leveler" is seen to improve considerably the handling qualities of the airplane. Notice that the initial, heavily damped oscillation is unaffected by the feedback to the ailerons. However, as would be

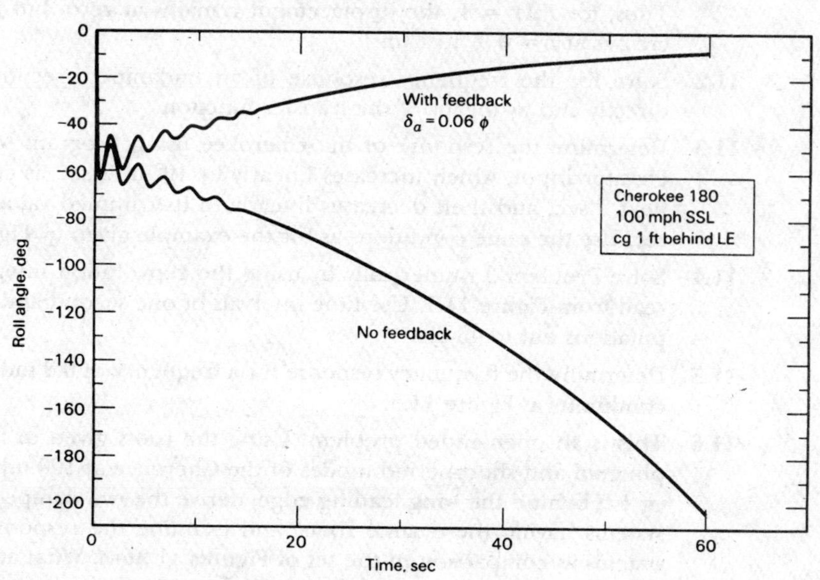

Figure 11.7 Effect of feedback on rolling motion.

expected from Figure 11.6, the perturbed roll angle tends quickly to zero with the feedback. Without it the airplane becomes inverted in approximately 50 sec.

SYSTEM AND PARAMETER IDENTIFICATION

System and parameter identifications is an important topic relating to aircraft stability and control and is receiving a lot of attention at this time. The topic pertains to the techniques involved in analyzing experimental flight test data in such a way as to obtain the equations of motion (system) and the stability derivatives (parameters). Unfortunately, the techniques are too complex and extensive to be treated here, but they are important and are therefore mentioned. As a start toward the study of these techniques, you are directed to Reference 11.4 published by the Advisory Group for Aerospace Research and Development (AGARD).

SUMMARY

This last chapter has presented a brief introduction to the subject of automatic stability and control. It has been shown that an airplane's handling qualities can be improved considerably by the use of feedback to its control system. Classical techniques only were covered involving the use of Laplace transforms and root locus methods. To continue further, the student should become knowledgeable in modern control, which applies optimization and other techniques through the use of linear algebra to the entire airplane state.

PROBLEMS

11.1 Calculate the impulsive admittance of a second-order, linear system without damping (mass on a spring), which is initially quiescent, by solving the differential equation directly. Repeat the solution using the transfer function. (Note: The initial conditions at $t = 0+$ are not all zero. $F\Delta t \approx m\Delta V$. Thus, for $F\Delta t = 1$, the displacement remains at zero, but the velocity increases at $t = 0+$ to $1/m$.)

11.2 Solve for the frequency response of an undamped, second-order system directly and by means of the transfer function.

11.3 Determine the response of the Cherokee using Program MOTION to an elevator input, which increases linearly by 10° in 2 sec, is constant for the next 2 sec, and then decreases linearly to its trimmed value in the next 2 sec. Use the same conditions as for the example given by Figure 11.1.

11.4 Solve Problem 3 numerically by using the convolution integral and points read from Figure 11.1. Use time intervals of one second and carry the computations out to 40 sec.

11.5 Determine the frequency response for a frequency of 0.2 rad/s for the same conditions as Figure 11.3.

11.6 This is an open-ended problem. Using the roots given in the text for the phugoid and short-period modes of the Cherokee at 100 mph SSL with the cg 1 ft behind the wing leading edge, derive the two damped, second-order systems having these same roots, and examine the response of these two systems in comparison to the set of Figures 11.3*a–d*. What are your observations?

REFERENCES

1. McLean, Donald, *Automatic Flight Control Systems,* Prentice Hall, New York, London, 1990.
2. McRuer, D., Ashkenas, I., Graham, D., *Aircraft Dynamics and Automatic Control,* Princeton University Press, NJ, 1973.
3. Stevens, B. L., and Lewis, F. L., *Aircraft Control and Simulation,* Wiley-Interscience Publications, John Wiley, New York, London, 1992.
4. Hamel, Peter, et al., *Rotorcraft System Identification,* AGARD Advisory Report 280, 1991.

APPENDIXES

APPENDIX A

THE SI SYSTEM

SI is an abbreviation for Système International and is a metric system of units that is being adopted internationally for all scientific and engineering work. This brief presentation presents only the aspects of the system pertinent to this text.

Names of International Units

Physical Quantity	Name of Unit	Symbol	
Length	Meter	m	
Mass	Kilogram	kg	
Time	Second	s	
Electric current	Ampere	A	
Temperature	Kelvin	K	
	DERIVED UNITS		
Area	Square meter	m^2	
Volume	Cubic meter	m^3	
Frequency	Hertz	Hz	(s^{-1})
Density	Kilogram per cubic meter	kg/m^3	
Velocity	Meter per second	m/s	
Angular velocity	Radian per second	rad/s	
Acceleration	Meter per second squared	m/s^2	
Angular acceleration	Radian per second squared	rad/s^2	
Force	Newton	N	$(kg \cdot m/s^2)$
Pressure	Newton per square meter[a]	N/m^2	
Kinematic viscosity	Square meter per second	m^2/s	
Dynamic viscosity	Newton-second per square meter	$N \cdot s/m^2$	
Work, energy, quantity of heat	Joule	J	$(N \cdot m)$
Power	Watt	W	(J/s)
Entropy	Joule per kelvin	J/K	
Specific heat	Joule per kilogram kelvin	$Jkg^{-1}K^{-1}$	
Thermal conductivity	Watt per meter kelvin	$Wm^{-1}K^{-1}$	

[a] 1 N/m^2 = 1 Pascal, denoted by Pa.

DEFINITIONS

newton (N) The newton is the force that gives a mass of 1 kg an acceleration of 1 m/s/s.

joule (J) The joule is the work done when the point of application of 1 N is displaced a distance of 1 m in the direction of force.

watt (W) The watt is the power that gives rise to the production of energy at the rate of 1 J/s.

Conversion Factors

Multiply	By	To Get
Pounds (lb)	4.448	Newtons (N)
Feet (ft)	0.3048	Meters (m)
Slugs	14.59	Kilograms (kg)
Slugs per cubic foot (slugs/ft^3)	515.4	Kilograms per cubic meter (kg/m^3)
Horsepower (hp)	0.7457	Kilowatts (kW)
Pounds per square inch (psi)	6895.0	Pascals (Pa)
Pounds per square foot (psf)	47.88	Pascals (Pa)
Miles per hour (mph)	0.4471	Meters per second (m/s)
Knots (kt)	0.5151	Meters per second (m/s)

MISCELLANEOUS

$$g = 32.2 \text{ ft/sec}^2 = 9.81 \text{ m/s}^2 \text{ at sea level}$$
$$°K = °C + 273.19$$
$$°R = °F + 459.7$$
$$°C = (°F - 32)(5/9)$$
Universal Gas Constant ($p = \rho RT$)
$$R = 1716 \text{ ft}^2/\text{s}^2/°R$$
$$= 287.0 \text{ m}^2/\text{s}^2/°K$$

The *slug* is a consistent unit of mass in the English system, which is obtained by dividing the weight of an object in pounds by the acceleration of gravity in ft/s^2.

PREFIXES

The names of multiples and submultiples of SI units may be formed by application of the following prefixes.

Factor By which Unit is Multiplied	Prefix	Symbol
10^{12}	tera	T
10^{9}	giga	G
10^{6}	mega	M
10^{3}	kilo	k
10^{2}	hecto	h
10	deka	da
10^{-1}	deci	d
10^{-2}	centi	c
10^{-3}	milli	m
10^{-6}	micro	μ
10^{-9}	nano	n
10^{-12}	pico	p
10^{-15}	femto	f
10^{-18}	atto	a

APPENDIX B

STANDARD ATMOSPHERE

There are many models of the atmosphere to be found in the literature. Most of them are indistinguishable below an altitude of 100,000 ft (30 km). The table given below is based on the data to be found in Reference 2.3 and is the basis for program ATMOS:

English Units

Altitude ft	Temperature degrees *R*	Pressure lb/ft^2	Density, slugs/ft^3	Speed of sound, ft/s	Kinematic viscosity ft^2/s
0	518.69	2116.2	2.3769E-03	1116.4	1.5723E-04
5,000	500.86	1760.9	2.0482E-03	1097.1	1.7755E-04
10,000	483.04	1455.6	1.7556E-03	1077.4	2.0132E-04
15,000	465.23	1194.8	1.4962E-03	1057.4	2.2927E-04
20,000	447.43	973.26	1.2673E-03	1036.9	2.6234E-04
25,000	429.64	786.33	1.0663E-03	1016.1	3.0168E-04
30,000	411.86	629.66	8.9068E-04	994.85	3.4884E-04
35,000	394.08	499.34	7.3820E-04	973.14	4.0575E-04
40,000	389.99	393.12	5.8727E-04	968.08	5.0560E-04
45,000	389.99	309.45	4.6227E-04	968.08	6.4231E-04
50,000	389.99	243.61	3.6391E-04	968.08	8.1591E-04
55,000	389.99	191.80	2.8652E-04	968.08	1.0363E-03
60,000	389.99	151.03	2.2561E-04	968.08	1.3161E-03
65,000	389.99	118.93	1.7767E-04	968.08	1.6712E-03
70,000	389.99	93.672	1.3993E-04	968.08	2.1219E-03
75,000	389.99	73.784	1.1022E-04	968.08	2.6938E-03

SI Units

Altitude m	Temperature degrees K	Pressure kPa	Density, kg/m^3	Speed of sound, m/s	Kinematic viscosity m^2/sec
0	299.16	101.325	1.2250	340.29	1.4607E-05
900	282.31	90.971	1.1226	336.82	1.5687E-05
1,800	276.46	81.494	1.0269	333.32	1.6869E-05
2,700	270.62	72.835	0.93765	329.77	1.8167E-05
3,600	264.77	64.939	0.85445	326.19	1.9595E-05
4,500	258.93	57.752	0.77704	322.57	2.1167E-05
5,400	253.09	51.226	0.70513	318.91	2.2903E-05
6,300	247.25	45.311	0.63845	315.21	2.4824E-05
7,200	241.41	39.963	0.57671	311.47	2.6953E-05
8,100	235.58	35.140	0.51967	307.68	2.9320E-05
9,000	229.74	30.800	0.46706	303.85	3.1957E-05
9,900	223.91	26.906	0.41864	299.97	3.4903E-05
10,800	218.08	23.422	0.37417	296.03	3.8202E-05
11,700	216.66	20.335	0.32699	295.07	4.3475E-05
12,600	216.66	17.654	0.28388	295.07	5.0078E-05
13,500	216.66	15.327	0.24646	295.07	5.7680E-05
14,400	216.66	13.308	0.21399	295.07	6.6434E-05
15,300	216.66	11.555	0.18580	295.07	7.6514E-05
16,200	216.66	10.033	0.16133	295.07	8.8119E-05
17,100	216.66	8.7119	0.14009	295.07	1.0148E-04
18,000	216.66	7.5652	0.12165	295.07	1.1686E-04
18,900	216.66	6.5696	0.10564	295.07	1.3457E-04

APPENDIX C
CHEROKEE 180 MEASURED OR ESTIMATED DATA

WING GEOMETRY

span	$b = 30$ ft
midspan chord	$c_0 = 5.25$ ft
taper ratio	$\lambda = 1.0$
LE sweepback	$\Lambda = 0.0$ deg
dihedral	$\Gamma = 7.5$ deg
twist	$\theta_T = -2.0$ deg
planform area	$S = 157.5$ ft^2
aspect ratio	$A = 5.71$

dimensionless stations

wing root	0.13
flap inboard	0.13
flap outboard	0.60
aileron inboard	0.60
aileron outboard	0.95

flap chord/wing chord $\dfrac{c_f}{c} = 0.2$

aileron chord/wing chord $\dfrac{c_a}{c} = 0.2$

airfoil

NACA $65_2 - 415$

lift curve slope	$C_{l\alpha} = 0.106$/deg
max lift/no flaps	$C_{l\,\max} = 1.45$
ac moment/no flaps	$C_{m_{ac}} = -0.07$

Horizontal Tail Geometry

span	$b_t = 10.0$ ft
midspan chord	$c_{0_t} = 2.5$ ft
taper ratio	$\lambda_t = 1.0$
LE sweepback	$\Lambda_t = 0.0$
dihedral	$\Gamma_t = 0.0$
twist	$\theta_{T_t} = 0.0$
airfoil	symmetrical
planform area	25 ft^2
aspect ratio	4.0
elevator chord/tail chord	$c_e/c_t = 0.2$

centerline distance from wing LE to tail LE = 13.96 ft

Vertical Tail Geometry

(Note: All dimensions and areas are actual values. Aspect ratio based on image of tail below root.)

span	$b_V = 4.47$
midspan chord	$c_{0_V} = 3.31$
taper ratio	$\lambda_V = 0.54$
LE sweepback	$\Lambda_V = 35.0°$
airfoil	symmetrical
planform area	11.4
aspect ratio	3.5

rudder chord/tail chord $\frac{c_r}{c} = 0.4$

centerline distance from wing LE to tail LE = 11.3 ft.

Fuselage

total length	23.9 ft
equivalent diameter	4.0 ft
nose length	5.0 ft
nose to wing LE	5.0 ft

Propeller

diameter	6.17 ft
rotational speed	2400 rpm
prop to wing LE	5.5 ft

Total Airplane

weight 2400 lb
mass moment of inertia about *x*-axis $I_X = 1070$ slugs
mass moment of inertia about *y*-axis $I_Y = 1249$ slugs
mass moment of inertia about *z*-axis $I_Z = 2312$ slugs
equivalent flat plate area $f = 5.0 \text{ ft}^2$
Oswald's efficiency factor $e = 0.6$
nominal conditions for most examples
  cg = 1.0 ft behind wing LE
  true airspeed = 100 mph
  altitude = 0 (SSL)

Calculated Results at Nominal Conditions

Longitudinal	Lateral
$X_u = -0.07151$	$Y_\beta = -15.71$
$X_\alpha = -27.64$	$L_\beta = -12.06$
$Z_u = -0.4390$	$L_\phi = -5.437$
$Z_\alpha = -258.0$	$N_\beta = 4.952$
$Z_{\dot{\alpha}} = -0.891$	$N_\psi = -0.492$
$M_\alpha = -17.31$	
$M_{\dot{\alpha}} = -0.646$	
$M_q = -1.927$	

APPENDIX D

SOLUTION OF NTH-ORDER LINEAR DIFFERENTIAL EQUATIONS WITH CONSTANT COEFFICIENTS

The solution of linear simultaneous differential equations with constant coefficients can be written as the sun of a transient solution and a steady-state solution. These are also referred to as homogeneous and particular solutions, respectively.

The steady-state, or particular, solution is the part of the total solution that satisfies exactly the differential equation, including the forcing function. The transient solution satisfies the differential equation, with the forcing function set equal to zero.

A linear differential equation of the *n*th order can be written in general form as

$$C_n \frac{d^n x}{dt^n} + C_{n-1} \frac{d^{n-1} x}{dt^{n-1}} + \cdots + C_1 \frac{dx}{dt} + C_0 x = f(t) \tag{D-1}$$

To specify the problem completely, initial conditions equal in number to the order of the equation must be given. This usually means specifying x and its derivative up to order $n - 1$ at $t = 0$.

The solution to the homogeneous equation $[f(t) = 0]$ is of the form

$$x = Ae^{\sigma t} \tag{D.2}$$

The *n*th derivative will be

$$\frac{d^n x}{dt^n} = A\sigma^n e^{\sigma t}$$

This equation substituted into Equation 9.46 for $f(t)$ equal to zero leads to an *n*th degree polynomial for σ known as the characteristic equation.

$$C_n\sigma^n + C_{n-1}\sigma^{n-1} + \ldots + C_2\sigma^2 + C_1\sigma + C_0 = 0 \tag{D.3}$$

There are n roots or values of σ that will satisfy this polynomial. Thus, the general solution for x will be

$$x = A_1 e^{\sigma_1 t} + A_2 e^{\sigma_2 t} + \ldots + A_n e^{\sigma_n t} \tag{D.4}$$

where $\sigma_1, \sigma_2, \ldots \sigma_n$ are the roots of Equation D.3. These roots may be positive or negative real numbers or complex numbers with positive or negative real parts. If a complex root exists, its conjugate must also exist. Complex roots always appear in pairs of the form

$$\sigma = a \pm ib$$

$A_1, A_2, \ldots, A_n$ are real or complex constants to be determined from the initial conditions.

Let us examine the behavior of a term containing a particular type of root. Consider the following items:

1. e^{at}
2. e^{-at}
3. $e^{(a+ib)t}$
4. $e^{(-a+ib)t}$

If a and b are positive real quantities, it is obvious that item 1 will become infinitely large as t approaches ∞. This is an unstable transient motion. Conversely, item 2 is stable. Item 3 can be written as

$$e^{at} e^{ibt}$$

Since

$$e^{ibt} = \cos bt + i \sin bt$$

the complex root with a positive real part represents an oscillatory motion having an amplitude that increases without limit as t approaches ∞. This situation is also unstable. Conversely, item 4 is a stable oscillatory root. Thus, to investigate the stability of the transient solution, one need only consider the roots of the polynomial given by Equation D.3.

The particular solution is referred to as the steady-state solution since, for a stable system, it remains while the transient solution vanishes as t approaches ∞. The particular solution is of the same form as $f(t)$ if $f(t)$ can be expressed as a polynomial in t or in the form of

$$f(t) = Ae^{i\omega t}$$

or

$$f(t) = A \cos \omega t + B \sin \omega t$$

For example, suppose $f(t)$ were given by

$$f(t) = At^3 + Be^{i\omega t} + C \cos \omega t$$

Then the steady-state solution for x would take the form

$$x = at^3 + bt^2 + ct + d + fe^{i\omega t} + g \cos \omega t$$

Notice that when $f(t)$ is a polynomial, x must include all orders of t up to the highest in $f(t)$, even though some of the lower-order terms may be missing in $f(t)$.

As a fairly simple (and well-used) example of the foregoing, consider the second-order damped system governed by the equation.

$$\ddot{x} + 2\zeta\omega_n \dot{x} + \omega_n^2 x = f(t) \tag{D.5}$$

The characteristic equation becomes

$$\sigma^2 + 2\zeta\omega_n \sigma + \omega_n^2 = 0$$

The roots of this equation are

$$\sigma_{1,2} = \omega_n(-\zeta \pm \sqrt{\zeta^2 - 1} \tag{D.6}$$

ζ is referred to as the critical damping ratio for reasons that are now obvious. When ζ is positive but less than unity, σ will be complex and x will be a damped oscillation. When ζ is greater than unity, σ will be a negative real number, and the motion will be aperiodic.

Suppose $f(t)$ is given by

$$f(t) = A + Be^{i\omega t} \tag{D.7}$$

The particular solution will be of this same form.

$$x = X_0 + X_1 e^{i\omega t}$$

Substituting this into the original equation gives

$$X_1(\omega_n^2 - \omega^2 + 2\zeta\omega\omega_n i)\, e^{i\omega t} + \omega_n^2 X_0 = A + Be^{i\omega t}$$

In order for this equation to be satisfied for all t, the coefficients of like terms involving t must identically satisfy the equation. Thus,

$$X_1 = \frac{B}{\omega_n^2 - \omega^2 + 2\zeta\omega\omega_n i}$$

$$X_0 = \frac{A}{\omega_n^2}$$

It is instructive to clear the denominator of X_l of the complex number, so that

$$X_1 = B\frac{\omega_n^2 - \omega^2 - 2\zeta\omega\omega_n i}{(\omega_n^2 - \omega^2)^2 + 4\zeta^2\omega^2\omega_n^2}$$

If we further denote the frequency ratio ω/ω_n by r, X_1 can be written as

$$X_1 = \frac{B}{\omega_n^2}\frac{1 - r^2 - 2\zeta ri}{(1 - r^2)^2 + 4\zeta^2 r^2}$$

B/ω_n^2 is simply the static value of X_l for $\omega = 0$. Thus, the ratio of X_l to its static value can be written as

$$\frac{\omega_n^2 X_1}{B} = \frac{1 - r^2}{(1 - r^2)^2 + 4\zeta^2 r^2} - i\frac{2\zeta r}{(1 - r^2)^2 + 4\zeta^2 r^2} \qquad \text{(D.8)}$$

The complete solution to Equation D.5, forced in the manner given by Equation 9.52, is thus

$$x = e^{-\zeta\omega_n t}(x_1 e^{i\sqrt{1-\zeta^2}\,\omega_n t} + x_2 e^{-i\sqrt{1-\zeta^2}\,\omega_n t}) + \frac{A}{\omega_n^2} + \frac{B}{\omega_n^2}\frac{1 - r^2 - 2\zeta ri}{(1 - r^2)^2 + 4\zeta^2 r^2} e^{i\omega t} \qquad \text{(D.9)}$$

x_1 and x_2 are constants to be determined by the values of x and $\dot{x}$ at $t = 0$.

The complex form for x and the forcing function may seem somewhat mysterious at first if one is not used to this type of mathematical treatment. Certainly there is nothing imaginary about the longitudinal motion of an airplane. The complex notation is simply a convenient mathematical device to show the amplitude and phase relationships between various terms in the system.

For example, consider Equation D.8. This can be written in the form

$$\frac{\omega_n^2 X_1}{B} = C_1(r, \zeta) - iC_2(r, \zeta)$$

so that the steady-state solution becomes

$$x = \frac{A}{\omega_n^2} + \frac{B}{\omega_n^2}[C_1(r, \zeta) - iC_2(r, \zeta)]\, e^{i\omega t} \qquad \text{(D.10)}$$

The complex quantity in the brackets can be further expressed as

$$[C_1(r, \zeta) - iC_2(r, \zeta)] = Re^{-i\phi}$$

where

$$R = \sqrt{C_1^2 + C_2^2}$$

$$\phi = \tan^{-1}\frac{C_2}{C_1}$$

Thus, Equation D.10 becomes

$$x = \frac{A}{\omega_n^2} + \frac{B}{\omega_n^2} Re^{i(\omega t - \phi)} \tag{D.11}$$

If $Be^{i\omega t}$ is pictured as a force vector of magnitude B rotating around the origin in the complex plane, the time-dependent part of x, $x - A/\omega_n^2$, can be pictured as a displacement vector of magnitude BR/ω_n^2 rotating at the same rate of ω but lagging behind the force vector by an angle of ϕ. This is illustrated in Figure D.1.

ω_n is known as the undamped natural frequency, since this is the frequency at which the force-free system will oscillate when perturbed and with $\zeta = 0$. It is interesting to examine Equation D.11 when the system is forced at its natural frequency. For this case, $r = 1.0$, so $C_1(r, \zeta) = 0$, and $C_2(r, \zeta) = (2\zeta)^{-1}$, and ϕ therefore equals $\pi/2$ or 90°. Thus, the displacement of a second-order damped system, when forced at its natural frequency, will always lag the force by 90°, regardless of the damping.

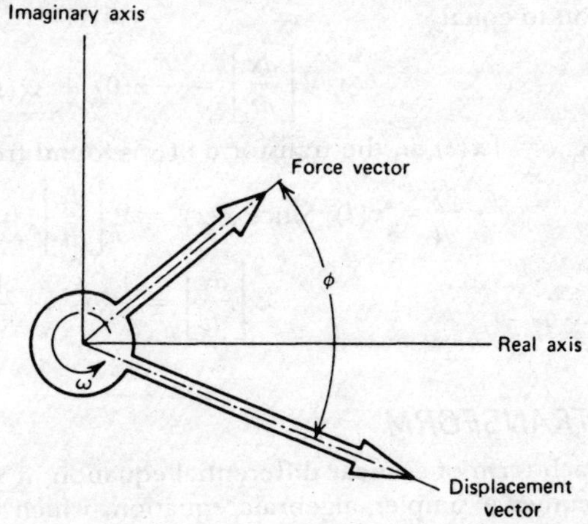

Figure D.1 Significance of complex representation for oscillatory vectors.

APPENDIX E

THE LAPLACE TRANSFORMATION

THE DIRECT TRANSFORM

The Laplace transform of a function, $x(t)$, is defined as

$$\bar{x}(s) = \mathfrak{L}[x(t)] = \int_0^\infty e^{-st}x(t)\,dt \tag{E.1}$$

The Laplace transform of the derivative of $x(t)$ can be found from the above definition to equal

$$\mathfrak{L}\left[\frac{dx}{dt}\right] = -x(0) + s\bar{x}(s) \tag{E.2}$$

Defining $y = \int x(t)\,dt$, the transform of y is found from the fact that

$\dfrac{dy}{dt} = x(t)$. Since $\bar{x}(s) = \mathfrak{L}\left[\dfrac{dy}{dt}\right]$, it follows that

$$\mathfrak{L}\left[\frac{dy}{dt}\right] = \frac{1}{s}\bar{x}(s) + \frac{1}{s}y(0) \tag{E.3}$$

THE INVERSE TRANSFORM

When each term of a linear differential equation in $x(t)$ is transformed, the equation becomes a simpler algebraic equation, which can be solved for $\bar{x}(s)$. The inverse transform must then be performed to obtain the solution for $x(t)$. Frequently, the transformed solution can be expressed as the ratio of two polynomials.

$$\bar{x}(s) = \frac{N(s)}{D(s)} \tag{E.4}$$

where the degree of $D(s)$ is higher than that of $N(s)$. In this case, the method of partial fractions can be used to separate each factor of $D(s)$ into a separate term. The inverse of each term is then found from a table of Laplace transforms. The solution for this case can also be solved by application of the Heaviside expansion. If $D(s) = (s - \sigma_1)(s - \sigma_2) \ldots (s - \sigma_n)$, then the inverse transform of Equation E.4 can be written as

$$x(t) = \sum_{l=1}^{n} \left\{\frac{(s - \sigma_l)N(s)}{D(s)}\right\}_{s=\sigma_l} e^{\sigma_l t}$$

Table of Laplace Transforms

$x(t)$	$x(s)$
1	$\frac{1}{s}$
t	$\frac{1}{s^2}$
$e^{\omega t}$	$\frac{1}{s - \omega}$
$\sin \omega t$	$\frac{\omega}{s^2 + \omega^2}$
$\cos \omega t$	$\frac{s}{s^2 + \omega^2}$
$\delta(t)$ (unit impulse)	1
$\delta(t - T)$ (unit impulse at T)	e^{-sT}
$x = 0$ for $t < T$ and 1 for $t > T$	$\frac{e^{-sT}}{s}$

APPENDIX F
AIRPLANE DATA

The following samples of airplane data (Figures F.1 to F.20 and Table F.1) are taken from *Aircraft Handling Qualities Data,* by Robert K. Heffley and Wayne F. Jewell, NASA CR-2144, December 1972. The figures include:

- Flight envelope.
- C_L and C_D as a function of M and altitude.
- Weight and moments of inertia.
- Three-view sketches to scale.

In addition, stability derivatives are tabulated for the Convair 880. The following definitions apply to notations on the sketches.

F.S. (fuselage station) = inches aft of reference transverse plane near the nose.
W.L. (waterline) = inches above horizontal reference plane.
B.B. (buttline) = inches to right of vertical plane of symmetry.

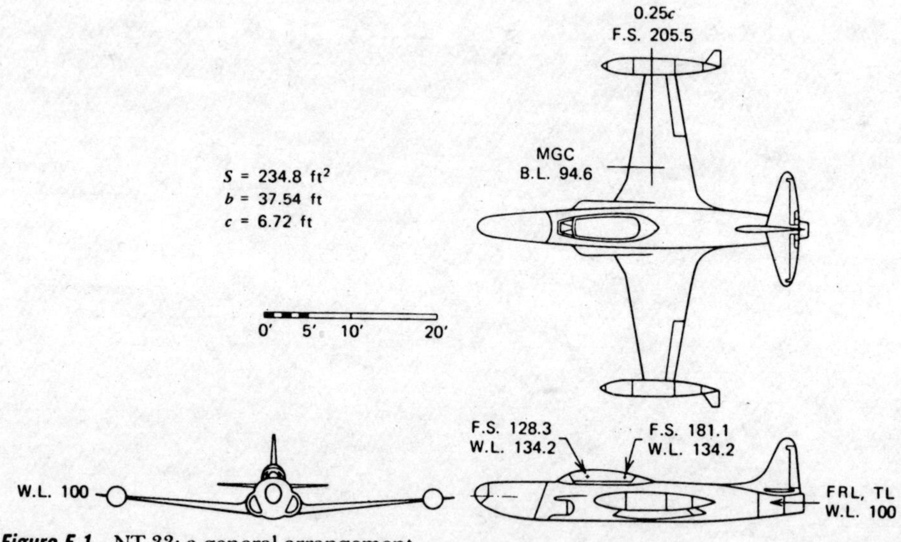

Figure F.1 NT-33; a general arrangement.

Nominal Configuration

230-gal tip tanks
60% internal fuel
$W = 13{,}700$ lb
cg at $0.263\bar{c}$, W.L. 100.2

$$\left.\begin{aligned} I_x &= 23{,}800 \text{ slug-ft}^2 \\ I_y &= 21{,}100 \text{ slug-ft}^2 \\ i_z &= 43{,}800 \text{ slug-ft}^2 \\ I_{xz} &= 480 \text{ slug-ft}^2 \end{aligned}\right\} \text{body axis}$$

Power Approach Configuration

230-gal tip tanks
25% internal fuel
Full flaps
Gear down
$1.4\ V_s$
$W = 11{,}800$ lb
cg at $0.260\bar{c}$, W.L. 100

$$\left.\begin{aligned} I_x &= 12{,}700 \text{ slug-ft}^2 \\ I_y &= 20{,}700 \text{ slug-ft}^2 \\ I_z &= 32{,}000 \text{ slug-ft}^2 \\ I_{xz} &= 480 \text{ slug-ft}^2 \end{aligned}\right\} \text{body axis}$$

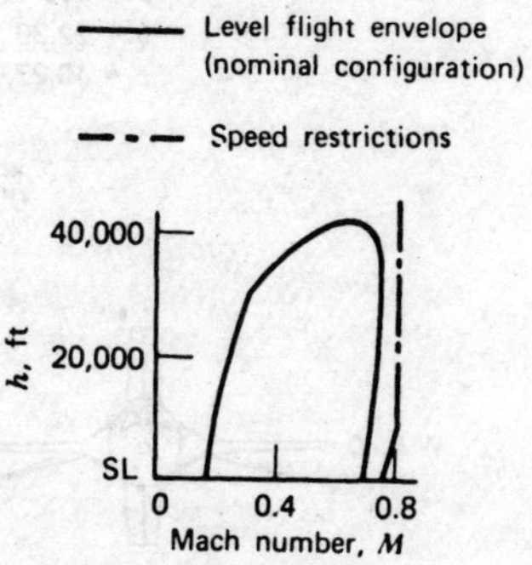

Figure F.2 NT-33A; flight envelope.

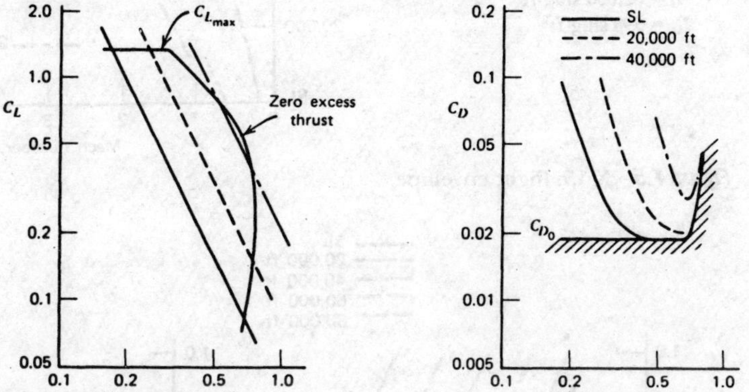

Figure F.3 NT-33A; 13,700 lb.

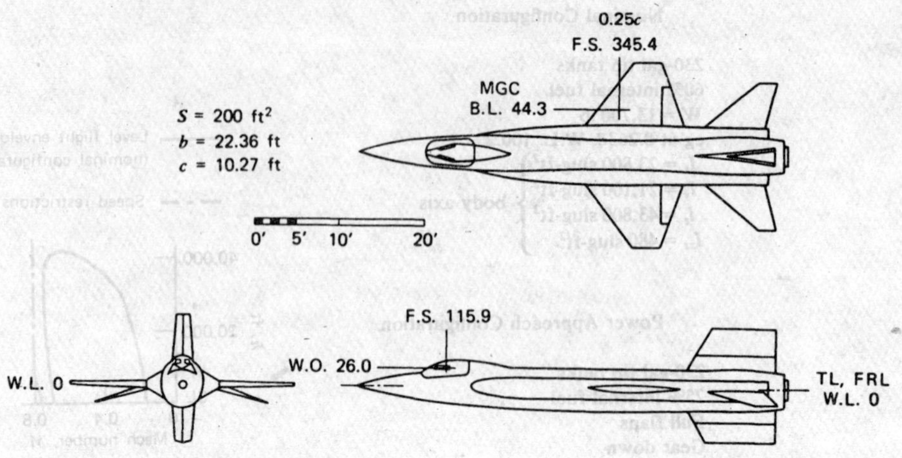

Figure F.4 X-15; a general arrangement.

Nominal Configuration

Zero fuel
Lower ventral on
Speed brakes retracted
$W = 15{,}560$ lb
cg at $0.22\bar{c}$
$i_x = 3650$ slug-ft²
$I_y = 80{,}000$ slug-ft²
$I_z = 82{,}000$ slug-ft²
$I_{xz} = 590$ slug-ft²

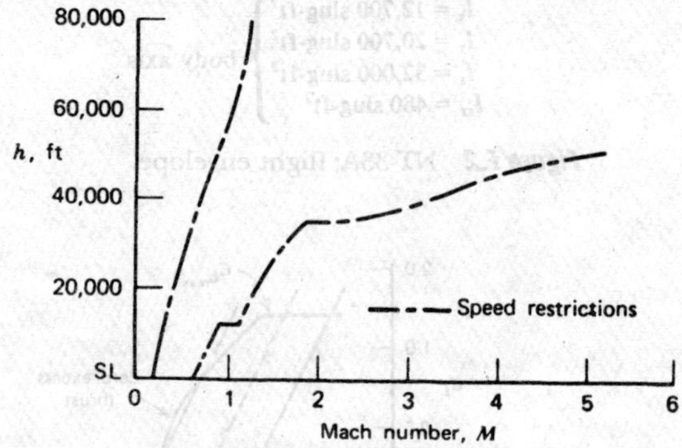

Figure F.5 X-15; flight envelope.

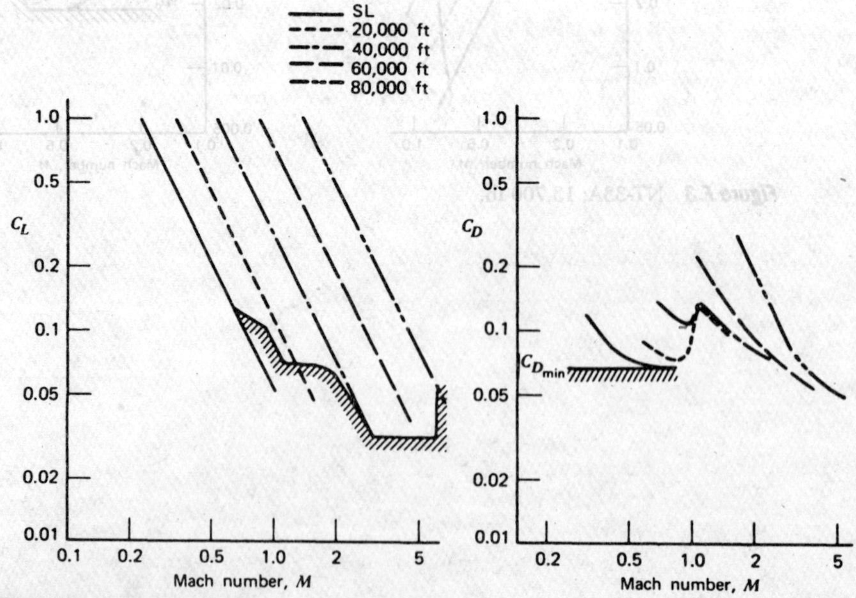

Figure F.6 X-15; 15,560 lb.

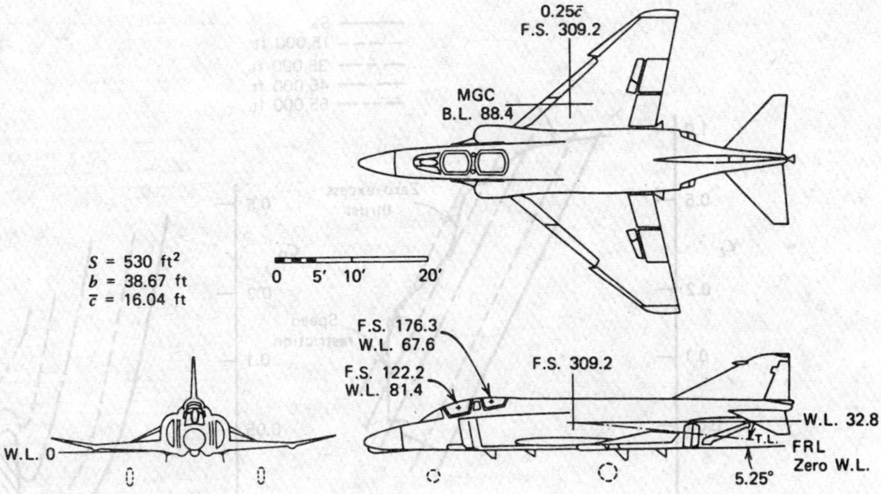

Figure F.7 F-4C; a general arrangement.

Nominal Configuration

4 AIM-7 missiles
60% internal fuel
$W = 38{,}924$ lb
cg at $0.289\bar{c}$, W.L. 27.65

$$\left.\begin{aligned} I_x &= 25{,}001 \text{ slug-ft}^2 \\ I_y &= 122{,}186 \text{ slug-ft}^2 \\ I_z &= 139{,}759 \text{ slug-ft}^2 \\ I_{xz} &= 2177 \text{ slug-ft}^2 \end{aligned}\right\} \text{body axis}$$

Power Approach Configuration

2 AIM-7 missiles aft
20% internal fuel
Full flaps, BLC
Gear down
19 units angle of attack
$W = 33{,}196$ lb
cg at $0.291\bar{c}$, W.L. 25.2

$$\left.\begin{aligned} I_x &= 23{,}668 \text{ slug-ft}^2 \\ I_y &= 117{,}500 \text{ slug-ft}^2 \\ I_z &= 133{,}723 \text{ slug-ft}^2 \\ I_{xy} &= 1573 \text{ slug-ft}^2 \end{aligned}\right\} \text{body axis}$$

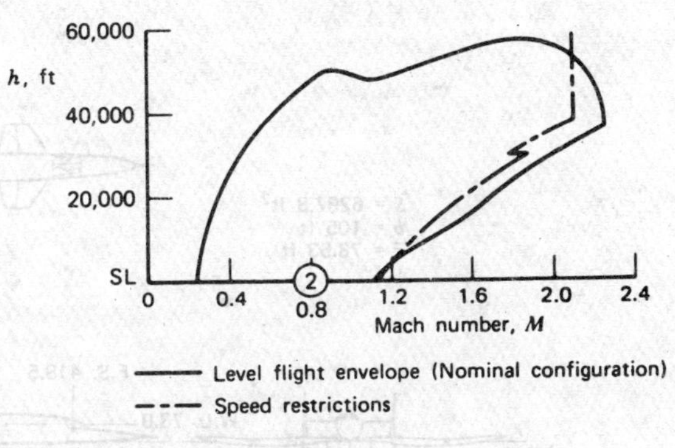

Figure F.8 F-4C; flight envelope.

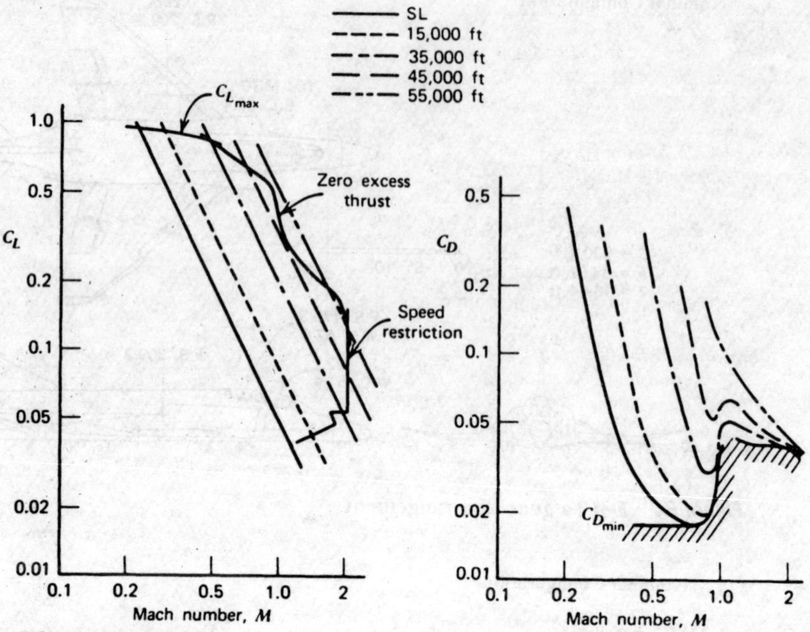

Figure F.9 F-4C; 38,942 lb.

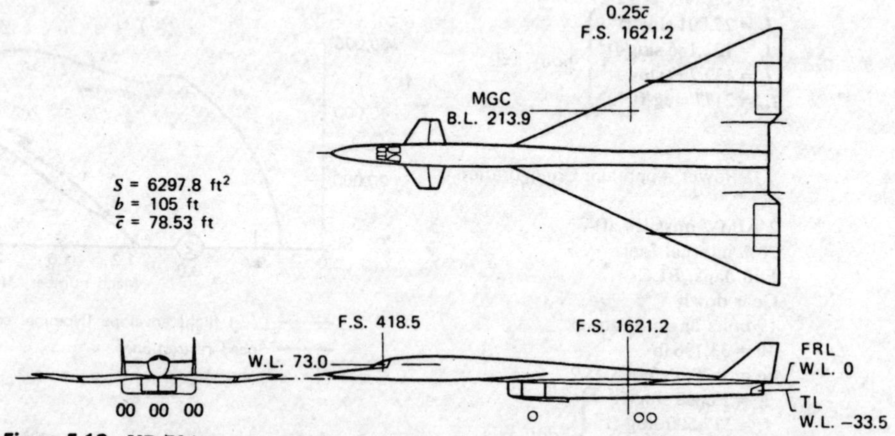

Figure F.10 XB-70A; a general arrangement.

Nominal Configuration

Tips folded according to flight condition
50% internal fuel
$W = 384{,}524$ lb
cg at $0.218\bar{c}$, W.L. −7.2

$$\left.\begin{aligned} I_x &= 1.8 \times 10^6 \text{ slug-ft}^2 \\ I_y &= 19.9 \times 10^6 \text{ slug-ft}^2 \\ I_z &= 22.1 \times 10^6 \text{ slug-ft}^2 \\ I_{xz} &= -0.88 \times 10^6 \text{ slug-ft}^2 \end{aligned}\right\} \text{body axis}$$

Power Approach Configuration

Tips extended
19% internal fuel
Canard flaps down
Gear down
$W = 300{,}000$ lb
cg at $0.235\bar{c}$

$$\left.\begin{aligned} I_x &= 1.45 \times 10^6 \text{ slug-ft}^2 \\ I_y &= 16 \times 10^6 \text{ slug-ft}^2 \\ I_z &= 17.2 \times 10^6 \text{ slug-ft}^2 \\ I_{xz} &= -0.6 \times 10^6 \text{ slug-ft}^2 \end{aligned}\right\} \text{body axis}$$

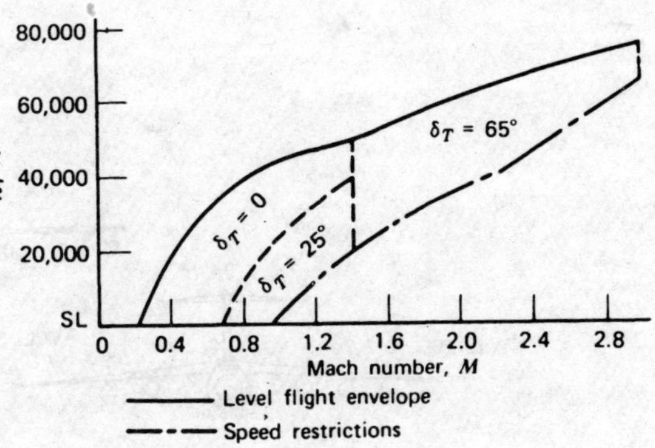

Figure F.11 XB-70A; flight envelope.

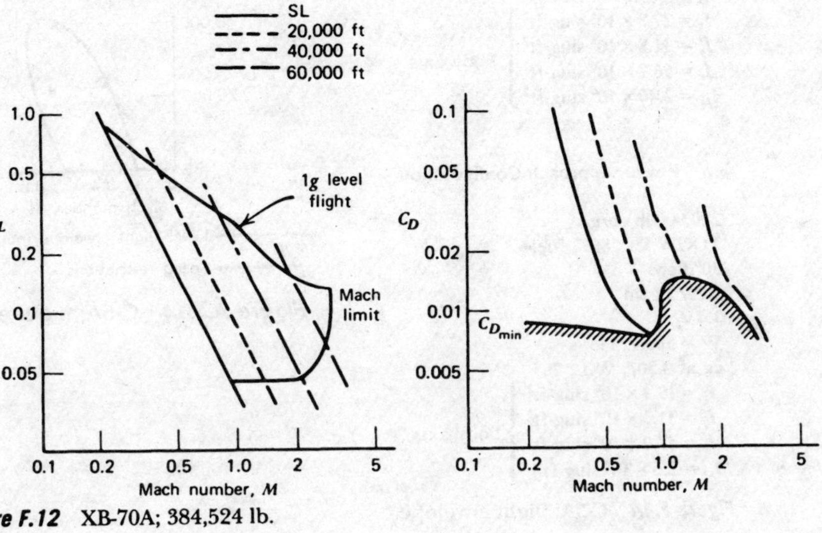

Figure F.12 XB-70A; 384,524 lb.

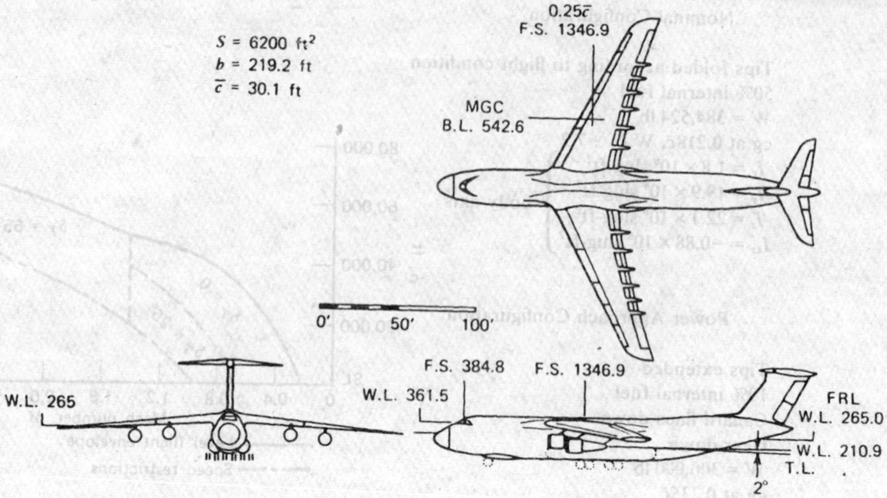

Figure F.13 C-5A; a general arrangement.

Nominal Configuration

220,000 lb cargo
TOGW less 40% fuel
$W = 654{,}362$ lb
cg at $0.30\bar{c}$, W.L. 265

$$\left.\begin{aligned} I_x &= 27.8 \times 10^6 \text{ slug-ft}^2 \\ I_y &= 31.8 \times 10^6 \text{ slug-ft}^2 \\ I_z &= 56.2 \times 10^6 \text{ slug-ft}^2 \\ I_{xz} &= 2.46 \times 10^6 \text{ slug-ft}^2 \end{aligned}\right\} \text{body axis}$$

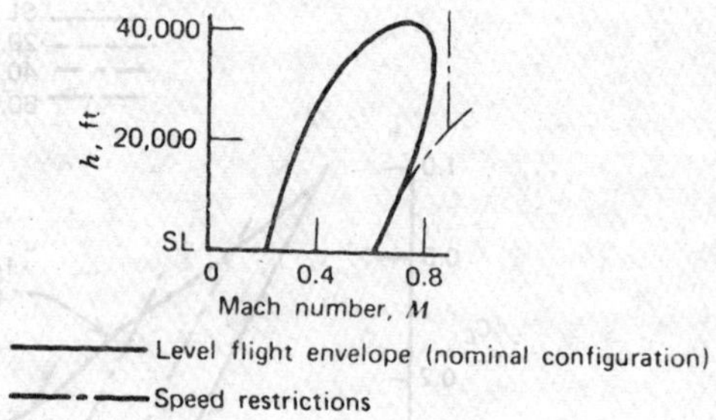

Figure A.3.14 C-5A; flight envelope.

Power Approach Configuration

220,000 lb cargo
TOGW less 80% fuel
30° flaps
Gear down
$1.4\,V_s$
$W = 580{,}723$ lb
cg at $0.30\bar{c}$, W.L. 265

$$\left.\begin{aligned} I_x &= 19.1 \times 10^6 \text{ slug-ft}^2 \\ I_y &= 31.3 \times 10^6 \text{ slug-ft}^2 \\ I_z &= 47.0 \times 10^6 \text{ slug-ft}^2 \\ I_{xz} &= 2.5 \times 10^6 \text{ slug-ft}^2 \end{aligned}\right\} \text{body axis}$$

Figure F.14 C-5A; flight envelope.

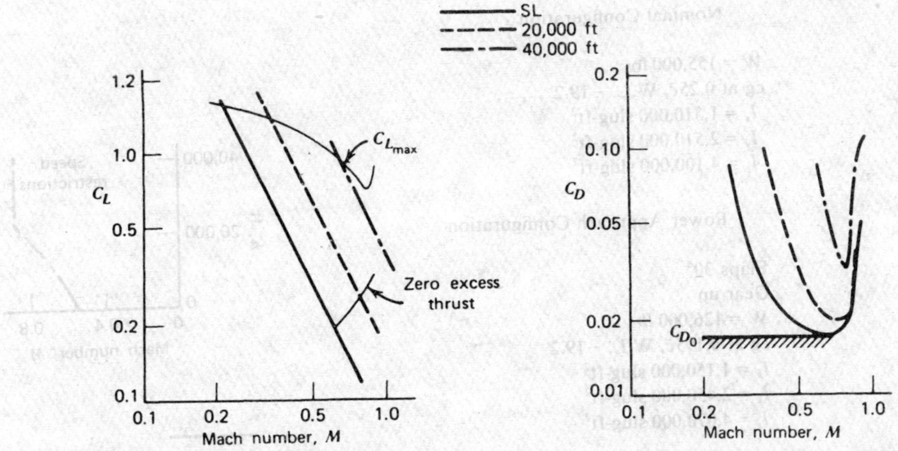

Figure F.15 C-5A; 654,362 lb.

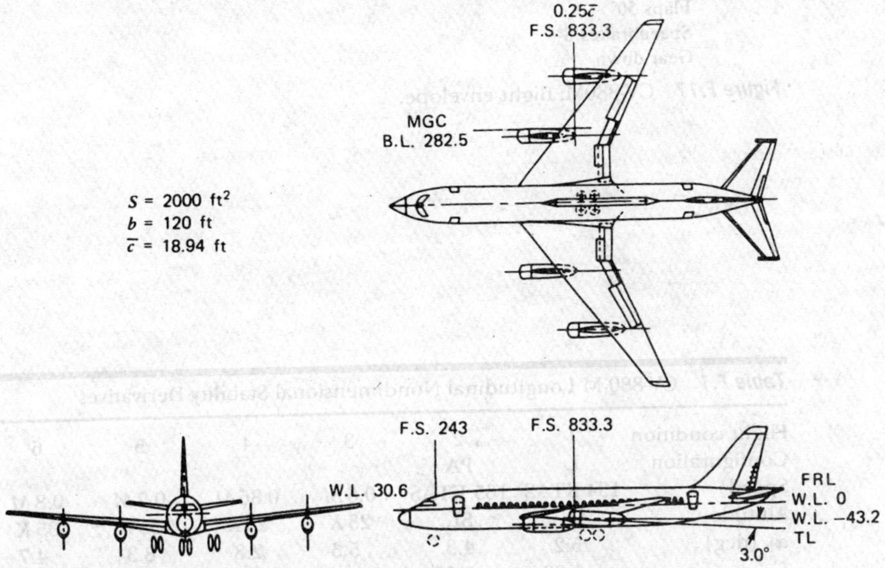

Figure F.16 CV-880M; a general arrangement.

Nominal Configuration

$W = 155{,}000$ lb
cg at $0.25\bar{c}$, W.L. -19.2
$I_x = 1{,}510{,}000$ slug-ft^2
$I_y = 2{,}510{,}000$ slug-ft^2
$I_z = 4{,}100{,}000$ slug-ft^2

Power Approach Configuration

Flaps 30°
Gear up
$W = 126{,}000$ *lb*
cg at $0.195\bar{c}$, W.L. -19.2
$I_x = 1{,}150{,}000$ slug-ft^2
$I_y = 2{,}450{,}000$ slug-ft^2
$I_z = 4{,}070{,}000$ slug-ft^2

Landing Configuration

Same as power approach except:
Flaps 50°
Speedbrakes 8°
Gear down

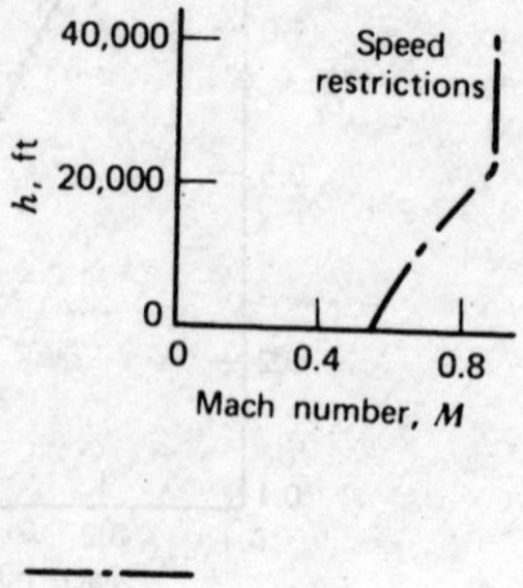

Figure F.17 CV-880M; flight envelope.

Table F.1 CV-880 M Longitudinal Nondimensional Stability Derivatives

Flight condition	1	2	3	4	5	6	7
Configuration	L	PA					
Speed	134 KTAS	165 KTAS	0.6 *M*	0.86 *M*	0.7 *M*	0.8 *M*	0.86 *M*
Altitude	SL	SL	23 *K*	23 *K*	35 *K*	35 *K*	35 *K*
α_0 (deg)	5.2	4.3	5.3	2.8	8.3	4.7	4.0
C_L	1.03	0.68	0.36	0.175	0.454	0.347	0.301
C_D	0.154	0.080	0.022	0.019	0.025	0.024	0.023
$C_{L\alpha}$ (1/rad)	4.66	4.52	4.28	4.41	4.62	4.8	4.9
$C_{D\alpha}$ (1/rad)	0.43	0.27	0.14	0.07	0.18	0.15	0.13
C_{m_α} (1/rad)	−0.381	−0.903	−0.522	−0.572	−0.568	−0.65	−0.74
$C_{L\alpha}$ (1/rad)	2.7	2.7	2.44	2.5	2.75	2.75	2.9
C_{L_q} (1/rad)	7.92	7.72	6.76	6.37	7.51	7.5	7.62
C_{m_α} (1/rad)	−4.17	−4.13	−4.16	−4.66	−4.4	−4.5	−4.6
C_{m_q} (1/rad)	−12.2	−12.1	−11.5	−11.8	−12.0	−12.0	−12.0
$C_{L\delta_e}$ (1/rad)	0.22	0.213	0.193	0.141	0.203	0.190	0.180
$C_{m\delta_e}$ (1/rad)	−0.657	−0.637	−0.586	−0.438	−0.618	−0.57	−0.532
$C_{h\delta_e}$ (1/rad)	−10.326	−0.328	−0.336	−0.278	−0.342	−0.31	−0.285
$C_{L\delta_{te}}$ (1/rad)	0.055	0.0532	0.482	0.0352	0.0508	0.047	0.0450
$C_{m\delta_{te}}$ (1/rad)	−0.164	−0.159	−0.146	−0.11	−0.155	−0.14	−0.134
$C_{h\delta_{te}}$ (1/rad)	−0.287	−0.285	−0.297	−0.343	−0.312	−0.335	−0.352

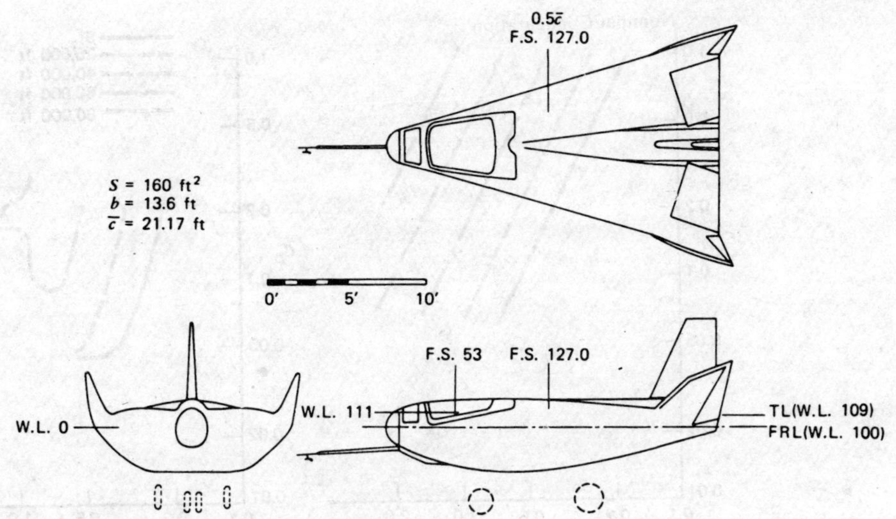

Figure F.18 HL-10; a general arrangement.

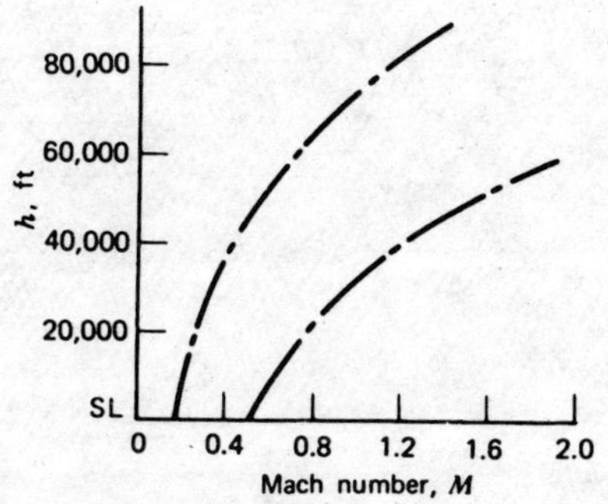

Nominal Configuration

Zero fuel (burnout)
Gear up
Transonic or subsonic configuration, depending on flight condition
$W = 6466$ lb
cg at $0.517\bar{c}$, W.L. 94.4
$I_x = 1353$ slug-ft^2
$I_y = 6413$ slug-ft^2
$I_z = 7407$ slug-ft^2
$I_{xz} = 399$ slug-ft^2
} body axis

Figure F.19 HL-10; flight envelope; nominal envelope extremes.

Configuration	Speed Brakes	Elevon Flaps	Tip-Fin Flaps
Subsonic	Zero	Zero	Zero
Transonic	8°	30°	30.5°/32.5°

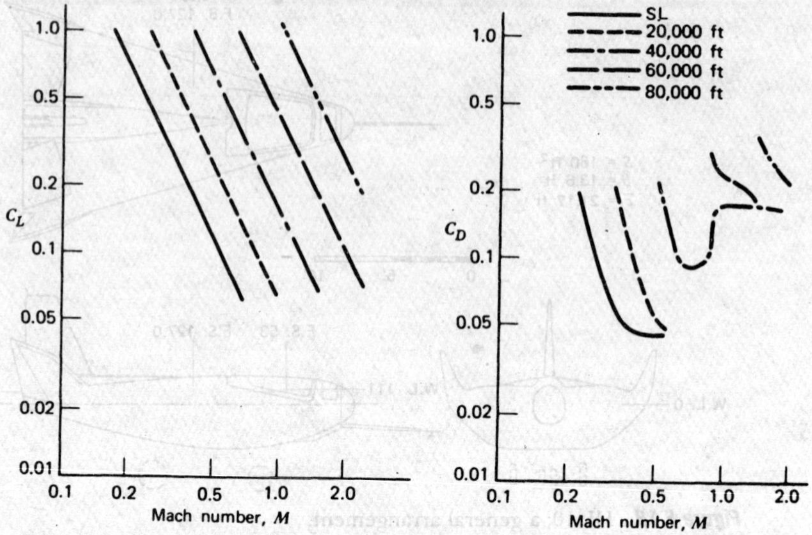

Figure F.20 HL-10; 6466 lb.

Table F.2 Stealth Aircraft

F-117 (see Figure 5.27)

Length:	65 ft 11 in
Wing span:	43 ft 4 in
Wing sweep:	67 deg
Wing area:	1140 ft^2
Height:	12 ft 5 in
Vertical tail sweep:	20 deg
Weight empty:	29,500 lb
Max gross weight:	52,500 lb
Max speed:	0.9 Mach @ 35,000 ft
Max alt.:	52,000 ft
Max range:	1250 mi/internal fuel
Powerplant: (two)	GE F404-GE-F1D2 (10,800 lb T class)
T.O. speed at combat weight:	165 kt
Ldg. speed at combat weight:	150 kt

F-22 (see Figure 5.27)

Length:	62 ft 1 in
Span:	44 ft 6 in
Height:	17 ft 9 in
Wing sweep:	42 deg
Max speed:	$> M = 2$
Wing area:	830 ft^2
Weight empty:	30,000 lb (est)
Max T.O. weight:	58,000 lb
Max level speed:	$M = 1.58$ (supercruise) (demonstrated) $M = 1.7$ @ 30,000 ft afterburning
Absolute ceiling:	50,000 ft
Max level speed at SL:	800 kt (target)

B-2 (see Figure 9.3)

Wing span:	172 ft
Length:	69 ft
Height:	17 ft
Planform area:	over 5000 ft^2
Weight empty:	100,000–110,000 lb
Normal T.O. weight:	371,330 lb
Max T.O. weight:	400,000 lb
Max wing loading:	80.0 psf
Approach speed:	140 kt
Service ceiling:	50,000 ft
Unrefueled range:	7,255 mi
Max fuel capacity:	180,000–200,000 lb
Powerplant (4):	GE F118-GE-110 (19,000 lb T class)
Material:	radar-absorbent honeycomb structure and skin

See Jane's *All the World's Aircraft*, 1991–92, for additional details on all three stealth airplanes.

APPENDIX G

NOMENCLATURE, ABBREVIATIONS, ACRONYMS

a	speed of sound
a_{sl}	speed of sound at sea level
a	slope of C_L or C_l with α
a	core radius of vortex
a_1	longitudinal flapping angle
A	coefficient of characteristic equation
A	aspect ratio
A	propeller disc area, πR^2
A	cross-sectional area or reference area
A	indicial admittance
A	propeller disc area
a	average acceleration for calculating ground roll distance
ac	aerodynamic center
ADF	automatic direction finder
AF	activity factor (Eq. 6.70)
AGARD	Advisory Group for Aerospace Research and Development
ASTOVL	advanced STOVL (usually means supersonic STOVL)
b	wing span
B	number of blades
B	coefficient of characteristic equation
B	factor multiplying blade radius to get effective radius
B	body force
b'	span of rolled-up trailing vortex system
bhp	brake horsepower
BSFC	brake specific fuel consumption
b_1	lateral flapping angle
c	chord
c	airfoil chord
C	coefficient of characteristic equation
c_0	midspan chord
c	mean aerodynamic chord
C_D	wing or body drag coefficient
C_{D_0}	drag coefficient at zero lift
C_d	average section drag coefficient
C_d	average section C_d
C_{Di}	induced drag coefficient
C_{D_v}	C_D based on body volume
c_f	flap chord
C_f	drag (skin friction) coefficient based on wetted area
C_F	total skin friction drag coefficient
CF	centrifugal force

cg	center-of-gravity
C_H	hinge moment coefficient
C_h	control surface hinge moment
C_l	airfoil lift coefficient
C_l	section lift coefficient
C_l	rolling moment coefficient
C_L	wing lift coefficient
C_{Ld}	propeller design lift coefficient
C_{l_0}	airfoil lift coefficient at zero angle of attack
C_{L_0}	wing lift coefficient at zero angle of attack
C_{L_0}	lift coefficient at zero angle of attack
C_{l_α}	derivative of C_l with respect to α
C_{l_a}	additional section C_l along wing span caused by α
C_{L_α}	derivative of C_L with respect to alpha
C_{l_b}	basic section C_l along wing for $C_L = 0$.
$\overline{C}_l$	average section C_l
$C_{L\delta}$	derivative of C_L with respect to elevator angle
$C_{L_{limit}}$	limit C_L for circulatory lift due to induced effects
$C_{l_{max}}$	maximum airfoil lift coefficient
$C_{L_{max}}$	maximum wing lift coefficient
C_M	pitching moment coefficient (usually about cg)
$C_{m_{1/4}}$	airfoil moment coefficient about quarter-chord point
$C_{m_{ac}}$	airfoil moment coefficient about the aerodynamic center
$C_{M_{ac}}$	moment coefficient about wing aerodynamic center
$C_{M\alpha}$	derivative of C_M with respect to α (usually about cg)
$C_{M\delta}$	derivative of C_M with respect to elevator angle
C_{mLE}	airfoil pitching moment coefficient about leading edge
M_A	pitching moment about midspan leading edge
C_μ	momentum (or blowing) coefficient for jet flap
C_n	normal force coefficient for an airfoil
C_N	yawing moment coefficient
cp	center of pressure
C_p	pressure coefficient, $(p - p_\infty)/q$
C_P	power coefficient
C_p	specific heat at constant pressure
C_p	pressure coefficient
c_t	tip chord
C_T	thrust coefficient
d	diameter
D	coefficient of characteristic equation
D	propeller diameter
D	drag
D_f	skin friction drag
DG	directional gyroscope
D_i	induced drag
DME	distance measuring equipment
e	Oswald's wing efficiency factor, $C_{D_i} = C_L^2/(\pi Ae)$
e	flapping hinge offset
E	coefficient of characteristic equation
EBF	externally blown flaps
EGT	exhaust gas temperature

EPR	engine pressure ratio
f	frequency
f	equivalent flat plate area; parasite area
F	correction factor to jet-flapped wing for aspect ratio
F	force
F	thrust for a turbinejet
F	Prandtl's tip-loss factor
F_0	static thrust for a turbojet
FAA	Federal Aviation Agency
F_R	ratio of fuel flow at part throttle to full-throttle flow
f_v	vertical equivalent flat plate area causing download
g	acceleration of gravity
G	control gearing
$G(s)$	transfer function
gph	gallons per hour
GPS	global positioning system
h	distance of rotor above cg
h	distance of cg from LE of MAC as fraction of MAC
h	impulsive admittance
h	altitude
h_1	dimensionless distance of tail ac from LE of MAC
h_n	neutral point location behind LE of MAC in terms of MAC
h_{nw}	equivalent to wing ac
h_{abs}	absolute ceiling
h_e	total airplane energy per unit weight
i	dimensionless mass moment of inertia
I	mass moment of inertia
I_F	mass moment of inertia of blade about flapping hinge
IGE	in-ground effect
i_{hs}	incidence angle of horizontal stabilizer
ILS	instrument landing system
IP	imaginary part
IR	infrared
i_t	incidence angle of horizontal tail, positive nose-down
J	propeller advance ratio, $V/(nD)$
K_β	coupling between rotor blade pitch angle and flapping
k_e	linkage factor for stabilator
l	a length of some kind
l	rolling moment
l_t	distance of tail ac behind cg
L	rolling moment about x-axis (positive right wing down)
L	lift
LE	leading edge
LEV	leading edge vortex
LFC	laminar flow control
LO	low observables
l_w	distance of wing ac ahead of cg
m	mass flow rate
m	doublet strength
m	airplane mass
m_j	mass flux in jet flap

M	pitching moment (usually about cg)
M	airplane mass
M	pitching moment about y-axis (positive nose up)
M	figure of merit for hovering rotor
M	Mach number
MAC	mean aerodynamic chord
M_C	blade moment about flapping hinge from centrifugal force
M_H	hub moment
M_I	blade inertial moment about flapping hinge
M_L	blade lift moment about flapping hinge
M_s	pitching moment about leading edge
n	load factor
n	angular velocity, revolutions per second
N	yawing moment
N	yawing moment about z-axis (positive nose right)
N	rpm
NACA	National Advisory Committee for Aeronautics (now NASA)
NASA	National Aeronautics and Space Administration
NavCom	navigation and communication
NLF	natural laminar flow
OGE	out-of-ground effect
OM	outer marker
$\bar{p}$	dimensionless pitch rate
p	pressure, usually static
p	pressure
p	pitch
p_c	pressure just ahead of compressor
p_2	pressure just behind compressor
P	power
P	angular velocity around body x-axis
P	angular velocity (roll rate) about airplane's x-axis
P	control force
P_A	power available
P_E	engine power
p_o	reservoir pressure
p_{sl}	static pressure at sea level
p_o	ambient static pressure
P_i	induced power
P_{i0}	induced power for static case
p_l	static pressure on lower surface
P_p	profile power required by profile drag of blades
P_{par}	parasite power required by parasite drag of aircraft
P_R	power required
P_s	excess specific power
p_u	static pressure on upper surface
P_{use}	useful power (TV)
$\bar{q}$	dimensionless roll rate
q	elemental source strength
q	dynamic pressure, $\rho V^2/2$
Q	angular momentum
Q	angular velocity around body y-axis

Q	torque
Q	total source strength
Q	angular velocity (pitch rate) about airplane's y-axis
Q	volume flux
Q	heat flux per unit weight of gas
r	radial coordinate
r	compression ratio
$\bar{r}$	dimensionless yaw rate
R	angular velocity around body z-axis
R	radius of cylinder
R	radius of curvature
R	angular velocity (yaw rate) about airplane's z-axis
R	range
R ≠ C	rate of climb
R	Reynolds' number (Eq. 2.14)
R	propeller radius
R	universal gas constant
RAM	radar-absorbent material
RCS	radar cross section
R_l	Reynolds number based on total length
RP	real part
Rx	local Reynolds number based on distance from LE
R_s	spin radius
s	Laplace variable
s	distance along the takeoff
S	Strouhal number, fd/V
S	wing planform area
S	distance along a path
S_f	flap planform area
SFC	specific fuel consumption
SSL	standard sea level
STOVL	short takeoff/vertical landing
s_{tran}	distance to flare (transition)
S_w	wetted area
t	maximum airfoil thickness
t	time
t^*	reference time to nondimensionalize equations of motion
T	thrust
T	period of oscillatory motion
T	temperature
T_c	constant temperature (11 to 23 km) (36,000 to 75,500 ft)
TDPF	tail damping power factor
t_e	maximum endurance
T_0	static thrust
T_R	ratio of thrust to maximum available thrust
TSFC	thrust specific fuel consumption
T_2	time-to-double-amplitude
$T_{1/2}$	time-to-damp to halve-amplitude
u	velocity component along x-axis, also perturbation
U	x component of velocity
U_{de}	derived gust velocity

UDF	unducted fan
USB	upper-surface blowing
v	velocity component along y-axis, also perturbation
V	airplane resultant velocity
V	y component of velocity
V/STOL	vertical/short takeoff and landing
V	local resultant velocity
V_o	free-stream velocity
V_A	approach velocity
V_1	critical engine failure speed
V_2	takeoff climb speed
V_{cal}	calibrated airspeed
V_E	resultant velocity at propeller section with w_a and w_t
V_E	velocity at exit of diffuser
V_H	horizontal tail volume
V_v	vertical tail volume
V_j	velocity of the jet for a jet flap
V_j	primary jet velocity
V_{LOF}	liftoff speed
V_M	body volume
V_{mc}	minimum control speed
V_μ	minimum unstick speed
V_n	velocity normal to surface
V_{opt}	speed for minimum drag
VOR	very high frequency, omnidirectional radio
V_R	takeoff rotation speed
V_R	resultant velocity at propeller blade section
V_s	stalling speed
V_{s_1}	one engine-out stall speed
v_θ	tangential component of velocity
V_T	tip speed due to rotation, ωR
VTOL	vertical takeoff and landing
V_w	headwind speed
w	velocity component along z-axis, also perturbation
w	downwash
w	induced velocity
W	gross weight
W	z component of velocity
w_a	axial component of induced velocity
W_E	weight empty
W_f	fuel weight
w_0	induced velocity for static case
w_0	impact velocity in vortex theory for propellers
WOT	wide-open throttle
w_t	tangential component of induced velocity
x	dimensionless distance along wing span
x	dimensionless radial distance along a propeller blade
x	dimensionless distance along airfoil chord
x	cartesian coordinate
x	distance of rotor axis behind cg
X	aerodynamic force in x direction

x'	earth-fixed x-axis
X_Λ	distance of wing ac behind midspan leading edge
x_{ac}	dimensionless location of aerodynamic center
x_{cp}	dimensionless location of center of pressure from LE
x_p	x location at which velocity is induced
X_w	distance of cg behind wing ac
y	spanwise distance along a wing span from the center
y	cartesian coordinate
Y	aerodynamic force in y direction
y'	earth-fixed y-axis
y_p	y location at which velocity is induced
z	distance of camber above chord line or an airfoil
z	cartesian coordinate
Z	aerodynamic force in z direction
z'	earth-fixed z-axis
z_{max}	maximum value of dimensionless camber
z_p	z location at which velocity is induced
Z_w	distance of cg above wing ac
z_v	distance of center of pressure of vertical tail above cg

GREEK NOMENCLATURE

α	rotor disc plane angle of attack
α	angle of attack
α	ratio of primary jet area to nozzle area
α_{0l}	angle of attack for zero lift of an airfoil
α_{0L}	angle of attack for zero lift of a wing
α_i	induced angle of attack
β	ratio of diffuser exit area to nozzle area
β	sideslip angle
β	$\sqrt{1 - M_\infty^2}$
β	flapping angle of rotor blade
δ	control surface angle
δ	jet flap angle
γ	ratio of specific heats
γ_j	equivalent distributed vortex strength of a jet flap
γ	resultant angle between leading edge and free-stream V
γ	strength of a vortex filament
γ	rotor blade Lock number
γ	strength of point vortex in two-dimensional flow
Γ	bound circulation about a wing section
Γ	total circulation around an airfoil
Γ	bound circulation
Γ	dihedral angle
Γ_0	midspan value of bound circulation
Γ_0	midspan value of Γ
Γ_c	circulation around airfoil having a jet flap
Γ_j	total circulation of the jet behind jet-flapped airfoil
$\Gamma(I)$	strength of horseshoe vortex in numerical wing models
δ_a	aileron angle
δ_e	elevator angle

δ_f	flap angle
δ_r	rudder angle
δ_t	tab angle
δ	boundary layer thickness
δ	ratio of ambient pressure to sea level value
δ	induced drag factor, $C_{D_i} = C_L^2/[\pi A(1 + \delta)]$
δ	flow turned by this angle through oblique shock wave
Δ	frequently denotes increment in quantity following it
δ^*	displacement thickness of boundary layer
ϵ	downwash angle at horizontal tail
ϵ	Oswald's efficiency factor, $C_{D_i} = C_L^2/(\pi A \epsilon)$
ϵ	drag-to-lift ratio
ϵ	downwash angle at tail caused by the wing
ϵ_α	derivative of ϵ with angle of attack
ϵ	wing twist
ϵ_T	total wing twist
ζ	critical damping ratio
η	real part of σ
η_t	ratio of q at tail to free-stream q
η	propeller efficiency
η	empirical correction to theoretical value of τ
η	propeller efficiency
η_i	ideal efficiency
θ	pitch angle
β	blade section pitch angle
θ	ratio of ambient temperature to sea level value
θ	angle of oblique shock wave
θ	climb angle
θ	Euler angle, about y-axis
θ	cylindrical coordinate
θ_0	rotor collective pitch
θ_1	lateral cyclic pitch
θ_2	longitudinal cyclic pitch
θ_c	climb angle
θ_c	constant θ (11–23 km) (36,000–75,500 ft)
θ_c	angle of climb
θ_D	descent angle
θ_T	total rotor blade twist
τ	flap effectiveness factor, $\Delta\alpha/\Delta\delta_f$
λ	propeller advance ratio, V/V_T
λ	ratio of rotor inflow to tip speed
λ	taper ratio
$\Lambda_{1/4}$	sweepback angle of quarter-chord line
Λ	leading edge sweepback angle
μ	dimensionless airplane mass
μ	dynamic viscosity
μ	coefficient of friction
μ	rotor advance ratio, V/V_T
ρ	air mass density
ρ_c	air mass density just ahead of compressor
ρ_o	reservoir mass density

σ	root of characteristic equation
σ	propeller solidity, blade area/disc area
σ	blade section solidity, $Bc/(\pi R)$
σ	ratio of air mass density to sea level value
τ	dimensionless blade weight moment
τ	dimensionless time
τ	time constant in first-order, differential equation
τ	a specific value of the time, t
τ_w	wall shear stress
ν	kinematic viscosity
ϕ	velocity potential
ϕ	angle of resultant flow at a propeller blade section
ϕ	thrust augmentation ratio
ϕ	Euler angle, about x-axis
ϕ_T	helix angle of vortex sheet at blade tip
ψ	Euler angle, about z-axis
ψ	azimuth angle around a rotor
ψ	stream function
ω	angular velocity, radians per second
ω	forcing frequency
ω	circular frequency (imaginary part of σ)
ω_n	natural frequency
Ω	rotational speed, radians per second
Ω	angular velocity about spin axis

SUBSCRIPTS/SUPERSCRIPTS

∞	free-stream
0	reservoir conditions
0	sea level
0	midspan
0	trim condition
1/4	quarter-chord line
1	immediately ahead of shock wave
2	immediately behind shock wave
ac	aerodynamic center
b	body (frequently the fuselage)
B	base
c	compressible
cr	critical
E	empty
ew	equivalent weight
F	fuel
fuse	fuselage
gr	ground-roll
i	induced
i	incompressible
LE	leading edge
LOF	liftoff
max	maximum
min	minimum

n	normal to leading edge
PL	payload
s	standard
s	coefficient based on slipstream q
t	horizontal tail
v	vertical tail
w	wing
w	wave
i	derivative with respect to incidence angle of tail
p	derivative with respect to dimensionless pitch rate
q	derivative with respect to dimensionless yaw rate
r	derivative with respect to dimensionless roll rate
x	referring to x-axis
y	referring to y-axis
z	referring to z-axis
*	throat of converging–diverging nozzle where $M = 1$
G	gravitational component
α	derivative with respect to α
β	derivative with respect to β
u	derivative with respect to perturbation velocity, u

Index